104.50
75I

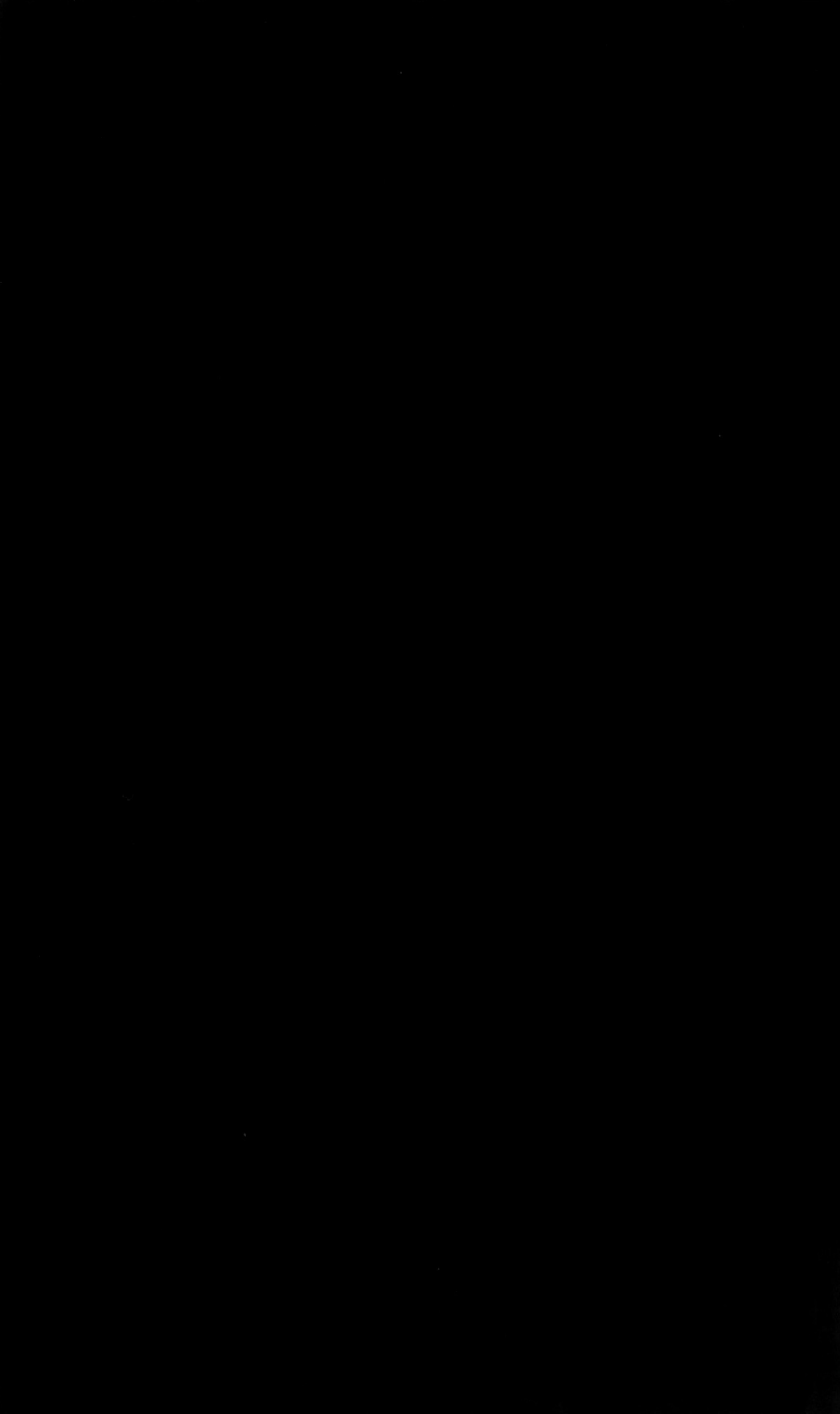

# BIOACTIVATION
# OF FOREIGN COMPOUNDS

Edited by

## M. W. Anders
*Department of Pharmacology*
*School of Medicine and Dentistry*
*University of Rochester*
*Rochester, New York*

1985

ACADEMIC PRESS, INC.
*(Harcourt Brace Jovanovich, Publishers)*
Orlando   San Diego   New York   London
Toronto   Montreal   Sydney   Tokyo

COPYRIGHT © 1985, BY ACADEMIC PRESS, INC.
ALL RIGHTS RESERVED.
NO PART OF THIS PUBLICATION MAY BE REPRODUCED OR
TRANSMITTED IN ANY FORM OR BY ANY MEANS, ELECTRONIC
OR MECHANICAL, INCLUDING PHOTOCOPY, RECORDING, OR ANY
INFORMATION STORAGE AND RETRIEVAL SYSTEM, WITHOUT
PERMISSION IN WRITING FROM THE PUBLISHER.

ACADEMIC PRESS, INC.
Orlando, Florida 32887

*United Kingdom Edition published by*
ACADEMIC PRESS INC. (LONDON) LTD.
24/28 Oval Road, London NW1 7DX

Library of Congress Cataloging in Publication Data

Main entry under title:

Bioactivation of foreign compounds.

  (Biochemical pharmacology and toxicology)
  Includes index.
  1. Xenobiotics--Metabolism.  2. Xenobiotics--Physiolog-
ical effect.  3. Xenobiotics--Toxicology.  4. Biotransformation (Metabolism)   I. Anders, M. W.  II. Series.
[DNLM: 1. Biotransformation.  2. Drugs--Metabolism.
3. Biochemistry.   QV 38 B61415]
QP529.B56   1985    612'.015    84-16772
ISBN 0-12-059480-3 (alk. paper)

PRINTED IN THE UNITED STATES OF AMERICA

85 86 87 88     9 8 7 6 5 4 3 2 1

# Contents

Contributors . . . . . . . . . . . . . . . . . . . . . . . . xi
Preface . . . . . . . . . . . . . . . . . . . . . . . . . . . xv

## PART I. INTRODUCTION

### 1. Some Historical Perspectives on the Metabolism of Xenobiotic Chemicals to Reactive Electrophiles
*Elizabeth C. Miller and James A. Miller*

| | | |
|---|---|---|
| I. | Introduction . . . . . . . . . . . . . . . . . . . . . | 3 |
| II. | Early Evidence for the Conversion of Chemical Carcinogens to Protein- and Nucleic Acid-Bound Derivatives . . . . . . . . | 4 |
| III. | Early Studies on the Metabolism of Foreign Chemicals by Enzymes of the Endoplasmic Reticulum and on the Inducibility of Some of These Enzymes . . . . . . . . . . . . . . . . . . . | 7 |
| IV. | Early Examples of Metabolic Activation: The Concepts of Pro-Drugs and Lethal Synthesis . . . . . . . . . . . . . . . . . . | 10 |
| V. | Experimental Demonstration That 2-Acetylaminofluorene Is a Procarcinogen . . . . . . . . . . . . . . . . . . . . . | 10 |
| VI. | Early Data on the Metabolic Formation of Electrophilic Aromatic N–O Derivatives . . . . . . . . . . . . . . . . . . . | 11 |
| VII. | Central Role of Electrophilic Reactants in Carcinogenesis, Mutagenesis, and Some Other Toxic Manifestations by Chemicals . | 15 |
| VIII. | Looking Ahead . . . . . . . . . . . . . . . . . . . . | 17 |
| | References . . . . . . . . . . . . . . . . . . . . . | 19 |

### 2. Pharmacokinetics of Biological Activation and Inactivation of Foreign Compounds
*James R. Gillette*

| | | |
|---|---|---|
| I. | Introduction . . . . . . . . . . . . . . . . . . . . . | 30 |

v

II. Fundamental Pharmacokinetic Concepts . . . . . . . . . . . . . 32
III. Concentration of the Parent Compound in Arterial Blood after Different Routes of Administration . . . . . . . . . . . . . . 48
IV. Concentration of the Parent Compound in Blood Exiting Various Organs in the Central Compartment after Different Routes of Administration . . . . . . . . . . . . . . . . . . . . . . . . 50
V. Kinetics of the Formation of Chemically Reactive Metabolites by Organs in the Central Compartment . . . . . . . . . . . . . 54
VI. General Comments . . . . . . . . . . . . . . . . . . . . . . . 69
References . . . . . . . . . . . . . . . . . . . . . . . . . . . 70

## 3. Cellular Defense Mechanisms against Reactive Metabolites
*Donald J. Reed*

I. Introduction . . . . . . . . . . . . . . . . . . . . . . . . . . 71
II. General Concepts of Protection . . . . . . . . . . . . . . . . 72
III. Types of Protection . . . . . . . . . . . . . . . . . . . . . . 75
IV. Cellular Aspects of Protection . . . . . . . . . . . . . . . . . 88
V. Comments . . . . . . . . . . . . . . . . . . . . . . . . . . . 96
References . . . . . . . . . . . . . . . . . . . . . . . . . . . 97

# PART II. BIOACTIVATION BY CHEMICAL CLASS

## 4. Alkanes
*James S. Bus*

I. Introduction . . . . . . . . . . . . . . . . . . . . . . . . . . 111
II. Alkane Toxicity . . . . . . . . . . . . . . . . . . . . . . . . 112
III. Alkane Bioactivation . . . . . . . . . . . . . . . . . . . . . 112
IV. Toxic Mechanisms of the $\gamma$-Diketones . . . . . . . . . . . . 116
References . . . . . . . . . . . . . . . . . . . . . . . . . . . 119

## 5. Alkenes and Alkynes
*Paul R. Ortiz de Montellano*

I. Introduction . . . . . . . . . . . . . . . . . . . . . . . . . . 121
II. Oxidative Chemistry of Olefins and Acetylenes . . . . . . . . 122
III. Metabolism of Olefins . . . . . . . . . . . . . . . . . . . . 126
IV. Metabolism of Acetylenes . . . . . . . . . . . . . . . . . . . 131
V. Destruction of Cytochrome $P$-450 by Olefins and Acetylenes . . . 135
VI. Mechanisms of $\pi$-Bond Oxidation and Cytochrome $P$-450 Destruction . . . . . . . . . . . . . . . . . . . . . . . . . . 140
VII. Olefin and Acetylene Toxicity . . . . . . . . . . . . . . . . . 144
References . . . . . . . . . . . . . . . . . . . . . . . . . . . 145

## 6. Benzene and Substituted Benzenes
*Laurence S. Kaminsky*

|  |  |
|---|---|
| I. Introduction | 157 |
| II. Chemical Properties | 158 |
| III. Reaction Mechanisms for Bioactivation | 159 |
| IV. Enzymology of Bioactivation | 170 |
| References | 170 |

## 7. Polycyclic Aromatic Hydrocarbons: Metabolic Activation to Ultimate Carcinogens
*Dhiren R. Thakker, Haruhiko Yagi, Wayne Levin, Alexander W. Wood, Allan H. Conney, and Donald M. Jerina*

|  |  |
|---|---|
| I. Introduction | 178 |
| II. Metabolic Pathways of Polycyclic Aromatic Hydrocarbons | 180 |
| III. Metabolism of Benzo[a]pyrene | 181 |
| IV. Biological Activity of Benzo[a]pyrene Derivatives | 195 |
| V. The Bay Region Theory | 203 |
| VI. Comparative Metabolism of Polycyclic Aromatic Hydrocarbons | 206 |
| VII. Stereoselectivity in the Metabolism of the Polycyclic Aromatic Hydrocarbons | 212 |
| VIII. Stereoselectivity in the Biological Activity of Polycyclic Aromatic Hydrocarbon Derivatives | 219 |
| IX. Summary | 221 |
| References | 222 |

## 8. Furans
*Leo T. Burka and Michael R. Boyd*

|  |  |
|---|---|
| I. Chemistry | 243 |
| II. Occurrence | 245 |
| III. Enzymology | 246 |
| IV. Reactive Intermediates | 247 |
| V. Fate of Reactive Intermediates | 252 |
| VI. Summary | 253 |
| References | 254 |

## 9. Phenols, Catechols, and Quinones
*Richard D. Irons and Tadashi Sawahata*

|  |  |
|---|---|
| I. Introduction | 259 |
| II. Phenols | 261 |
| III. Catechols and Hydroquinones | 266 |

## 10. Halogenated Alkanes
*M. W. Anders and Lance R. Pohl*

|     |     |     |
| --- | --- | --- |
| IV. | Quinones | 269 |
|     | References | 277 |

| | | |
|---|---|---|
| I. | Chemistry of the Halogen–Carbon Bond | 284 |
| II. | Oxidative Dehydrohalogenation Mechanism | 285 |
| III. | Oxygenation of Halocarbon Radicals: Reductive-Oxygenation Pathway of Metabolism | 290 |
| IV. | Cytochrome $P$-450–Dependent Reductive Reactions of Halogenated Hydrocarbons | 296 |
| V. | Glutathione-Dependent Metabolism of Halogenated Hydrocarbons | 302 |
|  | References | 306 |

## 11. Halogenated Alkenes and Alkynes
*Dietrich Henschler*

| | | |
|---|---|---|
| I. | Chemical Reactivity as a Basis for Predicting Biotransformation Pathways and Rates | 317 |
| II. | Halogenated Ethylenes | 318 |
| III. | Halogenated Allyl Compounds | 335 |
| IV. | Halogenated Alkynes | 339 |
|  | References | 341 |

## 12. Arylamines and Arylamides: Oxidation Mechanisms
*Sidney D. Nelson*

| | | |
|---|---|---|
| I. | Introduction | 349 |
| II. | Chemical Properties of Arylamines and Arylamides | 350 |
| III. | Reaction Mechanisms of N-Oxidation for Arylamines and Arylamides | 352 |
| IV. | Fate of Arylamine and Arylamide Oxidation Products | 363 |
| V. | Summary | 365 |
|  | References | 365 |

## 13. Arylhydroxylamines and Arylhydroxamic Acids: Conjugation Reactions
*Patrick E. Hanna and R. Bruce Banks*

| | | |
|---|---|---|
| I. | Introduction | 376 |
| II. | Bioactivation of Arylhydroxylamines and Arylhydroxamic Acids by Conjugation Reactions | 376 |

III. Fate of Reactive Intermediates Generated by Conjugation of Arylhydroxylamines and Arylhydroxamic Acids . . . . . . . . 392
References . . . . . . . . . . . . . . . . . . . . . . . . . 395

## 14. Nitrosamines

*Michael C. Archer and George E. Labuc*

I. Introduction . . . . . . . . . . . . . . . . . . . . . . . . . 403
II. Chemical Properties of Nitrosamines . . . . . . . . . . . . . . 404
III. Mechanisms in the Bioactivation of Nitrosamines . . . . . . . . 405
IV. Enzymology of the Bioactivation of Nitrosamines . . . . . . . . 417
V. Fate of Reactive Intermediates from Nitrosamines . . . . . . . . 419
References . . . . . . . . . . . . . . . . . . . . . . . . . 420

## 15. Hydrazines

*R. A. Prough and S. J. Moloney*

I. Introduction . . . . . . . . . . . . . . . . . . . . . . . . . 433
II. Chemical Properties of Hydrazines . . . . . . . . . . . . . . . 434
III. Bioactivation of Hydrazines . . . . . . . . . . . . . . . . . . 436
IV. Fates of Reactive Intermediates of Hydrazines . . . . . . . . . 441
V. Conclusion . . . . . . . . . . . . . . . . . . . . . . . . . 445
References . . . . . . . . . . . . . . . . . . . . . . . . . 446

## 16. Nitroimidazoles

*P. David Josephy and Ronald P. Mason*

I. Introduction . . . . . . . . . . . . . . . . . . . . . . . . . 451
II. Preparation of Aminoimidazoles . . . . . . . . . . . . . . . . 454
III. Zinc Reduction of Nitroimidazoles . . . . . . . . . . . . . . . 455
IV. Enzymatic Reduction of Nitroimidazoles: Diamagnetic Products . . 456
V. Free-Radical Intermediates in the Enzymatic Reduction of Nitro Compounds . . . . . . . . . . . . . . . . . . . . . . . . 458
VI. Catalytic, Radiolytic, and Electrochemical Reduction of Nitroimidazoles . . . . . . . . . . . . . . . . . . . . . . . 461
VII. Coreduction of Nitroimidazoles with Macromolecules . . . . . . 462
VIII. Nonenzymatic Reactions of the Nitro Anion Free Radical . . . . 462
IX. Other Free-Radical Metabolites of Nitro Compounds . . . . . . 469
X. Bacterial Metabolism of Nitroimidazoles . . . . . . . . . . . . 472
XI. Metabolism of Misonidazole by Mammalian Cells *in Vitro* . . . . 473
XII. Summary . . . . . . . . . . . . . . . . . . . . . . . . . . 475
References . . . . . . . . . . . . . . . . . . . . . . . . . 476

## 17. Nitriles
*Ahmed E. Ahmed, Mohammed Y. H. Farooqui, and Norman M. Trieff*

| | |
|---|---:|
| I. Introduction | 485 |
| II. Inorganic Cyanides | 486 |
| III. Nitriles | 490 |
| IV. Saturated Aliphatic Nitriles | 491 |
| V. Unsaturated Aliphatic Nitriles | 499 |
| VI. Alkylaryl Nitriles | 506 |
| VII. Aryl Nitriles | 508 |
| VIII. Summary | 509 |
| References | 510 |

## 18. Thiono-Sulfur Compounds
*Robert A. Neal*

| | |
|---|---:|
| I. Introduction | 519 |
| II. Mechanism of Metabolism of Parathion, Carbon Disulfide, and Thioacetamide | 523 |
| III. Summary | 537 |
| References | 538 |

Index . . . . . . . . . . . . . . . . . . . . . . . . . . . . . 541

# Contributors

Numbers in parentheses indicate the pages on which the authors' contributions begin.

*Ahmed E. Ahmed* (485), Departments of Pathology, Pharmacology, and Toxicology, University of Texas Medical Branch, Galveston, Texas 77550

*M. W. Anders* (283), Department of Pharmacology, School of Medicine and Dentistry, University of Rochester, Rochester, New York 14642

*Michael C. Archer* (403), Department of Medical Biophysics, University of Toronto, Ontario Cancer Institute, Toronto, Ontario, Canada M4X 1K9

*R. Bruce Banks* (375), Department of Chemistry, University of North Carolina, Greensboro, North Carolina 27412

*Michael R. Boyd* (243), Laboratory of Experimental Therapeutics and Metabolism, National Cancer Institute, National Institutes of Health, Bethesda, Maryland 20205

*Leo T. Burka*[1] (243), Laboratory of Experimental Therapeutics and Metabolism, National Cancer Institute, National Institutes of Health, Bethesda, Maryland 20205

*James S. Bus* (111), Department of Biochemical Toxicology, Chemical Industry Institute of Toxicology, Research Triangle Park, North Carolina 27709

*Allan H. Conney* (177), Department of Experimental Carcinogenesis and Metabolism, Hoffman-La Roche Inc., Nutley New Jersey 07110

*Mohammed Y. H. Farooqui* (485), Departments of Pathology, Pharmacology, and Toxicology, University of Texas Medical Branch, Galveston, Texas 77550

[1]Present address: Systematic Toxicology Branch, Toxicology Research and Testing Program, National Institute of Environmental Health Services, Research Triangle Park, North Carolina 27709.

*James R. Gillette* (29), Laboratory of Chemical Pharmacology, National Heart, Lung, and Blood Institute, National Institutes of Health, Bethesda, Maryland 20205

*Patrick E. Hanna* (375), Departments of Medicinal Chemistry and Pharmacology, University of Minnesota, Minneapolis, Minnesota 55455

*Dietrich Henschler* (317), Institut für Pharmakologie und Toxikologie, Universität Würzburg, D-8700 Würzburg, Federal Republic of Germany

*Richard D. Irons* (259), Chemical Industry Institute of Toxicology, Research Triangle Park, North Carolina 27709

*Donald M. Jerina* (177), Laboratory of Bioorganic Chemistry, National Institute of Arthritis, Diabetes, and Digestive and Kidney Diseases, National Institutes of Health, Bethesda, Maryland 20205

*P. David Josephy*[2] (451), Laboratory of Pulmonary Function and Toxicology, National Institute of Environmental Health Sciences, Research Triangle Park, North Carolina 27709

*Laurence S. Kaminsky* (157), Laboratory of Biochemical and Genetic Toxicology, New York State Department of Health, Albany, New York 12201

*George E. Labuc* (403), Department of Medical Biophysics, University of Toronto, Ontario Cancer Institute, Toronto, Ontario, Canada M4X 1K9

*Wayne Levin* (177), Department of Experimental Carcinogenesis and Metabolism, Hoffman-La Roche Inc., Nutley, New Jersey 07110

*Ronald P. Mason* (451), Laboratory of Molecular Biophysics, National Institute of Environmental Health Sciences, Research Triangle Park, North Carolina 27709

*Elizabeth C. Miller* (3), McArdle Laboratory for Cancer Research, University of Wisconsin School of Medicine, Madison, Wisconsin 53706

*James A. Miller* (3), McArdle Laboratory for Cancer Research, University of Wisconsin School of Medicine, Madison, Wisconsin 53706

*S. J. Moloney*[3] (433), Department of Biochemistry, The University of Texas Health Science Center, Dallas, Texas 75235

*Robert A. Neal* (519), Chemical Industry Institute of Toxicology, Research Triangle Park, North Carolina 27709

*Sidney D. Nelson* (349), Department of Medicinal Chemistry, University of Washington, Seattle, Washington 98195

*Paul R. Ortiz de Montellano* (121), Department of Pharmaceutical Chemistry, University of California, San Francisco, California 94143

[2]Present address: Department of Chemistry and Biochemistry, University of Guelph, Guelph, N1G 2W1 Ontario, Canada.
[3]Present address: Avon Products, Inc., Toxicology Department, Suffern, New York 10901.

CONTRIBUTORS

*Lance R. Pohl* (283), Laboratory of Chemical Pharmacology, National Heart, Lung, and Blood Institute, National Institutes of Health, Bethesda, Maryland 20205

*R. A. Prough* (433), Department of Biochemistry, The University of Texas Health Science Center, Dallas, Texas 75235

*Donald J. Reed* (71), Department of Biochemistry and Biophysics, Oregon State University, Corvallis, Oregon 97331

*Tadashi Sawahata* (259), Toxicology Laboratory, Toray Industries, Inc., Sonoyama, Otsu 520, Japan

*Dhiren R. Thakker*[a] (177), Laboratory of Bioorganic Chemistry, National Institute of Arthritis, Diabetes, and Digestive and Kidney Diseases, National Institutes of Health, Bethesda, Maryland 20205

*Norman M. Trieff* (485), Department of Preventive Medicine and Community Health, University of Texas Medical Branch, Galveston, Texas 77550

*Alexander W. Wood* (177), Department of Experimental Carcinogenesis and Metabolism, Hoffman-La Roche Inc., Nutley, New Jersey 07110

*Haruhiko Yagi* (177), Laboratory of Bioorganic Chemistry, National Institute of Arthritis, Diabetes, and Digestive and Kidney Diseases, National Institutes of Health, Bethesda, Maryland 20205

---

[a]Present address: Center for Drugs and Biologics, Food and Drug Administration, Bethesda, Maryland 20205.

# *Preface*

The elucidation of the understanding of the mechanisms by which chemicals produce toxic effects is a challenge to pharmacologists, toxicologists, and biochemists. Although significant gaps in our knowledge remain, work in many laboratories over the past two decades has shown that the metabolic alteration, or bioactivation, of chemicals—either to stable, but toxic, metabolites or to reactive electrophiles—is necessary for the elicitation of a toxic response.

This volume in the Biochemical Pharmacology and Toxicology series aims to summarize the body of knowledge on chemical bioactivation. The introductory chapters deal with historical developments and with factors that affect all chemicals. The emphasis of the remainder of the volume is on the mechanisms of bioactivation of chemical classes. These chapters provide information on biochemical reaction mechanisms and the fate of toxic metabolites. The enzymology of bioactivation enzymes has been treated briefly because this was the subject of earlier volumes in this series (*Enzymatic Basis of Detoxication,* Volumes I and II, edited by W. B. Jakoby, Academic Press, 1980).

The biochemical view presented in this work should enhance our ability to predict bioactivation mechanisms for new compounds and to make better judgments on the hazards posed by exposure to chemicals.

*M. W. Anders*

*Part 1*

# Introduction

Chapter 1

# Some Historical Perspectives on the Metabolism of Xenobiotic Chemicals to Reactive Electrophiles

ELIZABETH C. MILLER AND JAMES A. MILLER

*McArdle Laboratory for Cancer Research*
*University of Wisconsin School of Medicine*
*Madison, Wisconsin*

|      |                                                                                                                                                         |    |
|------|---------------------------------------------------------------------------------------------------------------------------------------------------------|----|
| I.   | Introduction                                                                                                                                            | 3  |
| II.  | Early Evidence for the Conversion of Chemical Carcinogens to Protein- and Nucleic Acid-Bound Derivatives                                                | 4  |
| III. | Early Studies on the Metabolism of Foreign Chemicals by Enzymes of the Endoplasmic Reticulum and on the Inducibility of Some of These Enzymes           | 7  |
| IV.  | Early Examples of Metabolic Activation: The Concepts of Pro-Drugs and Lethal Synthesis                                                                  | 10 |
| V.   | Experimental Demonstration That 2-Acetylaminofluorene Is a Procarcinogen                                                                                | 10 |
| VI.  | Early Data on the Metabolic Formation of Electrophilic Aromatic N–O Derivatives                                                                         | 11 |
| VII. | Central Role of Electrophilic Reactants in Carcinogenesis, Mutagenesis, and Some Other Toxic Manifestations by Chemicals                                | 15 |
| VIII.| Looking Ahead                                                                                                                                           | 17 |
|      | References                                                                                                                                              | 19 |

## I. Introduction

Our interest in and our work on the bioactivation of foreign compounds grew out of our graduate studies in biochemistry at the University of Wisconsin in the early 1940s. These studies introduced us to a fascinating set of xenobiotic chemicals, the newly discovered pure chemical carcinogens. Our graduate major advisor, Professor Carl Baumann, was collaborating with Professor Harold Rusch, the director of the then new McArdle Laboratory for Cancer Research, on the modulating effects of nutrition on cancer induction in rats and mice by some hepatocarcinogenic aminoazo dyes and certain carcinogenic polycyclic aromatic hydrocarbons. From participation in these studies we became interested in the mechanisms by which these agents so profoundly alter normal cells that they give rise to cancer cells. That interest and research direction have continued to the present.

The pathway of research is frequently circuitous and may encompass a broader field than that chosen for investigation. Thus, early investigations on the metabolism of the chemical carcinogens happened to open research areas of broad interest in pharmacology and toxicology.* Those early studies led to the recognition of what were later called the mixed-function oxygenase systems in the endoplasmic reticulum, provided the first observations on the induction of enzymes of the endoplasmic reticulum, demonstrated the activation of carcinogenic chemicals by their conversion to electrophilic derivatives that react covalently with cellular macromolecules, and furnished data on the probable importance of macromolecule-bound chemicals in the development of carcinogenic and other toxic lesions. The aim of this chapter is to put some of these earlier findings from a number of laboratories into perspective in relation to the expanding information that shows that many xenobiotic chemicals exert their biological effects through metabolism to electrophilic reactants that bind covalently to critical informational macromolecules of target cells.

## II. Early Evidence for the Conversion of Chemical Carcinogens to Protein- and Nucleic Acid-Bound Derivatives

The pioneer discoveries of the first pure chemical carcinogens, the polycyclic aromatic hydrocarbons[1] and the $p$-aminoazo dyes,[2-4] in the early 1930s encouraged analyses of the roles of these chemicals in the carcinogenic processes they caused. However, the tools that were available for the study of biochemical problems in the 1940s were, by current standards, very lim-

---

* In some cases the terminology used in the original literature has been changed to conform with current usage. The major examples are the nomenclature of the polycyclic aromatic hydrocarbons, the use of NADPH rather than TPNH as the abbreviation for reduced triphosphopyridine nucleotide, and the use of the terms microsomes and endoplasmic reticulum in place of small granules.

# I. METABOLISM OF XENOBIOTIC CHEMICALS

ited. For instance, detection of the polycyclic aromatic hydrocarbons depended on their absorption of ultraviolet light and on their strong fluorescence under such conditions. However, the instruments available for studies of these properties were relatively crude and not useful for study of the very small amounts of hydrocarbons in tissues of treated animals. The first commercial photoelectric spectrophotometers became available in the 1940s.[5] Paper chromatography was introduced in 1944,[6] and the first use of a $^{14}$C-labeled carcinogen was reported in 1948 by Heidelberger and Jones.[7]

Thus, the high extinction coefficient ($e_{max}$) of the hepatocarcinogen $N,N$-dimethyl-4-aminoazobenzene (DAB) at about 410 nm in organic solvents and its even higher extinction at 518 nm in strongly acidic solutions were important factors in our choice of DAB and structurally related dyes as carcinogens for study (Fig. 1).[8,9] Analysis with a photoelectric colorimeter of acid solutions of DAB permitted detection of as little as 1 μg. Furthermore, these dyes were readily chromotographed on small alumina columns. Early studies by Stevenson et al.[10] on the urinary metabolites of DAB showed that the dye or some of its metabolites, or both, were subject to cleavage of the azo linkage, N-demethylation, and ring hydroxylation in the rat in vivo.

These data and the dependence of the hepatocarcinogenicity for rats on the presence of at least one $N$-methyl group and the azo group[11,12] provided the base from which we tried to analyze the role of DAB and its congeners in hepatic tumor formation.

In the course of studies that required the precipitation of liver proteins from DAB-fed rats with trichloroacetic acid, it became evident that, in addition to DAB and its metabolites $N$-methyl-4-aminoazobenzene (MAB)

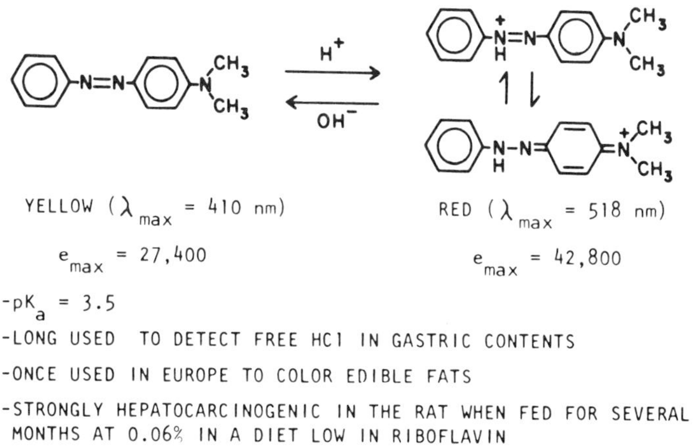

YELLOW ($\lambda_{max}$ = 410 nm)         RED ($\lambda_{max}$ = 518 nm)

$e_{max}$ = 27,400                           $e_{max}$ = 42,800

- $pK_a$ = 3.5
- LONG USED TO DETECT FREE HCl IN GASTRIC CONTENTS
- ONCE USED IN EUROPE TO COLOR EDIBLE FATS
- STRONGLY HEPATOCARCINOGENIC IN THE RAT WHEN FED FOR SEVERAL MONTHS AT 0.06% IN A DIET LOW IN RIBOFLAVIN

Fig. 1. Some properties of the rat hepatocarcinogen $N,N$-dimethyl-4-aminoazobenzene (DAB).

and 4-aminoazobenzene (AB) (Fig. 2), the liver contained aminoazo dyes that could not be separated from the protein without destruction of the latter and that retained the acid–base indicator properties of DAB.[13] This carcinogen derivative, evidently an aminoazo dye covalently bound to protein of the target tissue, appeared to provide an indicator of reactions that might be involved in the induction of tumors.

Early studies showed that the amount of bound dye in rat liver was approximately the same whether DAB or its equally carcinogenic metabolite MAB was administered, but that the level was much lower when the noncarcinogenic metabolite AB was given.[13] The amount of the bound dye increased for the first few weeks of dye feeding and then decreased or remained constant[13]; the time at which the maximum level of binding occurred was approximately inversely related to the carcinogenicities of a series of ring-methyl derivatives of DAB.[14] The largest amount of the bound dye was found in the liver, the target tissue, and in the liver of the rat, the major susceptible species.[13] Of great interest to us was the finding that liver tumors induced by DAB contained no detectable bound dye, even with continuous feeding of the dye.[13]

The small amounts of protein-bound dye (40–50 μg/liver) and the limited separation techniques for isolation of such a minor component in pure form in amounts suitable for the characterization techniques then available hindered structural characterization. However, it was determined that the dye was bound to the protein through the "diamine" portion of the dye, because reduction with tin in acid solution yielded amines, presumably aniline (from DAB or MAB) and a toluidine (from the dyes substituted with a methyl group in the prime ring), which could be steam distilled and then quantified by reaction with $N,N$-dimethyl-$p$-aminobenzaldehyde.[14] Studies with Sorof and Cohen[15] showed that there was specificity among the liver proteins for the binding of the carcinogenic aminoazo dyes, and Sorof and Cohen[16] noted that these dye-binding proteins of liver were present in greatly reduced amounts in the primary liver tumors induced by the dyes.

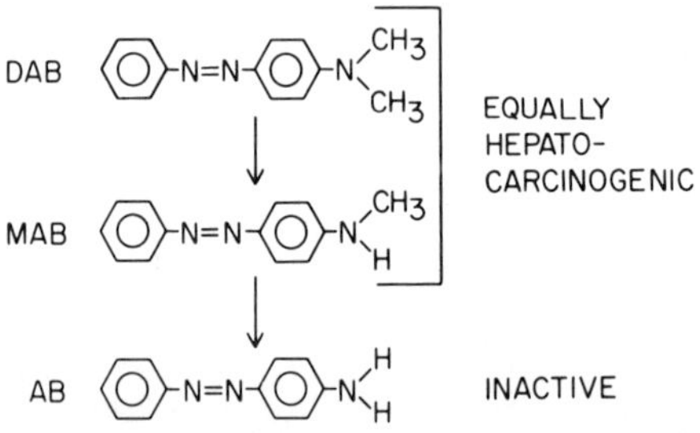

Fig. 2. The N-demethylation of DAB to MAB and AB in rat liver *in vivo* and *in vitro* and the carcinogenicities of these dyes in rat liver.[8,11,12,39]

# I. METABOLISM OF XENOBIOTIC CHEMICALS

The ability to be converted to protein-bound derivatives in the target tissues for carcinogenesis was not unique to the aminoazo dyes. Within a few years protein-bound fluorescent derivatives of benzo[a]pyrene were found in the skin of mice treated topically with the hydrocarbon.[17] At about the same time the incorporation of the hepatic carcinogen ethionine into liver proteins was noted by Levine and Tarver,[18] and shortly thereafter protein-bound derivatives of 2-[$^{14}$C]acetylaminofluorene and [$^{14}$C]dibenz[a,h]anthracene were detected in rat liver and mouse skin, respectively.[19-21]

The studies just described preceded the seminal paper of Watson and Crick[22] on the manner by which genetic information in DNA could be transferred from cells to their progeny cells and the later work of many investigators on DNA as a template for the synthesis of RNAs leading to protein synthesis. These advances triggered interest in the occurrence of DNA- and RNA-bound carcinogen residues in target cells. The first demonstration of nucleic acid-bound carcinogen residues in animal tissues was that of Wheeler and Skipper[23] with nitrogen mustard. By 1964 nucleic acid-bound derivatives of sulfur mustard, dimethylnitrosamine, ethionine, 2-acetylaminofluorene (AAF), DAB, and the polycyclic aromatic hydrocarbons were reported.[24] All of these demonstrations depended on the availability of $^{14}$C- or $^{3}$H-labeled carcinogens.

Since these early studies, protein- and nucleic acid-bound derivatives of carcinogens have been found in target tissues for members of nearly all classes of chemical carcinogens, and many of these derivatives have been characterized (reviewed in Ref. 25). The rapid development of sophisticated molecular biological techniques in the past few years is now facilitating detailed studies on the effects of carcinogen residues on the structure and function of the nucleic acids and proteins (see, for example, Refs. 26-32). The results of studies of this kind will surely provide basic data on the roles, at the molecular level, of chemical carcinogens in tumor induction. Current data suggest that these roles will include modifications of the DNA that result in changes in the structure or expression, or both, of cellular oncogenes (proto-oncogenes) in malignant cells induced by chemical carcinogens.[33,34]

## III. Early Studies on the Metabolism of Foreign Chemicals by Enzymes of the Endoplasmic Reticulum and on the Inducibility of Some of These Enzymes

The development of gentle systems for the disruption of mammalian cells[35] and of centrifugal methods for the preparation of enzymatically functional cellular organelles[36,37] facilitated approaches to the *in vitro* metabolism of xenobiotics. To the authors' knowledge and as noted by Trager,[38] the first of these studies was carried out by Mueller and Miller,[39] who showed in 1948 that N-demethylation and aromatic ring hydroxylation occurred on aerobic incubation of DAB with pyridine nucleotide-fortified rat liver homogenates. The enzymatic activity of rat liver for the reduction of the azo linkage was found primarily in the microsomal fraction (derived from the endoplasmic reticulum by homogenization and then called the small granule fraction) and was dependent on both NADPH and a flavin adenine dinucleotide-containing protein.[40] Mueller and Miller[41] further showed a dependence on oxygen and NADPH for the enzymatic N-demethylation of the *N*-methylaminoazo dyes by rat liver homogenates; formaldehyde was formed in stoichiometric amounts. The similar NADPH- and oxygen-dependent N-dealkylation of both *N*-dimethyl- and *N*-diethyl-4-aminoantipyrine by liver microsomes was subsequently shown by La Du *et al.* in 1955.[42] In that year Axelrod[43] extended the known substrate range of these enzymes by showing that a NADPH- and oxygen-dependent system in rabbit liver microsomes deaminated amphetamine with the formation of phenyl acetone and ammonia. The demonstration that hydroxylation of aromatic rings was catalyzed by the microsomal fraction was made by Mitoma *et al.*,[44] and the variety of substrates for this reaction was further extended by Booth and Boyland[45] and by Conney *et al.*[46] The studies of Mitoma *et al.*[44] further showed that more than one carbon atom of an aromatic ring might be a site for hydroxylation, and Parke and Williams[47] suggested that hydroxylation at ortho and para positions involved different enzymes. Other studies in a number of laboratories demonstrated microsomal NADPH-dependent oxidations that resulted in O-dealkylations, oxidations of the sulfur atoms of thioethers, and oxidation of aliphatic carbon atoms. Furthermore, each of these types of oxidations could occur with a wide range of substrates (reviewed in Ref. 48).

The requirements of these microsomal enzymes for both a reducing system and molecular oxygen placed them in the mixed-function oxygenase class of Mason.[49] This categorization implied that one oxygen atom was transferred to the substrate and that the other underwent reduction and was converted to water. Posner *et al.*[50] confirmed this mechanism for the hydroxylation of aniline by use of $^{18}O_2$. In 1958 Garfinkel[51] and Klingenberg[52] showed that these microsomal mixed-function oxygenases required a specific hemoprotein for their activity. This hemoprotein, which became known as cytochrome *P*-450 on the basis of its spectral properties, is now recognized as a family of hemoprotein monooxygenases with different substrate specificities.[53,54]

Research on the microsomal cytochrome *P*-450 mixed-function oxygenases increased explosively in the 1960s and early 1970s, and studies continuing to the present time have shown the versatility of these microsomal systems for the oxidation of aliphatic and aromatic carbon atoms, nitrogen atoms, and sulfur atoms contained in a wide variety of substrates and for dehalogenations.[55] Furthermore, although the activities are generally at higher levels in liver, the cytochrome P-450 oxygenases have now been observed in essentially all tissues studied.[56] Detailed studies, especially by Ziegler and colleagues,[57] have also revealed the occurrence of another microsomal oxygenase that oxidizes nitrogen and sulfur atoms in certain types of compounds. This flavoprotein enzyme requires NADPH and $O_2$, but it has no requirement for a hemoprotein. Another recently recognized system for the oxidation of xenobiotics is the prostaglandin endoperoxidase synthetase system of the endoplasmic reticulum of some tissues.[58-60]

Studies on the metabolism of the aminoazo dyes in our laboratory led to the unexpected finding that the hepatic activity for N-demethylation of aminoazo dyes was dependent on the nature of the diet; thus, the livers of mice fed certain crude diets had activities about twice those of mice on purified diets.[61] Studies on the effects of the administration of certain polycyclic aromatic hydrocarbons were prompted by the finding that these compounds inhibited hepatocarcinogenesis by aminoazo dyes.[62,63] Thus, a single dose of 3-methylcholanthrene caused rapid increases in the abilities of rat liver microsomes to remove by oxidation N-methyl groups of N-methyl-p-aminoazo dyes, to reduce the azo linkage of these dyes, and to hydroxylate benzo[a]pyrene.[46,64] Severalfold increases were observed within 3 to 24 hr after hydrocarbon administration; within hours to days after the maximum levels were reached, the enzymatic activities declined to normal. The administration of amino acid antagonists suggested that the increases in the oxygenase levels were the result of enhanced enzyme synthesis,[46,64] and this conclusion is now well documented.[65-69]

Shortly thereafter Remmer and Alsleben[70] and Conney and Burns[71] independently made the first observations on the stimulation of cytochrome P-450-catalyzed microsomal drug metabolism by administration of barbiturates. It was some time, however, before it was recognized that the barbiturates and the polycyclic aromatic hydrocarbons stimulate the synthesis of different forms of cytochrome P-450. When the induced synthesis of the cytochrome P-450 systems was reviewed by Conney in 1967,[72] it was already evident that the number of substrates for which metabolism could thus be increased was large and that the number and variety of inducing agents was also extensive. Furthermore, it is now evident that induced synthesis of other enzymes of the endoplasmic reticulum (e.g., epoxide hydrolase, uridine diphosphate glucuronosyltransferases) can also occur as a result of exposures to a variety of xenobiotic chemicals.[73,74]

The major effects of the enzymes in the endoplasmic reticulum on xenobiotic chemicals appear to be detoxifications, and this group of enzyme systems is frequently considered to have primarily a protective function for the host. However, the enzymes of the endoplasmic reticulum also convert a variety of chemicals to proximate or ultimate carcinogens and mutagens or to metabolites with other toxic manifestations, or both. Some examples are discussed later and in the subsequent chapters of this book.

## IV. Early Examples of Metabolic Activation: The Concepts of Prodrugs and Lethal Synthesis

In contrast to the majority of drugs, which are pharmacologically active as such, some agents require biotransformation to exert their pharmacological effects (Ref. 75, pp. 88–95; Ref. 76). Albert[77] introduced the term "prodrug" for these agents in 1958. Methenamide appears to have been the first drug to be designed as a prodrug; it was introduced in 1899 as a source of urinary formaldehyde for treatment of infections (Ref. 75, p. 88). In other cases (e.g., prontosil, phenylarsonic acids, phenacetin, aspirin, and parathion) it was discovered that the administered compound is in reality a prodrug after the efficacy of the original drug had been established (see Ref. 75, pp. 88–90). The case of parathion ($O,O$-diethyl-$O$-$p$-nitrophenyl thiophosphate) is of interest in the context of this chapter because it was apparently the first drug shown to be activated through catalysis by the mixed-function oxygenases of the endoplasmic reticulum.[78] Oxidative removal of the sulfur atom converts it to the corresponding phosphate, which is a direct-acting acylating agent and an inhibitor of acetylcholine esterase *in vitro*.[79,80]

Studies on the toxicity of fluoroacetate led Peters[81] to his concept of "lethal synthesis." Thus, Peters showed that the toxicity of fluoroacetate depends on its enzymatic conversion to fluorocitrate, which is a strong inhibitor of *cis*-aconitase (aconitate hydratase). These studies are of particular interest because, in addition to being a useful biochemical tool, fluoroacetate is a naturally occurring toxin in some plants of the genus *Dichapetalum* and has been responsible for major losses of South African cattle exposed to the plants during grazing.[81]

## V. Experimental Demonstration That 2-Acetylaminofluorene Is a Procarcinogen

The finding of protein- and nucleic acid-bound derivatives of chemical carcinogens and, with the exception of the carcinogenic alkylating agents, the lack of demonstrable reactivity of the carcinogens in the absence of tissue preparations suggested that reactive metabolites of chemical carcinogens were formed *in vivo*. The rough correlations between the carcinogenic

# 1. METABOLISM OF XENOBIOTIC CHEMICALS

activities of some carcinogens and the levels of their macromolecule-bound derivatives in the target tissues further suggested that reactive metabolites might be important in the carcinogenic process. One approach to searching for these reactive metabolites was to investigate the formation of the bound derivatives *in vitro*. Studies in our laboratory showed that incubation of AB, MAB, or DAB or their 3′-methyl derivatives with NADPH-fortified hepatic microsomes yielded protein-bound dyes,[82,83] but these bound dyes were different from those formed *in vivo* and thus their significance to protein-binding *in vivo* and to carcinogenesis was not evident.

A more fruitful lead came from studies on the *in vivo* metabolism of AAF administered to rats.[84,85] These studies led to the recognition of a new urinary metabolite, which was excreted as a conjugate cleavable by crude $\beta$-glucuronidase and which was found in increasing amounts with continued administration of AAF. The structure of the metabolite was deduced from its acidic character, its elementary analysis (which showed the addition of only one oxygen atom to the elementary formula of AAF), and its separability from all of the ring-hydroxy derivatives of AAF, most of which had been prepared and studied by the Weisburgers and their colleagues (reviewed in Ref. 86). Definitive characterization was provided by an unequivocal synthesis of *N*-hydroxy-AAF from 2-nitrofluorene and by demonstration of the identity of the synthetic and metabolic products.[84] On administration to rats[87] and other rodents,[88] *N*-hydroxy-AAF proved to be more active as a carcinogen than AAF; it thus became the first example of a proximate carcinogenic metabolite of a carcinogen. Subsequent studies with a variety of aromatic amines and amides similarly indicated that N-hydroxylation is the first step in their activation for carcinogenic activity.[89,90] *N*-Hydroxy metabolites formed by metabolic reduction also are presumed intermediates in the activation of aromatic nitro carcinogens and mutagens.[90-92]

N-Hydroxylation of AAF is catalyzed by a cytochrome *P*-450-dependent mixed-function oxygenase of the hepatic endoplasmic reticulum,[93-95] and this activity is induced by 3-methylcholanthrene.[94] However, administration of 3-methylcholanthrene to rats simultaneously with AAF resulted in an inhibition of carcinogenesis.[63,96] The balance between the induction of the deactivating ring-hydroxylation systems as compared to the induction of the activating N-hydroxylation system[94] and the inhibition of the N-hydroxylation of AAF by 3-methylcholanthrene[97] may both contribute to this *in vivo* result.

## VI. Early Data on the Metabolic Formation of Electrophilic Aromatic N–O Derivatives

In retrospect the concept of the electrophilicity of the reactive forms of chemical carcinogens was surprisingly slow in developing in studies on chemical carcinogenesis. By the early 1950s a number of alkylating agents were known to have carcinogenic activity,[98] but the carcinogenic activities of these compounds were relatively weak and they received only limited attention. In 1956 Magee and Barnes[99] reported the carcinogenicity of dimethylnitrosamine. Within a few years Magee and associates[100,101] showed that, like the direct methylating agents, an uncharacterized metabolite of dimethylnitrosamine that was formed in liver slices methylated histidine residues in proteins. Similarly, guanine residues in hepatic nucleic acids were methylated when dimethylnitrosamine was administered to rats. Nevertheless, to our knowledge the alkylating agents were not regarded as prototypes for the reactive forms of chemical carcinogens. The concept of electrophilic reactivity as a requirement for the active forms of chemical carcinogens came from direct observations, and the first of these came from studies on the $N$-hydroxy metabolites of $N$-aryl carcinogens.

The activation of hydroxylamines by protonation, already familiar to organic chemists, was applied to $N$-hydroxy-2-aminofluorene ($N$-hydroxy-AF) by Kriek,[102] who demonstrated in 1965 that it reacted with guanine residues in nucleic acids in weak acid. The formation of $N$-hydroxy-AF from $N$-hydroxy-AAF by a deacylase activity in liver microsomes was demonstrated by Irving[103] the next year. However, the degree of protonation at physiological pH of most aromatic hydroxylamines is low, and the significance of these proton-catalyzed reactions of aromatic hydroxylamines in amine carcinogenesis is still not clear.

A reinvestigation in the mid-1960s of the "free" dyes extracted from alkaline ethanolic digests of livers of dye-fed rats led to the demonstration with Scribner[104] that the "DAB" presumably obtained on administration of MAB was in fact 3-methylmercapto-MAB. This dye was inferred to be a degradation product of protein-bound 3-(methion-$S$-yl)-MAB residues in the livers of rats fed MAB or DAB. About the same time Poirier in our laboratory[105] synthesized the ester $N$-benzoyloxy-MAB as a substitute for $N$-hydroxy-MAB, for which a synthesis had not yet been devised. On incubation with methionine at neutrality in the absence of any tissue preparation, this ester yielded a polar derivative that decomposed to 3-methylmercapto-MAB.[105,106] Figure 3 shows this and other reactions of $N$-benzoyloxy-MAB as an electrophile and their relation to products derived from MAB *in vivo* in the rat liver.[105–112] Analogous studies showed that $N$-acetoxy-AAF had similar but stronger electrophilic reactivity,[106,113] while $N$-hydroxy-AAF, a nucleophile, was not reactive under these conditions. Likewise, 3-methylmercapto-AAF, which was identified as a reaction product of $N$-acetoxy-AAF with methionine,[106] was obtained on degradation of liver proteins from rats administered AAF or $N$-hydroxy-AAF.[114] Similarly, degradation of the DNA and RNA from the livers of rats given

# 1. METABOLISM OF XENOBIOTIC CHEMICALS

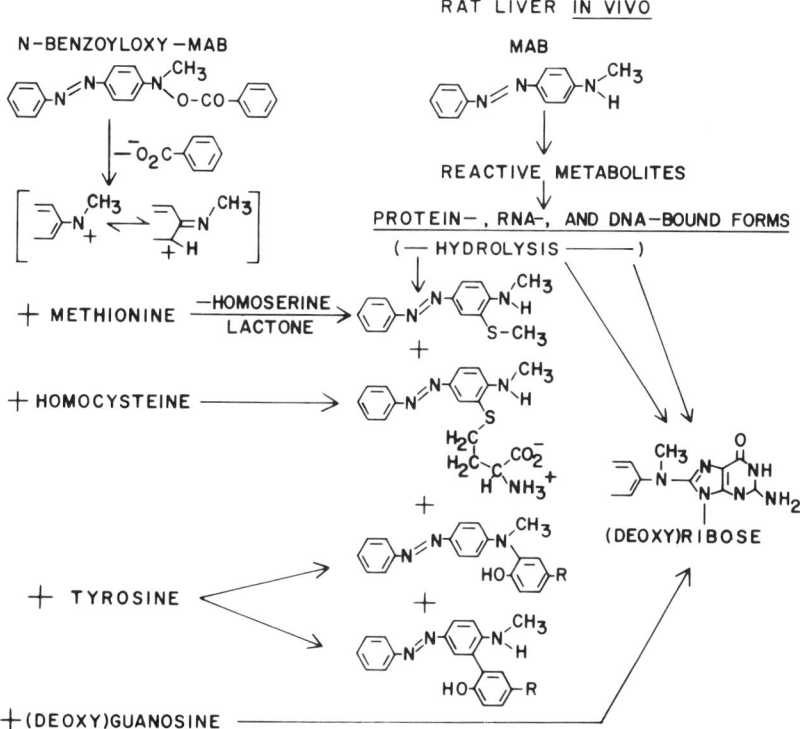

**Fig. 3.** The nonenzymatic reactions of N-benzoyloxy-MAB with methionine, tyrosine, and guanine residues.[105-110] The adducts formed in these reactions were shown by cochromatography to be identical with derivatives obtained from the liver protein or nucleic acids of rats administered MAB or DAB. Subsequently, 3-(deoxyguanosin-$N^2$-yl)-MAB[111] and 3-(deoxyadenosin-$N^6$-yl)-MAB[112] were also characterized as products formed by reaction of N-benzoyloxy-MAB with deoxyguanosine and deoxyadenosine *in vitro* and were obtained as hydrolysis products from the liver DNA of rats treated with MAB.

[9-$^{14}$C]AAF yielded products chromatographically indistinguishable from the N-(guan-8-yl)AAF derivatives obtained on reaction of N-acetoxy-AAF with deoxyguanosine and guanosine.[115-117]

These results, as well as the carcinogenicities of N-benzoyloxy-MAB and N-acetoxy-AAF in the subcutaneous tissue of the rat where MAB and AAF are not active,[105,117,118] stimulated searches for the enzymatic formation of reactive esters of N-hydroxy-AAF by tissue preparations. Studies from our laboratory[114,119] and that of King[120] soon provided evidence for the formation of an electrophilic derivative from N-hydroxy-AAF by liver cytosols supplemented with 3'-phosphoadenosine 5'-phosphosulfate (PAPS). In complementary studies the hepatic toxicity of N-hydroxy-AAF was reduced by the simultaneous administration of *p*-hydroxyacetanilide, and the hepatocarcinogenicity was inhibited by the administration of acetanilide. Both of these inhibitions were largely overcome by the administration of sodium sulfate.[121,122] These data suggested that the sulfuric acid ester of *N*-hydroxy-AAF is important in its biological activity.

However, the complexity of the situation became evident with continued study. By 1972, five enzymatic pathways for the conversion of *N*-hydroxy-AAF to electrophilic reactants had been demonstrated in rat liver (Fig. 4),[25] and the relative importance of each of these electrophilic reactants for carcinogenesis is not yet clear. In addition to the sulfotransferase reaction, these include the microsomal deacetylation of *N*-hydroxy-AAF to *N*-hydroxy-AF,[103] the formation of the *N-O*-glucuronide of *N*-hydroxy-AAF (a weak electrophilic reactant,[84,123,124] the one-electron oxidation of *N*-hydroxy-AAF by peroxidases and $H_2O_2$ to yield a free nitroxide radical and its electrophilic dismutation products *N*-acetoxy-AAF and 2-nitrosofluorene,[125] and a cytosolic acyltransferase reaction.[126,127] The latter reaction, in effect, transfers the *N*-acetyl group to the oxygen atom with the formation of *N*-acetoxy-AF, a strong electrophile. Similar reactions have also been demonstrated to occur with other *N*-hydroxyarylamides.[90] In addition, Tada and Tada[128] observed the activation of 4-hydroxyaminoquinoline 1-oxide through a seryl-tRNA synthetase-mediated reaction.

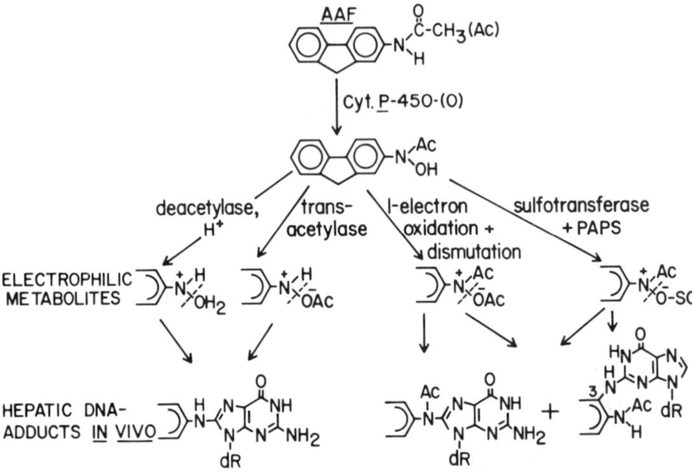

**Fig. 4.** Multiple pathways for the metabolism of AAF to strong electrophilic reactants in rat liver.[25] In addition to the reactions shown, *N*-hydroxy-AAF is metabolized to an *O*-glucuronide. The latter metabolite, although a weak electrophile, also reacts with nucleic acids to yield adducts of *N*-(guan-8-yl)-2-aminofluorene and its N-acetylated derivative. Cyt. *P*-450 stands for the complete system of cytochrome *P*-450 isozyme(s), cytochrome *P*-450 reductase, NADPH, and $O_2$.

## VII. Central Role of Electrophilic Reactants in Carcinogenesis, Mutagenesis, and Some Other Toxic Manifestations by Chemicals

By 1969 our studies on the reactivity of the esters of $N$-hydroxyamines and amides and a consideration of the literature on the known or postulated reactive forms of other chemical carcinogens led us to suggest that the active forms of most, if not all, chemical carcinogens are electrophilic reactants and that the critical reactions for carcinogenesis include those with nucleophilic groups in nucleic acids or proteins, or both.[129] By that time evidence had accrued for the metabolic formation of electrophilic alkylating species from alkylnitrosamides, alkylnitrosamines, cycasin (methyl azoxymethanol $\beta$-glucoside), alkylhydrazines, and aryl dialkyltriazenes (see Ref. 130). Pyrrolic esters were suggested as reactive intermediates in the carcinogenicity of the pyrrolizidine alkaloids,[131,132] and enzyme-catalyzed reactions had been demonstrated for the conversion of the polycyclic aromatic hydrocarbons to species that bound to nucleic acids and proteins.[133,134] Free radicals and carbonium ions had been suggested as metabolic products of carbon tetrachloride.[135,136] The electrophilic nature of metal ions was noted in relation to the carcinogenicity of some metal compounds.[129]

The deductions made in 1969 were greatly strengthened by subsequent findings. Extensive studies on a variety of polycyclic aromatic hydrocarbons point to the formation of bay region diol epoxides as major critical intermediates for their carcinogenicity,[137] but the possible importance of other electrophilic intermediates in some situations is also under study.[138,139] Epoxides also appear to be essential intermediates in the carcinogenicity of aflatoxins $B_1$ and $B_2$[140-143] and vinyl chloride.[144] Electrophilic esters were implicated from the structures of the hepatic DNA adducts as critical metabolic intermediates in the carcinogenicities of safrole and estragole,[145,146] and recent data have identified 1′-sulfooxysafrole as the major electrophilic ultimate carcinogenic metabolite of safrole in mouse liver.[147] Evidence has been presented for the formation of electrophilic derivatives of ethyl carbamate.[148,149] Carcinogenic and mutagenic metal ions reduce the fidelity of DNA polymerases.[150] Some examples of the wide variety of metabolic activation of chemical carcinogens to reactive electrophiles are shown in Fig. 5.

The mutagenic activities of the alkylating agents stimulated studies during the 1950s and 1960s of their electrophilic reactivities with biological constituents, expecially DNA.[151,152] The new finding that tissues could metabolize carcinogens *in vivo* or *in vitro* to electrophilic ultimate carcinogens implied that these electrophilic metabolites might be mutagenic. The first demonstration came from studies with Maher and Szybalski,[153] which showed that

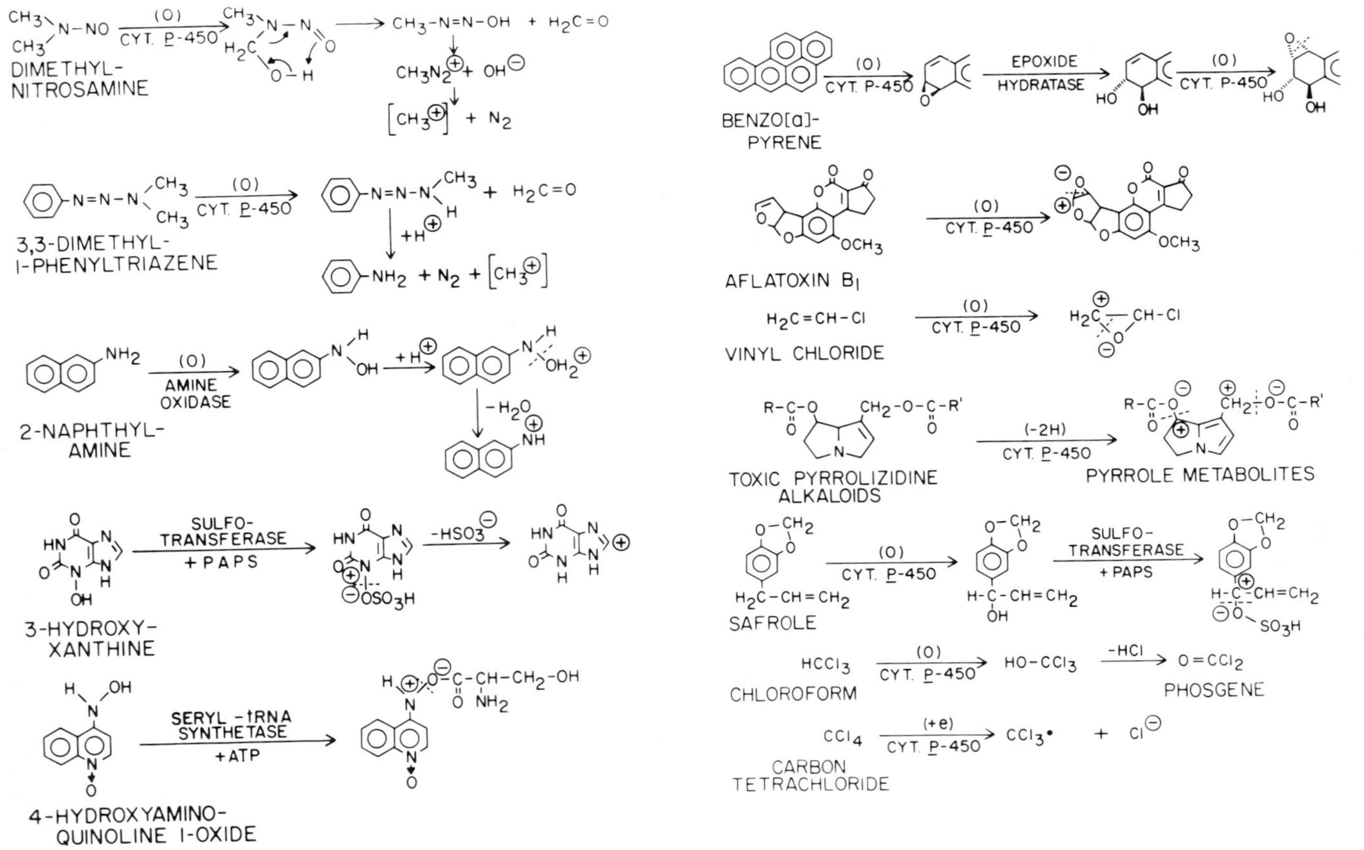


**Fig. 5.** The metabolic activation of members of a number of different classes of chemical carcinogens to reactive electrophiles.[25,144] In addition to the examples shown, some of these carcinogens are also metabolized to other electrophilic derivatives. Cyt. *P*-450 stands for the complete system of cytochrome *P*-450 isozyme(s), cytochrome *P*-450 reductase, NADPH, and $O_2$.

the hepatic metabolite AAF-$N$-sulfate ($N$-sulfoxy-AAF), as well as the synthetic esters $N$-acetoxy-AAF and $N$-benzoyloxy-MAB, is a strong mutagen in a *Bacillus subtilis*-transforming DNA system, while the related nonelectrophilic metabolites of AAF and MAB are without mutagenic effect. With the acceptance of the concept of metabolism of nonreactive chemicals to electrophilic reactants, the spectrum of chemicals for which mutagenic activity seemed possible increased considerably. These findings also suggested a new approach to examination of the potential mutagenic and carcinogenic activities of chemicals. As is now well known, *in vivo* host-mediated and *in vitro* tissue-mediated mutagenesis assays were developed for the detection of chemicals that are not mutagenic per se but can be metabolized to such derivatives.[154] These assays have demonstrated potential mutagenic activity for a wide variety of chemicals that are not mutagenic in assays that lack metabolic systems for activation to electrophilic metabolites.

In addition, metabolism to electrophilic reactants now appears to play an important role in some toxic manifestations other than carcinogenicity and mutagenicity. Examples are (1) the hepatotoxicities of acetaminophen, furosemide, halobenzenes, isoniazid, and iproniazid, (2) the bone marrow aplasia induced by benzene, (3) methemoglobinemia and hemolysis associated with aromatic amines, and (4) renal and pulmonary toxicity induced by derivatives of furan and thiophene.[155] These and further examples are discussed in detail in other chapters of this book, and electrophilic metabolites have been the focus of two international conferences on biological reactive intermediates.[156,157]

## VIII. Looking Ahead

The central theme of this chapter has been the development of the premise that in many cases the electrophilic reactivities of chemicals or their metabolites are essential to the biological events that they elicit. This unifying concept has facilitated the understanding of some phenomena, and it has had some predictive value. However, detailed understanding at the cellular and molecular levels of the adverse effects caused by electrophilic reactants is not yet available. One of the early approaches to the understanding of mutagenesis was the finding of Loveless[158] that the mutagenicity of ethylating and methylating agents was correlated with the concentrations of $O^6$-ethyl- and $O^6$-methylguanine residues that were formed in the DNA and not with the extents of reaction at N-7 of guanine residues. Detailed data, especially on mutagenesis by ethylating and methylating agents, now make it evident that, in addition to the natures of the initial lesions and their effects on the fidelity of DNA replication, the rates and fidelity of repair of the DNA of the cell have marked effects on the mutagenic responses.[154,159] Other parameters, such as the rate of cell and DNA replication (i.e., the time available for repair before replication occurs), may also affect the mutagenic response. Much more remains to be learned.

The present understanding of chemical carcinogenesis is more rudimentary (Fig. 6).[160-162] Because carcinogenesis is a multi-stage process,[160-163] it is essential to separate the activities of a chemical as an initiator of the malignant change (almost certainly dependent on changes in cellular DNA) from its role in tumor promotion, which seems at present not to result from alterations in DNA. Although much emphasis has been placed on mutagenesis as a prototype for initiation, some investigators are now probing the possibility that the critical lesions for initiation may be more complex than those generally studied as mutagenic lesions. Among the current suggestions are that rearrangements in the DNA may occur in response to reaction with electrophilic reactants and that these altered sequences may facilitate expression of cellular oncogenes.[33,34,161,164-166] It seems quite possible that major advances in understanding of the molecular mechanisms of chemical carcinogenesis lie just ahead.

It must also be kept in mind that, important as electrophilic reactions of

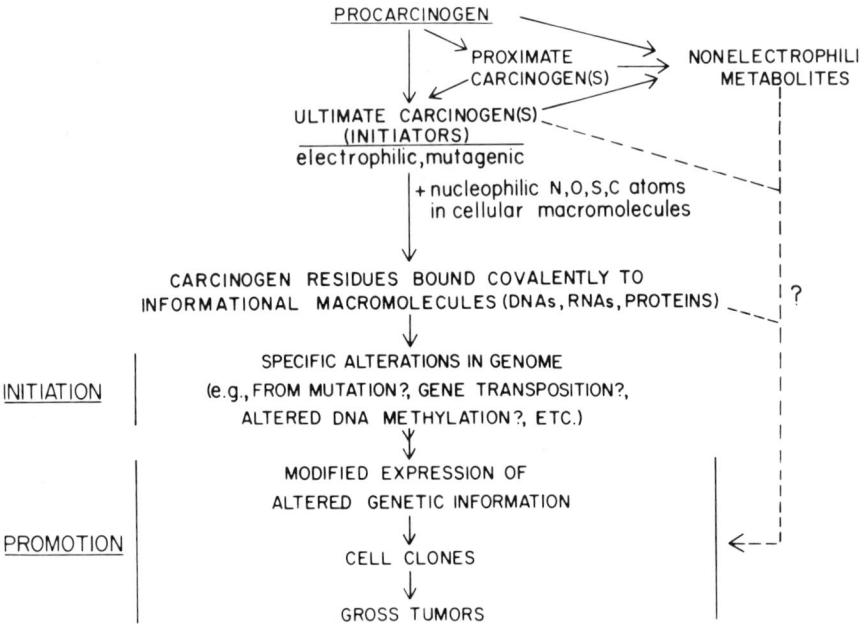

Fig. 6. A current view of molecular and cellular events in chemical carcinogenesis.

chemicals are in eliciting adverse biological effects, many toxic effects do not depend on reactions of electrophilic metabolites with target molecules. For instance, mutagenesis can be elicited by planar chemicals that, through noncovalent intercalation in DNA, cause frameshifts in the reading of DNA during its replication. Likewise, the biological effects of 2,3,7,8-tetrachloro-$p$-dioxin and its isosteres appear to provide important examples of effects mediated through noncovalent binding of the toxins to specific receptors.[167] 2,3,7,8-Tetrachloro-$p$-dioxin does not yield covalent adducts in the hepatic DNA of rats,[168] and the induction of hepatic tumors in rats exposed to this chemical for long periods now appears to be explicable in terms of its potent tumor-promoting activity.[169,170]

Toxicity is a broad term that encompasses a variety of adverse biological effects with widely different degrees of reversibility. The toxicities of electrophilic metabolites for cells are probably expressed through a range of molecular mechanisms, including depletion or inactivation of specific proteins, RNAs, or essential cofactors. At present it seems likely that, to considerable extent, the toxicities of individual chemicals or classes of chemicals are individual problems.

## Acknowledgments

We wish to acknowledge the essential contributions of our students and postdoctoral fellows at the McArdle Labortory to the research described in this chapter. Their work and ideas in our laboratory and their subsequent careers have been a constant inspiration and joy to us.

Our work in chemical carcinogenesis is presently supported by Grants CA-07175 and CA-22484 of the National Cancer Institute, USPHS.

## References

1. Cook, J. W., Hieger, I., Kennaway, E. L., and Mayneord, W. V. (1932). The production of cancer by pure hydrocarbons. Part I. *Proc. R. Soc. London, Ser. B* **111,** 455–484.
2. Kinosita, R. (1937). Studies on the cancerogenic chemical substances. *Trans. Jpn. Pathol. Soc.* **27,** 665–727.
3. Sasaki, T., and Yoshida, T. (1935). Experimentelle Erzeugung des Lebercarcinomas durch Fütterung mit o-Amidoazotoluol. *Virchow's Arch. Pathol. Anat. Physiol.* **295,** 175–200.
4. Yoshida, T. (1933). Über die serienweise Verfolgung der Veränderungen der Leber der experimentellen Hepatomerzeugung durch o-Aminoazotoluol. *Trans. Jpn. Pathol. Soc.* **23,** 636–638.
5. Hogness, T. R., and Potter, V. R. (1941). Spectrometric studies in relation to biology. *Annu. Rev. Biochem.* **10,** 509–530.
6. Consden, R., Gordon, A. H., and Martin, A. J. P. (1944). Qualitative analysis of proteins: A partition chromatographic method using paper. *Biochem. J.* **38,** 224–232.

7. Heidelberger, C., and Jones, H. B. (1948). The distribution of radioactivity in the mouse following administration of dibenzanthracene labeled in the 9 and 10 positions with carbon 14. *Cancer* **1**, 252–260.
8. Miller, J. A., Miller, E. C., and Baumann, C. A. (1945). On the methylation and demethylation of certain carcinogenic azo dyes in the rat. *Cancer Res.* **5**, 162–168.
9. Miller, J. A., and Baumann, C. A. (1945). The determination of p-dimethylaminoazobenzene, p-monomethylaminoazobenzene, and p-aminoazobenzene in tissue. *Cancer Res.* **5**, 157–161.
10. Stevenson, E. S., Dobriner, K., and Rhoads, C. P. (1942). The metabolism of dimethylaminoazobenzene (butter yellow) in rats. *Cancer Res.* **2**, 160–167.
11. Miller, J. A., and Baumann, C. A. (1945). The carcinogenicity of certain azo dyes related to p-dimethylaminoazobenzene. *Cancer Res.* **5**, 227–234.
12. Miller, J. A., and Miller, E. C. (1953). The carcinogenic aminoazo dyes. *Adv. Cancer Res.* **1**, 340–396.
13. Miller, E. C., and Miller, J. A. (1947). The presence and significance of bound aminoazo dyes in the livers of rats fed p-dimethylaminoazobenzene. *Cancer Res.* **7**, 468–480.
14. Miller, E. C., Miller, J. A., Sapp, R. W., and Weber, G. M. (1949). Studies on the protein-bound aminoazo dyes formed *in vivo* from 4-dimethylaminoazobenzene and its C-monomethyl derivatives. *Cancer Res.* **9**, 336–343.
15. Sorof, S., Cohen, P. P., Miller, E. C., and Miller, J. A. (1951). Studies on the soluble proteins from livers of rats fed aminoazo dyes. *Cancer Res.* **11**, 383–387.
16. Sorof, S., and Cohen, P. P. (1951). Electrophoretic and ultracentrifugal studies on the soluble proteins of various tumors and of livers from rats fed 4-dimethylaminoazobenzene. *Cancer Res.* **11**, 376–382.
17. Miller, E. C. (1951). Studies on the formation of protein-bound derivatives of 3,4-benzpyrene in the epidermal fraction of mouse skin. *Cancer Res.* **11**, 100–108.
18. Levine, M., and Tarver, H. (1951). Studies of ethionine. III. Incorporation of ethionine into rat proteins. *J. Biol. Chem.* **192**, 835–850.
19. Miller, E. C., and Miller, J. A. (1952). *In vivo* combinations between carcinogens and tissue constituents and their possible role in carcinogenesis. *Cancer Res.* **12**, 547–556.
20. Wiest, W. G., and Heidelberger, C. (1953). The interaction of carcinogenic hydrocarbons with tissue constituents. II. 1,2,5,6-dibenzanthracene-9,10-$C^{14}$ in skin. *Cancer Res.* **13**, 250–254.
21. Weisburger, E. K., Weisburger, J. H., and Morris, H. P. (1953). Studies on the metabolism of 2-acetylaminofluorene-9-$C^{14}$. *Arch. Biochem. Biophys.* **43**, 474–484.
22. Watson, J. D., and Crick, F. H. C. (1953). Molecular structure of nucleic acids: A structure for deoxyribose nucleic acid. *Nature (London)* **171**, 737–738.
23. Wheeler, G. P., and Skipper, H. E. (1957). Studies with mustards. III. *In vivo* fixation of $C^{14}$ from nitrogen mustard-$C^{14}H_3$ in nucleic acid fractions of animal tissues. *Arch. Biochem. Biophys.* **72**, 465–475.
24. Miller, E. C., and Miller, J. A. (1966). Mechanisms of chemical carcinogenesis: Nature of proximate carcinogens and interactions with macromolecules. *Pharmacol. Rev.* **18**, 805–838.
25. Miller, E. C., and Miller, J. A. (1981). Searches for ultimate chemical carcinogens and their reactions with cellular macromolecules. *Cancer* **47**, 2327–2345.
26. Beard, P., Kaneko, M., and Cerutti, P. (1981). N-Acetoxyacetylaminofluorene reacts preferentially with a control region of intracellular SV40 chromosomes. *Nature (London)* **291**, 84–85.
27. Kirschmeier, P., Gattoni-Celli, S., Dina, D., and Weinstein, I. B. (1982). Carcinogen- and radiation-transformed C3H 10T-1/2 cells contain RNAs homologous to the long

terminal repeat sequence of a murine leukemia virus. *Proc. Natl. Acad. Sci. U.S.A.* **79**, 2773–2777.
28. Lippke, J. A., Gordon, L. K., Brash, D. E., and Haseltine, W. E. (1981). Distribution of UV light-induced damage in a defined sequence of human DNA: Detection of alkaline-sensitive lesions at pyrimidine nucleoside–cytidine sequences. *Proc. Natl. Acad. Sci. U.S.A.* **78**, 3388–3392.
29. Fuchs, R. P. P., Schwartz, N., and Daune, M. P. (1981). Hot spots of frameshift mutations induced by the ultimate carcinogen N-acetoxy-N-2-acetylaminofluorene. *Nature (London)* **294**, 657–659.
30. Santella, R. M., Grunberger, D., Weinstein, I. B., and Rich, A. (1981). Induction of the Z conformation in poly(dG-dC) · poly(dG-dC) by binding of N-2-acetylaminofluorene to guanine residues. *Proc. Natl. Acad. Sci. U.S.A.* **78**, 1451–1455.
31. Swenson, D. H., Li, L. H., Hurley, L. H., Rokem, J. S., Petzold, G. L., Dayton, B. D., Wallace, T. L., Lin, A. H., and Krueger, W. C. (1982). Mechanism of interaction of CC-1065 (NSC 298223) with DNA. *Cancer Res.* **42**, 2821–2828.
32. Green, C. L., Loechler, E. L., Fowler, K. W., and Essigmann, J. M. (1984). Construction and characterization of extrachromosomal probes for mutagenesis by carcinogens: site-specific incorporation of $O^6$-methylguanine into viral and plasmid genomes. *Proc. Natl. Acad. Sci., U.S.A.* **81**, 13–17.
33. Balmain, A., and Pragnell, I. B. (1983). Mouse skin carcinomas induced *in vivo* by chemical carcinogens have a transforming Harvey-*ras* oncogene. *Nature (London)* **303**, 72–74.
34. Newbold, R. F., and Overell, R. W. (1983). Fibroblast immortality is a prerequisite for transformation by EJ c-Ha-ras oncogene. *Nature (London)* **304**, 648–651.
35. Potter, V. R., and Elvehjem, C. A. (1936). A modified method for the study of tissue oxidations. *J. Biol. Chem.* **114**, 495–504.
36. Claude, A. (1941). Particulate components of cytoplasm. *Cold Spring Harbor Symp. Quant. Biol.* **9**, 263–270.
37. Schneider, W. C. (1949). Intracellular distribution of enzymes. I. The distribution of succinic dehydrogenase, cytochrome oxidase, adenosine triphosphatase, and phosphorus compounds in normal rat tissues. *J. Biol. Chem.* **165**, 585–593.
38. Trager, W. F. (1980). Oxidative functionalization reactions. *Drugs Pharm. Sci.* **10**, Part A, 177–210.
39. Mueller, G. C., and Miller, J. A. (1948). The metabolism of 4-dimethylaminoazobenzene by rat liver homogenates. *J. Biol. Chem.* **176**, 535–544.
40. Mueller, G. C., and Miller, J. A. (1949). The reductive cleavage of 4-dimethylaminoazobenzene by rat liver: The intracellular distribution of the enzyme system and its requirement for triphosphopyridine nucleotide. *J. Biol. Chem.* **180**, 1125–1136.
41. Mueller, G. C., and Miller, J. A. (1953). The metabolism of methylated aminoazo dyes. II. Oxidative demethylation by rat liver homogenates. *J. Biol. Chem.* **202**, 579–587.
42. La Du, B. N., Gaudette, L., Trousof, N., and Brodie, B. B. (1955). Enzymatic dealkylation of aminopyrine (Pyramidon) and other alkylamines. *J. Biol. Chem.* **214**, 741–752.
43. Axelrod, J. (1955). The enzymatic deamination of amphetamine (Benzedrine). *J. Biol. Chem.* **214**, 753–763.
44. Mitoma, C., Posner, H. S., Reitz, H. C., and Udenfriend, S. (1956). Enzymatic hydroxylation of aromatic compounds. *Arch. Biochem. Biophys.* **61**, 431–441.
45. Booth, J., and Boyland, E. (1957). The biochemistry of aromatic amines. 3. Enzymic hydroxylation by rat-liver microsomes. *Biochem. J.* **66**, 73–78.
46. Conney, A. H., Miller, E. C., and Miller, J. A. (1957). Substrate-induced synthesis and other properties of benzpyrene hydroxylase in rat liver. *J. Biol. Chem.* **228**, 753–766.

47. Parke, D. V., and Williams, R. T. (1956). Species differences in the o- and p-hydroxylation of aniline. *Biochem. J.* **63**, 12P.
48. Brodie, B. B., Gillette, J. R., and La Du, B. N. (1958). Enzymatic metabolism of drugs and other foreign compounds. *Annu. Rev. Biochem.* **27**, 427–454.
49. Mason, H. S. (1957). Mechanisms of oxygen metabolism. *Science* **125**, 1185–1188.
50. Posner, H. S., Mitoma, C., Rothberg, S., and Udenfriend, S. (1961). Enzymic hydroxylation of aromatic compounds. III. Studies on the mechanism of microsomal hydroxylation. *Arch. Biochem. Biophys.* **94**, 280–290.
51. Garfinkel, D. (1958). Studies on pig liver microsomes. I. Enzymic and pigment composition of different microsomal fractions. *Arch. Biochem. Biophys.* **77**, 493–509.
52. Klingenberg, M. (1958). Pigments of rat liver microsomes. *Arch. Biochem. Biophys.* **75**, 376–386.
53. Coon, M. J., and Persson, A. V. (1980). Microsomal cytochrome *P*-450: A central catalyst in detoxication reactions. *In* "Enzymatic Basis of Detoxication" (W. B. Jakoby, ed.), Vol. 1, pp. 117–134. Academic Press, New York.
54. Lu, A. Y. H., and West, S. B. (1980). Multiplicity of mammalian microsomal cytochromes *P*-450. *Pharmacol. Rev.* **31**, 277–295.
55. Wislocki, P. G., Miwa, G. T., and Lu, A. Y. H. (1980). Reactions catalyzed by the cytochrome *P*-450 system. *In* "Enzymatic Basis of Detoxication" (W. B. Jakoby, ed.), Vol. 1, pp. 135–182. Academic Press, New York.
56. Vainio, H., and Hietanen, E. (1980). Role of extrahepatic metabolism. *Drugs Pharm. Sci.* **10**, Part A, 251–284.
57. Ziegler, D. M. (1980). Microsomal flavin-containing monooxygenase: Oxygenation of nucleophilic nitrogen and sulfur compounds. *In* "Enzymatic Basis of Detoxication" (W. B. Jakoby, ed.), Vol. 1, pp. 201–227. Academic Press, New York.
58. Eling, T., Boyd, J., and Sivarajah, K. (1982). Oxidation of chemical carcinogens by prostaglandin synthetase. *In* "Prostaglandins and Cancer: First International Conference" (T. J. Powles, R. S. Bockman, K. V. Honn, and P. Ramwell, eds.), pp. 113–122. Alan R. Liss, Inc., New York.
59. Marnett, L. J. (1981). Polycyclic aromatic hydrocarbon oxidation during prostaglandin biosynthesis. *Life Sci.* **29**, 531–546.
60. Zenser, T. V., Cohen, S. M., Mattammal, M. B., Murasaki, G., and Davis, B. B. (1982). Role of prostaglandin endoperoxide synthetase in benzidine and 5-nitrofuran-induced kidney and bladder carcinogenesis. *In* "Prostaglandins and Cancer: First International Conference" (T. J. Powles, R. S. Bockman, K. V. Hohn, and P. Ramwell, eds.), pp. 123–141. Alan R. Liss, Inc., New York.
61. Brown, R. R., Miller, J. A., and Miller, E. C. (1954). The metabolism of methylated aminoazo dyes. IV. Dietary factors enhancing demethylation *in vitro*. *J. Biol. Chem.* **209**, 211–222.
62. Richardson, H. L. Stier, A. R., and Borsos-Nachtnebel, E. (1952). Liver tumor inhibition and adrenal histologic responses in rats to which 3′-methyl-4-dimethylaminoazobenzene and 20-methylcholanthrene were simultaneously administered. *Cancer Res.* **12**, 356–361.
63. Miller, E. C., Miller, J. A., Brown, R. R., and MacDonald, J. C. (1958). On the protective action of certain polycyclic aromatic hydrocarbons against carcinogenesis by aminoazo dyes and 2-acetylaminofluorene. *Cancer Res.* **18**, 469–477.
64. Conney, A. H., Miller, E. C., and Miller, J. A. (1956). The metabolism of methylated aminoazo dyes. V. Evidence for induction of enzyme synthesis in the rat by 3-methylcholanthrene. *Cancer Res.* **16**, 450–459.
65. Conney, A. H., and Gilman, A. E. (1963). Puromycin inhibition of enzyme induction by 3-methylcholanthrene and phenobarbital. *J. Biol. Chem.* **238**, 3682–3685.

66. Gelboin, H. V., and Sokoloff, L. (1961). Effects of 3-methylcholanthrene and phenobarbital on amino acid incorporation in protein. *Science* **134**, 611–612.
67. Gelboin, H. V., and Blackburn, N. R. (1963). The stimulatory effect of 3-methylcholanthrene on microsomal amino acid incorporation and benzpyrene hydroxylase activity and its inhibition by actinomycin D. *Biochim. Biophys. Acta* **72**, 657–660.
68. Gelboin, H. V. (1964). Studies on the mechanism of methylcholanthrene induction of enzyme activities. II. Stimulation of microsomal and ribosomal amino acid incorporation: The effects of polyuridylic acid and actinomycin D. *Biochim. Biophys. Acta* **91**, 130–144.
69. von der Decken, A., and Hultin, T. (1960). Inductive effects of 3-methylcholanthrene on enzyme activities and amino acid incorporation capacity of rat liver microsomes. *Arch. Biochem. Biophys.* **90**, 201–207.
70. Remmer, H., and Allenben, B. (1958). Die Aktivierung der Entgiftung in den Lebermikrosomen während der Gewöhnung. *Klin. Wochenschr.* **36**, 332–333.
71. Conney, A. H., and Burns, J. J. (1959). Stimulatory effect of foreign compounds on ascorbic acid biosynthesis and on drug-metabolizing enzymes. *Nature (London)* **184**, 363–364.
72. Conney, A. H. (1967). Pharmacological implications of microsomal enzyme induction. *Pharmacol. Rev.* **19**, 317–366.
73. Kasper, C. B., and Henton, D. (1980). Glucuronidation. In "Enzymatic Basis of Detoxication" (W. B. Jakoby, ed.), Vol. 2, pp. 3–36. Academic Press, New York.
74. Oesch, F. (1980). Microsomal epoxide hydrolase. In "Enzymatic Basis of Detoxication" (W. B. Jakoby, ed.), Vol. 2, pp. 277–290. Academic Press, New York.
75. Albert, A. (1981). *Selective Toxicity*, 6th edition. Chapman & Hall, London.
76. Stella, V. (1975). Pro-drugs: An overview and definition. In "Pro-drugs as Novel Drug Delivery Systems" (T. Higuchi and V. Stella, eds.), pp. 1–115. Am. Chem. Soc., Washington, D.C.
77. Albert, A. (1958). Chemical aspects of selective toxicity. *Nature (London)* **182**, 421–423.
78. Murphy, S. P., and Du Bois, K. P. (1958). The influence of various factors on the enzymatic conversion of organic thiophosphates to anti-cholinesterase agents. *J. Pharmacol. Exp. Ther.* **124**, 194–202.
79. Diggle, W. M., and Gage, J. C. (1951). Cholinesterase inhibition by parathion *in vivo*. *Nature (London)* **168**, 998.
80. Gage, J. C. (1953). A cholinesterase inhibitor derived from $OO$-diethyl $O$-$p$-nitrophenyl, thiophosphate *in vivo*. *Biochem. J.* **54**, 426–430.
81. Peters, R. A. (1952). The puzzle for therapy in fluoroacetate poisoning. *Br. Med. J.* **2**, 1165–1170.
82. Gelboin, H. V., Miller, J. A., and Miller, E. C. (1958). The formation *in vitro* of protein-bound derivatives of aminoazo dyes by rat liver and its enhancement by benzpyrene pretreatment. *Biochim. Biophys. Acta* **27**, 655–656.
83. Gelboin, H. V., Miller, J. A., and Miller, E. C. (1959). The *in vitro* formation of protein-bound derivatives of aminoazo dyes by rat liver preparations. *Cancer Res.* **19**, 975–985.
84. Cramer, J. W., Miller, J. A., and Miller, E. C. (1960). N-Hydroxylation: A new metabolic reaction observed in the rat with the carcinogen 2-acetylaminofluorene. *J. Biol. Chem.* **235**, 885–888.
85. Miller, J. A., Cramer, J. W., and Miller, E. C. (1960). The N- and ring-hydroxylation of 2-acetylaminofluorene during carcinogenesis in the rat. *Cancer Res.* **20**, 950–962.
86. Weisburger, E. K., and Weisburger, J. H. (1958). Chemistry, carcinogenicity and metabolism of 2-fluorenamine and related compounds. *Adv. Cancer Res.* **5**, 331–431.
87. Miller, E. C., Miller, J. A., and Hartmann, H. A. (1961). *N*-Hydroxy-2-acetylamino-

fluorene: A metabolite of 2-acetylaminofluorene with increased carcinogenic activity in the rat. *Cancer Res.* **21,** 815–824.
88. Miller, E. C., Miller, J. A., and Enomoto, M. (1964). The comparative carcinogenicities of 2-acetylaminofluorene and its *N*-hydroxy metabolite in mice, hamsters, and guinea pigs. *Cancer Res.* **24,** 2018–2032.
89. Miller, J. A., and Miller, E. C. (1967). The activation of carcinogenic amines and amides by N-hydroxylation *in vivo*. *In* "Carcinogenesis: A Broad Critique" (M. Mandel, ed.), pp. 397–420. University of Texas, Austin.
90. Clayson, D. B., and Garner, R. C. (1976). Carcinogenic aromatic amines and related compounds. *ACS Monogr.* **173,** 366–461.
91. Tada, M., and Tada, M. (1976). Metabolic activation of 4-nitroquinoline 1-oxide and its binding to nucleic acid. *In* "Fundamentals of Cancer Prevention" (P. N. Magee, S. Takayama, T. Sugimura, and T. Matsushima, eds.), pp. 217–227. University Park Press, Baltimore, Maryland.
92. Howard, P. C., Heflich, R. H., Evans, F. E., and Beland, F. A. (1983). Formation of DNA adducts *in vitro* and in *Salmonella typhimurium* upon metabolic reduction of the environmental mutagen 1-nitropyrene. *Cancer Res.* **43,** 2052–2058.
93. Irving, C. C. (1964). Enzymatic N-hydroxylation of the carcinogen 2-acetylaminofluorene and the metabolism of *N*-hydroxy-2-acetylaminofluorene-9-$^{14}$C *in vitro*. *J. Biol. Chem.* **239,** 1589–1596.
94. Lotlikar, P. D., Enomoto, M., Miller, E. C., and Miller, J. A. (1967). Species variations in the N- and ring-hydroxylation of 2-acetaminofluorene and effects of 3-methylcholanthrene pretreatment. *Proc. Soc. Exp. Biol. Med.* **125,** 341–346.
95. Lotlikar, P. D., and Zaleski, K. (1975). Ring- and N-hydroxylation of 2-acetaminofluorene by rat liver reconstituted cytochrome *P*-450 enzyme system. *Biochem. J.* **150,** 561–564.
96. Miyaji, T., Moszkowski, L. I., Senoo, T., Ogata, M., Odo, T., Kawai, K., Sayama, Y., Ishida, H., and Matsuo, H. (1953). Inhibition of 2-acetylaminofluorene tumors in rats with simultaneously fed 20-methylcholanthrene; 9:10-dimethyl-1:2-benzanthracene and chrysene, and consideration of sex difference in tumor genesis with 2-acetylaminofluorene. *Gann* **44,** 281–283.
97. Razzouk, C., Mercier, M., and Roberfroid, M. (1980). Induction, activation, and inhibition of hamster and rat liver microsomal arylamide and arylamine *N*-hydroxylase. *Cancer Res.* **40,** 3540–3546.
98. Brookes, P., and Lawley, P. D. (1964). Alkylating agents. *Br. Med. Bull.* **20,** 91–95.
99. Magee, P. N., and Barnes, J. M. (1956). The production of malignant primary hepatic tumours in the rat by feeding dimethylnitrosamine. *Br. J. Cancer* **10,** 114–122.
100. Magee, P. N., and Hultin, T. (1962). Toxic liver injury and carcinogenesis. Methylation of proteins of rat-liver slices by dimethylnitrosamine *in vitro*. *Biochem. J.* **83,** 106–114.
101. Magee, P. N., and Farber, E. (1962). Toxic liver injury and carcinogenesis. Methylation of rat-liver nucleic acids by dimethylnitrosamine *in vivo*. *Biochem. J.* **83,** 114–124.
102. Kriek, E. (1965). On the interactions of *N*-2-fluorenylhydroxylamine with nucleic acids *in vitro*. *Biochem. Biophys. Res. Commun.* **20,** 793–799.
103. Irving, C. C. (1966). Enzymatic deacetylation of *N*-hydroxy-2-acetylaminofluorene by liver microsomes. *Cancer Res.* **26,** 1390–1396.
104. Scribner, J. D., Miller, J. A., and Miller, E. C. (1965). 3-Methylmercapto-*N*-methyl-4-aminoazobenzene: An alkaline-degradation product of a labile protein-bound dye in the livers of rats fed *N,N*-dimethyl-4-aminoazobenzene. *Biochem. Biophys. Res. Commun.* **20,** 560–565.
105. Poirier, L. A., Miller, J. A., Miller, E. C., and Sato, K. (1967). *N*-Benzoyloxy-*N*-methyl-

4-aminoazobenzene: Its carcinogenic activity in the rat and its reactions with proteins and nucleic acids and their constituents *in vitro*. *Cancer Res.* **27**, 1600–1613.
106. Lotlikar, P. D., Scribner, J. D., Miller, J. A., and Miller, E. C. (1966). Reaction of esters of aromatic *N*-hydroxy amines and amides with methionine *in vitro:* A model for *in vivo* binding of amine carcinogens to protein. *Life Sci.* **5**, 1263–1269.
107. Lin, J.-K., Miller, J. A., and Miller, E. C. (1968). Studies on structures of polar dyes derived from the liver proteins of rats fed *N*-methyl-4-aminoazobenzene. II. Identity of synthetic 3-(homocystein-*S*-yl)-*N*-methyl-4-aminoazobenzene with the major polar dye P2b. *Biochemistry* **7**, 1889–1895.
108. Lin, J.-K., Miller, J. A., and Miller, E. C. (1969). Studies on structures of polar dyes derived from the liver proteins of rats fed *N*-methyl-4-aminoazobenzene. III. Tyrosine and homocysteine sulfoxide polar dyes. *Biochemistry* **8**, 1573–1582.
109. Lin, J.-K., Schmall, B., Sharpe, I. D., Miura, I., Miller, J. A., and Miller, E. C. (1975). N-Substitution of C-8 in guanosine and deoxyguanosine by the carcinogen *N*-benzoyloxy-*N*-methyl-4-aminoazobenzene *in vitro*. *Cancer Res.* **35**, 832–843.
110. Lin, J.-K., Miller, J. A., and Miller, E. C. (1975). Structures of hepatic nucleic acid-bound dyes in rats given the carcinogen *N*-methyl-4-aminoazobenzene. *Cancer Res.* **35**, 844–850.
111. Beland, F. A., Tullis, D. L., Kadlubar, F. F., Straub, K. M., and Evans, F. E. (1980). Characterization of DNA adducts of the carcinogen *N*-methyl-4-aminoazobenzene *in vitro* and *in vivo*. *Chem.-Biol. Interact.* **31**, 1–17.
112. Tullis, D. L., Straub, K. M., and Kadlubar, F. F. (1981). A comparison of the carcinogen–DNA adducts formed in rat liver *in vivo* after administration of single or multiple doses of *N*-methyl-4-aminoazobenzene. *Chem.-Biol. Interact.* **38**, 15–27.
113. Miller, E. C., Juhl, U., and Miller, J. A. (1966). Nucleic acid guanine: Reaction with the carcinogen *N*-acetoxy-2-acetylaminofluorene. *Science* **153**, 1125–1127.
114. DeBaun, J. R., Miller, E. C., and Miller, J. A. (1970). *N*-Hydroxy-2-acetylaminofluorene sulfotransferase: Its probable role in carcinogenesis and protein–(methion-*S*-yl) binding in rat liver. *Cancer Res.* **30**, 577–595.
115. Kriek, E., Miller, J. A., Juhl, U., and Miller, E. C. (1967). 8(*N*-fluorenylacetamido)guanosine, an arylamidation reaction product of guanosine and the carcinogen *N*-acetoxy-*N*-fluorenylacetamide in neutral solution. *Biochemistry* **6**, 177–182.
116. Kriek, E. (1969). On the mechanism of action of carcinogenic aromatic amines. I. Binding of 2-acetylaminofluorene and *N*-hydroxy-2-acetylaminofluorene to rat liver nucleic acids *in vivo*. *Chem.-Biol. Interact.* **1**, 3–17.
117. Miller, J. A., and Miller, E. C. (1969). The metabolic activation of carcinogenic aromatic amines and amides. *Prog. Exp. Tumor Res.* **11**, 273–301.
118. Bartsch, H., Malaveille, C., Stich, H. F., Miller, E. C., and Miller, J. A. (1977). Comparative electrophilicity, mutagenicity, DNA repair induction activity, and carcinogenicity of some *N*- and *O*-acyl derivatives of *N*-hydroxy-2-aminofluorene. *Cancer Res.* **37**, 1461–1467.
119. DeBaun, J. R., Rowley, J. Y., Miller, E. C., and Miller, J. A. (1968). Sulfotransferase activation of *N*-hydroxy-2-acetylaminofluorene in rodent livers susceptible and resistant to this carcinogen. *Proc. Soc. Exp. Biol. Med.* **129**, 268–273.
120. King, C. M., and Phillips, B. (1968). Enzyme-catalyzed reactions of the carcinogen *N*-hydroxy-2-fluorenylacetamide with nucleic acid. *Science* **159**, 1351–1353.
121. DeBaun, J. R., Smith, J. Y. R., Miller, E. C., and Miller, J. A. (1970). Reactivity *in vivo* of the carcinogen *N*-hydroxy-2-acetylaminofluorene: Increase by sulfate ion. *Science* **167**, 184–186.
122. Weisburger, J. H., Yamamoto, R. S., Williams, G. M., Grantham, P. H., Matsushima, T.,

and Weisburger, E. K. (1972). On the sulfate ester of N-hydroxy-N-2-fluorenylacetamide as a key ultimate hepatocarcinogen in the rat. *Cancer Res.* **32**, 491–500.
123. Irving, C. C., Veazey, R. A., and Hill, J. T. (1969). Reaction of the glucuronide of the carcinogen N-hydroxy-2-acetylaminofluorene with nucleic acids. *Biochim. Biophys. Acta* **179**, 189–198.
124. Miller, E. C., Lotlikar, P. D., Miller, J. A., Butler, B. W., Irving, C. C., and Hill, J. T. (1968). Reactions *in vitro* of some tissue nucleophiles with the glucuronide of the carcinogen N-hydroxy-2-acetylaminofluorene. *Mol. Pharmacol.* **4**, 147–154.
125. Bartsch, H., and Hecker, E. (1971). On the metabolic activation of the carcinogen N-hydroxy-N-2-acetylaminofluorene. III. Oxidation with horseradish peroxidase to yield 2-nitrosofluorene and N-acetoxy-N-2-acetylaminofluorene. *Biochim. Biophys. Acta* **237**, 567–578.
126. Bartsch, H., Dworkin, M., Miller, J. A., and Miller, E. C. (1972). Electrophilic N-acetoxyaminoarenes derived from carcinogenic N-hydroxy-N-acetylaminoarenes by enzymatic deacetylation and transacetylation in liver. *Biochim. Biophys. Acta* **286**, 272–298.
127. King, C. M., and Allaben, W. T. (1980). Arylhydroxamic acid acyltransferase. *In* "Enzymatic Basis of Detoxication" (W. B. Jakoby, ed.), Vol. 2, pp. 187–197. Academic Press, New York.
128. Tada, M., and Tada, M. (1975). Seryl-tRNA synthetase and activation of the carcinogen 4-nitroquinoline 1-oxide. *Nature (London)* **255**, 510–512.
129. Miller, J. A., and Miller, E. C. (1969). Metabolic action of carcinogenic aromatic amines and amides via N-hydroxylation and N-hydroxy esterification and its relationship to ultimate carcinogens as electrophilic reactants. *Jerusalem Symp. Quantum Chem. Biochem.* **1**, pp. 237–261.
130. Magee, P. N., Montesano, R., and Preussmann, R. (1976). N-Nitroso compounds and related carcinogens. *ACS Monogr.* **173**, 491–625.
131. Mattocks, A. R. (1968). Toxicity of pyrrolizidine alkaloids. *Nature (London)* **217**, 723–728.
132. Culvenor, C. C. J., Downing, D. T., Edgar, J. A., and Jago, M. V. (1969). Pyrrolizidine alkaloids as alkylating and antimitotic agents. *Ann. N. Y. Acad. Sci.* **163**, 837–847.
133. Gelboin, H. V. (1969). A microsome-dependent binding of benzo[*a*]pyrene to DNA. *Cancer Res.* **29**, 1272–1276.
134. Grover, P. L., and Sims, P. (1969). Enzyme-catalyzed reactions of polycyclic hydrocarbons with deoxyribonucleic acid and protein *in vitro*. *Biochem. J.* **110**, 159–160.
135. Reynolds, E. S. (1967). Liver parenchymal cell injury. IV. Pattern of incorporation of carbon and chlorine from carbon tetrachloride into chemical constituents *in vivo*. *J. Pharmacol. Exp. Ther.* **155**, 117–126.
136. Butler, T. C. (1961). Reduction of carbon tetrachloride *in vivo* and reduction of carbon tetrachloride and chloroform *in vitro* by tissues and tissue constituents. *J. Pharmacol. Exp. Ther.* **134**, 311–319.
137. Jerina, D. M., Sayer, J. M., Thakker, D. R., Yagi, H., Levin, W., Wood, A. W., and Conney, A. H. (1980). Carcinogenicity of polycyclic aromatic hydrocarbons: The bay-region theory. *In* "Carcinogenesis: Fundamental Mechanisms and Environmental Effects" (B. Pullman, P.O.P. Ts'o, and H. Gelboin, eds.), pp. 1–12. Reidel Publ., Dordrecht, Netherlands.
138. Cavalieri, E., Roth, R., and Rogan, E. G. (1976). Metabolic activation of aromatic hydrocarbons by one-electron oxidation in relation to the mechanism of tumor initiation. *In* "Carcinogenesis" (R. I. Freudenthal and P. W. Jones, eds.), Vol. 1, pp. 181–190. Raven Press, New York.

139. Watabe, T., Ishizuka, T., Isobe, M., and Ozawa, N. (1982). A 7-hydroxymethyl sulfate ester as an active metabolite of 7,12-dimethylbenz[a]anthracene. *Science* **215**, 403–405.
140. Essigmann, J. M., Croy, R. G., Nazdan, A. M., Busby, W. F., Jr., Reinhold, V. N., Buchi, G., and Wogan, G. N. (1977). Structural identification of the major DNA adduct formed by aflatoxin $B_1$ *in vitro*. *Proc. Natl. Acad. Sci. U.S.A.* **74**, 1870–1874.
141. Martin, C. N., and Garner, R. C. (1977). Aflatoxin B-oxide generated by chemical or enzymic oxidation of aflatoxin $B_1$ causes guanine substitution in nucleic acids. *Nature (London)* **267**, 863–865.
142. Swenson, D. H., Lin, J.-K., Miller, E. C., and Miller, J. A. (1977). Aflatoxin $B_1$ 2,3-oxide as a probable intermediate in the covalent binding of aflatoxins $B_1$ and $B_2$ to rat liver DNA and ribosomal RNA *in vivo*. *Cancer Res.* **37**, 172–181.
143. Lin, J.-K., Miller, J. A., and Miller, E. C. (1977). 2,3-Dihydro-2-(guan-7-yl)-3-hydroxyaflatoxin $B_1$, a major acid hydrolysis product of aflatoxin $B_1$–DNA or –rRNA adducts formed in hepatic microsome-mediated reactions and in rat liver *in vivo*. *Cancer Res.* **37**, 4430–4438.
144. Zajdela, F., Croisy, A., Barbin, A., Malaveille, C., Tomatis, L., and Bartsch, H. (1980). Carcinogenicity of chloroethylene oxide, an ultimate reactive metabolite of vinyl chloride, and bis(chloromethyl)ether after subcutaneous administration and in initiation–promotion experiments in mice. *Cancer Res.* **40**, 352–356.
145. Phillips, D. H., Miller, J. A., Miller, E. C., and Adams, B. (1981). The $N^2$-atom of guanine and the $N^6$-atom of adenine residues as sites for covalent binding of metabolically activated 1′-hydroxysafrole to mouse-liver DNA *in vivo*. *Cancer Res.* **41**, 2664–2671.
146. Phillips, D. H., Miller, J. A., Miller, E. C., and Adams, B. (1981). Structures of the DNA adducts formed in mouse liver after administration of the proximate hepatocarcinogen 1′-hydroxyestragole. *Cancer Res.* **41**, 176–181.
147. Boberg, E. W., Miller, E. C., Miller, J. A., Poland, A., and Liem, A. (1983). Strong evidence from studies with brachymorphic mice and pentachlorophenol that 1′-sulfoöxysafrole is the major ultimate electrophilic and carcinogenic metabolite of 1′-hydroxysafrole in mouse liver. *Cancer Res.* **43**, 5163–5173.
148. Ribovich, M. L. Miller, J. A., Miller, E. C., and Timmins, L. G. (1982). Labeled 1,$N^6$-ethenoadenosine and 3,$N^4$-ethenocytidine in hepatic RNA of mice given ethyl-1,2-[$^3$H] or ethyl-1-[$^{14}$C]ethyl carbamate (urethan). *Carcinogenesis* **3**, 539–546.
149. Scherer, E., Steward, A. P., and Emmelot, P. (1980). Formation of precancerous islands in rat liver and modification of DNA by ethyl carbamate: Implications for its metabolism. *Dev. Toxicol. Environ. Sci.* **8**, 249–254.
150. Loeb, L. A., and Zakour, R. A. (1980). Metals and genetic miscoding. *In* "Nucleic Acid–Metal Ion Interactions" (T. G. Spiro, ed.), pp. 115–144. Wiley, New York.
151. Ross, W. C. J. (1962). "Biological Alkylating Agents." Butterworth, London.
152. Price, C. C., Gaucher, G. M., Koneru, P., Shibakawa, R., Sowa, J. R., and Yamaguchi, M. (1969). Mechanism of action of alkylating agents. *Ann. N. Y. Acad. Sci.* **163**, 593–598.
153. Maher, V. M., Miller, E. C., Miller, J. A., and Szybalski, W. (1968). Mutations and decreases in density of transforming DNA produced by derivatives of the carcinogens 2-acetylaminofluorene and N-methyl-4-aminoazobenzene. *Mol. Pharmacol.* **4**, 411–426.
154. Hollaender, A., and de Serres, F. J. eds. (1971–1982). Chemical mutagens. *In* "Principles and Methods for their Detection," Vols. 1–7. Plenum, New York.
155. Gillette, J. R., Mitchell, J. R., and Brodie, B. B. (1974). Biochemical mechanisms of drug toxicity. *Annu. Rev. Pharmacol.* **14**, 271–288.
156. Jollow, D. J., Kocsis, J. J., Snyder, R., and Vainio, H., eds. (1977). "Biological Reactive Intermediates." Plenum, New York.

157. Snyder, R., Parke, D. V., Kocsis, J. J., Jollow, D. J., Gibson, C. G., and Witmer, C. M., eds. (1981). "Biological Reactive Intermediates-II. Plenum, New York.
158. Loveless, A. (1969). Possible relevance of O-6 alkylation of deoxyguanosine to the mutagenicity and carcinogenicity of nitrosamines and nitrosamides. *Nature (London)* **223**, 206–207.
159. Singer, B., and Kuśmierek, J. T. (1982). Chemical mutagenesis. *Annu. Rev. Biochem.* **52**, 655–693.
160. Miller, E. C. (1978). Some current perspectives on chemical carcinogenesis in humans and experimental animals: Presidential address. *Cancer Res.* **38**, 1479–1496.
161. Weinstein, I. B., Yamasaki, H., Wigler, M., Lee, L.-S., Fisher, P. B., Jeffrey, A., and Grunberger, D. (1978). Molecular and cellular events associated with the action of initiating carcinogens and tumor promoters. *In* "Carcinogens, Identification and Mechanisms of Action" (A. C. Griffin and C. R. Shaw, eds.), pp. 399–418. Raven Press, New York.
162. Weinstein, I. B. (1981). Current concepts and controversies in chemical carcinogenesis. *J. Supramol. Struct. Cell. Biochem.* **17**, 99–120.
163. Slaga, T. J., Sivak, A., and Boutwell, R. K., eds. (1978). Carcinogenesis- Vol. 2. Mechanisms of Tumor Promotion and Cocarcinogenesis. Raven Press, New York.
164. Cairns, J. (1981). The origin of human cancers. *Nature (London)* **289**, 353–357.
165. Klein, G., and Klein, E. (1984). Oncogene activation and tumor progression. *Carcinogenesis* **5**, 429–435.
166. Bishop, J. M. (1983). Cellular oncogenes and retroviruses. *Annu. Rev. Biochem.* **52**, 301–354.
167. Poland, A., Greenlee, W. F., and Kende, A. S. (1979). Studies on the mechanism of action of chlorinated dibenzo-$p$-dioxins and related compounds. *Ann. N. Y. Acad. Sci.* **320**, 214–230.
168. Poland, A., and Glover, E. (1979). An estimate of the maximum *in vivo* covalent binding of 2,3,7,8-tetrachlorodibenzo-$p$-dioxin to rat liver protein, ribosomal RNA, and DNA. *Cancer Res.* **39**, 3341–3344.
169. Kociba, R. J., Keyes, D. G., Beyer, J. E., Carreon, R. M., Wade, C. E., Dittenber, D. A., and Humiston, C. G. (1978). Results of two-year chronic toxicity and oncogenicity study of 2,3,7,8-tetrachlorodibenzo-$p$-dioxin in rats. *Toxciol. Appl. Pharmacol.* **46**, 279–303.
170. Pitot, H. C., Goldsworthy, T., Campbell, H. A., and Poland, A. (1980). Quantitative evaluation of the promotion by 2,3,7,8-tetrachlorodibenzo-$p$-dioxin of hepatocarcinogenesis from diethylnitrosamine. *Cancer Res.* **40**, 3616–3620.

*Chapter 2*

# Pharmacokinetics of Biological Activation and Inactivation of Foreign Compounds

JAMES R. GILLETTE

*Laboratory of Chemical Pharmacology*
*National Heart, Lung, and Blood Institute*
*National Institutes of Health*
*Bethesda, Maryland*

|  |  |
|---|---|
| I. Introduction. | 30 |
| II. Fundamental Pharmacokinetic Concepts. | 32 |
|     A. Relationship between Intracellular and Extracellular Concentrations at Steady State. | 33 |
|     B. Two-Compartment Systems. | 36 |
|     C. Repeated Administration. | 41 |
|     D. Area under the Blood Concentration vs Time Curves. | 43 |
|     E. Range of Maximum and Minimum Concentrations within Cells. | 43 |
|     F. Organ Clearances and Total Body Clearances. | 44 |
| III. Concentration of the Parent Compound in Arterial Blood after Different Routes of Administration. | 48 |
| IV. Concentration of the Parent Compound in Blood Exiting Various Organs in the Central Compartment after Different Routes of Administration. | 50 |
| V. Kinetics of the Formation of Chemically Reactive Metabolites by Organs in the Central Compartment | 54 |
|     A. Short-Lived Chemically Reactive Metabolites. | 54 |
|     B. Intermediate-Lived Metabolites. | 57 |
|     C. Long-Lived Metabolites | 60 |
| VI. General Comments | 69 |
|     References | 70 |

## I. Introduction

It has long been recognized that the biological effects of a foreign compound may be caused not only by the foreign compound itself but also by one or more of its metabolites. Frequently the metabolite causes the same effects as the parent substance; sometimes the metabolite causes effects that differ from those of the parent substance. Occasionally, however, the parent substance is biologically inert, and thus the effects manifested after the administration of the parent compound are due solely to the metabolites.

In most instances investigators can prove that a metabolite evokes biological effects by administering the metabolite and comparing its effects with those of the parent compound. However, in some instances this approach fails, because the metabolite is poorly absorbed, is converted to the parent substance, or is very rapidly converted to other substances in the body by enzymatic or nonenzymatic reactions.

Although many biologically active metabolites exert their actions by combining reversibly with receptor sites, others react covalently with putative action sites, as described elsewhere in this book. Indeed, it has become increasingly apparent that many toxicities caused by foreign compounds are mediated through the formation of such chemically reactive metabolites. However, the term *chemically reactive metabolite* may be applied to substances having half-lives ranging from milliseconds to several hours. It is useful, therefore, to think of chemically reactive metabolites of foreign compounds as belonging to one or more of the categories.

Ultrashort-lived metabolites never leave the enzyme that catalyzes their formation. However, they are rapidly converted at the enzyme site to relatively stable metabolites and frequently react covalently with the enzyme and thereby inactivate the enzyme. Their formation is thus most easily detected by demonstrating an irreversible loss of enzyme activity that can be slowed by the presence of other substrates of the enzyme but not reversed by addition of an alternative substrate after inactivation of the enzyme has occurred. The addition of nucleophiles or other possible traps of the metabolite, however, should not affect the rate of inactivation, because such substances should react with the metabolite only after it has left the enzyme. The precursors of such metabolites have been called "suicide enzyme inhibitors" and, owing to their presumed enzyme specificity, are potentially useful as drugs in chemotherapy. Indeed several drugs currently used clinically are thought to act through the formation of ultrashort-lived metabolites.

Short-lived metabolites leave the enzymes that catalyze their formation, but not the cells in which they are formed. They either rapidly react covalently with cellular constituents such as proteins and nucleic acids, or are rapidly converted to other metabolites by enzymatic and nonenzymatic mechanisms. They may frequently be detected by measuring the covalent binding of radiolabeled material to macromolecules either *in vitro* or *in vivo*. However, unlike the covalent binding of ultrashort-lived metabolites, the covalent binding of these metabolites *in vitro* may be altered by addition or removal of various nucleophiles or enzymes that inactivate the reactive metabolite. Because these reactive metabolites are restricted to the cells in which they are formed, any toxicity that depends directly on the reaction of the chemically reactive metabolite with target substances within cells will be restricted to those cells that contain enzymes that convert the precursor to the active metabolite.

Intermediate-lived metabolites leave the cells where they are formed but rapidly enter other cells within the same organ or are inactivated in blood to such an extent that only trivial amounts enter cells in other organs of the body. Although the number of chemically reactive metabolites within this group will be small, members may theoretically be detected by measuring covalently bound material in blood initially and at various times after the withdrawal of the blood. With members in this class, there should be no change in the covalent binding to blood components after withdrawal of venous blood unless the reactive metabolite is formed in the blood. Presumably most of the toxicities caused by these metabolites would be restricted to organs containing the enzymes that catalyze the formation of the metabolites, but the metabolite may cause damage to cells within the organ that lack the activating enzyme. Thus, it is plausible that metabolites in this category that are generated solely in the liver may cause DNA damage in liver, but not in the ovaries or testes, which could result in heritable changes.

Long-lived metabolites leave the organ in which they are formed and are distributed to other organs in the body, but are not rapidly excreted into urine or bile. Members of this group of metabolites may frequently be detected by showing that inducers and inhibitors of the activating enzyme cause parallel effects on covalently bound material in hepatic and extrahepatic tissues. The active metabolite may also be detected by observing changes in covalent binding to blood components after withdrawal of the blood, or it may be isolated from the blood. Because the metabolite may be transported to most organs of the body, virtually any organ potentially may manifest toxicity caused by the metabolite. Thus, it is plausible that metabolites in this category could cause damage in sensitive cells of organs that do not generate the metabolite.

Ultralong-lived metabolites are eliminated from the body mainly by excretion into urine or bile. Those eliminated in urine thus may usually be detected in freshly voided urine. Because their concentrations are likely to be greater in bile or urine than in blood, the toxicities caused by this group of metabolites may occur predominantly in the bladder or in the intestines.

Because the classification depends on the half-life of the metabolite within cells and organ, it is important to realize that a given metabolite may be a short-lived metabolite in one tissue or situation and a long-lived metabolite in another tissue or situation. Moreover, a given precursor may be converted to several different chemically reactive metabolites by either parallel reactions or sequential reactions. Thus, one or more of the metabolites may belong to the short-lived metabolite category, whereas other metabolites may belong to the long-lived and ultralong-lived metabolite categories. In this context a short-lived metabolite may be inactivated to a stable metabolite that is transported to another organ, where it is either converted back to the original short-lived metabolite (as can happen with the hydrolysis of certain glucuronides) or converted to another short-lived metabolite. In this way, inducers that promote the formation of the initial short-lived metabolite in liver may increase the amount of covalently bound material in extrahepatic tissues. Thus, a short-lived metabolite can be misclassed as a long-lived metabolite.

## II. Fundamental Pharmacokinetic Concepts

Pharmacokinetic equations express the relationships between the many factors that govern the amounts of a parent drug and its metabolites in different pharmacokinetic compartments of the body, including those that contain cells that generate chemically reactive metabolites. Indeed, when properly assessed, they permit us to make inferences about factors that are not readily apparent. In order to gain a perspective of these relationships, therefore, it is necessary to understand some of the fundamental concepts of the pharmacokinetics of a parent compound and its metabolites.

As a first step, it is useful to consider the sequence of events that govern the concentration of a parent compound in arterial blood, because virtually all pharmacokinetic equations imply the use of arterial blood concentrations as a reference point.

If a parent compound were rapidly injected into the aorta of an animal, its concentration in aortic blood would initially oscillate as the blood perfuses the various organs of the body, returns to the heart and lung, and is recycled Within a very few minutes, however, the magnitude of the oscillations would diminish and the concentration of the compound in aortic blood would then decline in what frequently appears to be a biexponential or triexponential manner. If the concentration of the parent compound appears to decline biexponentially, the kinetics may be simulated by a two-compartment system, and if it appears to decline triexponentially, the kinetics may be simulated by a three-compartment system.

During the oscillation phase as well as during the other phases, many events occur, including the passage of the drug into and out of the cells of various organs. The amount of the parent drug within given sets of cells in the various organs increases, reaches a maximum, and then declines. At the time when the amount of drug in a given set of cells reaches a maximum, the rate at which the parent compound enters the cells will equal the rate at which it leaves the cells, both by transfer from the cells back into the blood and by its conversion to first-generation metabolites. At the time a maximum concentration of the parent compound is achieved within the cells, therefore, the reactions within the cells are said to be in steady state.

## A. RELATIONSHIP BETWEEN INTRACELLULAR AND EXTRACELLULAR CONCENTRATIONS AT STEADY STATE

Many of the factors that govern the steady-state concentration of a parent compound within cells are illustrated in Fig. 1. In the interest of simplicity, the presence of capillary walls and extracellular space separating the blood and the cytoplasm within the cells has been ignored. As shown in Fig. 1, the parent compound (D, drug) may enter cells from the blood within a zone of an organ by active transport (act-trans) of the unbound compound or by passive diffusion (dif) of the compound; usually the rate of passive diffusion is predominantly governed by the concentration of the nonionized form of the compound within the blood. After the compound enters the cells, it may be reversibly bound to receptor sites within the cells. The compound may leave the cells by active transport of the unbound compound back into the blood or into another fluid (not shown). In addition, the compound may leave the cells by metabolism (met) of the unbound compound or by passive diffusion mainly of the nonionized compound back into blood. The rate at which the foreign compound enters the cell may therefore be described by

$$\text{Rate}_{D(in)} = \text{Rate}_{D(act\text{-}trans,\,in)} + \text{Rate}_{D(dif,\,in)} \qquad (1a)$$

where

$$\text{Rate}_{D(act\text{-}trans,\,in)} = f_{fD}^{zone}[D]^{zone} CL_{D(act\text{-}trans,\,in)}^{cell(f)} \qquad (1b)$$

and

$$\text{Rate}_{D(dif,\,in)} = f_{fD}^{zone}[D]^{zone} f_{nD}^{zone} CL_{D(dif,\,in)}^{cell(n)} \qquad (1c)$$

## FREE OR TOTAL CONCENTRATION?

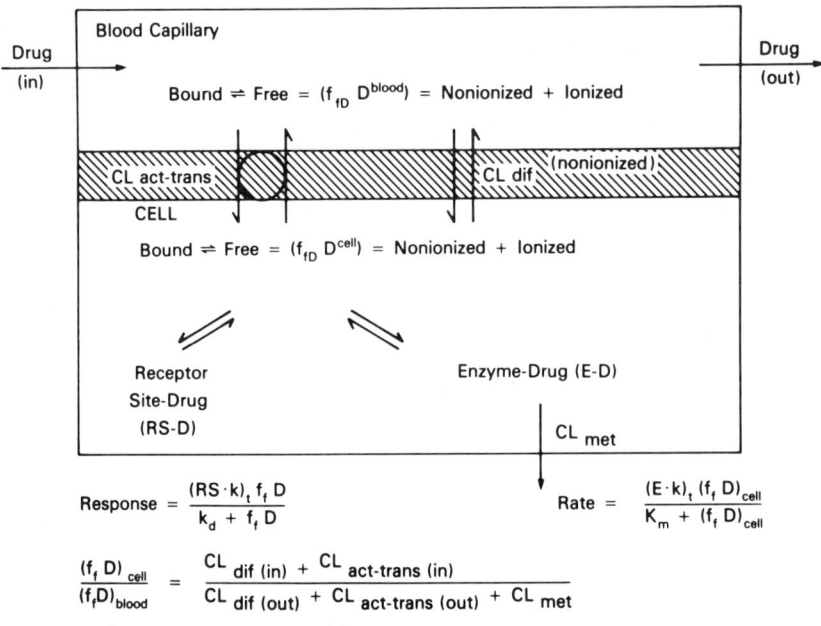

Fig. 1. Factors that govern the relationship between the steady-state concentration of a parent compound in cells and in blood.

whereas the rate at which the parent compound leaves the cell may be described by

$$\text{Rate}_{D(out)} = \text{Rate}_{D(act\text{-}trans,\ out)} + \text{Rate}_{D(dif,\ out)} + \text{Rate}_{D(met)} \quad (2a)$$

where

$$\text{Rate}_{D(act\text{-}trans,\ out)} = f_{fD}^{cell}[D]^{cell} CL_{D(act\text{-}trans,\ out)}^{cell(f)} \quad (2b)$$

$$\text{Rate}_{D(dif,\ out)} = f_{fD}^{cell}[D]^{cell} f_{nD}^{cell} CL_{D(dif,\ out)}^{cell(n)} \quad (2c)$$

$$\text{Rate}_{D(met)} = f_{fD}^{cell}[D]^{cell} CL_{D(met)}^{cell(f)} \quad (2d)$$

In these equations $f_{fD}$ is the unbound fraction of the compound and $f_{fD} f_{nD}$ is the nonionized fraction of the compound. $CL_D^{(f)}$ is the rate of the various processes divided by the concentration of the unbound compound, and $CL_{D(dif)}^{(n)}$ is the ratio of passive diffusion divided by the concentration of the nonionized form of the compound. Under steady-state conditions the rate

## 2. PHARMACOKINETICS OF FOREIGN COMPOUNDS

at which the foreign compound enters the cell must equal the rate at which it leaves the cell, that is, Eq. (1a) equals Eq. (2a). Thus, the ratio of the total concentration of the foreign compound in the cell to the total concentration of the compound in the blood serving the cell may be expressed as

$$\frac{[D]^{cell}}{[D]^{zone}} = \frac{f_{fD}^{zone}[CL_{D(act\text{-}trans, in)}^{cell(f)} + f_{nD}^{zone}CL_{D(dif, in)}^{cell(n)}]}{f_{fD}^{cell}[CL_{D(act\text{-}trans, out)}^{cell(f)} + f_{nD}^{cell}CL_{D(dif, out)}^{cell(n)} + CL_{D(met)}^{cell(f)}]}$$

$$= CL_{D(in)}^{zone}/CL_{D(out)}^{cell} = R_D^{cell} \tag{3a}$$

Thus, the ratio of the total concentration of the drug in the cell to that in the blood may be visualized as a partition ratio ($R_D^{cell}$) that depends on all of the factors that affect the concentration of the foreign compound. However, the ratio of the unbound concentrations in the cell and the blood will be independent of the reversible binding of the foreign compound to components in the blood and in the cell.

$$\frac{[D]^{cell(f)}}{[D]^{zone(f)}} = \frac{f_{fD}^{cell}[D]^{cell}}{f_{fD}^{zone}[D]^{zone}}$$

$$= \frac{CL_{D(act\text{-}trans, in)}^{cell(f)} + f_{nD}^{zone}CL_{D(dif, in)}^{cell(n)}}{CL_{D(act\text{-}trans, out)}^{cell(f)} + f_{nD}^{cell}CL_{D(dif, out)}^{cell(n)} + CL_{D(met)}^{cell(f)}} \tag{3b}$$

When the compound is not metabolized in the cell and when it enters and leaves the cell solely by passive diffusion of its nonionized form, Fick's Law of diffusion across thin membranes[1] predicts that at equilibrium, the concentration of the nonionized form outside and inside the cell would be the same. For this to occur, the clearance of the nonionized form into the cell must equal the clearance of the nonionized form out of the cell, that is,

$$CL_{D(dif, in)}^{cell(n)} = CL_{D(dif, out)}^{cell(n)}$$

The total concentration of the unbound compound, however, would depend on the ionized form as well as the nonionized form and, therefore, on the $pK_a$ of the compound and the intracellular and extracellular pH values. According to a modification of the Henderson–Hasselbalch equation, the fraction of the total unbound concentration that exists in the nonionized form ($f_n$) may be calculated from the relationships shown in Eqs. (4a) and (4b). For acids

$$f_{nD} = \frac{1}{1 + 10^{pH-pK_a}} \tag{4a}$$

and for bases,

$$f_{nD} = \frac{1}{1 + 10^{pK_a-pH}} \tag{4b}$$

For a more complete analysis of transfer mechanisms see Gillette and Pang.[2]

Simple though these relationships may be, they nevertheless illustrate several important principles in pharmacokinetics: (1) Although reversible binding of the parent compound to intracellular components may affect the time at which the maximum concentration of the unbound form of the parent compound is reached, it does not affect the relationship between the maximum intracellular concentration of the unbound form and the concentration of the unbound form of the compound in blood at that time, provided that there are not intervening compartments between the blood and the cell. (2) If the active-transport clearance either into the cell or out of the cell greatly exceeds the diffusional clearance of the unbound form, the relationships between the intracellular and extracellular concentrations of the unbound form will be obscure and probably highly variable because of the variability of the active-transport clearances in a given population of animals or humans. When such cells contain receptor sites that initiate a pharmacological response of the compound, attempts to correlate the blood concentrations with the intracellular responses are likely to be disappointing. (3) Although the diffusional clearance of the nonionized form is almost invariably greater than the diffusional clearance of the ionized form, the value of $f_n$ will be exceedingly small at pH values that markedly differ from the $pK_a$ of the compound. Under these conditions the rate of diffusion of the ionized form may exceed the rate of diffusion of the nonionized form. The ratios predicted by Eqs. (3a) and (3b) thus should be considered as maximal estimates. (4) The degree to which enzymes in cells alter the concentration of the compound depends not only on the activity of the enzymes, but also on the active-transport clearances and diffusional clearances of the compound out of the cells. Measurements of enzyme activity alone are not sufficient to determine the effects of metabolism on the intracellular concentration of the compound. It is, nevertheless, evident that a given set of activities of the enzymes will affect the intracellular concentration of a polar compound, which does not pass membranes readily, more than it will affect the intracellular concentration of a lipid-soluble compound. (5) Because the pH value of cells is usually somewhat lower than that of blood, the "ion-trapping" effect will be greater with bases than with acids. Indeed, the unbound concentration of acids will be lower in cells than in blood.

## B. TWO-COMPARTMENT SYSTEMS

After the maximum intracellular concentration of the compound is achieved, the concentration of the compound in blood declines and a concentration gradient is established, which governs the rate at which the compound leaves the cell. Thus, the ratio of the intracellular to the blood concentrations increases, until a maximum is reached, after which time the ratio of the rates of decline in the cells and blood remain constant provided that all processes follow first-order kinetics.

If the maximum concentration of the parent compound within a set of cells within an organ is achieved before the first blood sample for the assay of the drug in blood, the apparent volume ($V_{D_1}$) of the central compartment, as estimated from the relationship $V_{D_1} = \text{Dose}_D/[D]_{1(\text{initial})}$, will exceed the volume of the blood, because such cells would be a part of the central compartment. Indeed, it is frequently assumed by kineticists that the well-perfused organs, including the stomach, intestines, liver, lung, kidneys, various endocrine glands, and frequently the brain, are a part of the central compartment.

Because most foreign compounds are eliminated from the body almost entirely by some combination of these organs, the elimination of foreign compounds is almost invariably represented by a rate constant ($k_{10}$) emanating from the central compartment of either a two-compartment or a three-compartment model. The pharmacokinetic equation for a two-compartment model (Fig. 2) for total concentrations of a parent compound in arterial (art) blood is thus usually written as[3]:

$$[D]^{\text{art}} = \left[\frac{\text{Dose}_D^{\text{ia}}}{V_1}\right]\left[\frac{(k_{21} - \alpha)e^{-\alpha t}}{\beta - \alpha} + \frac{(k_{21} - \beta)e^{-\beta t}}{\alpha - \beta}\right] \quad (5)$$

and for the deep compartment as:

$$[D]^{\text{deep}} = \left[\frac{\text{Dose}_D^{\text{ia}} k_{12}}{V_2}\right]\left[\frac{e^{-\alpha t}}{\beta - \alpha} + \frac{e^{-\beta t}}{\alpha - \beta}\right] \quad (6)$$

where $V_1$ and $V_2$ are the apparent volumes of distribution of the compound

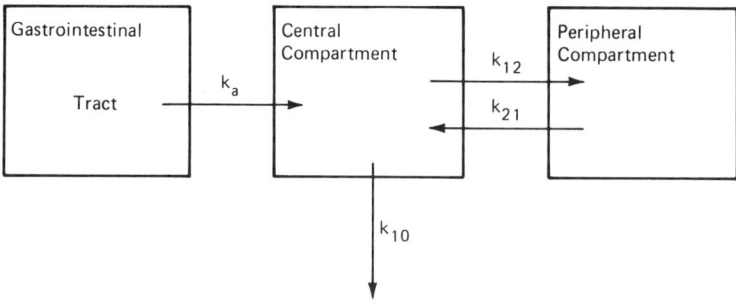

Fig. 2. Diagram of a two-compartment system with absorption of a compound from the gastrointestinal tract.

in the central and peripheral compartments, respectively, and $\text{Dose}_D^{ia}$ is the intraarterially administered dose of the compound. In these equations,

$$\alpha = 1/2[(k_{12} + k_{21} + k_{10}) + \sqrt{(k_{12} + k_{21} + k_{10})^2 - 4k_{21}k_{10}}] \quad (7a)$$

and

$$\beta = 1/2[(k_{12} + k_{21} + k_{10}) - \sqrt{(k_{12} + k_{21} + k_{10})^2 - 4k_{21}k_{10}}] \quad (7b)$$

The values of $\alpha$ and $\beta$ depend on all of the microconstants that describe the two-compartment model and are not particularly enlightening. However, other forms of the equations provide some insight into their meaning. For example, $\alpha$, may also be written as

$$\alpha = 1/2[(k_{12} + k_{21} + k_{10}) + \sqrt{k_{12}^2 + 2k_{12}(k_{21} + k_{10}) + (k_{21} + k_{10})^2 - 4k_{21}k_{10}}] \quad (8a)$$

or

$$\alpha = 1/2[(k_{12} + k_{21} + k_{10}) + \sqrt{k_{12}^2 + 2k_{12}(k_{21} + k_{10}) + (k_{21} - k_{10})^2}] \quad (8b)$$

or

$$\alpha = \frac{k_{12}}{2}\left[1 + \frac{k_{21}}{k_{12}} + \frac{k_{10}}{k_{12}} + \sqrt{1 + 2\left[\frac{k_{21}}{k_{12}} + \frac{k_{10}}{k_{12}}\right] + \left[\frac{k_{21}}{k_{12}} - \frac{k_{10}}{k_{12}}\right]^2}\right] \quad (8c)$$

Because $k_{10} = CL_{10}/V_1$ and $k_{12} = CL_{12}/V_1$, it follows that $k_{10}/k_{12} = CL_{10}/CL_{12}$. Moreover, if we set $CL_{21}$ equal to $CL_{12}$, it follows that $k_{21}/k_{12} = V_1/V_2$. Therefore

$$\alpha = \frac{k_{12}}{2}\left[1 + \frac{V_1}{V_2} + \frac{CL_{10}}{CL_{12}}\right.$$

$$\left. + \sqrt{1 + 2\left[\frac{V_1}{V_2} + \frac{CL_{10}}{CL_{12}}\right] + \left[\frac{V_1}{V_2} - \frac{CL_{10}}{CL_{12}}\right]^2}\right] \quad (8d)$$

This form of the equation provides insight into the importance of the relative volumes of distribution of the substance and the relative clearances emanating from the central compartment and of the transfer of the compound from the central compartment into the deep compartment. However, it should be stressed that the $V_2$ in this equation is defined as though there were no active-transport systems, no ion trapping, and no metabolism of the compound. It is not necessarily due solely to the actual volume of cells and the reversible binding of the compound to intracellular components.

When either $k_{21} \gg k_{10}$ of $k_{10} \gg k_{21}$, the equation for $\alpha$ degenerates to simple forms, because under these conditions $4k_{21}k_{10}$ will be negligible compared with $(k_{12} + k_{21} + k_{10})^2$. For example, when $k_{21} \gg k_{10}$, $\alpha = k_{21} + k_{12}$ and $\beta = k_{21}k_{10}/(k_{12} + k_{21})$. When $k_{10} \gg k_{21}$, $\alpha = k_{12} + k_{10}$ and $\beta = k_{21}k_{10}/(k_{12} + k_{10})$. The conditions under which these approximations are valid are readily recognized, because $\alpha$ will be much greater than $\beta$. Such approximations are especially useful in simplifying equations and estimating ranges of possible values.

In a two-compartment system, the time at which the compound reaches a maximum would be predicted by the relationship

$$t_{max} = \frac{[\ln(\alpha/\beta)]}{(\alpha - \beta)} \tag{9}$$

which may be inserted into Eq. (6). By a series of transformations, Eq. (6) becomes

$$[D]_{max}^{deep} = \left[\frac{\text{Dose}_D^{ia} k_{12}}{V_2 \alpha}\right] \left[\frac{\alpha}{\beta}\right]^{-\beta/(\alpha-\beta)} \tag{10}$$

Similarly,

$$[D]_{(at\, D_2\, max)}^{art} = \left[\frac{\text{Dose}_D^{ia} k_{21}}{V_1 \alpha}\right] \left[\frac{\alpha}{\beta}\right]^{-\beta/(\alpha-\beta)} \tag{11}$$

The ratio of Eq. (10) divided by Eq. (11) gives

$$\frac{[D]_{max}^{deep}}{[D]_{(at\, D_2\, max)}^{art}} = \frac{k_{12} V_1}{k_{21} V_2} = \frac{CL_{12}}{CL_{21}} \tag{12}$$

Equation (12) is analogous to Eq. (3a) in demonstrating that, at the time a maximum concentration is reached in the deep compartment, the ratio of the concentration in the deep compartment to that in blood depends on the ratio of the clearances into and out of the compartment.

It should be emphasized, however, that the deep compartment represents the mean set of similar, but not necessarily identical, compartments, which represent the distribution of a major portion of the compound into organs and tissues of the body. As such, it does not represent accurately any given set of cells in the body. Moreover, a small set of cells, which never contain more than a trivial proportion of the total amount of the compound in the body, may not be represented by either the central compartment or the peripheral compartment (in a two-compartment system) or compartments (in multiple-compartment systems). Thus, the maximum concentration of the compound in a given set of cells cannot be predicted with accuracy solely by studies of the arterial concentration of the compound. Indeed, a plot of the unbound concentrations in different sets of cells against time may appear

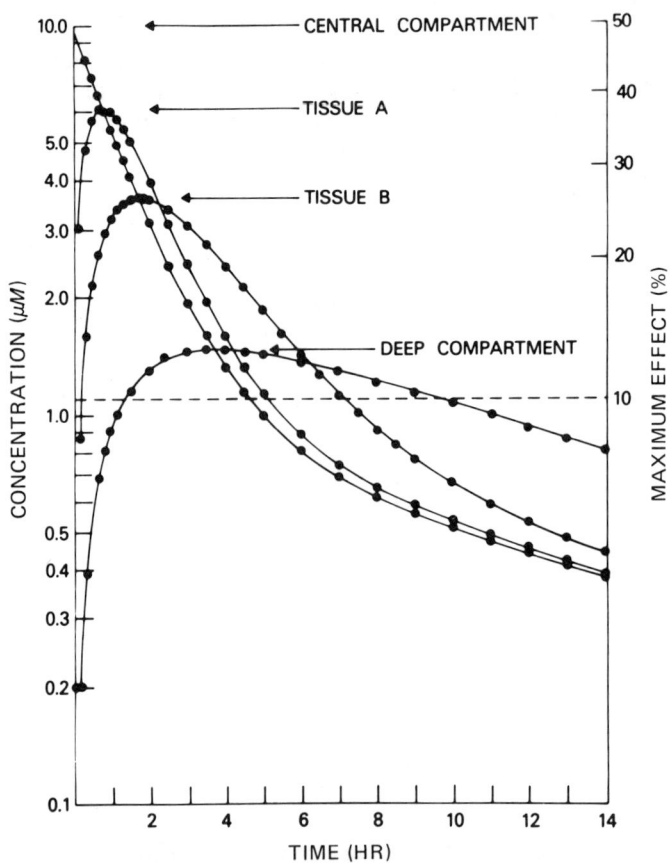

**Fig. 3.** A simulation of the concentrations of a hypothetical parent compound in various compartments after a single intraaortic administration of the compound. The system includes the central and deep compartments of a two-compartment system and two hypothetical small compartments. The scale on the right represents the fraction of receptor sites that theoretically should be occupied by the compound when the equilibrium constant of the receptor sites equals the initial concentration of the compound in the central compartment.

to be similar to the simulated concentration time curves shown in Fig. 3. Clearly, however, the maximum concentration of unbound drug in cells will never be greater than its highest concentration in blood (in the case of rapidly injected compounds, the initial concentration) unless an active transport, extensive ion trapping, or, in the case of the kidney, extensive water absorption from the glomerular filtrate occur.

## C. REPEATED ADMINISTRATION

When a compound is injected intraarterially in repeated dosages at a constant dosage interval ($\tau$), the equation for a two-compartment model becomes[3]

$$[D]^{art} = \left[\frac{Dose_D^{ia}}{V_1}\right]\left[\left[\frac{1-e^{-n\alpha\tau}}{1-e^{-\alpha\tau}}\right]\left[\frac{(k_{21}-\alpha)e^{-\alpha t}}{\beta-\alpha}\right]\right.$$
$$\left. +\left[\frac{1-e^{-n\beta}}{1-e^{-\beta\tau}}\right]\left[\frac{(k_{21}-\beta)e^{-\beta t}}{\alpha-\beta}\right]\right] \quad (13)$$

As $n$ approaches infinity, the system approaches a quasi-steady state in which the entire dose would be eliminated from the body during the dosage interval. Under such steady-state (ss) conditions the equation becomes

$$[D]^{art}_{ss} = \left[\frac{Dose_D^{ia}}{V_1}\right]\left[\left[\frac{1}{1-e^{-\alpha\tau}}\right]\left[\frac{(k_{21}-\alpha)e^{-\alpha t}}{\beta-\alpha}\right]\right.$$
$$\left. +\left[\frac{1}{1-e^{-\beta\tau}}\right]\left[\frac{(k_{21}-\beta)e^{-\beta t}}{\alpha-\beta}\right]\right] \quad (14)$$

Although the system theoretically can never actually achieve a steady state, the system is usually accepted as being in steady state when $n\beta\tau > 3.0$. Under these conditions the values of $e^{-n\beta\tau}$ would be less than 0.05. The steady-state concentration of the compound in the deep compartment during multidose administration may be expressed by similar equations:

$$[D]^{deep}_{ss} = \left[\frac{Dose_D^{ia} k_{12}}{V_2(\beta-\alpha)}\right]\left[\frac{e^{-\alpha t}}{[1-e^{-\alpha\tau}]} - \frac{e^{-\beta t}}{[1-e^{-\beta\tau}]}\right] \quad (15)$$

Under quasi-steady-state conditions, the ratio of the concentration of the compound in the deep compartment to that in the arterial blood at any given time during the dosage interval may be obtained by dividing Eq. (15) by Eq. (14). At the end of the dosage interval, however, the $\alpha$ phase may have been virtually completed, in which case the ratio of the equations becomes

$$\frac{[D]^{deep(\tau)}_{ss(min)}}{[D]^{art(\tau)}_{ss(min)}} = \frac{k_{12} V_1}{(k_{21}-\beta) V_2} \quad (16)$$

Thus, the minimum (min) concentration of unbound compound in the deep compartment at the end of the dosage interval will always be at least slightly higher than its minimum concentration in the arterial blood. This is also intuitively logical, because at the end of the dosage interval during steady state, a concentration gradient of unbound drug between the deep compartment and blood must be established in order for the compound to diffuse from the deep compartment to the blood. When $k_{21} \gg k_{10}$ or when $k_{10} \gg k_{21}$, estimation of the range of possible values for these minimum concentrations is possible.

$$\frac{CL_{12}}{CL_{21}} < \frac{[D]_{ss(min)}^{deep(\tau)}}{[D]_{ss(min)}^{art(\tau)}} < \frac{CL_{12} + CL_{10}}{CL_{21}} \qquad (17)$$

It should be evident from Eqs. (12) and (17) that the maximum and minimum concentrations of unbound compound in arterial blood frequently provide the range of possible unbound concentrations in the tissues represented by the deep compartment. This general rule may not be valid, however, when there are active-transport systems, differences between $f_n^{zone}$ and $f_n^{cells}$, or considerable metabolism in the cells representing a part of the deep compartment. With these restrictions in mind, it is also possible to demonstrate that the rule would also be true for cells representing compartments that are not a part of either the central compartment or the peripheral

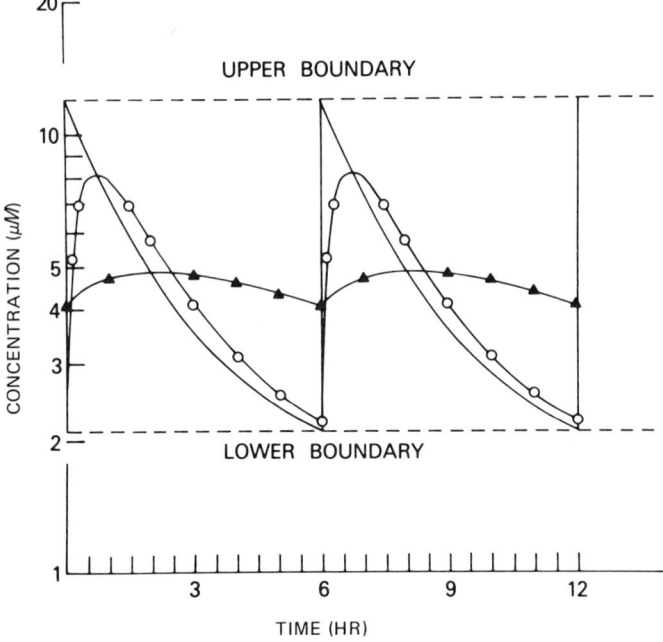

**Fig. 4.** A simulation of the concentrations of a hypothetical parent compound at a pseudo–steady state achieved during multidose administration intraaortically of the compound. The system includes the central compartment (———), the deep compartment (▲—▲), and a hypothetical small compartment (○—○).

## 2. PHARMACOKINETICS OF FOREIGN COMPOUNDS

compartment,[4] even though the presence of such compartments are not detectable by studying the decrease in arterial blood concentrations (Figs. 3 and 4).

### D. AREA UNDER THE BLOOD CONCENTRATION VS TIME CURVES

Integration of Eqs. (5) and (6) with respect to time gives the area under the concentration vs time curve ($AUC_D$) in the central and deep compartments after a single dose. Thus,

$$AUC_D^{art} = \frac{Dose_D^{ia}}{V_1 k_{10}} = \frac{Dose^{ia}}{CL_D^{body}} \tag{18}$$

$$AUC_D^{deep} = \frac{Dose_D^{ia} k_{12}}{V_2 k_{21} k_{10}} = \frac{Dose^{ia} CL_{12}}{CL_D^{body} CL_{21}} \tag{19}$$

Similarly, integration of Eqs. (14) and (15) with respect to time from $t = 0$ to $t = \tau$ gives the areas under the concentration vs time curves in the central and deep compartments under quasi-steady-state conditions.

$$AUC_D^{art(ss)} = \frac{Dose_D^{ia}}{V_1 k_{10}} = \frac{Dose_D^{ia}}{CL_D^{body}} \tag{20}$$

$$AUC_D^{deep(ss)} = \frac{Dose_D^{ia} k_{12}}{V_2 k_{21} k_{10}} = \frac{Dose^{ia} CL_{12}}{CL_D^{body} CL_{21}} \tag{21}$$

Comparison of Eq. (18) with Eq. (19) and comparison of Eq. (20) with Eq. (21) reveals that the ratio $AUC_D^{deep}/AUC_D^{art}$ will depend on the ratio of the clearances into and out of the deep compartment. Moreover, a comparison of Eq. (18) with Eq. (20) and a comparison of Eq. (19) with Eq. (21) reveals that the areas under the concentration vs time curves after a single dose should be the same as that during a dose interval in a system in steady state.

### E. RANGE OF MAXIMUM AND MINIMUM CONCENTRATIONS WITHIN CELLS

Dividing both $AUC_D^{art(ss)}$ and $AUC_D^{deep(ss)}$ by $\tau$ gives average steady-state concentrations in the arterial blood and the deep compartments, respectively. If lines were drawn on the curves during the dosage interval at these values, the maximum concentrations must be above the lines and the minimum concentrations must be below the lines. If other compartments were visualized, it would be possible to show that ratio of the areas under the concentration vs time curves in the compartment and blood would also be dependent on the relative clearances of the compound into and out of the compartments.

Thus, if the clearances of unbound compound into and out of a compartment were identical, the maximum concentration in the compartment must be somewhere between the maximum concentration of unbound compound in the arterial circulation and $AUC_{ss}^{art}/\tau$, regardless of the depth of the compartment. Similarly, the minimum concentration of the unbound drug in the compartment must be between $AUC_D^{art(ss)}/\tau$ and the minimum concentration of unbound compound within the arterial circulation. Accordingly, under the prescribed conditions, the range of possible concentrations of unbound compounds within the various cells in the body may be estimated.

## F. ORGAN CLEARANCES AND TOTAL BODY CLEARANCES

### 1. General

In the above equations, $CL_D^{body}$ represents the total rate of elimination of the parent compound by all organs in the central compartment at any given time divided by the total concentration of the compound in the arterial blood at that time. For some organs, such as the kidney, the compound enters the organ solely from the arterial blood, and the organ clearance for such organs may be portrayed rather simply. For such mono-input organs,

$$CL_D^{organ} = \frac{\text{Rate(in)} - \text{Rate(out)}}{[D]^{art}} = \frac{Q^{organ}[D]^{art} - Q^{organ}[D]^{out}}{[D]^{art}}$$

$$= \frac{Q^{organ}([D]^{art} - [D]^{out})}{[D]^{art}} = Q^{organ}\frac{(1 - [D]^{out})}{[D]^{art}} \quad (22a)$$

Setting

$$[D]^{out}/[D]^{art} = F_D^{organ} \quad \text{and} \quad 1 - F_D^{organ} = E_D^{organ}$$

we arrive at

$$CL_D^{organ} = Q^{organ}(1 - F_D^{organ}) = Q^{organ}E_D^{organ} \quad (22b)$$

where $F_D^{organ}$ is the organ availability of the compound, $E_D^{organ}$ is the organ extraction ratio of the compound; and, $Q$ is the blood flow rate through the organ.

However, some organs, notably the liver and the lungs, are perfused by both arterial blood and venous blood, some of which may have perfused other organs capable of eliminating the compound from the body. Thus, the rate of entrance of the compound into the organ is not necessarily given

# 2. PHARMACOKINETICS OF FOREIGN COMPOUNDS

by the total blood flow rate through the organ times the arterial blood concentration. Nevertheless, we may take advantage of the fact that the concentrations of the compound in all of the organs in the central compartment are almost instantaneously in steady state with the concentration of the compound in blood.

Thus, the "poly-input clearances" ($CL_D^{\text{pin organ}}$) may be represented by[5]

$$CL_D^{\text{pin organ}} = \sum (Q^{(i)} \Pi F_D^{\text{pre organ}(j)}) E_D^{\text{organ}} \qquad (23)$$

where $Q^{(i)}$ is the partitioned bloodflow through a set of organs leading from the aorta to the organ under investigation, and $F_D^{\text{pre organ}(j)}$ is the available fraction of the drug for a given organ ($j$) of the set of organs. The term $\Pi F_D^{\text{pre organ}(j)}$ would be the ratio of the concentration of drug in blood entering the organ to that of arterial blood if the organ were perfused by only one blood supply.

If the compound were eliminated from the body solely by the intestinal mucosa and the liver, the liver would be served by 3 blood supplies and the poly-input clearance of the liver ($CL_D^{\text{pin H}}$) would be

$$CL_D^{\text{pin H}} = (Q^{G-1} F_D^{G-1} + Q^{G-2} + Q^{HA}) E_D^H \qquad (24)$$

where $Q^{G-1}$ is the portion of the intestinal blood flow that perfuses the intestinal mucosa, $Q^{G-2}$ is the portion of the intestinal blood flow that bypasses the intestinal mucosa (only about 25% of the intestinal flow perfuses the intestinal mucosa), $Q^{HA}$ is the hepatic arterial blood flow, and $E_D^H$ is the hepatic extraction ratio of the compound.

The total body clearance in this situation would be

$$CL_D^{\text{body}} = CL_D^{G-1} + CL_D^{\text{pin H}} \qquad (25)$$

For more complex combinations of organs see Gillette.[5]

Because Eqs. 22–25 are based solely on the conservation of mass, they are valid regardless of the complexity of the various processes that govern the elimination of the compound by the organ.

## 2. Well-Stirred and Parallel-Tube Models

In recent years various attempts have been made to devise models that relate the organ extraction ratio to different combinations of the blood flow rate through an organ and the activities of various enzymes that metabolize the compound within the organ. Two models have gained popularity. In the simplest model, the "well-stirred model,"[6] the concentration of the parent compound and its metabolites in blood within all of the capillaries in the organ are assumed to be the same as their concentrations in the blood leaving the organ. The principal advantage of this model is the inherent simplicity of the mathematical equations, and it is used even though it ignores the fact that the concentrations of compounds are usually higher in the proximal end than in the distal end of the capillaries or sinusoids. In the other model, "the parallel-tube model,"[7] it is assumed that capillaries and sinusoids may be viewed as a set of identical tubes lined with cells containing the enzymes that metabolize the compound. In the development of this model, one visualizes that each tube may be subdivided into an infinite number of segments each of which may be viewed as a well-stirred model. The segments need not have the same length, but they must have the same ratio of the rate of metabolism of the compound divided by the rate of blood flow through the segment. When the compound is metabolized by several enzymes within the organ, it is assumed that all of the enzymes possess the same $K_m$ value, particularly when the model is applied to saturable systems. This model ignores the fact that residence times of the blood in the various capillaries may differ markedly and that the residence time of the blood within a given capillary may change with time even when the total blood flow through the organ remains constant. This model also ignores the fact that the capillaries are not exactly in parallel or of the same length. It further ignores the possibility that a compound and its metabolites may diffuse to a certain extent from one acinus to another within the organ. It is therefore obvious that neither of these simple models will invariably predict adequately the relationships between the total blood flow through the organ and the total activity of the enzymes within the organ. Nevertheless, they are useful in obtaining insights of the relationships that can govern the concentration of the compound and its metabolites within the different cells of the organ.

For a well-stirred model, the concentration of the parent compound may be predicted from these general equations:

$$[D]^{out} = \frac{[D]^{in} Q^{organ}}{Q^{organ} + R_D^{cell} CL_{D(met)}^{cell}} \quad (26a)$$

or

$$F_D^{organ} = \frac{Q^{organ}}{Q^{organ} + R_D^{cell} CL_{D(met)}^{cell}} \quad (26b)$$

where $R_D^{cell}$ is a partition ratio of the compound in cells that lack active-transport systems, that is,

$$R_D^{cell} = CL_{D(dif, in)}^{cell} / [CL_{D(dif, out)}^{cell} + CL_{D(met)}^{cell}] \quad (26c)$$

When the clearances are defined as rates divided by concentrations of unbound compound, however,

$$R_D^{cell} = f_{fD}^{zone} CL_{D(dif, in)}^{cell(f)} / f_{fD}^{cell} [CL_{D(dif, out)}^{cell(f)} + CL_{D(met)}^{cell(f)}] \quad (26d)$$

## 2. PHARMACOKINETICS OF FOREIGN COMPOUNDS

and since

$$CL_{D(met)}^{cell} = f_{fD}^{cell} CL_{D(met)}^{cell(f)}$$

then

$$R_D^{cell} CL_{D(met)}^{cell} = f_{fD}^{zone} CL_{D(dif, in)}^{cell(f)} CL_{D(met)}^{cell(f)} / CL_{D(dif, out)}^{cell(f)} + CL_{D(met)}^{cell(f)}] \quad (26e)$$

Note that Eq. (26e) does not contain $f_{fD}^{cell(f)}$, which confirms the view that the value of $F_D^{organ}$ is independent of reversible binding of the compound to intracellular components. Also note that when

$$CL_{D(met)}^{cell(f)} \ll CL_{D(dif, out)}^{cell(f)}$$

and when

$$CL_{D(dif, out)}^{cell(f)} = CL_{D(dif, in)}^{cell(f)}$$

then

$$R_D^{cell} CL_{D(met)}^{cell} = f_{fD}^{zone} CL_{D(met)}^{cell(f)} \quad (27)$$

These approximations are usually assumed to be valid in most discussions of the well-stirred model.[6,8]

Because the organ clearance of a compound may be expressed as the rate of metabolism of the compound divided by the arterial concentration, the organ clearance of a mono-input organ may be expressed by

$$CL_D^{organ} = \frac{[D]^{cell} CL_{D(met)}^{cell}}{[D]^{in}} = \frac{Q^{organ} R_D^{cell} CL_{D(met)}^{cell}}{Q^{organ} + R_D^{cell} CL_{D(met)}^{cell}} \quad (28)$$

and the organ extraction ratio may be written as

$$E_D^{organ} = \frac{R_D^{cell} CL_{D(met)}^{cell}}{Q^{organ} + R_D^{cell} CL_{D(met)}^{cell}} \quad (29)$$

The equation for the parallel-tube model may be expressed as a series of any number of segments (seg) of well-stirred compartments. From Eq. (26b) we may write

$$F_D^{organ} = [F_D^{seq}]^n = \left[ \frac{Q^{organ}}{Q^{organ} + R_D^{cells} CL_{D(met)}^{cells}} \right]^n \quad (30)$$

but if the cells in the organ were identical,

$$\sum [R_D^{cells} CL_{D(met)}^{cells}]^{seg} = CL_{D(met)}^{organ}/n.$$

Thus,

$$F_D^{organ} = \left[ \frac{Q^{organ}}{Q^{organ} + [CL_{D(met)}^{organ}/n]} \right]^n \quad (31)$$

But the limit of Eq. (28) as $n$ approaches infinity would be

$$F_D^{\text{organ}} = e^{-CL_{D(\text{met})}^{\text{organ}}/Q^{\text{organ}}} \qquad (32)$$

Therefore,

$$E_D^{\text{organ}} = 1 - e^{-CL_{D(\text{met})}^{\text{organ}}/Q^{\text{organ}}} \qquad (33)$$

and

$$CL_D^{\text{organ}} = Q^{\text{organ}}[1 - e^{-CL_{D(\text{met})}^{\text{organ}}/Q^{\text{organ}}}] \qquad (34)$$

Occasionally, the investigator wishes to visualize the concentration of the compound within a segment located between the proximal and distal ends of the tube. This concentration in blood in zone (i), $[D]_{F_D}^{\text{in zone}(i)}$, may be expressed as a fraction of the total length of the tube, $f_{\text{length}}^{\text{zone}(i)}$. Thus, Eq. (29) becomes

$$F_D^{\text{zone}(i)} = e^{-(i/n)CL_{D(\text{met})}^{\text{organ}}/Q^{\text{organ}}} \qquad (35)$$

where $i$ is the number of the $t$th zone in an organ containing $n$ zones. In these equations,

$$CL_{D(\text{met})}^{\text{cells}} = \sum f_{fD}^{\text{cell}} V_{\text{max}}^{\text{enzyme,cell}}/K_m^{\text{enzymes}} \qquad (36)$$

where $V_{\text{max}}^{\text{enzyme, cell}}$ is the theoretical maximal rate, expressed as amount per time, at which a given enzyme within a cell at an infinite concentration of unbound compound catalyzes the metabolism of the compound, and $K_m$ is the concentration of the unbound compound within the cell at which the rate will be one-half the maximal rate. It is usually assumed that the concentration of unbound substance within the cell is always negligible compared with the $K_m$ of the enzyme. It should be noted, however, that a given enzyme may catalyze the formation of several different metabolites of the compound.

## III. Concentration of the Parent Compound in Arterial Blood after Different Routes of Administration

Because compounds are seldom, if ever, injected directly into the aorta, Eq. (5) is seldom valid. Even after rapid intravenous injection of the compound, the equation for the concentration of the compound in arterial blood frequently requires some modification, especially when the compound is rapidly eliminated by the lung.

For visualization of the effects of different routes of administration on the concentration of the parent compound in arterial blood, it is useful to assume that the compound is absorbed from the site of administration by a first-order process. It is also useful to assume that rates of transfer of the compound between the cells and into organ blood for all of the organs through which the compound must pass on its way to the arterial blood reach steady-state conditions almost instantaneously. After oral (po) administration of the compound, the fraction of the dose of the parent compound reaching the aorta will depend on the fraction of the dose that is absorbed and on the fraction of the absorbed dose that escapes metabolism and other processes of elimination by the intestinal mucosa, liver, and lung. Under these conditions, the two-compartment model predicts that the concentration of the compound in the arterial blood may be estimated from this equation

$$[D]^{art} = \left[\frac{\text{Dose}_D^{po} k_a F_D^{G-1} F_D^H F_D^L}{V_1}\right]\left[\frac{(k_{21} - \alpha)e^{-\alpha t}}{(g - \alpha)(\beta - \alpha)} + \frac{(k_{21} - \beta)e^{-\beta t}}{(g - \beta)(\alpha - \beta)}\right.$$
$$\left. + \frac{(k_{21} - g)e^{-gt}}{(\alpha - g)(\beta - g)}\right] \tag{37}$$

where $F_D^{G-1}$, $F_D^H$, and $F_D^L$ are the organ available fractions of the compound for the intestinal mucosa, the liver, and the lung, respectively, $k_a$ is the rate constant for the absorption of the compound, and $g$ is the total rate constant of disappearance of the compound from the site of administration. A ratio of $k_a/g$ is the fraction of the dose that is absorbed. All other symbols are the same as those used in Eq. (5).

The area under the concentration vs time curve for this equation would be

$$AUC_D^{art} = \frac{\text{Dose } F_D^A F_D^{G-1} F_D^H F_D^L}{CL_D^{body}} \tag{38}$$

where $CL_D^{body}$ is the sum of poly-input and mono-input organ clearances and $F_D^A = k_a/g$. Similarly, the concentration of the compound in the deep compartment may be predicted from the equation:

$$[D]^{deep} = \left[\frac{\text{Dose}_D^{po} k_a k_{12} F_D^{G-1} F_D^H F_D^L}{V_2}\right]\left[\frac{e^{-\alpha t}}{(g - \alpha)(\beta - \alpha)}\right.$$
$$\left. + \frac{e^{-\beta t}}{(g - \beta)(\alpha - \beta)} + \frac{e^{-gt}}{(\alpha - g)(\beta - g)}\right] \tag{39}$$

and the area under the concentration vs time curve in the deep compartment will be

$$AUC_D^{deep} = \frac{\text{Dose}_D^{po} F_D^A F_D^{G-1} F_D^H F_D^L CL_{12}}{CL_D^{body} CL_{21}} \tag{40}$$

A ratio of the areas under the arterial blood concentration vs time curves after the administration of the same dose of a compound orally and intraortically should therefore provide an estimate of the total fraction of the dose of the compound administered orally that reaches the arterial circulation, that is,

$$\frac{AUC_D^{art}/Dose_D^{po}}{AUC_D^{art}/Dose_D^{ia}} = F_D^A F_D^{G-1} F_D^H F_D^L \qquad (41)$$

The equations for arterial concentrations of compounds after administration by other routes will be identical to Eqs. (34)–(38), except that the organ availability terms will be different. For example, if the compound were administered subcutaneously or intramuscularly, the coefficients would contain $F_D^L$, but not $F_D^{G-1}$ or $F_D^H$, because the compound would pass through the lung, but not the gastric mucosa or the liver, during the first pass from the site of administration to arterial blood. Moreover, $k_a$ would presumably equal $g$, and $F_D^A$ in Eqs. (38), (40), and (41) would be 1.0.

After intravenous administration of the compound, $e^{-gt}$ would equal zero and $k_a = k_g \gg \alpha$ and $\beta$. Thus, Eqs. (37) and (39) would lack the third exponential term, $k_a F_D^{G-1}$, $F_D^H$, $(g - \alpha)$, $(g - \beta)$, but would still retain $F_D^L$.

## IV. Concentration of the Parent Compound in Blood Exiting Various Organs in the Central Compartment after Different Routes of Administration

After the intraaortic administration of a compound, its concentration in blood exiting an organ within the central compartment will be proportional to the concentration of the compound in arterial blood (art). For mono-input organs such as the kidney ($K$),

$$[D]_{out}^{K(cycle)} = [D]^{art} F_D^K \qquad (42)$$

where $[D]^{art}$ at any given time may be estimated from Eq. (5) and $F_D^K$ may be estimated from either Eq. (26b) or Eq. (32). For poly-input (pin) organs, such as the liver (H), the concentration of the parent compound exiting the organ may be governed by a more complex availability term $F_D^{pin\ organ}$. For example

$$[D]_{out}^{H(cycle)} = [D]^{art} F_D^{pin\ H} = \frac{[D]^{art}[Q^{G-1}F_D^{G-1} + Q^{G-2} + Q^{HA}]F_D^H}{Q^{G-1} + Q^{G-2} + Q^{HA}} \qquad (43)$$

where $[D]^{art}$ may also be expressed by Eq. (37).

## 2. PHARMACOKINETICS OF FOREIGN COMPOUNDS

After oral administration of the compound, its concentration in blood exiting the kidney would still be proportional to its concentration in arterial blood, but, in this case, the concentration in arterial blood would be estimated by Eq. (37).

The concentration of the compound exiting the liver after oral administration, however, depends on the relative contribution of the compound as it is being absorbed on its way to the arterial circulation (first-pass portion) and of the compound that reenters the liver (cycle portion)(Figs. 5 and 6). Thus, the equation for the total concentration of the compound leaving the organ may be partitioned according to the kinetic pathway of the compound. For example,

$$[D]_{out}^{H} = [D]_{out}^{H(cycle)} + [D]_{out}^{H(first\text{-}pass)} \tag{44}$$

When it is assumed that the amount of the compound in the intestines and in the liver reaches a steady state with that in the blood almost instantaneously, the cycle portion will be proportional to that in the arterial circulation. Indeed, the equation for the cycle portion would be the same as Eq. (43) except that the equation for $[D]^{art}$ would be Eq. (37). The equation for the first-pass portion depends on the rate of absorption of the compound and on the blood flow through the liver, but is obviously independent of the parameters that govern the $\alpha$ phase and the $\beta$ phase, which describe the concentration of the compound in arterial blood. When it is assumed that the absorption of the compound follows first-order kinetics, the equation for the contribution of the first-pass portion to the total concentration of the compound leaving the liver would be

$$[D]_{out}^{H(first\text{-}pass)} = \frac{Dose_D^{po} F_D^{G-1} F_D^{H} k_a e^{-gt}}{Q^{G-1} + Q^{G-2} + Q^{HA}} \tag{45}$$

The total concentration of the compound leaving the liver would be the sum of the concentrations obtained from Eqs. (43) and (45).

In the well-stirred model, these equations also provide the concentrations of the compound in the blood within all the capillaries (or sinusoids) of the organ. If the concentration of the compound in a set of contingous homoge-

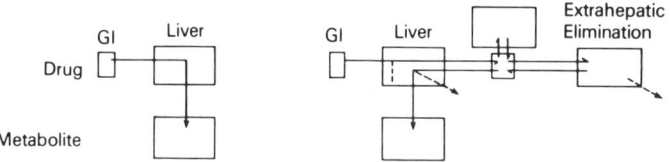

**Fig. 5.** Partitioned kinetic pathways describing the concentration and metabolism of a parent compound after oral administration of the compound.

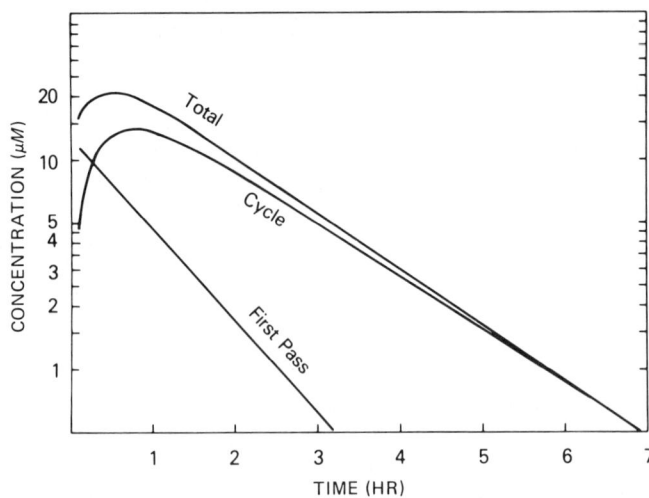

**Fig. 6.** A simulation of the partitioned concentrations of a parent compound due to the partitioned kinetic pathways of the parent compound after its oral administration. The partitioned kinetic pathways are illustrated in Fig. 5. The equations used are those for a well-stirred model, described in the text.

neous zones in a parallel-tube model is desired, one may use these equivalent terms:

$$F_D^{organ} = F_D^{zone(1)} F_D^{zone(2)} = \left[e^{-CL_{D(met)}^{zone(1)}/Q}\right]\left[e^{-CL_{D(met)}^{zone(2)}/Q}\right] \quad (46)$$

The average concentration of the compound within any given zone of the tube may be expressed by

$$[D]^{zone} = [D]_{in}^{zone}(F_D^{zone} - 1)/\ln F_D^{zone} \quad (47)$$

The simulation shown in Fig. 6 is based on the partitioned Eqs. (43) and (45) for the concentration of parent compound in blood leaving the liver after oral administration of the compound. Such simulations are enlightening, because they dramatize the relative contributions of the various kinetic pathways of the compound at any given time. In the simulation shown in Fig. 6, for example, the maximum concentration of the compound did not occur almost instantaneously because the rate constant for the absorption of the compound from the site of administration was set at a small value. Indeed, at the time the maximal total concentration occurred, the contribution of the first-pass portion was rather small. By contrast, if the rate constant of absorption were considerably larger, the maximum concentration in

## 2. PHARMACOKINETICS OF FOREIGN COMPOUNDS

the liver would have occurred almost immediately after the administration of the compound, but the concentration would have declined rapidly as the compound was absorbed until the concentration was predominantly due to the cycle portion [Eq. (43)] of the partitioned Eq. (44). Under these conditions, marked changes in the elimination of the substance by the kidney and other organs that are not involved in the first-pass portion of the concentration would not be expected to have a significant effect on the maximal concentration of the compound in liver, but may have an effect on the duration of action of the compound in liver.

Area under the concentration vs time curves may also be partitioned according to different kinetic pathways that govern the concentration of the compound in first-pass organs after different routes of administration. For example, after oral administration,

$$AUC_D^H = AUC_D^{H(\text{first-pass})} + AUC_D^{H(\text{cycle})} \tag{48a}$$

where

$$AUC_{D(\text{out})}^{H(\text{cycle})} = AUC_D^{\text{art}} \left[ \frac{(Q^{G-1}F_D^{G-1} + Q^{G-2} + Q^{HA})F_D^H}{Q^{G-1} + Q^{G-2} + Q^{HA}} \right]$$

$$= AUC_D^{\text{art}} F_D^{\text{pin H}} F_D^H \tag{48b}$$

in which $AUC_D^{\text{art}}$ may be obtained from Eq. (35). Moreover,

$$AUC_{D(\text{out})}^{H(\text{first-pass})} = \frac{\text{Dose}_D^{po} F_D^A F_D^{G-1} F_D^H}{Q^{G-1} + Q^{G-2} + Q^{HA}} \tag{48c}$$

As will be shown later, these equations for the areas under the concentration time curves are components of the equations for the areas under the concentration vs time curves for the formation of metabolites by the liver. They are therefore useful in dramatizing the effects of different routes of administration on the exposure of an organ to the parent compound and its metabolites. For example, if the compound were eliminated from the body solely by the liver and the kidney (i.e., both $F^{G-1}$ and $F^L$ equal 1.0), we may calculate that the ratio of the areas under the concentration vs time curves after intravenous (iv) and oral administration of the same dose of a compound should be given by

$$\frac{AUC_D^H/\text{Dose}_D^{po}}{AUC_D^H/\text{Dose}_D^{iv}} = 1 + [CL_D^K/Q^H] \tag{49}$$

Inspection of Eq. (49) reveals that the ratio should range from 1.0 (when $CL_D^K$ is negligible) to somewhat less than 2.0 (when $CL_D^K$ is limited by the renal blood flow).[9] Thus, when $CL_D^K$ is negligible compared with the hepatic blood flow, the average concentration of the compound in liver will be independent of the route of administration of the parent compound.

## V. Kinetics of the Formation of Chemically Reactive Metabolites by Organs in the Central Compartment

This chapter has thus far been devoted to the development of the equations that permit the calculation of the concentration of the parent compound in blood within a zone of an organ. It has also provided many of the mathematical tools required to estimate the rate of formation of chemically reactive metabolites.

### A. SHORT-LIVED CHEMICALLY REACTIVE METABOLITES

Figure 7 illustrates the factors that govern the rate of synthesis and the steady-state concentration of a metabolite within cells. As a parent compound enters the zone, a portion of it enters a group of cells containing enzymes that form a given metabolite. By definition, a short-lived metabolite does not leave cells in which it is formed, and we are not, for the moment, concerned with either the rate at which the metabolite enters the blood or the rate at which the metabolite enters the zone from other sources.

There is no reason for believing, however, that all cells in the zone contain enzymes in exactly the same amounts or in exactly the same proportions. Differences in the amounts of such enzymes may be represented, in part, by visualizing that the parent compound is metabolized by two sets of homogeneous cells in the zone; one of the sets contains enzymes that generate the short-lived metabolite [cell(gen)], and the other set does not [cell(nongen)]. It may further be visualized that the rates of metabolism of the parent compound and the metabolite within the two groups of cells may differ. The equation for the zonal availability of the compound—that is, the ratio of the concentration of the compound in blood within the zone divided by the concentration in the preceding zone—would be

$$F_D^{zone} = \frac{Q^{zone}}{Q^{zone} + R_D^{cell(gen)}CL_{D(met)}^{cell(gen)} + R_D^{cell(nongen)}CL_{D(met)}^{cell(nongen)}} \quad (50)$$

where $F_D^{zone}$ may be used in Eqs. (42), (43), (45)–(47), (48b), or (48c) to relate the concentration of the compound in the blood within the zone or organ to that which enters the zone or organ.

The steady-state concentration of the compound within the cells that

## 2. PHARMACOKINETICS OF FOREIGN COMPOUNDS

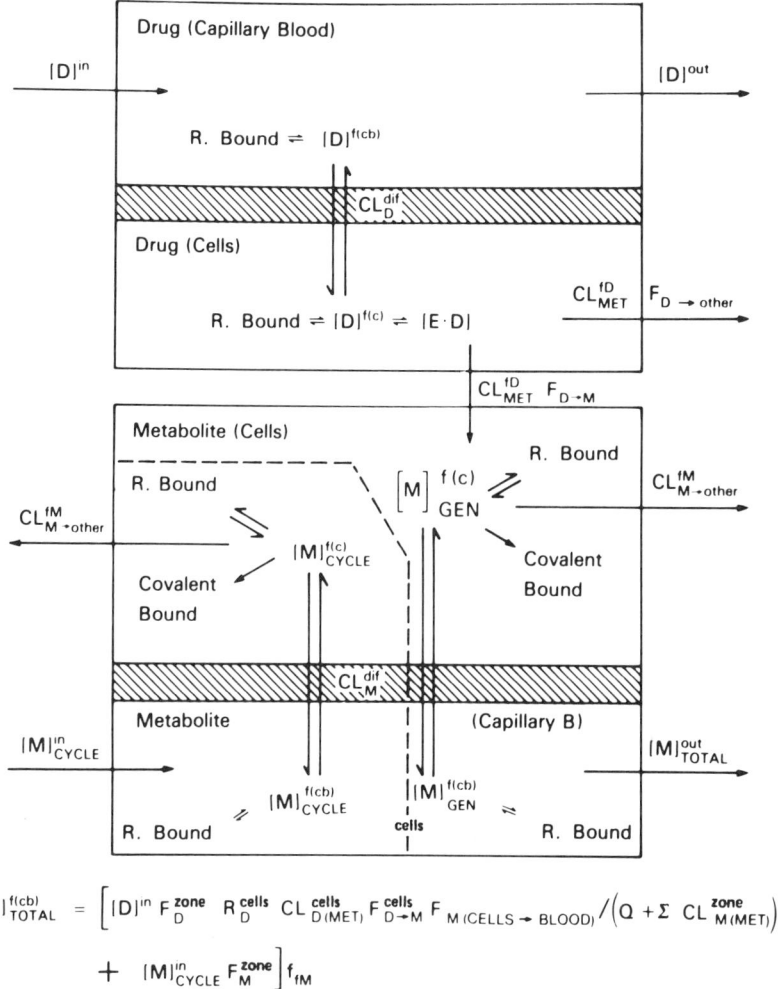

$$[M]_{TOTAL}^{f(cb)} = \left[ [D]^{in} F_D^{zone} R_D^{cells} CL_{D(MET)}^{cells} F_{D \to M}^{cells} F_{M(CELLS \to BLOOD)} / \left( Q + \Sigma \, CL_{M(MET)}^{zone} \right) \right.$$
$$\left. + [M]_{CYCLE}^{in} F_M^{zone} \right] f_{fM}$$

**Fig. 7.** Factors that govern the concentration of a metabolite within cells and the contiguous blood within an organ.

generate the reactive metabolite may be expressed by

$$[D]^{cells(gen)} = [D]^{zone} R_D^{cells(gen)} \tag{51a}$$

where $R_D^{cells(gen)}$ is the partition ratio that relates the concentration of the compound in the cells to that in zonal blood and is equal to

$$CL_{D(dif, in)}^{cells(gen)} / [CL_{D(dif, out)}^{cells(gen)} + CL_{D(met)}^{cells(gen)}] \tag{51b}$$

The steady-state rate of conversion of the parent compound to all its first-generation metabolites within the cells that form the chemically reactive metabolite may be expressed by

$$\text{Rate}_{D(met)}^{cells(gen)} = [D]^{cells(gen)} CL_{D(met)}^{cells(gen)} \tag{52}$$

The rate of formation of the chemically reactive metabolite may be expressed as a fraction of the total rate of metabolism within the cells. Thus,

$$\text{Rate}_{D \to M}^{cells(gen)} = \text{Rate}_{D(met)}^{cells(gen)} F_{D \to M}^{cells(gen)}$$

where

$$F_{D \to M}^{cells(gen)} = CL_{D \to M}^{cells(gen)} / CL_{D(met)}^{cells(gen)} \tag{53}$$

Because by definition all of the short-lived chemically reactive metabolite must be inactivated within the cells in which it is generated, the rate of formation of any second-generation metabolite may be expressed as a fraction of the rate of formation of the short-lived chemically reactive metabolite. For example, when the chemically reactive metabolite is covalently bound to protein (cbp), then the rate of formation of the chemically reactive metabolite may be expressed by

$$\text{Rate}_{M \to cbp}^{cells(gen)} = \text{Rate}_{D \to M}^{cells(gen)} F_{M \to cbp}^{cells(gen)}$$

where

$$F_{M \to cbp} = CL_{M \to cbp}^{cells(gen)} / CL_{D \to M}^{cells(gen)} \tag{54}$$

Sequential substitution of Eqs. (51)–(53) into Eqs. (54) thus gives

$$\text{Rate}_{M \to cbp}^{cells(gen\,zone)} = [D]^{zone} R_D^{cells(gen)} CL_{D(met)}^{cells(gen)} F_{D \to M}^{cells(gen)} F_{M \to cbp}^{cells(gen)} \tag{55}$$

Thus, the rate of covalent binding of the reactive metabolite should be directly proportional to the concentration of the parent compound in the blood within the organ containing cells that generate the chemically reactive metabolite, provided that the various fractions do not change with time. In turn, the concentration of the parent compound within the zone of a well-stirred model is governed by the partitioned concentration represented by Eqs. (37), (43), and (45). It is therefore evident that simulations such as those shown in Fig. 6 should be useful in relating the rates of covalent binding to the concentrations of the parent compound in blood.

Moreover, we may also relate the areas under the rate vs time curve of the covalent binding to the area under the concentration vs time curve of the parent compound. For example,

$$AUC_{rate(M \to cpb)}^{zone} = AUC_D^{zone} R_D^{cells(gen)} CL_{D(met)}^{cells(gen)} F_{D \to M}^{cells(gen)} F_{M \to cbp}^{cells(gen)} \tag{56}$$

## 2. PHARMACOKINETICS OF FOREIGN COMPOUNDS

which gives the amount of covalently bound material in the zone that occurs from the time of the administration until all of the parent compound is elminated from the body. Again, the $AUC_D^{zone}$ and therefore $AUC_{rate(M\to cbp)}^{zone}$ may be partitioned according to the first-pass and cycle portions of the kinetic pathways of the parent compound. Thus, the relative importance of these pathways in contributing to the total amount of covalent binding of the reactive metabolite to protein may be assessed.

### B. INTERMEDIATE-LIVED METABOLITES

By definition, members of this category of chemically reactive metabolites leave the cells, but never the organ, in which they are formed. Therefore, only a portion of the chemically reactive metabolite is inactivated in the cells as it is being formed. This portion may be called the zonal first-pass portion. The rest of the chemically reactive metabolite leaves the cells and enters the blood within the zone. One portion of the chemically reactive metabolite reenters the cells in which it was generated (the zonal cycle portion). A second portion enters the set of cells that are unable to generate the chemically reactive metabolite. A third portion reacts with components in blood, and a fourth portion is swept downstream into the next zone.

Thus, the rate of formation of the chemically reactive metabolite must equal the rate at which the metabolite is eliminated from the zone by these processes.

$$\text{Rate}_{D\to M}^{\text{cells(gen)}} = \text{Rate}_{M(\text{met})}^{\text{cells(gen, zonal first-pass)}} + \text{Rate}_{M(\text{met})}^{\text{cells(gen, zonal cycle)}}$$
$$+ \text{Rate}_{M(\text{met})}^{\text{cells(nongen)}} + \text{Rate}_{M(\text{met})}^{\text{zonal}} + Q^{\text{zone}}[M]^{\text{zone}} \quad (57)$$

As with the short-lived chemically reactive metabolite, the covalent binding resulting from the zonal first-pass portion of metabolism of the metabolite—the first term in Eq. (54)—may be represented by

$$\text{Rate}_{M\to cbp}^{\text{cells(gen, zonal first-pass)}} = [D]_{\text{in}}^{\text{zone}} F_D^{\text{zone}} R_D^{\text{cells(gen)}} CL_{D(\text{met})}^{\text{cells(gen)}}$$
$$\times F_{D\to M}^{\text{cells(gen)}} F_{M(\text{met})}^{\text{cells(gen)}} F_{M\to cbp}^{\text{cells(gen)}} \quad (58)$$

but, in this case,

$$F_{M\to cbp} = CL_{M\to cbp}^{\text{cells(gen)}} / CL_{M(\text{met})}^{\text{cells(gen)}} \quad (59)$$

The solution of the other parts of Eq. (57) requires an equation for the steady-state concentration of the metabolite in blood within the zone.

$$[M]^{\text{zone(from cells, gen)}} / [D]_{\text{in}}^{\text{zone}} = \frac{F_D^{\text{zone}} R_D^{\text{cells(gen)}} CL_{D(\text{met})}^{\text{cells(gen)}} F_{D\to M}^{\text{cells(gen)}} F_{M(\text{cells}\to\text{blood})}^{\text{cells(gen)}}}{Q^{\text{zone}} + CL_{M(\text{met})}^{\text{zone}}}$$
$$(60)$$

where

$$F_{M(\text{cells}\to\text{blood})} = \frac{CL^{\text{cells(gen)}}_{M(\text{cells}\to\text{blood})}}{[CL^{\text{cells(gen)}}_{M(\text{cells}\to\text{blood})} + CL^{\text{cells(gen)}}_{M(\text{met})}]} \quad (61a)$$

and

$$CL^{\text{zone}}_{M(\text{met})} = CL^{\text{zone(blood)}}_{M(\text{met})}$$
$$+ R^{\text{cells(gen)}}_{M} CL^{\text{cells(gen)}}_{M(\text{met})} + R^{\text{cells(nongen)}}_{M} CL^{\text{cells(nongen)}}_{M(\text{met})} \quad (61b)$$

The rate of formation of covalently bound metabolite by the zonal cycle pathway within the cells that generate the metabolite may be expressed by the following equation:

$$\text{Rate}^{\text{cells(gen, zonal cycle)}}_{M\to\text{cbp}} = [M]^{\text{zone(from cells, gen)}} R^{\text{cells(gen)}}_{M} CL^{\text{cells(gen)}}_{M(\text{met})} F^{\text{cells(gen)}}_{M\to\text{cbp}} \quad (62)$$

The rate of formation of covalently bound metabolite within the cells that do not generate the metabolite may be expressed by the following equation

$$\text{Rate}^{\text{cells(nongen) from cell(gen)}}_{M\to\text{cbp}}$$
$$= [M]^{\text{zone(from cells, gen)}} R^{\text{cells(nongen)}}_{M} CL^{\text{cells(nongen)}}_{M(\text{met})} F^{\text{cells(nongen)}}_{M\to\text{cbp}} \quad (63)$$

The rate of covalent binding of the metabolite to components in blood within the zone may be expressed by the equation

$$\text{Rate}^{\text{zone(from cells, gen)}}_{R\to\text{cbp}} = [M]^{\text{zone(from cells, gen)}} CL^{\text{zone}}_{M\to\text{cbp}} \quad (64)$$

The rate at which the metabolite, generated in the zone, leaves the zone and enters other zones may be expressed by

$$\text{Rate}^{\text{zone(from cells, gen)}}_{M(i\to i+1)} = [M]^{\text{zone(from cells, gen)}} Q^{\text{zone}} \quad (65)$$

When the zone under consideration is the last zone of a parallel-tube model, then, by definition of an intermediate-lived metabolite, the reactive metabolite leaving the zone via the blood will become metabolized in the blood before it reaches the next organ. The rate at which this portion becomes covalently bound to venous blood components (ven) may be expressed as a fraction of the rate at which it left the organ, that is,

$$\text{Rate}^{\text{ven}}_{M\to\text{cbp}} = \text{Rate}^{\text{zone(from cells, gen)}}_{M(i\to i+1)} F^{\text{organ(ven)}}_{M\to\text{cbp}} \quad (66)$$

When the zone under consideration is not the last zone of a parallel-tube model, then Eq. (65) represents the portion of the metabolite formed in the zone that is swept by the blood into the next zone.

When the zone under consideration is not the first zone of a parallel-tube model, some of the metabolite formed in prior zones may also be swept into the zone under consideration.

The relative activities of the enzymes that metabolize the parent compound and the metabolite or the relative proportion of the cells that contain enzymes that catalyze the formation of the metabolite in a segment are seldom, if ever, identical for all segments of the tube. It is perhaps realistic, therefore, to describe the organ availability as a set of mathematical terms representing several different zones within an organ. For example, in a four-zone system,

$$[M]_{out}^{organ}/[D]_{in}^{organ} = [[M]_{out}^{zone\,1}/[D]_{in}^{zone\,1}]F_M^{zone\,2}F_M^{zone\,3}F_M^{zone\,4}$$
$$+ F_D^{zone\,1}[[M]_{out}^{zone\,2}/[D]_{in}^{zone\,2}]F_M^{zone\,3}F_M^{zone\,4}$$
$$+ F_D^{zone\,1}F_D^{zone\,2}[[M]_{out}^{zone\,3}/[D]_{in}^{zone\,3}]F_M^{zone\,4}$$
$$+ F_D^{zone\,1}F_D^{zone\,2}F_D^{zone\,3}[[M]_{out}^{zone\,4}/[D]_{in}^{zone\,4}] \quad (67)$$

In this set of terms,

$$F_D^{zone} = \frac{Q^{zone}}{Q^{zone} + CL_{D(met)}^{zone}} \quad (68)$$

where

$$CL_{D(met)}^{zone} = F_D^{cells(gen)}CL_{D(met)}^{cells(gen)} + F_D^{cells(nongen)}CL_{D(met)}^{cells(nongen)}$$

Moreover,

$$F_M^{zone} = \frac{Q^{zone}}{Q^{zone} + CL_{M(met)}^{zone}} \quad (69)$$

where

$$CL_{M(met)}^{zone} = CL_{M(met)}^{zone(blood)} + R_M^{cells(gen)}CL_{M(met)}^{cells(gen)} + R_M^{cells(nongen)}CL_{M(met)}^{cells(nongen)}$$

Equation (60) describes the value of the terms $[M]_{out}^{zone}/[D]_{in}^{zone}$.

If all of the zones were identical, the equations for the parallel-tube model used by Pang and Gillette[10] might be used. However, as pointed out by these authors, the assumption that all zones within the organ are identical is probably invalid.

In a sequential system, such as one that includes the intestinal mucosa, portal venous blood, liver, hepatic venous blood, pulmonary arterial blood, lungs, and pulmonary venous blood, the set of terms represented by Eq. (67) may be expanded to include the portions of the metabolite that are formed in and escape from poly-input organs. For example, the set of terms may include terms representing the intestinal mucosa, liver, and lung, and the portions of the metabolite that are inactivated in the blood within the portal vein, the hepatic vein, the pulmonary artery, and the pulmonary vein.

From Eq. (67) it is possible to visualize the concentration of the metabolite in any intermediate zone within the organ, which may be used to calculate the rates of covalent binding in cells that generate the metabolite, in cells that do not generate the metabolite, and to components in blood within the zone.

Inspection of these equations for sequential zone models reveals that changes in the blood flow may alter the relative rates of covalent binding in the different zones in unpredictable ways. When the organ extraction of the parent compound is high at a low blood flow rate, an increase in blood flow should decrease the extraction ratio and increase the concentration of the parent compound in the distal portion of the tube. The increase in blood flow will therefore increase the rate of formation of metabolites catalyzed by the enzymes in the cells within the distal portion. On the other hand, regardless of the extraction ratio of the parent compound, an increase in the blood flow will increase the rate at which the metabolite in the blood within a zone is swept downstream. An increase in blood flow rate may therefore not only increase the organ availability of the metabolite, but also increase the covalent binding and the formation of second-generation metabolites in the distal zones of the organ. When the enzymes that metabolize the first-generation metabolites to metabolites that differ from those formed by enzymes in the proximal zones of the organ, the pattern of second-generation metabolites may be changed by increasing the blood flow rate through the organ, even though [D] never approaches the $K_m$ values of the various enzymes. Such interzonal differences in the distribution of the different enzymes may be frequently detected by comparing the pattern of metabolites formed after normal and retrogarde perfusion of the organ.[11]

## C. LONG-LIVED METABOLITES

By definition, members of the long-lived category of chemically reactive metabolites not only leave the cells in which they are generated, but also leave the organ containing the cells, enter the systemic circulation, and are cycled to virtually every organ in the body. The rate of covalent binding of a chemically reactive metabolite in cells that generate the metabolite within an organ depends not only on the factors that govern the rates of covalent binding of intermediate-lived metabolites, which include the concentration of parent drug entering the organ, but also on the factors that govern the concentration of the metabolite in systemic blood, which include the concentration of the metabolite in blood entering the organ. However, the symbols have been changed to differentiate the terms associated with the parent compound for those associated with the metabolite in Eq. (70).

## 2. PHARMACOKINETICS OF FOREIGN COMPOUNDS

### 1. Partitioned Equations for the Concentration of the Parent Compound Entering a First-Pass Organ

Figure 8 illustrates the kinetic pathways by which a parent compound and its primary metabolite might contribute to the total concentration of the metabolite as it enters a first-pass organ.

*D First-Pass.* As the parent compound is absorbed from the site of administration, the portion [D]$^{\text{first-pass}}$ of the total concentration of the parent compound entering the first-pass, mono-input organ may be described by an equation analogous to Eq. (45), that is,

$$[D]_{\text{in}}^{\text{first-pass}} = \frac{\text{Dose}_D(D_a)(\Pi F_D^{\text{pre organ}})e^{-D_g t}}{Q^{\text{organ(first-pass)}}} \quad (70)$$

$D_a$ is the rate constant of absorption and $D_g$ is the rate constant of elimination of the parent compound from the site of administration. The fraction of the dose that would be absorbed ultimately would be $(D_a)/(D_g)$.

*D Cycle.* A portion of the parent compound may escape the first-pass organs and enter the systemic circulation in the cycle portion. The equation for this portion of the parent compound in arterial blood would be

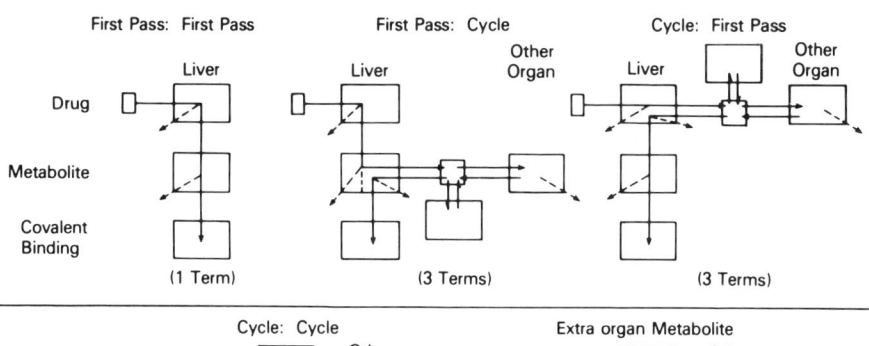

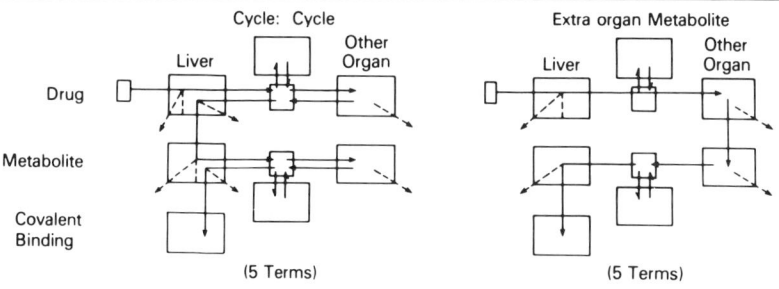

**Fig. 8.** Partitioned kinetic pathways describing the concentration and metabolism of a first-generation metabolite after oral administration of the parent compound.

$$[D]^{art(cycle)} = \frac{Dose_D(D_a)[\Pi F_D^{organ(first\text{-}pass)}][A_D e^{-D_\alpha t} + B_D e^{-D_\beta t} + G_D e^{-D_g t}]}{V_{D_1}} \quad (71)$$

where

$$A_D = \frac{D_{21} - D_\alpha}{(D_g - D_\alpha)(D_\beta - D_\alpha)}$$

$$B_D = \frac{D_{21} - D_\beta}{(D_g - D_\beta)(D_\alpha - D_\beta)}$$

$$G_D = \frac{D_{21} - D_g}{(D_\alpha - D_g)(D_\beta - D_g)}$$

which is the same as Eq. (37), but again note the change in symbols.

Moreover, $\Pi F_D^{organ(first\text{-}pass)}$ is the mathematical product of the organ availabilities of all the first-pass organs through which the drug must pass from the site of administration to the arterial blood. $D_{21}$ is the rate constant for the passage of the parent compound out of the deep compartment. $D_\alpha$ and $D_\beta$ are the hybrid constants for the $\alpha$ phase and $\beta$ phase of the parent compound.

## 2. Partioned Equations for a First-Generation Metabolite within a First-Pass Organ

For the first-generation metabolite, a more complex set of kinetic pathways must be visualized.

*M First-Pass: First-Pass.* As the parent compound is absorbed from its site of administration, it passes through a first-pass organ in which a portion of the parent drug is converted to a given metabolite, but a portion of the metabolite is converted to second-generation metabolites (including that portion that combines with tissue macromolecules) before it leaves the organ.

The equation for the concentration of the parent compound entering the organ by this kinetic pathway would be Eq. (70).

*M Cycle: First-Pass.* After the parent compound reaches the arterial circulation, it enters both first-pass and nonfirst-pass organs in which a portion of the parent compound is converted to a given metabolite, but a portion of the metabolite is converted to second-generation metbolites (including that portion that combines with tissue macromolecules) before it leaves the organ. The equation for the concentration of the parent compound in arterial blood for this kinetic pathway would be Eq. (71), and the equation for the parent compound entering the organ would be $[D]^{art} F_D^{pin\,organ}$.

For both of these kinetic pathways, the metabolite is derived directly from

## 2. PHARMACOKINETICS OF FOREIGN COMPOUNDS

the parent compound. For any given zone within the organ, the concentration of the metabolite in blood may be described by Eq. (67). The rate of covalent binding of the metabolite formed in previous zones of the organ may be described by the following:

$$\text{Rate}_{M\to cbp}^{cells(i)(M \text{ from prezones})} = [M]_{in}^{zone(i)} R_M^{cells(i)} CL_{M(met)}^{cells(i)} F_{M\to cbp}^{cells(i)} \quad (72)$$

And the rates of covalent binding from the metabolite generated within the zone may be described by Eqs. (58)–(64).

$$\text{Rate}_{M\to cbp}^{zone(blood)} = [M]_{in}^{zone} CL_{M\to cbp}^{zone(blood)} \quad (73)$$

The total rate of formation of a chemically reactive metabolite within cells and the total rate of covalent binding of the metabolite may be calculated from the appropriate use of these equations.

As shown in Fig. 8, the portion of the first-generation metabolite that reaches the arterial blood may be partitioned into several kinetic pathways.

*M First-Pass: Cycle.* As the parent compound is absorbed from the site of administration, it enters a first-pass organ in which portion is converted to the metabolite before the parent compound escapes the organ; this portion of the parent compound never reaches the arterial blood and represents the first-pass metabolism of the parent compound. A portion of the metabolite, however, escapes the organ and enters the arterial blood. The equation that describes the concentrations of the metabolite derived from first-pass metabolism of the parent compound that enters the arterial circulation contains an exponential terms that represent the $M_\alpha$ phase and $M_\beta$ phase of the metabolite. Thus,

$$[M]^{art(first\text{-}pass\,cycle)}$$
$$= \{[\text{Dose}_D(D_a)]\Pi F_D^{pre\,organ}[[M]_{out}^{organ}/[D]_{in}^{organ}]^{single\,pass}\Pi F_M^{post\,organ}]$$
$$[A_M e^{-M_\alpha t} + B_M e^{-M_\beta t} + C_M e^{-D_g t}]\}/V_{M_1} \quad (74)$$

where

$$A = \frac{M_{21} - M_\alpha}{(D_g - M_\alpha)(M_\beta - M_\alpha)}$$

$$B = \frac{M_{21} - M_\beta}{(D_g - M_\beta)(M_\alpha - M_\beta)}$$

$$C = \frac{M_{21} - D_g}{(M_\beta - D_g)(M_\beta - D_g)}$$

in which $([M]_{out}/[D]_{in})^{single\,pass}$ would be the ratio of the concentration of the metabolite that escapes the organ to the concentration of the parent compound that enters the organ in a single-pass, steady-state perfusion of the organ [Eq. (67)]. The term, $\Pi F_M^{\text{post organ}}$ is the mathematical product of the organ availabilities of the metabolite after it leaves the organ [see Eq. (26b) or (32)]. $M_{21}$ is the rate constant for the passage of the metabolite from its deep compartment to its central compartment, and $M_\alpha$ and $M_\beta$ are the hybrid rate constants for the $\alpha$ phase and $\beta$ phase of the metabolite.

As with Eq. (39), Eq. (74) may be converted to the equation for intravenous injections by assuming that $D_g$ is much greater than all other constants and by making appropriate adjustments to the first-pass availability of the parent compound.

*M Cycle: Cycle.* After the parent compound is administered and enters the first-pass organs, a portion enters the arterial blood and is distributed to the deep compartment of the parent compound. It then reenters the first-pass organ, and a portion of it is converted to the first-generation metabolite. A portion of the metabolite escapes the organ and enters arterial blood and then the deep compartment of the metabolite. The equation for this kinetic pathway contains an exponential term representing the absorption of the parent compound, two exponential terms representing the $\alpha$ pase and $\beta$ phase of the parent compound, and two exponential terms representing the $\alpha$ phase and $\beta$ phase of the metabolite. The concentration of the metabolite in arterial blood due to this kinetic pathway may be represented as

$$[M]^{\text{art(cycle: cycle)}} = \{\text{Dose}_D(D_a)[[\Pi F_D^{\text{first-pass organ}}]Q^{\text{organ}}F_D^{\text{pin organ}}$$
$$[M]_{\text{out}}^{\text{organ}}/[D]_{\text{in}}^{\text{organ}}]^{\text{single pass}}[\Pi F_M^{\text{post organ}}]$$
$$[Ae^{-D_\alpha t} + Be^{-D_\beta t} + Ce^{-M_\alpha t} + De^{-M_\beta t} + Ee^{-D_g t}]\}/V_{D_1}V_{M_1} \quad (75)$$

where

$$A = \frac{(D_{21} - D_\alpha)(M_{21} - D_\alpha)}{(D_\beta - D_\alpha)(M_\alpha - D_\alpha)(M_\beta - D_\alpha)(D_g - D_\alpha)}$$

$$B = \frac{(D_{21} - D_\beta)(M_{21} - D_\beta)}{(D_\alpha - D_\beta)(M_\alpha - D_\beta)(M_\beta - D_\beta)(D_g - D_\beta)}$$

$$C = \frac{(D_{21} - M_\alpha)(M_{21} - M_\alpha)}{(D_\alpha - M_\alpha)(D_\beta - M_\alpha)(M_\beta - M_\alpha)(D_g - M_\alpha)}$$

$$D = \frac{(D_{21} - M_\beta)(M_{21} - M_\beta)}{(D_\alpha - M_\beta)(D_\beta - M_\beta)(M_\alpha - M_\beta)(D_g - M_\beta)}$$

$$E = \frac{(D_{21} - D_g)(M_{21} - D_g)}{(D_\alpha - D_g)(D_\beta - D_g)(M_\alpha - D_g)(M_\beta - D_g)}$$

## 2. PHARMACOKINETICS OF FOREIGN COMPOUNDS

In this partitioned equation, $F_D^{\text{first-pass organ}}$ is the fraction of the absorbed dose of the parent compound that reaches aortic blood. $Q^{\text{organ}} F^{\text{pin organ}} / V_{D(1)}$ is the rate constant at which the parent compound is delivered from arterial blood to the organ under consideration.

$$[[M]_{\text{out}}^{\text{organ}} / [D]_{\text{in}}^{\text{organ}}]^{\text{single pass}}$$

is the term that describes the portion of the parent compound in blood entering the organ that is converted to the metabolite and escapes the organ in a single-pass, steady-state perfusion, and $\Pi F_M^{\text{post organ}}$ is the portion of the metabolite leaving the organ that ultimately reaches the aortic blood. The other parameters have been defined in Eqs. (71) and (74).

When the metabolite is generated and escapes from a set of sequential organs, an expanded set of terms—similar to that represented by Eq. (67)—is required to describe the total amount of metabolite generated in the set of sequential organs and entering the aortic blood.

As with Eq. (39), Eq. (75) may be converted to an equation for the concentrations of the metabolite after intravenous administration of the parent compound by assuming that $D_g$ is much greater than the other kinetic constants and by making appropriate adjustments for the various first-pass organ availabilites.

*M Non first-Pass: Cycle.* After the parent compound reaches the aortic blood, it may reach other organs in the central compartment where it may be converted to the metabolite. A portion of the metabolite escapes the organ and eventually enters the aortic blood. For each of such organs, the partitional equation would be the same as Eq. (75), except that the values for

$$Q^{\text{organ}} F_D^{\text{pin organ}} [[M]_{\text{out}}^{\text{organ}} / [D]_{\text{in}}^{\text{organ}}]^{\text{single pass}} \Pi F_M^{\text{post organ}}$$

would represent the organ under consideration rather than a first-pass organ.

For each of the partitioned equations representing cycled metabolite, the concentration of the metabolite entering the organ under consideration may be represented by

$$[M]_{\text{in}}^{\text{organ}} = [M]^{\text{art}} F_M^{\text{pin organ}} \tag{76}$$

and for the zone under consideration,

$$[M]_{\text{in}}^{\text{zone}} = [M]_{\text{in}}^{\text{organ}} \Pi F_M^{\text{pre zones}} \tag{77}$$

The rates of covalent binding of the cycled metabolite for each of the kinetic pathways within the zone may then be expressed by Eq. (72) or (73).

Simulations based on these parititioned equations are especially revealing, because they dramatize the relative contributions of the various kinetic pathways to the total concentration of the metabolite within a given zone of any given organ within the central compartment at any given time. For example, a given set of parameters may result in the partitioned concentrations of a metabolite in blood within the liver shown in Fig. 9. With this set of parameters, the concentration of the metabolite is initially due to the M first-pass: first-pass, kinetic pathways, whereas later it is due predominantly to the M cycle: cycle, kinetic pathway. Moreover, the contributions of the various pathways depend on different patameters. For example, an increase in the activities of the enzymes that generate the first-generation metabolite in liver would increase the magnitude of the contributions of the M first-pass: first-pass and the M first-pass: cycle pathways, but would not affect the time course of these pathways. By contrast, an increase in the enzyme in liver would decrease both the magnitude and the time course of the M cycle: first-pass, M cycle: cycle, and the M extrahepatic: cycle pathways. An increase in the elimination clearance of the parent compound in liver by mechanisms that do not lead to the first-generation metabolite would decrease the magnitude of all of the partitioned equations, but would not affect the time course of the M first-pass: first-pass and M first-pass: cycle path-

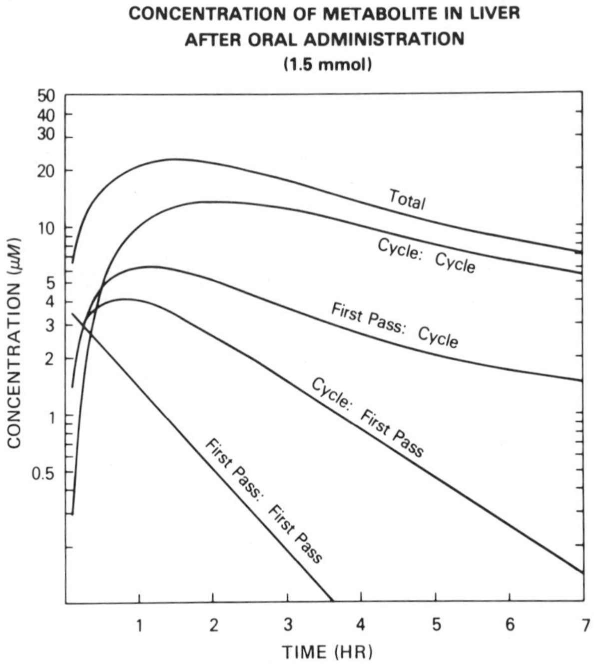

**Fig. 9.** Simulations of the partitioned concentrations of a first-generation metabolite arising from the partitioned kinetic pathways of the metabolite. The partitioned kinetic pathways are illustrated in Fig. 8. The equations used for the simulations are those for a well-stirred model, described in the text.

## 2. PHARMACOKINETICS OF FOREIGN COMPOUNDS

ways. An increase in the elimination clearance of the parent compound in an extrahepatic organ by mechanisms that do not generate the metabolite would decrease the magnitude and the time course of the M cycle: first-pass and M cycle: cycle partitioned kinetic pathways, but would have no influence on the M first-pass: first-pass or M first-pass: cycle pathways. By similar reasoning, the investigator may analyze the effects of changes in the elimination clearances by enzymes and other processes in various organs on the contributions of the various partitioned kinetic pathways and may also investigate the influence of changes in the rate constants of adsorption and the elimination clearances from the site of administration.

Such simulations also serve as a first step in predicting whether total concentrations of the parent compound or the metabolite might reach levels that approach or exceed the $K_m$ values of various enzymes that catalyze the formation or elimination of the metabolite and the predominant kinetic pathways at these times. For example, in some instances the $K_m$ values may be exceeded only during the rapid absorption of the parent compound, whereas in other instances they may be exceeded only during the highest concentrations of the D-cycle kinetic pathways.

### 3. Partitioned Area under the Concentration vs Time Curves

Each of the partitioned equations may be integrated to estimate the contribution each makes to the estimate of the total area under the concentration vs time curves for the parent compound and the metabolite.

Integration of Eq. (70) gives

$$AUC_{D(in)}^{\text{first pass}} = \frac{\text{Dose}_D(D_a)[\Pi F_D^{\text{pre organ}}]}{D_g Q^{\text{organ}}} \quad (78)$$

Integration of Eq. (71) gives

$$AUC_D^{\text{art(cycle)}} = \frac{\text{Dose}_D(D_a)\Pi F_D^{\text{organ(first-pass)}}}{D_g CL_D^{\text{body}}} \quad (79)$$

and

$$AUC_{D(in)}^{\text{organ(cycle)}} = AUC_D^{\text{art(cycle)}} F_D^{\text{pin organ}} \quad (80)$$

With these equations and Eqs. (61)–(63) and (67)–(69), as appropriate, it is possible to obtain area under the rate vs time curves for the covalent binding for the M first-pass: first-pass and M cycle: first-pass portion of the total covalent binding of the reactive metabolites in cells within a given zone of an organ. The area under the rate vs time curve gives the total amount of covalently bound material that would occur after all the metabolite had left the body and all of the covalent bound material remained unchanged in the cells.

Integration of Eq. (74) gives

$$AUC_M^{\text{art(first-pass cycle)}} = \frac{\text{Dose}_D^{po}(D_a) F_D^{\text{pre organ}} [[M]_{\text{out}}^{\text{organ}}/[D]_{\text{in}}^{\text{organ}}]^{\text{single pass}} \prod F_M^{\text{post organ}}}{D_g CL_M^{\text{body}}} \quad (81)$$

and integration of Eq. (75) gives

$$AUC_M^{\text{art(cycle, cycle)}} = \frac{\text{Dose}_D^{po}(D_a)[\prod F_D^{\text{first-pass organ}}] Q_D^{\text{organ}} [[M]_{\text{out}}^{\text{organ}}/[D]_{\text{in}}^{\text{organ}}]^{\text{single pass}} \prod F_M^{\text{post organ}}}{(D_g) CL_D^{\text{body}} CL_M^{\text{body}}} \quad (82)$$

Equations for the areas under the rate vs time curves may be calculated for the cycled metabolite by substituting these equations into the following equation:

$$AUC_{\text{rate}(M \to \text{cbp})}^{\text{zone}} = AUC_M^{\text{art(cycle)}} F_M^{\text{pin organ}} [\prod_{o \to i} F_M^{\text{zone}(i)}] X \quad (83)$$

where $X$ is $\text{rate}_{M \to \text{cbp}}/[M]^{\text{zone}(i)}$ described in Eqs. (62) or (63).

The total amount of covalent binding expected to occur within a given zone of a given organ would be the sum of the areas under the rate vs time curves of the individual partitioned kinetic pathways.

The primary purpose of partitioning the equations, however, is to dramatize the role of each of the kinetic pathways in the total covalent binding and to what extent alterations in enzyme activity in different organs affect each of the partitioned areas.

When the parent compound in hepatic cells is rapidly converted to a chemically reactive metabolite, which, in turn, is rapidly metabolized within hepatic cells, the M first-pass: first-pass pathway will account for most of the covalent binding of the metabolite within hepatocytes after oral administration of the parent compound. Consequently, regardless of the activities of the enzymes that metabolize either the parent compound or the metabolite in extrahepatic organs, changes in the activities of these enzymes may not appreciably alter the covalent binding of the metabolite within hepatic cells after oral administration of the parent compound, because the contributions of the cycle pathways will have been negligible. When the parent compound is administered intravenously, however, the M first-pass: first-pass pathway will not occur, and the covalent binding of the metabolite within hepatocytes will be due solely to the D-cycle pathways.

When the parent compound is eliminated from the body solely in the liver, the total amount of covalent binding of the metabolite within the liver will be independent of the route of administration of the parent compound regardless of the interorgan distribution of the enzymes that inactivate the metabolites [see Eq. (49)]. However, when the parent compound is eliminated in extrahepatic organs by pathways that did not form the chemically reactive metabolite, increases in the activities of the enzymes that eliminate the parent compound in the extrahepatic organs will decrease the covalent binding of the chemically reactive metabolite within hepatocytes to the extent that such increases in enzyme activity preferentially decrease the $AUC_D^H$ after intravenous administration [see Eq. (49)].

Other analyses may be made by visualizing the effects of changes in the organ availabilities of either the parent compound or the metabolite in various zones of the liver and other organs. Such analyses dramatically illustrate how these changes would alter the pattern of covalent binding within various zones of a given organ or between organs.

## VI. General Comments

This chapter has provided most of the basic mathematical tools for constructing models that describe the concentrations of a foreign compound and its first-generation metabolites in cells within various zones of different organs. In developing these mathematical tools, it has been assumed that all processes follow first-order kinetics and that the concentrations of the compound and its metabolites within the cells of vital organs reach a steady state relative to their concentrations in blood virtually instantaneously.

Although the mathematical tools provide insights to the basic framework of the pharmacokinetics of chemically reactive metabolites within various cells of the body, it is recognized that the assumption of first-order kinetics for all processes will not always be valid and that the models should be modified in those cases in which first-order kinetics are not applicable. Although dose-dependent and concentration-dependent clearances of the various processes are clearly important, they are beyond the objectives of this chapter. Even when such dose-dependent clearances occur, however, ranges that define the limits within which dose-dependent processes exert their influence may be established with first-order systems. For example, the steady-state intrinsic clearance of an enzyme [i.e., $V_m/(K_m + f_{fD}[D])$] that catalyzes a reaction must be between zero and $V_m/K_m$ of the enzyme, and the range of influence of that enzyme must be between these two clearances. When the enzyme accounts for only a small proportion of the total body clearance of the substances, such ranges of influence will be relatively small and may be of little pharmacological or toxicological significance. The partitioned equations representing different kinetic pathways of a compound, however, may illustrate that such small amounts of enzyme may, nevertheless, have a profound influence in first-pass portions of the equations. The partitioning of the equations for linear models provides insights into the results expected to occur in nonlinear models.

## References

1. Riggs, D. S. (1963). "The Mathematical Approach to Physiological Problems." Williams & Wilkins, Baltimore, Maryland.
2. Gillette, J. R., and Pang, K. S. (1980). Drug absorption, distribution and elimination. *In* The Basis of Medicinal Chemistry (M. E. Wolff, ed.), 4th ed., Part 1, pp. 55–106. Wiley, New York.
3. Gibaldi, M., and Perrier, D. (1975). "Pharmacokinetics." Dekker, New York.
4. Gillette, J. R. (1984). Solvable and unsolvable problems in extrapolation of toxicological data between animal species and strains. *In* "Drug Metabolism and Drug Toxicity" (J. R. Mitchell and M. G. Horning, eds.), pp. 237–260. Raven Press, New York.
5. Gillette, J. R. (1982). Sequential organ first-pass effects: Simple methods for constructing compartmental pharmacokinetic models from physiological models of drug disposition by several organs. *J. Pharm. Sci.* **71**, 673–677.
6. Rowland, M., Benet, L. Z., and Graham, G. G. (1973). Clearance concepts in pharmacokinetics. *J. Pharmacokinet. Biopharm.* **1**, 123–126.
7. Winkler, K., Keiding, S., and Tygstrup, N. (1973). Clearance as a quantitative measure of liver function. *In* "The Liver: Quantitative Aspects of Structure and Function" (G. Paumgartner and R. Presig, eds.), pp. 144–155. Karger, Basel.
8. Wilkinson, G. R., and Shand, D. G. (1975). Commentary: A physiological approach to hepatic drug clearance. *Clin. Pharmacol. Ther.* **18**, 377–390.
9. Gillette, J. R., Saul, W. F., and Maling, H. M. (1981). A new principle for estimating hepatic blood flow rates. *Pharmacology* **23**, 237–246.
10. Pang, K. S., and Gillette, J. R. (1978). Kinetics of metabolite formation and elimination in the perfused rat liver preparation: Differences between the elimination of preformed acetaminophen and acetaminophen formed from phenacetin. *J. Pharmacol. Exp. Ther.* **207**, 178–194.
11. Pang, K. S., and Terrell, J. A. (1981). Retrograde perfusion to probe the heterogeneous distribution of hepatic drug-metabolizing enzymes in rats. *J. Pharmacol. Exp. Ther.* **216**, 339–346.

*Chapter 3*

# Cellular Defense Mechanisms Against Reactive Metabolites

DONALD J. REED

*Department of Biochemistry and Biophysics*
*Oregon State University*
*Corvallis, Oregon*

|     |     |     |
| --- | --- | --- |
| I.  | Introduction | 71 |
| II. | General Concepts of Protection | 72 |
|     | A. Protection against Electrophiles | 72 |
|     | B. Protection against Free Radicals | 74 |
| III.| Types of Protection | 75 |
|     | A. Water | 75 |
|     | B. Endogenous Thiols | 77 |
|     | C. Exogenous Thiols | 81 |
|     | D. Vitamins and Antioxidants | 85 |
|     | E. Covalent Binding of Reactive Intermediates | 86 |
| IV. | Cellular Aspects of Protection | 88 |
|     | A. Compartmentation | 88 |
|     | B. Enzyme Induction | 90 |
|     | C. Redox Cycling | 93 |
| V.  | Comments | 96 |
|     | References | 97 |

## I. Introduction

As so elegantly described in Chapter 1 by Miller and Miller, the bioactivation of foreign compounds yields reactive intermediates and metabolites that are capable of cellular interactions. These interactions may be enzymatic or nonenzymatic, and the formation of electrophiles and free radicals by bioactivation processes is well documented, as illustrated in subsequent chapters with 15 classes of chemicals. Thus, enzymatic detoxication enzymes and the metabolic pathways that describe the reactions catalyzed now represent well-defined processes of foreign compound metabolism. Not so well delineated are the dynamics of these processes, such as competition within detoxication pathways, determination of pool sizes, and rates of turnover of precursors for conjugate formation.

Genetic and somatic alterations of cellular constituents occur during the detoxication of reactive metabolites. The extent of these alterations is related to the status of the cellular defense systems that protect cellular constituents and stabilize their functions. This chapter will examine the general nature of reactive metabolites and their secondary reactive products and will outline the types of events involved in cellular protection. Pertinent to "built-in" protection at the cellular level during chemical intoxication is the status of naturally occurring substances. Additionally, compartmentation of cellular protective systems must be considered, because its role is beginning to be placed in better perspective with regard to cell viability during chemical intoxication.

Cell viability and cell survival are dependent on various inherent protective features. The protection of vital cellular constituents appears to be dependent on the structural integrity of the cell, the compartmentation of both functions and constituents, and the presence of certain enzymes. These enzymes include glutathione $S$-transferases, epoxide hydrolases, superoxide dismutases (SOD), catalase, glutathione peroxidases (GSH-Px), and glutathione reductase. Essential low molecular weight constituents include water, thiols (particularly glutathione), vitamins C and E, and the vitamin A precursor, $\beta$-carotene; an absolute requirement for any of these agents for cellular protection has not been ascertained.

## II. General Concepts of Protection

### A. PROTECTION AGAINST ELECTROPHILES

Cellular defense against reactive intermediates can be defined or expressed on the basis of organic reagents in which the reactive intermediates are agents that possess a low electron density. These agents attack the cellular constituents at positions of high electron density and are therefore referred to as electrophilic reagents or electrophiles. Such reactive electrophiles, with their electron-deficient centers, are subject to attack by anions (negatively charged) such as hydroxide, or by an electron-rich center, such as the nitrogen atom in amines. These electron-rich entities are referred to as nucleo-

Table I

Nucleophilic Constants[a]

| Nucleophile | Nucleophilicity constant($n$) |
|---|---|
| $H_2O$ | 0.01 |
| $SO_4^{2-}$ | 2.5 |
| $CH_3COO^-$ | 2.76 |
| $Cl^-$ | 2.99 |
| $HPO_4^{2-}$ | 3.52 |
| $HCO_3^-$ | 3.8 |
| $Br^-$ | 4.02 |
| $(NH_2)_2CS$ | 4.1 |
| $C_6H_5NH_2$ | 4.35[b] |
| $SCN^-$ | 4.80 |
| $I^-$ | 4.93 |
| $SO_3^{2-}$ | 5.1 |
| $CN^-$ | 5.1 |
| $S_2O_3^{2-}$ | 6.35 |

[a] Koskikallio.[3a]
[b] Koskikallio.[3b]

philic reagents or nucleophiles. This electrophile–nucleophile dichotomy has been described as a unique case of the acid–base concept.[1] A Lewis acid becomes an electrophile by accepting electrons, and a Lewis base becomes a nucleophile by providing electrons. Electrophiles and nucleophiles can also show a relationship with oxidizing and reducing agents, even when covalent bonding does not accompany changes in electron states between two such reagents.

Swain and Scott[2] described a linear free-energy relationship for nucleophilic substitution reactions. The relative strength of a nucleophilic site is correlated with the ratio of the alkylation reaction rate of that nucleophilic site to the rate with water. The nucleophilic strength approximates the propensity of electrons at the reaction site to participate in covalent bond-forming activities.[3] Analysis of alkylation reactions with a variety of reaction sites over a range of nucleophilic strengths can indicate the sensitivity of the alkylating agent (electrophile) toward differences in nucleophilic strength, which is designated as $n$ for such reagents (Table I).[3a,3b] A greater selectivity of an alkylating agent is indicated by a higher substrate $s$ value, which can be related to the degree of $S_N2$ character of the reaction. Methyl bromide was chosen as a standard substrate for which $s = 1.00$.[2] The substrate $s$ values for $N$-methyl-$N$-nitrosoureas (MNU) and $N$-ethyl-$N$-nitrosourea (ENU) have been estimated to be 0.42 and 0.26, respectively.[4] Thus, MNU reacts more selectively than ENU and indicates that the reactive entity

**Table II**
Thiol Groups in Cysteinyl Derivatives[a]

| Compound | Percentage in the Thiolate form at pH 7.5 |
|---|---|
| L-Cysteinyl ester | 16 |
| L-Cysteinyl glycine | 11 |
| L-Cysteine | 6 |
| Glutathione | 1 |

[a] Benesch and Benesch.[4a]

is not a carbonium ion but rather an intermediate occurring ahead of the reaction chain.[4]

As a general rule, in terms of cellular protection, the higher the nucleophilic constant, the greater the effectiveness in protection if the nucleophile is functioning as an electrophile scavenger. The ability of thiols to protect depends greatly on the percentage of the thiol group present as an anion. Cysteine has a sixfold greater ionization than glutathione at pH 7.5 (Table II),[4a] but the relative concentration of glutathione (GSH) to cysteine in liver is about 50 to 1, and therefore the overall total protection is greater with glutathione because of its high concentration in most tissues.

## B. PROTECTION AGAINST FREE RADICALS

Radicals are involved in certain bioactivation processes as reactants. They are less susceptible to variations in electron density than the substrate being attacked. Reactions of this type are highly susceptible to the presence of small amounts of agents that either liberate or scavenge radicals.[1] The importance of agents such as ferrous iron for formation and vitamin E for scavenging of radicals continues to be a subject of lively polemics among toxicologists.

Reduced species of dioxygen (including those that are free radicals) are endogenous to cells and tissues. However, chemical intoxication by certain foreign chemicals can result in a magnification of the cellular content of these reactive intermediates in the pathway of the four-electron conversion of dioxygen to water.

Cellular constituents that protect against free radical-mediated cellular damage reside in the subcellular compartments. These subcellular organelle compartments are important, but attention must be paid to the distribution of both prooxidants and antioxidants between the aqueous and the lipid phases. Additionally, oxidation–reduction (redox) potentials vary with the degree of complexation with metal ions, macromolecules, and so on. Thus,

# 3. CELLULAR DEFENSE MECHANISMS

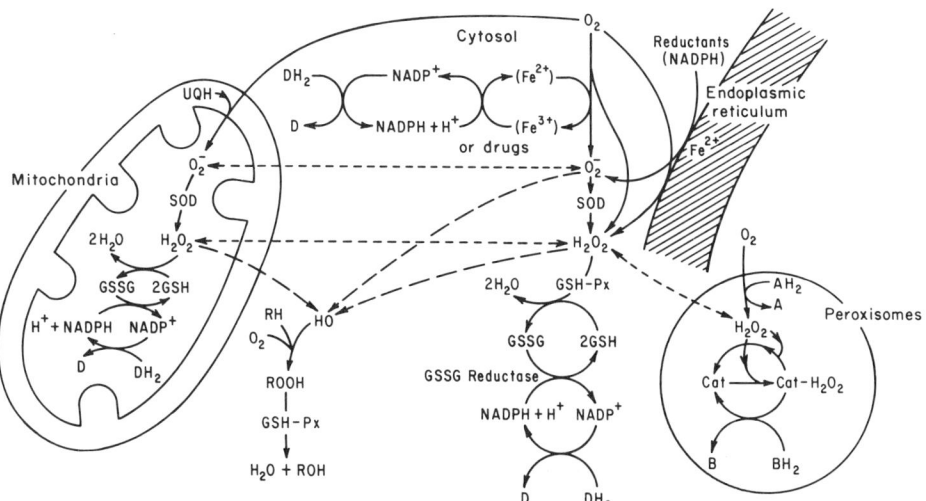

**Fig. 1.** Schematic diagram of cellular protection against oxygen-mediated toxicity. Modified from Sies et al.[4b] UQH: semiquinone of ubiquinone. D, DH, $AH_2$, B, $BH_2$: various cellular substrates involved in enzymatic redox reactions.

the general nature of these reactions and the protective systems are influenced by the particular milieu of the cellular interior.

Specific enzymes have a major responsibility for cellular protection against oxygen-mediated toxicity. The glutathione peroxidase–reductase redox cycle enzymes, catalase (Cat), and superoxide dismutases (SOD) catalyze reactions that cause rapid decreases in the concentration of reduced oxygen intermediates (Fig. 1).[4b] Glutathione is a substrate for the glutathione peroxidase–reductase redox cycle (an enzymatic reaction), but in nonenzymatic reactions the cell primarily utilizes $\beta$-carotene, $\alpha$-tocopherol, ascorbic acid, and possibly uric acid as important cellular protective agents.

Correlation of cellular damage by reduced oxygen species with lipid peroxidation processes is a major thrust of current research and therefore will be discussed again in this chapter.

## III. Types of Protection

### A. WATER

Water at intracellular concentrations up to 55 $M$ participates in cellular protection by providing the hydroxide nucleophile at a physiological pH of 7.4.

The formation of corresponding alcohols of carbon electrophiles, including the 2-chloroethyl carbonium ion,[5] is evidence of such protection. Nonenzymatic competitive reactions with halide ions or oxygen-, nitrogen-, or sulfur-containing anions in proportion to their nucleophilicity and reaction rate constants are described in the early work of Swain and Scott.[2]

Water has a vital role in cellular protection as a substrate for epoxide hydrolases, which convert epoxides to the corresponding diols. Although this role for water can be generally thought of as a protection mechanism, the conversion of certain arene oxides to the corresponding *trans*-dihydrodiols is a bioactivation step. For example, the 7,8-epoxide of benzo[*a*]pyrene, upon conversion to 7,8-*trans*-dihydrodiol, serves as the precursor of 7,8-*trans*-dihydrodiol 9,10-epoxide of benzo[*a*]pyrene, which is a potent mutagen and carcinogen.[6] Thus, polycyclic aromatic hydrocarbon epoxides present a most interesting challenge to the cellular protection systems. The structure of the first ultimate carcinogen identified for any member of the polycyclic hydrocarbon class is shown in Fig. 2.

The initial 7,8-epoxide undergoes "detoxication" with water as the reagent to form the corresponding *trans*-dihydrodiol. Failure of an adequate cellular elimination of this product can result in formation of the ultimate carcinogen with a 9,10-epoxide ring, which is in the bay region. The four possible metabolites are optically active isomers of the 7,8-diol 9,10-epoxide. The (+)-(7$R$,8$S$)-diol (9$S$,10$R$)-epoxide is the isomer responsible for nearly all of the tumorigenic activity attributed to benzo[*a*]pyrene.[7]

But why is one isomer much more carcinogenic than the other three isomers? One can suggest several important reasons. The active isomer has an epoxide group that forms part of a bay region of the hydrocarbon. Although this isomer is predicted to have high chemical reactivity,[6] it shares the common features of all four isomers; that is, it is a part of the bay region and has a sterically hampered enzymatic detoxication of the epoxide function.[8]

More than 10 polycyclic aromatic hydrocarbons have been examined as candidates for proximate and ultimate carcinogens, and all the reactive intermediates found conformed to the bay region theory (for a review, see Ref. 9).

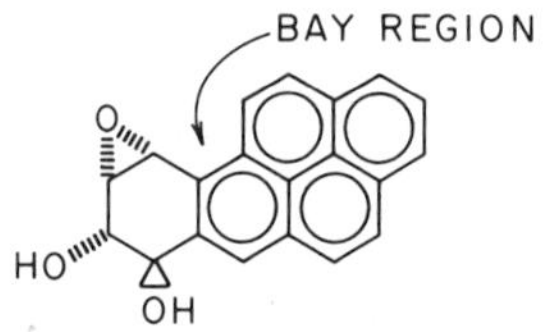

Fig. 2. The structure of the first ultimate carcinogen identified for any member of the polycyclic hydrocarbon class: (+)-(7$R$,8$S$)-dihydroxy-(9$S$,10$R$)-epoxy-7,8,9,10-tetrahydrobenzo[*a*]pyrene.

Protection by epoxide hydrolase activity can vary markedly as a result of differences in the efficiency of catalysis with different substrates. Microsomal epoxide hydrolase catalysis of the hydration of epoxides is thought to occur via a general base mechanism with the nucleophilic attack of water or hydroxide ion on the oxirane carbon.[10-12] Efficiency of catalysis is illustrated with the hydration of phenanthrene 9,10-oxide. The enzyme rate over that of hydroxide ion ($10^8$) and spontaneous water reactions ($10^{11}$) is $10^3$ times more efficient than hydronium ion catalysis.[12]

Strong support for the concept of hampered detoxication rates for the ultimate carcinogens comes from a study of compounds that cause destruction of the reactive epoxide group. A naturally occurring plant phenol with a high affinity for the diol epoxide through $\pi-\pi$ interactions (between the aromatic regions of the diol epoxide and the blocking agent) was found, which also has acidic and nucleophilic groups that participate in the destruction of the epoxide group. This compound, ellagic acid, reacts rapidly with benzo[a]pyrene-7,8-diol 9,10-epoxide to form a cis adduct that markedly inhibits mutagenicity in vitro.[13,14] Catalysis rates of epoxide hydrolases and glutathione S-transferases vary widely because of the nature of the polycyclic aromatic hydrocarbon epoxide isomers (see review in Ref. 15). Therefore, it was not surprising that glutathione was shown to conjugate with (±)-benzo[a]pyrene-7,8-dihydrodiol and (±)-benzo[a]pyrene-7,8-diol 9,10-epoxide when Jernström et al.[16] used isolated hepatocytes prepared from 3-methylcholanthrene-treated rats. More important, the intracellular level of glutathione must be maintained to prevent the reactive intermediates from interacting with the DNA in intact hepatocytes.[16]

## B. ENDOGENOUS THIOLS

Glutathione (L-$\gamma$-glutamyl-L-cysteinylglycine) accounts for about 90% of the intracellular nonprotein thiol content. In the hepatocyte, the major subcellular compartments of glutathione are the cytoplasm (85–90%) and the mitochondrial matrix (10–15%). The glutathione concentration of the latter appears as high as, or higher than, that in the cytoplasm.[17] Glutathione is resistant to typical proteases because of the $\gamma$-glutamyl linkage. Hydrolytic degradation of glutathione and its conjugates is mainly an extracellular process catalyzed by a specific enzyme, $\gamma$-glutamyltransferase. Thus, glutathione is an ideal cellular constituent for cellular protection systems.

The intracellular presence of a foreign chemical often requires that glutathione functions in a dual role for maintenance of cellular protection. Formation of glutathione conjugates has been shown with many foreign chemicals (for reviews, see Refs. 18 and 19). As early as 1879, the excretion of foreign chemicals as N-acetylcysteine derivatives was recognized.[20] The origin of the cysteine moiety remained obscure for over half a century, however, because the structure of glutathione was not established until 1929 by Hopkins[21] and independently by Kendall et al.[22]

The role of glutathione as the main intracellular interceptor of reactive electrophilic compounds is becoming well established. Early studies that demonstrated liver necrosis as a result of giving bromobenzene or acetaminophen led investigators to the determination that reactive intermediates can react enzymatically and nonenzymatically with glutathione at rates sufficient to deplete the pool of glutathione in the liver. Because the hepatotoxicity of these and other, similar foreign chemicals can be mitigated by pretreatment with cysteine, acetylcysteine, or methionine, it follows that glutathione biosynthesis, rather than direct interaction of these thiols, provides the majority of the protective effects.

Protection provided by glutathione against chemical damage is altered by various factors. The liver glutathione content is altered by diurnal or circadian variations as well as by fasting. Fasting of rats for 1 or 2 days decreases the level of liver glutathione to about 50 to 70% of the level in fed animals.[23-25] Diurnal variation of the liver glutathione content results in higher glutathione concentrations at night and early morning and lower concentrations in the late afternoon, with a maximum variation of 25 to 30%.[26-28]

Both fasting and depletion of glutathione with agents such as diethyl maleate cause a loss of protection and an enhanced hepatotoxicity of several chemicals. In an early study, Davis and Whipple[29] compared the effects of fasting and various diets on chloroform hepatotoxicity and noted that fasting increased toxicity. Other agents showing increased hepatotoxicity after fasting include carbon tetrachloride,[30-39] 1,1-dichloroethylene,[37] bromobenzene,[36,38,39] acetaminophen,[36,38,39] and thioacetamide.[36] Loss of glutathione protection has been demonstrated for acetaminophen, bromobenzene, carbon tetrachloride, and allyl alcohol[40,41] after diethyl maleate was used to deplete glutathione concentrations.

An explanation for the decrease in liver glutathione during fasting is apparent in the studies of Meredith and Reed[17] and Reed and Ellis,[42] which support the role of glutathione as a physiological reservoir for cysteine, as proposed by Tateishi et al.[43] As such, glutathione efflux from the liver appears to account for essentially all liver glutathione turnover except that associated with glutathione disulfide (GSSG) and glutathione conjugate efflux. In vivo inactivation of $\gamma$-glutamyltransferase activity causes extensive glutathionuria.[42] The glutathionuria prevents the normal reutilization of the cysteinyl moiety of glutathione after extracellular hydrolysis of glutathione, which is catalyzed by extracellular $\gamma$-glutamyltransferase in the kidney.[44] For example, after AT-125 treatment, fed animals showed a decrease in liver glutathione content similar to that exhibited after fasting, except that AT-125 treatment caused the urinary glutathione content to increase from less than 50 $\mu M$ to greater than 15 m$M$.[42] Thus, liver glutathione appears to efflux into the plasma, and a major portion of the sulfur of effluxed glutathione returns to the liver as cysteine to assist in maintenance of a high glutathione content. During fasting, the lack of dietary amino acid intake causes a decrease in total sulfur amino acids available to the liver as well as to other organs. One of the functions of liver glutathione is to provide the extracellular cysteine and cystine needed by many extrahepatic cells for biosynthesis of intracellular glutathione. These cells show a complete or nearly complete absence of the cystathionine pathway.[45]

Maintenance of intracellular glutathione appears totally dependent on *in situ* biosynthesis of glutathione. Rapid glutathione depletion and resynthesis, after 2-chloroethanol treatment of rats, require rapid replacement of the cysteine pool by either the influx of cysteine or cystine or the conversion of methionine and serine to cysteine via the cystathionine pathway.[46] Thus far, however, only hepatocytes have been demonstrated to utilize the cystathionine pathway to an appreciable extent during glutathione depletion and resynthesis.[47]

In the liver, methionine metabolism is a dynamic cycle with a continual flow between methionine and homocysteine, and the only outlet for the intermediates is the irreversible utilization of homocysteine to form cystathionine.[48] In fed rats, hepatic methionine sulfur has a half-life of about 5 min, and 40–60% of the tissue homocysteine is diverted to cystathionine formation at each pass through the cycle. During dietary protein restriction the half-life of methionine sulfur increases and the proportion of homocysteine diverted to cystathionine decreases, thereby conserving the tissue methionine content.[48] The interplay between dietary cystine and methionine for the maintenance of liver cysteine, and in turn glutathione, is a topic of current active investigation.

Glutathione *S*-transferases have a high specificity for glutathione, which is bound with a high affinity (~0.3 m$M$) to provide an enzyme-bound glutathione thiolate anion (GS$^-$) at an estimated intracellular concentration in the liver ranging up to about 0.1 m$M$. The combined effects of substrate affinities and the high concentration of substrate-bound nucleophilic anions provide an effective scavenger system for many reactive electrophiles, which undergo glutathione conjugate formation. Reactive chemicals, particularly hydrophilic electrophiles, are effectively sequestered by glutathione *S*-transferases. These transferases are present in liver in high concentrations; approximately 10% of the rat hepatic cytosolic protein is glutathione *S*-transferases.[49]

Because many oxygen-containing compounds exist in cells, one might think that sulfur compounds such as glutathione are not needed to react with electrophiles. However, as a general rule, sulfur compounds have been assumed to be about 1000 times better nucleophiles than oxygen compounds.[50] Ketterer et al.[51] have characterized glutathione and its thiolate anion (GS⁻) as strong nucleophiles that react best with "soft" electrophilic centers. When the electrophilic center is soft, the nucleophile causes polarization of the electrophile, and a rapid reaction ensues. In contrast, a "hard" electrophilic center polarizes a weak nucleophile, so that there is much less dependency on the nucleophile, and often competing reactions with water or other compounds result. The efficacy of glutathione against most carbon-centered electrophiles, which have hard electrophilic centers, is not understood.

An interesting comparison of reactivity toward acrylonitrile has been made between $HS-CH_2CH_2COOH$ ($\beta$-mercaptopropionic acid) and $NH_2-CH_2CH_2COOH$ ($\beta$-alanine), because the thiol and amino groups of these compounds have identical $pK_a$ values (10.05 – 10.06).[52] Steric environments for these compounds are also comparable. Therefore, the 280-fold greater reactivity of the sulfur group compared to the amino group must be explained on the basis of differences in reactivities, which depend on polarizabilities of nonbonded electrons on nitrogen and sulfur, charge distributions in ground and transition states, and solvation factors[52] (Fig. 3).

In the comparison of the rate of reaction of the nonprotonated amino

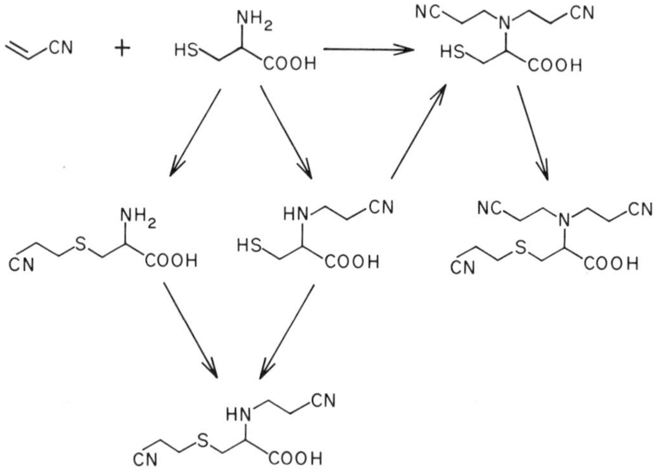

Fig. 3. A schematic diagram of the possible reaction pathways of aminothiols with $\alpha,\beta$-unsaturated compounds.

groups or the thiolate anion with the double bond of acrylonitrile, both functional groups must approach the double bond almost at right angles to the plane of the molecule of acrylonitrile. However, the amino group forces electron relocation that is energetically less favorable than the sulfur atom electron distribution.[52]

The influence of pH on the rates at which thiols react with electrophiles can be indicated by examples. The ratio of rate constants for the reaction of methyl iodide with thiophenol anion ($C_6H_5S^-$) and thiophenol ($C_6H_5SH$) is $1.6810^4$ in methanol at 25°C.[53]

The thiol group of glutathione has a $pK_a$ of 8.56, and the ratio of rate constants was $3 \times 10^3$ when the reaction of glutathione with acrylonitrile was determined at pH 10.5 ($GS^-$ only) and 5.3 glutathione (GSH) only.

At physiological pH, 99% of the glutathione exists as glutathione and 1% as the anion ($GS^-$). Because the reaction rate with electrophiles is about 3 orders of magnitude greater for the anion, it is not surprising that glutathione S-transferases catalyze glutathione conjugation reactions by binding glutathione, promoting the formation of the anion, as well as by binding lipophilic electrophiles.

## C. EXOGENOUS THIOLS

Exogenous thiols and other sulfur-containing compounds have been utilized to mimic or assist glutathione in detoxifying reactive intermediates. In general, administration of sulfur-containing compounds provides the greatest protective effects when such compounds enhance the synthesis of cysteine and, in turn, glutathione.

The complexity of protective mechanisms for exogenous thiols can be illustrated with N-acetylcysteine treatment for an overdose of acetaminophen. N-Acetylcysteine either prevents an acute depletion of glutathione or allows repletion by hepatocytes.[54,55] Presumably these effects occur through deacetylation of N-acetylcysteine[56] and cysteine incorporation into glutathione, but direct evidence has not been reported. Other therapeutic effects are possible: (1) N-acetylcysteine may inhibit the oxidation transformation of acetaminophen to its reactive metabolite; (2) N-acetylcysteine may be metabolized to inorganic sulfate, thereby increasing the availability of the limiting precursor in sulfate conjugate formation; and (3) N-acetylcysteine or cysteine may react nonenzymatically with reactive intermediates.

The relative formation of acetaminophen glutathione, sulfate, and glucuronide conjugates *in vivo* remains under active investigation.[57,58] In the presence of N-acetylcysteine, isolated mouse hepatocytes, when incubated with acetaminophen, decrease formation of acetaminophen sulfate and increase formation of acetaminophen glutathione compared to incubation without $N$-acetylcysteine.[55] Hinson[59] has reviewed the reactive metabolites of phenacetin and acetaminophen and has described evidence suggesting that $N$-acetyl-$p$-benzoquinoneimine is a reactive metabolite.

Protection by thiols against free radicals can be described by either a hydrogen donation reaction:

$$R\cdot + R'SH \rightarrow RH + R'S\cdot$$

or a charge transfer reaction:

$$R^{\ddagger} + R'SH \rightarrow R + R'S\cdot + H^+$$

The latter reaction has been examined[60] after oxidation of 2'-deoxyguanosine 5'-monophosphate by $\cdot OH$ or $Br^{\bar{}}$ to form $dGMP^{\ddot{+}}$ with the reaction:

$$dGMP^{\ddot{+}} + RSH \rightarrow dGMP + RS\cdot + H^+$$

The "rate constants" for thiols are estimated in Table III. These rate constants appear comparable to the relative reaction rates in the protection against electrophiles by thiols.

The thiyl radical $RS\cdot$ is probably the most prominent sulfur-centered radical within cells and is formed either by hydrogen atom abstraction from thiols by other radicals ($R\cdot + R'SH \rightarrow RH + R'S\cdot$), by reduction ($RSSR + R'_{aq} \rightarrow RS\cdot + RS^- + R'_{aq}$) or displacement reactions of disulfides ($RSSR + PO_3\cdot^{2-} \rightarrow RS\cdot + RSPO_3^{2-}$). Sulfur-centered radical anion formation results from the stabilization of the thiyl radical ($RS\cdot + RS^- \rightarrow RSSR^{\bar{}}$). Radical cation formation can occur with disulfides by a one-electron oxidation to yield $RSSR^{\ddot{+}}$. Because sulfur-centered radicals undergo many types of radical reactions, including rearrangement, displacement, fragmentation, addition, abstraction, and redox reactions, the potential for multiple cellular effects is obvious.

An interesting example of radical generation and rapid elimination of such radicals is found with 4-dimethylaminophenol (DMAP). Eyer and Lengfelder[61] reported that in the presence of an oxidant, potassium ferricyanide, the transient 4-phenoxyl radical was generated and reached a steady-state concentration under the conditions they used in about 40 msec. Addition of equimolar concentrations of glutathione and DMAP caused an immediate disappearance of the phenoxyl radicals and the formation of glutathione disulfide. These workers also observed that the rate of ferricyanide reduction by glutathione was only $10^{-4}$ that of ferricyanide oxidation of DMAP.[61] Interestingly, DMAP is relatively stable in oxygenated solutions as a result of the rapid reduction of the phenoxyl radical to

# 3. CELLULAR DEFENSE MECHANISMS

**Table III**
Estimated Rate Constants for Thiols

| Thiol (pH ≈ 7.4) | $10^{-4} k$ {[OH(dGMP) product] + RSH} (dm$^3$ mol$^{-1}$ sec$^{-1}$) |
|---|---|
| Cysteamine | 20 |
| Cysteine | 6 |
| Dithiothreitol | 1.3 |
| Glutathione | 0.6 |
| N-Acetylcysteine | 0.4 |

DMAP. The rate of this reaction appears to exceed greatly the rate of the dismutation reaction of $O_2^-$.[61]

Aminothiols, thiocarbamides, mercaptoimidazoles, and dithiocarbamates are similar in their ability to deplete liver glutathione. Flavin-containing monooxygenase catalyzes the oxidation of the aminothiol cysteamine to cystamine by the following scheme[62]:

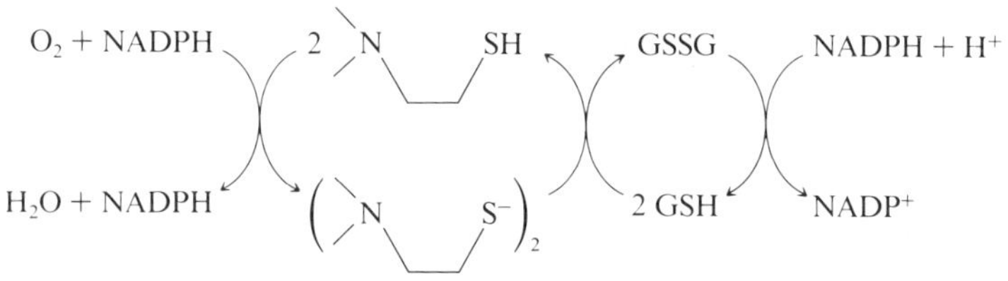

$$\begin{array}{c} O_2 + NADPH \\ \searrow \\ H_2O + NADP^+ \end{array} \underset{\longleftarrow}{\overset{2 \; \text{N-SH}}{\rightleftharpoons}} (\text{N-S}^-)_2 \underset{\longleftarrow}{\overset{GSSG}{\rightleftharpoons}} 2 \; GSH \quad \underset{\longleftarrow}{\overset{NADPH + H^+}{\rightleftharpoons}} NADP^+$$

Depletion of NADPH decreases the rate of glutathione disulfide reduction, and glutathione disulfide accumulation in the liver is accompanied by glutathione disulfide efflux from the liver. A time- and dose-dependent depletion of glutathione occurs with a maximum decrease in liver glutathione concentrations about 2 hr after animals are given 120 mg cysteamine/kg body weight.[63] The glutathione disulfide lost from the liver enters the circulation and is rapidly degraded hydrolytically by extracellular γ-glutamyltransferase in the kidneys to yield cystine.[44] Cystine is utilized by the kidneys for glutathione biosynthesis.[64] Glutathione and cystine can undergo a rapid thiol–disulfide interchange reaction as follows:

Cystine + glutathione → cysteine + cysteinylglutathione

Cysteinylglutathione is an excellent substrate for γ-glutamyltransferase resulting in the overall conversion of effluxed glutathione to extracellular cysteine. This may be important in the cysteine requirement for biosynthesis of glutathione in certain cell types lacking the cystathionine pathway for cysteine biosynthesis.[47] The observed depletion of liver glutathione by substrates for flavin-containing monoamine oxidase indicates why protection by agents such as cysteamine is complex and may reflect alterations in extracellular thiol–disulfide status. These alterations are as important as the perturbation of the intracellular thiol–disulfide homeostasis during protection. An important aspect of oxidative challenge by exogenous thiols such as cysteamine is the effect of thiol–disulfide interchange reactions. Rate constants and equilibrium constants for thiol–disulfide interchange reactions involving glutathione disulfide have been reported for a variety of reducing thiols.[65] The rate of reduction follows a Brønsted relation, with a large contribution to the equilibrium constants being derived from the relative $pK_a$ values of the reducing thiols and the thiols derived from the disulfides. Interestingly, the rate of reduction does not necessarily correlate with the equilibrium constant for the reduction of a disulfide by a thiol.[65] A $S_N 2$ displacement reaction characterizes such thiolate–disulfide interchanges.

$$RSH \rightleftharpoons RS^- + H^+$$

$$RS^- + GSSG \rightleftharpoons GSSR + GS^-$$

$$RS^- + GSSR \rightleftharpoons RSSR + GS^-$$

$$GS^- + H^+ \rightleftharpoons GSH$$

The susceptibility of strong reducing thiols to autoxidation, including protein thiols,

$$2\ RSH + O_2 \rightarrow RSSR + H_2O_2$$

$$RSSR + GS^- \rightleftharpoons RSSG + RS^-$$

$$RSSG + GS^- \rightleftharpoons RSH + GSSG$$

and hydrogen peroxide production mean that protection of intracellular thiols is a complex process.

Thiols are generally given a protective role in the prevention of lipid peroxide formation. However, thiols may promote lipid peroxidation under certain *in vitro* conditions. Recent studies have suggested that certain thiols may reduce iron directly, and, in turn, $Fe^{2+}$ may undergo autoxidation without an $O_2^-$ intermediate:

$$2\ Fe^{2+} + O_2 + 2\ H^+ \rightarrow 2\ Fe^{3+} + H_2O_2$$

Therefore, a thiol-dependent lipid peroxidation may occur by a direct re-

# 3. CELLULAR DEFENSE MECHANISMS

duction of iron by thiols followed by initiation of

$$RS^- + Fe^{3+} \rightarrow Fe^{2+} + RS\cdot$$

lipid peroxidation by the thiyl radical that is independent of superoxide anion radical.[66]

## D. VITAMINS AND ANTIOXIDANTS

Vitamin E (Fig. 4) is an efficient inhibitor of lipid peroxidation *in vivo* (for a review, see Ref. 67). Autoxidation, in simple terms, is a chain reaction and is described in the following reaction scheme of Burton and Ingold.[68] RH represents the organic substrate and initiation:

| | | | |
|---|---|---|---|
| *Initiation:* | $RH \rightarrow R\cdot$ or $ROO\cdot$ | | (1) |
| *Propagation:* | $R\cdot + O_2 \rightarrow ROO\cdot$ | | (2) |
| | $ROO\cdot + HR' \rightarrow ROOH + R'\cdot$ | | (3a) |
| | $ROO\cdot + R'H \rightarrow ROOR'\cdot H (\equiv R\cdot)$ | | (3b) |
| *Termination:* | $ROO\cdot + ROO\cdot \rightarrow$ nonradical products | | (4) |

with $ROO\cdot$ the peroxy radical as the product.

Several individual tocopherols constitute vitamin E, and the relative and the absolute antioxidant effectiveness *in vitro* has only recently been clarified.[68] These chain-breaking phenolic antioxidants (ArOH) shorten the oxidation chain. Whereas chain termination by reaction (4) is suppressed, termination may occur by reactions (5) and (6) with $n$ being the stoichiometric factor for the antioxidant.

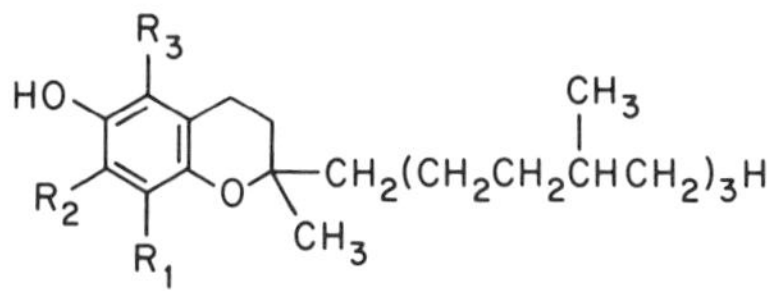

$\alpha$-Tocopherol ($\alpha$-T): $R_1 = R_2 = R_3 = CH_3$

$\beta$-Tocopherol ($\beta$-T): $R_1 = R_3 = CH_3; R_2 = H$

$\gamma$-Tocopherol ($\gamma$-T): $R_1 = R_2 = CH_3; R_3 = H$

$\delta$-Tocopherol ($\delta$-T): $R_1 = CH_3; R_2 = R_3 = H$

**Fig. 4.** Vitamin E.

$$\text{ROO} \cdot + \text{ArOH} \rightarrow \text{ROOH} + \text{ArO} \cdot \qquad (5)$$

$$(n-1)\text{ROO} \cdot + \text{ArO} \cdot \rightarrow \text{nonradical products} \qquad (6)$$

In an inhibited reaction in which all ArO· are destroyed by reaction (6), the rate of autoxidation is described by the following equation:

$$\frac{-d[O_2]}{dt} = \frac{k_3[\text{RH}]R_i}{nk_5[\text{ArOH}]}$$

where $R_1$ is the rate of chain initiation.

The abstraction by peroxyl radicals of the phenolic hydrogens from these tocopherols is described by the rate constant $k_5$. The values of $k_5$ for $\alpha$-, $\beta$-, $\gamma$-, and $\delta$-tocopherols are 23.5, 16.6, 15.9, and $6.5 \times 10^5$ $M^{-1}$ sec$^{-1}$, respectively, at 30°C.[68] Each tocopherol was found to react with two peroxyl radicals, and all the tocopherols appear to be exceptionally good chain-breaking antioxidants *in vitro*. Furthermore, the data of Burton and Ingold[68] are in agreement with *in vivo* tests of the relative biological activities of these tocopherols(T): $\alpha$-T > $\beta$-T > $\gamma$-T > $\delta$-T.[69]

Rate constants have been reported for the reaction of superoxide radicals with ferricytochrome $c$ ($k = 2.6 \pm 0.2 \times 10^5$ $M^{-1}$ sec$^{-1}$ at pH 9.0) and ascorbate ($k = 1.52 \pm 0.1 \times 10^5$ $M^{-1}$ sec$^{-1}$ at pH 9.9).[70] It has been suggested that the facile oxidation of ascorbate by $O_2^-$ probably occurs via hydrogen atom transfer.[71]

In aqueous media, reduced paraquat (methyl viologen; MV$^+$) combines with oxygen to give a stoichiometric yield of $O_2^-$. In an aprotic solvent such as dimethylformamide, MV$^+$ and $O_2^-$ combine in a 1:1 stoichiometry to form irreversibly a peroxy zwitterion adduct [MV$^+$ $O_2^-$] with the O—O group at the 2-position.[72] Rapid decomposition via ring rupture and oxidative reactions yields a multitude of products.[72]

Interestingly, no evidence has been obtained to indicate that $O_2^-$ acts as an initiator of radical chain reactions.[71] The powerful nucleophilic properties of $O_2^-$ in aprotic solvents do not exist in aqueous media.[71]

### E. COVALENT BINDING OF REACTIVE INTERMEDIATES

During the past two decades, it has become generally accepted that chemically inert foreign compounds can be transformed to chemically reactive intermediates that react covalently with cellular macromolecules as well as with low molecular weight nucleophiles.

The Millers pioneered in establishing the transformation processes of the endoplasmic reticulum, and they advanced our understanding of covalent interactions as related to specific chemical carcinogens.[73]

A threshold dose of certain chemicals above which covalent alkylation

## 3. CELLULAR DEFENSE MECHANISMS

becomes much greater has been discussed in detail by Gillette.[74,75] The threshold dose response for bromobenzene has been explained on the basis that the concentrations of glutathione in liver are decreased after the administration of toxic doses of this agent until the rate of formation of glutathione conjugate is limited by the availability of glutathione.[75] Jollow et al.[76] showed that the rate of covalent binding of bromobenzene metabolites to liver macromolecules was markedly increased after glutathione was depleted. Glutathione protects the liver against the toxic effects of bromobenzene epoxide by conversion to the glutathione conjugate.

Furosemide shows negligible covalent binding and causes no necrosis below a threshold dose of about 150 mg/kg.[77] In contrast to bromobenzene and acetaminophen, furosemide does not deplete the liver of glutathione or other protective nucleophiles. A lack of a delay in the covalent binding after administration supports these observations.[75]

Chemically reactive metabolites of chemicals bind to cellular proteins, including endoplasmic reticulum,[78,79] nuclear proteins,[80-82] and cytosolic proteins.[83-85]

There is some evidence that binding to certain proteins may occur preferentially rather than randomly. For example, after administration of either 3-methylcholanthrene or benzo[a]pyrene *in vivo,* ligandin is the preferred target protein in rat liver cytosol.[80,85-88]

Protection against certain substrates containing olefinic bonds is afforded by cytochrome $P$-450 monooxygenases, especially the isozymes that are inducible by phenobarbital treatment. This protection results from the selective destruction of the enzyme prosthetic heme group and the formation of "green" porphyrin by N-alkylation.[89]

Interestingly, 2,3,7,8-tetrachlorodibenzo-$p$-dioxin (TCDD), an extraordinarily potent toxin, binds to hepatic DNA or RNA at 6 and 12 pmol/mol of nucleotide residue.[90] Proteins are alkylated at a level of 60 pmol TCDD per mole amino acid residue. This level of binding is 5 or 6 orders of magnitude less than most carcinogens, which covalently bind to protein RNA and DNA to the extent of $10^{-4}$ to $10^{-6}$ mole of carcinogen per mole of amino acid or nucleotide residue.[91] Poland and Glover[90] concluded that at a maximum level of covalent binding of TCDD to DNA, binding is about 1 molecule of TCDD per DNA equivalent to 35 cells. Therefore, they conclude that covalent binding of TCDD is not likely to have a role in TCDD-induced carcinogenesis.

Alkylation of DNA *in vivo* by electrophiles may include a significant fraction of "apparent" alkylation due to extensive alkylation of DNA precursor nucleotides. Cellular DNA precursor pools of deoxyribonucleotides are generally more susceptible to methylation than are residues within the DNA duplex.[92] The N-1 position of adenosine is at least 13,000 times more susceptible to methylation by $N$-methyl-$N$-nitrosourea than the same site in DNA.[93] Even though alkylated deoxynucleotides can be incorporated into DNA, the alkylation of DNA precursor nucleotides may provide a type of protection against DNA damage. Further work is needed to evaluate the effects of products resulting from alkylation of DNA precursor nucleotides on cell viability.

Covalent binding to DNA has been proposed as a quantitative indicator for genotoxicity.[94]

3,5-Di-*tert*-butyl-4-hydroxytoluene (BHT) undergoes a cytochrome $P$-450-catalyzed conversion to a quinone methide.[95] The specific binding to protein, but not to nucleic acids, has led to the detection of a conjugate between the 4-methyl group and the cysteine sulfhydryl group of microsomal proteins. These workers have concluded that *in vivo* BHT can conjugate not only to glutathione and cysteine, but also to protein sulfhydryls, and that the latter reaction may produce BHT-mediated damage leading to the observed latent toxicity of this compound.[96]

## IV. Cellular Aspects of Protection

### A. COMPARTMENTATION

The concentration and nature of nonprotein thiols and protein sulfhydryl groups has been a topic of long-standing interest. Vital cellular functions of protein sulfhydryls are numerous and cannot be enumerated here. The dynamic nature and relatively high concentration of protein sulfhydryl groups, which may exceed the concentration of total nonprotein thiols (6–10 m$M$ in liver), are closely related to the status and functions of nonprotein thiols. Protection of protein thiols from alterations (including the homeostatic oxidation state) involves the total cellular thiol:disulfide potential.[97] Subcellular distribution and concentrations of both nonprotein thiols and protein sulfhydryls may relate closely to their respective functions. More than 90% of the total nonprotein thiols present in cells is glutathione. The presence of a discrete mitochondrial pool of glutathione was proposed by Vignais and Vignais.[98] Mitochondria retained glutathione during experiments involving nonaqueous media.[99,100] Isolated rat mitochondria contain about 10% of the total hepatic glutathione content, and about 90% of the glutathione is present in the reduced form.[99] Wahlländer *et al.*[101] reported the mitochondrial glutathione content to be 13% of the total liver content by a nonaqueous extraction procedure. Meredith and Reed[17] suggested that, on the basis of compartment water space,[101,102] the mitochondrion maintains a higher glutathione concentration than the cytoplasm, 10 m$M$ vs 7 m$M$, respectively. The apparent impermeability of the inner membrane of the mitochondrion led to the speculation that mitochondria maintain intramitochondrial glutathione by *in situ* synthesis.[101] Higashi *et al.*[103] have suggested that liver glutathione is a two-compartment physiological reservoir of L-cysteine. A labile compartment serves as a cysteine reservoir and as a more stable compartment, which is not readily available even during starvation. Cho *et al.*[104] have provided confirmatory evidence in that fasted and refed rats maintain a constant concentration of plasma cystine.

Studies on glutathione biosynthesis demonstrate separate pools of glutathione in the cytosol and the mitochondria with *in vivo* turnover half-lives of 2 and 30 hr, respectively.[17] Short-term starvation depletes the cytosolic pool, but not the mitochondrial pool.

Differential depletion of cytosolic and mitochondrial glutathione of freshly isolated hepatocytes with glutathione-depleting agents has permitted an evaluation of the protective role of intracellular glutathione. Chemical intoxication observed with ethacrynic acid,[17] acetaminophen,[105] or bromobenzene[106] suggests that the short-term depletion of cytosolic glutathione contents does not cause a significant loss of cell viability. When depletion of cytosolic glutathione contents was accompanied by partial depletion of mitochondrial glutathione, a rapid increase in loss of cell viability was observed. Much more needs to be understood about the consequences of depletion of the pools of intracellular glutathione.

Epoxides of aromatic and olefinic carbon–carbon bonds are electrophilic compounds, which are capable of alkylating cellular constituents. Studies utilizing intact organs or isolated organ cell suspensions demonstrate clearly that both the glutathione *S*-transferases and the epoxide hydrolases have a major role in the metabolism of epoxides. Different subcellular localizations of these enzymes place the glutathione *S*-transferases in the cytosol and epoxide hydrolase in the endoplasmic reticulum or the cytoplasm, depending on the substrate (see review, Ref. 15). An example of the dual role of these enzymes for cellular protection is the prevention of styrene 7,8-oxide-induced hepatotoxicity.[107]

Hill and Burk[108] have speculated that vitamin E–deficient hepatocytes are more suspectible to oxidative stress than normal hepatocytes. Also, other oxidant defenses were unable to prevent the lipid peroxidation that occurred under the incubation conditions employed in their experiments. The loss of cell viability during incubation was not accompanied by depletion of glutathione, which indicates that some aspect of membrane fragility may relate to a specific function of vitamin E. Compartmentation may have an important role in these observations.

## B. ENZYME INDUCTION

Administration of certain "antioxidants" is an important aspect of cellular protection for the prevention of induction of experimental tumors by a variety of chemical carcinogens. Wattenberg[109-112] made an extensive study of this phenomenon followed by reports by Weisburger's laboratory.[113-115]

Theories advanced to explain these effects of antioxidants have been discussed by Benson et al.[116] and include (a) direct interaction of the carcinogen or of its activated metabolic products with the antioxidant, (b) enhanced activities of enzymes that inactivate the proximate or ultimate carcinogens, thereby limiting the damaging interactions with critical macromolecular components and also preventing the initiation process of cellular transformation, (c) blockage or competition against specific metabolic activation processes required to convert procarcinogens to their reactive intermediates, and (d) increased efficiency in repair of damaged DNA.

Addition of either 2(3)-*tert*-butyl-4-hydroxyanisole (BHA) or 1,2-dihydro-6-ethoxy-2,2,4-trimethylquinoline (ethoxyquin) to the diet greatly decreases the levels of mutagenic metabolites of benzo[*a*]pyrene in CD-1 mice.[117] These observations have led to the finding that liver cytosol from BHA-fed mice and rats exhibits glutathione *S*-transferase activities that are enhanced more than 10-fold.[116] Increased enzyme protein included multiple glutathione *S*-transferase species.[116]

Studies by Nakagawa et al. show that butylated hydroxytoluene (BHT) given orally increases the total glutathione content in rat liver and the activities of glutathione *S*-transferase and glutathione disulfide reductase but has no effect on glutathione peroxidase.[118] The isomers of BHA, 2-*tert*-butyl-4-hydroxyanisole (2-BHA) and 3-*tert*-butyl-4-hydroxyanisole (3-BHA), have been compared in their induction properties in mouse liver, and both isomers increased the acid-soluble thiol content and increased the activities of glutathione *S*-transferases and epoxide hydrolase.[119] The 3-BHA induction was more than three times higher than that of 2-BHA in the liver. In the forestomach, the induction by the isomers was reversed.[119]

Dietary administration of BHA to mice also enhances dicoumaryl-inhibited hepatic NADPH: quinone reductase [NAD(P)H dehydrogenase(quinone); NAD(P)H (quinone acceptor) oxidoreductase, EC 1.6.99.2] activity 10-fold.[120] Severalfold increases were observed in kidney, lung, and mucosa of the upper small intestine. The protective effects of BHA may be a result of increased conversion of quinones to readily excreted conjugates.[120]

There is a growing body of evidence that *vic*-diol epoxides, reactive intermediates of polycyclic hydrocarbon, are largely responsible for most of the mutagenic and carcinogenic effects induced by polycyclic

## 3. CELLULAR DEFENSE MECHANISMS

hydrocarbons.[121-132] Experiments with *trans*-7,8-dihydro-7,8-dihydroxybenzo-[*a*]pyrene and 7$\beta$,8$\alpha$-dihydroxy-9$\alpha$,10$\alpha$-epoxy-7,8,9,10-tetrahydrobenzo[*a*]pyrene indicate some inactivation is glutathione dependent[16,80,133-138] but not dependent on microsomal epoxide hydrolase.[138-141]

Earlier studies on the induction of phase II enzymes by phenobarbital and 3-methylcholanthrene,[142-144] TCDD (2,3,7,8-tetrachlorodibenzo-*p*-dioxin),[142] and 3,4-benzo[*a*]pyrene[143] indicated less induction than that caused by antioxidants, such as BHA.[116]

Using a mutagenic test system, Glatt *et al.*[145] have reported the inactivation of a polycyclic aromatic diol epoxide by dihydrodiol dehydrogenase activity. Dihydrodiol dehydrogenase, after purification to apparent homogeneity from rat liver cytosol,[146] converted dihydrodiols to catechols with $NADP^+$ as a cosubstrate. The mutagenicity of *r*-8, *t*-8, *t*-9-dihydroxy-*t*-10,11-oxy-8,9,10,11-tetrahydrobenz[*a*]anthracene (BA-8,9-diol 10,11-oxide) was not decreased by the presence of large amounts of highly purified microsomal or cytosolic epoxide hydrolase.[145] Because highly purified dihydrodiol dehydrogenase inactivated this diol epoxide, the authors suggest that dihydrodiol dehydrogenase is probably more important than either microsomal or cytosolic epoxide hydrolase in the inactivation of diol epoxides.[145]

Phenobarbital induction followed by isolation of liver poly(A)$^+$RNA and translation experiments led Pickett *et al.*[147] to conclude that the level of translatable glutathione *S*-transferase B mRNA is elevated about three- to fourfold by phenobarbital administration. Felton *et al.*[148] have shown that induction of hepatic glutathione *S*-transferase activity by polycyclic aromatic compounds does not correlate with the Ah receptor, which controls the inducible aryl hydrocarbon (benzo[*a*]pyrene) hydroxylase activity. *trans*-Stilbene oxide has different inducing properties than either phenobarbital or 3-methylcholanthrene.[149,150] Treatment of rats with *trans*-stilbene oxide increases the epoxide hydrolase activity of liver microsomes to more than 700% of control values.[151,152] Cytoplasmic glutathione *S*-transferase activity increases 300-400%, leading to the conclusion that this agent induces phase II enzymes considerably more than the cytochrome *P*-450 system, which is about doubled.[152] The structural requirements of stilbene oxide induction have been examined. A comparison of induction in rats by *cis*- and *trans*-stilbene, *cis*- and *trans*-stilbene oxide, benzoin, and benzil led to the conclusion that *trans*-stilbene oxide is the actual inducing agent and that benzil is more selective as an inducer of epoxide hydrolase than is *trans*-stilbene oxide.[153]

Certain genetic differences in cellular protection mechanisms can be described as specific genetic deficiencies. For example, in a review by Dutton,[154] the Gunn rat is characterized by the absence of monoglucuronidation of bilirubin and a decreased glucuronidation of thyroxine and phenol red in comparison to other rat strains, but the species has an unimpaired ability to conjugate morphine and edrophon.

The discovery that selenium, as selenocysteine,[155] is an essential element of glutathione peroxide has provided much insight into the role of selenium in oxidative processes. Glutathione peroxidase is the only identified selenium-containing mammalian enzyme.[156] The structure and function of this enzyme have been reviewed.[157-159]

Selenium-deficient rat liver has a non-selenium-dependent glutathione peroxidase activity that is associated with one or more of the glutathione S-transferases.[160,161] Organic hydroperoxides, such as cumene and tent-butyl hydroperoxide, but not hydrogen peroxide, are reduced by these enzymes.

A membrane-bound glutathione peroxidase-like activity from mouse liver and cardiac mitochondrial membrane has been described.[162] This enzyme activity has been distinguished from the cytosol and mitochondrial matrix selenium-dependent glutathione peroxidases. Three features are important: (1) the enzyme activity is membrane bound and is not affected by selenium deficiency; (2) it does not appear to be a glutathione S-transferase[162]; and (3) the enzyme appears to be located in mitochondrial intermembrane space.

The relationship between selenium and glutathione in cells is understood in terms of the essential role of selenium in the function of glutathione peroxidase.[155] Hill and Burk[108] reported that when selenium-deficient rats were used as a source of freshly isolated hepatocytes, the hepatocytes demonstrated an altered glutathione status after incubation. The glutathione concentration in selenium-deficient hepatocytes rose to 1.4 times that of control hepatocytes. The selenium-deficient cells released twice as much glutathione into the incubation medium as did the control cells. Plasma glutathione concentrations in selenium-deficient rats were twofold greater than in control rats, indicating that increased glutathione synthesis and release occurs *in vivo* during selenium deficiency.[108] Selenium deficiency protects against acetaminophen hepatotoxicity,[163] but not against inorganic mercury poisoning.[164] Clearly, much is yet to be understood about the intra- and extracellular levels and chemical forms of thiols that influence the toxicity of foreign chemicals.

A soluble, heat-labile factor, which is not glutathione peroxidase, can confer glutathione-dependent inhibition of lipid peroxidation to both microsomal and mitochondria preparations from rat liver.[165] The relationship of this protection system to the protective functions of vitamin E remain to be established.

## 3. CELLULAR DEFENSE MECHANISMS

### C. REDOX CYCLING

A second role for glutathione is in a powerful protection system against oxidative damage that is mediated largely by metabolites of molecular oxygen. Toxic effects associated with oxygen metabolism and subsequent redox cycling and lipid peroxidation appear to be a part of aerobic life. The relationship of very important events to reactive intermediates of chemicals has been reviewed by Kappus and Sies.[166] We are just beginning to understand and assess the energy required by the consumption of reducing equivalents that result from redox cycling.

The combined effects of protection by superoxide dismutase, glutathione peroxidase, and glutathione reductase are

$$4 O_2^{\bar{\cdot}} + 4 H^+ \rightarrow 2 H_2O_2 + 2 O_2$$
$$2 H_2O_2 \rightarrow 2 H_2O + O_2$$
$$\text{Net:} \quad 4 O_2^{\bar{\cdot}} + 4 H^+ \rightarrow 2 H_2O + 3 O_2$$

Because superoxide anion radicals can migrate across artificial lipid bilayer membranes at temperatures above the lipid phase transition,[167] they may traverse membranes in general, although their crossing membranes of erythrocytes[168] and granulocytes[169] is thought to occur via anion channels. This point could be important in that most quinones are located in membranes and the concentration of oxygen in the lipid plasma membrane is eight times that in the aqueous medium.[170] Thus, superoxide anion radicals may migrate in both directions across membranes from their site of formation. The rapid reaction of semiquinones with oxygen indicates that semiquinones would not diffuse far in the presence of oxygen.[171] Lack of protection by superoxide dismutase may indicate some superoxide anion production outside of cells.[171]

Protection from reduced oxygen species may occur in membranes by specific membrane-associated proteins that are capable of membrane protection and can be shown to limit the lipid peroxidation of membrane-associated polyunsaturated lipids.

An excellent example of the complexity of redox cycling concerning the reduction of nitro compounds has been discussed by Biaglow.[172] Reduction of nitro groups to nitro radical anions is catalyzed by reductases, including NADPH cytochrome $P$-450 reductase.[173] The fate of such radicals depends on many factors, including oxygen and glutathione concentrations. Oxygen and glutathione compete for the nitro radical anion electron and, in turn, form superoxide anion radical and GS·, which then forms glutathione disulfide. Decreasing the concentration of either oxygen or glutathione appears to increase macromolecule damage.

Such damage has been shown to be coincident with radiation damage and currently is being investigated extensively for therapeutic application in cancer treatment with nitro compounds such as misonidazole.[174]

Protection against quinones appears to involve several aspects of cellular function including oxygen reduction. Quinones can undergo either two-electron reductions to corresponding hydroquinones or one-electron reductions to the corresponding semiquinone radicals.[175] The main cytotoxic effects of quinones are thought to be mediated through one-electron reduction to the semiquinone radical.[176] This radical is known to form the superoxide anion radical by a one-electron reduction of molecular oxygen.[177]

It has been suggested that, whereas the rate of NADPH formation is not limiting for monooxygenase activity, it may be rate limiting for quinone-stimulated superoxide formation.[171] Simple quinones stimulate the formation of $O_2^-$ by isolated rat hepatocytes at rates up to 15 nmol/min per $10^6$ cells. Destruction of $O_2^-$ and water formation would require the consumption of 15 nmol/min per $10^6$ cells of intracellular glutathione or nearly a complete turnover of glutathione to glutathione disulfide and reduction back to glutathione in 2 min. An equal quantity (7.5 nmol) of NADPH must be furnished. Sies et al.[4b,178] have calculated the maximum rate of NADPH production to be equivalent to 15 nmol/min per $10^6$ cells.

Sies et al.[178] have reviewed the metabolism of organic hydroperoxides and concluded that the enzymatic reduction of organic hydroperoxides is the result of the activities of two glutathione-requiring enzymes, glutathione peroxidase [EC 1.11.1.9] and glutathione transferase [EC 2.5.1.1.8]. Additional protection against hydroperoxides is afforded by endogenous "antioxidants," including α-tocopherol, ascorbic acid, and β-carotene. Synthetic antioxidants such as butylated hydroxytoluene (BHT) are thought to mimic α-tocopherol in the termination of free-radical reaction sequences.

During oxidative stress, the rate of cellular generation of NADPH from $NADP^+$ appears to be rate limiting for monooxygenase reactions. In the intact liver, cytochrome P-450-dependent drug metabolism is decreased when an organic hydroperoxide is reduced to the corresponding alcohol by glutathione peroxidase, and in turn, NADPH-reducing equivalents are consumed in the conversion of newly generated glutathione disulfide to glutathione.[179]

A serious question has been raised concerning the role of hydrogen peroxide in microsomal lipid peroxidation. Hydrogen peroxide reacts with reduced transition metals, especially ferrous iron, to generate the most oxidative oxygen species, the highly reactive hydroxyl radical. This radical is thought to predominate in initiation of microsomal lipid peroxidation. Morehouse et al.[180] have concluded, however, that hydroxyl radical generated from hydrogen peroxide does not initiate microsomal lipid peroxidation.

## 3. CELLULAR DEFENSE MECHANISMS

NADPH-dependent lipid peroxidation, which is initiated by NADPH-cytochrome $P$-450 reductase, may occur by hydroxyl radical formation through an iron-catalyzed, Haber–Weiss reaction.[181] Such lipid peroxidation may also arise from the formation of a reactive ADP–Fe–oxygen complex.[182] Possible events include the reduction of perferryl ion to the ferryl ion for initiation of lipid peroxidation. Superoxide anion may be the reductant:

$$ADP-Fe^{3+} + O_2^- \rightarrow ADP-Fe^{2+}O_2$$

followed by

$$ADP-Fe^{2+}O_2 \xrightarrow[2\,H^+]{2\,e^-} ADP-[FeO]^{2+} + H_2O$$

In contrast, NADPH-cytochrome $P$-450 reductase-dependent lipid peroxidation with EDTA-Fe may occur through a hydroxyl ion-dependent mechanism rather than through an ADP ferrous ion–oxygen type of complex.[183]

The effect of redox cycling can now be extended to calcium. Hydroperoxides can modulate the redox state of pyridine nucleotides and the calcium balance in rat liver mitochondria via the participation of the glutathione redox cycle involving glutathione peroxidase and glutathione reductase.[184] It has been proposed that the redox state of mitochondrial pyridine nucleotides can be controlled, in part, by glutathione peroxidase and glutathione reductase and that oxygen metabolites are a factor in the distribution of calcium between mitochondria and extramitochondrial space.[184]

A calcium-dependent process has been described as the common final pathway of toxic chemical cell death. Using primary cultures of adult rat hepatocytes, Schanne et al.[185] demonstrated that toxic cell death caused by 10 different chemicals was dependent on extracellular calcium. Similarly, Chenery et al.[186] showed that cultured hepatocytes exposed to $CCl_4$ and the calcium ionophore A23187 were dependent on extracellular calcium for the expression of toxicity. These investigators proposed a two-step mechanism for toxic cell death. The first step is the disruption of the integrity of the plasma membrane by widely differing mechanisms followed by a common final step: the influx of extracellular calcium across the damaged plasma membrane.

Recent experiments cast doubt on the commonality of this pathway for cell death. Studies by Acosta and Sorenson[187] have demonstrated that the toxicity of cadmium chloride is accelerated in cultured hepatocytes incubated in calcium-free media. These results are in direct conflict with those reported by Schanne et al.[185] and Chenery et al.[186] In addition, Smith et al.[188] have demonstrated in freshly isolated hepatocytes that three different liver cell toxins ($CCl_4$, bromobenzene, and ethylmethanesulfonate) are far more toxic to hepatocytes in the absence of extracellular calcium than in its presence. These studies suggest that certain aspects of toxic cell injury are not dependent on extracellular calcium.

The reasons for variability in cellular responses to extracellular calcium and chemical toxicants are presently unknown. One can only speculate that differences in media conditions play an important role. Reed and Fariss[189] have shown that the entry of extracellular calcium is not a prerequisite for cell death in isolated rat hepatocytes.

It is interesting to speculate that the susceptibility to cell injury afforded cells in calcium-free media may be the result of inadequate intracellular calcium concentrations, in that A23187 permits the influx of extracellular calcium but does not cause cellular damage.[189] Indeed, these studies support the contention that the entry of extracellular calcium is not the cause, but more likely the result, of cell death.

A puzzling result of the extracellular calcium studies is the accelerated toxic cell injury afforded hepatocytes incubated in calcium-free media. This phenomenon has been observed with numerous compounds, including ADR-BCNU, $CCl_4$, bromobenzene, and EMS, which deplete glutathione concentrations.[188] Babson et al.[190] have demonstrated that ADR-BCNU toxicity is dependent on the concentration of intracellular glutathione. That is, cell death occurs once the level of intracellular glutathione falls below 20% of its initial value. Accelerated cell death of hepatocytes incubated in the absence of calcium resulted from an enhanced loss of intracellular glutathione.[189] This accelerated loss of glutathione has recently been shown to occur regardless of the toxin used.[192]

Subsequent developments suggest that chemically induced cell injury results from changes in intracellular calcium homeostasis as opposed to the influx of extracellular calcium. These studies have demonstrated the depression of calcium sequestration in liver microsomes and mitochondria after treatment with a variety of hepatotoxins. Furthermore, this loss in calcium retention appears to be related to alterations in the status of glutathione and reducing equivalents.

## V. Comments

Bioactivation of chemicals involves metabolic processes that are reasonably well understood in terms of the formation of reactive intermediates and some of the detoxication events following bioactivation. Emerging from studies about this important aspect of metabolism and related toxic events are the types of processes that lead to cellular damage, loss of organ function, and, ultimately, death of the organism, if the exposure to a particular chemical is sufficiently severe. Moreover, evolution has equipped most cell types with an appropriate defense system. As described in this chapter, typical defense processes against either radical or ionic-reactive intermediates of organic substances or radical oxygen intermediates are based, in part, on a high concentration of a low molecular weight, thiol-containing tripeptide, glutathione. The ability of glutathione to serve as the reducing agent for radicals and as a good nucleophile for electron-deficient ionic intermediates appears to occur both enzymatically and nonenzymatically. Enzymatic reactions are characterized by two main features: (1) the ability to enhance greatly the nucleophilicity of glutathione by its conversion to the thiolate anion, as caused by the various glutathione $S$-transferases, and (2) the ability to provide access to cellular reducing equivalents via glutathione and NADPH for the elimination of hydroperoxides and hydrogen peroxide. Redox cycling with the conversion of molecular oxygen to water via hydrogen peroxide is beginning to be understood in terms of the energy demand that may be imposed as a result of electron reduction events.

A major question needing further clarification is the possible consequences of alterations in the levels of protective agents. For example, does an inverse relationship between protection and damage exist, which can be exacerbated by events that lower the intracellular concentration of protective agents such as glutathione? Whether short- or long-term effects or even genetic effects occur, a better understanding of the manner in which protective mechanisms function is clearly needed.

Covalent adduct formation with macromolecules is interesting from the standpoint of a protective process. Do such adducts pose a hazard during their elimination, because they release uncommon amino acids or nucleoside adducts that may compete in further metabolic events? These and other questions make cellular defense mechanisms an intriguing area of future research.

## References

1. Sykes, P. (1975). "A Guideline to Mechanisms in Organic Chemistry." Wiley, New York.
2. Swain, C. G., and Scott, C. B. (1953). Quantitative correlation of reaction rates. Comparison of hydroxide ion with other nucleophilic reagents toward alkyl halides, esters, epoxides and acyl halides. *J. Am. Chem. Soc.* **75,** 141-147.
3. Jensen, D. E. (1978). Reaction of DNA with alkylating agents. Differential alkylation of poly[dA-dT] by methylnitrosourea and ethylnitrosourea. *Biochemistry* **17,** 5108-5113.
3a. Koskikallio, J. (1969a). Nucleophilic reactivity. Part II. Kinetics of reactions of methyl perchlorate with nucleophiles in water. *Acta Chem. Scand.* **23,** 1477-1489.
3b. Koskikallio, J. (1969b). Nucleophilic reactivity. Part III. Kinetics of reactions of methyl perchlorate with amines in water and methanol. *Acta Chem. Scand.* **23,** 1490-1494.

4. Veleminsky, J., Osterman-Golkar, S., and Ehrenberg, L. (1970). Reaction rates and biological action of N-methyl- and N-ethyl-N-nitrosourea. *Mutat. Res.* **10**, 169–174.
4a. Benesch, R. E., and Benesch, R. (1955). The acid strength of the -SH group in cysteine and related compounds. *J. Am. Chem. Soc.* **77**, 5877–5881.
4b. Sies, H., Wefers, H., Graf, P., and Akerboom, T. P. M. (1983). Hepatic hydroperoxide metabolism: Studies on redox cycling and generation of $H_2O_2$. *In* "Isolation, Characterization, and Use of Hepatocytes" (R. A. Harris and N. W. Cornell, eds.), pp. 341–348. Am. Elsevier, New York.
5. Reed, D. J., May, H. E., Boose, R. B., Gregory, K. M., and Beilstein, M. A. (1975). 2-Chloroethanol formation as evidence for a 2-chloroethyl alkylating intermediate during chemical degradation of 1-(2-chloroethyl)-3-cyclohexyl-1-nitrosourea and 1-(2-chloroethyl)-3-(*trans*-4-methylcyclohexyl)-1-nitrosourea. *Cancer Res.* **35**, 568–576.
6. Jerina, D. M. (1983). The 1982 Bernard B. Brodie Award Lecture. Metabolism of aromatic hydrocarbons by the cytochrome P-450 system and epoxide hydrolase. *Drug Metab. Drug Dispos.* **11**, 1–4.
7. Buening, M. K., Wislocki, P. G., Levin, W., Yagi, H., Thakker, D. R., Akagi, H., Koreeda, M., Jerina, D. M., and Conney, A. H. (1978). Tumorigenicity of the optical enantiomers of the diastereomeric benzo[*a*]pyrene-7,8-diol 9,10-epoxides in newborn mice: Exceptional activity of (+)-(7$R$,8$S$)-dihydroxy-(9$S$,10$R$)-epoxy-7,8,9,10-tetrahydrobenzo[*a*]pyrene. *Proc. Natl. Acad. Sci. U.S.A.* **75**, 5358–5361.
8. Harvey, R. G. (1981). Activated metabolites of carcinogenic hydrocarbons. *Acc. Chem. Res.* **14**, 218–226.
9. Conney, A. H. (1982). Induction of microsomal enzymes by foreign chemicals and carcinogenesis by polycyclic aromatic hydrocarbons: G. H. A. Clowes Memorial Lecture. *Cancer Res.* **42**, 4875–4917.
10. Hanzlik, R. P., Edelman, M., Michaely, W. J., and Scott, G. (1976). Enzymatic hydration of [$^{18}$O] epoxides. Role of nucleophilic mechanisms. *J. Am. Chem. Soc.* **98**, 1952–1955.
11. Dansette, P. M., Makedonska, V. B., and Jerina, D. M. (1978). Mechanism for catalysis for the hydration of substituted styrene oxides by hepatic epoxide hydrase. *Arch. Biochem. Biophys.* **187**, 290–298.
12. Armstrong, R. N., Levin, W., and Jerina, D. M. (1980). Hepatic microsomal epoxide hydrolase—Mechanistic studies of the hydration of K-region arene oxides. *J. Biol. Chem.* **255**, 4698–4705.
13. Sayer, J. M., Yagi, H., Wood, A. W., Conney, A. H., and Jerina, D. M. (1982). Extremely facile reaction between the ultimate carcinogen benzo[*a*]pyrene-7,8-diol 9,10-epoxide and ellagic acid. *J. Am. Chem. Soc.* **104**, 5562–5564.
14. Wood, A. W., Huang, M.-T., Chang, R. L., Newmark, H. L., Lehr, R. E., Yagi, H., Sayer, J. M., Jerina, D. M., and Conney, A. H. (1982). Inhibition of the mutagenicity of bay-region diol epoxides of polycyclic aromatic hydrocarbons by naturally occurring plant phenols—Exceptional activity of ellagic acid. *Proc. Natl. Acad. Sci. U.S.A.* **79**, 5513–5517.
15. Hernandez, O., and Bend J. R. (1982). Metabolism of epoxides. *In* "Metabolic Basis of Detoxication" (W. B. Jakoby, J. R. Bend, and J. Caldwell, eds.), Chapter 11, pp. 207–228. Academic Press, New York.
16. Jernström, B., Babson, J. R., Moldéus, P., Holmgren, A., and Reed, D. J. (1982). Glutathione conjugation and DNA binding of (+)-7$\beta$,8$\alpha$-dihydroxy-9$\alpha$,10$\alpha$-epoxy-7,8, 9,10-tetrahydrobenzo[*a*]pyrene in isolated rat hepatocytes. *Carcinogenesis* **3**, 861–866.
17. Meredith, M. J., and Reed, D. J. (1982). Status of the mitochondrial pool of glutathione in the isolated hepatocyte. *J. Biol. Chem.* **257**, 3447–3453.

18. Boyland, E., and Chasseaud, L. F. (1969). Role of glutathione and glutathione S-transferases in mercapturic acid biosynthesis. *Adv. Enzymol.* **32,** 173.
19. Chasseaud, L. F. (1979). Role of glutathione and glutathione S-transferases in the metabolism of chemical carcinogens and other electrophilic agents. *Adv. Cancer Res.* **29,** 175–274.
20. Baumann, E., and Preusse, C. (1879). Über Bromphenylmercaptursäure. *Ber. Dtsch. Chem. Ges.* **12,** 806.
21. Hopkins, F. G. (1929). On glutathione: A reinvestigation. *J. Biol. Chem.* **84,** 269–320.
22. Kendall, E. C., McKenzie, B. F., and Mason, H. L. (1929). A study of glutathione. I. Its preparation in crystalline form and its identification. *J. Biol. Chem.* **84,** 657–674.
23. Leaf, G., and Neuberger, A. (1947). The effect of diet on the glutathione content of the liver. *Biochem. J.* **41,** 280–287.
24. Maruyama, E., Kojima, J., Higashi, T., and Sakamoto, Y. (1968). Effect of diet on liver glutathione and glutathione reductase. *J. Biochem. (Tokyo)* **63,** 398–399.
25. Tateishi, N., Higashi, T., Shinya, S., Naruse, A., and Sakamoto, Y. (1974). Studies on the regulation of glutathione level in rat liver. *J. Biochem. (Tokyo)* **75,** 93–103.
26. Beck, L. V., Rieks, V. D., and Duncan, B. (1958). Diurnal variation in mouse and rat liver sulfhydryl. *Proc. Soc. Exp. Biol. Med.* **97,** 229–231.
27. Calcutt, G., and Ting, M. (1969). Diurnal variations in rat tissue disulphide levels. *Naturwissenschaften* **56,** 419–420.
28. Boyd, S. C., Sasame, H. A., and Boyd, M. R. (1979). High concentrations of glutathione in glandular stomach—Possible implications for carcinogenesis. *Science* **205,** 1010–1012.
29. Davis, N. C., and Whipple, W. (1919). The influence of fasting and various diets on the liver injury effected by chloroform anesthesia. *Arch. Intern. Med.* **23,** 612–633.
30. Davis, N. C. (1924). The influence of diet upon liver injury produced by carbon tetrachloride. *J. Med. Res.* **44,** 601–614.
31. Campbell, R., and Kosterlitz, H. (1948). The effects of short-term changes in dietary protein on response of liver to carbon tetrachloride injury. *Br. J. Exp. Pathol.* **29,** 149–159.
32. Krishnan, N., and Stenger, R. J. (1966). Effects of starvation on the hepatotoxicity of carbon tetrachloride; a light and electron microscopic study. *Am. J. Pathol.* **49,** 239–255.
33. McLean, A. E. M., and McLean, E. K. (1966). The effect of diet and 1,1,1-trichloro-2,2-bis (p-chlorophenyl)ethane (DDT) on microsomal hydroxylating enzymes and on sensitivity of rats to carbon tetrachloride poisoning. *Biochem. J.* **100,** 564–571.
34. Highman, B., Cyr, W. H., and Streett, R. P., Jr. (1973). Effect of x-irradiation and fasting on hepatotoxicity of carbon tetrachloride in rats. *Radiat. Res.* **54,** 444–452.
35. Diaz Gomez, M. I., DeCastro, C. R., DeFerreyra, C., D'Acosta, N. DeFenos, O. M., and Castro, J. A. (1975). Mechanistic studies on carbon tetrachloride hepatotoxicity in fasted and fed rats. *Toxicol. Appl. Pharmacol.* **32,** 101–108.
36. Strubelt, O., Dost-Kempf, E., Siegers, C.-P., Younes, M., Völpel, M., Preuss, U., and Dreckmann, J. G. (1981). The influence of fasting on the susceptibility of mice to hepatotoxic injury. *Toxicol. Appl. Pharmacol.* **60,** 66–77.
37. Jaeger, R. J., Connolly, R. B., and Murphy, S. D. (1974). Effect of 18-hr fast and glutathione depletion on 1,1-dichloroethylene-induced hepatotoxicity and lethality in rats. *Exp. Mol. Pathol.* **20,** 187–198.
38. Pessayre, D., Dolder, A., Artigou, J.-Y., Wandscheer, J.-C., Descatoire, V., Degott, C., and Benhamou, J. P. (1979). Effect of fasting on metabolic-mediated hepatotoxicity in the rat. *Gastroenterology* **77,** 264–271.
39. Pessayre, D., Wandscheer, J.-C., Cobert, B., Level, R., Degott, C., Batt, A.M., Martin, N.,

and Benhamou, J. P. (1980). Additive effects of inducers and fasting on acetaminophen hepatotoxicity. *Biochem. Pharmacol.* **29,** 2219–2223.
40. Mitchell, J. R., Jollow, D. J., Potter, W. Z., Gillette, J. R., and Brodie, B. B. (1973). Acetaminophen-induced hepatic necrosis. I. Role of drug metabolism. *J. Pharmacol. Exp. Ther.* **187,** 185–194.
41. Siegers, C.-P., Schutt, A., and Strubelt, O. (1977). Influence of some hepatotoxic agents of hepatic glutathione levels in mice. *Proc. Eur. Soc. Toxicol.* **18,** 160–162.
42. Reed, D. J., and Ellis, W. W. (1982). Influence of $\gamma$-glutamyltranspeptidase inactivation on the status of extracellular glutathione and glutathione conjugates. *Adv. Exp. Med. Biol.* **136A–136B,** 75–86.
43. Tateishi, N., Higashi, T., Naruse, A., Nakashimo, K., Shiozaki, H., and Sakamoto, Y. (1977). Rat liver glutathione; possible role as a reservoir of cysteine. *J. Nutr.* **107,** 51–60.
44. McIntyre, T. M., and Curthoys, N. P. (1980). The interorgan metabolism of glutathione. *Int. J. Biochem.* **12,** 545–551.
45. Reed, D.J., Brodie, A. E., and Meredith, M. J. (1983). Cellular heterogeneity in the status and functions of cysteine and glutathione. *In* "Functions of Glutathione: Biochemical, Physiological, Toxicological, and Clinical Aspects" (A. Larsson *et al.*, eds.), pp. 39–49. Raven Press, New York.
46. White, I. N. H. (1976). The role of liver glutathione in the acute toxicity of retrorsine to rats. *Chem.-Biol. Interact.* **13,** 333–342.
47. Reed, D. J. (1983). Regulation and function of glutathione in cells. *In* "Radioprotectors and Anticarcinogens" (O. F. Nygaard and M. G. Simic, eds.), pp. 153–168. Academic Press, New York.
48. Finkelstein, J. D. (1974). 1. Methionine metabolism in mammals. Biochemical basis for homocystinuria. *Metab., Clin. Exp.* **23,** 387–398.
49. Jakoby, W. B. (1978). The glutathione *S*-transferases: A group of multifunctional detoxication proteins. *Adv. Enzymol.* **46,** 383–414.
50. Janssen, M. J. (1972). Nucleophilicity of organic sulfur compounds. *In* "Sulfur in Organic and Inorganic Chemistry" (A. Senning, ed.), Vol. 3, Chapter 29, pp. 355–378. Dekker, New York.
51. Ketterer, B., Coles, B., and Meyer, D. J. (1983). The role of glutathione in detoxication. *Environ. Health Perspect.* **49,** 56–69.
52. Friedman, M., Cavins, J. F., and Wall, J. S. (1965). Relative nucleophilic reactivities of amino groups and mercaptide ions in addition reactions with $\alpha,\beta$-unsaturated compounds. *J. Am. Chem. Soc.* **87,** 3672–3682.
53. Pearson, R. G., Sobel, H., and Songstad, J. (1968). Nucleophilic reactivity constants toward methyl iodide and *trans*-[Pt(py)$_2$Cl$_2$]. *J. Am. Chem. Soc.* **90,** 319–326.
54. Thor, H., Moldéus, P., and Orrenius, S. (1979). Metabolic activation and hepatotoxicity —Effect of cysteine, *N*-acetylcysteine, and methionine on glutathione biosynthesis and bromobenzene toxicity in isolated rat hepatocytes. *Arch. Biochem. Biophys.* **192,** 405–413.
55. Massey, T. E., and Racz, W. J. (1981). Effects of *N*-acetylcysteine on metabolism, covalent binding, and toxicity of acetaminophen in isolated mouse hepatocytes. *Toxicol. Appl. Pharmacol.* **60,** 220–228.
56. Sheffner, A. L., Medler, E. M., Bailey, K. R., Gallo, D. G., Mueller, A. J., and Sarett, H. P. (1966). Metabolic studies with acetylcysteine. *Biochem. Pharmacol.* **15,** 1523–1535.
57. Galinski, R. E., and Levy, G. (1979). Effect of *N*-acetylcysteine on the pharmacokinetics of acetaminophen in rats. *Life Sci.* **25,** 693–700.
58. Piperno, E., Mosher, A. H., Berssenbruegge, D. A., Winkler, J. D., and Smith, R. B.

(1978). Pathophysiology of acetaminophen overdosage toxicity: Implications for management. *Pediatrics* **62**, 880–889.
59. Hinson, J. A. (1983). Reactive metabolites of phenacetin and acetaminophen: A review. *Environ. Health Perspect.* **49**, 71–79.
60. Willson, R. L. (1983). Free radical repair mechanisms and the interactions of glutathione and vitamins C and E. *In* "Radioprotectors and Anticarcinogens" (O. F. Nygaard and M. G. Simic, eds.), pp. 1–22. Academic Press. New York.
61. Eyer, P., and Lengfelder, E. (1984). Radical formation during autooxidation of 4-dimethylaminophenol and some properties of the reaction products. *Biochem. Pharmacol.* **33**, 1005–1013.
62. Ziegler, D.M., Poulsen, L.L., and Richerson, R.B. (1983). Oxidative metabolism of sulfur-containing radioprotective agents. *In* "Radioprotectors and Anticarcinogens" (O.F. Nygaard and M. G. Simic, eds.), pp. 191–202. Academic Press, New York.
63. Lauterburg, B. H., and Mitchell, J. R. (1981). Cysteamine, a modulator of the intracellular thiol–disulfide ratio. *Hepatology* **1**, 523.
64. Moldéus, P., Ormstad, K., and Reed, D. J. (1981). Turnover of cellular glutathione in isolated rat-kidney-cells—Role of cystine and methionine. *Eur. J. Biochem.* **116**, 13–16.
65. Szajewski, R. P., and Whitesides, G. M. (1980). Rate constants and equilibrium constants for thiol–disulfide interchange reactions involving oxidized glutathione. *J. Am. Chem. Soc.* **102**, 2010–2026.
66. Tien, M., Morehouse, L. A., Bucher, J. R., and Aust, S. (1982). The multiple effects of ethylenediaminetetraacetate in several model lipid peroxidation systems. *Arch. Biochem. Biophys.* **218**, 450–458.
67. Tappel, A. L. (1979). Measure of and protection from vivo lipid peroxidation. *In* "Biochemical and Clinical Aspects of Oxygen" (W. A. Caughey, ed.), pp. 679–698. Academic Press, New York.
68. Burton, G. W., and Ingold, K. U. (1981). Autoxidation of biological molecules. I. The antioxidant activity of vitamin E and related chain-breaking phenolic antioxidants in vitro. *J. Am. Chem. Soc.* **103**, 6472–6477.
69. Century, B., and Horwitt, M. (1965). Biological availability of various forms of vitamin E with respect to different indices of deficiency. *Fed. Proc., Fed. Am. Soc. Exp. Biol.* **24**, 906–911.
70. Bielski, B. H. J., and Richter, H. W. (1977). Study of superoxide radical chemistry by stopped-flow radiolysis and radiation-induced oxygen consumption. *J. Am. Chem. Soc.* **99**, 3019–3023.
71. Sawyer, D. T., and Valentine, J. S. (1981). How super is superoxide? *Acc. Chem. Res.* **14**, 393–400.
72. Nanni, E. J., Jr., Angelis, C. T., Dickson, J., and Sawyer, D. T. (1981). Oxygen activation by radical coupling between superoxide ion and reduced methyl viologen. *J. Am. Chem. Soc.* **103**, 4268–4270.
73. Miller, E. C., and Miller, J. A. (1966). Mechanisms of chemical carcinogenesis: Nature of proximate carcinogens and interactions with macromolecules. *Pharmacol. Rev.* **18**, 805–838.
74. Gillette, J. R. (1974a). Commentary: A perspective of the role of chemically reactive metabolites of foreign compounds in toxicity. I. Correlation of changes in covalent binding of reactive metabolites with changes in the incidence and severity of toxicity. *Biochem. Pharmacol.* **23**, 2785–2794.
75. Gillette, J. R. (1974b). A perspective on the role of chemically reactive metabolites of

foreign compounds in toxicity. II. Alterations in the kinetics of covalent binding. *Biochem. Pharmacol.* **23**, 2927–2938.
76. Jollow, D. J., Mitchell, J. R., Zampaglione, N., and Gillette, J. R. (1974). Bromobenzene-induced liver necrosis. Protective role of glutathione and evidence for 3,4-bromobenzene oxide as the hepatotoxic metabolite. *Pharmacology* **11**, 151–169.
77. Weihe, M., Potter, W. Z., Nelson, W. L., Jollow, D. J., and Mitchell, J. R. (1974). Mechanism of dose threshold for furosemide hepatotoxicity. *Toxicol. Appl. Pharmacol.* **29**, 90–91.
78. Uehleke, H. (1974). The model system of microsomal drug activation and covalent binding to endoplasmic proteins. *Proc. Eur. Soc. Study Drug Toxic.* **15**, 119–129.
79. Raha, C. R., Gallagher, C. H., Shubik, P., and Peratt, S. (1976). Covalent binding to protein of the K-region oxide of benzo[a]pyrene formed by microsome incubation. *J. Natl. Cancer. Inst. (U.S.)* **57**, 33–38.
80. Ketterer, B. (1980). Interactions between carcinogens and proteins. *Br. Med. Bull.* **36**, 71–78.
81. MacLeod, M. C., Kootstra, A., Mansfield, B. K., Slaga, T. J., and Selkirk, J. K. (1980). Specificity in interaction of benzo[a]pyrene with nuclear macromolecules: Implication of derivatives of two dihydrodiols in protein binding. *Proc. Natl. Acad. Sci. U.S.A.* **77**, 6396–6400.
82. Stout, D. L., Hemminki, K., and Becker, F. F. (1980). Covalent binding of 2-acetylaminofluorene, 2-amino-fluorene, and $N$-hydroxy-2-acetylaminofluorene to rat liver nuclear DNA and protein in vivo and in vitro. *Cancer Res.* **40**, 3579–3584.
83. Ohmi, N., Bhargava, M., and Arias, I. M. (1981). Binding of 3′-methyl-$N,N$-dimethyl-4-aminoazobenzene metabolites to rat liver cytosol proteins and ligandin subunits. *Cancer Res.* **41**, 3461–3464.
84. Reeve, V. E., Gallagher, C. H., and Raha, C. R. (1981). The water-soluble and protein-bound metabolites of benzo[a]pyrene formed by rat liver. *Biochem. Pharmacol.* **30**, 749–755.
85. Jakoby, W. B., and Keen, J. H. (1977). A triple-threat in detoxification: The glutathione S-transferases. *Trends Biochem. Sci.* **2**, 229–231.
86. Ketterer, B., Tipping, E., Beale, D., and Meuwissen, J. A. T. P. (1976). Ligandin, glutathione and carcinogen binding. In "Glutathione: Metabolism and Function" (I. M. Arias and W. B. Jakoby, eds.), pp. 243–258. Raven Press, New York.
87. Singer, S., and Litwack, G. (1971). Identity of corticosteroid binder 1 with the macromolecule binding 3-methylcholanthrene in liver cytosol in vivo. *Cancer Res.* **31**, 1364–1368.
88. Schelin, C., Tunek, A., and Jergil, B. (1983). Covalent binding of benzo[a]pyrene to rat liver cytosolic proteins and its effect on the binding to microsomal proteins. *Biochem. Pharmacol.* **32**, 1501–1506.
89. Ortiz de Montellano, P. R., Kunze, K. L., and Mico, B. A. (1980). Destruction of cytochrome $P$-450 by olefins: N-alkylation of prosthetic heme. *Mol. Pharmacol.* **18**, 602–605.
90. Poland, A., and Glover, E. (1979). An estimate of the maximum in vivo covalent binding of 2,3,7,8-tetrachlorodibenzo-$p$-dioxin to rat liver protein, ribosomal RNA and DNA. *Cancer Res.* **39**, 3341–3344.
91. Farber, E. (1968). Biochemistry of carcinogenesis. *Cancer Res.* **28**, 1859–1869.
92. Topal, M. D., and Baker, M. S. (1982). DNA precursor pool: A significant target for $N$-methyl-$N$-nitrosourea in C3H/10T$^1$/$_2$ clone 8 cells. *Proc. Natl. Acad. Sci. U.S.A.* **79**, 2211–2215.
93. Topal, M. D., Hutchinson, C. A., III, and Baker, M. S. (1982). DNA precursors in

chemical mutagenesis: A novel application of DNA sequencing. *Nature (London)* **298,** 863–865.
94. Lutz, W. K. (1979). *In vivo* covalent binding of organic chemicals to DNA as a quantitative indicator in the process of chemical carcinogenesis. *Mutat. Res.* **65,** 289–356.
95. Nakagawa, Y., Hiraga, K., and Suga, T. (1983). On the mechanism of covalent binding of butylated hydroxytoluene to microsomal protein. *Biochem. Pharmacol.* **32,** 1417–1421.
96. Kehrer, J. P., and Witschi, H. (1980). Effects of drug metabolism inhibitors on butylated hydroxytoluene-induced pulmonary toxicity in mice. *Toxicol. Appl. Pharmacol.* **53,** 333–342.
97. Ziegler, D. M., Duffel, M. W., and Poulsen, L. L. (1980). Studies on the nature and regulation of the cellular thiol: Disulphide potential. *Ciba Found. Symp.* **72,** 191–204.
98. Vignais, P. M., and Vignais, P. V. (1973). Fuscin, an inhibitor of mitochondrial SH-dependent transport-linked functions. *Biochim. Biophys. Acta* **325,** 357–374.
99. Jocelyn, P. C., and Kamminga, A. (1974). The non-protein thiol of rat liver mitochondria. *Biochim. Biophys. Acta* **343,** 356–362.
100. Jocelyn, P. C. (1975). Some properties of mitochondrial glutathione. *Biochim. Biophys. Acta* **396,** 427–436.
101. Wahlländer, A., Soboll, S., Sies, H., Linke, I., and Muller, M. (1979). Hepatic mitochondrial and cytosolic glutathione content and the sub-cellular distribution of GSH $S$-transferases. *FEBS Lett.* **97,** 138–140.
102. Elbers, R., Heldt, H. W., Schmucker, P., Soboll, S., and Wiese, H. (1974). Measurement of the ATP/ADP ratio in mitochondria and in the extramitochondrial compartment by fractionation of freeze-stopped liver tissue in non-aqueous media. *Hoppe-Seyler's Z. Physiol. Chem.* **355,** 378–393.
103. Higashi, T., Tateishi, N., Naruse, A., and Sakamoto, Y. (1977). A novel physiological role of liver glutathione as a reservoir of L-cysteine. *J. Biochem. (Tokyo)* **82,** 117–124.
104. Cho, E. S., Sahyoun, N., and Stegink, L. D. (1981). Tissue glutathione as a cysteine reservoir during fasting and refeeding of rats. *J. Nutr.* **111,** 914–922.
105. Meredith, M. J., and Reed, D. J. (1983). Depletion *in vitro* of mitochondrial glutathione in rat hepatocytes and enhancement of lipid peroxidation by adriamycin and 1,3-bis(2-chloroethyl)-1-nitrosourea (BCNU). *Biochem. Pharmacol.* **32,** 1383–1388.
106. Meredith, M. J., and Reed, D. J., unpublished.
107. Van Anda, J., Bend, J. R., and Fouts, J. R. (1978). Effect of diethyl maleate pre-treatment on metabolism and toxicity of [$^{14}$C]styrene oxide in the isolated perfused rat liver. *Pharmacologist* **20,** 200.
108. Hill, K. E., and Burk, R. F. (1982). Effect of selenium deficiency and vitamin E deficiency on glutathione metabolism in isolated rat hepatocytes. *J. Biol. Chem.* **257,** 10668–10672.
109. Wattenberg, L. W. (1972a). Inhibition of carcinogenic effects of diethylnitrosamine and 4-nitroquinoline-$N$-oxide by antioxidants. *Fed. Proc., Fed. Am. Soc. Exp. Biol.* **31,** 633.
110. Wattenberg, L. W. (1972b). Inhibition of carcinogenic and toxic effects of polycyclic hydrocarbons by phenolic antioxidants and ethoxyquin. *J. Natl. Cancer Inst. (U.S.)* **48,** 1425–1430.
111. Wattenberg, L. W. (1974). Inhibition of carcinogenic and toxic effects of polycyclic hydrocarbons by several sulfur-containing compounds. *J. Natl. Cancer Inst. (U.S.)* **52,** 1583–1587.
112. Wattenberg, L. W., Lam, L. K. T., Speier, J. L., Loub, W. D., and Borchert, P. (1977). Inhibitors of chemical carcinogenesis. *Cold Spring Harbor Conf. Cell Proliferation* **4,** 785–799.

113. Weisburger, E. K., Evarts, R. P., and Wenk, M. L. (1977). Inhibitory effect of butylated hydroxytoluene (BHT) on intestinal carcinogenesis in rats by azoxymethane. *Food Cosmet. Toxicol.* **15,** 139–141.
114. Grantham, P. H., Weisburger, J. H., and Weisburger, E. K. (1973). Effect of the antioxidant butylated hydroxytoluene (BHT) on the metabolism of the carcinogens N-2-fluorenylacetamide and N-hydroxy-N-2-fluorenylacetamide. *Food Cosmet. Toxicol.* **11,** 209–217.
115. Ulland, B. M., Weisburger, J. H., Yamamoto, R. S., and Weisburger, E. K. (1973). Antioxidants and carcinogenesis: Butylated hydroxytoluene, but not diphenyl-p-phenylenediamine, inhibits cancer induction by N-2-fluorenylacetamide and by N-hydroxy-N-2-fluorenylacetamide in rats. *Food Cosmet. Toxicol.* **11,** 199–207.
116. Benson, A. M., Batzinger, R. P., Ou, S.-Y. L., Bueding, E., Cha, Y.-N., and Talalay, P. (1978). Elevation of hepatic glutathione S-transferase activities and protection against mutagenic metabolites of benzo[a]pyrene by dietary antioxidants. *Cancer Res.* **38,** 4486–4495.
117. Batzinger, R. P., Ou, S.-Y. L., and Beuding, E. (1978). Anti-mutagenic effects of 2(3)-tert-butyl-4-hydroxyanisole and of antimicrobial agents. *Cancer Res.* **38,** 4476–4483.
118. Nakagawa, Y., Hiraga, K., and Suga, T. (1981). Effects of butylated hydroxytoluene (BHT) on the level of glutathione and the activity of glutathione S-transferase in rat liver. *J. Pharmacobio-Dyn.* **4,** 823–826.
119. Lam, L. K., Sparnins, V. L., Hochalter, J. B., and Wattenberg, L. W. (1981). Effects of 2- and 3-tert-butyl-4-hydroxyanisole on glutathione S-transferase and epoxide hydrolase activities and sulfhydryl levels in liver and forestomach of mice. *Cancer Res.* **41,** 3940–3943.
120. Benson, A. M., Hunkeler, M. J., and Talalay, P. (1980). Increase of NAD(P)H : quinone reductase by dietary antioxidants: Possible role in protection against carcinogenesis and toxicity. *Proc. Natl. Acad. Sci. U.S.A.* **77,** 5216–5220.
121. Wislocki, P. G., Wood, A. W., Chang, R. L., Levin, W., Yagi, H., Hernandez, O., Jerina, D. M., and Conney, A. H. (1976). Toxicity of a diol epoxide derived from benzo[a]pyrene. *Biochem. Biophys. Res. Commun.* **68,** 1006–1012.
122. Huberman, E., Sachs, L., Yang, S. K., and Gelboin, H. V. (1976). Identification of mutagenic metabolites of benzo[a]pyrene in mammalian cells. *Proc. Natl. Acad. Sci. U.S.A.* **73,** 607–611.
123. Newbold, R. F., and Brookes, P. (1976). Exceptional mutagenicity of a benzo[a]pyrene diol epoxide in cultured mammalian cells. *Nature (London)* **261,** 52–54.
124. Boyland, E., and Sims, P. (1967). Carcinogenic activities in mice of compounds related to benz[a]-anthracene. *Int. J. Cancer* **2,** 500–504.
125. Flesher, J. W., Harvey, R. G., and Sydnor, K. L. (1976) Oncogenicity of K-region epoxides of benzo[a]pyrene and 7,12-dimethylbenz[a]anthracene. *Int. J. Cancer* **18,** 351–353.
126. Slaga, T. J., Viaje, A., Berry, D. L., Bracken, W. M., Buty, S. G., and Scribner, J. D. (1976). Skin tumor initiating ability of benzo[a]pyrene 4,5-epoxides, 7,8-epoxides and 7,8-diol 9,10-epoxides, and 7,8 diols. *Cancer Lett.* **2,** 115–122.
127. Sims, P., Grover, P. L., Swaisland, A., Pal, K., and Hewer, A. (1974). Metabolic activation of benzo[a]pyrene proceeds by a diol epoxide. *Nature (London)* **252,** 326–327.
128. Bigger, C. A. H., Tomaszweski, J. E., and Dipple, A. (1980). Differences between products of binding of 7,12-dimethylbenz[a]anthracene to DNA in mouse skin and in a rat liver microsomal system. *Biochem. Biophys. Res. Commun.* **80,** 229–235.
129. Hecht, S. S., LaVoie, E., Mazzarese, R., Amin, S., Bedenko, V., and Hoffman, D. (1978).

1,2-Dihydro-1,2-dihydroxy-5-methylchrysene, a major activated metabolite of the environmental carcinogen 5-methylchrysene. *Cancer Res.* **38**, 2191–2194.
130. Levin, W., Wood, A. W., Lu, A. Y. H., Ryan, D., West, S., Conney, A. H., Thakker, D. R., Yagi, H., and Jerina, D. M. (1977). Role of purified cytochrome *P*-448 and epoxide hydrase in the activation and detoxification of benzo[*a*]pyrene. *ACS Symp. Ser.* **44**, 99–126.
131. Cooper, C. S., MacNicoll, A. D., Ribeiro, O., Gervasi, P. G., Hewer, A., Walsh, C., Pal, K., Grover, P. L., and Sims, P. (1980). Involvement of a non-bay-region diol epoxide in the metabolic activation of benz[*a*]-anthracene in hamster-embryo cells. *Cancer Lett.* **9**, 53–59.
132. MacNicoll, A. D., Cooper, C. S., Ribeiro, O., Pal, K., Hewer, A., Grover, P. L., and Sims, P. (1981). The metabolic activation of benz[*a*]anthracene in three biological systems. *Cancer Lett.* **11**, 243–249.
133. Burke, M. D., Vadi, H., Jernström, B., and Orrenius, S. (1977). Metabolism of benzo[*a*]pyrene with isolated hepatocytes and the formation and degradation of DNA-binding derivatives. *J. Biol. Chem.* **252**, 6424–6436.
134. Glatt, H. R., and Oesch F. (1977). Inactivation of electrophilic metabolites by glutathione *S*-transferase and limitation of the system due to subcellular localization. *Arch. Toxicol.* **39**, 87.
135. Shen, A. L., Fahl, W. E., and Jefcoate, C. R. (1980). Metabolism of benzo[*a*]pyrene by isolated hepatocytes and factors affecting covalent binding of benzo[*a*]-pyrene metabolites to DNA in hepatocytes and microsomal systems. *Arch. Biochem. Biophys.* **204**, 511–523.
136. Hesse, S., Jernstöm, B., Martinez, M., Guenthner, O., and Orrenius, S. (1980). Inhibition of binding benzo[*a*]pyrene metabolites to nuclear DNA by glutathione and glutathione *S*-transferase B. *Biochem. Biophys. Res. Commun.* **94**, 612–617.
137. Guenthner, T. M., Jernstöm, B., and Orrenius, S. (1980). On the effect of cellular nucleophiles on the binding of metabolites of 7,8-dihydroxy-7,8-dihydrobenzo[*a*]pyrene and 9-hydroxybenzo[*a*]pyrene to nuclear DNA. *Carcinogenesis (N.Y.)* **1**, 407–418.
138. Glatt, H. R., Billings, R., Platt, K. L., and Oesch, F. (1981). Improvement of the correlation of bacterial mutagenicity with carcinogenicity of benzo[*a*]pyrene and four of its major metabolites by activation with intact liver cells instead of cell homogenate. *Cancer Res.* **41**, 270–277.
139. Wood, A. W., Levin, W., Lu, A. Y. H., Yagi, H., Hernandez, O., Jerina, D. M., and Conney, A. H. (1976). Metabolism of benzo[*a*]pyrene and benzo[*a*]pyrene derivatives to mutagenic products by highly purified hepatic microsomal enzymes. *J. Biol. Chem.* **251**, 4882–4890.
140. Wood, A. W., Chang, R. L., Levin, W., Lehr, R. E., Schaefer-Ridder, M., Karle, J. M., Jerina, D. M., and Conney, A. H. (1977). Mutagenicity and cytotoxicity of benz[*a*]anthracene diol epoxides and tetrahydroepoxides; experimental activity of the bay regions 1,2-epoxides. *Proc. Natl. Acad. Sci. U.S.A.* **74**, 2746–2750.
141. Bentley, P., Oesch, F., and Glatt, H. R. (1977). Dual role of epoxide hydratase in both activation and inactivation of benzo[*a*]pyrene. *Arch. Toxicol.* **39**, 65–75.
142. Baars, A. J., Jansen, M., and Breimer, D. D. (1978). The influence of phenobarbital, 3-methylcholanthrene and 2,3,7,8-tetrachlorodibenzo-*p*-dioxin on glutathione *S*-transferase activity of rat liver cytosol. *Biochem. Pharmacol.* **27**, 2487–2494.
143. Clifton, G., and Kaplowitz, N. (1978). Effect of dietary phenobarbital, 3,4-benzo[*a*]pyrene and 3-methylcholanthrene on hepatic, intestinal and renal glutathione *S*-transferase activities in the rat. *Biochem. Pharmacol.* **27**, 1284–1287.

144. Hales, B. F., and Neims, A. H. (1977). Induction of rat hepatic glutathione S-transferase B by phenobarbital and 3-methylcholanthrene. *Biochem. Pharmacol.* **26**, 555-556.
145. Glatt, H. R., Cooper, C. S., Grover, P. L., Sims, P., Bentley, P., Merdes, M., Waechter, F., Vogel, K., Guenthner, T. M., and Oesch, F. (1982). Inactivation of a diol epoxide by dihydrodiol dehydrogenase but not by two epoxide hydrolases. *Science* **215**, 1507-1509.
146. Vogel, K., Bentley, P., Platt, K. L., and Oesch, F. (1980). Rat liver cytoplasmic dihydrodiol dehydrogenase. Purification to apparent homogeneity and properties. *J. Biol. Chem.* **255**, 9621-9625.
147. Pickett, C. B., Wells, W., Lu A. Y. H., and Hales, B. F. (1981). Induction of translationally active rat liver glutathione S-transferase B messenger RNA by phenobarbital. *Biochem. Biophys. Res. Commun.* **99**, 1002-1010.
148. Felton, J. F., Ketley, J. N., Jakoby, W. B., Aitio, A., Bend, J. R., and Nebert, D. W. (1980). Hepatic glutathione transferase (EC 2.5.1.18) activity induced by polycyclic aromatic compounds: Lack of correlation with the murine *Ah* locus. *Mol. Pharmacol.* **18**, 559-564.
149. Conney, A. H. (1967). Pharmacological implications of microsomal enzyme induction. *Pharmacol. Rev.* **19**, 317-366.
150. Estabrook, R. W., and Lindenlaub, E., eds. (1979). "The Induction of Drug Metabolism." Verlag-Stuttgart, New York.
151. Schmassman, H. U., and Oesch, H. (1978). *trans*-Stilbene oxide: A selective inducer of rat liver epoxide hydratase. *Mol. Pharmacol.* **14**, 834-847.
152. Guthenberg C., Morgenstern, R., DePierre, J. W., and Mannervik, B. (1980). Induction of glutathione S-transferase A, S-transferase B, and S-transferase C in rat-liver cytosol by *trans*-stilbene oxide. *Biochim. Biophys. Acta* **631**, 1-10.
153. Seidegard J., DePierre, J. W., Morgenstern, R., Pilotti, A., and Ernster, L. (1981). Induction of drug-metabolizing systems and related enzymes with metabolites and structural analogues of stilbene. *Biochim. Biophys. Acta* **672**, 65-78.
154. Dutton, G. J. (1980). "Glucuronidation of Drugs and other Compounds." CRC Press, Boca Raton, Florida.
155. Rotruck, J. T., Pope, A. L., Ganther, H. E., Swanson, A. B., Hafeman, D. G., and Hoekstra, W. G. (1973). Selenium: Biochemical role as a component of glutathione peroxidase. *Science* **179**, 588-591.
156. Stadtman, T. C. (1980). Selenium-dependent enzymes. *Annu. Rev. Biochem.* **49**, 93-110.
157. Flohé, L., Günzler, W. A., and Loschen, G. (1979). The glutathione peroxidase reaction: A key to understand the selenium requirement of mammals. *In* "Trace Metals in Health and Disease" (N. Karasch, ed.), pp. 263-266. Raven Press, New York.
158. Wendel, A. (1980). Glutathione peroxidase. *In* "Enzymatic Basis of Detoxication" (W. B. Jakoby, ed.), Vol. 1, pp. 333-353. Academic Press, New York.
159. Sunde, R. A., and Hoekstra, W. G. (1980). Structure, synthesis and function of glutathione peroxidase. *Nutr. Rev.* **38**, 265-273.
160. Lawrence, R. A., and Burk, R. F. (1978). Species, tissue and subcellular distribution of non-Se-dependent glutathione peroxidase activity. *J. Nutr.* **108**, 211-215.
161. Prohaska, J. R., and Ganther, H. E. (1977). Glutathione peroxidase activity of glutathione S-transferases purified from rat liver. *Biochem. Biophys. Res. Commun.* **76**, 437-445.
162. Katki, A. G., and Myers, C. E. (1980). Membrane-bound glutathione peroxidase-like activity in mitochondria. *Biochem. Biophys. Res. Commun.* **96**, 85-91.
163. Burk, R. F., and Lane, J. M. (1979). Ethane production and liver necrosis in rats after administration of drugs and other chemicals. *Toxicol. Appl. Pharmacol.* **50**, 467-478.

164. Burk, R. F., Jordan, H. E., and Kiker, K. W. (1977). Some effects of selenium status on inorganic mercury metabolism in the rat. *Toxicol. Appl. Pharmacol.* **40**, 71–82.
165. McCay, P. B., Gibson, D. D., and Hornbrook, K. R. (1981). Glutathione-dependent inhibition of lipid peroxidation by a soluble, heat-labile factor not glutathione peroxidase. *Fed. Proc., Fed. Am. Soc. Exp. Biol.* **40**, 199–205.
166. Kappus, H., and Sies, H. (1981). Toxic drug effects associated with oxygen metabolism — redox cycling and lipid peroxidation. *Experientia* **37**, 1233–1241.
167. Rumyantseva, G. V., Weiner, L. M., Molin, Y. N., and Budker, V. G. (1979). Permeation of liposome membrane by superoxide radical. *FEBS Lett.* **108**, 477–480.
168. Lynch, R. E., and Fridovich, I. (1978). Permeation of erythrocyte stroma by superoxide radical. *J. Biol. Chem.* **253**, 4697–4699.
169. Gennaro, R., and Romeo, D. (1979). The release of superoxide anion from granulocytes: Effect of inhibitors of anion permeability. *Biochem. Biophys. Res. Commun.* **88**, 44–99.
170. Pryor, W. A. (1973). Free radical reactions and their importance in biochemical systems. *Fed. Proc., Fed. Am. Soc. Exp. Biol.* **32**, 1862–1868.
171. Powis, G., Svingen, B. A., and Appel, P. (1981). Quinone-stimulated superoxide formation by subcellular fractions, isolated hepatocytes, and other cells. *Mol. Pharmacol.* **20**, 387–394.
172. Biaglow, J. E. (1981). Cellular electron transfer and radical mechanisms for drug metabolism. *Radiat. Res.* **86**, 212–242.
173. Lu, A. Y. H. (1982). Personal communication.
174. Bump, E. A., Taylor, Y. C., and Brown, M. (1983). Role of glutathione in the hypoxic cell cytotoxicity of misonidazole. *Cancer Res.* **43**, 997–1002.
175. Iyanagi, T., and Yamazaki, I. (1970). One-electron-transfer reactions in biochemical systems. V. Difference in the mechanism of quinone reduction by the NADH dehydrogenase and the NAD(P)H dehydrogenase (DT-diaphorase). *Biochim. Biophys. Acta* **216**, 282–294.
176. Bachur, N. R., Gordon, S. L., and Gee, M. V. (1978). A general mechanism for microsomal activation of quinone anticancer agents to free radicals. *Cancer Res.* **38**, 1745–1750.
177. Fridovich, I. (1976). I. Oxygen radicals, hydrogen peroxide, and oxygen toxicity. In "Free Radicals in Biology" (W. A. Pryor, ed.), Vol. 1, pp. 249–276. Academic Press, New York.
178. Sies, H., Wendel, A., and Bors, W. (1982). Metabolism of organic hydroperoxides. In "Metabolic Basis of Detoxication" (W. B. Jakoby, J. R. Bend, and J. Caldwell, eds.), Chapter 16, pp. 307–321. Academic Press, New York.
179. Gerstenecker, C., and Sies, H. (1980). Restriction of hexobarbital metabolism by *t*-butylhydroperoxide in perfused rat liver. *Biochem. Pharmacol.* **29**, 3112–3113.
180. Morehouse, L. A., Tien, M., Bucher, J. R., and Aust, S. D. (1983). Effect of hydrogen peroxide on the initiation of microsomal lipid peroxidation. *Biochem. Pharmacol.* **32**, 123–127.
181. Fong, K., McCay, P. B., Poyer J. L., Keele, B. B., and Mirsa, H. (1973). Evidence that peroxidation of lysosomal membranes is initiated by hydroxyl free radicals produced during flavin enzyme activity. *J. Biol. Chem.* **248**, 7792–7797.
182. Tien, M., Svingen, B. A., and Aust, S. D. (1981). Superoxide-dependent lipid peroxidation. *Fed. Proc., Fed. Am. Soc. Exp. Biol.* **40**, 179–182.
183. Bucher, J. R., Tien, M., Morehouse, L. A., and Aust, S. D. (1982). Redox cycling and lipid peroxidation: The central role of iron chelates. *Fundam. Appl. Toxicol.* **3**, 222–226.
184. Lötscher, H. R., Winterhalter, K. H., Carafoli, E., and Richter, C. (1979). Hydroperox-

ides can modulate the redox state of pyridine nucleotides and the calcium balance in rat liver mitochondria. *Proc. Natl. Acad. Sci. U.S.A.* **76,** 4340–4344.
185. Schanne, F. A. X., Kane, A. B., Young, E. E., and Farber, J. L. (1979). Rapid killing of single neurons by irradiation of intracellularly injected dye. *Science* **206,** 700–702.
186. Chenery, R., George, M., and Krishna, G. (1981). The effect of ionophore A23187 and calcium on carbon tetrachloride-induced toxicity in cultured rat hepatocytes. *Toxicol. Appl. Pharmacol.* **60,** 241–252.
187. Acosta, D., and Sorenson, E. M. B. (1983). Role of calcium in cytotoxic injury of cultured hepatocytes. *Ann. N. Y. Acad. Sci.* **407,** 78–92.
188. Smith, M. T., Thor, H., and Orrenius, S. (1981). Toxic injury to isolate hepatocytes is not dependent on extracellular calcium. *Science* **213,** 1257–1259.
189. Reed, D. J., and Fariss, M. W. (1984). Glutathione depletion and susceptibility. *Pharmacol. Rev.* **36,** 25S–33S.
190. Babson, J. R., Abell, N. S., and Reed, D. J. (1981). Protective role of the glutathione redox cycle against adriamycin-mediated toxicity in isolated hepatocytes. *Biochem. Pharmacol.* **30,** 2299–2304.
191. Fariss, M. W., and Reed, D. J., unpublished data.

*Part II*

# Bioactivation by Chemical Class

Chapter 4

# Alkanes

JAMES S. BUS

*Department of Biochemical Toxicology*
*Chemical Industry Institute of Toxicology*
*Research Triangle Park, North Carolina*

|      |                                                           |     |
|------|-----------------------------------------------------------|-----|
| I.   | Introduction                                              | 111 |
| II.  | Alkane Toxicity                                           | 112 |
|      | A. Neurotoxicity                                          | 112 |
|      | B. Testicular Toxicity                                    | 112 |
| III. | Alkane Bioactivation                                      | 112 |
|      | A. *n*-Hexane and Methyl *n*-Butyl Ketone Metabolism      | 112 |
|      | B. The Neurotoxic Metabolite, 2,5-Hexanedione             | 113 |
|      | C. The $\gamma$-Diketone Hypothesis                       | 115 |
| IV.  | Toxic Mechanisms of the $\gamma$-Diketones                | 116 |
|      | A. Inhibition of Glycolysis                               | 116 |
|      | B. Derivatization of Amino Groups of Axon Proteins        | 117 |
|      | C. Inhibition of Sterol Synthesis                         | 118 |
|      | References                                                | 119 |

## I. Introduction

The alkanes are a homologous series of compounds with the general molecular formula, $C_nH_{2n+2}$. Although $C_1$–$C_5$ alkanes are toxic to humans and animals, the toxicity has not been linked with any general or specific bioactivation process. In recent years, however, it has been recognized that the six-carbon straight-chain alkane, *n*-hexane, produces both neurotoxicity and testicular toxicity. The toxic action of this alkane, however, clearly requires metabolic activation.

## II. Alkane Toxicity

### A. NEUROTOXICITY

The neurotoxicity of *n*-hexane has been classified as a central–peripheral distal axonopathy.[1] This description replaced the earlier, less accurate characterization of the nerve lesion as a "dying-back" neuropathy limited to the peripheral nervous system.

The axonal degeneration observed with *n*-hexane exposure is characterized by development of multifocal axonal swellings in the distal ends of long and large-diameter axons of both the central and peripheral nervous systems.[1,2] The swellings occur on the proximal sides of the distal nodes of Ranvier and are accompanied by an apparent displacement of the myelin sheath. An additional early feature of the neuropathy is the accumulation of densely packed masses of 10-nm neurofilaments in the axonal swellings.[3]

The neurotoxicity of *n*-hexane and related alkanes is similar to that seen with a variety of other known neurotoxins including acrylamide, carbon disulfide, and the organophosphate, triorthocresylphosphate.[2] In contrast to some of these agents, however, repeated exposure to *n*-hexane is required for development of neurotoxicity.

### B. TESTICULAR TOXICITY

A second toxic response that has been reported after *n*-hexane exposure is atrophy of the germinal epithelium of the testes.[4] A comparative time course for development of testicular vs the axonal toxicity has not been described. Nonetheless, *in vivo* studies with the putative neurotoxic metabolite of *n*-hexane suggested a direct toxic effect on the testis, with no disturbance of central nervous system gonadotropin control systems.[5]

## III. Alkane Bioactivation

### A. *n*-HEXANE AND METHYL *n*-BUTYL KETONE METABOLISM

Reports linking occupational exposure to *n*-hexane[6] or methyl *n*-butyl ketone (2-hexanone)[7] to development of pathologically similar neuropathies provided the first hint that bioactivation of alkanes was a possible requirement for expression of neurotoxicity. A subsequent metabolic study conducted by DiVincenzo *et al.*[8] using guinea pigs confirmed that *n*-hexane and methyl n-butyl ketone were metabolized to at least two common metabolites, 5-hydroxy-2-hexanone and 2,5-hexanedione. Identification of these metabolites indicated that n-hexane was sequentially oxidized at the C-2 and C-5 ($\omega$-1) positions. The hydroxylation of n-hexane is apparently catalyzed by the microsomal monooxygenase system, in that in vitro microsomal metabolism of n-hexane yields 2-hexanol as an oxidized metabolite.[9,10] The primary site of oxidation was the 2-position, although 1- and 3-hexanol were also formed to a lesser extent.[9,10]

DiVincenzo et al.[11] have outlined a proposed pathway for methyl n-butyl ketone metabolism in rats (Fig. 1).[11] Inhalation studies with rats have demonstrated that n-hexane is also extensively metabolized to methyl n-butyl ketone,[12,13] indicating that this metabolic route may account for a large fraction of n-hexane metabolism (Fig. 1). With [1-$^{14}$C]methyl n-butyl ketone, DiVincenzo et al.[11] found that expiration of $^{14}CO_2$ represented the largest portion of excreted radioactivity. They proposed that $\alpha$ oxidation to 2-ketohexanoic acid with decarboxylation to pentanoic acid represented the primary source of the $^{14}CO_2$. This assumption was based on the observation that only a small amount of labeled 2-aminohexanoic acid was detected in the urine, indicating decarboxylation as the preferred pathway. Decarboxylation was further suggested in that urinary acetic acid was unlabeled, and there was little incorporation of radiolabel into tissue neutral lipids. Thus, methyl n-butyl ketone appeared to have little potential to enter intermediary metabolism via metabolism to acetate. A second potential source of $^{14}CO_2$, decarboxylation of 2-keto-5-hydroxyhexanoic acid, was apparently a minor metabolic step, because only a small amount of $\gamma$-valerolactone was detected in the urine.

The $\omega$-1 oxidation of [1-$^{14}$C]methyl n-butyl ketone was found by DiVincenzo et al.[11] to be the second major metabolic route for this agent. Approximately 35% of the urinary radioactivity, which represented 35–40% of the administered dose, was identified as free or glucuronide or sulfate conjugates of 5-hydroxy-2-hexanone and 2,5-hexanedione. In addition, the peak concentrations of these metabolites in sera after oral administration of methyl n-butyl ketone were equal to or higher than the parent compound, and their elimination time from sera was longer than that of the precursors.

## B. THE NEUROTOXIC METABOLITE, 2,5-HEXANEDIONE

Several lines of evidence have developed implicating 2,5-hexanedione as the likely metabolite responsible for the neurotoxic activity of n-hexane and methyl n-butyl ketone. The observation that the nerve lesion produced in

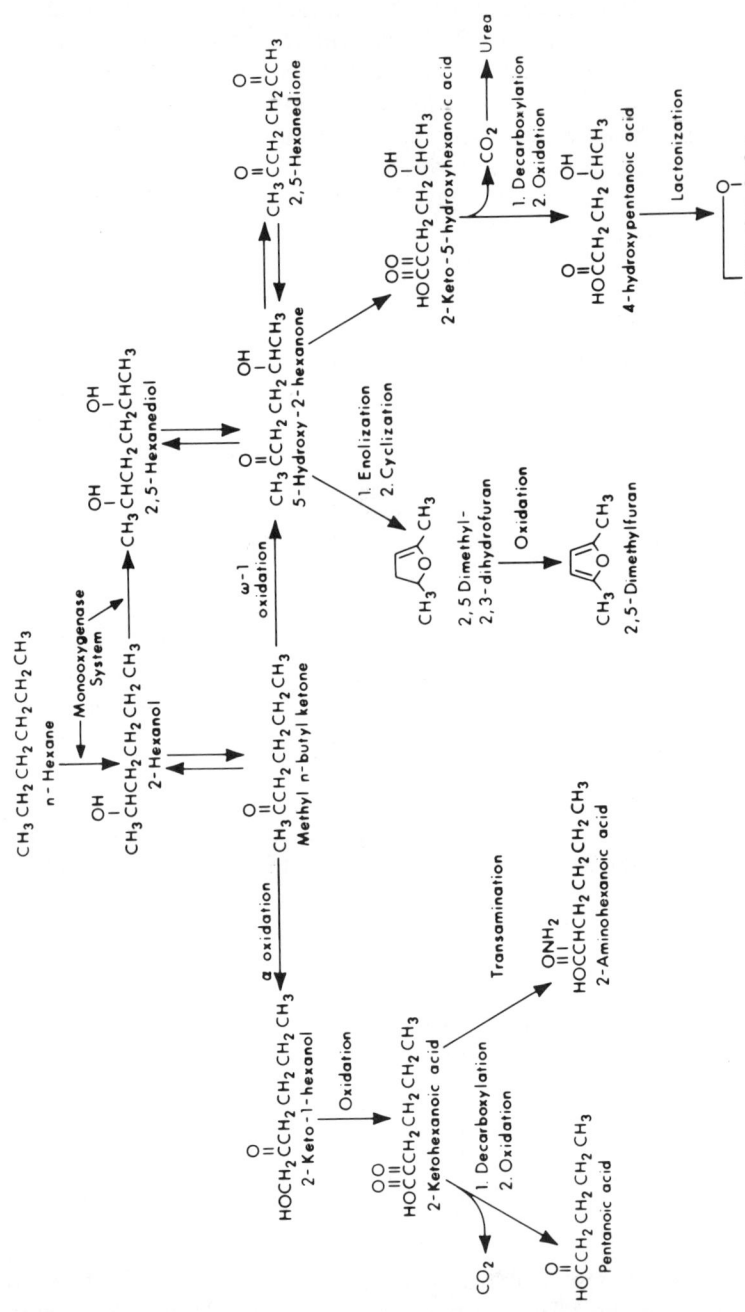

Fig. 1. Proposed metabolism of *n*-hexane to toxic metabolites. Modified from DiVincenzo et al.[11]

rats by 2,5-hexanedione treatment is indistinguishable from that of the other two compounds[3,14,15] strongly supported 2,5-hexanedione as the proximate neurotoxic metabolite. Furthermore, the relative neurotoxic potential of n-hexane and its metabolites, as determined by the time to clinical appearance of neurotoxicity, has been described as follows in rats given equimolar doses of each agent: 2,5-hexanedione > 5-hydroxy-2-hexanone > 2,5-hexanediol > methyl n-butyl ketone > 2-hexanol > n-hexane.[4] Most importantly, the neurotoxicity ranking obtained from the clinical parameters correlated directly with both the peak serum concentration and the total area under the curve for 2,5-hexanedione. In another study evaluating the disposition of n-hexane in rats after inhalation exposure, 2,5-hexanedione was retained in a target tissue, sciatic nerve, relative to nontarget tissues such as liver and blood.[12] The prolonged excretion of 2,5-hexanedione has also been noted in rats given methyl n-butyl ketone.[11] These investigations, therefore, have clearly implicated 2,5-hexanedione as the neurotoxic metabolite of n-hexane.

## C. THE $\gamma$-DIKETONE HYPOTHESIS

The delineation of 2,5-hexanedione as the putative neurotoxic metabolite of n-hexane has raised the question whether formation of a $\gamma$-diketone metabolite ($RCOCH_2CH_2COR$) is a necessary prerequisite for neurotoxic activity of structurally similar compounds. This appears to be so. Treatment of cats with methyl isobutyl ketone or methyl ethyl ketone, which cannot be metabolized to $\gamma$-diketone metabolites, produced no neurotoxicity.[16] In a subsequent study, Spencer et al.[17] found histopathological evidence of neurotoxicity in rats given 2,5-hexanedione or 2,5-hexanediol in the drinking water for 12 weeks. However, neurotoxicity was not observed in rats similarly treated with 2,4-hexanedione, 2,3-hexanedione, or 1,6-hexanediol. Analogs of shorter chain lengths, acetone, 1,4-butanediol, and glutaraldehyde, were also without neurotoxic activity. A common feature of all of these analogs is their inability to form a $\gamma$-diketone metabolite.

The neurotoxic activity of the $\gamma$-diketone functionality was further seen when the toxicity of straight-chain analogs containing greater than six carbons was characterized. 2,5-Heptanedione[18,19] and 3,6-octanedione[18] were neurotoxic in rats, while 2,6-heptanedione was not.[18] It is interesting to note that 2- and 3-heptanone also were not neurotoxic,[17,19] even though 3-heptanone was metabolized to 2,5-heptanedione.[19] However, when rats were exposed to equal concentrations of methyl n-butyl ketone or 3-heptanone, the serum concentration of 2,5-hexanedione was approximately 20-fold greater than the corresponding concentration of 2,5-heptanedione. Thus, the neurotoxic potential of the seven-carbon or greater alkanes appears to be limited as a result of their relatively minor capacity for metabolism to $\gamma$-diketones. This supposition is consistent with the results of a comparative toxicity study in rats of $n$-pentane, $n$-hexane, and $n$-heptane.[20] Despite exposure to a high concentration (3000 ppm) of the agents for 12 hr/day for up to 16 weeks, only $n$-hexane produced neurotoxicity.

## IV. Toxic Mechanisms of the $\gamma$-Diketones

### A. INHIBITION OF GLYCOLYSIS

A detailed evaluation of the pathogenesis of the nerve lesion produced by neurotoxic hexacarbons, both *in vivo* and *in vitro,* led Spencer et al.[21] to conclude that these agents exerted their primary toxic effects directly on the axon itself and not on the nerve cell perikaryon. The observation that the activity of the energy-requiring fast axonal transport system was decreased by methyl $n$-butyl ketone treatment[22] further suggested that 2,5-hexanedione might disrupt essential energy production processes within the axon.[21]

In a series of biochemical investigations, the potential for 2,5-hexanedione to inhibit key enzymes of glycolysis was evaluated. 2,5-Hexanedione or methyl $n$-butyl ketone inhibited both glyceraldehyde 3-phosphate dehydrogenase (GAPDH) and fructose 6-phosphate kinase (PFK) activity when incubated at millimolar concentrations with purified enzyme or homogenates of rat brain or nerve.[23,24] Coincubation with dithiothreitol prevented the enzyme inhibition in both cases, suggesting that these agents may interfere with essential sulfhydryl groups of the enzymes. This conclusion was supported by the observation that lactate dehydrogenase, a non-sulfhydryl-dependent enzyme, was not inhibited by either compound. The inhibitory effects were also consistent with the $\gamma$-diketone hypothesis, because 1,6-hexanediol, acetone,[23] and 2,4-hexanedione[24] were without effect on enzyme activity. A later study conducted by Howland et al.[25] confirmed that 2,5-hexanedione inhibited rat brain GAPDH *in vitro,* although the calculated $K_I$ was 47.9 m$M$. The high $K_I$ for 2,5-hexanedione contrasted with the $K_I$ of 0.41 m$M$ for acrylamide, which produces neurotoxicity similar to the hexacarbons. 2,5-Hexanedione also inhibited brain enolase, another glycolytic enzyme, with an estimated $K_I$ of 67.7 m$M$.

Although the previous studies demonstrated the ability of 2,5-hexanedione to inhibit key glycolytic enzymes *in vitro,* the concentrations required were approximately 100-fold greater than that found in sciatic nerves of rats exposed to 1000 ppm $n$-hexane for 6 hr or given 1% 2,5-hexanedione in the drinking water for 6 weeks.[26] In addition, no indication of an inhibition of glycolysis was obtained when lipogenesis from [$^{14}$C]glucose was examined in sciatic nerves isolated from rats treated with 2,5-hexanedione for 6 weeks.[27] Graham and Abou-Donia[28] also questioned the importance of GAPDH inhibition as a factor in the pathogenesis of neuropathy, because the concentration of 2,5-hexanedione required for *in vitro* enzyme inhibition was over $10^4$ times greater than that required for the sulfhydryl reagents, $N$-ethylmaleimide and $p$-chloromercuribenzoate. Thus, the relationship of the *in vitro* inhibition of glycolytic enzymes to the *in vivo* development of hexacarbon neuropathy remains to be clarified.

## B. DERIVATIZATION OF AMINO GROUPS OF AXON PROTEINS

In the evaluation of GAPDH inhibition by 2,5-hexanedione, Graham and Abou-Donia[28] observed a weak irreversible inhibition of the enzyme by 2,5-hexanedione and a weak reversible inhibition by acetone. These investigators proposed that the irreversible inhibition was due to formation of a stable conjugated diimine or Schiff base with one or two amino groups on proteins. The nonneurotoxic diketones, 2,3-, 2,4-, or 2,6-diketones, would be incapable of forming a stable conjugated imine and therefore would not irreversibly bind to protein. Such a phenomenon could account for the specific toxicity of the $\gamma$-diketones. In a subsequent study, it was demonstrated that incubation of 2,5-hexanedione with bovine serum albumin for 14 to 20 days produced a chromophore consistent with formation of conjugated imines.[29] It was suggested from these data that $\gamma$-diketones may form inter- and intramolecular protein crosslinks among critical proteins of the axon and, in particular, the 10-nm neurofilaments. The covalently crosslinked neurofilaments, which move slowly down the axon, would be too large to pass the nodes of Ranvier. This would account for the early observation in hexacarbon neurotoxicity of neurofilament accumulation on the proximal sides of the nodes.

An alternative to the formation of conjugated diimines was subsequently suggested by DeCaprio *et al.*,[30] who found evidence for formation of a pyrrole derivative as a product of the reaction of 2,5-hexanedione with lysine groups of protein (Fig. 2). Graham *et al.*[31] subsequently identified, by NMR spectroscopy and gas chromatography–mass spectrometry, a pyrrole derivative formed by the reaction of 2,5-hexanedione with the model amine, ethanolamine. When the pyrrole derivative of ethanolamine was allowed to stand in solution or subjected to thin-layer chromatography, an orange chromophore was produced. The chromophore was similar to the conjugated diimine proposed earlier,[29] which suggested that an oxidized pyrrole,

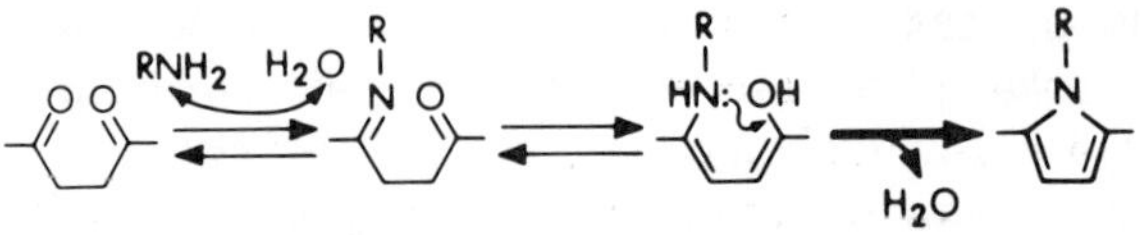

**Fig. 2.** Proposed formation of pyrrole derivatives from the reaction of 2,5-hexanedione with lysine amino groups of protein.

rather than a conjugated diimine, was responsible for the orange chromophore. Evidence for pyrrole formation was further supported in that attempts to establish proof of structure of conjugated diimine formation from reaction of 2,5-hexanedione with primary amines were unsuccessful. Incubations of 2,5-hexanedione with bovine serum albumin also resulted in a specific loss of lysinyl residues, and incubation with ovalbumin yielded evidence, based on electrophoretic analysis, of trimer, tetramer, and higher order polymer formation. These observations, therefore, were consistent with derivatization of lysinyl residues and subsequent oxidation of the pyrrole resulting in inter- and intramolecular protein crosslinks. Although these investigations have all been conducted *in vitro,* preliminary evidence has been obtained for formation of pyrrole derivatives in central nervous system proteins isolated from White Leghorn hens treated for 20 weeks with 2,5-hexanedione.[32]

## C. INHIBITION OF STEROL SYNTHESIS

Sciatic nerves isolated from rats fed 1% 2,5-hexanedione in the drinking water for 6 weeks had a significantly depressed incorporation of [$^{14}$C] acetate into several lipid fractions including total sterols, squalene, ubiquinone, and triacylglycerols.[33] Incorporation of [$^{14}$C]acetate into phospholipids, free fatty acids, and cholesteryl esters was not affected by the treatment. A comparison of [$^{14}$C]acetate and [$^{3}$H]mevalonolactone incorporation into nerve sterols indicated that 3-hydroxy-3-methylglutaryl-CoA reductase, the regulatory enzyme of cholesteryl synthesis, was the likely site of inhibition in sterol synthesis.[26] The inhibition of sterologenesis was not a direct or immediate effect of *n*-hexane or its metabolites, in that sciatic nerve sterologenesis was not altered by incubation with 2,5-hexanedione or by exposure of rats to 1000 ppm *n*-hexane for 6 hr.

The inhibition of sterol synthesis induced by 2,5-hexanedione may have several detrimental effects on nerve structure and function. Cholesterol constitutes approximately 20% of the lipid in the neurofilaments, which are one of the earliest axonal elements affected in hexacarbon neurotoxicity.[3] In addition, a decreased membrane cholesterol content may alter the fluidity of the axon membrane. Consistent with this, Couri and Nachtman[34] have reported changes in sciatic nerve membrane microviscosity in 2,5-hexanedione-treated rats, which preceded clinical signs of neurotoxicity. The precise role of the inhibition of sterol synthesis in the development of the neurotoxicity, however, remains to be clarified.

## References

1. Spencer, P. S., and Schaumburg, H. H. (1977). Central-peripheral distal axonopathy—the pathology of dying-back polyneuropathies. *Prog. Neuropathol.* **3**, 253–295.
2. Spencer, P. S., Schaumburg, H. H., Sabri, M. I., and Veronesi, B. (1980). The enlarging view of hexacarbon neurotoxicity. *CRC Crit. Rev. Toxicol.* **7**, 279–356.
3. Schaumburg, H. H., and Spencer, P. S. (1976). Degeneration in central and peripheral nervous systems produced by pure *n*-hexane: An experimental study. *Brain* **99**, 183–192.
4. Krasavage, W. J., O'Donoghue, J. L., DiVincenzo, G. D., and Terhaar, C. J. (1980). The relative neurotoxicity of methyl *n*-butyl ketone, *n*-hexane and their metabolites. *Toxicol. Appl. Pharmacol.* **52**, 433–441.
5. Chapin, R. E., Norton, R. M., Popp, J. A., and Bus, J. S. (1982). The effects of 2,5-hexanedione on reproductive hormones and testicular enzyme activities in the F-344 rat. *Toxicol. Appl. Pharmacol.* **62**, 262–272.
6. Herskowitz, A., Ishii, N., and Schaumburg, H. H. (1971). *n*-Hexane neuropathy: A syndrome occurring as a result of industrial exposure. *N. Engl. J. Med.* **285**, 82–85.
7. Billmaier, D., Yee, H. T., Allen, N., Craft, R., Williams, N., Epstein, F., and Fontaine, R. (1974). Peripheral neuropathy in a coated fabric plant. *J. Occup. Med.* **16**, 665–671.
8. DiVincenzo, G. D., Kaplan, C. J., and Dedinas, J. (1976). Characterization of the metabolites of methyl *n*-butyl ketone, methyl isobutyl ketone, and methyl ethyl ketone in guinea pig serum and their clearance. *Toxicol. Appl. Pharmacol.* **36**, 511–522.
9. Krämer, A., Staudinger, H., and Ullrich, V. (1974). Effect of *n*-hexane inhalation on the monooxygenase system in mice liver microsomes. *Chem.-Biol. Interact.* **8**, 11–18.
10. Nilsen, O. G., Toftgard, R., Eng, L., and Gustafsson, J.-A. (1981). Regio-selectivity of purified forms of rabbit liver microsomal cytochrome *P*-450 in the metabolism of benzo[*a*]pyrene, *n*-hexane and 7-ethoxyresorufin. *Acta Pharmacol. Toxicol.* **48**, 369–376.
11. DiVincenzo, G. D., Hamilton, M. L., Kaplan, C. J., and Dedinas, J. (1977). Metabolic fate and disposition of $^{14}$C-labeled methyl *n*-butyl ketone in the rat. *Toxicol. Appl. Pharmacol.* **41**, 547–560.
12. Bus, J. S., White, E. L., Gillies, P. G., and Barrow, C. S. (1981). Tissue distribution of *n*-hexane, methyl *n*-butyl ketone, and 2,5-hexanedione in rats after single or repeated inhalation exposure to *n*-hexane. *Drug Metab. Dispos.* **9**, 386–387.
13. Baker, T. S., and Rickert, D. E. (1981). Dose-dependent uptake, distribution, and elimination of inhaled *n*-hexane in the Fischer-344 rat. *Toxicol. Appl. Pharmacol.* **61**, 414–422.
14. Spencer, P. S., and Schaumburg, H. H. (1975). Experimental neuropathy produced by 2,5-hexanedione—a major metabolite of the neurotoxic industrial solvent methyl *n*-butyl ketone. *J. Neurol., Neurosurg, Psychiatry* **38**, 771–775.
15. Spencer, P. S., Schaumburg, H. H., Raleigh, R. L., and TerHaar, C. J. (1975). Nervous system degeneration produced by the industrial solvent methyl *n*-butyl ketone. *Arch. Neurol. (Chicago)* **32**, 219–222.
16. Spencer, P. S., and Schaumburg, H. H. (1976). Feline nervous system response to chronic intoxication with commercial grades of methyl *n*-butyl ketone, methyl isobutyl ketone, and methyl ethyl ketone. *Toxicol. Appl. Pharmacol.* **37**, 301–311.

17. Spencer, P. S., Bischoff, M. C., and Schaumburg, H. H. (1978). On the specific molecular configuration of neurotoxic aliphatic hexacarbon compounds causing central-peripheral distal axonopathy. *Toxicol. Appl. Pharmacol.* **44,** 17-28.
18. O'Donoghue, J. L., and Krasavage, W. J. (1979). Hexacarbon neuropathy: A γ-diketone neuropathy. *J. Neuropathol. Exp. Neurol.* **38,** 333.
19. Katz, G. V., O'Donoghue, J. L., DiVincenzo, G. D., and TerHarr, C. J. (1980). Comparative neurotoxicity and metabolism of ethyl $n$-butyl ketone and methyl $n$-butyl ketone in rats. *Toxicol. Appl. Pharmacol.* **52,** 153-158.
20. Takeuchi, Y., Ono, Y., Hisanaga, N., Kitoh, J., and Sugiura, Y. (1980). A comparative study on the neurotoxicity of $n$-pentane, $n$-hexane, and $n$-heptane in the rats. *Br. J. Ind. Med.* **37,** 241-247.
21. Spencer, P. S., Sabri, M. I., Schaumburg, H. H., and Moore, C. L. (1978). Does a defect of energy metabolism in the nerve fiber underlie axonal degeneration in polyneuropathies? *Ann. Neurol.* **5,** 501-507.
22. Mendell, J. R., Sahenk, Z., Saida, K., Weiss, H. S., Savage, R., and Couri, D. (1977). Alterations of fast axonal transport in experimental methyl $n$-butyl ketone neuropathy. *Brain Res.* **133,** 107-118.
23. Sabri, M. J., Moore, C. L., and Spencer, P. S. (1979). Studies on the biochemical basis of distal axonopathies. I. Inhibition of glycolysis by neurotoxic hexacarbon compounds. *J. Neurochem.* **32,** 683-689.
24. Sabri, M. I., Ederle, K., Holdsworth, C. E., and Spencer, P. S. (1979). Studies on the biochemical basis of distal axonopathies. II. Specific inhibition of fructose 6-phosphate kinase by 2,5-hexanedione and methyl $n$-butyl ketone. *Neurotoxicology (Park Forest South, Ill.)* **1,** 285-297.
25. Howland, R. D., Vyas, I. L., Lowndes, H. E., and Argentieri, T. M. (1980). The etiology of toxic peripheral neuropathies: *In vitro* effects of acrylamide and 2,5-hexanedione on brain enolase and glycolytic enzymes. *Brain Res.* **202,** 131-142.
26. Gillies, P. J., Norton, R. M., White, E. L., and Bus, J. S. (1980). Inhibition of sciatic nerve sterologenesis in hexacarbon-induced distal axonopathy in the rat. *Toxicol. Appl. Pharmacol.* **54,** 217-222.
27. Gillies, P. J., Norton, R. M., and Bus, J. S. (1981). Inhibition of sterologenesis but not glycolysis in 2,5-hexanedione-induced distal axonopathy in the rat. *Toxicol. Appl. Pharmacol.* **59,** 287-292.
28. Graham, D. G., and Abou-Donia, M. B. (1980). Studies of the molecular pathogenesis of hexane neuropathy. I. Evaluation of the inhibition of glyceraldehyde 3-phosphate dehydrogenase by 2,5-hexanedione. *J. Toxicol. Environ. Health* **6,** 621-631.
29. Graham, D. G. (1980). Hexane neuropathy: A proposal for pathogenesis of a hazard of occupational exposure and inhalant abuse. *Chem.-Biol. Interact.* **32,** 339-345.
30. DeCaprio, A. P., Olajos, E. J., and Weber, P. (1982). Covalent binding of a neurotoxic $n$-hexane metabolite: Conversion of primary amines to substituted pyrrole adducts by 2,5-hexanedione. *Toxicol. Appl. Pharmacol.* **65,** 440-450.
31. Graham, D. G., Anthony, D. C., Boekelheide, K., Maschmann, N. A., Richards, R. G., Wolfram, J. W., and Shaw, B. R. (1982). Studies on the molecular pathogenesis of hexane neuropathy. II. Evidence that pyrrole derivatization of lysinyl residues leads to protein cross linking. *Toxicol. Appl. Pharmacol.* **64,** 415-422.
32. DeCaprio, A. P., Weber, P., and Abraham, R. (1982). *In vivo* formation of substituted pyrrole adducts and the neurotoxicity of 2,5-hexanedione in the hen. *Toxicologist* **2,** 156.
33. Gillis, P. J., Norton, R. M., and Bus, J. S. (1980). Effect of 2,5-hexanedione on lipid biosynthesis in sciatic nerve and brain of the rat. *Toxicol. Appl. Pharmacol.* **54,** 210-216.
34. Couri, D., and Nachtman, J. P. (1979). Biochemical and biophysical studies of 2,5-hexanedione neuropathy. *Neurotoxicology (Park Forest South, Ill.)* **1,** 269-283.

Chapter 5

# Alkenes and Alkynes

PAUL R. ORTIZ DE MONTELLANO
Department of Pharmaceutical Chemistry
School of Pharmacy
University of California
San Francisco, California

I. Introduction . . . . . . . . . . . . . . . . . . . . 121
II. Oxidative Chemistry of Olefins and Acetylenes . . . . . . 122
III. Metabolism of Olefins . . . . . . . . . . . . . . . 126
IV. Metabolism of Acetylenes . . . . . . . . . . . . . . 131
V. Destruction of Cytochrome $P$-450 by Olefins and Acetylenes . 135
VI. Mechanisms of $\pi$-Bond Oxidation and Cytochrome $P$-450 Destruction . . . . . . . . . . . . . . . . . . . . 140
VII. Olefin and Acetylene Toxicity . . . . . . . . . . . . 144
References . . . . . . . . . . . . . . . . . . . . 145

## I. Introduction

The focal concern of this chapter is the mammalian metabolism of discrete olefinic and acetylenic functions. The metabolic fate of more complex conjugated functionalities is consequently dealt with only where it illuminates the intrinsic metabolic properties of carbon–carbon $\pi$ bonds. This distinction between isolated and conjugated $\pi$ bonds is not entirely arbitrary, because the chemical properties of unsaturated carbon–carbon bonds can be markedly altered by conjugation. The high electrophilic reactivity of $\alpha,\beta$-unsaturated carbonyl functions, for instance, reflects more the properties of the carbonyl group than it does the inherent reactivity of the carbon–

carbon double bond. The metabolism of olefins and acetylenes is examined here from a mechanistic perspective, an approach that, in view of limited space, allows only a cursory discussion of the further transformations and toxicity of the primary oxidative metabolites.

## II. Oxidative Chemistry of Olefins and Acetylenes

The $\pi$ bond of an olefin is weaker, and consequently more reactive, than the $\sigma$ bonds to its vinylic protons. Homolytic rupture of a carbon–hydrogen bond in ethylene, for example, requires 103 kcal/mol (Table I),[1-6] whereas rotation of one methylene group relative to the other, a motion that breaks the $\pi$ bond, requires a maximum of 65 kcal/mol.[7] Homolytic cleavage of the carbon–hydrogen bond in a terminal acetylene is also more difficult than addition to one of the acetylenic $\pi$ bonds. Furthermore, it is evident from the short length and high strength of the bonds in an acetylene (Table I) that it will generally react less readily with an electron-deficient species than will a comparable olefin. The higher reactivity of olefins in such reactions is also predicted by their lower gas phase ionization and solution electrolytic potentials, two measures of the ease with which an electron can be removed from the $\pi$ bond. The predicted difference in reactivity is borne out by the relative rates of reaction of olefins and acetylenes with radicals[8,9] and electrophilic reagents, like 4-chlorobenzenesulfenyl chloride,[5,6,10] and by the fact that electrochemical oxidation of olefins, but not acetylenes, is of preparative utility.[1,11,12]

The presence of a heteroatomic substituent, which can supply electrons to the $\pi$ bond, increases the reactivity of olefins and acetylenes. The electronic asymmetry introduced by such a substituent, however, makes one of the two carbons in the $\pi$ bond substantially more reactive than the other. An illustration of both effects is provided by the facile electrochemical dimerization of enol ethers in methanol, a process believed to involve coupling of an enol radical cation with a second unactivated substrate molecule (Scheme 1).[6,13,14] Only tail-to-tail dimers are produced in the reaction, because both radicals and cations are stabilized by electron donation from vicinal heteroatoms. Radicals and cations are not stabilized to the same degree by conjugation to electron-rich centers, however. The higher electron deficiency and, consequently, higher stabilization by electron donation of a

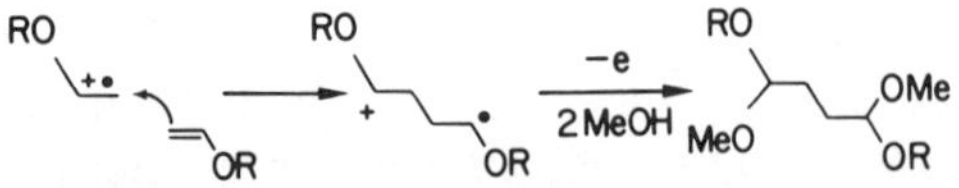

Scheme 1

## 5. ALKENES AND ALKYNES

Table I

Physical and Chemical Properties of Olefins and Acetylenes

| Property | Ethylene | 1-Butene | Acetylene | 1-Butyne | Reference |
|---|---|---|---|---|---|
| Ionization potential (eV) | 10.51 | 9.58 | 11.41 | 10.18 | 1 |
| $E_{1/2}$ (V) | 3.47 | 2.61 | 4.30 | 3.17 | 1 |
| Bond length (Å) | | | | | 2 |
| C–H | 1.083 | —[a] | 1.059 | — | |
| C–C | 1.338 | — | 1.206 | — | |
| Dissociation energy, C–H bond (kcal/mol) | | | | | |
| Relative rates (25°C) | 103 | — | 128 | — | 3 |
| Hydrogen radical addition | 8.6 | | | | |
| 4-Chlorobenzenesulfenyl chloride | | — | 1.0 | — | 4 |
| addition | $2.82 \times 10^5$ | — | 1 | 429 | 5, 6 |

[a] Dash indicates that no data are available.

cation is evident in the fact that homolysis of a carbon–hydrogen bond in methanol to give the hydroxymethyl *radical* is only 5 kcal/mol easier than formation of the ethyl radical from ethane,[15] whereas the methoxymethyl *cation* is more stable than the ethyl cation by 29 kcal/mol.[16,17] A practical consequence of the higher stabilization of cations than radicals is that, although the preferred regiochemistry of addition to the π bond is predicted to be the same in both cases, the regiochemical specificity is much higher in the case of cations. The addition of radicals to vinyl fluoride[18-20] and enol ethers[21,22] occurs preferentially at the β-carbon, for example, but the proportion of addition to the β- over the α-carbons has been shown with vinyl fluoride not to exceed a value of 50. The reaction of enol ethers with cations, on the other hand, proceeds exclusively at the β-carbon.[23]

Increasing the electron affinity of a carbon atom makes homolytic cleavage of a carbon–hydrogen bond more difficult but facilitates dissociation of the hydrogen as a proton. Acetylenes are consequently more acidic than olefins[24,25]: the p$K_a$ of ethylene is 36.5, whereas the corresponding values for acetylene and phenylacetylene are 25 and 18.5, respectively. The extremely low acidity of vinylic protons makes their participation in biological acid–base reactions highly unlikely. The analogous reactions of acetylenic protons are still difficult but not biologically impossible.

The chemical oxidation of olefins to epoxides occurs by concerted and nonconcerted mechanisms. A concerted mechanism, one in which both carbon–oxygen bonds are formed in a synchronous manner even if not necessarily at the same rate, is usually attributed to epoxidation reactions in which the olefin stereochemistry is completely retained. Stereochemical retention does not, however, rule out nonconcerted pathways involving intermediates in which rotation about the carbon–carbon axis is slow relative to closure of the epoxide ring. The stereochemical fidelity observed in the epoxidation of olefins by organic peracids, in view of the enormous number and variety of structures that have been examined, nevertheless leaves little doubt that this particular reaction proceeds by a concerted mechanism.[26] Nonconcerted mechanisms are commonly observed in metal-catalyzed epoxidation reactions.[27,28] Of particular interest in this regard are epoxidation reactions catalyzed by metalloporphyrins, because they are directly relevant to the biological epoxidation reactions promoted by cytochrome $P$-450 enzymes. The development of appropriate model systems has been facilitated by the finding that peroxides and oxygen donors, like phenyliodosobenzene, can replace the NADPH and oxygen requirement of the normal enzymatic process.[29] Chlorodimethylferriprotoporphyrin IX, the diester of the prosthetic heme moiety in cytochrome $P$-450, catalyzes the epoxidation of *cis*- and *trans*-stilbene in the presence of phenyliodosobenzene.[30] The stereochemistry of the parent stilbene is retained in this reaction, but is not retained in the analogous epoxidation of the stilbene isomers catalyzed by a manganese tetraphenylporphyrin complex.[30a] The loss of stereochemistry in the latter instance has been rationalized in terms of the oxomanganese radical intermediate shown in Scheme 2 (the porphyrin is

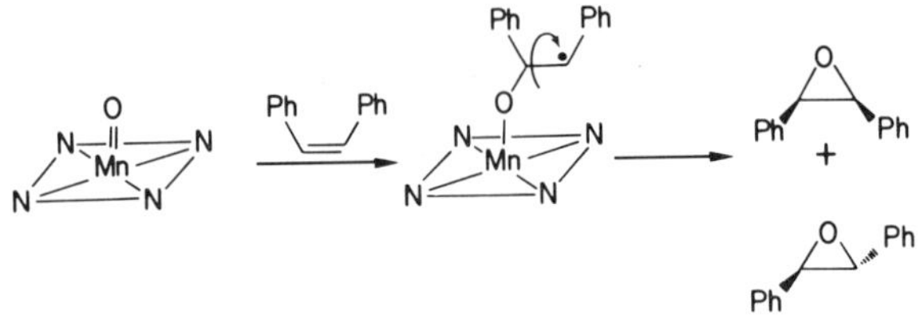

Scheme 2

indicated by the four nitrogens). The delicate balance between concerted and nonconcerted mechanisms is delineated by the contrasting observation that the olefin stereochemistry is preserved in the epoxidation by related metalloporphyrins of (a) *cis*- and *trans*-2-hexene in the presence of *tert*-butylperoxide[31] and (b) the isomers of 2-octene in the presence of oxygen and a catalytic hydrogenation system.[32]

One mechanism for retention of stereochemistry in a nonconcerted mechanism is provided by the proposal that a metalloorganic intermediate intervenes in the epoxidation of olefins by chromyl chloride.[33] The postulated

## 5. ALKENES AND ALKYNES

intermediate (Scheme 3), obtained by nucleophilic attack of the π bond on the electron-deficient metal, is converted to the epoxide by an internal ligand transfer of the metal-bound carbon to the oxygen. This reaction is predicted to proceed with retention of the olefin stereochemistry. The formation of analogous metallocyclic intermediates has been suggested as a general mechanism,[33] although no evidence is now available to support the formation of such metallocyclic species in metalloporphyrin-catalyzed epoxidation reactions.

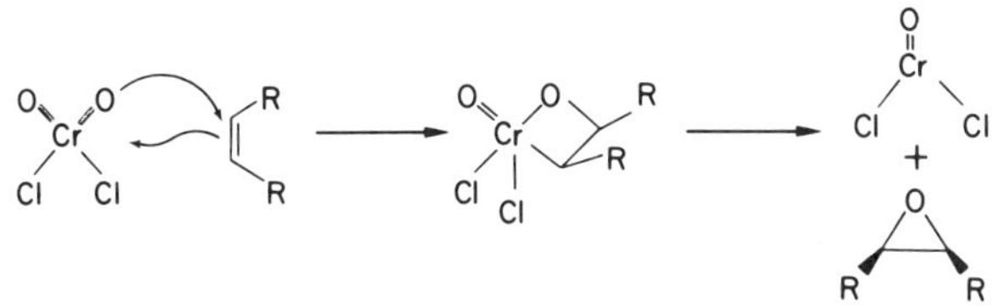

Scheme 3

The epoxidation of an acetylene, in principle, yields an unsaturated epoxide, or oxirene. Oxirenes are so exceedingly unstable, however, that the only examples so far directly observed are perfluorinated analogs trapped in an argon matrix.[34] Theoretical calculations indicate that the rearrangement of oxirenes to ketocarbenes, acyclic dipolar species, and, eventually, ketenes is an energetically favored process with negligible reaction barriers (Scheme 4).[35,36] These theoretical predictions are borne out by the known oxidative chemistry of acetylenes.[37-42] An example pertinent to the biological oxida-

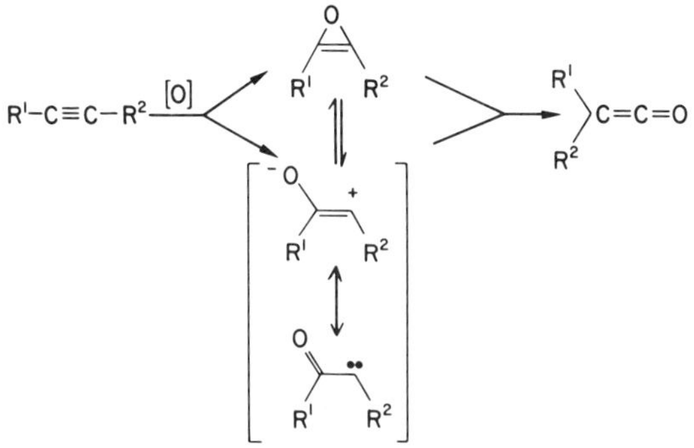

Scheme 4

tion of a substrate discussed in Section IV is provided by the demonstration that the oxidation of [1-$^2$H]biphenylacetylene with *m*-chloroperbenzoic acid in methanol gives methyl [2-$^2$H]biphenylacetate as the only product (Scheme 5).[42,43]

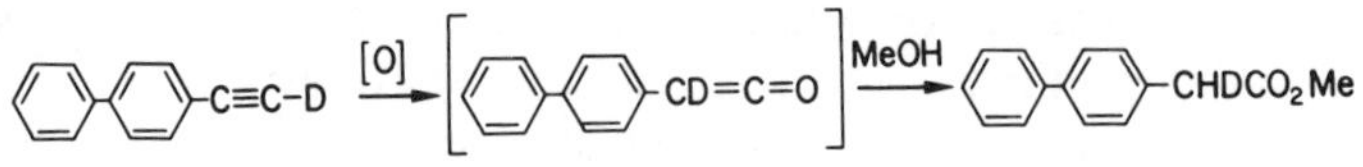

$$\text{Ph-Ph-C≡C-D} \xrightarrow{[O]} \left[\text{Ph-Ph-CD=C=O}\right] \xrightarrow{\text{MeOH}} \text{Ph-Ph-CHDCO}_2\text{Me}$$

**Scheme 5**

## III. Metabolism of Olefins

The only general pathway for the metabolism of an olefin function is its oxidation to an epoxide. The chemical reactivity of epoxides, and the presence of efficient enzymes for their further metabolism, however, frequently result in the experimental observation, not of the epoxides themselves, but of secondary products. The enzymes responsible for the metabolism of epoxides are epoxide hydrolases[44,45] and glutathione transferases,[46,47] and the products generated by these enzymes are, respectively, *vic*-diols and glutathione conjugates (Scheme 6). A partial list of the simple olefins that have been shown to undergo metabolic conversion to *vic*-diols and glutathione conjugates is given in Table II.[48–65] Although the glutathione conju-

**Scheme 6**

gates are obtained *in vitro*, the metabolites excreted *in vivo* are usually the mercapturic acid derivatives obtained by enzymatic removal of the glycine and glutamic acid residues from the tripeptide moiety followed by acetylation of the cysteinyl amino group in the conjugate.[47]

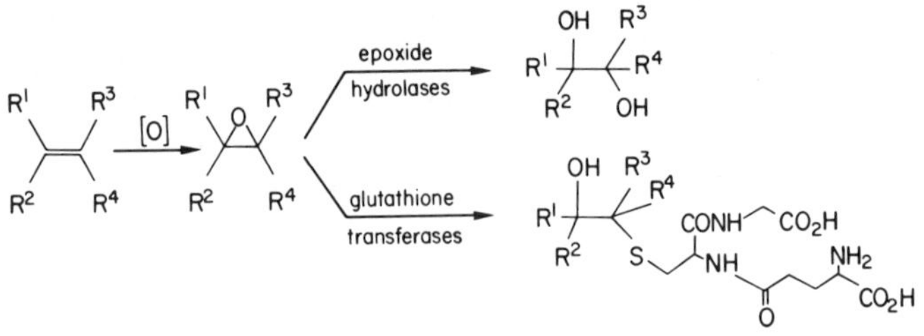

The epoxide metabolites themselves can be isolated when they are poor substrates for epoxide hydrolases and glutathione transferases or *in vitro* when the epoxide hydrolases are deliberately inhibited. The isolation in 1953 of the epoxide (**1**) of heptachlor from the fatty tissue of dogs fed the parent compound,[66] the first demonstration of a biological epoxidation, reflects the fortuitous circumstance that this sterically hindered epoxide is resistant to further metabolism.[67] A number of epoxides, some of which can be formed enzymatically from the corresponding olefins, are poor enough substrates that they have found utility as inhibitors of microsomal epoxide hydrolases.[49,68] The observation that 4-octene is converted to 4,5-octane-

## 5. ALKENES AND ALKYNES

**Table II**
Metabolism of Olefins to Diols and Glutathione Conjugates

| Olefin | Diol | Reference Diol | Reference Glutathione conjugate |
|---|---|---|---|
| 1-Octene | 1,2-Octanediol | 48, 49 | —[a] |
| 4-Octene | 4,5-Octanediol | 48 | — |
| trans-Stilbene | meso-1,2-Diphenyl-1,2-ethanediol | 50 | — |
| cis-Stilbene | threo-1,2-Diphenyl-1,2-ethanediol | 51 | — |
| Oleic acid | threo-9,10-Dihydroxystearic acid | 52 | — |
| Styrene | Phenyl-1,2-ethanediol | 53, 54 | 55 |
| Cyclohexene | trans-1,2-Cyclohexanediol | 54, 56 | 57, 58 |
| Indene | trans-1,2-Indanediol | 59, 60 | 61 |
| Secobarbital | Secobarbital diol | 62, 63 | — |
| 2-Isopropyl-4-pentenamide | 4,5-Dihydroxy-2-isopropylpentanamide | 64 | — |
| p-Methylstyrene | p-Methylphenyl-1,2-ethanediol | 65 | 65 |

[a] Dash indicates that no data are available.

diol by rat hepatic microsomes, but to 4,5-oxidooctane when the incubation also includes 20 mM 1,2-oxidooctane,[48] provided early support for the view that epoxides are mandatory precursors of *vic*-diol metabolites. More effective inhibitors of epoxide hydrolases than 1,2-oxidooctane, including oxidocyclohexane and 3,3,3-trichloro-1,2-oxidopropane, are now commonly used. The development of sensitive analytical techniques has made possible, in more recent times, the detection of epoxide metabolites excreted in the urine, even when these are not particularly resistant to further metabolism.[63,69-71] A partial list of the epoxide metabolites, which have been isolated and characterized either *in vivo* or *in vitro*, is given in Table III.[48,51,54,61,63,69-78] A comparison of Tables II and III provides several examples of the role of epoxides as precursors of *vic*-diols and glutathione conjugates.

The epoxidation of olefins in animals is catalyzed by microsomal cytochrome *P*-450 enzymes. One of the few known exceptions is the oxidation of the terminal double bond of squalene (8) in the biosynthesis of cholesterol,[78] a reaction catalyzed by a substrate-specific flavoprotein.[79,80] Epoxidation reactions catalyzed by nonheme enzymes are not uncommon in unicellular organisms, however.[81,82] The role of cytochrome *P*-450 in mammalian metabolism of unsaturated hydrocarbons has been most extensively documented in the case of aromatic hydrocarbons (see Chapters 6 and 7, this volume).[83-86] Ancillary evidence for the analogous role of cytochrome *P*-450 in olefin epoxidation is provided by the requirement for NADPH and oxygen,[48,52,59] by inhibition of the reaction by carbon monoxide and other cytochrome *P*-450 inhibitors,[87-89] by alterations in epoxidation activity caused by inducers of cytochrome *P*-450,[59,64,87,88] and by the epoxidation of olefins by purified, reconstituted enzymes.[90]

Olefin epoxidation is a stereospecific process that proceeds with retention of the olefin stereochemistry. The metabolism of *cis*-stilbene and oleic acid gives *threo*-1,2-diphenyl-1,2-ethanediol[51] and *threo*-9,10-dihydroxystearic acid,[52] respectively, the products expected from enzymatic hydrolysis of the *cis*-epoxides. A more definitive test of the stereospecificity of the epoxidation process is provided by the recent demonstration that the stereochemistry of *trans*-[1-$^2$H]-1-octene is retained in the epoxide metabolite (Scheme 7).[91] This latter experiment is important, because it shows that hydrophobic bonding of substituents at the two ends of a $\pi$ bond, an interaction that could suppress the rotation required for stereochemical inversion, is not responsible for the observed retention of stereochemistry.

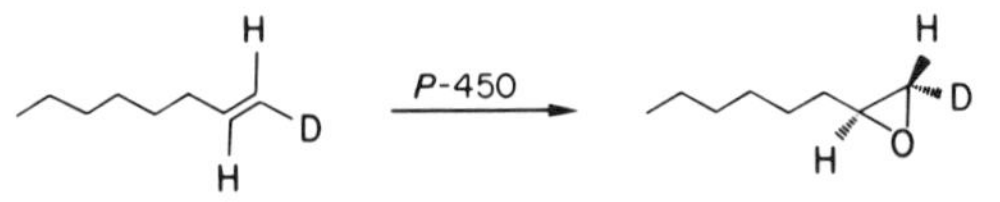

Scheme 7

## 5. ALKENES AND ALKYNES

**Table III**
Isolated Epoxide Metabolites

| Olefin | Metabolite | Reference |
|---|---|---|
| 1-Octene | 1,2-Oxidooctane | 48 |
| 4-Octene | 4,5-Oxidooctane | 48 |
| 1-Hexadecene | 1,2-Oxidohexadecane | 72 |
| Cyclohexene | 1,2-Oxidocyclohexane | 54 |
| Styrene | 1-Phenyl-1,2-oxidoethane | 54 |
| cis-Stilbene | cis-1,2-Diphenyl-1,2-oxidoethane | 51 |
| Indene | 1,2-Oxidoindane | 61 |
| 4-Vinylcyclohexene | (2'-Cyclohexenyl)-1,2-oxidoethane | 73 |
| Hexobarbital | Hexobarbital epoxide (**2**) | 70 |
| Secobarbital | Secobarbital epoxide (**3**) | 69 |
| Allyl alcohol | Glycidol | 74 |
| $\Delta^9$-Tetrahydrocannabinol | 1,2-Oxidotetrahydrocannabinol (**4**) | 75 |
| Aldrin | Dieldrin (**5**) | 76, 77 |
| Quinine | 10,11-Oxidoquinine (**6**) | 63 |
| Alclophenac | Alclophenac epoxide (**7**) | 71 |
| Squalene (**8**) | 2,3-Oxidosqualene | 78 |

The activated oxygen species involved in cytochrome $P$-450-catalyzed oxidations can be transferred not only to the $\pi$ bond of an olefin but also to other oxidizable functions of the same substrate. As a result of this internal competition between reactive sites, the less substituted a carbon–carbon $\pi$ bond, the more important epoxidation is likely to be. This is illustrated by the finding that the rate of epoxidation of 1-octene is four times higher than that of 4-octene, a difference compensated for by the formation of an octenol metabolite in the case of 4-octene.[48] In practical terms, epoxidation is seldom a significant pathway in the metabolism of trisubstituted or tetrasubstituted olefins. A relevant example is provided by the metabolism of geraniol, a trisubstituted olefinic substrate related to squalene (**8**). The observation of only allylic hydroxylation with geraniol (Scheme 8)[92] underscores the possible advantage of a nonheme enzyme for the biosynthetic epoxidation of squalene.[79,80]

**Scheme 8**

The steric effect of substituents, and their competition with the double bond for the activated oxygen species, obscure the electronic effect of such substituents on the intrinsic reactivity of the $\pi$ bond. Increasing substitution, for example, is known to increase the rate of chemical epoxidation of olefins by peracids, because carbon substituents stabilize the development of positive charge more effectively than hydrogen atoms.[48] The rate of epoxidation of benzene, a system relatively insulated from complications introduced by steric effects or internal competition for the reactive oxidant, is reduced by halide substitution.[93] Halogenated ethylenes are metabolized more slowly *in vivo* than ethylene,[94] a result that is not easily translated into a mechanistic argument because of the ambiguities inherent in the *in vivo* approach. A systematic study of the rates of epoxidation of para-substituted styrenes has, in contrast, failed to detect a correlation between the rates and the electronic properties of the para substituents.[95] It should be noted that related study of the benzylic hydroxylation of 1,3-diphenylpropanes bearing para substituents on one of the two phenyl rings has provided evidence that electron-withdrawing substituents reduce the rate of hydroxylation.[96] The evidence thus suggests, but not without some contradiction, that epoxidation of a $\pi$ bond is facilitated by electron release from the $\pi$-bond substituents.

A few instances have been reported of the metabolism of isolated $\pi$ bonds to products other than epoxides, diols, and glutathione conjugates. The conversion of olefins to aldehydes or ketones, a reaction that may involve rearrangement of epoxide intermediates but may also arise by a still undefined direct pathway, is one of these alternative metabolic routes. The ketone derivatives of the double bonds in novonal (**9**)[93] and brallobarbital

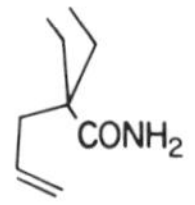

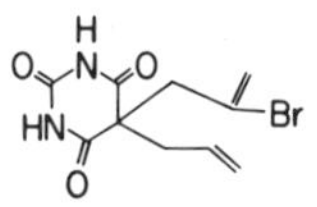

(**10**)[97] have been found, albeit in very low yield, among the urinary metabolites of these drugs. The *in vivo* conversion of styrene to phenylacetic acid (Scheme 9) has been shown by deuterium-labeling studies to involve a 1,2-shift of one of the terminal protons.[98] The finding that such carbonyl

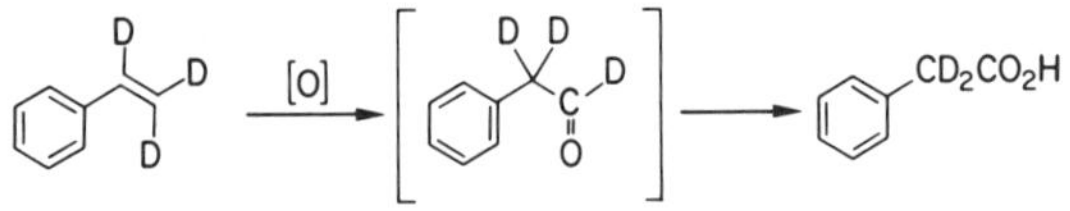

Scheme 9

derivatives have only been observed among urinary metabolites suggests that they may originate by acid-catalyzed ring opening of the epoxides, a reaction for which chemical precedent exists (Scheme 10).[99] The study with deuterated styrene specifically excludes insertion of oxygen into the $\pi$ bond between the terminal carbon and the associated vinylic proton. The forma-

## 5. ALKENES AND ALKYNES

tion of a carbonyl derivative during the metabolism of a $\pi$ bond has only been unequivocally demonstrated *in vitro* in the case of halogenated ethylenes. The best documented example is the conversion of trichloroethylene to trichloroacetaldehyde (Scheme 10),[100] a reaction recently suggested to involve rearrangement of the chloride during oxygen transfer to the $\pi$ bond (see Section VI),[101,102] but which could also involve rearrangement of the epoxide metabolite.[103]

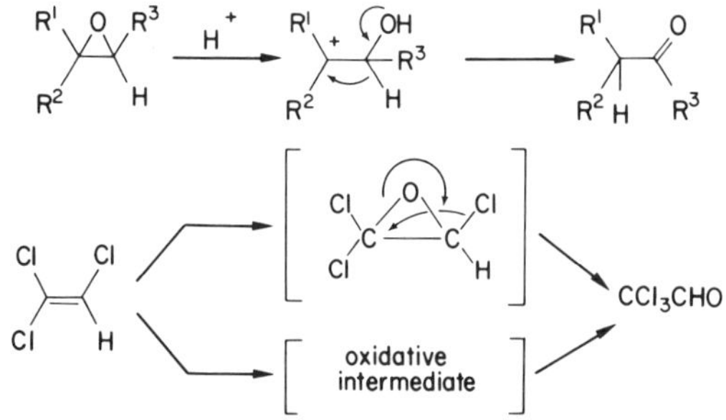

**Scheme 10**

A second type of transformation for which isolated examples exist is the conversion of a $\pi$ bond to the corresponding alcohol function. This is illustrated in Scheme 11 by a transformation of tamoxifen (**11**) observed, so far, only in monkeys.[104] This transformation may involve hydration of the $\pi$ bond or reduction of the epoxide metabolite, a reaction recently reported for 24,25-oxidolanosterol.[105]

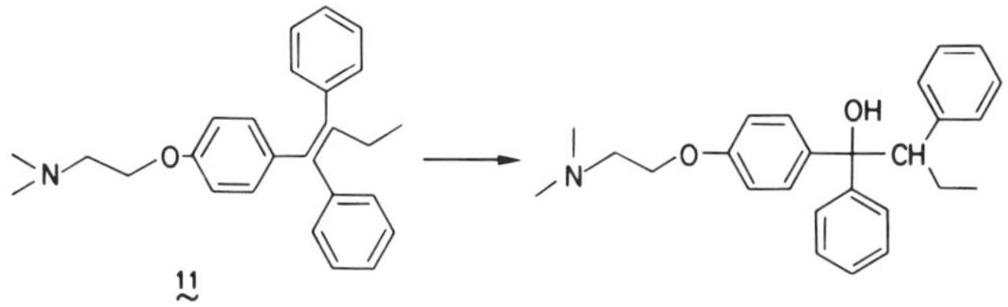

**Scheme 11**

### IV. Metabolism of Acetylenes

The metabolism of acetylenes has received relatively little attention. The older literature records the *in vivo* conversion of 3-(2-nitrophenyl)propynoic acid[106] and 2-ethinylnitrobenzene[107] to 3-hydroxyindole (Scheme 12), and the excretion of phenylacetylene in rabbit urine as 2-phenylacetic acid.[53] It is not possible to determine from these and more recent[108] *in vivo* studies,

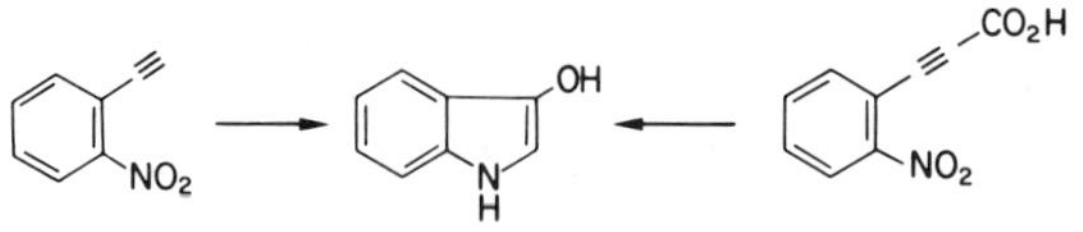

Scheme 12

however, whether the acetylene group has undergone a hydrolytic or oxidative transformation. The demonstration that 4'-ethinyl-2-fluorobiphenyl[109] and biphenylacetylene[110,111] are converted, both *in vivo* and *in vitro*, to the corresponding biphenylacetic acids (Scheme 13), however, provides credible evidence for the involvement of a cytochrome *P*-450-catalyzed oxidative process.

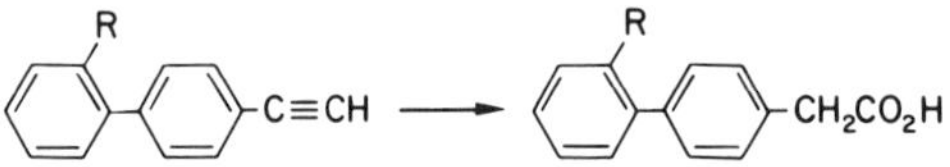

R = H, F

Scheme 13

Two alternative mechanisms can be formulated for oxidative conversion of a terminal acetylene to an acetic acid derivative, both of which involve hydration of a ketene intermediate as the final step (Scheme 14). The ketene in the first mechanism, however, is obtained by insertion of oxygen into the acetylenic carbon–hydrogen bond, giving a hydroxyacetylene that tautomerizes to the ketene, whereas the ketene in the second mechanism arises by 1,2-shift of the acetylenic hydrogen atom during oxidation of a $\pi$ bond in the acetylene. The two mechanisms, one triggered by carbon hydroxylation and the other by $\pi$-bond oxidation, can be readily distinguished. The former predicts that the acetylenic proton will be lost by exchange with the medium, while the latter predicts that the proton will be retained, but on the carbon adjacent to that on which it was originally located (Scheme 14). The

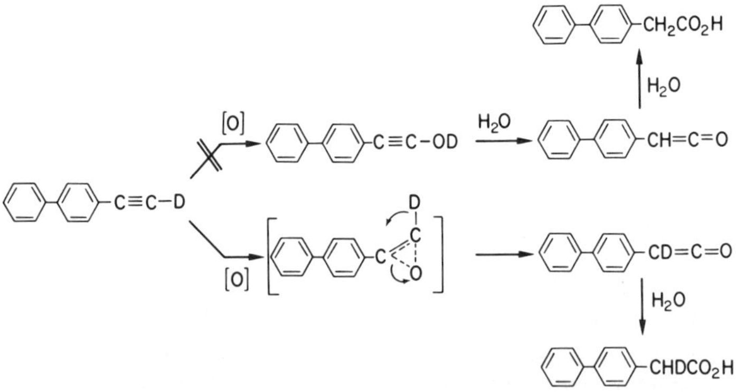

Scheme 14

# 5. ALKENES AND ALKYNES

oxidation of [1-$^2$H]biphenylacetylene by rat hepatic microsomes gives exclusively the product in which the deuterium is retained on the vicinal carbon, a result completely analogous to that obtained when the oxidation is performed chemically with *m*-chloroperbenzoic acid (Scheme 5).[42,43] The dominance of π-bond oxidation is not unexpected, of course, in view of the high homolytic bond strength of the acetylenic carbon–hydrogen bond (see Section II). The observation that the rates of metabolism of 4′-ethinyl-2-fluorobiphenyl- and biphenylacetylene are subject to an isotope effect when deuterium replaces the acetylenic proton has nevertheless been used to argue against a π-bond oxidative mechanism.[112] The isotope effects observed ($k_H/k_D$ of 1.42 and 1.95, respectively), however, are readily explained by a π-bond oxidation mechanism if the hydrogen shift is synchronous with rate-limiting π-bond oxidation, or if, as discussed in the next section, the hydrogen shift determines the rate of inactivation of cytochrome *P*-450.

The metabolism of one other set of acetylenic compounds, the 17α-ethinyl sterols, has been extensively investigated because of the widespread use of these substances as contraceptives. Abdel-Aziz and Williams first reported the urinary excretion by rabbits[113] and humans,[114] after treatment with 17α-ethinylestradiol (**12**), of a D-ring-expanded metabolite in which the acetylene group is no longer present (**13**). A metabolic scheme involving oxidation of the acetylenic group was postulated by these authors to explain the formation of the unusual metabolite (Scheme 15). The same D-ring expansion has been observed more recently with norgestrel (**14**).[115,116] The results with norgestrel include evidence for the presence in the urine of an unstable intermediate, the analog of the keto acid in Scheme 15, which

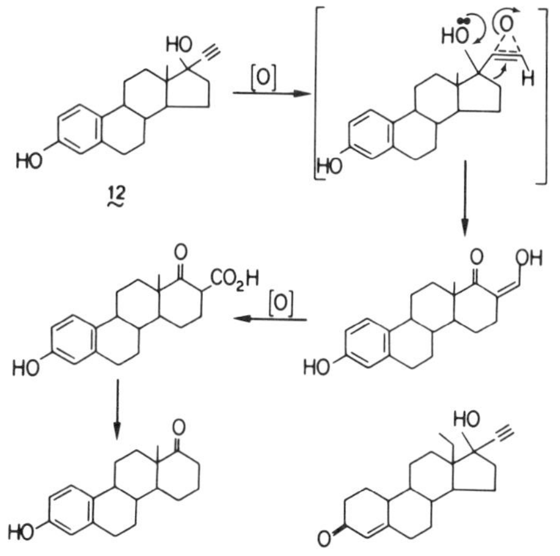

Scheme 15

decarboxylates in acidic solution to the final D-homoketone metabolite.[116] Metabolism of the ethinyl group in sterols thus also involves oxidation of the $\pi$ bond, except that the oxidation is accompanied by shift of a carbon from the *vic*-sterol skeleton rather than of the terminal acetylenic proton. The carbon shift is favored in these sterols, because the positive charge that develops during $\pi$-bond oxidation can be stabilized by electron donation from the 17-hydroxyl group if the ring expansion occurs in concert with the oxidative reaction.

A second metabolic pathway involving the acetylenic group in the 17α-ethinyl sterols has been noted. The 17-keto derivatives formally obtained

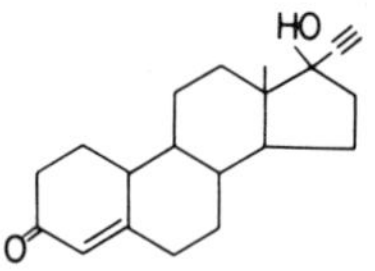

**15**

by loss of the ethinyl group have been found among the urinary metabolites of norethisterone (**15**)[117] and 17α-ethinylestradiol (**12**).[118,119] Little is known about the mechanism of this transformation other than that it requires NADPH and is inhibited by SKF-525A, two indications of a cytochrome *P*-450-dependent process.[119] The most attractive, if still unconfirmed, mechanism for this ethinyl group loss (Scheme 16) is that rearrangement of the terminal hydrogen during oxidation of the acetylenic $\pi$ bond competes with the carbon shift that leads to the D-ring-expanded products. The acetic acid moiety obtained by rearrangement of the hydrogen atom, or an esterified derivative of it if the ketene is immediately trapped by a nucleophile other than water, is then eliminated with formation of the 17-keto function by a retro-aldol type of reaction.[120]

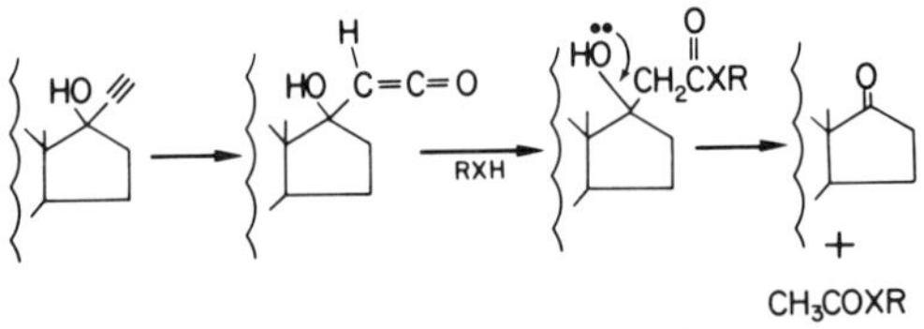

**Scheme 16**

The carbon glucuronide (**16**) of ethchlorvynol (**17**) has been isolated from the urine of rabbits treated with the parent drug.[121] This unusual metabolite presumably reflects the relatively high acidity of the acetylenic proton. Although this is the only known example of this transformation, the possibility that other examples have escaped detection is high. The most common

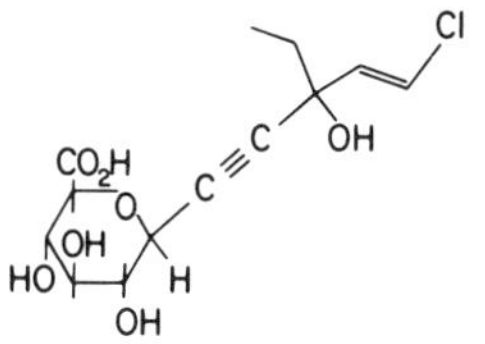

# 5. ALKENES AND ALKYNES

[Structures 16 and 17]

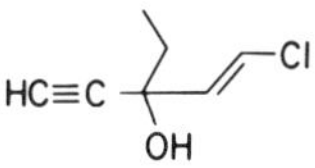

approach for the characterization of glucuronide metabolites, hydrolytic release of the parent compounds with glucuronidases, is ineffective in the present situation because carbon glucuronides are not susceptible to the action of these enzymes.[121]

## V. Destruction of Cytochrome P-450 by Olefins and Acetylenes

The perturbation of heme biosynthesis associated with administration of 2-isopropyl-4-pentenamide (18), commonly referred to as allylisopropyl-

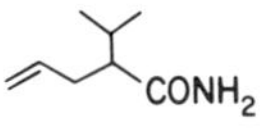

[Structures 18, 19, 20]

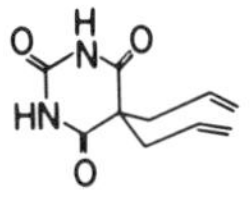

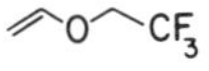

acetamide (AIA), and related homoallylic amides has been known for several decades.[122,123] The observation in the late 1960s that administration of AIA causes a rapid decline in the cytochrome P-450 content of hepatic microsomes implied a possible connection between the effect of AIA on heme biosynthesis and cytochrome P-450 content.[124,125] Evidence for such a connection was provided by De Matteis,[126] who also established that the *in vivo* loss of cytochrome P-450 reflects destruction of preexisting enzyme by a process intimately tied to the olefinic terminus in the substrate.[127,128] Levin *et al.*[129,130] concurrently established the importance of the olefin moiety and furthermore demonstrated that enzyme loss occurs *in vitro* if the incubation with the unsaturated substrates includes the cofactors required for catalytic turnover of the enzyme. Subsequent work has led to the realization that the ability to destroy cytochrome P-450 is not restricted to a narrow group of homoallylic amides like AIA and secobarbital but is, in fact, an inherent feature of the oxidative metabolism of isolated $\pi$ bonds. Cytochrome P-450 is effectively destroyed by simple olefins like 1-heptene, 4-ethyl-1-hexene, and, indeed, even by ethylene itself.[131] A list of some of the known destructive substrates is given in Table IV.[127,129,131,132] The structure–activity data

**Table IV**

Examples of Cytochrome P-450 Destruction by Olefins Where the Role of the π Bond Has Been Confirmed

| Compound | Reference In vitro | In vivo |
|---|---|---|
| 2-Isopropyl-4-pentenamide (**18**) | 131 | 127 |
| Secobarbital | 129 | 129 |
| 2-Phenyl-4-pentenamide | —[a] | 127 |
| 5,5-Diallylbarbituric acid (**19**) | — | 127, 129 |
| Aprobarbital | — | 129 |
| 1-Heptene | 131 | — |
| 4-Ethyl-1-hexene | 131 | — |
| Ethylene | 131 | — |
| Fluroxene (**20**) | 132 | — |

[a] Dash indicates that no data are available.

now available suggest that enzyme destruction, at its maximum with terminal olefins, is sharply reduced or suppressed by the presence of substituents on the π bond.[131] Terminal olefins are known, nevertheless, that do not detectably destroy cytochrome P-450.[131,133] The absence of destruction by a terminal olefin is likely, at least in some instances, to reflect the absence of significant π-bond oxidation. Insufficient information is available at this time, however, to allow differentiation of the possible explanations for lack of activity, or to define a reliable relationship between structure and activity.

The ability of acetylenes to inhibit oxidative xenobiotic metabolism underlies the incorporation of the acetylenic function into insecticide synergists.[134,135] The inhibitory properties of the synergists were shown in 1971 to be associated with an *in vivo* decrease in cytochrome P-450 concentration.[135] Ethynyl sterols are also known to inhibit drug metabolism, although the specific role of the ethynyl function is less readily discerned because the sterols can interfere with metabolic processes by indirect hormonal mechanisms.[136,137] The demonstration in 1977 that norethisterone (**15**) and ethinyl estradiol (**12**) cause the *in vitro* destruction of cytochrome P-450, that catalytic turnover of the enzyme is a prerequisite of the destructive process, and that the destructive activity resides in the triple bond, established at least a partial role for the ethynyl function in the inhibition of drug metabolism by 17-ethynyl sterols.[138] The generality of the destruction of cytochrome P-450 during oxidative metabolism of acetylenic functions has received strong confirmation.[42,139-142] Of some three dozen acetylenic substrates, including acetylene itself, only a small number have not been found to cause detectable

# 5. ALKENES AND ALKYNES

cytochrome *P*-450 destruction. The inactive compounds are ethinamate (**21**), pargyline (**22**), barban (**23**), and tremorine (**24**).[133,141] Destructive

activity is highest in terminal acetylenes, as it is in olefins, but in the case of acetylenes, destructive activity is not necessarily lost when the triple bond is not at a terminal position.[140,141] The destruction caused by internal acetylenes, however, differs from that mediated by their terminal counterparts in that only the latter give rise to isolable prosthetic heme adducts (see later discussion).[140] It should also be mentioned that cytochrome *P*-450 is analogously destroyed during the metabolism of allenes,[143] although nothing is known of the metabolic fate of this functionality.

The destruction of cytochrome *P*-450 by AIA (**18**) is the result of a suicidal interaction of the enzyme with the substrate.[144] Evidence for this is provided by adherence to pseudo-first-order kinetics, by the requirement for oxygen and NADPH, by the observation that only cytochrome *P*-450 isozymes that metabolize the substrate are vulnerable to inactivation, and by the fact that electron flow to the enzyme must remain unhampered for destruction to occur.[144,145] Analogous, if less complete evidence indicates that the destruction caused by other unsaturated substrates also results from an autocatalytic, suicidal process. The statistical number of substrate molecules oxidized by an enzyme molecule before it is inactivated has been measured in the case of AIA and fluroxene (**20**). A value of approximately 200–240 has been obtained for this partition ratio when AIA is incubated with microsomes from phenobarbital-treated rats,[144,146] or with a purified, reconstituted enzyme preparation from such microsomes.[145,146] In the case of fluroxene, a ratio of approximately 296 has been obtained in a microsomal system.[147] The structural and mechanistic parameters that govern the partition ratio are not known.

The key to the mechanism of inactivation of cytochrome *P*-450 by unsaturated functionalities is provided by the parallel, equimolar loss of microsomal heme[128,139,140,145–151] and by the simultaneous formation in many, but not all, instances of abnormal heme derivatives. As already noted, abnor-

Table V

Structure of the N-Alkylated Protoporphyrin IX Derivatives

| Substrate | N-alkyl group (R) | Alkylated pyrrole ring | Reference |
|---|---|---|---|
| Ethylene | —CH$_2$CH$_2$OH | D | 153, 156 |
| Propene | —CH$_2$CHOHCH$_3$ | D | 156 |
| Octene | —CH$_2$CHOH(CH$_2$)$_5$CH$_3$ | D | 156 |
| Acetylene | —CH$_2$CHO | Several | 154 |
| Fluroxene (20) | —CH$_2$CHO | Several | 154 |
| Vinyl fluoride | —CH$_2$CHO | Several | 154 |
| Propyne | —CH$_2$COCH$_3$ | A | 142 |
| Octyne | —CH$_2$CO(CH$_2$)$_5$CH$_3$ | A | 156 |
| Ethchlorvynol (17) | —CH$_2$COC(CH$_2$CH$_3$)(OH)(CH=CHCl) | A,B,C,D | 155 |

mal heme derivatives have not been found with allenes[143] or with internal acetylenes.[140] Ethylene[131] and norethisterone[140,151] have been shown to cause the formation of abnormal heme derivatives *in vitro*, although this facet of the process has usually been investigated *in vivo*. The evidence indicates that the abnormal porphyrins, which are isolated after removal of the iron atom, derive specifically from the prosthetic heme moiety of inactivated cytochrome P-450 enzymes,[152] a result that implicates prosthetic heme modification as the cause of enzyme loss. The structures of the porphyrins derived from the prosthetic heme after treatment with ethylene,[153] fluroxene,[154] vinyl fluoride,[154] acetylene,[154] propyne,[142] ethchlorvynol,[155] propene,[156] octene,[156] and octyne[156] have recently been elucidated. The porphyrin in each instance is an N-alkylated protoporphyrin IX in which the N-alkyl moiety is directly derived from the administered olefin or acetylene. A summary of the actual structures is provided in Table V. It is

## 5. ALKENES AND ALKYNES 139

evident from the structural data that the *N*-alkyl group formally represents the addition of an oxygen to the internal carbon and of the porphyrin nitrogen to the terminal carbon of the $\pi$ bond involved in the reaction. The result of this addition reaction in the case of an acetylene is an enol, which readily tautomerizes to the more stable carbonyl structure. If a leaving group is present on the carbon to which the oxygen is added, as it is in vinyl fluoride and fluoroxene, the group is eliminated in the observed product. These reactions are briefly outlined in Scheme 17. The "N" in the scheme represents a porphyrin nitrogen, and the "O" a catalytically activated form of atomic oxygen. The oxygen has been shown in one study to derive, in fact, from molecular oxygen.[156]

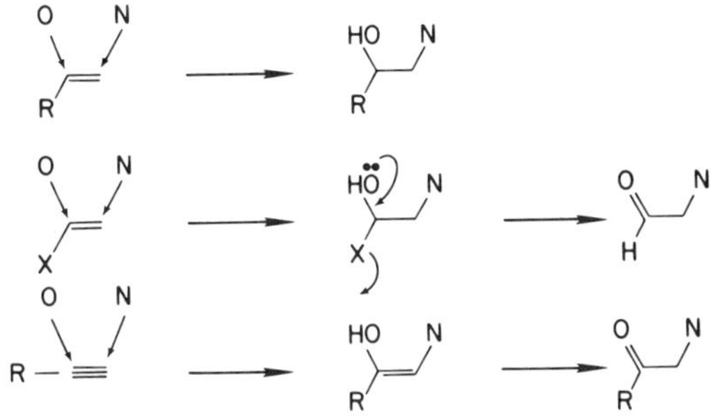

**Scheme 17**

The autocatalytic nature of the alkylation process, the origin of the oxygen atom in the *N*-alkyl group, and the structures of the heme adducts are all consistent with a reaction sequence initiated by oxygen transfer from the enzyme to the $\pi$ bond. Two other points of relevance to the heme alkylation process should be mentioned before the mechanism of $\pi$-bond oxidation is taken up (next section). A study of the alkylation of prosthetic heme by *trans*-1-[1-$^2$H]octene has established that the oxygen and the porphyrin nitrogen add to the same side of the double bond in a stereospecific manner.[91] It is also now apparent that different substrates result in the specific alkylation of different nitrogens in the protoporphyrin IX structure. The specificity of the process, summarized in Table V, is underscored by the finding that ethylene, propene, and octene essentially alkylate only the nitrogen of pyrrole ring D, whereas propyne and octyne predominantly alkylate the nitrogen of pyrrole ring A. The alkylation regiochemistry has led to formulation of an active-site topology in which the hydrocarbon chain is bound in a lipophilic region over pyrrole ring C and the protein imposes a steric or electronic barrier over pyrrole ring B.[156]

## VI. Mechanisms of π-Bond Oxidation and Cytochrome P-450 Destruction

The broad features of the catalytic cycle of cytochrome P-450, defined over the last few years, have been reviewed.[157,158] It is now generally believed that the enzyme reductively cleaves the bond between the two atoms of a molecule of oxygen before it transfers one of those atoms to its substrate. The iron atom of the prosthetic heme effects this oxygen activation by a mechanism that is still obscure. One possibility is that the iron-bound dioxygen molecule is acylated before oxygen–oxygen bond cleavage.[159] A second possibility, on the basis of model studies, is that the dioxygen molecule closes to an intermediate with iron bound to both ends of the molecule before heterolytic or homolytic rupture of the oxygen–oxygen bond.[160] In any case, the iron–oxygen complex generated by one of these mechanisms or by an alternative mechanism is considered to be the species that actually reacts with the substrate. The distribution of the electrons in the reactive oxoiron complex, which determines the properties of the subsequent reaction with substrates, has not yet been determined. Four reasonable alternatives can be formulated for the electronic distribution if the protein matrix is not involved in electron exchange processes and a somewhat larger number if this arbitrary constraint is not imposed. The four electronic configurations, in the absence of protein involvement, are given in Scheme 18. The four nitrogens of the porphyrin and the thiolate ligand from the protein are shown. Formal charges and iron oxidation states, including the porphyrin radical cation present in the last formulation, are indicated. It is evident from the available alternatives that substrate oxidation by one-electron steps, or by a single two-electron reaction, can be reasonably envisioned. The oxygen atom in formulation B, for example, would be expected to react

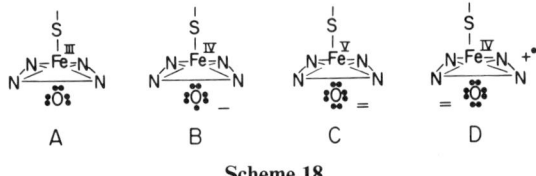

**Scheme 18**

initially as a one-electron acceptor, whereas the oxygen in formulation A, depending on the orbitals occupied by the electrons, could participate in stepwise or concerted transfer of the two electrons.

The electronic and chemical properties of the catalytic iron–oxygen complex will determine the mechanism by which a given carbon–carbon π bond is oxidized. One of the possibilities, as in chemical epoxidation, is that the two carbon–oxygen bonds are formed simultaneously without the interven-

## 5. ALKENES AND ALKYNES

tion of a discrete intermediate (Scheme 19, path a). A second possibility is that the oxygen adds as an electrophile to the $\pi$ bond, generating a transient carbon cation, which subsequently collapses to the epoxide (Scheme 19, path b). The oxygen could also add to the $\pi$ bond as a radical species, in which case the initial intermediate would be a radical rather than a cation (Scheme 19, path c). This radical could directly collapse to the epoxide, but could also first undergo transfer of the unpaired electron to the heme center to generate the cation of path b. Formation of the carbon–oxygen bond in the radical intermediate of path c could, in fact, be preceded by the analogous transfer of one of the $\pi$-bond electrons to the iron–oxygen center (Scheme 19, path d). Finally, the possibility must be considered that the intermediate is not an acyclic species but rather, in analogy with that postulated for the chromyl chloride oxidation of olefins,[33] a four-membered metallocycle obtained by nucleophilic attack of the $\pi$ bond on the electron-deficient iron atom (Scheme 19, path e).

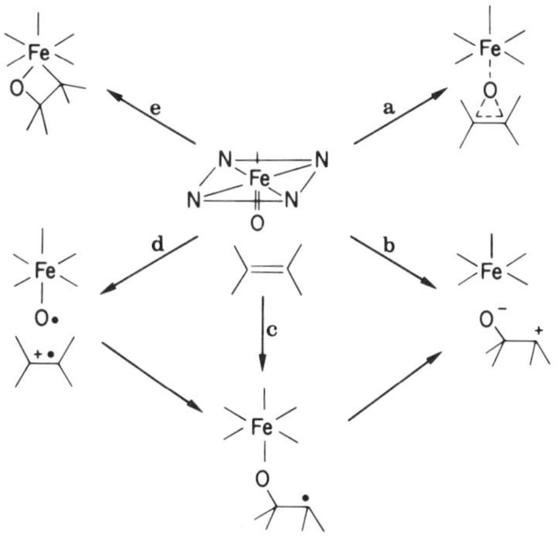

**Scheme 19**

It has been expedient, until recently, to assume that the epoxidation of olefins is mediated by a concerted mechanism (Scheme 19, path a). Experimental support for a concerted mechanism is provided by the consistent observation that the stereochemistry of the olefin is conserved during the enzymatic reaction.[51,52] A concerted oxidation sequence is also consistent with the more extensive data on hydroxylation reactions catalyzed by cytochrome $P$-450, particularly with the small isotope effects[161] and retention of configuration[162] associated with such reactions. The view that hydroxylation reactions occur by concerted insertion of oxygen into carbon–

hydrogen bonds, however, has been severely shaken by the observation of large intramolecular isotope effects[163,164] and by the demonstration that deuterium is stereochemically scrambled during the hydroxylation of labeled norbornane.[163] The premise that no intermediates are formed during the enzymatic epoxidation of an olefin is also inconsistent with the finding that olefin epoxidation and prosthetic heme alkylation are divergent consequences of the same catalytic process. The trivial explanation for the association of these two processes, that prosthetic heme is alkylated by the epoxide metabolite rather than by an intermediate in the oxidative sequence, is definitively excluded by a variety of observations. The epoxides of several active olefins are reversibly bound in the active site, but do not cause prosthetic heme alkylation.[131,150,165] An epoxide has been shown, in one instance, to inhibit competitively the destruction of the enzyme by a related olefin.[133,144] The prosthetic heme adducts obtained with fluroxene and vinyl fluoride (Table V) require addition of the porphyrin nitrogen to the unsubstituted end of the double bond,[154] a reaction orientation opposite to that expected for the reaction of a nucleophile with an epoxide bearing an oxygen or fluorine substituent.[166] Molecular orbital calculations on the addition of nucleophiles to the epoxide of fluroxene do, in fact, predict that the addition should occur on the trifluoroethoxy-substituted terminus.[167] Finally, the demonstration that the oxygen and the porphyrin nitrogen add to the $\pi$ bond of 1-octene from the same side is incompatible with alkylation by the epoxide, because nucleophilic attack on an epoxide results formally in addition to the opposite faces of the $\pi$ bond.[91] The catalytic involvement of cytochrome $P$-450 in its own destruction,[144,146] in view of the destructive impotence of epoxide metabolites, demands that heme alkylation occur before completion of oxygen transfer from the enzyme to the $\pi$ bond. Olefin epoxidation, therefore, does not appear to be a concerted process, although neither the point at which epoxidation and heme alkylation diverge, nor the factors that determine into which of these two alternatives a given catalytic event is channeled, are now known.

The nature of the intermediate responsible for alkylation of the prosthetic heme group has not been unambiguously defined. The regiochemistry of the reaction with fluroxene and vinyl fluoride, however, does not support the involvement of a cationic intermediate such as that in path b (Scheme 19). The presence of these heteroatoms overwhelmingly favors localization of the positive charge on the carbon to which they are attached, a reaction orientation that does not lead to the isolated heme adducts. The wrong orientation is also favored for the radical addition of path c (Scheme 19) in the presence of an oxygen or fluorine substituent, but the difference in the thermodynamic stability of the two orientations is predicted in the case of *radicals* to be sufficiently small that the minor fraction of catalytic events leading to the disfavored orientation could still account for the observed efficiency of the destructive process. The epoxide: alkylation partition ratios of 200 to 300 observed with AIA[144-146] and fluroxene[147] require that the disfavored orientation of radical addition occur with at least this frequency, a requirement compatible with the finding that radicals add to vinyl fluoride only 3–50 times more frequently at the substituted than at the unsubstituted end.[18] The only mechanisms in which the observed regiochemistry of the addition of heme to the $\pi$ bond is preferred are paths d and e (Scheme 19). The mechanism of prosthetic heme alkylation is subject to the further constraints imposed by the observations that oxygen and nitrogen add in a cis fashion to the $\pi$ bond and that the olefin stereochemistry is retained both in the epoxide metabolite and in the heme alkylation product. Nonconcerted mechanisms are not excluded by these constraints, however, if rotation about the carbon–carbon bond in an intermediate such as that of path c is suppressed by electronic or steric interactions of the terminal carbon in the intermediate with the metalloporphyrin center. The finding that different substrates alkylate different heme nitrogen atoms with remarkable specificity does, in fact, require the existence of strong steric or electronic interactions. The awkward, sterically hindered approach required to form the metallocyclic intermediate of path e, on the other hand, argues against the viability of this mechanism, although it is consistent with all the experimental data. In sum, the available evidence suggests, but does not prove, that a radical-like intermediate formed in the catalytic step commits the enzyme to oxidation of the $\pi$ bond, although further evidence is required to elaborate the details of the coupled mechanisms of $\pi$-bond oxidation and prosthetic heme alkylation.

The oxidation of a $\pi$ bond in an acetylene is more difficult than the oxidation of the analogous bond in an olefin, but the two reactions are otherwise similar because the $\pi$ bonds in an acetylene are orthogonal and therefore react as independent entities. The epoxide of an olefin, as already mentioned, is a much more stable structure than the unsaturated epoxide expected from an acetylene. It is very difficult, because oxirenes have never been isolated, to determine whether the destruction of cytochrome P-450 caused by acetylenes reflects reactions of the oxirene metabolites or of intermediates in the oxidative sequence. The mechanistic alternatives formulated for the reaction of olefins (Scheme 19) are equally applicable to the reaction of acetylenes if, of course, the extra double bond is incorporated into the substrate molecule. The extra double bond, however, requires that one further type of intermediate be considered. As shown in Scheme 4, the ring-opened dipolar species in equilibrium with an oxirene can equally well be written as an $\alpha$-ketocarbene. An iron–carbene complex (as shown on top of next page) is therefore conceivable as an intermediate in the reaction of acetylenes with cytochrome P-450. The little evidence now available on the mechanism of oxidation of acetylene π bonds does not permit differentiation of the possible alternatives.

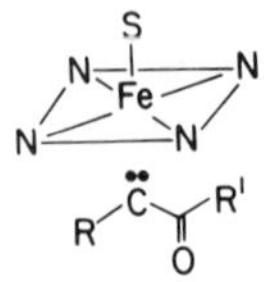

## VII. Olefin and Acetylene Toxicity

Unactivated carbon–carbon π bonds do not react chemically with biological entities under physiological conditions. These functionalities can be activated, however, either by conjugation with groups that accept π electrons or by enzymatic oxidation. The conjugation of an olefin or acetylene moiety with a carbonyl function, an example of the first mode of activation, results in electrophilic reactivity of biological significance.[168] The metabolic formation of such conjugated functionalities has not been reviewed here, because it lies outside the main focus of the chapter. The oxidation of allyl alcohol to acrolein (Scheme 20), a toxic alkylating agent, nevertheless illustrates one of the mechanisms by which a reactive, conjugated functionality can be metabolically unmasked.[169] The lethal synthesis of conjugated acetylenic or olefinic moieties by enzymatically catalyzed oxidation or rear-

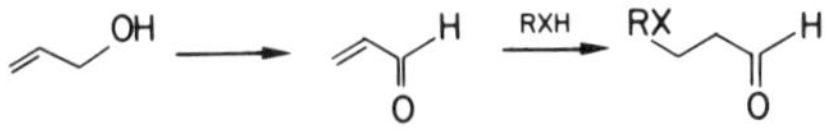

Scheme 20

rangement reactions has, in fact, been exploited in the design of suicide substrates for a number of enzymes.[170]

The toxicity of epoxides, the primary metabolites of olefins, is determined by their inherent reactivity, their susceptibility to detoxication by enzymatic hydrolysis or glutathione conjugation, and their ability to reach and irreparably damage critical cellular targets. The epoxides of even simple hydrocarbon epoxides are sufficiently reactive to alkylate biological macromolecules, although this intrinsic activity can be greatly enhanced by conjugation, heteroatomic substitution, or other factors. High intrinsic reactivity of the epoxide group appears to be a more important determinant of acute cellular toxicity than of mutagenicity or carcinogenicity, although the relationships between reactivity, enzymatic detoxication, and pharmacokinetic targeting are not simple. This is illustrated by the finding that, although the epoxides of propene,[171,172] styrene,[172] and AIA[146] are able to alkylate DNA *in vitro*, the epoxide of propene is both a mutagen[173,174] and a carcinogen,[175] that of

# 5. ALKENES AND ALKYNES

styrene is a mutagen[173,174] but is, at best, a weak carcinogen,[176,177] and that from AIA appears not even to be a mutagen.[133] The more potent carcinogens nevertheless usually incorporate a highly reactive epoxide moiety. Among these are the 7,8-dihydrodiol 9,10-epoxide of benzo[$a$]pyrene (**25**)[178] and the 2,3-epoxide of aflatoxin $B_1$ (**26**).[179] The direct necrotic activity of

**25**      **26**      **27**

highly reactive epoxides is expressed in the cellular toxicity of fluroxene (**20**) and divinyl ether (**27**).[180,181] The toxicity of aromatic and halogen-substituted $\pi$ bonds is discussed more explicitly in Chapters 6 and 11 of this volume.

Little is known about the potential toxicity of acetylene group metabolites because, until recently, it was doubted that triple bonds underwent oxidative metabolism. Acetylene groups also occur less frequently than olefin moieties in biologically relevant structures. The relationship between the metabolism of the 17$\alpha$-ethinyl function in sterol hormone analogs and the possible weak carcinogenicity of these compounds,[182] for example, is difficult to assess. If reactive species are released during the oxidative metabolism of triple bonds, it is likely that these will be the ketene products obtained by rearrangement[42,43] rather than the oxirene metabolites themselves. The rapid intramolecular rearrangement of oxirenes, and their high reactivity, makes diffusion of the oxirenes out of the active site of the oxidative enzymes unlikely.

## Acknowledgments

The author thanks Drs. Bruce A. Mico, Garold S. Yost, Kent L. Kunze, Hal S. Beilan, Bonnie L. K. Mangold, and Conrad Wheeler for their important contributions to the work reported from his laboratory, and Dr. Kunze in addition for suggestions pertinent to this manuscript. The studies in the author's laboratory were supported by the National Institutes of Health.

## References

1. Utley, J. H. P., and Lines, R. (1978). The electrochemistry of the carbon–carbon triple bond. *In* "The Chemistry of the Carbon–Carbon Triple Bond" (S. Patai, ed.), pp. 739–753. Wiley, New York.

2. Sutton, L. E. (1958). "Table of Interatomic Distances." Chemical Society, London.
3. Wyatt, J. R., and Stafford, F. E. (1972). Mass spectrometric determination of the heat of formation of ethynyl radical, $C_2H$, and of some related species. *J. Phys. Chem.* **76**, 1913–1918.
4. Cowfer, J. A., Keil, D. G., Michael, J. V., and Yeh, C. (1971). Absolute rate constants for the reactions of hydrogen atoms with olefins. *J. Phys. Chem.* **75**, 1584–1592.
5. Schmid, G. H., Modro, A., Lenz, F., Garratt, D. G., and Yates, K. (1976). Effect of acetylene structure on the rates and products of addition of 4-chlorobenzenesulfenyl chloride. *J. Org. Chem.* **41**, 2331–2336.
6. Yates, K., Schmid, G. H., Regulski, T. W., Garratt, D. G., Leung, H. W., and McDonald, R. (1973). Relative ease of formation of carbonium ions and vinyl cations in electrophilic additions. *J. Am. Chem. Soc.* **95**, 160–165.
7. Rabinovitch, B. S., and Looney, F. S. (1955). Nitric oxide-catalyzed cis–trans isomerization of dideuteroethylene. *J. Chem. Phys.* **23**, 2439–2440.
8. Nagase, S., and Kern, C. W. (1980). *Ab initio* study of radical reactions. Relative reactivity of olefin and acetylenic bonds in addition reactions. *J. Am. Chem. Soc.* **102**, 4513–4515.
9. Holt, P. M., and Kerr, J. A. (1977). Kinetics of gas phase addition reactions of methyl radicals. I. Addition to ethylene, acetylene, and benzene. *Int. J. Chem. Kinet.* **9**, 185–200.
10. Melloni, G., Modena, G., and Tonellato, U. (1981). Relative reactivities of carbon–carbon double and triple bonds toward electrophiles. *Acc. Chem. Res.* **14**, 227–233.
11. Schäfer, H. J. (1981). Anodic and cathodic C–C bond formation. *Angew. Chem., Int. Ed. Engl.* **20**, 911–934.
12. Bard, A. J., Ledwith, A., and Shine, H. J. (1976). Formation, properties, and reactions of cation radicals in solution. *Adv. Phys. Org. Chem.* **13**, 155–278.
13. Koch, D., Schäfer, H., and Steckhan, E. (1974). Anodische Dimerisierung von Enoläthern. *Chem. Ber.* **107**, 3640–3657.
14. Belleau, B., and Au-Young, Y. K. (1969). Electrochemical methoxylation of vinyl ethers: A novel anodic dimerization reaction. *Can. J. Chem.* **47**, 2117–2118.
15. Benson, S. W. (1965). Bond energies. *J. Chem. Educ.* **42**, 502–518.
16. Taft, R. W., Martin, R. H., and Lampe, F. W. (1965). Stabilization energies of substituted methyl cations. The effect of strong demand on the resonance order. *J. Am. Chem. Soc.* **87**, 2490–2492.
17. Martin, R. H., Lampe, F. W., and Taft, R. W. (1966). An electron-impact study of ionization and dissociation in methoxy- and halogen-substituted methanes. *J. Am. Chem. Soc.* **88**, 1353–1357.
18. Tedder, J. M., and Walton, J. C. (1976). The kinetics and orientation of free-radical addition to olefins. *Acc. Chem. Res.* **9**, 183–191.
19. Minisci, F. (1975). Free-radical additions to olefins in the presence of redox systems. *Acc. Chem. Res.* **8**, 165–171.
20. Chen, K. S., Krusic, P. J., Meakin, P., and Kochi, J. K. (1974). Electron spin resonance studies of fluoroalkyl radicals in solution. I. Structures, conformations, and barriers to hindered internal rotation. *J. Phys. Chem.* **78**, 2014–2030.
21. Tarrant, P., and Stump, E. C. (1964). Free radical additions involving fluorine compounds. VII. The addition of perhaloalkanes to vinyl ethyl ether and vinyl 2,2,2-trifluoroethyl ether. *J. Org. Chem.* **29**, 1198–1202.
22. Edge, D. J., Gilbert, B. C., Norman, R. O. C., and West, P. R. (1971). Electron spin resonance studies. Part XXVIII. Oxidation of enols, enol ethers, and related compounds with the hydroxyl and the amino radical. *J. Chem. Soc. B* pp. 189–196.

23. Fischer, P. (1980). Enol ethers—structure, synthesis and reactions. *In* "The Chemistry of Ethers, Crown Ethers, Hydroxyl Groups and Their Sulphur Analogues" (S. Patai, ed.), Part 2, pp. 761–820. Wiley, New York.
24. Hopkinson, A. C. (1978). Acidity, hydrogen bonding, and complex formation. *In* "The Chemistry of the Carbon–Carbon Triple Bond" (S. Patai, ed.), pp. 75–136. Wiley, New York.
25. Cram, D. J. (1965). "Fundamentals of Carbanion Chemistry." Academic Press, New York.
26. Berti, G. (1973). Stereochemical aspects of the synthesis of 1,2-epoxides. *Top. Stereochem.* **7,** 95–251.
27. Groves, J. T. (1980). Mechanisms of metal-catalyzed oxygen insertion. *In* "Metal Ion Activation of Dioxygen" (T. G. Spiro, ed.), pp. 125–162. Wiley, New York.
28. Mimoun, H. (1981). Activation of molecular oxygen and selective oxidation of olefins catalyzed by Group VIII transition metal complexes. *Pure Appl. Chem.* **53,** 2389–2399.
29. O'Brien, P. J. (1978). Hydroperoxides and superoxides in microsomal oxidations. *Pharmacol. Ther., Part A* **2,** 517–536.
30. Groves, J. T., Nemo, T. E., and Myers, R. S. (1979). Hydroxylation and epoxidation catalyzed by iron–porphine complexes. Oxygen transfer from iodosylbenzene. *J. Am. Chem. Soc.* **101,** 1032–1033.
30a. Groves, J. T., Kruper, W. J., and Haushalter, R. C. (1980). Hydrocarbon oxidations with oxometalloporphinates. Isolation and reactions of a (porphinato)manganese (V) complex. *J. Am. Chem. Soc.* **102,** 6375–6377.
31. Ledon, H. J., Durbut, P., and Varescon, F. (1981). Selective epoxidation of olefins by molybdenum porphyrin-catalyzed peroxy-bond heterolysis. *J. Am. Chem. Soc.* **103,** 3601–3603.
32. Tabushi, I., and Yazaki, A. (1981). *P*-450-Type dioxygen activation using $H_2$/colloidal Pt as an effective electron donor. *J. Am. Chem. Soc.* **103,** 7371–7373.
33. Sharpless, K. B., Teranishi, A. Y., and Bäckvall, J. E. (1977). Chromyl chloride oxidation of olefins. Possible role of organometallic intermediates in the oxidations of olefins by oxy transition metal species. *J. Am. Chem. Soc.* **99,** 3120–3128.
34. Torres, M., Bourdelande, J. L., Clement, A., and Strausz, O. P. (1983). Argon matrix isolation of bis(trifluoromethyl)oxirene, perfluoromethylethyloxirene, and their isomeric ketocarbenes. *J. Am. Chem. Soc.* **105,** 1698–1700.
35. Tanaka, K., and Yoshimine, M. (1980). An *ab initio* study on ketene, hydroxyacetylene, formylmethylene, oxirene, and their rearrangement paths. *J. Am. Chem. Soc.* **102,** 7655–7662.
36. Strausz, D. P., Gosavi, R. K., Denes, A. S., and Czizmadia, I. G. (1976). Mechanism of the Wolff rearrangement. 6. *Ab initio* molecular orbital calculations on the thermodynamic and kinetic stability of the oxirene molecule. *J. Am. Chem. Soc.* **98,** 4784–4786.
37. Stille, J. K., and Whitehurst, D. D. (1964). Oxirene. An intermediate in the peroxyacid oxidation of acetylenes. *J. Am. Chem. Soc.* **86,** 4871–4876.
38. Ciabattoni, J., Campbell, R. A., Renner, C. A., and Concannon, P. W. (1970). The peracid oxidation of acetylenes. 1,2-methyl migration, cyclopropane formation, and stereoselective 1,5- and 1,6-transannular insertion. *J. Am. Chem. Soc.* **92,** 3826–3828.
39. McDonald, R. N., and Schwab, P. A. (1964). Strained ring systems. I. Peroxidation studies with certain acetylenes. The relevance of oxirene intermediates. *J. Am. Chem. Soc.* **86,** 4866–4871.
40. Concannon, P. W., and Ciabattoni, J. (1973). Peroxy acid oxidation of cycloalkynes and the decomposition of 2-diazocycloalkanones. *J. Am. Chem. Soc.* **95,** 3284–3289.
41. Ogata, Y., Sawaki, Y., and Ohno, T. (1982). Mechanism for oxidation of phenylacety-

lenes with peroxymonophosphoric acid. Oxirene as an intermediate inconvertible to ketocarbene. *J. Am. Chem. Soc.* **104**, 216–219.
42. Ortiz de Montellano, P. R., and Kunze, K. L. (1980). Occurrence of a 1,2-shift during enzymatic and chemical oxidation of a terminal acetylene. *J. Am. Chem. Soc.* **102**, 7373–7375.
43. Ortiz de Montellano, P. R., and Kunze, K. L. (1981). Shift of the acetylenic hydrogen during chemical and enzymatic oxidation of the biphenylacetylene triple bond. *Arch. Biochem. Biophys.* **209**, 710–712.
44. Oesch, F. (1972). Mammalian epoxide hydrases: Inducible enzymes catalyzing the inactivation of carcinogenic and cytotoxic metabolites derived from aromatic and olefinic compounds. *Xenobiotica* **3**, 305–340.
45. Ota, K., and Hammock, B. D. (1980). Cytosolic and microsomal epoxide hydrolases: Differential properties in mammalian liver. *Science* **207**, 1479–1481.
46. Jakoby, W. B., and Habig, W. H. (1980). Glutathione transferases. *In* "Enzymatic Basis of Detoxication" (W. B. Jakoby, ed.), pp. 63–94. Academic Press, New York.
47. Chasseaud, L. F. (1979). The role of glutathione and glutathione *S*-transferases in the metabolism of chemical carcinogens and other electrophilic agents. *Adv. Cancer Res.* **29**, 176–274.
48. Maynert, E. W., Foreman, R. L., and Watabe, T. (1970). Epoxides as obligatory intermediates in the metabolism of olefins to glycols. *J. Biol. Chem.* **245**, 5234–5238.
49. Watabe, T., and Akamatsu, K. (1974). Enzymatic hydrolysis of mono-*n*-alkyl-substituted ethylene oxides and their inhibitory effects on hepatic microsomal epoxide hydrolase. *Chem. Pharm. Bull.* **22**, 2155–2158.
50. Watabe, T., and Akamatsu, K. (1972). Stereoselective hydrolysis of acyclic olefin oxides to glycols by hepatic microsomal epoxide hydrolase. *Biochim. Biophys. Acta* **279**, 297–305.
51. Watabe, T., and Akamatsu, K. (1974). Microsomal epoxidation of *cis*-stilbene: Decrease in epoxidase activity related to lipid peroxidation. *Biochem. Pharmacol.* **23**, 1079–1085.
52. Watabe, T., Ueno, Y., and Imazumi, J. (1971). Conversion of oleic acid into *three*-dihydroxystearic acid by rat liver microsomes. *Biochem. Pharmacol.* **20**, 912–913.
53. El Masri, A. M., Smith, J. N., and Williams, R. T. (1958). The metabolism of alkylbenzenes: Phenylacetylene and phenylethylene (styrene). *Biochem. J.* **68**, 199–204.
54. Leibman, K. C., and Ortiz, E. (1970). Epoxide intermediates in microsomal oxidation of olefins to glycols. *J. Pharmacol. Exp. Ther.* **173**, 242–246.
55. Seutter-Berlage, F., Delbressine, L. P. C., Smeets, F. L. M., and Ketelaars, H. C. J. (1978). Identification of three sulphur-containing urinary metabolites of styrene in the rat. *Xenobiotica* **8**, 413–418.
56. Leibman, K. C., and Ortiz, E. (1971). Oxidation of cycloalkenes in liver microsomes. *Biochem. Pharmacol.* **20**, 232–236.
57. James, S. P., Waring, R. H., and White, D. A. (1967). Some metabolites of bromocyclopentane, bromocyclohexane, and bromocycloheptane. *Biochem. J.* **103**, 25p.
58. van Bladeren, P. J., Breimer, D. D., Seghers, C. J. R., Vermeulen, N. P. E., van der Gen, A., and Cauvet, J. (1981). Dose-dependent stereoselectivity in the formation of mercapturic acids from cyclohexene oxide by the rat. *Drug. Metab. Dispos.* **9**, 207–211.
59. Leibman, K. C., and Ortiz, E. (1968). Oxidation of indene in liver microsomes. *Mol. Pharmacol.* **4**, 201–207.
60. Brooks, C. J. W., and Young, L. (1956). The metabolic conversion of indene into *cis*- and *trans*-indane-1,2-diol. *Biochem. J.* **63**, 264–269.
61. Francis, T. J. R., Bick, R. J., Callaghan, P., and Hopkins, R. P. (1975). The role of

1,2-epoxyindene in the metabolism of indene by rat liver fractions. *Biochem. Soc. Trans.* **3**, 1244–1246.
62. Waddell, W. J. (1965). The metabolic fate of 5-allyl-5-(1-methylbutyl) barbituric acid (secobarbital). *J. Pharmacol. Exp. Ther.* **149**, 23–28.
63. Liddle, C., Graham, G. G., Christopher, R. K., Bhuwapathanapun, S., and Duffield, A. M. (1981). Identification of new urinary metabolites in man of quinine using methane chemical ionization gas chromatography–mass spectrometry. *Xenobiotica* **11**, 81–87.
64. Smith, A. (1976). The metabolism of 2-allyl-2-isopropylacetamide *in vivo* and in the isolated perfused rat liver. *Biochem. Pharmacol.* **25**, 2429–2442.
65. Bergemalm-Rynell, K., and Steen, G. (1982). Urinary metabolites of vinyltoluene in the rat. *Toxicol. Appl. Pharmacol.* **62**, 19–31.
66. Davidow, B., and Radomski, J. L. (1953). Isolation of an epoxide metabolite from fat tissues of dogs fed heptachlor. *J. Pharmacol. Exp. Ther.* **107**, 259–265.
67. Brooks, G. T., Harrison, A., and Lewis, S. E. (1970). Cyclodiene epoxide ring hydration by microsomes from mammalian liver and houseflies. *Biochem. Pharmacol.* **19**, 255–273.
68. Oesch, F., Kaubisch, N., Jerina, D. M., and Daly, J. W. (1971). Hepatic epoxide hydrase. Structure–activity relationships for substrates and inhibitors. *Biochemistry* **10**, 4858–4866.
69. Harvey, D. J., Glazener, L., Johnson, D. B., Butler, C. M., and Horning, M. G. (1977). Comparative metabolism of four allylic barbiturates and hexobarbital by the rat and guinea pig. *Drug. Metab. Dispos.* **5**, 527–546.
70. Vermeulen, N. P. E., Bakker, B. H., Schultink, J., van der Gen, A., and Breimer, D. D. (1979). The epoxide-diol pathway in the metabolism of hexobarbital in rat and man. *Xenobiotica* **9**, 289–299.
71. Slack, J. A., and Ford-Hutchinson, A. W. (1980). Determination of a urinary epoxide metabolite of alclofenac in man. *Drug Metab. Dispos.* **8**, 84–86.
72. Watabe, T, and Yamada, N. (1975). The biotransformation of 1-hexadecene to carcinogenic 1,2-epoxyhexadecane by hepatic microsomes. *Biochem. Pharmacol.* **24**, 1051–1053.
73. Watabe, T., Hiratsuka, A., Isobe, M., and Ozawa, N. (1980). Metabolism of *d*-limonene by hepatic microsomes to non-mutagenic epoxides toward *Salmonella typhimurium*. *Biochem. Pharmacol.* **29**, 1068–1071.
74. Patel, J. M., Wood, J. C., and Leibman, K. C. (1980). The biotransformation of allyl alcohol and acrolein in rat liver and lung preparations. *Drug Metab. Dispos.* **8**, 305–308.
75. Gurny, O., Maynard, D. E., Pitcher, R. G., and Kierstead, R. W. (1972). Metabolism of $(-)$-$\Delta^9$- and $(-)$-$\Delta^8$-tetrahydrocannabinol by monkey liver. *J. Am. Chem. Soc.* **94**, 7928–7929.
76. Korte, F., and Kochen, W. (1966). Insektizide im Stoffwechsel. XII. Isolierung und Identifizierung von Metaboliten des Aldrin-$^{14}$C aus dem Urin von Kaninchen. *Med. Pharmacol. Exp.* **15**, 409–414.
77. Brooks, G. T. (1969). The metabolism of diene-organochlorine (cyclodiene) insecticides. *Residue Rev.* **27**, 81–138.
78. Corey, E. J., Russey, W. E., and Ortiz de Montellano, P. R. (1966). 2,3-oxidosqualene, an intermediate in the biological synthesis of sterols from squalene. *J. Am. Chem. Soc.* **88**, 4750–4751.
79. Ono, T., and Bloch, K. (1975). Solubilization with partial characterization of rat liver squalene epoxidase. *J. Biol. Chem.* **250**, 1571–1579.
80. Ono, T., Takahashi, K., Odani, S., Konno, H., and Imai, Y. (1980). Purification of

squalene epoxidase from rat liver microsomes. *Biochem. Biophys. Res. Commun.* **96**, 522–528.
81. Ruettinger, R. T., Griffith, G. R., and Coon, M. J. (1977). Characterization of the ω-hydroxylase of *Pseudomonas oleovorans* as a nonheme iron protein. *Arch. Biochem. Biophys.* **183**, 528–537.
82. May, S. W., and Abbott, B. J. (1973). Enzymatic epoxidation. Comparison between the epoxidation and hydroxylation reactions catalyzed by the ω-hydroxylation system of *Pseudomonas oleovorans*. *J. Biol. Chem.* **248**, 1725–1730.
83. Wiebel, F. J., Selkirk, J. K., Gelboin, H. V., Haugen, D. A., van der Hoeven, T. A., and Coon, M. J. (1975). Position-specific oxygenation of benzo[a]pyrene by different forms of purified cytochrome *P*-450 from rabbit liver. *Proc. Natl. Acad. Sci. U.S.A.* **72**, 3917–3920.
84. Deutsch, J., Leutz, J. C., Yang, S. K., Gelboin, H. V., Chiang, Y. L., Vatsis, K. P., and Coon, M. J. (1978). Regio- and stereoselectivity of various forms of purified cytochrome *P*-450 in the metabolism of benzo[a]pyrene and (−)-*trans*-7,8-dihydroxy-7,8-dihydrobenzo[a]pyrene as shown by product formation and binding to DNA. *Proc. Natl. Acad. Sci. U.S.A.* **75**, 3123–3127.
85. Thomas, P. E., Lu, A. Y. H., Ryan, D., West, S. B., Kawalek, J., and Levin, W. (1976). Multiple forms of rat liver cytochrome *P*-450. Immunochemical evidence with antibody against cytochrome *P*-448. *J. Biol. Chem.* **251**, 1385–1391.
86. Harvey, R. G. (1981). Activated metabolites of carcinogenic hydrocarbons. *Acc. Chem. Res.* **14**, 218–226.
87. Costa, A. K., and Ivanetich, K. M. (1980). Tetrachloroethylene metabolism by the hepatic microsomal cytochrome *P*-450 system. *Biochem. Pharmacol.* **29**, 2863–2869.
88. Costa, A. K., Katz, I. D., and Ivanetich, K. M. (1980). Trichloroethylene: Its interaction with hepatic microsomal cytochrome *P*-450 *in vitro*. *Biochem. Pharmacol.* **29**, 433–439.
89. Campbell, T. C., and Hayes, J. R. (1976). The role of aflatoxin metabolism in its toxic lesion. *Toxicol. Appl. Pharmacol.* **35**, 199–222.
90. Guengerich, F. P., and Watanabe, P. G. (1979). Metabolism of [14]C- and [36]Cl-labeled vinyl chloride *in vivo* and *in vitro*. *Biochem. Pharmacol.* **28**, 589–596.
91. Ortiz de Montellano, P. R., Mangold, B. L. K., Wheeler, C., Kunze, K. L., and Reich, N. O. (1983). Stereochemistry of cytochrome *P*-450-catalyzed epoxidation and prosthetic heme alkylation. *J. Biol. Chem.* **258**, 4202–4207.
92. Licht, H. J., and Coscia, C. J. (1978). Cytochrome *P*-450 LM$_2$ mediated hydroxylation of monoterpene alcohols. *Biochemistry* **17**, 5638–5646.
93. Uehleke, H., and Brinkschulte-Freitas, M. (1978). Kinetics and metabolism of 2,2-diethylallylacetamide in dog and man. *Naunyn-Schmiedeberg's Arch. Pharmacol.* **302**, 11–18.
94. Bolt, H. M., Laib, R. J., and Filser, J. G. (1982). Reactive metabolites and carcinogenicity of halogenated ethylenes. *Biochem. Pharmacol.* **31**, 1–4.
95. Hanzlik, R. P., Shearer, G. O., Hamburg, A., and Gillesse, T. (1978). Metabolism *in vitro* of para-substituted styrenes. Kinetic observation of substituent effects. *Biochem. Pharmacol.* **27**, 1435–1439.
96. Hjelmeland, L.M., Aronow, L., and Trudell, J. R. (1977). Intramolecular determination of substituent effects in the hydroxylations catalyzed by cytochrome *P*-450. *Mol. Pharmacol.* **13**, 634–639.
97. Keding, H., and Schmidt, G. (1969). Isolierung und Identifizierung von Brallobarbital-Metaboliten. *Arzneim.-Forsch.* **19**, 342–347.
98. Delbressine, L. P. C., Ketelaars, H. C. J., Seuter-Berlage, F., and Smeets, F. L. M. (1980).

Phenaceturic acid, a new urinary metabolite of styrene in the rat. *Xenobiotica* **10,** 337-342.
99. Pocker, Y., and Ronald, B. P. (1980). The region of mechanistic transition in acid-catalyzed epoxide ring opening. A mechanistic switch mediated by salt in aqueous media. *J. Am. Chem. Soc.* **102,** 5311-5316.
100. Daniel, J. W. (1963). The metabolism of $^{36}$Cl-labeled trichloroethylene and tetrachloroethylene in the rat. *Biochem. Pharmacol.* **12,** 795-802.
101. Henschler, D., Hoos, W. R., Fetz, H., Dallmeier, E., and Metzler, M. (1978). Reactions of trichloroethylene epoxide in aqueous systems. *Biochem. Pharmacol.* **28,** 543-548.
102. Miller, R. E., and Guengerich, F. P. (1982). Oxidation of trichloroethylene by liver microsomal cytochrome *P*-450: Evidence for chlorine migration in a transition state not involving trichloroethylene oxide. *Biochemistry* **21,** 1090-1097.
103. McDonald, R. N. (1971). Rearrangements of $\alpha$-halo epoxides and related $\alpha$-substituted epoxides. *In* "Mechanisms of Molecular Rearrangements" (B. S. Thyagarajan, ed.), Vol. 3, pp. 67-107. Wiley (Interscience), New York.
104. Fromson, J. M., Pearson, S., and Bramah, S. (1973). The metabolism of tamoxifen (I. C. I. 46,474). Part I. In laboratory animals. *Xenobiotica* **3,** 693-709.
105. Steckbeck, S. R., Nelson, J. A., and Spencer, T. A. (1982). Enzymic reduction of an epoxide to an alcohol. *J. Am. Chem. Soc.* **104,** 893-895.
106. Hoppe-Seyler, G. (1882). Über das physiologische Verhalten der Orthonitrophenylpropiolsäure. *Z. Physiol. Chem.* **7,** 178-182.
107. Böhm, F. (1939). Mechanism of indoxyl formation *in vivo* from certain *o*-nitrophenyl derivatives. *Hoppe-Seyler's Z. Physiol. Chem.* **261,** 35-42.
108. Yih, R. Y., and Swithenbank, C. (1971). Identification of metabolites of *N*-(1,1-dimethylpropynyl)-3,5-dichlorobenzamide in rat and cow urine and rat feces. *J. Agric. Food Chem.* **19,** 320-324.
109. Sullivan, H. R., Roffey, P., and McMahon, R. E. (1979). Biotransformation of 4'-ethynyl-2-fluorobiphenyl in the rat. *In Vitro* and *In Vivo* Studies. *Drug. Metab. Dispos.* **7,** 76-80.
110. Wade, A., Symons, A. M., Martin, L., and Parke, D. V. (1980). The metabolic oxidation of the ethynyl group in 4-ethynylbiphenyl *in vitro*. *Biochem. J.* **188,** 867-872.
111. Wade, A., Symons, A. M., Martin, L., and Parke, D. V. (1979). Metabolic oxidation of the ethynyl group in 4-ethynylbiphenyl. *Biochem. J.* **184,** 509-517.
112. McMahon, R. E., Turner, J. C., Whitaker, G. W., and Sullivan, H. R. (1981). Deuterium isotope effect in the biotransformation of 4-ethynyl-biphenyls to 4-biphenylacetic acids by rat hepatic microsomes. *Biochem. Biophys. Res. Commun.* **99,** 662-667.
113. Abdel-Aziz, M. T., and Williams, K. I. H. (1969). Metabolism of 17$\alpha$-ethynylestradiol and its 3-methyl ether by the rabbit; an *in vivo* D-homoannulation. *Steroids* **13,** 809-820.
114. Abdel-Aziz, M. T., and Williams, K. I. H. (1970). Metabolism of radioactive 17$\alpha$-ethynylestradiol by women. *Steroids* **15,** 695-710.
115. Sisenwine, S. F., Kimmel, H. B., Liu, A. L., and Ruelius, H. W. (1973). The metabolic disposition of norgestrel in female rhesus monkeys. *Drug. Metab. Dispos.* **7,** 1-6.
116. Sisenwine, S. F., Kimmel, H. B., Liu, A. L., and Ruelius, H. W. (1973). Urinary metabolites of D,L-norgestrel in women. *Acta Endocrinol. (Copenhagen)* **73,** 91-104.
117. Palmer, K. H., Feierabend, J. F., Baggett, B., and Wall, M. E. (1969). Metabolic removal of a 17$\alpha$-ethynyl group from the antifertility steroid, norethindrone. *J. Pharmacol. Exp. Ther.* **167,** 217-222.
118. Williams, M. C., Helton, E. D., and Goldzieher, J. W. (1975). The urinary metabolites of

17α-ethynylestradiol-9α,11-³H in women. Chromatographic profiling and identification of ethynyl and non-ethynyl compounds. *Steroids* **25**, 229–246.
119. Helton, E. D., Williams, M. C., and Goldzieher, J. W. (1977). Oxidative metabolism and de-ethynylation of 17α-ethynylestradiol by baboon liver microsomes. *Steroids* **30**, 71–83.
120. House, H. O. (1972). "Modern Synthetic Reactions," 2nd ed., pp. 629–733. Benjamin, Menlo Park, California.
121. Abolin, C. R., Tozer, T. N., Craig, J. C., and Gruenke, L. D. (1980). C-Glucuronidation of the acetylenic moiety of ethchlorvynol in the rabbit. *Science* **209**, 703–704.
122. Schmid, R., and Schwartz, S. (1952). Experimental porphyria. III. Hepatic type produced by sedormid. *Proc. Soc. Exp. Biol. Med.* **81**, 685–689.
123. Goldberg, A., Rimington, C., and Fenton, J. C. B. (1954–1955). Experimentally produced porphyria in animals. *Proc. R. Soc. London, Ser. B* **143**, 257–279.
124. Wada, O., Yano, Y., Urata, G., and Nakao, K. (1968). Behavior of hepatic microsomal cytochromes after treatment of mice with drugs known to disturb porphyrin metabolism in liver. *Biochem. Pharmacol.* **17**, 595–603.
125. Waterfield, M. D., Del Favero, A., and Gray, C. H. (1969). Effect of 1,4-dihydro-3,5-dicarbethoxycollidine on hepatic microsomal haem, cytochrome $b_5$ and cytochrome $P$-450 in rabbits and mice. *Biochim. Biophys. Acta* **184**, 470–473.
126. De Matteis, F. (1970). Rapid loss of cytochrome $P$-450 and haem caused in the liver microsomes by the porphyrogenic agent 2-allyl-2-isopropylacetamide. *FEBS Lett.* **6**, 343–345.
127. Abbritti, G., and De Matteis, F. (1971-1972). Decreased levels of cytochrome $P$-450 and catalase in hepatic porphyria caused by substituted acetamides and barbiturates. Importance of the allyl group in the molecule of the active drugs. *Chem.-Biol. Interact.* **4**, 281–286.
128. De Matteis, F. (1971). Loss of haem in rat liver caused by the porphyrogenic agent 2-allyl-2-isopropylacetamide. *Biochem. J.* **124**, 767–777.
129. Levin, W., Sernatinger, E., Jacobson, M., and Kuntzman, R. (1972). Destruction of cytochrome $P$-450 by secobarbital and other barbiturates containing allyl groups. *Science* **176**, 1341–1343.
130. Levin, W., Lu, A. Y. H., Jacobson, M., and Kuntzman, R. (1973). Lipid peroxidation and the degradation of cytochrome $P$-450 heme. *Arch. Biochem. Biophys.* **158**, 842–852.
131. Ortiz de Montellano, P. R., and Mico, B. A. (1980). Destruction of cytochrome $P$-450 by ethylene and other olefins. *Mol. Pharmacol.* **18**, 128–135.
132. Murphy, M. J., Dunbar, D. A., Guengerich, F. P., and Kaminsky, L. S., (1981). Destruction of highly purified cytochrome $P$-450 associated with metabolism of fluorinated ether anesthetics. *Arch. Biochem. Biophys.* **212**, 360–369.
133. Ortiz de Montellano, P. R., Mico, B. A., Beilan, H. S., and Kunze, K. L. (1981). Olefins as suicide inhibitors of cytochrome $P$-450. *In* "Molecular Basis of Drug Action" (T. Singer and R. Ondarza, eds.), pp. 151–166. Am. Elsevier, New York.
134. Fellig, J., Barnes, J. R., Rachlin, A. I., O'Brien, J. P., and Focella, A. (1970). Substituted phenyl 2-propynyl ethers as carbamate synergists. *J. Agric. Food Chem.* **18**, 78–80.
135. Skrinjaric-Spoljar, M., Matthews, H. B., Engel, J. L., and Casida, J. E. (1971). Response of hepatic microsomal mixed-function oxidases to various types of insecticide chemical synergists administered to mice. *Biochem. Pharmacol.* **20**, 1607–1618.
136. Mackinnon, M., Sutherland, E., and Simon, F. R. (1977). Effects of ethinyl estradiol on hepatic microsomal proteins and the turnover of cytochrome $P$-450. *J. Lab. Clin. Med.* **90**, 1096–1106.

137. Breckenridge, A. M., Back, D. J., and Orme, M. (1979). Interactions between oral contraceptives and other drugs. *Pharmacol. Ther.* **7**, 617-626.
138. White, I. N. H., and Müller-Eberhard, U. (1977). Decreased liver cytochrome *P*-450 in rats caused by norethindrone or ethynyloestradiol. *Biochem. J.* **166**, 57-64.
139. White, I. N. H. (1978). Metabolic activation of acetylenic substituents to derivatives in the rat causing the loss of hepatic cytochrome *P*-450 and haem. *Biochem. J.* **174**, 853-861.
140. Ortiz de Montellano, P. R., and Kunze, K. L. (1980). Self-catalyzed inactivation of hepatic cytochrome *P*-450 by ethynyl substrates. *J. Biol. Chem.* **255**, 5578-5585.
141. White, I. N. H. (1980). Structure-activity relationships in the destruction of cytochrome *P*-450 mediated by certain ethynyl-substituted compounds in rats. *Biochem. Pharmacol.* **29**, 3253-3255.
142. Ortiz de Montellano, P. R., and Kunze, K. L. (1981). Cytochrome *P*-450 inactivation: Structure of the prosthetic heme adduct with propyne. *Biochemistry* **20**, 7266-7271.
143. Ortiz de Montellano, P. R., and Kunze, K. L. (1980). Inactivation of hepatic cytochrome *P*-450 by allenic substrates. *Biochem. Biophys. Res. Commun.* **94**, 443-449.
144. Ortiz de Montellano, P. R., and Mico, B. A. (1981). Destruction of cytochrome *P*-450 by allylisopropylacetamide is a suicidal process. *Arch. Biochem. Biophys.* **206**, 43-50.
145. Ortiz de Montellano, P. R., Mico, B. A., Mathews, J. M., Kunze, K. L., Miwa, G. T., and Lu, A. Y. H. (1981). Selective inactivation of cytochrome *P*-450 isozymes by suicide substrates. *Arch. Biochem. Biophys.* **210**, 717-728.
146. Loosemore, M. J., Wogan, G. N., and Walsh, C. (1981). Determination of partition ratios for allylisopropylacetamide during suicidal processing by a phenobarbital-induced cytochrome *P*-450 isozyme from rat liver. *J. Biol.Chem.* **256**, 8705-8712.
147. Marsh, J. A., Bradshaw, J. J., Sapeika, G. A., Lucas, S. A., Kaminsky, L. S., and Ivanetich, K. M. (1977). Further investigations of the metabolism of fluroxene and the degradation of cytochrome *P*-450 *in vitro*. *Biochem. Pharmacol.* **26**, 1601-1606.
148. Guengerich, F. P., and Strickland, T. W. (1977). Metabolism of vinyl chloride: Destruction of the heme of highly purified liver microsomal cytochrome *P*-450 by a metabolite. *Mol. Pharmacol.* **13**, 993-1004.
149. Levin, W., Jacobson, M., and Kuntzman, R. (1972). Incorporation of radioactive δ-aminolevulinic acid into microsomal cytochrome *P*-450: Selective breakdown of the hemoprotein by allylisopropylacetamide and carbon tetrachloride. *Arch. Biochem. Biophys.* **148**, 262-269.
150. Bradshaw, J. J., Ziman, M. R., and Ivanetich, K. M. (1978). The degradation of different forms of cytochrome *P*-450 *in vivo* by fluroxene and allyl-*iso*-propylacetamide. *Biochem. Biophys. Res. Commun.* **85**, 859-866.
151. White, I. N. H. (1981). Destruction of liver haem by norethindrone. Conversion into green pigments. *Biochem. J.* **196**, 575-583.
152. Correia, M. A., Farrell, G. C., Olson, S., Wong, J. S., Schmid, R., Ortiz de Montellano, P. R., Beilan, H. S., Kunze, K. L., and Mico, B. A. (1981). Cytochrome *P*-450 heme moiety. The specific target in drug-induced heme alkylation. *J. Biol. Chem.* **256**, 5466-5470.
153. Ortiz de Montellano, P. R., Beilan, H. S., Kunze, K. L., and Mico, B. A. (1981). Destruction of cytochrome *P*-450 by ethylene. Structure of the resulting prosthetic heme adduct. *J. Biol. Chem.* **256**, 4395-4399.
154. Ortiz de Montellano, P. R., Kunze, K. L., Beilan, H. S., and Wheeler, C. (1982). Destruction of cytochrome *P*-450 by vinyl fluoride, fluroxene, and acetylene. Evidence for a radical intermediate in olefin oxidation. *Biochemistry* **21**, 1331-1339.

155. Ortiz de Montellano, P. R., Beilan, H. S., and Mathews, J. M. (1982). Alkylation of the prosthetic heme in cytochrome *P*-450 during oxidative metabolism of the sedative-hypnotic ethchlorvynol. *J. Med. Chem.* **25**, 1174–1179.
156. Kunze, K. L., Mangold, B. L. K., Wheeler, C., Beilan, H. S., and Ortiz de Montellano, P. R. (1982). The cytochrome *P*-450 active site. Regiospecificity of prosthetic heme alkylation by olefins and acetylenes. *J. Biol. Chem.* **258**, 4202–4207.
157. Gander, J. E., and Mannering, G. J. (1980). Kinetics of hepatic cytochrome *P*-450 dependent monooxygenase systems. *Pharmacol. Ther.* **10**, 191–221.
158. White, R. E., and Coon, M. J. (1980). Oxygen activation by cytochrome *P*-450. *Annu. Rev. Biochem.* **49**, 315–356.
159. Sligar, S. G., Kennedy, K. A., and Pearson, D. C. (1980). Chemical mechanisms for cytochrome *P*-450 hydroxylation: Evidence for acylation of heme-bound dioxygen. *Proc. Natl. Acad. Sci. U.S.A.* **77**, 1240–1244.
160. Welborn, C. H., Dolphin, D., and James, B. R. (1981). One-electron electrochemical reduction of a Ferrous porphyrin dioxygen complex. *J. Am. Chem. Soc.* **103**, 2869–2871.
161. Björkhem, I. (1977). Rate-limiting step in microsomal cytochrome *P*-450-catalyzed hydroxylations. *Pharmacol. Ther., Part A* **1**, 327–348.
162. Hamilton, G. A. (1974). Chemical models and mechanisms for oxygenases. *In* "Molecular Mechanisms of Oxygen Activation" (O. Hayaishi, ed.), pp. 405–451. Academic Press, New York.
163. Groves, J. T., McClusky, G. A., White, R. E., and Coon, M. J. (1978). Aliphatic hydroxylation by highly purified liver microsomal cytochrome *P*-450. Evidence for a carbon radical intermediate. *Biochem. Biophys. Res. Commun.* **81**, 154–160.
164. Hjelmeland, L. M., Aronow, L., and Trudell, J. R. (1977). Intramolecular determination of primary kinetic isotope effects in hydroxylations catalyzed by cytochrome *P*-450. *Biochem. Biophys. Res. Commun.* **76**, 541–549.
165. Ortiz de Montellano, P. R., Yost, G. S., Mico, B. A., Dinizo, S. E., Correia, M. A., and Kambara, H. (1979). Destruction of cytochrome *P*-450 by 2-isopropyl-4-pentenamide and methyl 2-isopropyl-4-pentenoate: Mass spectrometric characterization of prosthetic heme adducts and nonparticipation of epoxide metabolites. *Arch. Biochem. Biophys.* **197**, 524–533.
166. Parker, R. E., and Isaacs, N. S. (1959). Mechanisms of epoxide reactions. *Chem. Rev.* **59**, 737–799.
167. Adams, S. M., Murphy, M. J., and Kaminsky, L. S. (1981). Molecular orbital studies of the metabolism of fluroxene and analogous fluorinated ether anesthetics. *Mol. Pharmacol.* **20**, 423–428.
168. Fujita, E., and Nagao, Y. (1977). Tumor inhibitors having potential for interaction with mercapto enzymes and/or coenzymes. *Bioorg. Chem.* **6**, 287–309.
169. Reid, W. D. (1972). Mechanism of allyl alcohol-induced hepatic necrosis. *Experientia* **28**, 1058–1061.
170. Walsh, C. (1977). Recent developments in suicide substrates and other active site-directed inactivating agents of specific target enzymes. *Horiz. Biochem. Biophys.* **3**, 36–81.
171. Lawley, P. D., and Jarman, M. (1972). Alkylation by propylene oxide of deoxyribonucleic acid, adenine, guanosine, and deoxyguanylic acid. *Biochem. J.* **126**, 893–900.
172. Hemminki, K., Paasivirta, J., Kurkirinne, T., and Virkki, L. (1980). Alkylation products of DNA bases by simple epoxides. *Chem.-Biol. Interact.* **30**, 259–270.
173. Voogd, C. E., van der Stel, J. J., and Jacobs, J. J. A. A. (1981). The mutagenic action of aliphatic epoxides. *Mutat. Res.* **89**, 269–282.

# 5. ALKENES AND ALKYNES

174. Wade, D. R., Airy, S. C., and Sinsheimer, J. E. (1978). Mutagenicity of aliphatic epoxides. *Mutat. Res.* **58**, 217–223.
175. International Agency for Research on Cancer (1976). "IARC Monographs on the Evaluation of Carcinogenic Risk of Chemicals," Vol. 11, pp. 191–199. IARC, Lyon, France.
176. International Agency for Research on Cancer (1976). "IARC Monographs on the Evaluation of Carcinogenic Risk of Chemicals," Vol. 11, pp. 201–208. IARC, Lyon, France.
177. National Cancer Institute (1979). "Bioassay of Styrene for Possible Carcinogenicity," Tech. Rep. 185. Nat. Cancer Inst., Bethesda, Maryland.
178. Weinstein, I. B., Jeffrey, A. M., Jennette, K. W., Blobstein, S. H., Harvey, R. G., Harris, C., Autrup, H., Kasai, H., and Nakanishi, K. (1976). Benzo[a]pyrene diol epoxides as intermediates in nucleic acid binding *in vitro* and *in vivo*. *Science* **193**, 592–595.
179. Campbell, T. C., and Hayes, J. R. (1976). The role of aflatoxin metabolism in its toxic lesion. *Toxicol. Appl. Pharmacol.* **35**, 199–222.
180. Harrison, G. G., Ivanetich, K., Kaminsky, L., and Halsey, M. J. (1976). Fluroxene (2,2,2-trifluoroethyl vinyl ether) toxicity: A chemical aspect. *Anesth. Analg. (Cleveland)* **55**, 529–533.
181. Orth, O. S., Slocum, H. C., Stutzman, J. W., and Meek, W. J. (1940). Vinethene as an anesthetic agent. *Anesthesiology* **1**, 246–260.
182. Klatskin, G. (1977). Hepatic tumors: Possible relationship to use of oral contraceptives. Gastroenterology **73**, 386–394.

*Chapter 6*

# Benzene and Substituted Benzenes

LAURENCE S. KAMINSKY

*Laboratory for Biochemical
and Genetic Toxicology
New York State Department of Health
Albany, New York*

| | |
|---|---:|
| I. Introduction | 157 |
| II. Chemical Properties | 158 |
|    A. Benzene and Substituted Benzenes | 158 |
|    B. Biphenyl and Substituted Biphenyls | 158 |
| III. Reaction Mechanisms for Bioactivation | 159 |
|    A. Bioactivation of Benzene and Substituted Benzenes to Arene Oxides | 159 |
|    B. Reactions of Benzene and Substituted Benzene Arene Oxides | 163 |
| IV. Enzymology of Bioactivation | 170 |
| References | 170 |

## I. Introduction

Benzene and many substituted benzenes exhibit a variety of profound toxic effects. In many cases, at least in laboratory animals, this toxicity is a consequence of a bioactivation of the phenyl ring.[1-3] In humans, epidemiological studies have implicated benzene in myeloblastic leukemia,[4,5] aleukemia,[6] chronic myeloid leukemia,[4,7] and, possibly, lymphatic leukemia[8] and Hodgkin's disease.[9] Benzene is not a mutagen in the Ames test,[10,11] but does increase the frequency of sister chromatid exchanges in murine bone marrow cells by a mechanism that requires bioactivation.[1]

The toxicity of biphenyls, and particularly halogenated biphenyls, has been extensively investigated.[12] Polychlorinated biphenyls (PCBs) produce hepatocellular carcinomas in rats[13] and hepatomas in mice,[14] and polybrominated biphenyls (PBBs), given at high doses, cause liver tumors in rats.[15] 4-Chloro-[16] and 4-bromobiphenyl[17] are mutagenic in the Ames test, but only after bioactivation by hepatic microsomal systems.

This chapter will review current knowledge of the bioactivation of benzene and substituted benzenes. The role of substituent groups in modifying the bioactivation of benzene will be addressed, but not the bioactivation of the substituent groups themselves or of fused aromatic systems, which will be reviewed elsewhere in this volume.

Emphasis will be placed on biphenyl and substituted biphenyls because of the ubiquity of halogenated, and particularly chlorinated, isomers and congeners of biphenyl in the environment, because they pose health risks, and because of the unusual influence of one phenyl ring on the reactivity of the other in these molecules. Furthermore, many of the aspects of bioactivation of polycyclic aromatic systems are relevant to that of benzene and are discussed in Chapter 7.

## II. Chemical Properties

### A. BENZENE AND SUBSTITUTED BENZENES

The development of an understanding of the aromatic character of benzene, from the time of its first isolation by Faraday in 1825 through the incorporation of molecular orbital theories, has been clearly documented by Chandler and Craig.[18] Benzene is thermochemically more stable (by 36 kcal/mol) and less reactive than if it had three conjugated double bonds. The reactivity of the benzene nucleus is affected by substituents. Electron-withdrawing substituents deactivate the ring to electrophilic substitution, with the meta positions being least affected, and electron-releasing substituents facilitate electrophilic substitution at ortho and para positions.[19] Conjugative interactions with substituents can reverse the influence of the substituents. Nucleophilic substitution occurs at substituted benzene carbons and only rarely at hydrogen-substituted carbons.[20] Benzene can undergo homolytic substitution, an example of which is hydroxylation.[21]

### B. BIPHENYL AND SUBSTITUTED BIPHENYLS

The relative orientations of the phenyl rings in biphenyl affect the bioactivation of the molecule by determining which form of enzyme will catalyze

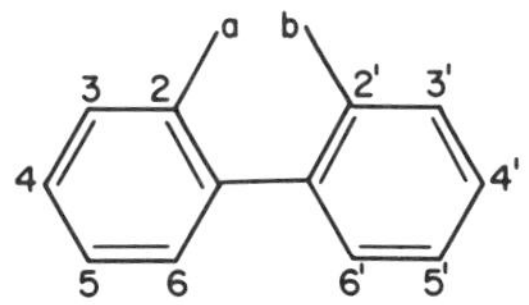

## 6. BENZENE AND SUBSTITUTED BENZENES

Fig. 1. Biphenyl numbering system: a and b represent ortho substituents.

the activation.[22-24] X-ray diffraction analysis of crystalline biphenyl reveals an essentially planar molecule in which the ortho hydrogen–hydrogen repulsive interactions are insufficient to induce nonplanarity.[25] Bulky ortho substituents force the two rings out of coplanarity, and if the sum of the bond lengths C–a and C–b (Fig. 1) exceeds 2.90 Å, the molecule can be resolved into two enantiomers.[26] In the case of substitution by a particular atom or group of atoms, the greater the number of ortho substituents the greater the probability of the biphenyl being noncoplanar. Thus, for a series with increasing ortho fluoro substitution, from 2-fluorobiphenyl through 2,6,2′,6′-tetrafluorobiphenyl, the barriers to internal rotation increase from 8 to 49 kcal/mol.[27] Ortho substituents on the biphenyl nucleus can thus affect the bioactivation of the molecule through their influence on its stereochemistry.[22] Apart from a role in determining the enzyme involved in the bioactivation, stereochemical factors can affect bioactivation electronically. In the coplanar state, biphenyl has additional resonance forms, which do not contribute to the noncoplanar state and which could, for example, stabilize certain carbonium ions of bioactivated intermediates.[22,28] Resonance effects in biphenyls also alter the electronic influences of substituents on the phenyl ring from those operative in benzene,[28] particularly ortho to the substituent. $^{13}C$ Nuclear magnetic resonance studies have indicated that substituent electronic effects on one ring of PCBs can be transmitted to the other ring.[28] The net atomic charges on the carbon atoms of a series of dichlorobiphenyls, calculated by a semiempirical molecular orbital method, support this conclusion, but indicate that the transmitted effects are minor.[22]

### III. Reaction Mechanisms for Bioactivation

#### A. BIOACTIVATION OF BENZENE AND SUBSTITUTED BENZENES TO ARENE OXIDES

The hypothesis by Boyland in 1950[29] that carcinogenic aromatic hydrocarbons could be bioactivated by formation of arene oxides provided an impetus for the investigation of the bioactivation of benzene, although there was "very little evidence to support the hypothesis." Support for the bioactivation of benzene to benzene oxide was provided when previously identified *in vivo* metabolites of benzene were formed from synthetic benzene oxide in *in vitro* systems.[30] Rabbit liver microsomes yielded *trans*-1,2-dihydroxybenzene and phenol, and a rat liver soluble system, in the presence of glutathione, yielded *S*-(1,2-dihydro-2-hydroxyphenyl)glutathione from synthetic benzene oxide (Fig. 2.)[30]

It is now generally accepted that arene oxides are the principal activated intermediates of benzene and substituted benzene bioactivation,[31] notwithstanding the failure to isolate metabolically formed benzene oxide from benzene.[32] Subsequent studies have, however, indicated that benzene oxide, although an important reactive intermediate, is not always the ultimate reactive and toxic intermediate of benzene metabolism.[32,33] Incubation of [$^{14}$C]benzene with rat liver microsomes and with NADPH as an electron source yielded metabolites that were irreversibly bound, primarily to the microsomal protein. In the presence of glutathione or cysteine, most of the binding was prevented and formation of water-soluble metabolites was increased.[32] However, in this microsomal system, synthetic benzene oxide did not bind to a similar extent, and added phenol bound more extensively than benzene. This result implies that a secondary metabolite of benzene, probably a metabolite of phenol, is the ultimate reactive metabolite of benzene in this system. It is capable of binding to cellular macromolecules, and is presumably the ultimate source of toxicity.[32] Addition of

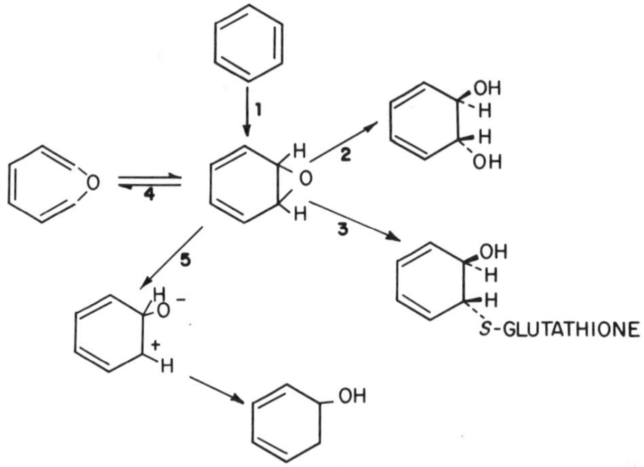

Fig. 2. (1) The bioactivation of benzene to the arene oxide; (2) hydration of the arene oxide to the *trans*-dihydrodiol catalyzed by epoxide hydrolase; (3) spontaneous or glutathione *S*-epoxide transferase-catalyzed addition of glutathione; (4) tautomerism of benzene to the oxepin; (5) spontaneous isomerization of the arene oxide to the phenol via a carbonium ion intermediate.

superoxide dismutase to the system inhibited the binding by greater than 50% and caused an accumulation of hydroquinone, indicating that the superoxide anion plays a role in the bioactivation of benzene.[34] Similar results were observed with 2,2'-dichlorobiphenyl, where the majority of the protein-bound metabolite arose from the activation of the arene oxide-derived phenolic metabolite rather than from the arene oxide itself.[35] This was supported by *in vivo* studies on benzene metabolite binding to bone marrow, which indicated that it was the benzene metabolites hydroquinone and 1,2,4-benzenetriol that were autoxidized to the activated cytotoxic compounds.[33] In contrast, neither phenol nor hydroquinone was capable of mimicking the effects of benzene in reducing the cellularity of bone marrow in mouse tibia.[36] The further bioactivation of benzene metabolites such as phenol are reviewed in Chapter 9.

The arene oxide of benzene is unstable and has a half-life in water at pH 7.0 of approximately 2 min.[31] However, its formation in biological systems can be confirmed when, in the presence of epoxide hydrolase, benzenedihydrodiol is formed (Fig. 2). In rat hepatic microsomes, with large added quantities of purified epoxide hydrolase, benzene yielded the dihydrodiol and phenol in a ratio of approximately 1:5, indicating that a considerable portion of benzene metabolism proceeds via the epoxide intermediate.[32] Apart from the formation of dihydrodiols, several other criteria can be used to demonstrate bioactivation of benzenes to arene oxides: formation of glutathione conjugates of the corresponding phenols (Fig. 2), appearance of precursors of mercapturic acids in the urine, and the occurrence of "NIH shifts" in substituted benzenes (see Section III,B,2) during biotransformation to phenols.[31] There is a report that a 3,4-oxide of 2,5,2',5'-tetrachlorobiphenyl was isolated from a hepatic microsomal preparation from phenobarbital (PB)-induced rats. The arene oxide required NADPH for its formation, and, in the presence of an epoxide hydrolase inhibitor, isolation of the arene oxide was facilitated.[37] This report should be accepted with some skepticism, because the same authors subsequently reported that hydroxylation of this PCB occurred primarily directly and not via an arene oxide.[38] The involvement of a direct 3-hydroxylation mechanism during microsomal metabolism of this substrate was also suggested, because the intermediacy of the 3,4-oxide in the major metabolic pathways was precluded based on metabolite patterns.[39] In a direct confirmation of arene oxide formation from a benzene derivative, the 2,3- and 3,4-oxides of chlorobenzene have been trapped in a metabolizing system of rat hepatic microsomes containing NADPH and unlabeled carrier chlorobenzene oxides.[40]

The bioactivation of benzene and its derivatives to arene oxides is catalyzed by the microsomal mixed-function oxidase system, with cytochromes *P*-450 as the terminal oxidases. Cytochrome *P*-450 occurs in an unknown number of isozymic forms, which are differentially inducible by variety of xenobiotics (see Section IV).[41] It has not been clearly established what form of activated oxygen is associated with cytochrome $P$-450-catalyzed epoxidation of benzene or whether, in fact, the different isozymes generate different forms of activated oxygen.[42] It was determined at an early stage that the oxygen atom incorporated into the substrate is derived from dioxygen.[43] The dioxygen complexes with the ferrocytochrome $P$-450–substrate complex, and an electron is introduced into the complex.[42] Model studies have indicated that the axial heme ligand in this complex is thiolate.[44] The bound oxygen molecule can undergo heterolytic or homolytic scission. In the former case a possible product is $-S-Fe(V)O^{2-}$, which would be resonance stabilized with a resonance form being $-S \cdot Fe(III)O^-$.[42] Homolytic cleavage could lead to a thiyl ferric hydroxide complex, $-S \cdot Fe(III)OH^-$. For aliphatic hyroxylations, the activated oxygen is proposed to abstract a hydrogen from the substrate,[42] but no mechanism for the epoxidation of aromatic compounds by cytochrome $P$-450 has been experimentally determined. A singlet oxygen atom could be transferred to the aromatic double bond to produce an arene oxide in one step. Addition of a triplet oxygen atom could form a non-ring-closed, tetrahedral intermediate, which could ultimately generate a phenol by (1) ring closure to the arene oxide, and (2) rearrangement to the keto tautomer of the phenol, or (3) by rearrangement directly to the phenol. Because the majority of bioactivations of benzenes occur with isotope effects (see later), the direct-addition mechanism is favored.[44] In a model system, $Mn(V)=O$ epoxidation of an olefin proceeded via addition to the double bond to give a freely rotating free-radical intermediate, which ring-closed to yield the epoxide.[45] Such a system may provide some insight into the mechanism of arene oxide formation.

The metabolism of [4-$^3$H]acetanilide to 4-hydroxyacetanilide is catalyzed by rat liver microsomes or a variety of reconstituted purified cytochrome $P$-450 systems.[46] The reaction, which proceeds via an arene oxide, occurs in the same manner whether $NADPH/O_2$ or cumene hydroperoxide provides the activated oxygen.

Cytochrome $P$-450 isozymes differ in the extent to which they catalyze the epoxidation of benzene and its derivatives, and they also exhibit differing regioselectivities for epoxidation. Regioselective differences are based on observed differences in phenolic metabolites, which are formed from the same substrate but catalyzed by different isozymes. It is presumed that different phenol metabolites imply that different sites were epoxidized, because it is unlikely that the cytochrome isozymes differentially affect the direction of ring opening of the intermediate epoxide. The most convincing evidence for differences in isozymic regioselectivities of epoxidation has been obtained with reconstituted systems of purified cytochrome $P$-450 and

NADPH–cytochrome $P$-450 reductase. A reconstituted system with the cytochrome $P$-450 isozyme from PB-induced rats yielded primarily 4-hydroxybiphenyl from biphenyl: with the isozyme from 3-methylcholanthrene (3-MC)-induced rats, the metabolites were 2- and 4-hydroxybiphenyl in a ratio of 0.44.[47] Thus, the PB-induced isozyme catalyzes the formation of the 3,4-oxide of biphenyl and the 3-MC-induced isozyme the 2,3- and 3,4-oxides. An even more marked regioselectivity difference was noted with dichlorobiphenyls (with both substituents on the same ring) when metabolized in the reconstituted system. Cytochrome $P$-450 purified from PB-induced rats epoxidized unchlorinated rings, while the isozyme from $\beta$-naphthoflavone (BNF)-induced rats epoxidized chlorinated rings.[22]

The differing regioselectivities of the isozymes of cytochrome $P$-450 can have major consequences for toxicity associated with the bioactivation. Thus, the hepatotoxicity of bromobenzene is enhanced in PB-treated animals and diminished in 3-MC- or BNF-treated animals.[48] The explanation, based on results with purified cytochromes $P$-450 from PB-induced animals, which catalyze 3,4-epoxidation of bromobenzene, and from BNF-induced animals, which catalyze 2,3-epoxidation,[49] is that the 3,4-epoxide is hepatotoxic and the 2,3-epoxide is not.[50] However, the various cytochrome $P$-450 isozymes may be associated with differing toxicities of benzene and its derivatives, not only because of differences in regioselectivities of activation to arene oxides, but also because of differences in secondary metabolism of metabolites. Studies with DBA/2 and C57BL/6 mice, in which benzene was more toxic to the former than the latter strain, have demonstrated the importance of this possibility.[51]

Clear evidence that different isozymes of cytochrome $P$-450 metabolize and thus bioactivate benzene derivatives to different extents was provided with a reconstituted system containing purified hepatic isozymes from either PB- or BNF-induced rats.[22] A series of 10 dichlorobiphenyls (DCBs) were used as substrates, and total metabolites were quantified. For DCBs with two $o$-chloro substituents, the PB-induced isozyme was approximately 100-fold more effective as a catalyst; for DCBs with no $o$-chloro substituents, the BNF-induced isozyme was approximately 50-fold more effective; for DCBs with one $o$-chloro substituent, the two isozymes were approximately equivalent.[22]

## B. REACTIONS OF BENZENE AND SUBSTITUTED BENZENE ARENE OXIDES

In biological systems, the activated arene oxides of benzene and its derivatives can undergo a number of enzyme- and non-enzyme-catalyzed reactions (Fig. 2). The fate of arene oxides is a function of the nature, number, and relative positions of substituents on the benzene ring and their effect on the stability of the epoxide, the availability of conjugating systems, the susceptibility to catalytic hydration, and the specific biological environment. Benzene oxide tautomerizes to an oxepin (Fig. 2), but no metabolites of this tautomer have been observed.[52] The metabolism of benzene has been reviewed.[53]

1. Hydration

Arene oxides can undergo hydrations catalyzed by epoxide hydrolase[54] (see Section IV) to the *trans*-1,2-dihydrodiol (Fig. 2). It has been estimated in experimental animals, however, that only 1.6% of the benzene administered is converted to the dihydrodiol.[53] No systematic assessment of the role of benzene substituents in modifying the susceptibility of benzene arene oxides to hydration by purified epoxide hydrolase is available. Benzene oxide is a poor substrate for microsomal epoxide hydrolase relative to napthalene 1,2-oxide,[30,55] but toluene 3,4-oxide[56] and *p*-xylene 3,4-oxide are better substrates.[57] The methyl group in the toluene oxide destabilizes the oxirane ring, but in the xylene oxide the epoxide ring is trisubstituted, which sterically hinders enzymatic hydration. These disparate reasons for preventing hydration of epoxides indicate the complexity of predicting whether an activated arene oxide will be a substrate.

A study of bromobenzene in two strains of mice, where the relative rates of *p*-bromophenol and *o*-bromophenol formation were compared and where dihydrodiol formation rates from the intermediate arene oxides were determined, apparently indicated that the 2,3-oxide was a better substrate for epoxide hydrolase than was the 3,4-oxide of bromobenzene.[48]

The extent to which arene oxides are hydrated and thus detoxified by epoxide hydrolase is a function of their rate of formation, which depends on the cytochrome *P*-450 isozyme responsible for the formation of the arene oxide, and their interaction with epoxide hydrolase. In an *in vitro* reconstituted system containing BNF-induced cytochrome *P*-450, 2,3-dichlorobiphenyl as the substrate, and epoxide hydrolase, 76% of the total metabolites was dihydrodiol and 24% phenolic. In contrast, with a PB-induced cytochrome *P*-450 isozyme in an otherwise identical system, only 33% of the total metabolites was dihydrodiol.[58] These results, together with data on naphthalene as substrate,[58,59] indicate that the BNF-induced isozyme associates more firmly with epoxide hydrolase than does the PB-induced isozyme. This difference in functional interaction could have major consequences for chemical carcinogenicity and toxicology of benzene and its derivatives.

## 2. Acid-Catalyzed and Spontaneous Arene Oxide Ring Opening

Arene oxides of benzene and its derivatives can undergo spontaneous or acid-catalyzed isomerization to phenols (Fig. 2). This isomerization, should it proceed through to the phenol, essentially affords a route for deactivation of activated benzenes, unless the phenol is susceptible to further bioactivation to more reactive and potentially toxic metabolites.[32,33] However, intermediates of some isomerizations are, under certain conditions, capable of interacting with cellular macromolecules with consequent toxic effects.

The isomerization of arene oxides to phenols is associated with the NIH shift, an intramolecular migration of benzene ring substituents to an adjacent carbon atom.[60] Kinetic studies of the spontaneous isomerization under basic conditions and the acid-catalyzed isomerization of benzene oxide and a number of methyl-substituted benzene oxides have indicated that the reactions are not concerted.[61] The stepwise reactions incorporate the development of a carbonium ion in the transition state after heterolytic cleavage of the C–O bond, and this step is rate determining for the overall isomerization.[44,61] Carbonium ion formation precedes the substituent shift. On the basis of entropic considerations, it was initially suggested that the arene oxide ring spontaneously isomerizes in a concerted reaction to a dienone, which enolizes to the phenol.[62] The keto form is, however, now accepted as being on the major pathway of isomerization of arene oxides to phenols. The overall pathway is shown in Fig. 3.

NIH shifts occur with retention of the migrating substituent. The migration and retention of deuterium from a series of $p$-deuterium-substituted benzene derivatives has been reported.[52] The greatest migration (65%) occurred with the —$OCH_3$-substituted benzene and the least (1%) with —$NHSO_2C_6H_5$-substituted benzene. In general, nonionizable substituents para to the deuterium show high migration and retention of deuterium, and ionizable substituents retain less of the para deuterium substituents. For 4,4′-dichlorobiphenyl in a reconstituted system of purified cytochrome $P$-450, 60% of the chloro substituent migrated to form 3,4′-dichloro-4-biphenylol.[52] Formation of a phenol from an arene oxide proceeds without an isotope effect, when deuterated arene oxides are compared with their protium congeners.[44] Because arene oxides are formed without cleavage of C–H bonds, their formation also proceeds without a significant isotope effect.[44] The absence of an isotope effect during formation of phenols from benzene and substituted benzenes has consequently been used as an indicator of arene oxide intermediates.[44] Phenol formation from bromobenzene,

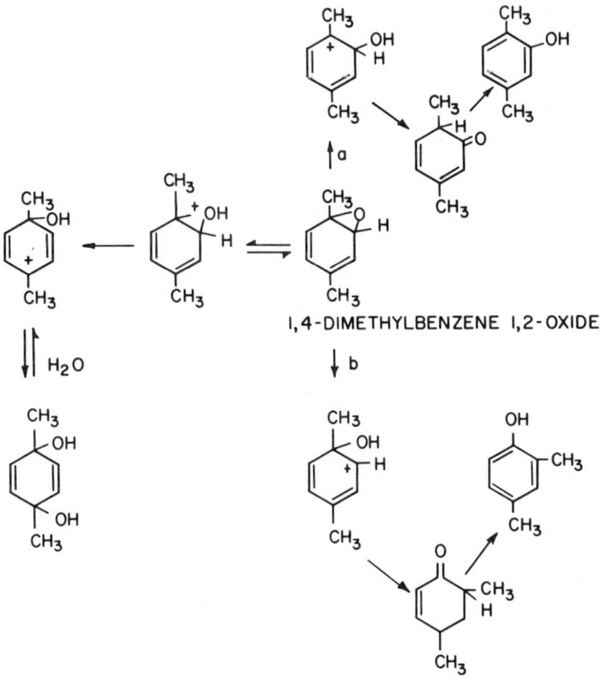

Fig. 3. Spontaneous and acid-catalyzed isomerization of 1,4-dimethylbenzene 1,2-oxide. Spontaneous isomerization: (a) without migration of a methyl group; (b) with migration of a methyl group (NIH shift).

anisole, acetanilide, benzonitrile, biphenyl, benzene, *o*- and *p*-xylene, toluene, and mesitylene all proceeded without an isotope effect when ring-deuterated substrates were used. In contrast, nitrobenzene and methyl phenyl sulfone exhibited significant primary isotope effects associated with the formation of phenols in *in vitro* systems. It was tentatively concluded that meta hydroxylation of these two compounds proceeds via direct insertion of oxygen into the C–H bond rather than via arene oxide formation.[44] Other examples of a direct insertion mechanism have been reported for biphenyls and chlorobenzene.[38,63] In the latter case, because neither the 2,3- or 3,4-oxides of chlorobenzene isomerized to *m*-chlorophenol, a metabolite of chlorobenzene isolated from the microsomes and reconstituted purified metabolizing systems, it was apparent that meta hydroxylation occurred via direct insertion.[63] Formation of phenols by direct-insertion mechanisms would not result in bioactivation of the benzene, except by providing a suitable substrate for further bioactivation.

## 6. BENZENE AND SUBSTITUTED BENZENES

Hydroxylation of [7-$^2$H]warfarin with rat hepatic microsomes occurs with retention of deuterium, and there is an isotope effect.[64] These results were interpreted to indicate that direct insertion, abstraction, or direct epoxidation were unlikely mechanisms for this reaction. The simplest mechanism compatible with all the data was addition of a triplet oxygen atom to yield a tetrahedral adduct.[64]

A factor that governs the direction of ring opening of the arene oxide is the relative stability of the two possible carbonium ion intermediates. This is clearly demonstrated with toluene 2,3-oxide and toluene 1,2-oxide, which isomerize exclusively to 2-hydroxytoluene, probably because these pathways permit the intermediate to form the relatively stable tertiary alkylic carbonium ion.[65,66] In general, alkyl substituents on the oxirane ring of arene oxides stabilize the ring against isomerization, and alkyl substituents on nonoxirane carbons of the arene oxide destabilize the ring.[61,65] Ring opening of asymmetrical oxides is favored in the direction that leads to carbonium ions, which can be delocalized to resonance states with a greater number of tertiary ions.[52] Electron-withdrawing substituents on the phenyl ring stabilize an arene oxide ring, thus facilitating its epoxide hydrolase-catalyzed hydration.[61]

For biphenyls, the degree of coplanarity of the two phenyl rings can affect the relative stabilities of the two carbonium ions arising from an arene oxide. The 3,4-oxide of 2,5,2′,5′-tetrachlorobiphenyl undergoes spontaneous isomerization to 4- and 3-hydroxybiphenyl in a 4:1 ratio.[38] Semiempirical molecular orbital calculations (MNDO) indicated that, at phenyl ring twist angles of 90°, 67.5°, and 45°, the energy barriers to ring opening to the 4-OH isomer were 3.1, 2.8, and 2.0 kcal/mol, respectively; for ring opening to the 3-OH isomer, the values were 3.3, 3.5, and 3.6 kcal/mol, respectively. Ratios for 4-OH to 3-OH formation calculated from these values were 1.25, 3.70, and 14.29. These results indicated the influence of the phenyl ring twist angle on the stability of arene oxide rings and suggested that the twist angle for this biphenyl is approximately 67.5°.[67] Similar calculations indicated that chloro substituents on a ring of biphenyls did not affect the stability of an arene oxide on the other ring, apart from an effect of ortho substituents on the phenyl ring twist angle.[22]

The 1,2-oxide of 1,4-dimethylbenzene provides an interesting example of the NIH shift and of the role of benzene ring substituents on the fate of the arene oxide.[68,69] The oxide spontaneously isomerizes at pH values above 6.0 to two phenols, with the NIH-shift form predominating in a ratio of 87:13 (Fig. 3). As the pH was decreased, this ratio was altered until both products were produced in approximately equivalent amounts at pH 4. At low pH, a dihydroxylated metabolite was formed, which was postulated to arise from reaction of the delocalized and resonance-stabilized carbonium ion formed by isomerization of the arene oxide with water. This reaction demonstrated the ease of nucleophilic reactions with the bioactivated benzene. The diol was converted to the two phenolic metabolites spontaneously, thus completing a novel mechanism for isomerization of an arene oxide to a phenol.[68,69]

The possibility that arene oxides of benzene and its substituents could isomerize to ring-opened carbonium ions, which would have the potential to react with nucleophilic cell components to produce toxic effects, was examined by theoretical calculations.[70] Intermediate Neglect of Differential Overlap semiempirical calculations were performed on benzene oxide, 3-methylbenzene oxide, 4-methylbenzene oxide, 4,5-dimethylbenzene oxide, 1-methylbenzene oxide, and 1,4-dimethylbenzene oxide. The calculated activation energies for isomerization of the three oxides of toluene were lowest for the experimentally determined preferred pathways, indicating that the calculations were valid approximations. Experimentally determined rates of acid-catalyzed isomerization of the oxides were compared with a variety of calculated parameters. Only the transition state energy difference — that is, the energy difference between the ground-state reactant and the ring-opened carbonium ion relative to that of benzene oxide — correlated with the experimental data.[70] This correlation supports the kinetic data, which indicate the formation of carbonium ion intermediates and point to a role for carbonium ion stability in determining the fate of the epoxide. In this study, the stabilities of the oxides of the benzene derivatives were lower than those of the oxides of the polycyclic aromatic compounds investigated.

### 3. Conjugation

Although most benzene metabolites are excreted as conjugates, conjugation usually occurs after isomerization of the arene oxide to the phenol.[53] There is, however, some deactivation of arene oxides by direct conjugation with glutathione.[52] This conjugation, which ultimately yields a premercapturic acid, can occur spontaneously or can be catalyzed by glutathione S-epoxide transferase, but the spontaneous reaction probably predominates *in vivo*.[71] It was originally demonstrated[30] that benzene oxide and glutathione, when incubated with a rat liver postmicrosomal supernatant fraction, yielded the addition product shown in Fig. 2. The nonenzymatic addition reaction of arene oxides and glutathione is facilitated by strong electron-withdrawing substituents on the benzene, which is suggestive of nucleophilic attack on the oxide rather than on an intermediate carbonium ion.[52]

## 4. Covalent Binding of Activated Benzene and Benzene Derivatives to Cellular Macromolecules

It is generally accepted that the majority of toxic effects of benzene and its derivatives arise from the bioactivation of the compounds to arene oxides, which interact with cellular macromolecules to form covalent adducts. The extent to which this occurs is dependent on factors already discussed, which either deactivate the activated benzene or permit it to react with any macromolecules within its range of transport. The variety of alternatives makes predictions of the fate of an arene oxide difficult. This is demonstrated by a study of the mutagenic effects of a series of oxiranes of benzene derivatives and hydrogenated congeners tested with four *Salmonella* strains in the Ames test.[72] No bioactivating system was used in these studies. Benzene oxide was inactive, a probable consequence of its lack of stability, its rapid isomerization to phenol, and, less likely, its partial tautomerism to the oxepin.[73] The dioxide of benzene was weakly mutagenic, and, although no halogenated derivatives of benzene oxide were tested, bromo substituents on cyclohexane oxides increased their mutagenicity.[72]

There are numerous reports on the covalent binding of benzene derivatives. A commercial mixture of PCBs became bound to macromolecules in rat, mouse, or rabbit liver microsomal preparations after bioactivation, but not to rat kidney or lung microsomes.[74] Induction of cytochromes *P*-450 increased the binding, and PB induction was more effective than 3-MC induction in liver; in kidney microsomes, PB induction was ineffective, but 3-MC induction increased binding.[74] Clearly, the cytochrome *P*-450 isozyme composition affects the binding of the activated compounds. In Chinese hamster ovary cells, bioactivated 4-chlorobiphenyl becomes covalently bound to protein (85% of total binding), DNA, and RNA; the specific binding to DNA is 3.5 times that to protein.[75] Benzene itself covalently binds to DNA after bioactivation,[76] and *in vivo* in rats binds to both liver and bone marrow.[77] A new technique, involving dialysis in the presence of detergent, has been used to demonstrate that, after bioactivation, bromobenzene binds to macromolecules in primary hepatocytes or microsomes.[78] There is, however, no clear relationship between the toxicity and covalent binding of benzene derivatives. The hepatotoxicities of a series of substituted benzenes decreases in the order: *o*-bromobenzonitrile, bromobenzene, *o*-bromotoluene, and *o*-bromobenzotrifluoride. However, the extent of *in vitro* covalent binding of these compounds in rat microsomal preparations did not follow this order, and *o*-bromotoluene was the most extensively bound.[79]

The nucleophilic attack on benzene oxide by a variety of nitrogen, oxygen, and sulfur nucleophiles has been extensively investigated.[80] The attack by water on the oxide of 1,4-dimethylbenzene was described previously. Additions of oxygen nucleophiles to benzene oxides had previously been detected only in the case of alkyl-substituted oxirane rings.[80] Alkoxides, added very slowly to benzene oxide, and hydrogen peroxide, in large excess under basic conditions, yielded the 1,2-adduct, which on reduction yielded the dihydrodiol of benzene.

Only fairly polarizable nitrogen nucleophiles added to benzene oxide. Ammonia and amino compounds did not react, whereas azide added readily to both 1,2- and 1,6-positions in a ratio of 3:2.[80] The sulfur nucleophile, sodium thiophenoxide, added to 1,2-positions of benzene oxide, as did thioethanol and thioacetate. The reactions tested in this study were highly stereospecific, which is indicative of nucleophilic ring opening rather than the trapping of intermediate ring-opened carbonium ions. However, carbonium ions involved in tight ion pairs could also explain the stereospecificity. This study[80] highlights the complexity of the problem of determining what specific interactions occur between cellular macromolecules and activated benzenes.

The newer and previously discussed results indicate that at least some of the toxic effects of benzene derivatives arise from a secondary activation of phenolic metabolites, and this may possibly be a widespread phenomenon. Certainly many of the published studies have not tested for this possibility, and further studies are required to resolve this question.

## IV. Enzymology of Bioactivation

The enzyme system involved in the bioactivation of benzene and its derivatives in mammalian systems is the microsomal mixed-function oxidase system. The system comprises NADPH, NADPH–cytochrome $P$-450 reductase, a FAD- and FMN- linked protein, and an as yet unknown number of isozymes of cytochrome $P$-450. This system has been extensively reviewed.[41,81-83]

Epoxide hydrolase, the enzyme involved in the detoxification of arene oxides by hydration to dihydrodiols, has also been extensively reviewed.[84-86]

## References

1. Tice, R. R., Costa, D. L., and Drew, R. T. (1980). Cytogenic effects of inhaled benzene in murine bone marrow: Induction of sister chromatid exchanges, chromosomal aberrations, and cellular proliferation inhibition in DBA/2 mice. *Proc. Natl. Acad. Sci. U.S.A.* **77**, 2148–2152.
2. Sammett, D., Lee, E. W., Kocsis, J. J., and Snyder, R. (1979). Partial hepatectomy reduces both metabolism and toxicity of benzene. *J. Toxicol. Environ. Health* **5**, 785–792.

3. Snyder, R., and Kocsis, J. J. (1975). Current concepts of chronic benzene toxicity. *CRC Crit. Rev. Toxicol.* **3**, 265–288.
4. Vigliani, E. C., and Saita, G. (1964). Benzene and leukemia. *N. Engl. J. Med.* **271**, 872–876.
5. Forni, A., and Moreo, L. (1967). Cytogenetic studies in a case of benzene leukaemia. *Eur. J. Cancer* **3**, 251–255.
6. Saita, G. (1973). Benzene-induced hypoplastic anaemia and leukaemias. *In* "Blood Disorders Due to Drugs and Other Agents" (R. H. Girwood, ed.), pp. 127–146. Exerpta Medica, Amsterdam.
7. Selleyei, M., and Kelemen, E. (1971). Chromosome study in a case of granulocytic leukaemia with "Pelgerisation" 7 years after benzene pancytopenia. *Eur. J. Cancer* **7**, 83–85.
8. Girard, R., Tolot, F., and Bourret, J. (1971). Malignant haemopathies and benzene poisoning. *Med. Lav.* **62**, 71–76.
9. Aksoy, M., Erdem, S., Dincol, T., Hepyuksel, T., and Dincol, G. (1974). Chronic exposure to benzene as a possible contributory etiologic factor in Hodgkins' disease. *Blut* **28**, 293–298.
10. Lyon, J. P. (1976). Mutagenicity studies with benzene. *Diss. Abstr. Int. B* **36**, 5537.
11. Lyon, J. P. (1975). Mutagenicity studies with benzene. Ph.D. Thesis, University of California.
12. McConnell, E. E. (1980). Acute and chronic toxicity, carcinogenesis, reproduction, teratogenesis and mutagenesis in animals. *In* "Halogenated Biphenyls, Terphenyls, Naphthalenes, Dibenzodioxins and Related Products" (R. D. Kimbrough, ed.), pp. 109–150. Elsevier/North-Holland Biomedical Press, Amsterdam.
13. Kimbrough, R. D., Squire, R. A., Linder, R. E., Strandberg, J. D., Montali, R. J., and Burse, V. W. (1975). Induction of liver tumours in Sherman strain female rats by polychlorinated biphenyl Aroclor 1260. *J. Natl. Cancer Inst. (U.S.)* **55**, 1453–1459.
14. Kimbrough, R. D., and Linder, R. E. (1974). Induction of adenofibrosis and hepatomas of the liver of BALB/cd mice by polychlorinated biphenyl (Aroclor 1254). *J. Natl. Cancer Inst. (U.S.)* **53**, 547–552.
15. Kimbrough, R. D., Burse, V. W., and Liddle, J. A. (1978). Persistent liver lesions in rats after a single oral dose of polybrominated biphenyls (Firemaster FF-1) and concomitant PBB tissue levels. *Environ. Health Perspect.* **23**, 265–273.
16. Wyndham, C., Devenish, J., and Safe, S. (1976). The in vitro metabolism, macromolecular binding and bacterial mutagenicity of 4-chlorobiphenyl, a model PCB substrate. *Res. Commun. Chem. Pathol. Pharmacol.* **15**, 563–570.
17. Kohli, J., Wyndham, C., Smylie, M., and Safe, S. (1978). Metabolism of bromobiphenyls. *Biochem. Pharmacol.* **27**, 1245–1249.
18. Chandler, G. S., and Craig, D. P. (1971). Aromatic character and the benzene nucleus. *In* "Rodd's Chemistry of Carbon Compounds" (S. Coffey, ed.), Vol 3, Part A, pp. 5–42. Elsevier, Amsterdam.
19. De La Mare, P. B. D. (1971). Electrophilic aromatic substitution. *In* "Rodd's Chemistry of Carbon Compounds" (S. Coffey, ed.), Vol. 2, Part A, pp. 45–88. Elsevier, Amsterdam.
20. Chapman, N. B. (1971). Nucleophilic aromatic substitution. *In* "Rodd's Chemistry of Carbon Compounds" (S. Coffey, ed.), Vol. 3, Part A, pp. 89–111. Elsevier, Amsterdam.
21. Hey. D. H., and Williams, G. H. (1971). Homolytic aromatic substitution. *In* "Rodd's Chemistry of Carbon Compounds" (S. Coffey, ed.), Vol. 3 Part A, pp. 113–129. Elsevier, Amsterdam.
22. Kaminsky, L. S., Kennedy, M. W., Adams, S. M., and Guengerich, F. P. (1982). Metabolism of dichlorobiphenyls by highly purified isozymes of rat liver cytochrome $P$-450. *Biochemistry* **20**, 7379–7384.

23. Kennedy, M. W., Carpentier, N. K., Dymerski, P. P., and Kaminsky, L. S. (1981). Metabolism of dichlorobiphenyls by hepatic microsomal cytochrome P-450. *Biochem. Pharmacol.* **30**, 577–588.
24. Kennedy, M. W., Carpentier, N. K., Dymerski, P. P., Adams, S. M., and Kaminsky, L. S. (1980). Metabolism of monochlorobiphenyls by hepatic microsomal cytochrome P-450. *Biochem. Pharmacol.* **29**, 727–736.
25. Robertson, G. B. (1961). Crystal and molecular structure of diphenyl. *Nature (London)* **191**, 593–594.
26. Newman, M. S. (1956). "Steric Effects in Organic Chemistry." Wiley, New York.
27. Farbrot, E. M., and Shanke, P. N. (1970). A theoretical study of the equilibrium conformation and the barrier to internal rotation in some fluorobiphenyls. *Acta Chem. Scand.* **24**, 3645–3654.
28. Wilson, N. K. (1975). Carbon-13 nuclear magnetic resonance. $^{13}C$ Shieldings and spin-lattice relaxation times in chlorinated biphenyls. *J. Am. Chem. Soc.* **97**, 3573–3579.
29. Boyland, E. (1950). The biological significance of metabolism of polycyclic compounds. *In* "Biological Oxidation of Aromatic Rings" (R. T. Williams, ed.), pp. 40–54. Cambridge Univ. Press, London and New York.
30. Jerina, D., Daly, J., Witkop, B., Zaltzman-Nirenberg, P., and Udenfriend, S. (1968). Role of the arene oxide–oxepin system in the metabolism of aromatic substrates. *Arch. Biochem. Biophys.* **128**, 176–183.
31. Jerina, D. M., and Daly, J. W. (1974). Arene oxides: A new aspect of drug metabolism. *Science* **185**, 573–582.
32. Tunek, A., Platt, K. L., Bentley, P., and Oesch, F. (1978). Microsomal metabolism of benzene to species irreversibly binding to microsomal protein and effects of modifications of this metabolism. *Mol. Pharmacol.* **14**, 920–929.
33. Greenlee, W. F., Sun, J. D., and Bus, J. S. (1981). A proposed mechanism of benzene toxicity: Formation of reactive intermediates from polyphenol metabolites. *Toxicol. Appl. Pharmacol.* **59**, 187–195.
34. Tunek, A., Platt, K. L., Przybylski, M., and Oesch, F. (1980). Multistep metabolic activation of benzene. Effect of superoxide dismutase on covalent binding to microsomal macromolecules and identification of glutathione conjugates using high-pressure liquid chromatography and field desorption mass spectrometry. *Chem.-Biol. Interact.* **33**, 1–17.
35. Hesse, S., Mezger, M., and Wolff, T. (1978). Activation of [$^{14}C$]chlorobiphenyls to protein binding metabolites by rat liver microsomes. *Chem.-Biol. Interact.* **20**, 355–365.
36. Tunek, A., Olofsson, T., and Berlin, M. (1981). Toxic effects of benzene metabolites on granulopoietic stem cells and bone marrow cellularity in mice. *Toxicol. Appl. Pharmacol.* **59**, 149–156.
37. Forque, S. T., Preston, B. D., Hargraves, W. A., Reich, I. L., and Allen, J. R. (1979). Direct evidence that an arene oxide is a metabolic intermediate of 2,2′,5,5′-tetrachlorobiphenyl. *Biochem. Biophys. Res. Commun.* **91**, 475–483.
38. Preston, B. D., and Allen, J. R. (1981). 2,2′,5,5′-tetrachlorobiphenyl: Isolation and identification of metabolites generated by rat liver microsomes. *Drug Metab. Dispos.* **8**, 197–204.
39. Preston, B. D., Miller, J. A., and Miller, E. C. (1983). Nonarene oxide aromatic ring hydroxylation of 2,2′,5,5′,-tetrachlorobiphenyl as the major metabolic pathway catalyzed by phenobarbital-induced rat liver microsomes. *J. Biol. Chem.* **258**, 8304–8311.
40. Selander, H. G., Jerina, D. M., Piccolo, D. E., and Berchtold, G. A. (1975). Synthesis of 3- and 4-chlorobenzene oxides. Unexpected trapping results during metabolism of [$^{14}C$]chlorobenzene by hepatic microsomes. *J. Am. Chem. Soc.* **97**, 4428–4430.

41. Guengerich, F. P. (1979). Isolation and purification of cytochrome *P*-450, and the existence of multiple forms. *Pharmacol. Ther.* **6**, 99–121.
42. White, R. E., and Coon, M. J. (1980). Oxygen activation by cytochrome *P*-450. *Annu. Rev. Biochem.* **49**, 315–356.
43. Hyano, M., Lindberg, M. C., Dorfman, R. I., Hancock, J. E. H., and Doring, W. von E. (1955). On the mechanism of the C-11$\beta$-hydroxylation of steroids; A study with $^{18}H_2O$ and $^{18}O_2$. *Arch. Biochem. Biophys.* **59**, 529–532.
44. Tomaszewski, J. E., Jerina, D. M., and Daly, J. W. (1975). Deuterium isotope effects during formation of phenols by hepatic monooxygenases. Evidence for an alternative to the arene oxide pathway. *Biochemistry* **3**, 2024–2031.
45. Groves, J. T., Kruper, W. J., Jr., and Haushalter, R. C. (1980). Hydrocarbon oxidations with oxometalloporphinates. Isolation and reactions of a (porphinate)manganese (V) complex. *J. Am. Chem. Soc.* **102**, 6375–6377.
46. Rahimutula, A. D., O'Brian, P. J., Seigfried, H. E., and Jerina, D. M. (1978). The mechanism of action of cytochrome *P*-450. Occurrence of the "NIH shift" during hydroperoxide-dependent aromatic hydroxylations. *Eur. J. Biochem.* **89**, 133–141.
47. Burke, M. D., and Mayer, R. T. (1975). Inherent specificities of purified cytochromes *P*-450 and *P*-448 toward biphenyl hydroxylation and ethoxy resorufin deethylation. *Drug Metab. Dispos.* **3**, 245–252.
48. Lau, S., and Zannoni, V. G. (1979). Hepatic microsomal epoxidation of bromobenzene to phenols and its toxicological implication. *Toxicol. Appl. Pharmacol.* **50**, 309–318.
49. Lau, S., and Zannoni, V. G. (1981). Bromobenzene metabolism in the rabbit. Specific forms of cytochrome *P*-450 involved in 2,3- and 3,4-epoxidation. *Mol. Pharmacol.* **20**, 234–235.
50. Lau, S., Abrams, G. D., and Zannoni, V. G. (1980). Metabolic activation and detoxification of bromobenzene leading to cytotoxicity. *J. Pharmacol. Exp. Ther.* **214**, 703–708.
51. Longacre, S. L., Kocsis, J. J., and Snyder, R. (1981). Influence of strain differences in mice on the metabolism and toxicity of benzene. *Toxicol. Appl. Pharmacol.* **60**, 398–409.
52. Daly, J. W., Jerina, D. M., and Witkop, B. (1972). Arene oxides and the NIH shift: The metabolism, toxicity and carcinogenicity of aromatic compounds. *Experientia* **28**, 1129–1149.
53. Rusch, G. M., Leong, B. K. J., and Laskin, S. (1977). Benzene metabolism. *In* "Benzene Toxicity: A Critical Evaluation" (S. Laskin and B. Goldstein, eds.), pp. 23–36. Hemisphere Publ. Corp., Washington, D.C.
54. Lu, A. Y. H., and Miwa, G. T. (1980). Molecular properties and biological functions of microsomal epoxide hydrase. *Annu. Rev. Pharmacol. Toxicol.* **20**, 513–531.
55. Oesch, F., Jerina, D. M., and Daly, J. W. (1971). Substrate specificity of hepatic epoxide hydrase in microsomes and in a purified preparation: Evidence for homologous enzymes. *Arch. Biochem. Biophys.* **144**, 253–261.
56. Jerina, D. M., Daly, J. W., and Witkop, B. (1968). The role of arene oxide–oxepin systems in the metabolism of aromatic substrates. II. Synthesis of 3,4-toluene-4-$^2$H oxide and subsequent "NIH shift" to 4-hydroxytoluene-3-$^2$H. *J. Am. Chem. Soc.* **90**, 6523–6525.
57. Oesch, F., Kaubisch, N., Jerina, D. M., and Daly, J. W. (1971). Hepatic epoxide hydrase. Structure–activity relationships for substrates and inhibitors. *Biochemistry* **10**, 4858–4866.
58. Kaminsky, L. S., Kennedy, M. W., and Guengerich, F. P. (1981). Differences in the functional interaction of two purified cytochrome *P*-450 isozymes with epoxide hydrolase. *J. Biol. Chem.* **256**, 6359–6362.
59. Oesch, F., Jerina, D. M., Daly, J. W., Lu, A. Y. H., Kuntzman, R., and Conney, A. H.

(1972). A reconstituted enzyme system that converts naphthalene to *trans*-1,2-dihydroxy-1,2-dihydronaphthalene via naphthalene-1,2-oxide: Presence of epoxide hydrase in cytochrome *P*-450 and *P*-448 fractions. *Arch. Biochem. Biophys.* **153,** 62–67.
60. Guroff, G., Daly, J. W., Jerina, D. M., Renson, J., Witkop, B., and Udenfriend, S. (1967). Hydroxylation-induced migration: The NIH shift. *Science* **157,** 1524–1530.
61. Kasperek, G. J., Bruice, T. C., Yagi, H., and Jerina, D. M. (1972). Differentiation between the concerted and stepwise mechanisms for aromatization (NIH shift) of arene oxides. *J. Chem. Soc., Chem. Commun.* pp. 784–785.
62. Kasperek, G. J., and Bruice, T. C. (1972). The mechanism of the aromatization of arene oxides. *J. Am. Chem. Soc.* **94,** 198–202.
63. Selander, H. G., Jerina, D. M., and Daly, J. W. (1975). Metabolism of chlorobenzene with hepatic microsomes and solubilized cytochrome *P*-450 systems. *Arch. Biochem. Biophys.* **168,** 309–321.
64. Bush, E. D., and Trager, W. F. (1982). Evidence against an abstraction or direct insertion mechanism for cytochrome *P*-450 catalyzed meta hydroxylations. *Biochem. Biophys. Res. Commun.* **104,** 626–632.
65. Jerina, D. M., Kaubisch, N., and Daly, J. W. (1971). Arene oxides as intermediates in the metabolism of aromatic substrates: Alkyl and oxygen migrations during isomerization of alkylated arene oxides. *Proc. Natl. Acad. Sci. U.S.A.* **68,** 2545–2548.
66. Guengerich, F. P., Fu, P. P., MacDonald, T. L., Kaminsky, L. S., and Adams, S. M. (1983). Applications of theoretical chemistry in prediction of metabolism. *TIPS* **4,** 443–446.
67. Chao, H. S. I., Berchtold, G. A., Boyd, D. R., Dynak, J. N., Tomaszewski, J. E., Yagi, H., and Jerina, D. M. (1981). Migration and retention of deuterium on aromatization of toluene 1,2-oxide and 2,3-oxide to *o*-cresol. *J. Org. Chem.* **46,** 1948–1950.
68. Kasperek, G. J., Bruice, T. C., Yagi, H., Kaubisch, N., and Jerina, D. M. (1972). Solvolytic chemistry of 1,4-dimethylbenzene oxide. A new and novel mechanism for the NIH shift. *J. Am. Chem. Soc.* **94,** 7876–7882.
69. Yagi, H., Jerina, D. M., Kasperek, G. J., and Bruice, T. C. (1972). A novel mechanism for the NIH shift. *Proc. Natl. Acad. Sci. U.S.A.* **69,** 1985–1986.
70. Marsh, M., and Jerina, D. M. (1978). Calculated properties of arene oxides of biological interest. 1. Molecular orbital examination and simple models. *J. Med. Chem.* **21,** 1298–1301.
71. Jeffrey, A. M., and Jerina, D. M. (1975). Novel rearrangement during dehydration of nucleophile adducts of arene oxides. A reappraisal of premercapturic acid structures. *J. Am. Chem. Soc.* **97,** 4427–4428.
72. Jung, R., Beermann, D., Glatt, H. R., and Oesch, F. (1981). Mutagenicity of structurally related oxiranes derivatives of benzene and its hydrogenated congeners. *Mutat. Res.* **81,** 11–19.
73. Vogel, E., and Gunther, H. (1976). Benzoloxidoxepin-valenztautomerie. *Angew. Chem.* **79,** 429–446.
74. Shimada, T., and Sato, R. (1978). Covalent binding in vitro of polychlorinated biphenyls to microsomal macromolecules. *Biochem. Pharmacol.* **27,** 585–593.
75. Wong, A., Basrur, P., and Safe, S. (1979). The metabolically mediated DNA damage and subsequent DNA repair by 4-chlorobiphenyl in Chinese hamster ovary cells. *Res. Commun. Chem. Pathol. Pharmacol.* **24,** 543–550.
76. Lutz, W. K., and Schlatter, C. H. (1977). Mechanism of the carcinogenic action of benzene: Irreversible binding to rat liver DNA. *Chem.-Biol. Interact.* **18,** 241–245.
77. Snyder, R., Lee, E. W., and Kocsis, J. J. (1978). Binding of labeled benzene metabolites to mouse liver and bone marrow. *Res. Commun. Chem. Pathol. Pharmacol.* **20,** 191–194.

78. Sun, J. D., and Dent, J. G. (1980). A new method for measuring covalent binding of chemicals to cellular macromolecules. *Chem.-Biol. Interact.* **32,** 41–61.
79. Wiley, R. A., Hanzlik, R. P., and Gillesse, T. (1979). Effect of substituents on in vitro metabolism and covalent binding of substituted bromobenzenes. *Toxicol. Appl. Pharmacol.* **49,** 249–255.
80. Jeffrey, A. M., Yeh, H. J. C., Jerina, D. M., DeMarinis, R. M., Foster, C. H., Piccolo, D. E., and Berchtold, G. A. (1974). Stereochemical course in reactions between nucleophiles and arene oxides. *J. Am. Chem. Soc.* **96,** 6929–6937.
81. Paine, A. (1981). Hepatic cytochrome $P$-450. *Essays Biochem.* **17,** 85–126.
82. Lu, A. Y. H., and West, S. B. (1980). Multiplicity of mammalian microsomal cytochromes $P$-450. *Pharmacol. Rev.* **32,** 277–295.
83. Johnson, E. F. (1979). Multiple forms of cytochrome $P$-450; Criteria and significance. *Rev. Biochem. Toxicol.* **I,** 1–26.
84. Guengerich, F. P. (1982). Epoxide hydrolase: Properties and metabolic rates. *Rev. Biochem. Toxicol.* **4,** 5–30.
85. Oesch, F. (1979). Epoxide hydratase. *Prog. Drug Metab.* **3,** 253–301.
86. Lu, A. Y. H., and Miwa, G. T. (1980). Molecular properties and biological functions of microsomal epoxide hydrase. *Annu. Rev. Pharmacol. Toxicol.* **20,** 513–531.

Chapter 7

# Polycyclic Aromatic Hydrocarbons: Metabolic Activation to Ultimate Carcinogens

Dhiren R. Thakker, Haruhiko Yagi, Wayne Levin,* Alexander W. Wood,* Allan H. Conney,* and Donald M. Jerina

*Laboratory of Bioorganic Chemistry*
*National Institute of Arthritis, Diabetes, and Digestive and Kidney Diseases*
*National Institutes of Health*
*Bethesda, Maryland*

*\*Department of Experimental Carcinogenesis and Metabolism*
*Hoffmann–La Roche Inc.*
*Nutley, New Jersey*

|       |                                                                   |     |
| ----- | ----------------------------------------------------------------- | --- |
| I.    | Introduction                                                      | 178 |
| II.   | Metabolic Pathways of Polycyclic Aromatic Hydrocarbons            | 180 |
| III.  | Metabolism of Benzo[*a*]pyrene                                    | 181 |
|       | A. General                                                        | 181 |
|       | B. Arene Oxides                                                   | 182 |
|       | C. Dihydrodiols                                                   | 185 |
|       | D. Phenols and Quinones                                           | 193 |
|       | E. Glutathione Conjugates                                         | 194 |
|       | F. Sulfates and Glucuronides                                      | 195 |
| IV.   | Biological Activity of Benzo[*a*]pyrene Derivatives               | 195 |
|       | A. Covalent Binding to Cellular Macromolecules                    | 196 |
|       | B. Mutagenicity of Benzo[*a*]pyrene Derivatives                   | 197 |
|       | C. Tumorigenicity of Benzo[*a*]pyrene Derivatives                 | 200 |
| V.    | The Bay Region Theory                                             | 203 |
| VI.   | Comparative Metabolism of Polycyclic Aromatic Hydrocarbons        | 206 |
|       | A. Overall Rates of Metabolism                                    | 206 |
|       | B. Comparison of Metabolite Profiles for a Series of Polycyclic Aromatic Hydrocarbons | 208 |

C. Effect of Substituents on the Metabolism and Biological Activity of Polycyclic Aromatic Hydrocarbons . . . . . . 210
VII. Stereoselectivity in the Metabolism of the Polycyclic Aromatic Hydrocarbons. . . . . . . . . . . . . . . . . 212
A. A Model for the Catalytic Site of Cytochrome P-450c . . 212
B. Stereoselectivity in the Metabolic Formation of Dihydrodiols from Polycyclic Aromatic Hydrocarbons Supports the Predictions of the Binding-Site Model . . . . . . . . 214
C. Regio- and Stereoselectivity in the Metabolism of Dihydrodiols of Polycyclic Aromatic Hydrocarbons . . . . . . 216
VIII. Stereoselectivity in the Biological Activity of Polycyclic Aromatic Hydrocarbon Derivatives . . . . . . . . . . 219
IX. Summary . . . . . . . . . . . . . . . . . . 221
References . . . . . . . . . . . . . . . . . 222

## I. Introduction

Polycyclic aromatic hydrocarbons (PAHs)* were first implicated as carcinogenic constituents of coal tar in the early part of the twentieth century,[1] soon after it had been demonstrated that coal tar was carcinogenic to rabbits[2] and mice.[3] A few years later, it was found that a pure pentacyclic aromatic hydrocarbon, dibenz[a,h]anthracene, caused cancer in experimental animals,[4] although this was not the active component in coal tar. Shortly thereafter, benzo[a]pyrene (B[a]P) was isolated from coal tar by Cook et al.[5] and was shown to be highly carcinogenic. Since then, a large number of PAHs have been identified as mutagenic, carcinogenic, and toxic components of our environment. For the pentacyclic hydrocarbons alone, at least 20 different compounds are possible by various arrangements of benzene rings. Not only do these compounds have different physicochemical properties, but their biological activity is also diverse. For example, B[a]P (Fig. 1) is highly tumorigenic to mice, whereas its isomer benzo[e]pyrene (B[e]P) is practically inactive at a severalfold higher dose.[6-10] Thus, the similarity in structure but diversity in biological activity make the PAHs an ideal class of compounds for the study of structure–activity correlations.

* Abbreviations used are as follows: PAHs, polycyclic aromatic hydrocarbons; B[a]P, benzo[a]pyrene; B[e]P, benzo[e]pyrene; B[a]P 4,5-dihydrodiol, trans-4,5-dihydroxy-4,5-dihydro-B[a]P; other dihydrodiols of B[a]P and of other PAHs are similarly abbreviated; B[a]P 7,8-diol 9,10-epoxide-1, (±)-7β,8α-dihydroxy-9β,10β-epoxy-7,8,9,10-tetrahydro-B[a]P; B[a]P 7,8-diol 9,10-epoxide-2, (±)-7β,8α-dihydrodroxy-9α,10α-epoxy-7,8,9,10-tetrahydro-B[a]P; other diol epoxides are similarly abbreviated, so that diol epoxide-1 isomers always have cis stereochemistry between the benzylic hydroxyl group and oxirane oxygen, and diol epoxide-2 isomers have trans stereochemistry between these two groups; B[a]P $H_4$-7,8-epoxide, 7,8-epoxy-7,8,9,10-tetrahydro-B[a]P; B[a]P $H_4$-9,10-epoxide is similarly abbreviated.

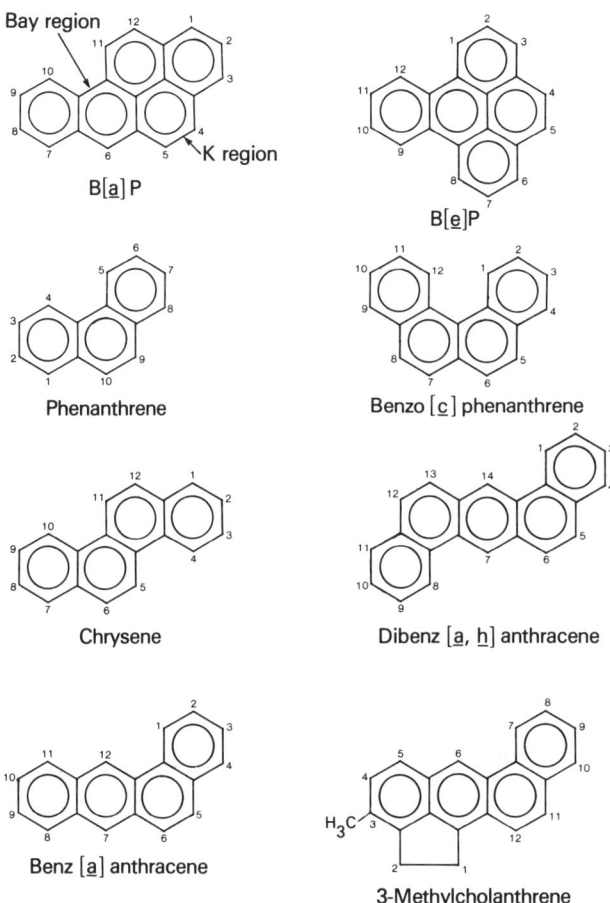

Fig. 1. Structures and numbering of several polycyclic aromatic hydrocarbons.

One of the best-known attempts to correlate physicochemical properties of PAHs with their carcinogenic potency is the Pullmans' application of Hückel-type quantum chemical calculations.[11] The Pullmans attempted to predict the carcinogenic potency of PAHs on the basis of calculated reactivity parameters. This and other structure–activity correlations (refs. 12 13, and references therein), which took into account only the physicochemical properties of the parent hydrocarbons, met with limited success, but stimulated considerable research in the area. Overwhelming evidence has been amassed in recent years indicating that PAHs, like other chemical carcinogens,[14] exert their carcinogenicity on oxidative metabolism to reactive intermediates.[15-21] The interrelationships of metabolic factors and physicochemical properties of the PAHs, which influence their biological activity, are the subject of this chapter.

## II. Metabolic Pathways of Polycyclic Aromatic Hydrocarbons

Certain monooxygenases, the cytochromes P-450, catalyze the initial oxidations of the PAHs. The cytochromes P-450 are localized predominantly in endoplasmic reticulum (cf. ref. 22) and, to a smaller extent, in nuclear membranes of many cell types.[23,24] Oxidation of the PAHs by molecular oxygen under catalysis by cytochromes P-450 requires the participation of an NADPH–cytochrome c (or P-450) reductase. Early studies established that the PAHs are metabolized primarily to phenols and dihydrodiols.[25-27] Later, arene oxides[28] were proposed as possible precursors of dihydrodiols by Boyland.[15] Initial attempts to establish that stable K-region oxides of phenanthrene, benz[a]anthracene, and dibenz[a,h]anthracene were precursors of dihydrodiols were inconclusive.[29,30] The first unequivocal evidence for the formation of an arene oxide as a primary oxidative metabolite was provided by Jerina et al.[31] in their studies with naphthalene. Arene oxides are now established as the principal primary oxidative metabolites of the PAHs, which are subsequently converted to *trans*-dihydrodiols by epoxide hydrolase, to glutathione conjugates by cytosolic glutathione S-transferase,[32] or to phenols by spontaneous isomerization (Fig. 2).[18] The isomerization to phenols is accompanied by migration and retention of substituents from the carbon atom bearing the final phenolic hydroxyl group (NIH shift) (Fig. 2).[33,34] The occurrence of the NIH shift in monooxygenase-catalyzed formation of phenols constitutes important evidence that such aromatic hy-

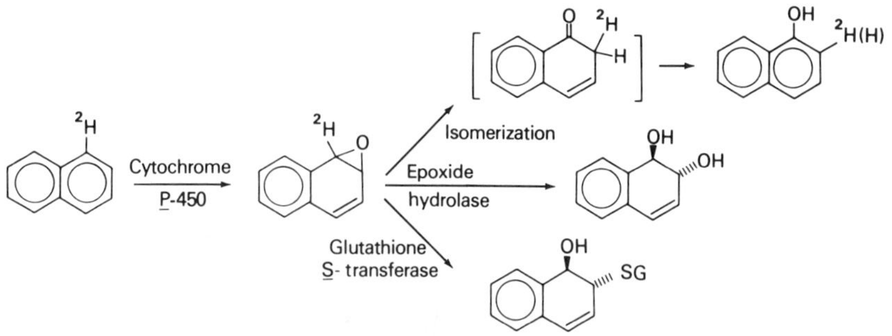

Fig. 2. Major pathways for the disposition of a typical arene oxide. Although the formation of dihydrodiols and glutathione conjugates is enzyme catalyzed, the NIH-shift pathway for the formation of phenols as well as the addition of glutathione occur spontaneously.

droxylations proceed via initial epoxidation of the aromatic substrates. Occurrence of the NIH shift is now considered one of the criteria for mixed-function hydroxylases that act on aromatic hydrocarbons (Fig. 2).

Dihydrodiols and phenols are, in turn, metabolized to a variety of metabolites. They are conjugated by microsomal UDPglucuronosyltransferases to form glucuronides (refs. 35–41, and references therein) and by cytosolic sulfotransferase to form sulfates.[42-48] The dihydrodiols are oxidized by monooxygenases to diol epoxides,[49-63] diol phenols,[53,64,65] bisdihydrodiols,[61,63,65,66] and quinones.[65] Dihydrodiols are also metabolized to catechols by diol dehydrogenases.[67-70] Phenols are oxidized to polyphenolic compounds,[71-73] as well as quinones.[71,74] Thus, the PAHs are metabolized to a wide range of metabolites, some of which represent metabolic detoxication of the parent compound, whereas others are highly toxic, mutagenic, or carcinogenic and are responsible for the adverse biological effects of the PAHs.

## III. Metabolism of Benzo[a]pyrene

### A. GENERAL

Benzo[a]pyrene, a prevalent environmental carcinogen,[75] is the most extensively studied PAH. As with other PAHs, early studies identified several phenols, dihydrodiols, and quinones as metabolites of B[a]P.[26,76-79] However, most of the details concerning the metabolic formation of isomeric dihydrodiols and phenols, as well as other metabolites of B[a]P, have become available only in the past 10 years through the application of high-performance liquid chromatography (HPLC).[80,81] In vitro studies with rat liver microsomes have shown that B[a]P is metabolized to 4,5-, 7,8-, and 9,10-dihydrodiols, to 1-, 3-, 7-, and 9-hydroxy-B[a]P, and to 1,6-, 3,6-, and 6,12-quinones[77,80-85] (Fig. 3). Many of these metabolites are also formed by other mammalian and nonmammalian tissues and organs, including rat lung,[46,86] rat intestinal microsomes,[87] rat liver nuclei,[88-90] mouse liver and lung microsomes,[85,91] mouse skin organ culture,[92] mouse epidermis,[93] hamster embryo cells,[94] hamster lung,[46] hamster embryo cultures,[51] hamster liver and lung microsomes,[95] microsomes from Rhesus monkey liver and lung,[96] human tissues and cells,[97-100] cultured tracheobronchial tissue from mice, rats, hamsters, bovines, and humans,[101] fish liver microsomes,[102] and the filamentous fungus *Cunninghamella elegans*.[103] Several other metabolites of B[a]P have also been reported. For example, 6-hydroxymethyl-B[a]P was claimed to be a metabolite of B[a]P,[104,105] although this metabolite has not been detected in subsequent studies. 6-Oxy-B[a]P has been reported as

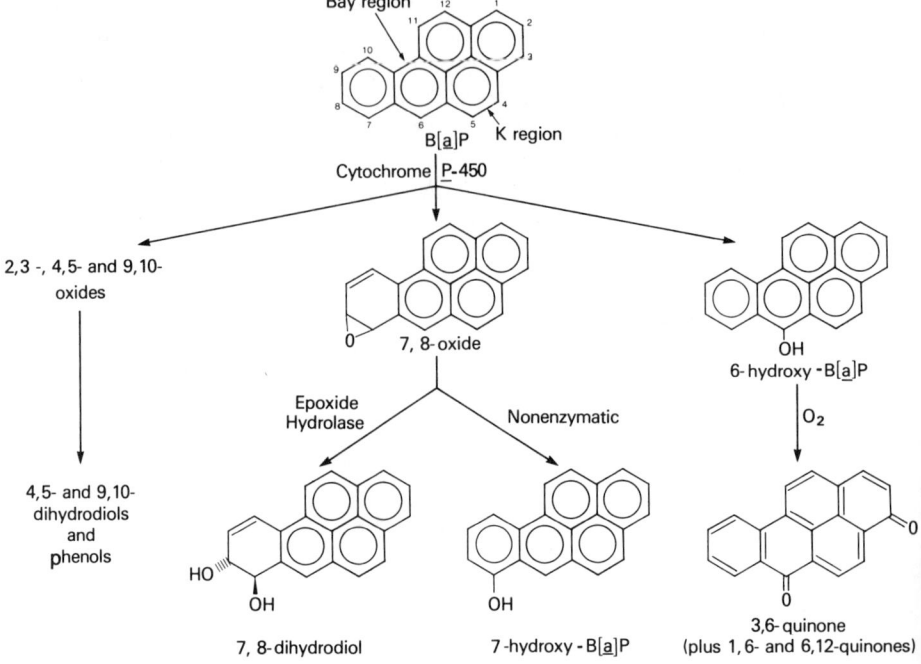

**Fig. 3.** Microsomal metabolism of B[*a*]P. The scheme indicates the major oxidative metabolites formed by cytochromes *P*-450 and epoxide hydrolase.

an intermediate in the formation of quinones.[106-108] B[*a*]P 11,12-dihydrodiol, the most polar dihydrodiol of B[*a*]P[85] has been reported to be a metabolite of B[*a*]P formed by rat skin *in vivo*.[109] The microsomal enzyme UDPglucuronosyltransferase catalyzes glucuronide formation from dihydrodiols and phenols of B[*a*]P. In the presence of cytosolic enzymes, glutathione and sulfate conjugates are also formed. These conjugates are highly polar metabolic products, which facilitate elimination of the hydrocarbon derivatives from the body.

## B. ARENE OXIDES

Arene oxides[28] appear to be the principal primary oxidative metabolites of B[*a*]P[80] formed by cytochrome *P*-450-dependent monooxygenases. Isolation of the K-region 4,5-oxide as a metabolite of B[*a*]P provided direct evidence for its formation.[19,83,110] The non-K-region 7,8- and 9,10-oxides have not been isolated as metabolites because of their instability.[111,112] However, studies with systems containing highly purified cytochrome *P*-450 and epoxide hydrolase, the enzyme that catalyzes hydrolysis of arene oxides to trans-dihydrodiols,[113,114] have clearly established that B[a]P 7,8- and 9,10-oxides are metabolites of B[a]P.[80] These studies have shown that B[a]P is metabolized to phenols and quinones by cytochrome P-450 in the absence of epoxide hydrolase, whereas in the presence of epoxide hydrolase, B[a]P 4,5-, 7,8-, and 9,10-dihydrodiols are formed at the expense of some of the phenols. These results provide unequivocal evidence for the formation of arene oxides as primary metabolites of B[a]P. Studies with 3-deuterio-B[a]P showed that the microsomal metabolism of the substrate yielded 3-hydroxy-B[a]P with 29% retention of deuterium. Although the magnitude of deuterium retention is curiously low, the results implicate B[a]P 2,3-oxide as an intermediate in the formation of 3-hydroxy-B[a]P.[115]

The absolute configurations and the enantiomeric compositions of B[a]P 4,5-oxide[110] and B[a]P 7,8-oxide,[116,117] formed by cytochrome P-450c and rat liver microsomes, respectively, have been determined. The nonenzymatic trans-addition of glutathione to enzymatically synthesized [$^{14}$C]B[a]P 4,5-oxide gave a pair of conjugates that cochromatographed with the glutathione conjugates derived from (+)-B[a]P (4$S$,5$R$)-oxide (Structures 1 and 2, Fig. 4), thus providing evidence for the (4$S$,5$R$)-absolute configuration of the enzymatically formed 4,5-oxide.[110,118] In these experiments, cytochrome P-450c exhibited > 97% preference for oxidation from the bottom face of the prochiral B[a]P molecule (cf. Fig. 13). The synthesis and assignment of absolute stereochemistry to the enantiomers of B[a]P 7,8-oxide by X-ray crystallography[116] have allowed the examination of stereoselectivity of rat liver microsomes in the formation of the 7,8-oxide from B[a]P.[117] Mi-

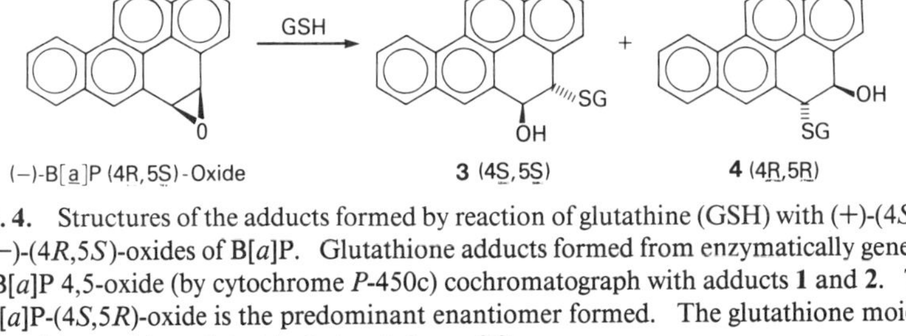

Fig. 4. Structures of the adducts formed by reaction of glutathine (GSH) with (+)-(4$S$,5$R$)- and (−)-(4$R$,5$S$)-oxides of B[a]P. Glutathione adducts formed from enzymatically generated [$^{14}$C]B[a]P 4,5-oxide (by cytochrome P-450c) cochromatograph with adducts 1 and 2. Thus, (+)-B[a]P-(4$S$,5$R$)-oxide is the predominant enantiomer formed. The glutathione moiety in the glutathione adducts 1 to 4 is designated as −SG.

crosomal epoxide hydrolase from rat liver converts (+)-(7R,8S)-B[a]P 7,8-oxide to the (−)-(7R,8R)-dihydrodiol, whereas the (−)-(7S,8R)-oxide is converted to the (+)-(7S,8S)-dihydrodiol[117] (Fig. 5). These findings are consistent with attack by solvent water at C-8 of either enantiomer as determined by [18]O-labeled water studies.[74,119] These results, coupled with the findings that the 7,8-dihydrodiol formed by rat liver microsomes from B[a]P is predominantly the (−)-enantiomer (≥96%)[119] with (R,R)-absolute configuration,[120,121] established that the metabolically formed 7,8-oxide has (7R,8S) configuration. [Yang et al.[74] failed to detect any of the (+)-enantiomer in the B[a]P 7,8-dihydrodiol formed by rat liver microsomes, and thus reported the dihydrodiol to be an optically pure (−)-enantiomer.] However, the high enantiomeric purity of the 7,8-dihydrodiol could also have resulted from formation of 7,8-oxide of low enantiomeric purity, followed by preferential hydration of (+)-B[a]P 7,8-oxide at low substrate conversion.[117] Under conditions simulating the formation of the 7,8-dihydrodiol from B[a]P, racemic 7,8-oxide gave practically racemic dihydrodiol,[117] thus providing unequivocal evidence for the formation of (7R,8S)-

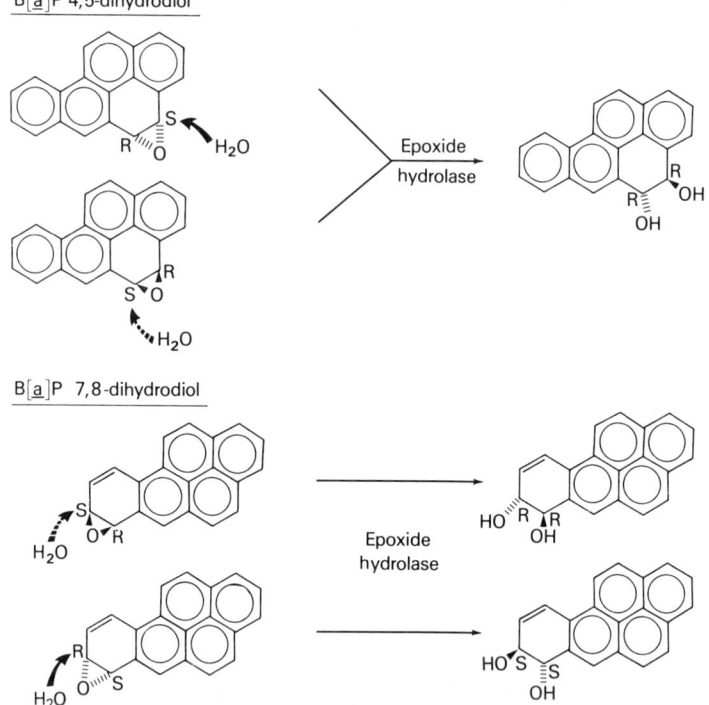

Fig. 5. Regio- and stereoselectivity of epoxide hydrolase in the metabolic formation of B[a]P 4,5- and 7,8-dihydrodiols.

B[a]P 7,8-oxide of high enantiomeric purity from B[a]P by rat liver microsomes.

Although hydrolysis and conjugation appear to be major enzymatic pathways by which the arene oxides are metabolized, reduction of B[a]P arene oxides to the parent hydrocarbon by rat liver microsomes in oxygen-deficient environments and in the presence of trichloropropene oxide (TCPO), an inhibitor of epoxide hydrolase, has been reported.[122-126] This reaction appears to be catalyzed by cytochrome $P$-450 and requires NADPH. The reduction of B[a]P 4,5-oxide was enhanced by four- to six-fold per nanomole of cytochrome $P$-450 by pretreatment of the rats with 3-methylcholanthrene. In the presence of oxygen the oxide reductase activity was only 2.5[126] to 7%[125] of that observed in the absence of oxygen. Although Sugiura et al.[125] found that the reductase activity in rat liver microsomes was enhanced by 105,000 $g$ supernatant only in the presence of oxygen, Wrighton et al.[126] observed stimulation of microsomal reductase activity both in the presence and in the absence of oxygen when washed microsomes were used at low protein concentrations. Evidence for a role for cytochrome $P$-450 in the reduction of B[a]P 4,5-oxide is provided in the observation that a purified, reconstituted cytochrome $P$-450 system catalyzes the reaction.[124] However, Wrighton et al.[126] were unable to observe reductase activity by a purified, reconstituted cytochrome $P$-450 system in the absence of cytosolic proteins.

The demonstration of reduction of B[a]P 7,8-oxide by rat liver microsomes was claimed by Sugiura et al.,[125] who carried out the reaction in the presence of reduced riboflavin. The rate of oxidation of reduced riboflavin was assumed to equal the rate of reduction of B[a]P 7,8-oxide. Because no evidence was presented to demonstrate that B[a]P 7,8-oxide was reduced by rat liver microsomes to B[a]P, these results must be considered highly speculative.

The role of the reduction of arene oxides of the PAHs to their parent hydrocarbons in the overall metabolism of these substrates is uncertain. Because the $K_m$ and $V_{max}$ for reduction of B[a]P 4,5-oxide are 100 $\mu M$ and 3 nmol/min/mg proteins, respectively,[125] and the $K_m$ and $V_{max}$ for hydration of the same substrate are $\leq 1$ $\mu M$ and 387 nmol/min/mg protein, respectively,[114] it is unlikely that the reduction pathway is of physiological significance, at least in this case.

## C. DIHYDRODIOLS

### 1. Metabolic Formation

The further metabolism of arene oxides of B[a]P to *trans*-dihydrodiols by epoxide hydrolase and to glutathione conjugates by glutathione *S*-transferases has been extensively investigated. Studies with benzene oxide provided the first unequivocal evidence for a microsomal enzyme, now called epoxide hydrolase, which catalyzes the addition of water to arene oxides.[113] trans-Dihydrodiols are formed exclusively.[31,68,74,80,113,127] Epoxide hydrolase exhibits a great deal of regio- and stereoselectivity in the metabolism of B[a]P arene oxides. Both the 7,8-oxide[74,119] (Fig. 5) and the 9,10-oxide[74] are hydrolyzed by attack of water exclusively at the allylic positions. However, this is not true for naphthalene 1,2-oxide. Recent studies[128a] show that the minor (−)-(1S,2R)-enantiomer of naphthalene 1,2-oxide formed by cytochrome P-450c (23% of total 1,2-oxide) undergoes epoxide hydrolase-catalyzed hydrolysis by attack of water at both the benzylic (40%) and the allylic (60%) positions. The major (+)-(1R,2S)-enantiomer (73% of total 1,2-oxide formed) is hydrolyzed by attack of water exclusively at the allylic position.[128a] Studies with the enantiomers of B[a]P 7,8-oxide showed that (+)-(7R,8S)- and (−)-(7S,8R)-oxides were metabolized by microsomal epoxide hydrolase to (−)-(7R,8R)- and (+)-(7S,8S)-dihydrodiols, respectively[117]; however, when the racemic oxide was used as a substrate, the (−)-(7R,8R)-dihydrodiol was formed in 80% enantiomeric excess at low substrate conversion.[74,117] The studies by Levin et al.[117] established that the (+)-enantiomer of the 7,8-oxide was preferentially metabolized by rat liver epoxide hydrolase. Interestingly, at saturating substrate concentrations, the (−)-enantiomer of the 7,8-oxide was metabolized by epoxide hydrolase at a rate approximately four times greater than the (+)-enantiomer, whereas the racemic substrate was metabolized at a rate equal to the (+)-enantiomer. Hence, the (+)-enantiomer appeared to have a much higher affinity for the enzyme than the (−)-enantiomer. Further studies with purified epoxide hydrolase have confirmed that the $K_{m_{app}}$ for the (+)-enantiomer is 10- to 15-fold lower than that for the (−)-enantiomer.[117] Thus, in the metabolism of the racemic 7,8-oxide, the high product stereoselectivity of rat liver epoxide hydrolase appears to reside in the much higher affinity of the enzyme for the (+)-enantiomer of the oxide.

Racemic B[a]P 4,5-oxide is metabolized by rat liver epoxide hydrolase to B[a]P 4,5-dihydrodiol of high (92%)[119] to moderate (42%)[74] enantiomeric purity. The unusually low enantiomeric purity of the product in the latter study may be in error, because subsequent studies with (+)- and (−)-enantiomers of B[a]P 4,5-oxide showed that both rat liver microsomes and purified epoxide hydrolase converted these enantiomers to the 4,5-dihydrodiol with 100 and 70% enantiomeric purity, respectively.[129] Further studies showed that the predominant enantiomer of the 4,5-dihydrodiol formed has (R,R)-absolute configuration and a positive $[\alpha]_D$ in tetrahydrofuran.[118] Determination of the position of the enzyme-catalyzed incorporation of $^{18}O$ from $^{18}O$-labeled water into the (+)- and (−)-enantiomers of B[a]P 4,5-oxide was achieved by trifluoroacetylation of the product diol, thermal decomposition of the bis ester to a 50:50 mixture of 4- and 5-trifluoroacetoxy-B[a]P, and gas chromatographic–mass spectrometric analysis of the trifluoroace-

**Table I**

Enantiomeric Composition and Absolute Configuration of the *trans*-Dihydrodiols of Benzo[a]pyrene (B[a]P) Formed by Liver Microsomes from Control and Treated Long–Evans Rats[a]

| Dihydrodiol | Treatment | Enantiomeric composition (%) | |
|---|---|---|---|
| | | (R,R) | (S,S) |
| B[a]P 4,5- | 3-Methylcholanthrene | 96 | 4 |
| B[a]P 9,10- | 3-Methylcholanthrene | 96 | 4 |
| B[a]P 7,8- | 3-Methylcholanthrene | 96 | 4 |
| | Phenobarbital | 92 | 8 |
| | None | 93 | 7 |

[a] Data obtained from Thakker et al.[53] and Jerina et al.[138] Absolute configuration of B[a]P 4,5-, 7,8-, and 9,10-dihydrodiols was reported by Kedzierski et al.,[118] Yagi et al.,[121] and Yagi and Jerina,[138a] respectively.

tates.[110] The results of the study established that preferential attack of water occurs at the 4-position of the (+)-oxide and the 5-position of the (−)-oxide, giving (+)-(4R,5R)-dihydrodiol in either instance (Fig. 5). Thus, epoxide hydrolase exhibits a high degree of stereoselectivity in directing the attack of water almost exclusively at the carbon atom with S-configuration (Fig. 5), as has been the case for several other epoxide hydrolase substrates.[130-132] The results of Armstrong et al.[110] are in contrast to the claim by Yang et al.[74] that both enantiomers of B[a]P 4,5-oxide are hydrated by cleavage of $C_4-O$ bonds. The conclusions of Yang et al. were derived from experiments that included hydration of the oxide in $^{18}O$-labeled water, dehydration of the *trans*-4,5-dihydrodiol to phenols, identification of 5-hydroxy-B[a]P as the only dehydration product by HPLC, and determination of $^{18}O$ in the phenols. The failure of Yang et al.[74] to recognize (1) that both 4- and 5-hydroxy-B[a]P are formed on acid-catalyzed dehydration of B[a]P 4,5-dihydrodiol (because of their inability to separate the 4- and 5-hydroxy-B[a]P) and (2) that large losses of label from the phenols occur during acid-catalyzed dehydration of 4,5-diol,[119] may provide the basis for their incorrect finding.

The overall metabolic conversion of B[a]P to the 4,5-, 7,8-, and 9,10-dihydrodiols by microsomes from 3-methylcholanthrene-treated rats occurs with high overall stereoselectivity[74,119] (Table I). In the case of B[a]P 7,8-dihydrodiol, a proximate carcinogen of B[a]P,[133-137] prior treatment of the animals with inducers has little effect on the enantiomeric composition of the dihydrodiol (Table I).[138] Formation of highly enantiomerically enriched 4,5- and 7,8-dihydrodiols is the result of high stereoselectivity of both cytochrome $P$-450[110,117] and epoxide hydrolase.[117,119,129]

## 2. Further Metabolism of Dihydrodiols

a. BENZO[a]PYRENE 7,8-DIHYDRODIOL. The first indication that B[a]P 7,8-dihydrodiol is further metabolized by rat liver microsomes to a reactive metabolite came from a study by Borgen et al.,[139] which demonstrated that microsomal metabolites of B[a]P 7,8-dihydrodiol became bound covalently to DNA to about a 10-fold greater extent than did B[a]P or other metabolites of B[a]P. Sims et al.[49] provided evidence that a 7,8-diol 9,10-epoxide was the reactive metabolite responsible for the DNA binding. Subsequently, synthesis of a diol epoxide by a method analogous to that used by Sims et al.[49] was reported.[140] In both studies, however, the stereochemistry of the product diol epoxides was not elucidated. Concurrently, Yagi et al.[141] reported the synthesis of the bay region 7,8-diol 9,10-epoxide-1 with cis stereochemistry between the benzylic hydroxyl group and the oxirane ring, and of the diastereomer (7,8-diol 9,10-epoxide-2), in which these groups are trans. Synthesis of both diastereomers and the elucidation of stereochemistry were presented at the Symposium on Biological Reactive Intermediates (July 1975) and at the Battelle Symposium on Polynuclear Aromatic Hydrocarbons (October 1975), and were later published in the proceedings of these symposia.[142,143] The synthesis of both diastereomers by practically an identical method was subsequently reported by Beland and Harvey.[144] The role of stereochemical factors in the synthesis of the diastereomeric diol epoxides has been discussed in detail by Yagi et al.[145] The high mutagenicity of racemic diol epoxide-1 (cf. Fig. 6) toward *Salmonella typhimurium* strains TA 98 and TA 100 and toward Chinese hamster V79 cells[146] was also demonstrated. Later, both diastereomeric diol epoxides (cf. Fig. 6) were tested for their mutagenic potency toward *S. typhimurium* strains TA 98, TA 100, and TA 1538,[147] and toward Chinese hamster V79 cells,[147,148] and were found to be highly mutagenic, with the diol epoxide-1 showing higher mutagenic activity toward bacterial cells and the diol epoxide-2 being more mutagenic toward the mammalian cells. Huberman et al.[51] also found the diastereomer 2 to be highly mutagenic toward Chinese hamster V79 cells; however, mutagenicity and toxicity of diol epoxide-1 reported in this study was much less than that reported by Wislocki et al.[146] and by Wood et al.[147] Previously, Malaveille et al.[149] had reported mutagenic activity of a B[a]P 7,8-diol 9,10-epoxide toward strain TA 100 of *S. typhimurium*. Although the purity and relative stereo-chemistry of the compound were not defined, the method of synthesis[49] would be expected to yield diol epoxide-2.[145] The mutagenic activity of this diol epoxide[149] was less than 10% of the activity of diol epoxide-2 reported elsewhere.[146,147] Tumorigenicity studies in newborn mice[134,136] provided the most direct and strongest evidence that B[a]P 7,8-diol 9,10-epoxide-2 is an ultimate carcinogen derived from B[a]P.

Metabolism studies with synthetic racemic B[a]P 7,8-dihydrodiol by rat

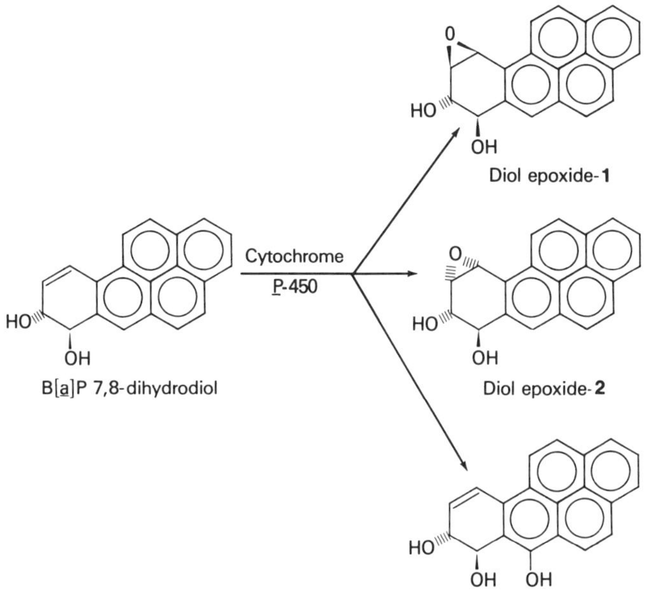

# 7. POLYCYCLIC AROMATIC HYDROCARBONS

**Fig. 6.** Metabolism of racemic B[a]P 7,8-dihydrodiol by rat liver cytochromes P-450. The structure of the diol phenol is tentative. It is formed in large amounts (21–37% of total metabolites) by liver microsomes from control and phenobarbital-treated rats, but not by liver microsomes from 3-methyl-cholanthrene-treated rats (4–6%). Absolute configuration is not implied.

liver microsomes[50] and with metabolically formed 7,8-dihydrodiol by mammalian cells[51] provided direct evidence that the highly mutagenic 7,8-diol 9,10-epoxides were the principal metabolites of these substrates (Fig. 6). The racemic dihydrodiol was metabolized to both diastereomers of the diol epoxides in approximately equal proportions by rat liver microsomes.[50] The bay region diol epoxides* constituted 59–95% of the total metabolites of B[a]P 7,8-dihydrodiol formed by microsomes from control and induced rats (Table II). Microsomes from 3-methylcholanthrene-treated rats, which contain approximately 70% of their total cytochromes P-450 as cytochrome P-450c,[150,151] showed the highest degree of regioselectivity in that diol epoxides constitute over 90% of the total metabolites. Purified cytochrome P-450c showed a similarly high regioselectivity (bay region diol epoxides constitute 92% of the total metabolites). Microsomes from control and

---

* The term bay region is used to describe the sterically hindered region between positions 10 and 11 of B[a]P. All the PAHS shown in Fig. 1 contain one or more bay regions. The bay region of benzo[c]phenanthrene is even more hindered than a typical bay region.

Table II

Regio- and Stereoselectivity on Metabolism of the (−)-(R,R)- and (+)-(S,S)-Enantiomers of Benzo[a]pyrene 7,8-Dihydrodiol to Bay Region Diol Epoxides by Rat Liver Microsomes and by a Purified Monooxygenase System Reconstituted with Cytochrome P-450c.[a]

| Substrate | Enzyme preparation[b] | Bay region diol epoxides (% of total metabolites) | Relative amounts of | | | |
|---|---|---|---|---|---|---|
| | | | (+)-Diol epoxide-1 | (−)-Diol epoxide-1 | (+)-Diol epoxide-2 | (−)-Diol epoxide-2 |
| (−)-(7R,8R)-B[a]P 7,8-dihydrodiol | Microsomes | | | | | |
| | Control | 76 | — | 30 | 70 | — |
| | Phenobarbital | 66 | — | 29 | 71 | — |
| | 3-Methylcholanthrene | 95 | — | 14 | 86 | — |
| | Reconstituted system (cytochrome P-450c) | 93 | — | 18 | 82 | — |
| (+)-(7S,8S)-B[a]P 7,8-dihydrodiol | Microsomes | | | | | |
| | Control | 59 | 86 | — | — | 14 |
| | Phenobarbital | 61 | 84 | — | — | 16 |
| | 3-Methylcholanthrene | 90 | 97 | — | — | 3 |
| | Reconstituted system (cytochrome P-450c) | 80 | 97 | — | — | 3 |

[a] Data obtained from Thakker et al.[53]
[b] Microsomes were obtained from livers of control, phenobarbital-treated, or 3-methylcholanthrene-treated immature male rats of Long–Evans strain.

phenobarbital-treated rats formed significant amounts (23–37%) of a metabolite tentatively identifed as the 6-hydroxy derivative of the 7,8-dihydrodiol[53] (Fig. 6).

Studies with the resolved enantiomers of B[a]P 7,8-dihydrodiol[53] or with biosynthetic diol, which was highly enriched in (−)-enantiomer,[53,152] revealed that metabolic conversion to 7,8-diol 9,10-epoxides was highly stereoselective (Table II). The (+)-(7$S$,8$S$)-enantiomer was predominantly metabolized by liver microsomes from 3-methylcholanthrene-treated rats to the diol epoxide isomer **1**, in which the benzylic hydroxyl group and oxirane oxygen are cis to each other.[53] In contrast, the (−)-(7$R$,8$R$)-enantiomer was predominantly converted by these microsomes to diol epoxide isomer **2**, in which the benzylic hydroxyl group and oxirane oxygen are trans to each other.[53,152] These results indicate that the oxygen is added to the double bond on the same face of the dihydrodiol regardless of the configuration of the hydroxyl groups. Subsequently, similar selectivity was exhibited by the cytochromes $P$-450 in microsomes from 3-methylcholanthrene-treated rats on oxidation of the bay region double bond of dihydrodiols of phenanthrene, chrysene, and benz[a]anthracene.[54,59,60,63,65] Metabolism of the (+)- and (−)-enantiomers of B[a]P 7,8-dihydrodiol by purified cytochromes $P$-450 from rabbit liver is also stereoselective[153]; however, the rates of metabolism (per nanomole of cytochrome $P$-450) are almost two orders of magnitude lower than those observed with rat liver cytochrome $P$-450c.[21] Unlike mammalian systems, the filamentous fungus *C. elegans* selectively oxidizes racemic B[a]P 7,8-dihydrodiol to the 7,8-diol 9,10-epoxide-**2**.[154] It is not known whether both enantiomers of the 7,8-dihydrodiol are metabolized to the same diastereomer of the diol epoxide or only one of the two enantiomers is a substrate for the fungal enzymes.

Benzo[a]pyrene 7,8-dihydrodiol is cooxidized to bay region 7,8-diol 9,10-epoxides during the metabolism of arachidonic acid by another microsomal enzyme, prostaglandin synthetase.[155–158] In contrast to the metabolism by cytochrome $P$-450 monooxygenases,[50,152] racemic 7,8-dihydrodiol is metabolized only to isomer **2** of the 7,8-diol 9,10-epoxide by prostaglandin synthetase.[157,159] Microsomal prostaglandin synthetase oxidizes B[a]P almost exclusively at the 6-position,[160,161] hence, the only way this enzyme can participate in the metabolic activation of B[a]P is by activating the 7,8-dihydrodiol formed by cytochrome $P$-450 and epoxide hydrolase. Whether or not this pathway contributes to the tumorigenicity of B[a]P is unknown.

b. BENZO[a]PYRENE 9,10-DIHYDRODIOL. The bay region 9,10-dihydrodiol of B[a]P is metabolized by liver microsomes from 3-methylcholanthrene-treated rats predominantly to a 1- or a 3-hydroxy derivative of the dihydrodiol, or both[64] (Fig. 7). Originally, this metabolite was mistakenly characterized as the catechol 9,10-hydroxy-B[a]P.[162] Very little metabolism occurred at the 7,8-olefinic double bond to form diol epoxides. These results are in sharp contrast with the results on the metabolism of B[a]P 7,8-dihydrodiol, which was oxidized predominantly (>90%) at the 9,10-double bond to form diol epoxides by the same enzymes.[50] The filamentous fungus C. elegans, however, produces 9,10-diol 7,8-epoxides from racemic B[a]P 9,10-dihydrodiol.[163] The percentage of total metabolism at the 7,8-position is not known, because unlabeled substrate was used in this study. Lack of optical activity in the unreacted substrate suggested that both enantiomers of B[a]P 9,10-dihydrodiol are metabolized at approximately equal rates by the fungal enzymes.[163]

It was suggested that steric bulk or increased polarity, or both, due to the pseudodiaxial hydroxyl groups of the bay region 9,10-dihydrodiol, retarded oxidation of the 7,8-double bond in the 9,10-dihydrodiol by rat liver cytochromes P-450.[64] Subsequently, metabolism of bay region dihydrodiols of chrysene,[55] benz[a]anthracene,[52,63] B[e]P,[58,66] and dibenz[a,h]anthracene[56] has been reported. In studies where quantification of the metabolites was reported,[55,58,63,66] the metabolism at the olefinic double bond in the bay region dihydrodiols by liver microsomes from 3-methylcholanthrene-treated rats to form diol epoxides ranged from low (B[e]P 9,10-dihydrodiol) to moderate (benz[a]anthracene 1,2-dihydrodiol) to high (chrysene 3,4-dihydrodiol) (Fig. 8).

Studies with the pseudodiaxial diol B[e]P 9,10-dihydrodiol showed that hamster liver microsomes produced much higher amounts of diol epoxides than did rat liver microsomes.[58] Moreover, 7,8-benzoflavone, when present in the incubation medium, enhanced the formation of 9,10-diol 11,12-epoxides from B[e]P 9,10-dihydrodiol with liver microsomes from humans and rabbits.[58] Similar enhancement in the presence of 7,8-benzoflavone was also reported in the formation of 9,10-diol 7,8-epoxides from B[a]P 9,10-dihydrodiol by human liver microsomes.[58] Thus, the regioselectivity in the metabolism of bay region dihydrodiols to diol epoxides appears to be

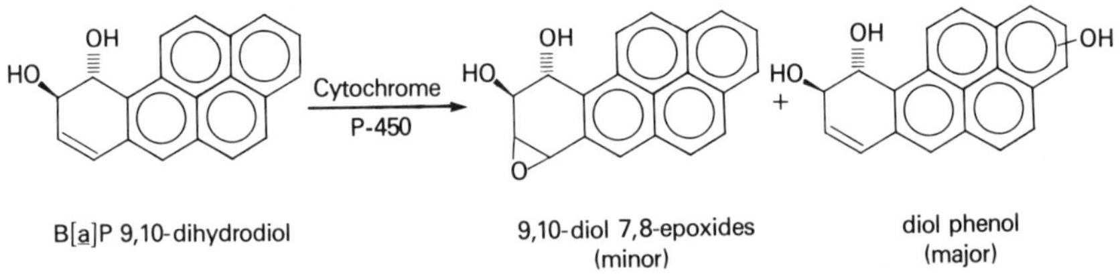

B[a]P 9,10-dihydrodiol → Cytochrome P-450 → 9,10-diol 7,8-epoxides (minor) + diol phenol (major)

Fig. 7. Metabolism of B[a]P 9,10-dihydrodiol by rat liver microsomes. Absolute configuration is not implied.

# 7. POLYCYCLIC AROMATIC HYDROCARBONS

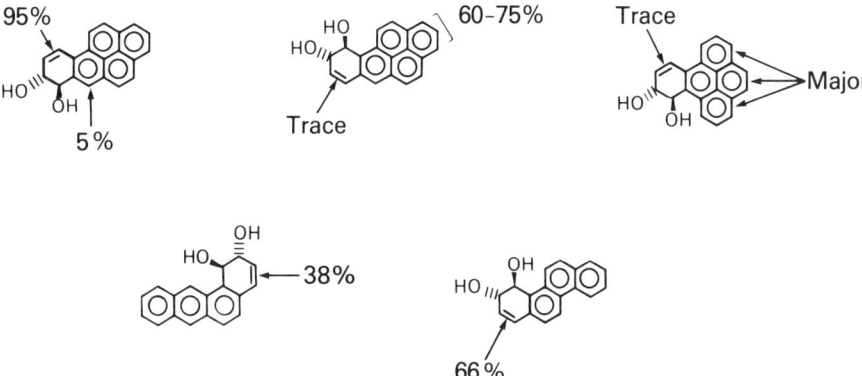

**Fig. 8.** Regioselectivity of cytochromes P-450 in liver microsomes from 3-methylcholanthrene-treated rats in the metabolic formation of diol epoxides from bay region dihydrodiols of several PAHs. B[a]P 7,8-dihydrodiol is included for comparison. Absolute configuration is not implied.

dependent on several factors, such as the substrate, species, source of microsomes, and the presence of cosubstrates in the incubation medium.

## D. PHENOLS AND QUINONES

Phenols and quinones constitute a high percentage of the metabolites of B[a]P formed by rat liver microsomes or by a purified monooxygenase system reconstituted with cytochrome P-450.[77,80-82,164] 3-Hydroxy-B[a]P and 9-hydroxy-B[a]P are the two major phenolic metabolites of B[a]P formed by rat liver microsomes.[82] In addition, small amounts of 1-hydroxy-B[a]P and 7-hydroxy-B[a]P have also been identifed as metabolites of B[a]P.[77,84] 6-Hydroxy-B[a]P has not been isolated as a metabolite of B[a]P, presumably because of its extreme instability.[71,107] However, the presence of 1,6-, 3,6-, and 6,12-quinones among the metabolites of B[a]P is indicative of the formation of this phenol.[107,165] The phenols of B[a]P are presumably formed by isomerization of arene oxides with the possible exception of 6-hydroxy-B[a]P. Formation of 6-hydroxy-B[a]P via an arene oxide would require that B[a]P 5a,6-oxide or B[a]P 6,6a-oxide is an intermediate. Although the formation of such trisubstituted arene oxides may be possible, direct insertion of oxygen into the C-H bond at C-6 may be a more attractive mechanism to explain the formation of 6-hydroxy-B[a]P. Evidence for direct oxygen insertion in the hydroxylation of certain benzene derivatives was provided by studies on deuterium isotope effects ($k_H/k_D = 1.30-1.75$).[166] With specifically tritium-labeled B[a]P at C-6, practically no primary isotope effect was observed[21] during the metabolism of the substrate by rat liver microsomes. The absence of an isotope effect is consistent with either the arene oxide pathway or an insertion reaction in which little carbon–tritium bond breaking occurs in the transition state. Thus, the mechanism for 6-hydroxylation of B[a]P remains unknown.

Phenols formed metabolically are further metabolized along several pathways. 3-Hydroxy-B[a]P is oxidatively metabolized by rat liver microsomes to products that include the 3,6-quinone as the major metabolite.[71,74] Both 3- and 9-hydroxy-B[a]P, on further metabolism, yield products that bind covalently to DNA.[72,167,168] Capdevila et al.[73] have used difference spectrophotometry and HPLC to show that these phenols are converted to diphenols under steady-state conditions. For example, 3-hydroxy-B[a]P is metabolized to the 3,6-diphenol, which autoxidizes to B[a]P-3,6-quinone.[169] Another product of 3-hydroxy-B[a]P, the 3,9-diphenol, is also formed by oxidative metabolism of 9-hydroxy-B[a]P. In addition, 9-hydroxy-B[a]P is metabolized to an arene oxide, presumably its 4,5-oxide,[167,170] which may be converted to a dihydrodiol phenol.[170-174] Wood et al.[175] evaluated the metabolic activation of all 12 isomeric phenols of B[a]P by cytochrome P-450 to mutagenic products toward bacteria. The 1-, 2-, 3-, 6-, 9-, and 12-hydroxy-B[a]P were activated to mutagenic products, but less so than B[a]P. Lubet et al.[176] had found that 9-hydroxy-B[a]P was metabolized to mutagenic products toward bacterial cells by microsomes from 5,6-benzoflavone-treated rats and that trichloropropene oxide, an epoxide hydrolase inhibitor, had little effect on the metabolic activation of this compound.

### E. GLUTATHIONE CONJUGATES

The glutathione S-transferases are a group of enzymes capable of further metabolizing arene oxides of polycyclic aromatic hydrocarbons to nontoxic products (cf. refs. 177, 178). Among the four arene oxides of B[a]P examined as substrates for rat liver glutathione S-transferases, the 11,12-oxide was the best substrate and was metabolized at a 1.8-, 8-, and 60-fold greater rate than were the 4,5-, 7,8-, and 9,10-oxides, respectively. Interestingly, the two K-region oxides, which are considerably more stable than the two non-K-region oxides,[179] are much better substrates for glutathione S-transferases. Preliminary studies with homogeneous glutathione S-transferases A, B, C, D, E, and AA from rat liver[177] indicated that transferases A and C were most active toward B[a]P 4,5-oxide.[180,181] B[a]P 4,5-oxide is also a substrate for human transferases.[180] Hernandez et al.[182] have shown that enzymatic conjugation of racemic B[a]P 4,5-oxide with glutathione is highly stereoselective. The $^{13}$C NMR spectrum of the glutathione conjugates of [4,5-$^{13}$C]B[a]P 4,5-oxide produced enzymatically by a purified glutathione S-transferase from little skate liver indicated that both C-4 and C-5 were attacked by glutathione in the racemic substrate to the extent that only one of the diastereomers of each of the two positional isomers was formed. A chromatographic method showed that rat liver enzymes also formed the same two diastereomers.[182] Armstrong et al.[110] subsequently established that the two isomers, which constituted the major enzymatic products, had (4S,5S)-absolute configuration. Studies with phenanthrene 9,10-oxide, pyrene 4,5-oxide, (+)- and (−)-benz[a]anthracene 5,6-oxide, and (+)- and (−)-B[a]P 4,5-oxides[183] have shown that glutathione S-transferase isozyme C from rat liver selectively directs the attack of glutathione at the carbon atom with (R)-absolute configuration, resulting in the formation of (S,S) isomers of glutathione conjugates (>96%) in each case. This is in contrast to the specificity shown by epoxide hydrolase.

## F. SULFATES AND GLUCURONIDES

The phenols, quinones, and dihydrodiols of B[a]P are converted to polar sulfate conjugates by cytosolic sulfotransferases,[46,47] and to water-soluble glucuronides by microsomal UDPglucuronyltransferases.[38-40] These enzymatic conversions appear to represent detoxication mechanisms whereby metabolites of B[a]P are converted to water-soluble compounds, which are readily excreted. In certain instances, however, formation of conjugates may represent metabolic activation of the substrate. For example, the sulfate ester of 7-hydroxymethyl-12-methylbenz[a]anthracene is mutagenic to S. typhimurium strain TA 98.[184] Glucuronide formation may also lead to similar activation of the substrate.[185] During enzymatic hydrolysis of the glucuronide of 3-hydroxy-B[a]P, a B[a]P derivative is formed that binds to DNA to a far greater extent than does the 3-hydroxy-B[a]P or its glucuronide.[185]

## IV. Biological Activity of Benzo[a]pyrene Derivatives

In attempts to establish the nature of the ultimate carcinogens derived from B[a]P, a large number of derivatives of the hydrocarbon have been evaluated in the past decade for their ability to bind covalently to cellular macromolecules, for their mutagenic activity toward bacterial and mammalian cells, and for their tumorigenicity toward rodents. As a result of these extensive studies, B[a]P 7,8-diol- 9,10-epoxide-2 has been identified as the ultimate carcinogen of B[a]P. These studies are briefly reviewed in the following sections.

## A. COVALENT BINDING TO CELLULAR MACROMOLECULES

Since the first demonstration by Miller[16] that topical application of B[a]P on mouse skin resulted in small amounts of products that were bound covalently to proteins, covalent interactions of B[a]P derivatives with cellular macromolecules have been the subject of extensive investigation. These studies have provided a great deal of insight into the metabolic activation of B[a]P. The report by Borgen et al.[139] that the 7,8-dihydrodiol of B[a]P covalently binds to DNA on metabolism by rat liver microsomes to a much greater extent than does B[a]P and its other metabolites provided the first clue that B[a]P is activated by these enzymes via the 7,8-dihydrodiol to highly reactive molecules. Numerous subsequent studies have indicated that B[a]P 7,8-diol 9,10-epoxides are the metabolites of B[a]P that react with cellular macromolecules to form covalent adducts.[49,120,186-193] Studies in which polyguanylic acid was modified by B[a]P 7,8-diol 9,10-epoxide-1[191,194] and by B[a]P 7,8-diol 9,10-epoxide-2[191,195] showed that the major covalent adducts are formed by attack of the $N^2$-amino group of guanine at C-10 of the diol epoxides, resulting in the formation of trans adducts in both cases (Fig. 9). Small, but significant amounts of cis adducts were also formed from the diol epoxide-1[191,194] and from the diol epoxide-2.[191] Only the trans adduct, however, along with some unidentified products, was iso-

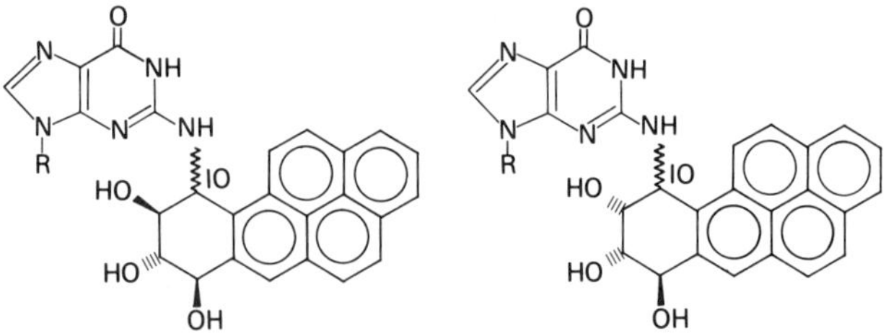

Derived from diol epoxide - 1     Derived from diol epoxide - 2

Fig. 9. Structures of nucleoside adducts formed by reaction of B[a]P 7,8-diol 9,10-epoxide-1 and B[a]P 7,8-diol 9,10-epoxide-2 with polyguanylic acid. The same adducts were formed in major accounts from B[a]P *in vivo*.

lated by Jeffrey et al.[195] after reaction of B[a]P 7,8-diol 9,10-epoxide-2 with polyguanylic acid. It was later shown that when B[a]P was applied on mouse skin, (+)-B[a]P diol epoxide-1 (7$S$,8$R$,9$S$,10$R$) and (+)-diol epoxide-2 (7$R$,8$S$,9$S$,10$R$) were produced *in vivo* from the (+)-(7$S$,8$S$)- and (−)-(7$R$,8$R$)-dihydrodiols, respectively, and subsequently reacted with the 2-amino group of guanine in RNA by cis- and trans-addition.[191,193] When bovine bronchial explants were exposed to B[a]P, only (+)-diol epoxide-2 was detected bound to the 2-amino group of guanine in RNA and DNA.[120,190] Evidence has also been presented for the formation of DNA and RNA adducts from both B[a]P 7,8-diol 9,10-epoxides-1 and -2 in bovine and human bronchial explants,[196] as well as in hamster kidney cells and in mouse embryo fibroblasts.[197,198] Interestingly, Baird and Diamond[94] reported that the ratio of DNA-bound B[a]P diol epoxides-1 and -2 in cultured hamster embryo cells is dependent on the time at which DNA is analyzed. Adducts from isomer 1 predominate at 4 to 6 hr and those from isomer 2 predominate at 72 hr after treatment with B[a]P. This has been attributed, in part, to different rates of repair involving the adducts from the two diastereomers of the diol epoxides. A higher rate of excision of the covalent DNA adducts formed by B[a]P 7,9-diol 9,10-epoxide-1 compared with those formed by B[a]P 7,8-diol 9,10-epoxide-2 in mouse embryo fibroblasts had been reported earlier by Shinohara and Cerutti.[197] In addition to the 2-amino group of guanine, other sites on nucleic acids have been reported to be modified by the bay region diol epoxides of B[a]P. For example, studies with polyguanylic acid,[191] as well as with superhelical ColE1 DNA,[199] indicated that B[a]P 7,8-diol 9,10-epoxides react with phosphate of the phosphodiester backbone of nucleic acids. Evidence for reaction of the B[a]P 7,8-diol 9,10-epoxide-2 at the $N^7$ position of guanine has also been presented[200]; however, covalent adducts involving the $N^7$ position of guanine could not be detected *in vivo* after treatment of mouse embryo cells or Chinese hamster V79 cells with B[a]P or B[a]P 7,8-diol 9,10-epoxide-2, respectively.[201] Evidence also exists for the covalent interactions of B[a]P 7,8-diol 9,10-epoxides with deoxy-adenosine and deoxycytidine residues,[188,190,202,203] although no information is available regarding the structure of these covalent adducts.

## B. MUTAGENICITY OF BENZO[a]PYRENE DERIVATIVES

As is the case with other chemical carcinogens, the mutagenicity of the derivatives of B[a]P and other PAHs has been extensively studied in an attempt to identify potentially carcinogenic compounds. Moreover, the well-characterized mutagenicity test systems allow systematic studies, which result in improved understanding of how B[a]P and other PAHs are metabolically activated to genotoxic products. The two test systems most extensively used for the study of PAH derivatives are a bacterial reverse-mutation system developed by Ames,[204,205] which utilizes different histidine-dependent strains of *S. typhimurium*, and a mammalian forward-mutation system, which utilizes the Chinese hamster V79 cell line.[206] Neither the bacterial[204] nor the mammalian cells[207] contain detectable levels of the monooxygenase system responsible for metabolic activation of the PAH. Hence they allow evaluation of intrinsic mutagenicity of the PAH derivatives. Metabolic activation and detoxication of the hydrocarbons and their derivatives can also be examined by carrying out the tests in the presence of added enzymes.

The intrinsic mutagenicity of more than 30 derivatives of B[a]P has been determined in *S. typhimurium* strains TA 1538, TA 98, and TA 100 and in Chinese hamster V79 cells.[146-148, 208] A comparison of the relative mutagenicity of all these derivatives has been reported by Levin *et al.*[209] The B[a]P derivatives that were compared include the 4,5-, 7,8-, 9,10-, and 11,12-oxides, the four corresponding *trans*-dihydrodiols, all 12 possible phenols, the 1,6-, 3,6-, 6,12-, 4,5-, 7,8-, and 9,10-quinones, two tetrahydroepoxides (B[a]P $H_4$-7,8- and B[a]P $H_4$-9,10-epoxides), and five diol epoxides (B[a]P 7,8-diol 9,10-epoxides-1 and -2, B[a]P 9,10-diol 7,8-epoxides-1 and -2, and B[a]P 7,10-diol 8,9-epoxide) (Table III). In *S. typhimurium* strains TA 98 and TA 100, B[a]P 7,8-diol 9,10-epoxide-1 was the most active derivative, followed by B[a]P $H_4$-9,10-epoxide, B[a]P 7,8-diol 9,10-epoxide-2, B[a]P 4,5-oxide, and B[a]P $H_4$-7,8-oxide. All other derivatives except 6-hydroxy-B[a]P (5% of the activity of most active compound tested in strain TA 98) had less than 2% of the activity of the most active compound tested. In strain TA 1538 of *S. typhimurium*, the $H_4$-9,10-epoxide was the most active derivative, followed by B[a]P 4,5-oxide, B[a]P 7,8-diol 9,10-epoxide-1, B[a]P $H_4$-7,8-epoxide, and B[a]P 7,8-diol 9,10-epoxide-2. 6-Hydroxy-B[a]P was 5% as active as the most mutagenic derivative in strains TA 1538 and TA 98. 1-Hydroxy-B[a]P, 3-hydroxy-B[a]P, and 12-hydroxy-B[a]P showed low but significant mutagenic activity. Low intrinsic mutagenicity of 1-, 3-, 4-, 7-, and 9-hydroxy-B[a]P in bacterial systems has also been observed in other studies.[208,210] In Chinese hamster V79 cells, B[a]P 7,8-diol 9,10-epoxide-2 was the most active derivative, followed by B[a]P 7,8-diol 9,10-epoxide-1 and B[a]P $H_4$-9,10-epoxide, which were equipotent. Thus, the B[a]P 7,8-diol 9,10-epoxides emerged from these studies as the most mutagenic metabolites of B[a]P. Many B[a]P derivatives have been evaluated for mutagenic activity on metabolic activation with a monooxygenase system, which was reconstituted with highly purified cytochrome

## Table III
Intrinsic Mutagenicity of Benzo[a]pyrene Derivatives toward *Salmonella typhimurium* Strains TA 1538, TA 98, and TA 100 and toward Chinese Hamster V79 Cells[a]

| B[a]P derivative | Relative (%) activity[b] | | | |
|---|---|---|---|---|
|  | TA 1538 | TA 98 | TA 100 | V79 |
| 7,8-Diol 9,10-epoxide-1 | 40 | 100 | 100 | 40 |
| 7,8-Diol 9,10-epoxide-2 | 15 | 35 | 65 | 100 |
| 9,10-Diol 7,8-epoxide-1 | — | 2 | 4 | <0.1 |
| 9,10-Diol 7,8-epoxide-2 | — | 11 | 1 | 0.2 |
| 7,10-Diol 9,10-epoxide | — | <0.1 | 0.4 | <0.1 |
| $H_4$-9,10-epoxide | 100 | 95 | 90 | 40 |
| $H_4$-7,8-epoxide | 20 | 10 | 2 | 0.2 |
| 4,5-Oxide | 60 | 20 | 6 | 1 |
| 6-Hydroxy-B[a]P | 5 | 5 | 0.6 | 0.3 |

[a] The comparisons of relative activity are from data obtained in several different experiments. However, in each experiment B[a]P 4,5-oxide or one of the 7,8-diol 9,10-epoxides was used as a positive control (cf. ref. 209).

[b] The 7,8-, 9,10-, and 11,12-oxides, the 4,5-, 7,8-, 9,10-, and 11,12-dihydrodiols, the 1,6-, 3,6-, 6,12-, 4,5-, 7,8-, and 11,12-quinones, and all the possible phenols of B[a]P except 6-hydroxy-B[a]P had 2% or less mutagenic activity of the most active derivative in each of the four test systems.

P-450c.[175] These studies showed that only B[a]P 7,8-dihydrodiol was activated to metabolites more mutagenic to *S. typhimurium* strains TA 98 and TA 1538 than those of B[a]P. B[a]P 7,8-oxide, the precursor of the 7,8-dihydrodiol, requires both epoxide hydrolase and the monooxygenase system for metabolic activation to mutagens. Thus, both B[a]P 7,8-oxide and 7,8-dihydrodiol are proximate mutagens derived from B[a]P, which are further metabolically activated to ultimate mutagens, the bay region 7,8-diol 9,10-epoxides. 1-, 2-, 3-, 6-, 9-, and 12-hydroxy B[a]P were metabolically activated to mutagens to a much lesser extent than was B[a]P. In another study,[211] 1-, 2-, and 3-hydroxy-B[a]P exhibited equal or higher mutagenicity than B[a]P in strain TA 98 or TA 100, or both, on metabolic activation by liver homogenate (S9 fraction) from rats treated with polychlorinated biphenyls. In the same study, 6-methyl-B[a]P and 6-hydroxymethyl-B[a]P were 1.5- to 3-fold more mutagenic than B[a]P on activation by S9 fraction. In general, these latter results do not correlate well with the tumorigenic activity of the compounds tested.

## C. TUMORIGENICITY OF BENZO[a]PYRENE DERIVATIVES

The studies on the mutagenicity, covalent binding, and metabolism of PAHs and their derivatives provided evidence that bay region 7,8-diol 9,10-epoxides of B[a]P were excellent candidates for the ultimate carcinogenic metabolites of B[a]P. However, definitive identification of an ultimate carcinogen could only come from tumor studies. Hence a large number of known and potential metabolites of B[a]P were tested as carcinogens in mice in three different tumor models in our laboratories. The results of these studies are briefly reviewed next.

### 1. Chronic Studies

Benzo[a]pyrene, four arene oxides (B[a]P 4,5-, 7,8-, 9,10-, and 11,12-oxides), the four corresponding dihydrodiols, and the diastereomeric B[a]P 7,8-diol 9,10-epoxide-1 and -2 have all been tested by the topical application of 0.1 or 0.4 $\mu$mol of each compound once every 2 weeks to the backs of C57BL/6J mice for 60 weeks.[64,112,133,135,212] Among the arene oxides, only B[a]P 7,8-oxide caused skin tumors.[112] However, at an equimolar dose, the 7,8-oxide was only 20% as active as B[a]P. Among the dihydrodiols, only the 7,8-dihydrodiol was tumorigenic. B[a]P 7,8-dihydrodiol was equipotent with B[a]P at the 0.1-$\mu$mol dose, but was threefold more active than B[a]P at the 0.025-$\mu$mol dose. The tetrahydro-7,8-epoxide and 7,8-$H_4$-diol of B[a]P, in which the 9,10-double bonds were saturated, were nontumorigenic. These results suggested that the carcinogenicity of B[a]P 7,8-oxide and B[a]P 7,8-dihydrodiol is due to their metabolic activation to 7,8-diol 9,10-epoxides. The 7,8-diol 9,10-epoxides had little, if any, carcinogenic activity in the test system, presumably because of their high chemical reactivity, short biological half-lives, and consequent inability to reach the critical cellular targets. Evaluation of the carcinogenicity of all 12 isomeric phenols of B[a]P[212,213] revealed that 2-hydroxy-B[a]P was highly carcinogenic and 11-hydroxy-B[a]P was weakly tumorigenic. 2-Hydroxy-B[a]P has not yet been identified as a metabolite of B[a]P, and hence, the contribution of this phenol to the overall carcinogenicity of B[a]P is unknown.

### 2. Initiation Promotion Studies

In this test system, a single dose of the PAH carcinogen is applied to the skin of female CD-1 mice, followed by repetitive treatment with the tumor promoter 12-O-tetradecanoylphorbol 13-acetate. In Table IV, tumor initiating activities of B[a]P and its derivatives are compared.[138,214,215] These results established that all of the highly tumorigenic derivatives of B[a]P

## Table IV

Skin Tumor-Initiating Activity of Benzo[a]pyrene Derivatives in Charles River CD-1 Mice[a,b,c]

| Initiator | Mice with tumors (%)[d] | Tumors per mouse | Relative activity[e] |
|---|---|---|---|
| B[a]P | 90–97 | 5.3 | 100 |
|  |  | 7.4–7.5 |  |
| 4,5-Oxide | 24 | 0.24 | 5 |
| 7,8-Oxide | 89 | 2.5 | 48 |
| 9,10-Oxide | 15 | 0.15 | 3 |
| 11,12-Oxide | 38 | 0.45 | 8 |
| 4,5-Dihydrodiol | 30 | 0.30 | 4 |
| 7,8-Dihydrodiol | 94 | 6.5 | 88 |
| 9,10-Dihydrodiol | 13 | 0.10 | 1 |
| 2-Hydroxy | 85 | 6.0 | 80 |
| 11-Hydroxy | 80 | 2.1 | 28 |
| 1-, 3-, 4-, 5-, 6-, 7-, 8-, 9-, 10-, 12-Hydroxy | 3–30 | 0.03–0.33 | 0.3–4 |
| 1,6-, 3,6-, 6,12-Quinones | 21–29 | 0.20–0.30 | 3–4 |
| 7,8-Diol 9,10-epoxide-**1** | 7 | 0.07 | 1 |
| 7,8-Diol 9,10-epoxide-**2** | 69 | 1.5 | 28 |
| 6-Methyl | 74 | 1.6 | 22 |
| 6-Hydroxymethyl | 57 | 1.0 | 14 |

[a] Results from Slaga et al.[137] The comparisons are derived from results obtained in several different experiments. However, in each experiments B[a]P was used as a positive control.

[b] All compounds were applied once, followed 1 week later by twice-weekly applications of 10 μg of tetradecanoylphorbol 13-acetate. The arene oxides and diol epoxides of B[a]P were tested at an initiating dose of 200 nmol. All other compounds were tested at an initiating dose of 400 nmol. B[a]P was tested at both doses. At the 200-nmol dose B[a]P initiated tumors in 92% of the animals and produced 5.3 tumors per mouse; at the 400-nmol dose B[a]P initiated tumors in 90 to 97% of the animals and produced 7.4–7.5 tumors per mouse.

[c] Of the initial 30 animals, 27–30 survived in all cases at the thirtieth week of promotion.

[d] Percentage of surviving mice with papillomas.

[e] Relative activity was calculated with respect to B[a]P. The number of tumors per mouse, calculated as total number of papillomas divided by total number of surviving mice, is used as an index of tumor-initiating activity in these comparisons.

identified in the chronic studies were also good tumor initiators. Tumor-initiating activities of B[a]P 7,8-oxide and B[a]P 7,8-dihydrodiol were higher than those of other arene oxides and dihydrodiols, respectively. Although B[a]P 7,8-dihydrodiol was almost as active as B[a]P (Table IV), Chourou-

## Table V
### Tumorigenic Activity of Benzo[a]pyrene Derivatives in Newborn Mice[a]

| Compound | Dose (nmol) | Mice with pulmonary adenomas (%) | Number of adenomas per mouse |
|---|---|---|---|
| Experiment 1 | | | |
| Control | — | 1.2 | 0.13 |
| B[a]P | 28 | 19 | 0.24 |
| 7,8-Dihydrodiol | 28 | 66 | 1.77 |
| 7,8-Diol 9,10-epoxide-1 | 28 | 10 | 0.14 |
| 7,8-Diol 9,10-epoxide-2 | 28 | 86 | 4.42 |
| Experiment 2 | | | |
| Control | — | 8 | 0.08 |
| B[a]P | 1400 | 93 | 10.0 |
| 4,5-Oxide | 1400 | 10 | 0.10 |
| 7,8-Oxide | 1400 | 72 | 2.08 |
| 9,10-Oxide | 1400 | 0 | 0 |
| 11,12-Oxide | 1400 | 20 | 0.32 |
| Experiment 3 | | | |
| Control | — | 8 | 0.08 |
| B[a]P | 1400 | 81 | 6.40 |
| 2-Hydroxy-B[a]P | 1400 | 98 | 24.0 |
| 6-Hydroxy-B[a]P | 1400 | 11 | 0.11 |

[a] Results from Kapitulnik et al.[136] (experiment 1), Wislocki et al.[218] (experiment 2), and Chang et al.[219] (experiment 3). Male and female mice were utilized, and the results are presented as the average of both.

linkov et al.[216] found only 50–66% of the tumor-initiating activity of B[a]P. 2-Hydroxy-B[a]P and 11-hydroxy-B[a]P had 80 and 28% of the tumor-initiating activity of B[a]P, respectively, as determined by the number of papillomas produced per mouse at the same dose. 6-Methyl-B[a]P and 6-hydroxymethyl-B[a]P had low but significant tumor-initiating activity. B[a]P 7,8-diol 9,10-epoxide-2 had about 28% of the tumor-initiating activity of B[a]P, whereas the 7,8-diol 9,10-epoxide isomer **1** was inactive. The (+)-enantiomer of diol epoxide-**2**, when applied in fractionated doses (six applications), was equipotent with B[a]P.[217] As prime candidates for ultimate carcinogenic metabolites of B[a]P, the bay region diol epoxides are expected to be more tumorigenic than the parent hydrocarbons. As in the chronic studies, however, the low tumorigenicity of the diol epoxides possibly resulted from the short biological half-lives of these compounds and the lower accessibility of the target cells.

## 3. Tumorigenicity of Benzo[a]pyrene Derivatives in Newborn Mice

In this tumor model, the PAH derivatives are administered intraperitoneally to Swiss–Webster mice within 24 hr of birth (one-seventh of the total dose) and at 8 (two-sevenths of the total dose) and 15 (four-sevenths of the total dose) days of age. The animals are killed at 6 to 7 months of age and examined for lung tumors and for tumors in other organs. The tumorigenic activities of B[a]P and selected derivatives are shown in Table V.[134,136,218,219] In this tumor model, B[a]P 7,8-diol 9,10-epoxide-**2** was greater than 20-fold more tumorigenic than B[a]P and 2.5-fold more tumorigenic than the 7,8-dihydrodiol at the 28-nmol total dose. B[a]P 7,8-diol 9,10-epoxide-**1** was nontumorigenic in this test system. These results established, for the first time, that B[a]P 7,8-diol 9,10-epoxide-**2** was the ultimate carcinogenic metabolite of B[a]P[134,136] and provided the first demonstration of an ultimate carcinogen from any member of the PAH class of carcinogens. Two other derivatives of B[a]P found to be tumorigenic in this tumor model were B[a]P 7,8-oxide and 2-hydroxy-B[a]P. B[a]P 7,8-oxide had about 20% of the activity of B[a]P at a dose of 1.4 μmol, whereas 2-hydroxy-B[a]P was about fourfold more active than B[a]P at the same dose.

## V. The Bay Region Theory

In 1976, increasing evidence suggested the B[a]P 7,8-diol 9,10-epoxides as the ultimate mutagenic and carcinogenic metabolites of B[a]P. This raised two important questions: (1) Do diol epoxides represent a general class of ultimate carcinogenic metabolites for the PAHs, and if so, (2) do the structural features of B[a]P 7,8-diol 9,10-epoxides provide a clue to the structures of highly reactive and carcinogenic diol epoxides of other PAHs? These questions led to the proposal that the critical structural feature in B[a]P 7,8-diol 9,10-epoxides responsible for their high chemical reactivity and biological activity was the presence of an epoxide group on the saturated angular benzo ring in the bay region of the molecule (see Fig. 1 for the definition of a bay region).[220,221] This hypothesis, which is now generally known as the bay region theory, predicted that for the benzo-ring diol epoxides of a given PAH, those compounds in which the epoxide group formed part of a bay region would have the highest chemical reactivity and the highest biological activity. For the carcinogenic PAHs, such bay region diol epoxides should be prime candidates as ultimate carcinogens, if formed as metabolites.

Both experimental evidence and theoretical considerations led to the formulation of the bay region theory. In addition to the high biological activity of bay region diol epoxides of B[a]P, existing data on methyl- and fluoro-substituted PAHs implied the importance of oxidative metabolism on the angular benzo ring for metabolic activation of the hydrocarbons. Substitution in the benzo ring of benz[a]anthracene by a methyl group[222] or substitution of fluorine in the benzo ring of 7-methyl benz[a]anthracene[223,224] resulted in less tumorigenic derivatives. It was argued that substitution on benzo rings inhibited metabolic formation of bay region diol epoxides and enhanced detoxication by metabolism at other sites.[220] Quatum chemical calculations[221,225] predicted that formation of a carbocation at a bay region benzylic position, by opening of an oxirane ring on a saturated angular benzo ring (Fig. 10), would be energetically more favorable than formation of a carbocation at other benzylic positions. This prediction was derived from perturbational molecular orbital calculations of carbocation stabilization ($\Delta E_{\text{deloc.}}$).

In the 7 years following the formulation of the bay region theory, bay region diol epoxides were either proven or implicated to be ultimate mutagenic or carcinogenic metabolites, or both, of at least 13 other PAHs in addition to B[a]P. These include benz[a]anthracene,[226-232] dibenz[a,h]-anthracene,[233,234] chrysene,[235-241] benzo[e]pyrene,[66,242,243] benzo[c]phenanthrene,[244-246] dibenzo[a,h]pyrene and dibenzo[a,i]-pyrene,[247-248] 3-methylcholanthrene,[249-256] 7-methylbenz[a]anthracene,[257-261] 7,12-dimethylbenz[a]anthracene,[253,262-270] 5-methylchrysene,[271-273] 15,16-dihydro-11-methylcyclopenta[a]phenanthrene-17-one,[274] and benzo[b]fluoranthene.[275] Structures of the bay region diol epoxides of these hydrocarbons are shown in Fig. 11. Studies on many of these hydrocarbons, which led to

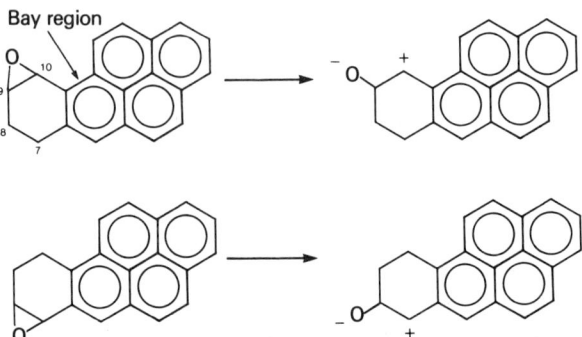

Fig. 10. Benzylic carbocations formed from bay region and non-bay region epoxides of B[a]P.

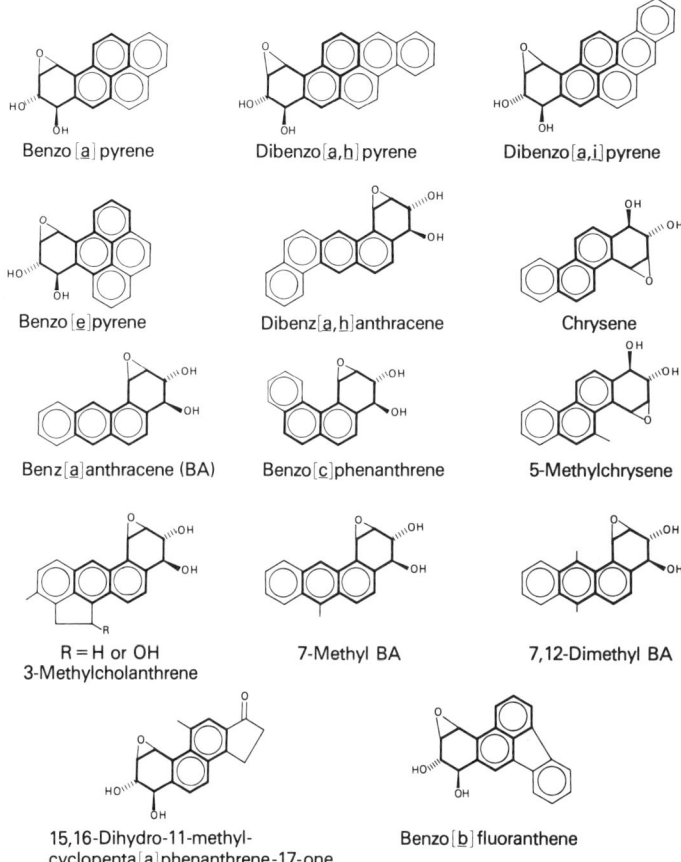

**Fig. 11.** Structures of bay region diol epoxides that are known or probable ultimate mutagens or carcinogens, or both. The phenanthrene portion of each of these diol epoxides, which forms the bay region, is emphasized. Absolute configuration is not implied. Bay region diol epoxides of benzopyrenes, dibenzopyrenes, benz[a]anthracene, chrysene, and benzo[c]phenanthrene have been synthesized, and their biological activity has been evaluated. For other hydrocarbons, bay region diol epoxides have been implied as ultimate mutagens or carcinogens, or both, from studies with their precursor dihydrodiols.

the identification or implication of bay region diol epoxides as ultimate carcinogens, have been reviewed in detail elsewhere.[20,21] Notably, all the bay region diol epoxides of the PAHs thus far tested are very poor substrates — if substrates at all — for epoxide hydrolase.[50,147,175,227,237,244,247] No effective enzymatic detoxication mechanism for these compounds has been demonstrated.

## VI. Comparative Metabolism of Polycyclic Aromatic Hydrocarbons

### A. OVERALL RATES OF METABOLISM

Oxidative metabolism of the PAHs to bay region diol epoxides is the major pathway by which PAHs are activated to mutagens and carcinogens. Hence regio- and stereoselectivity of the monooxygenases and epoxide hydrolase, the two enzymes essential for metabolic activation of the PAHs, toward PAH substrates play a pivotal role in the bioactivation of this class of carcinogens. Comparison of the overall rates of metabolism of several PAHs by rat liver microsomes from control and treated rats is given in Table VI,[54,80,246,276-281] and the metabolite profiles of these PAHs produced by microsomes from 3-methylcholanthrene-treated rats are given in Table VII.

Among the unsubstituted PAHs examined, benzo[c]phenanthrene, phenanthrene, and benz[a]anthracene are metabolized at the highest rates, the two benzo-pyrenes at intermediate rates, and chrysene and dibenz[a,h]anthracene at the lowest rates. There appears to be some correlation between the aqueous solubility of these PAHs (Table VI) and the

Table VI

Effect of Inducers on the Rate of Metabolism of Polycyclic Aromatic Hydrocarbons (PAHs)[a]

| PAH | Rate of metabolism (nmol/nmol cytochrome $P$-450/min) by liver microsomes from rats | | | Aqueous solubility[b] ($\mu$mol/liter) |
|---|---|---|---|---|
| | Control | Phenobarbital-treated | 3-Methylcholan-threne-treated | |
| B[a]P | 0.8 | 0.6 | 3.7 | 0.01–0.02 |
| B[e]P | 0.5 | 1.5 | 3.7 | 0.01 |
| Phenanthrene | 4.2 | 4.1 | 7.6 | 9–15 |
| Chrysene | 0.4 | 0.2 | 1.5 | 0.008 |
| Dibenz[a,h]anthracene | 0.6 | 0.2 | 1.5 | 0.002 |
| Benzo[c]phenanthrene | 3.9 | 4.2 | 7.8 | 0.5 |
| Benz[a]anthracene | 1.5 | 1.4 | 6.9 | 0.04–0.05 |
| 3-Methylcholanthrene | 1.0 | 0.8 | 2.4 | N.D.[c] |
| 6-Fluoro-B[a]P | 0.4 | 0.3 | 4.1 | N.D. |

[a] Data for B[a]P,[80] B[e]P,[276] phenanthrene and chrysene,[54] dibenz[a,h]anthracene,[277] benzo[c]phenanthrene,[278] benz[a]anthracene,[279] 3-methylcholanthrene,[280] and 6-fluoro-B[a]P[281] were taken from the indicated references.

[b] Aqueous solubilities of the PAHs are obtained from Jerina et al.[246]

[c] N.D., not determined.

## Table VII

Comparative Metabolism of Polycyclic Aromatic Hydrocarbons (PAHs) by Microsomes from 3-Methylcholanthrene-Treated Rats[a]

| PAH | Bay region dihydrodiol | K-Region dihydrodiol | Dihydrodiol with bay region double bond | Metabolites (%) Phenols | Quinones | Unknown metabolites |
|---|---|---|---|---|---|---|
| B[a]P | 22 | 10 | 13 | 30 | 25 | — |
| B[e]P | 3[b] | 34 | 3[b] | 29 | N.D.[c] | 34 |
| Phenanthrene | 9 | 69 | 18 | 4 | — | — |
| Chrysene | 47 | 4 | 26 | 13 | N.D. | 10 |
| Dibenz[a,h]anthracene | 13 | 2 | 24 | 33 | N.D. | 28 |
| Benzo[c]phenanthrene | N.D. | 89 | 9 | 2 | — | — |
| Benz[a]anthracene[d] | N.D. | 42 | 2 | 4 | — | — |
| 3-Methylcholanthrene[e] | N.D. | <1 | N.D. | N.D. | N.D. | 40 |
| 6-Fluoro-B[a]P | 20 | 8 | 28 | 31 | 13 | — |

[a] Data obtained from sources indicated in the legend to Table VI.
[b] B[a]P is a symmetrical molecule with two identical bay regions. The bay region dihydrodiol is also the dihydrodiol with bay region double bond.
[c] N.D., not determined.
[d] Benz[a]anthracene was also metabolized to benzo-ring 8,9-dihydrodiol (49% of total metabolites) and the 10,11-dihydrodiol (3% of total metabolites).
[e] 60% of total metabolites of 3-methylcholanthrene were formed by benzylic oxidation at C-1 and C-2. Very little metabolism occurred in the aromatic moiety of the molecule.

overall rate of their metabolism.[246] Phenanthrene, benzo[c]phenanthrene, and benz[a]anthracene are more soluble and are metabolized at higher rates than chrysene and dibenz[a,h]anthracene. B[a]P and B[e]P are intermediate in both their solubility and their overall rates of metabolism. In addition to solubility, steric and electronic factors are likely to contribute to the overall rate at which the PAHs are metabolized by monooxygenases. Systematic studies to identify such factors and establish a quantitative correlation between specific physicochemical properties of the PAHs and their rates of metabolism are unavailable.

Treatment of rats with 3-methylcholanthrene results in a 2- to 10-fold increase in the rate at which the PAHs are metabolized by liver microsomes (Table VI). Among the PAHs studied, the rate enhancement is lowest for phenanthrene (2-fold) and highest for 6-fluoro-B[a]P (10-fold). Both phenanthrene and benzo[c]phenanthrene are metabolized by liver microsomes from control rats at rates much higher than any of the other PAHs listed in Table VI. Treatment of the animals with phenobarbital causes little change or a small reduction in the rate of metabolism of the PAH, when expressed as nanomoles of substrate metabolized per nanomole of cytochrome $P$-450, although total metabolism of the PAH per milligram of protein increases after treatment of rats with phenobarbital. The increased rate of metabolism of PAH after treatment of rats with 3-methylcholanthrene is predominantly due to the induction of cytochrome $P$-450c, the isozyme of cytochrome $P$-450 with the highest catalytic activity toward PAHs.[282] This was shown by experiments in which antibody prepared against this isozyme inhibited at least 85% of the metabolism of B[a]P by microsomes from 3-methylcholanthrene-treated rats.[21,282] In contrast, this antibody inhibited metabolism of B[a]P by microsomes from control and phenobarbital-treated rats only to the extent of 20 to 25%.

## B. COMPARISON OF METABOLITE PROFILES FOR A SERIES OF POLYCYCLIC AROMATIC HYDROCARBONS

The percentages of metabolites formed from several PAHs by liver microsomes from 3-methylcholanthrene-treated rats are summarized in Table VII. Dihydrodiols and phenols are the major classes of metabolites formed from the PAH with the exception of 3-methylcholanthrene, which is metabolized predominantly at the benzylic C-1 and C-2 positions. Significant amounts of quinones are formed from B[a]P and 6-fluoro-B[a]P. B[e]P, chrysene, dibenz[a,h]anthracene, and 3-methylcholanthrene produce significant amounts of presently uncharacterized metabolites, some of which may arise as a result of secondary metabolism. For these PAHs, quinones have not been identified. Dihydrodiols account for as little as <1% (3-methylcholanthrene) to as much as 89% (benzo[c]phenanthrene) of the total metabolites of the PAH (Table VII). The K-region dihydrodiols are the major metabolites of the two best PAH substrates for cytochrome P-450 (Table VII), benzo[c]phenanthrene and phenanthrene, and are also formed in large amounts from benz[a]anthracene and B[e]P. K-Region dihydrodiols are formed by cytochrome P-450 only in trace quantities from the two poorest PAH substrates, chrysene and dibenz[a,h]anthracene (Table VII), and from 3-methylcholanthrene, which is predominantly metabolized at the benzylic C-1 and C-2 positions. Bay region dihydrodiols are formed in the highest percentages from B[a]P and 6-fluoro-B[a]P, in intermediate percentages from chrysene and dibenz[a,h]anthracene and in the lowest percentage from B[e]P; they are only trace metabolites from benz[a]anthracene, benzo[c]phenanthrene, and 3-methylcholanthrene. Dihydrodiols with a bay region double bond are formed in small percentages from B[e]P, benz[a]anthracene, and 3-methylcholanthrene and in moderate to high percentages from the other PAHs listed in Table VII. The percentage of phenols formed is dependent on the stability of the precursor arene oxides toward spontaneous isomerization and toward enzymatic hydration by epoxide hydrolase. Phenanthrene, benz[a]anthracene, and benzo[c]phenanthrene are metabolized to a very small percentage of phenols (2–4% of total metabolites). In contrast, 13–33% of total metabolites of B[a]P, 6-fluoro-B[a]P, B[e]P, chrysene, and dibenz[a,h]anthracene are phenols. Only for the PAHs such as phenanthrene, benz[a]anthracene, and benzo[c]phenanthrene, which are metabolized to only trace quantities of phenols and quinones, does the distribution of dihydrodiols accurately reflect the regioselectivity of the cytochromes P-450.

The profiles of metabolites of the PAHs formed by microsomes from 3-methylcholanthrene-treated rats (Table VII) are, as expected, similar to the profiles formed by a monooxygenase system reconstituted with highly purified cytochrome P-450c, the predominant isozyme of cytochrome P-450 in these microsomes,[150,151] and epoxide hydrolase.[21,62] The data obtained with the purified, reconstituted system on metabolism of the polycyclic hydrocarbons prove that a single isozyme of cytochrome P-450[283] is capable of oxidizing the PAHs at multiple sites (cf. refs. 21,62). In contrast to microsomes from 3-methylcholanthrene-treated rats, the microsomes from control rats contain little cytochrome P-450c (~2% of total as cytochrome P-450c), as determined by immunoquantification studies.[151] Metabolic profiles of a series of PAHs produced by microsomes from control and phenobarbital-treated rats are very similar to those produced by microsomes from 3-methylcholanthrene-treated rats.[21,62] These results indicate marked similarity in the regioselectivity of several of the isozymes of cytochrome $P$-450 toward the PAH substrates.

## C. EFFECT OF SUBSTITUENTS ON THE METABOLISM AND BIOLOGICAL ACTIVITY OF POLYCYCLIC AROMATIC HYDROCARBONS

3-Methylcholanthrene, an alkyl-substituted PAH, was included in Table VII, along with other nonsubstituted PAHs, to demonstrate the dramatic effect of alkyl substituents on the metabolite profile of a PAH. 3-Methylcholanthrene is metabolized at about one-third the rate of its unsubstituted analog benz[a]anthracene.[279,280] Furthermore, the alkyl substituents are oxidized in preference to the aromatic moiety under conditions of linearity with respect to time and protein.[280] Less than 1% of the total metabolites of 3-methylcholanthrene are dihydrodiols. In contrast, dihydrodiols account for 96% of the total metabolites of benz[a]anthracene. Whether this is attributable to steric factors or to selection of cytochrome $P$-450c for benzylic oxidation in preference to aromatic oxidation is not clear. When 7,12-dimethylbenz[a]anthracene is metabolized by liver microsomes from 3-methylcholanthrene-treated rats, approximately 40% of total metabolites are formed by benzylic oxidation.[52] Dihydrodiols constitute ~50% of the total metabolites. The 8,9-dihydrodiol is the major metabolite, accounting for greater than 30% of total metabolism.[52] When 5-methylchrysene is metabolized by rat liver homogenates, benzylic hydroxylation accounts for about 20% of the total metabolites.[284] Interestingly, 1-hydroxy and 2-hydroxy derivatives of 3-methylcholanthrene are metabolized by liver microsomes from 3-methylcholanthrene-treated rats at very different rates. Under identical conditions, the 1-hydroxy derivative is metabolized at an approximately 10-fold greater rate than the 2-hydroxy derivative.[280] Although the parent hydrocarbon is not metabolized to detectable levels of the 9,10-dihydrodiol, which has a bay region double bond, 1-hydroxy-3-methylcholanthrene is metabolized to a pair of diastereomeric 9,10-dihydrodiols (Fig. 12), which account for about 11% of the total metabolites.[250] Hydroxylation at C-1 appears to be an important first step in the metabolic activation of this hydrocarbon to ultimate carcinogens, the bay region diol epoxides.[249-252] Benzylic hydroxylation also appears to play some role in the metabolic activation of 7,12-dimethylbenz[a]anthracene.[263,269,285] The role of benzylic hydroxylation is not entirely clear, however, because the 3,4-dihydrodiol of 7,12-dimethylbenz[a]anthracene, with a bay region double bone, was found to be several fold more active as a tumor initiator (in

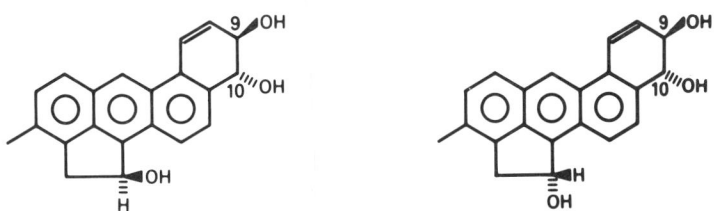

Fig. 12. Structures of the diastereomeric 9,10-dihydrodiols formed from racemic 1-hydroxy-3-methylcholanthrene by liver microsomes from 3-methylcholanthrene-treated rats. Absolute configuration is not implied.

CD-1 mice) than the corresponding 3,4-dihydrodiol of 7-hydroxymethyl-12-methyl-benz[a]anthracene.[286]

A 6-fluoro substituent in B[a]P, although not present on the critical benzo ring, reduces the tumor-initiating activity of B[a]P.[281,287] Hence, metabolism of 6-fluoro-B[a]P has been examined[281] (Table VII). Fluorination at the 6-position only slightly reduces the rate of metabolism of B[a]P. The overall distribution of the metabolites remains the same with two exceptions: The percentage of the 7,8-dihydrodiol with 9,10-bay region double bond is increased by greater than twofold, and the percentage of quinones is slightly decreased. Interestingly, quinone formation is accompanied by loss of fluorine. The 6-fluorinated 7,8- and the 4,5-dihydrodiols both have their hydroxyl groups in a pseudodiaxial conformation,[281,288] in contrast to the pseudodiequatorial conformation for these hydroxyl groups in the corresponding dihydrodiols of B[a]P. The 6-fluoro-7,8-dihydrodiol with pseudodiaxial conformation is metabolized to a somewhat lesser extent to bay region diol epoxides than is B[a]P 7,8-dihydrodiol.[281,281a] Moreover, both of the diastereomeric diol epoxides formed from the fluorodiol would be expected to have their hydroxyl groups in the pseudodiaxial conformation, and bay region diol epoxides with pseudodiaxial hydroxyl groups have not been found to be highly tumorigenic. Hence, metabolism studies have provided a basis for the "peri-effect,"[220] that is, the inhibitory effect on tumorigenicity of substituents peri to the benzo ring. Metabolism studies of 7-, 8-, 9-, and 10-fluoro-B[a]P[288] established that substitution of hydrogen by fluorine completely inhibited the formation of dihydrodiols at sites where oxygenated substituents would have been introduced at fluorinated positions. The lack of tumor-initiating activity of 7-, 8-, 9-, and 10-fluoro-B[a]P[288] can be attributed primarily to the inhibition of the metabolic formation of bay region 7,8-diol 9,10-epoxides.

A nonbenzo-ring, bay region methyl substituent generally enhances the tumorigenicity of the PAH. This has been demonstrated for several PAHs, including 7,12-dimethylbenz[a]anthracene,[289,290] 5-methyl- and 5,11-dimethylchrysene,[284,291] 11-methylcyclopentaphenanthrene-17-one,[292] 11-methyl B[a]P,[293] 1,4- and 4,10-dimethylphenanthrene,[294,295] and 7,12-dimethyldibenz[a,h]anthracene, 7,14-dimethyldibenz[a,j]anthracene, and 3,6-dimethylcholanthrene.[296,297] Steric strain and resultant deformation from planarity have been considered potential factors responsible for the enhanced tumorigenic activity of 7,12-dimethylbenz[a]anthracene and other PAHs with nonbenzo-ring, bay region methyl substituents. The high tumorigenicity of the bay region diol epoxides of benzo[c]phenanthrene,[245,246] an unsubstituted but nonetheless highly hindered and nonplanar PAH, provide the initial support for this concept. To explore the role of steric strain in the bay region, the tumorigenic activity of 3,6-dimethylcholanthrene and 7,11,12-trimethylbenz[a]anthracene has been examined.[297] 3,6-Dimethyl-cholanthrene was found to be two- to threefold more potent as a tumor initiator on mouse skin than 3-methylcholanthrene at doses of 15 and 45 nmol. At a lower dose (5 nmol), 3-methylcholanthrene was inactive, and 3,6-dimethylcholanthrene produced an average of 1.27 tumors per mouse and a 57% tumor incidence. 7,11,12-Trimethylbenz[a]anthracene was tested to determine if a further increase in the steric strain in the bay region of 7,12-dimethylbenz[a]anthracene would produce a more potent carcinogen.[297] The results indicated that the 11-methyl group caused at least a 20-fold decrease in the tumor-initiating activity relative to 7,12-dimethylbenz[a]anthracene; 3.3 tumors/mouse at a 5-nmol dose of 7,12-dimethylbenz[a]anthracene vs 1.1 tumors/mouse at a 45-nmol dose of 7,11,12-trimethylbenz[a]anthracene. In view of a much lower tumorigenicity of the severely hindered and strained 7,11,12-trimethylbenz[a]anthracene, the enhancement in the tumorigenic activity caused by nonbenzo-ring, bay region substitution by methyl groups remains unexplained.

## VII. Stereoselectivity in the Metabolism of the Polycyclic Aromatic Hydrocarbons

### A. A MODEL FOR THE CATALYTIC SITE OF CYTOCHROME P-450c

The results discussed in Section VI indicate that cytochrome $P$-450c shows broad substrate selectivity and regioselectivity in the metabolism of PAHs. Formation of highly enantiomerically enriched B[a]P 4,5- and 7,8-oxides by liver microsomes from 3-methylcholanthrene-treated rats, however, in which the (4S,5R)-oxide and the (7R,8S)-oxide enantiomers predominated over the other enantiomers by more than 20 to 1 (see Section III,B), demonstrated the high stereoselectivity of cytochrome P-450c. High stereoselectivity by cytochrome P-450c is also apparent in the metabolic formation of B[a]P 7,8-diol 9,10-epoxides from the (+)- and the (−)-enantiomers of B[a]P 7,8-dihydrodiol. (+)-B[a]P (7S,8S)-dihydrodiol and (−)-B[a]P (7R,8R)-dihydrodiol are metabolized by microsomes from 3-methylcholanthrene-treated rats to B[a]P 7,8-diol 9,10-epoxide-1 and B[a]P 7,8-diol 9,10-epoxide-2, respectively, in 95 and 85% diastereomeric excess (see Section III,C,2, Table II).

These results led to the formulation of a model for the binding site of cytochrome P-450c.[298] Two main considerations went into the formulation of this site model: (1) The steric constraints of the site must be such that it best accommodates the substrates in the orientations that result in epoxidation with the preferred stereoselectivity (Fig. 13), and (2) the overall shape of the site must be such that it represents a composite of all substrates, when the double bonds being epoxidized are aligned (Fig. 13). This hypothetical composite molecule represents a minimal boundary for the hydrophobic cleft or depression of the binding site of cytochrome P-450c[298] (Fig. 13).

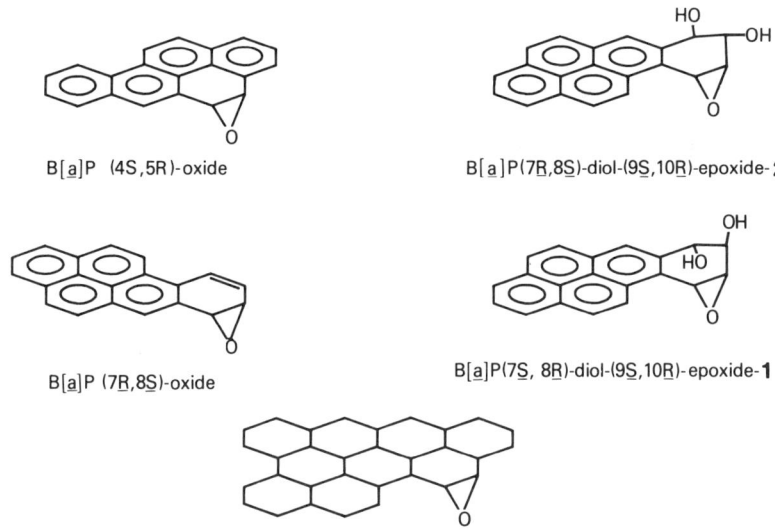

Fig. 13. A model for the minimum area of the binding site of cytochrome P-450c (298).

## B. STEREOSELECTIVITY IN THE METABOLIC FORMATION OF DIHYDRODIOLS FROM POLYCYCLIC AROMATIC HYDROCARBONS SUPPORTS THE PREDICTIONS OF THE BINDING-SITE MODEL

The proposed binding-site model for cytochrome P-450c predicts that enzymatically formed arene oxides on benzo rings of the PAH, which fit into the hypothetical cavity, must have (R)- and (S)-absolute configuration at their benzylic and allylic centers, respectively. Epoxide hydrolase-catalyzed hydrolysis of benzo-ring arene oxides occurs by inversion of the allylic centers[68,74,119,128,129] to form dihydrodiols, and hence, the (R,R)-enantiomers predominate. Stereoselectivity in the formation of dihydrodiols has been examined for benz[a]anthracene,[299] phenanthrene,[54] chrysene,[54] benzo[c]phenanthrene,[278] and the 9,10-dihydrodiol of B[a]P.[138a] These studies (Table VIII)[54,55,277,278,299,300] have established that all of the benzo-ring dihydrodiols (Table VIII) formed contain the (R,R)-enantiomer to the extent of 90 to 98%, which is consistent with the predictions of the model. Although the absolute configuration of dibenz[a,h]anthracene 3,4-dihydrodiol has not been determined, the metabolically formed dihydrodiol is formed with high enantiomeric purity[277] (Table VIII). The predominant enantiomer should have (R,R)-absolute configuration, according to the predictions of the model.

The regiospecificity of epoxide hydrolase toward K-region arene oxides is more complex than it is for the benzo-ring arene oxides. For example, enzymatic hydration of the (+)-(4S,5R)- and (−)-(4R,5S)-oxides of B[a]P occurs by inversion at the S center to form predominantly the (+)-(4R,5R)-dihydrodiol in each case.[129] In contrast, the hydration of benz[a]anthracene 5,6-oxide is much less specific. In this case, the (+)-(5S,6R)-oxide is converted predominantly to the (+)-(5R,6R)-dihydrodiol, but the (−)-(5R,6S)-oxide is hydrated to form 5,6-dihydrodiol of low enantiomeric purity with only a slight excess (7%) of the (−)-(5S,6S)-enantiomer.[129] Hence absolute configuration and enantiomeric composition of the metabolically formed K-region dihydrodiols provide little information about the stereoselectivity of cytochrome P-450 in the metabolism of benz[a]anthracene at the K region. However, studies with homogeneous cytochrome P-450 have shown that benz[a]anthracene is oxidized to form predominantly the (+)-(5S,6R)-oxide.[301] These results confirm the prediction of the model for oxidation of benz[a]anthracene at the K-region 5,6- as well as at the 8,9-position of the hydrocarbon. The K-region dihydrodiol of benzo[c]phenanthrene is of low enantiomeric purity (Table VIII), although the model predicts that the precursor arene oxide should be highly enantiomerically enriched and that the predominant enantiomer should have (5S,6R)-abso-

**Table VIII**

Enantiomeric Composition and Absolute Configuration of Dihydrodiols of Polycyclic Aromatic Hydrocarbons Formed by Liver Microsomes from 3-Methylcholanthrene-Treated Rats

| Dihydrodiol | Enantiomeric composition (%)[a] | | Enantiomeric purity (%) |
|---|---|---|---|
| | (R,R) | (S,S) | |
| B[a]P | | | |
| 4,5- | 96 (−) | 4 (+) | 92 |
| 7,8- | 96 (−) | 4 (+) | 92 |
| 9,10- | 96 (−) | 4 (+) | 92 |
| Benz[a]anthracene | | | |
| 3,4- | —[b] (−) | —[b] (+) | —[b] |
| 5,6- | 81 (+) | 19 (−) | 62 |
| 8,9- | 98 (−) | 2 (+) | 96 |
| 10,11- | 98 (−) | 2 (+) | 96 |
| Phenanthrene | | | |
| 1,2- | 96.5 (−) | 3.5 (+) | 93 |
| 3,4- | 98.5 (−) | 1.5 (+) | 97 |
| 9,10- | 42 (+) | 58 (−) | 16 |
| Chrysene | | | |
| 1,2- | 90 (−) | 10 (+) | 80 |
| 3,4- | 98.5 (−) | 1.5 (+) | 97 |
| Benzo[c]phenanthrene | | | |
| 3,4- | 94 (−) | 6 (+) | 88 |
| 5,6- | 75[c,d] | 25 | 50[d] |
| Dibenz[a,h]anthracene 3,4- | 80[c] | 20 | 60 |

[a] Data for dihydrodiols of B[a]P from Table I are included for comparison. The data for dihydrodiols of benz[a]anthracene,[299] phenanthrene and chrysene,[54,55] benzo[c]phenanthrene,[278,300] and dibenz[a,h]anthracene[277] were obtained from indicated references. The sign of $[\alpha]_D$ for the (R,R)-enantiomers has been indicated in the parentheses.

[b] A sufficient quantity of benz[a]anthracene 3,4-dihydrodiol could not be isolated to determine its configuration when liver microsomes from 3-methylcholanthrene-treated rats were used.

[c] The absolute configuration of benzo[c]phenanthrene 5,6-dihydrodiol and dibenz[a,h]anthracene 3,4-dihydrodiol has not been determined.

[d] The benzo[c]phenanthrene 5,6-dihydrodiol was reported to be formed by liver microsomes from 3-methylcholanthrene-treated rats as a racemic mixture[278]. However, our recent studies indicate that the 5,6-dihydrodial is formed with ~50% enantiomeric purity.

lute configuration. The elucidation of stereoselectivity of cytochrome P-450c and epoxide hydrolase in the formation of the K-region dihydrodiol of benzo[c]phenanthrene is under study.

## C. REGIO- AND STEREOSELECTIVITY IN THE METABOLISM OF DIHYDRODIOLS OF POLYCYCLIC AROMATIC HYDROCARBONS

High stereoselectivity in the metabolism of dihydrodiols to diastereomeric bay region diol epoxides was first demonstrated in the case of B[a]P. The (−)-(R,R)-enantiomer of B[a]P 7,8-dihydrodiol was metabolized predominantly to (+)-7,8-diol 9,10-epoxide-2 by liver microsomes from 3-methylcholanthrene-treated rats[53,152] and by a highly purified system reconstituted with cytochrome P-450c[53] (Table IX). In contrast, the (+)-(S,S)-enantiomer of B[a]P 7,8-dihydrodiol was converted predominantly to (+)-7,8-diol 9,10-epoxide-1 by both the microsomal enzymes and the purified and reconstituted system[53] (Table IX). Thus, for both enantiomers of the B[a]P 7,8-dihydrodiol, oxygen is added to the same heterotopic face of the molecule.[302] These results also contributed to the formulation of the model for the binding site of cytochrome P-450c[298] (see Section VII,A). The model makes a clear prediction regarding the stereochemical course preferred in the metabolism of benzo-ring dihydrodiols of other PAHs to bay region diol epoxides. According to the predictions of the model, (S,S)-dihydrodiols with bay region double bonds should be metabolized by cytochrome P-450c to diol epoxide-1 diastereomers, and (R,R)-dihydrodiols should be metabolized to diol epoxide-2 diastereomers. This prediction implies that cytochrome P-450c ignores the presence of the hydroxyl groups in either enantiomer and that oxygen is added from the same diastereoheterotopic face of the dihydrodiol. A summary of the results on the metabolism of enantiomeric benzo-ring dihydrodiols of several PAHs to bay region diol epoxides is presented in Table IX. The diol epoxides formed from enantiomeric dihydrodiols of benz[a]anthracene, phenanthrene, chrysene, and benzo[c]phenanthrene consisted of the predicted diastereomer to the extent of 77 to 99%. The diol epoxides formed, if any, from the (+)-(3S,4S)-enantiomer of benz[a]anthracene 3,4-dihydrodiol were not sufficient to allow quantification.

Metabolism of B[a]P 7,8-dihydrodiol by liver microsomes from 3-methylcholanthrene-treated rats occurs with high regioselectivity, so that 90–95% of total metabolites formed are bay region diol epoxides (Table IX). The percentage of bay region diol epoxides formed is lower when dihydrodiols with bay region double bonds of phenanthrene (78% of total metabolites) and chrysene (38–66% of total metabolites) are metabolized and is lowest when dihydrodiols of benz[a]anthracene (13% of total metabolites) and benzo[c]phenanthrene (5–8% of total metabolites) are metabolized. The 3,4-dihydrodiols of benz[a]anthracene and benzo[c]phenanthrene are metabolized to form bisdihydrodiols as their major

### Table IX
Comparative Metabolism of Dihydrodiols with Bay Region Double Bonds to Bay Region Diol Epoxides by Liver Microsomes from 3-Methylcholanthrene-Treated Rats[a]

| Dihydrodiol | Bay region diol epoxides (% of total metabolites) | Relative proportions of | |
| --- | --- | --- | --- |
| | | Diol epoxide-1 (%) | Diol epoxide-2 (%) |
| (−)-(7$R$,8$R$)-B[$a$]P | 95 | 14 | 86 |
| (+)-(7$S$,8$S$)-B[$a$]P | 90 | 96 | 4 |
| (−)-(3$R$,4$R$)-Benz[$a$]anthracene | 13 | <2 | >98 |
| (+)-(3$S$,4$S$)-Benz[$a$]anthracene | — | — | — |
| (−)-(1$R$,2$R$)-Phenanthrene[b] | 78 | 15 | 85 |
| (+)-(1$S$,2$S$)-Phenanthrene | —[b] | 85 | 15 |
| (−)-(1$R$,2$R$)-Chrysene | 66 | 5 | 95 |
| (+)-(1$S$,2$S$)-Chrysene | 38 | 86 | 14 |
| (−)-(3$R$,4$R$)-Benzo[$c$]phenanthrene | 8 | 23 | 77 |
| (+)-(3$S$,4$S$)-Benzo[$c$]phenanthrene | 5 | >99 | <1 |

[a] Data for dihydrodiols of B[$a$]P,[53] benz[$a$]anthracene,[65] phenanthrene,[54,60] chrysene,[59] and benzo[$c$]phenanthrene[61] were obtained from indicated sources.

[b] Quantification of bay region diol epoxides as percentage of total metabolites was obtained from Nordqvist et al.[54] The radiolabeled substrate used in that study was obtained biosynthetically and consisted almost exclusively (97%) of the (−)-($R$,$R$)-enantiomer. Studies with (+)- and (−)-enantiomers of phenanthrene 1,2-dihydrodiol were carried out with unlabeled substrate,[60] and thus no quantification was possible.

metabolites[62,65]; they are probably formed by initial oxidation to dihydrodiol arene oxides and subsequent enzymatic hydrolysis to the bisdihydrodiols. Small amounts of diol phenols are also formed from benzo[$c$]phenanthrene 3,4-dihydrodiols. Aside from forming only small amounts of diol epoxides from benz[$a$]anthracene 3,4-dihydrodiol, the cytochromes $P$-450 were also found to have a new catalytic activity toward this substrate. Benz[$a$]anthracene 3,4-dihydrodiol was metabolized by microsomal enzymes to the 3,4-quinone[65] (Fig. 14). The role of cytochrome $P$-450 was established by showing that antibodies to cytochrome $P$-450c and to NADPH−cytochrome $c$ reductase inhibited quinone formation.[65] A possible mechanism for formation of the quinone is initial oxidation of the 3,4-dihydrodiol to the catechol and subsequent autoxidation or enzymatic oxidation to the quinone (see ref. 65, and references therein). This was the first demonstration of a dehydrogenase-like activity of cytochrome $P$-450 toward a dihydrodiol substrate. Preliminary experiments showed the quinone to be nonmutagenic to bacteria.

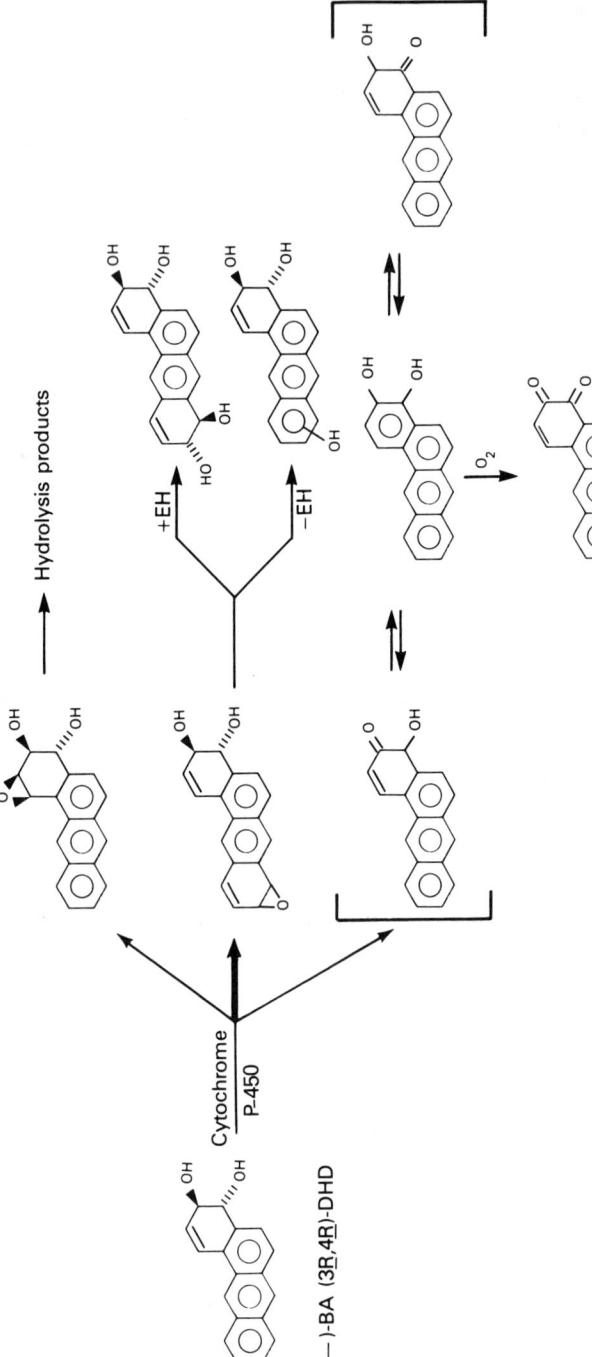

Fig. 14. Metabolism of (−)-(3R,4R)-benz[a]anthracene 3,4-dihydrodiol [(−)-BA (3R,4R)-DHD] by rat liver microsomes. The position and relative stereochemistry of the arene oxide on the dihydrodiol is arbitrary. A similar profile of metabolites is obtained from the (+)-(S,S)-enantiomer, except that the bay region diol epoxide is not detectable as a metabolite. EH, Epoxide hydrolase.

## VIII. Stereoselectivity in the Biological Activity of Polycyclic Aromatic Hydrocarbon Derivatives

Mutagenicity and tumorigenicity studies have shown that the biological activity of PAH metabolites is also highly dependent on their relative and

Table X

Skin Tumor-Initiating Activities of Enantiomers of Dihydrodiols with Bay Region Double Bonds and of Enantiomeric Bay Region Diol Epoxides of Benzo[a]pyrene, Benz[a]anthracene, and Chrysene in Female Charles River CD-1 mice[a]

| Initiator | Dose ($\mu$mol) | Mice with tumors (%) | Tumors per mouse |
|---|---|---|---|
| B[a]P | 0.1 | 77 | 2.60 |
| (−)-(7R,8R)-Dihydrodiol | 0.1 | 77 | 3.80 |
| (+)-(7S,8S)-Dihydrodiol | 0.1 | 23 | 0.43 |
| (−)-Diol epoxide-1 | 0.1 | 10 | 0.10 |
| (+)-Diol epoxide-1 | 0.1 | 17 | 0.17 |
| (−)-Diol epoxide-2 | 0.1 | 6 | 0.06 |
| (+)-Diol epoxide-2 | 0.1 | 47 | 1.10 |
| Benz[a]anthracene | 0.4 | 7 | 0.07 |
|  | 2.0 | 21 | 0.40 |
| (−)-(3R,4R)-Dihydrodiol | 0.1 | 43 | 1.00 |
|  | 0.4 | 50 | 1.80 |
| (+)-(3S,4S)-Dihydrodiol | 0.1 | 3 | 0.07 |
|  | 0.4 | 20 | 0.40 |
| Chrysene | 0.4 | 43 | 0.77 |
|  | 1.2 | 43 | 1.07 |
| (−)-(1R,2R)-Dihydrodiol | 0.4 | 67 | 1.47 |
|  | 1.2 | 83 | 2.77 |
| (+)-(1S,2S)-Dihydrodiol | 0.4 | 3 | 0.03 |
|  | 1.2 | 23 | 0.40 |
| (−)-Diol epoxide-1 | 0.4 | 7 | 0.07 |
|  | 1.2 | 13 | 0.13 |
| (+)-Diol epoxide-1 | 0.4 | 13 | 0.13 |
|  | 1.2 | 21 | 0.24 |
| (−)-Diol epoxide-2 | 0.4 | 10 | 0.13 |
|  | 1.2 | 23 | 0.33 |
| (+)-Diol epoxide-2 | 0.4 | 31 | 0.52 |
|  | 1.2 | 60 | 1.47 |

[a] All compounds were applied at indicated doses and were followed 1 week later by twice-weekly applications of 10 $\mu$g of tetradecanoylphorbol 13-acetate. The experiments lasted from 20 to 25 weeks. Data for 7,8-dihydrodiols of B[a]P,[135,303] diol epoxides of B[a]P,[217] dihydrodiols of benz[a]anthracene,[229] and dihydrodiols and diol epoxides of chrysene[241] were obtained from the indicated sources.

### Table XI

Pulmonary Tumors in Newborn Swiss–Webster Mice Treated with the Enantiomeric Dihydrodiols with Bay Region Double Bonds and/or Bay Region Diol Epoxides from Benzo[a]pyrene, Benz[a]anthracene, and Chrysene[a]

| Compound | Dose (μmol) | Mice with tumors (%) | Tumors per mouse[b] |
|---|---|---|---|
| B[a]P | 0.14 | 74 | 4.13 |
| (−)-(7R,8R)-Dihydrodiol | 0.14 | 98 | 9.28 |
|  | 0.70 | 100 | 32.20 |
| (+)-(7S,8S)-Dihydrodiol | 0.14 | 16 | 0.16 |
|  | 0.70 | 54 | 2.34 |
| (−)-Diol epoxide-1 | 0.007 | 13 | 0.15 |
|  | 0.014 | 22 | 0.25 |
| (+)-Diol epoxide-1 | 0.007 | 14 | 0.15 |
|  | 0.014 | 15 | 0.34 |
| (−)-Diol epoxide-2 | 0.007 | 9 | 0.09 |
|  | 0.014 | 12 | 0.13 |
| (+)-Diol epoxide-2 | 0.007 | 71 | 1.72 |
|  | 0.014 | 100 | 7.67 |
| Benz[a]anthracene | 0.28 | 16 | 0.22 |
| (−)-(3R,4R)-Dihydrodiol | 0.28 | 72 | 1.88 |
| (+)-(3S,4S)-Dihydrodiol | 0.28 | 9 | 0.09 |
| Chrysene | 1.4 | 13 | 0.16 |
| (−)-(1R,2R)-Dihydrodiol | 1.4 | 92 | 9.01 |
| (+)-(1S,2S)-Dihydrodiol | 1.4 | 18 | 0.37 |
| (−)-Diol epoxide-1 | 0.7 | 22 | 0.24 |
| (+)-Diol epoxide-1 | 0.7 | 19 | 0.20 |
| (−)-Diol epoxide-2 | 0.7 | 11 | 0.13 |
| (+)-Diol epoxide-2 | 0.7 | 90 | 5.45 |

[a] Data for dihydrodiols[304] and diol epoxides of B[a]P,[306] dihydrodiols of benz[a]anthracene,[232] and dihydrodiols and diol epoxides of chrysene[241] were obtained from the indicated sources. Swiss–Webster mice were injected intraperitoneally with one-seventh, two-sevenths, and four-sevenths of the indicated doses of the compounds within 24 hr of birth and at 8 and 15 days of age, respectively. Mice were sacrificed at 17 to 41 weeks of age depending on the hydrocarbon being examined.

[b] Tumors per mouse represent average tumors per mouse in both male and female mice.

absolute configuration. The tumorigenicity data for isomeric bay region-diol epoxides of B[a]P and for (+)-and (−)-enantiomers of B[a]P 7,8-dihydrodiol are shown in Tables X and XI. The (−)-(7R,8R)-dihydrodiol has 8- to 9-fold higher tumor-initiating activity (Table X) than the (+)-(7S,8S)-enantiomer on mouse skin[303] and also produced greater than 14-fold more lung tumors per mouse when tested in newborn mice[304] (Table XI). Among the four isomeric diol epoxides, the (+)-7,8-diol 9,10-epoxide-2 was severalfold more mutagenic to Chinese hamster V79 cells[305] and was the only isomer with high tumorigenic activity on mouse skin[217] and in newborn mice.[306] Thus, the (+)-7,8-diol 9,10-epoxide-2 was established as the ultimate carcinogen of this hydrocarbon. Subsequently, studies with chrysene derivatives also showed that the (−)-(R,R)-enantiomer of the 1,2-dihydrodiol was more tumorigenic than the (+)-(S,S)-enantiomer, and that the (+)-1,2-diol 3,4-epoxide-2 was the most tumorigenic diol epoxide among the four possible isomers (Tables X and XI).[241] The (−)-(3R,4R)-dihydrodiol of benz[a]anthracene, the precursor for (+)-3,4-diol 1,2-epoxide-2,[65] was highly tumorigenic in the newborn mouse at a dose that produced no tumors when the (+)-(3S,4S)-enantiomer was tested[232] (Table X). Similarly, the (−)-enantiomer was at least fourfold more potent than the (+)-enantiomer as a tumor initiator on mouse skin[229] (Table IX). Thus, for three different hydrocarbons, the bay region diol epoxides with the same absolute configuration [(R,S)-diol (S,R)-epoxide] are much more tumorigenic than the other isomers. When their bay regions are aligned, all three diol epoxides are superimposable. Further studies are in progress to determine whether this configuration is essential for the high tumorigenicity of other bay region diol epoxides.

## IX. Summary

The PAHs must be metabolically activated to elicit their mutagenic and carcinogenic activities. Oxidative metabolism of PAH leads to the formation of polar metabolites, some of which are noncarcinogenic and are eliminated directly or after conjugation. These metabolic pathways represent the detoxication of PAHs. The same oxidative enzymes also produce reactive molecules capable of eliciting mutagenic or carcinogenic responses, or both. Bay region diol epoxides are the only known ultimate mutagenic and carcinogenic metabolites of the PAHs. To date, bay region diol epoxides have been shown to be, or implicated as, ultimate carcinogens for over a dozen PAHs. These results confirm the predictions of the bay region theory.

The extent to which a PAH is metabolized to its bay region diol epoxides or to any other carcinogenic metabolites plays a major role in the carcinogenicity of the PAHs. Relative and absolute configuration of the dihydrodiols and bay region diol epoxides also plays a pivotal role in the expression of biological activity. Only a specific enantiomer of diol epoxide-2, the (R,S)-diol (S,R)-epoxide, has high tumorigenic activity for the three hydrocarbons examined thus far. Several possible mechanisms can be postulated to explain the high degree of stereoselectivity observed in the tumorigenic activity of diol epoxides: (a) covalent interactions of enantiomeric diol epoxides with chiral target molecules, (b) selective excision of the covalent adducts, leading to repair of the modified DNA, (c) stereoselective reversible interactions with carrier proteins, and (d) selective covalent reactions with chiral nucleophiles, leading to the inactivation of diol epoxides. For the three PAHs examined thus far, the most tumorigenic isomer of the diol epoxide is also the predominant isomer formed metabolically. Another aspect of stereochemistry that bears on the carcinogenicity of PAH is the conformation of hydroxyl groups in bay region diol epoxides. Bay region diol epoxides with pseudodiaxial hydroxyl groups are nontumorigenic or only weakly so. Conceivably, adverse steric interactions due to the axial hydroxyl groups could hinder the reaction at the critical site on the target molecules. In addition to these factors, several other physicochemical properties of bay region diol epoxides, such as stability, solubility, size, and shape, may also prove to be important in the expression of their carcinogenic activity.

Examination of stereoselectivity in the metabolic transformations of PAHs to dihydrodiols and diol epoxides has led to the formulation of a model for the active site of cytochrome $P$-450c. This model has, thus far, explained and predicted satisfactorily the absolute configuration of arene oxides, dihydrodiols, and diol epoxides formed from PAHs. An increased understanding of the metabolic transformations of PAHs to their ultimate carcinogens and of the various factors that influence the carcinogenic potency of the ultimate carcinogens should provide us with better predictions of the carcinogenic potential of untested polycyclic hydrocarbons.

## Acknowledgments

The authors gratefully acknowledge the excellent help of Mrs. Dorothy Dougherty, Ms. Colleen LePore, and Ms. Judy Thomas in the preparation of this manuscript.

## References

1. Bloch, B., and Dreifuss, W. (1921). Über die experimentelle Erzeugung von Carcinomen mit Lymphdrüzen und Lungenmetastasen durch Teerbestandteile. *Schweiz. Med. Wochenschr.* **51,** 1033-1037.
2. Yamagiwa, K., and Ichikawa, K. (1915). Experimentelle Studie über die Pathogenese der Epithelialgeschwulste. *Mitt. Med. Fak. Tokio* **15,** 295-344.
3. Tsutsui, H. (1918). Über das künstlich erzeugte Cancroid bei der Maus. *Gann* **12,** 17-21.

4. Hieger, I. (1930). The spectra of cancer-producing tars and oils and of related substances. *Biochem. J.* **24,** 505–511.
5. Cook, J. W., Hewett, C. L., and Hieger, I. (1933). The isolation of a cancer-producing hydrocarbon from coal tar. *J. Chem. Soc.* pp. 395–405.
6. Hartwell, J. L., and Shubika, P. (1967). "Survey of Compounds which have been Tested for Carcinogenic Activity." Thompson, Rockville, Maryland.
7. Van Duuren, B. L., Sivak, A., Langseth, L., Goldschmidt, B. M., and Segal, A. (1968). Initiators and promoters in tobacco carcinogenesis. *Natl. Cancer Inst. Monogr.* **28,** 173–180.
8. Dipple, A. (1976). Polynuclear aromatic carcinogens. *ACS Monogr.* **173,** 245–314.
9. Slaga, T. J., Becker, L., Bracken, W. M., and Weeks, C. E. (1979). The effects of weak or non-carcinogenic polycyclic hydrocarbons on 7,12-dimethyl benz[a]anthracene and benzo[a]pyrene skin tumor initiation. *Cancer Lett.* **7,** 51–59.
10. Buening, M. K., Levin, W., Wood, A. W., Chang, R. L., Lehr, R. G., Taylor, C., Yagi, H., Jerina, D. M., and Conney, A. H. (1980). Tumorigenic activity of benzo[e]pyrene derivatives on mouse skin and in newborn mice. *Cancer Res.* **40,** 203–206.
11. Pullman, A., and Pullman, B. (1955). Electronic structure and carcinogenic activity of aromatic molecules. *Adv. Cancer Res.* **3,** 117–169.
12. Herndon, W. C. (1974). Theory of carcinogenic activity of aromatic hydrocarbons. *Trans. N.Y. Acad. Sci.* [2] **36,** 200–217.
13. Scribner, J. D. (1975). Molecular orbital theory in carcinogenesis research. *J. Natl. Cancer Inst. (U.S.)* **55,** 1035–1038.
14. Miller, J. A. (1970). Carcinogenesis by chemicals: An overview. *Cancer Res.* **30,** 559–586.
15. Boyland, C. (1950). The biological significance of metabolism of polycyclic compounds. *Symp. Biochem. Soc.* **5,** 40–54.
16. Miller, E. C. (1951). Studies on the formation of protein-bound derivatives of 3,4-benzpyrene in epidermal reaction of mouse skin. *Cancer Res.* **11,** 100–108.
17. Heidelberger, C. (1973). Chemical carcinogenesis in culture. *Adv. Cancer Res.* **18,** 317–366.
18. Jerina, D. M., and Daly, J. W. (1974). Arene oxides: A new aspect of drug metabolism. *Science* **185,** 573–582.
19. Sims, P., and Grover, P. L. (1974). Epoxides in polycyclic aromatic hydrocarbon metabolism and carcinogenesis. *Adv. Cancer Res.* **20,** 165–274.
20. Nordqvist, M., Thakker, D. R., Yagi, H., Lehr, R. E., Wood, A. W., Levin, W., Conney, A. H., and Jerina, D. M. (1980). Evidence in support of the bay region theory as a basis for the carcinogenic activity of polycyclic aromatic hydrocarbons. *In* "Molecular Basis of Environmental Toxicity" (R. S. Bhatnagar, ed.), pp. 329–357. Ann Arbor Sci. Publ., Ann Arbor, Michigan.
21. Levin, W., Wood, A., Chang, R., Ryan, D., Thomas, P., Yagi, H., Thakker, D., Wyas, K., Boyd, C., Chu, S.-Y., Conney, A., and Jerina, D. (1982). Oxidative metabolism of polycyclic aromatic hydrocarbons to ultimate carcinogens. *Drug Metab. Rev.* **13,** 555–580.
22. Conney, A. H. (1967). Pharmacological implications of microsomal enzyme induction. *Pharmacol. Rev.* **19,** 317–366.
23. Kasper, C. (1971). Biochemical distinctions between the nuclear and microsomal membranes from rat hepatocytes. *J. Biol. Chem.* **246,** 577–581.
24. Khandwala, A. S., and Kasper, C. B. (1973). Preferential induction of aryl hydrocarbon hydroxylase activity in rat liver nuclear envelope by 3-methylcholanthrene. *Biochem. Biophys. Res. Commun.* **54,** 1241–1246.

25. Boyland, C., and Levi, A. A. (1935). Metabolism of polycyclic compounds. I. Production of dihydroxydihydroanthracene from anthracene. *Biochem. J.* **29**, 2679–2683.
26. Berenblum, Z., and Schoental, R. (1946). The metabolism of 3,4-benzpyrene into 8- and 10-benzpyrenols in the animal body. *Cancer Res.* **6**, 699–706.
27. Booth, J., and Boyland, E. (1949). Metabolism of polycyclic compounds. Formation of 1,2-dihydroxy-1,2-dihydronaphthalenes. *Biochem. J.* **44**, 361–365.
28. Boyd, D. R., and Jerina, D. M., (1984). Arene Oxides and oxepins. *In* "Small Ring Heterocycles" (A. Hassner, ed.), Part 3, pp. 197–282. Wiley, New York.
29. Boyland, C., and Sims, P. (1965). Metabolism of polycyclic compounds. The metabolism of 9,10-epoxy-9,10-dihydrophenanthrene in rats. *Biochem. J.* **95**, 788–792.
30. Boyland, E., and Sims, P. (1965). The metabolism of benz[a]anthracene and dibenz[a,h]anthracene and their 5,6-epoxy-5,6-dihydro derivatives by rat liver homogenates. *Biochem. J.* **97**, 7–16.
31. Jerina, D. M., Daly, J. W., Witkop, B., Zaltzman-Nirenberg, P., and Udenfriend, S. (1970). 1,2-Naphthalene oxide as an intermediate in the microsomal hydroxylation of naphthalene. *Biochemistry* **9**, 147–156.
32. Jeffrey, A. M., and Jerina, D. M. (1975). Novel rearrangements during dehydration of nucleophile adducts of arene oxides: A reappraisal of premercapturic acid structures. *J. Am. Chem. Soc.* **97**, 4427–4428.
33. Guroff, G., Daly, J. W., Jerina, D. M., Renson, J., Witkop, B., and Udenfriend, S. (1967). Hydroxylation-induced migrations: The NIH shift. *Science* **158**, 1524–1530.
34. Daly, J. W., Jerina, D. M., and Witkop, B. (1972). Arene oxides and the NIH shift: The metabolism, toxicity and carcinogenicity of aromatic compounds. *Experientia* **28**, 1129–1149.
35. Dutton, G. J., and Storey, I. D. E. (1954). Uridine compounds in glucuronic acid metabolism. *Biochem. J.* **57**, 275–283.
36. Dutton, G. J., and Storey, I. D. E. (1962). Glucuronide-joining enzymes. *In* "Methods in Enzymology" (S. P. Colowick and N. O. Kaplan, eds.), Vol. 5, pp. 159–164. Academic Press, New York.
37. Dahm, K., and Breuer, H. (1966). Biogenese von Östriol-16α-monoglucuronid und Östriol-17β-monoglucuronid. *Acta Endocrinol. (Copenhagen)* **52**, 43–53.
38. Nemoto, N., and Gelboin, H. V. (1976). Enzymatic conjugation of benzo[a]pyrene oxides, phenols and dihydrodiols with UDP-glucuronic acid. *Biochem. Pharmacol.* **25**, 1221–1226.
39. Lind, C., Vadi, H., and Ernster, L. (1978). Metabolism of benzo[a]pyrene 3,6-quinones and 3-hydroxybenzo[a]pyrene in liver microsomes from 3-methylcholanthrene-treated rats. *Arch. Biochem. Biophys.* **190**, 97–108.
40. Fahl, W., Shen, A. L., and Jefcoate, C. R. (1978). UDP-Glucuronosyl transferase and the conjugation of benzo[a]pyrene metabolites for DNA. *Biochem. Biophys. Res. Commun.* **85**, 891–899.
41. Dutton, G. J. (1980). "Glucuronidation of Drugs and Other Compounds." CRC Press, Boca Raton, Florida.
42. Gregory, J. D., and Lipmann, F. (1957). The transfer of sulfate among phenolic compounds with 3′,5′-diphosphoadenosine as coenzyme. *J. Biol. Chem.* **229**, 1081–1090.
43. Roy, A. B. (1960). The enzymatic synthesis of aryl sulphamates. *Biochem. J.* **74**, 49–56.
44. Singer, S. S., and Sylvester, S. (1976). Enzymatic sulfation of steroids. *Endocrinology* **98**, 963–974.
45. Farooqui, A. A., Rabel, G., and Mandel, P. (1977). Sulphatide metabolism in brain. *Life Sci.* **20**, 569–584.
46. Cohen, G. M., Haws, S. H., Moore, B. P., and Bridges, J. W. (1976). Benzo[a]pyrene-3-yl

hydrogen sulfate, a major ethyl acetate-extractable metabolite of benzo[a]pyrene in human, hamster and rat lung cultures. *Biochem. Pharmacol.* **25**, 2561–2570.
47. Cohen, G. M., Moore, B. P., and Bridges, J. W. (1977). Organic solvent-soluble sulfate ester conjugates of monohydroxybenzo[a]pyrenes. *Biochem. Pharmacol.* **26**, 551–553.
48. Nemoto, N., Takayama, S., and Gelboin, H. V. (1977). Enzymatic conversion of benzo[a]pyrene phenols, dihydrodiols and quinones to sulfate conjugates. *Biochem. Pharmacol.* **26**, 1825–1829.
49. Sims, P., Grover, P. L., Swaisland, A., Pal, K., and Hewer, A. (1974). Metabolic activation of benzo[a]pyrene proceeds by a diol epoxide. *Nature (London)* **252**, 326–328.
50. Thakker, D. R., Yagi, H., Lu, A. Y. H., Levin, W., Conney, A. H., and Jerina, D. M. (1976). Metabolism of benzo[a]pyrene: Conversion of (±)-*trans*-7,8-dihydroxy-7,8-dihydrobenzo[a]pyrene to the highly mutagenic 7,8-diol 9,10-epoxides. *Proc. Natl. Acad. Sci. U.S.A.* **73**, 2281–2285.
51. Huberman, G., Sachs, L., Yang, S. K., and Gelboin, H. V. (1976). Identification of mutagenic metabolites of benzo[a]pyrene in mammalian cells. *Proc. Natl. Acad. Sci. U.S.A.* **73**, 607–611.
52. Yang, S. K., Chou, M. W., Wislocki, P. G., and Lu, A. Y. H. (1980). Metabolism of 7,12-dimethylbenz[a]anthracene: Quantitation of metabolite formation in rat liver microsomes and a reconstituted enzyme system containing highly purified cytochrome *P*-450 or *P*-448. *In* "Polynuclear Aromatic Hydrocarbons: Chemistry and Biological Effects" (A. Bjorseth and A. J. Dennis, eds.), pp. 733–752. Battelle Press, Columbus, Ohio.
53. Thakker, D. R., Yagi, H., Akagi, H., Koreeda, M., Lu, A. Y. H., Levin, W., Wood, A. W., Conney, A. H., and Jerina, D. M. (1977). Metabolism of benzo[a]pyrene. VI. Stereoselective metabolism of benzo[a]pyrene and benzo[a]pyrene 7,8-dihydrodiol to diol epoxides. *Chem.-Biol. Interact.* **16**, 281–300.
54. Nordqvist, M., Thakker, D. R., Vyas, K. P., Yagi, H., Levin, W., Ryan, D. E., Thomas, P. E., Conney, A. H., and Jerina, D. M. (1981). Metabolism of chrysene and phenanthrene to bay region diol epoxides by rat liver enzymes. *Mol. Pharmacol.* **19**, 168–178.
55. Vyas, K. P., Yagi, H., Levin, W., Conney, A. H., and Jerina, D. M. (1981). Metabolism of (−)-*trans*-(3R,4R)-dihydroxy-3,4-dihydrochrysene to diol epoxides of liver microsomes. *Biochem. Biophys. Res. Commun.* **98**, 961–969.
56. Chou, M. W., Fu, P. P., and Yang, S. K. (1981). Metabolic conversion of dibenz[a,h]anthracene (±)-*trans*-1,2-dihydrodiol and chrysene (±)-*trans*-3,4-dihydrodiol to *vicinal*-dihydrodiol epoxides. *Proc. Natl. Acad. Sci. U.S.A.* **78**, 4270–4273.
57. Yang, S. K., and Chou, M. W. (1980). Metabolism of the bay region *trans*-1,2-dihydrodiol of benz[a]anthracene in rat liver microsomes occurs primarily at the 3,4-double bond. *Carcinogenesis (N.Y.)* **1**, 803–806.
58. Thakker, D. R., Levin, W., Buening, M., Yagi, H., Lehr, R. E., Wood, A. W., Conney, A. H., and Jerina, D. M. (1981). Species-specific enhancement by 7,8-benzoflavone of hepatic microsomal metabolism of benzo[e]pyrene 9,10-dihydrodiol to bay region diol epoxides. *Cancer Res.* **41**, 1389–1396.
59. Vyas, K. P., Levin, W., Yagi, H., Thakker, D. R., Ryan, D. E., Thomas, P. E., Conney, A. H., and Jerina, D. M. (1982). Stereoselective metabolism of the (+)- and (−)-enantiomers of *trans*-1,2-dihydroxy-1,2-dihydrochrysene to bay region 1,2-diol 3,4-epoxide diastereomers by rat liver enzymes. *Mol. Pharmacol.* **22**, 182–189.
60. Vyas, K. P., Thakker, D. R., Levin, W., Yagi, H., Conney, A. H., and Jerina, D. M. (1982). Stereoselective metabolism of the optical isomers of *trans*-1,2-dihydroxy-1,2-dihydrophenanthrene to bay region diol epoxides by rat liver microsomes. *Chem.-Biol. Interact.* **38**, 203–213.

61. Thakker, D. R., Ittah, Y., Levin, W., Croisy-Delcey, M., Conney, A. H., and Jerina, D. M. (1982). Metabolism of benzo[c]phenanthrene by rat liver microsomes and by a highly purified and reconstituted cytochrome P-450 system. *Proc. Int. Cancer Congr. 13th, 1982.* p. 164.
62. Thakker, D. R., Levin, W., Yagi, H., Conney, A. H., and Jerina, D. M. (1982). Regio- and stereoselectivity of hepatic cytochrome P-450 toward polycyclic aromatic hydrocarbon substrates. *In* "Advances in Experimental Medicine and Biology: Biological Reactive Intermediates IIA" (R. Synder, D. V. Parke, J. J. Kocsis, D. J. Jollow, C. G. Gibson, and C. M. Witmer, eds.), pp. 525–539. Plenum, New York.
63. Vyas, K. P., van Bladeren, P. J., Thakker, D. R., Yagi, H., Sayer, J. M., Levin, W., and Jerina, D. M. (1983). Regioselectivity and stereoselectivity in the metabolism of *trans*-1,2-dihydroxy-1,2-dihydrobenz[a]anthracene. *Mol. Pharmacol.* **24,** 115–123.
64. Thakker, D. R., Yagi, H., Lehr, R. E., Levin, W., Buening, M., Lu, A. Y. H., Chang, R. L., Wood, A. W., Conney, A. H., and Jerina, D. M. (1978). Metabolism of *trans*-9,10-dihydroxy-9,10-dihydrobenzo[a]pyrene occurs primarily by arylhydroxylation rather than formation of a diol epoxide. *Mol. Pharmacol.* **14,** 502–513.
65. Thakker, D. R., Levin, W., Yagi, H., Tada, M., Ryan, D. E., Thomas, P. E., Conney, A. H., and Jerina, D. M. (1982). Stereoselective metabolism of the (+)- and (−)-enantiomers of *trans*-3,4-dihydroxy-3,4-dihydrobenzo[a]anthracene by rat liver microsomes and by a purified and reconstituted cytochrome P-450 system. *J. Biol. Chem.* **257,** 5103–5110.
66. Wood, A. W., Levin, W., Thakker, D. R., Yagi, H., Chang, R. L., Ryan, D. E., Thomas, P. E., Dansette, P. M., Whittaker, N., Turujman, S., Lehr, R. E., Kumar, S., Jerina, D. M., and Conney, A. H. (1979). Biological activity of benzo[e]pyrene: An assessment based on mutagenic activities and metabolic profiles of the polycyclic hydrocarbon and its derivatives. *J. Biol. Chem.* **254,** 4408–4415.
67. Ayengar, P. K., Hayaishi, V., Nakajima, M., and Tomida, J. (1959). Enzymatic aromatization of 3,5-cyclohexadiene 1,2-diol. *Biochim. Biophys. Acta* **33,** 111–119.
68. Jerina, D. M., Ziffer, H., and Daly, J. W. (1970). The role of the arene oxide–oxepin system in the metabolism of aromatic substrates. IV. Stereochemical considerations of dihydrodiol formation and dehydrogenation. *J. Am. Chem. Soc.* **92,** 1056–1061
69. Vogel, K., Bentley, P., Platt, K. L., and Oesch, F. (1980). Rat liver cytoplasmic dihydrodiol dehydrogenase purification to apparent homogeneity and properties. *J. Biol. Chem.* **255,** 9621–9625.
70. Glatt, H. R., Vogel, K., Bentley, P., and Oesch, F. (1979). Reduction of benzo[a]pyrene mutagenicity by dihydrodiol dehydrogenase. *Nature (London)* **277,** 319–320.
71. Wiebel, F. J. (1975). Metabolism of monohydroxybenzo[a]pyrenes by rat liver microsomes and mammalian cells in culture. *Arch. Biochem. Biophys.* **168,** 609–621.
72. Capdevila, J., Janstrom, B., Vadi, H., and Orrenius, S. (1975). Cytochrome P-450-linked activation of 3-hydroxybenzo[a]pyrene. *Biochem. Biophys. Res. Commun.* **65,** 894–900.
73. Capdevila, J., Estabrook, R. W., and Prough, R. A. (1978). The microsomal metabolism of benzo[a]pyrene phenols. *Biochem. Biophys. Res. Commun.* **82,** 518–525.
74. Yang, S. K., Roller, P. P., and Gelboin, H. V. (1977). Enzymatic mechanism of benzo[a]pyrene conversion to phenols and diols and an improved high-pressure liquid chromatographic separation of benzo[a]pyrene derivatives. *Biochemistry* **16,** 3680–3687.
75. Committee on Biologic Effects of Atmospheric Pollutants (1972). "Particulate Polycyclic Organic Matter." Nat. Acad. Sci., Washington, D.C.
76. Cook, J. W., Ludwiczak, R. S., and Schoental, R. (1950). Polycyclic aromatic hydrocar-

bons. Part XXXVI. Synthesis of three metabolic oxidation products of 3,4-benzpyrene. *J. Chem. Soc.* pp. 1112–1121.
77. Conney, A. H., Miller, E. C., and Miller, J. A. (1957). Substrate-induced synthesis and other properties of benzo[a]pyrene hydroxylase in rat liver. *J. Biol. Chem.* **228,** 753–766.
78. Sims, P. (1967). The metabolism of benzo[a]pyrene by rat liver homogenates. *Biochem. Pharmacol.* **16,** 613–618.
79. Sims, P. (1970). Qualitative and quantitative studies on the metabolism of a series of aromatic hydrocarbons by rat liver preparations. *Biochem. Pharmacol.* **19,** 795–818.
80. Holder, G., Yagi, H., Dansette, P., Jerina, D. M., Levin, W., Lu, A. Y. H., and Conney, A. H. (1974). Effects of inducers and epoxide hydrase on the metabolism of benzo[a]pyrene by liver microsomes and a reconstituted system: Analysis by high-pressure liquid chromatography. *Proc. Natl. Acad. Sci. U.S.A.* **71,** 4356–4360.
81. Selkirk, J. K., Croy, R. G., and Gelboin, H. V. (1974). Benzo[a]pyrene metabolites: Efficient and rapid separation by high-pressure liquid chromatography. *Science* **184,** 169–171.
82. Holder, G. Yagi, H., Levin, W., Lu, A. Y. H., and Jerina, D. M. (1975). Metabolism of benzo[a]pyrene. III. An evaluation of the fluorescence assay. *Biochem. Biophys. Res. Commun.* **65,** 1363–1370.
83. Selkirk, J. K., Croy, R. G., and Gelboin, H. V. (1975). Isolation and characterization of benzo[a]pyrene metabolites. *Arch. Biochem. Biophys.* **168,** 322–326.
84. Selkirk, J. K., Croy, R. G., and Gelboin, H. V. (1976). High-pressure liquid chromatographic separation of 10 benzo[a]pyrene phenols and the identification of 1-phenol and 7-phenol as new metabolites. *Cancer Res.* **36,** 922–926.
85. Seifried, H. E., Birkett, D. J., Levin, W., Lu, A. Y. H., Conney, A. H., and Jerina, D. M. (1977). Metabolism of benzo[a]pyrene: Effect of 3-methylcholanthrene pretreatment on metabolism by microsomes from lungs of genetically "responsive" and "nonresponsive" mice. *Arch. Biochem. Biophys.* **178,** 256–263.
86. Grover, P. L., Hewer, A., and Sims, P. (1974). Metabolism of polycyclic hydrocarbons by rat lung preparations. *Biochem. Pharmacol.* **23,** 323–332.
87. Stohs, S. J., Grafström, R. C., Burke, M. D., Molders, P. W., and Orrenius, S. G. (1976). The isolation of rat intestinal microsomes with stable cytochrome $P$-450 and their metabolism of benzo[a]pyrene. *Arch. Biochem. Biophys.* **177,** 105–116.
88. Bresnick, E., Stoming, T. A., Vaught, J. B., Thakker, D. R., and Jerina, D. M. (1977). Nuclear metabolism of benzo[a]pyrene and of ($\pm$)-*trans*-7,8-dihydroxy-7,8-dihydrobenzo[a]pyrene. Comparative chromatographic analysis of alkylated DNA. *Arch. Biochem. Biophys.* **183,** 31–37.
89. Jenstrom, B., Vadi, H., and Orrenius, S. (1976). Function of isolated rat liver microsomes and nuclei of benzo[a]pyrene metabolites that bind to DNA. *Cancer Res.* **36,** 4107–4113.
90. Pezzuto, J. M., Yang, C. S., Yang, S. K., McCourt, D. W., and Gelboin, H. V. (1978). Metabolism of benzo[a]pyrene and (−)-*trans*-7,8-dihydroxy-7,8-dihydrobenzo[a]pyrene by rat liver nuclei and microsomes. *Cancer Res.* **38,** 1241–1245.
91. Holder, G. H., Yagi, H., Jerina, D. M., Levin, W., Lu, A. Y. H, and Conney, A. H. (1975). Metabolism of benzo[a]pyrene. Effect of substrate concentration and 3-methylcholanthrene pretreatment on hepatic metabolism by microsomes from rats and mice. *Arch. Biochem. Biophys.* **170,** 557–566.
92. MacNicoll, A. D., Grover, P. L., and Sims, P. (1980). The metabolism of a series of polycyclic hydrocarbons by mouse skin maintained in short-term organ culture. *Chem.-Biol. Interact.* **29,** 169–188.
93. Berry, D. L., Brecken, W. R., Slaga, T. J., Wilson, N. M., Butty, S. G., and Jordan, M. R.

(1977). Benzo[a]pyrene metabolism in mouse epidermis: Analysis by high-pressure liquid chromatography and DNA binding. *Chem.-Biol. Interact.* **18**, 129–142.
94. Baird, W. M., and Diamond, L. (1977). The nature of benzo[a]pyrene adducts formed in hamster embryo cells depends on the length of time of exposure to benzo[a]pyrene. *Biochem. Biophys. Res. Commun.* **77**, 162–167.
95. Wang, E., Rasmussen, R. C., and Crucker, T. T. (1976). Metabolism of benzo[a]pyrene by microsomes from tissues of pregnant and fetal hamster. *Life Sci.* **15**, 1291–1300.
96. Hudley, S. G., and Frundenthal, R. I. (1977). High-pressure liquid chromatographic analysis of benzo[a]pyrene metabolism by microsomal enzymes from rhesus liver and lung. *Cancer Res.* **37**, 244–249.
97. Harris, C. C., Autrup, H., and Stoner, G. (1978). Metabolism of benzo[a]pyrene in cultured human tissues and cells. *In* "Polycyclic Aromatic Hydrocarbons and Cancer" (H. V. Gelboin and P.O.P. Ts'o, eds.), Vol. 2, pp. 331–342. Academic Press, New York.
98. Booth, J., Keysall, G. R., Pat, K., and Sims, P. (1974). The metabolism of polycyclic hydrocarbons by cultured human lymphocytes. *FEBS Lett.* **43**, 341–344.
99. Baird, W. M., and Diamond, L. (1978). Metabolism and DNA binding of polycyclic aromatic hydrocarbons by human diploid fibroblasts. *Int. J. Cancer* **22**, 189–195.
100. Marshall, M. V., McLemare, T. L., Martin, R. R., Jenkins, W. T., Snodgrass, D. R., Carson, M. A., Arnold, M. S., Wray, N. P., and Griffin, A. C. (1979). Patterns of benzo[a]pyrene metabolism in normal human pulmonary alveolar macrophages. *Cancer Lett.* **8**, 103–109.
101. Autrup, H., Wejald, F. C., Jeffrey, A. M., Tate, H., Schwartz, R. D., Trump, B. F., and Harris, C. C. (1980). Metabolism of benzo[a]pyrene by cultured tracheo-bronchoid tissues from mice, rats, hamsters, bovines and humans. *Int. J. Cancer* **25**, 293–300.
102. Varanasi, U., and Gumm, D. J. (1981). *In vivo* metabolism of naphthalene and benzo[a]pyrene by flat-fish. *In* "Chemical Analysis and Biological Fate: Polynuclear Aromatic Hydrocarbons" (M. Cooke and A. J. Dennis, eds.), pp. 367–376. Battelle Press, Columbus, Ohio.
103. Cerniglia, C. G., and Gibson, D. T. (1979). Oxidation of benzo[a]pyrene by the filamentous fungus *Cunninghamella elegans*. *J. Biol. Chem.* **254**, 12174–12180.
104. Flesher, J. W., and Sydnor, K. L. (1973). Possible role of 6-hydroxymethylbenzo[a]pyrene as proximate carcinogen of benzo[a]pyrene and 6-methylbenzo[a]pyrene. *Int. J. Cancer* **11**, 433–437.
105. Sloan, N. H. (1975). α-Naphthoflavone activation of 6-hydroxymethylbenzo[a]pyrene synthetase. *Cancer Res.* **35**, 3731–3734.
106. Lasko, S., Caspary, W., Lorentzen, R., and Ts'o, P.O.P. (1975). Enzymatic formation of 6-oxobenzo[a]pyrene radical in rat liver homogenates from carcinogenic benzo[a]pyrene. *Biochemistry* **14**, 3978–3984.
107. Lorentzen, R. J., Caspary, W. S., Lasker, S. A., and Ts'o, P.O.P. (1975). The autoxidation of 6-hydroxybenzo[a]pyrene and 6-oxobenzo[a]pyrene radical reactive metabolites of benzo[a]pyrene. *Biochemistry* **14**, 3970–3977.
108. Nagata, C., Tagashira, Y., and Kodama, M. (1974). Metabolic activation of benzo[a]pyrene: Significance of free radical. *In* "Chemical Carcinogenesis" (P.O.P. Ts'o and J. A. DiPolo, eds.), pp. 87–111. Dekker, New York.
109. Weston. A., Grover, P. L., and Sims, P. (1982). Formation of the 11,12-diol as a metabolite of benzo[a]pyrene by rat skin *in vivo*. *Biochem. Biophys. Res. Commun.* **105**, 935–941.
110. Armstrong, R. N., Levin, W., Ryan, D. E., Thomas, P. E., Mah, M. D., and Jerina, D. M. (1981). Stereoselectivity of rat liver cytochrome $P$-450c on formation of benzo[a]pyrene 4,5-oxide. *Biochem. Biophys. Res. Commun.* **100**, 1077–1084.

111. Wood, A. W., Goode, R. L., Chang, R. L., Levin, W., Conney, A. H., Yagi, H., Dansette, P. M., and Jerina, D. M. (1975). Mutagenic and cytotoxic activity of benzo[a]pyrene 4,5-, 7,8-, and 9,10-oxides and the six corresponding phenols. *Proc. Natl. Acad. Sci. U.S.A.* **72**, 3176–3180.
112. Levin, W., Wood, A. W., Yagi, H., Dansette, P. M., Jerina, D. M., and Conney, A. H. (1976). Carcinogenicity of benzo[a]pyrene 4,5-, 7,8-, and 9,10-oxides on mouse skin. *Proc. Natl. Acad. Sci. U.S.A.* **73**, 243–247.
113. Jerina, D. M., Daly, J. W., Witkop, B., Zaltzman-Nirenberg, P., and Udenfriend, S. (1968). Role of the arene oxide–oxepin system in the metabolism of aromatic substrates. I. *In vitro* conversion of benzene oxide to a premercapturic acid and a dihydrodiol. *Arch. Biochem. Biophys.* **128**, 176–183.
114. Lu, A. Y. H., Jerina, D. M., and Levin, W. (1977). Liver microsomal epoxide hydrase: Hydration of alkene and arene oxides by membrane-bound and purified enzymes. *J. Biol. Chem.* **252**, 3715–3723.
115. Yang, S. K., Roller, P. P., Fu, P. P., Harvey, R. G., and Gelboin, H. V. (1977). Evidence for a 2,3-epoxide as an intermediate in the microsomal metabolism of benzo[a]pyrene to 3-hydroxybenzo[a]pyrene. *Biochem. Biophys. Res. Commun.* **77**, 1176–1182.
116. Boyd, D. R., Gadaginamath, G. S., Kher, A., Malone, J. F., Yagi, H., and Jerina, D. M. (1980). (+)- and (−)-Benzo[a]pyrene 7,8-oxide: Synthesis, absolute stereochemistry, and stereochemical correlation with other mammalian metabolites of benzo[a]pyrene. *J. Chem. Soc., Perkin Trans.* pp. 2112–2116.
117. Levin, W., Buening, M. K., Wood, A. W., Change, R. L., Kedzierski, B., Thakker, D. R., Boyd, D. R., Gadaginamath, G. S., Armstrong, R. N., Yagi, H., Karle, J. M., Slaga, T. J., Jerina, D. M., and Conney, A. H. (1980). An enantiomeric interaction in the metabolism and tumorigenicity of (+)- and (−)-benzo[a]pyrene 7,8-oxide. *J. Biol. Chem.* **255**, 9067–9074.
118. Kedzierski, B., Thakker, D. R., Armstrong, R. N., and Jerina, D. M. (1980). Absolute configuration of the K-region 4,5-dihydrodiols and 4,5-oxide of benzo[a]pyrene. *Tetrahedron Lett.* **22**, 405–408.
119. Thakker, D. R., Yagi, H., Levin, W., Lu, A. Y. H., Conney, A. H., and Jerina, D. M. (1977). Stereospecificity of microsomal and purified epoxide hydrase from rat liver: Hydration of arene oxides of polycyclic hydrocarbons. *J. Biol. Chem.* **252**, 6328–6334.
120. Nakanishi, K., Kasai, H., Cho, H., Harvey, R. G., Jeffrey, A. M., Jennette, K. W., and Weinstein, I. B. (1977). Absolute configuration of a ribonucleic acid adduct formed *in vivo* by metabolism of benzo[a]pyrene. *J. Am. Chem. Soc.* **99**, 258–260.
121. Yagi, H., Akagi, H., Thakker, D. R., Mah, H. D., Koreeda, M., and Jerina, D. M. (1977). Absolute stereochemistry of the highly mutagenic 7,8-diol 9,10-epoxides derived from the potent carcinogen *trans*-7,8-dihydroxy 7,8-dihydrobenzo[a]pyrene. *J. Am. Chem. Soc.* **99**, 2358–2359.
122. Booth, J., Hewer, A., Kepett, G. R., and Sims, P. (1975). Enzymatic reduction of aromatic hydrocarbon epoxides by the microsomal fraction of rat liver. *Xenobiotica* **50**, 197–203.
123. Kato, R., Iwasaki, K., Shiraga, T., and Noguchi, M. (1976). Evidence for the involvement of cytochrome *P*-450 in reduction of benzo[a]pyrene 4,5-oxide by rat liver microsomes. *Biochem. Biophys. Res. Commun.* **70**, 681–687.
124. Yamazoe, Y., Sugiura, M., Kamataki, T., and Kato, R. (1978). Reconstitution of benzo[a]pyrene 4,5-oxide reductase activity by purified cytochrome *P*-450. *FEBS Lett.* **88**, 337–340.
125. Sugiura, M., Yamazoe, Y., Kamataki, T., and Kato, R. (1980). Reduction of epoxy

derivatives of benzo[a]pyrene by microsomal cytochrome P-450. *Cancer Res.* **40**, 2910–2914.

126. Wrighton, S. A., Fahl, W. E., Shinnick, F. L., Jr., and Jefcoate, C. R. (1982). Characteristic of microsomal reduction of benzo[a]pyrene 4,5-oxide. *Chem.-Biol. Interact.* **40**, 345–356.

127. Jerina, D. M., Dansette, P. M., Lu, A. Y. H., and Levin, W. (1977). Hepatic microsomal epoxide hydrase: A sensitive radiometric assay for hydration of arene oxides of carcinogenic aromatic hydrocarbons. *Mol. Pharmacol.* **13**, 342–351.

128. Holtzman, J., Gillette, J. R., and Milne, G. W. A. (1967). The metabolic products of naphthalene in mammalian systems. *J. Am. Chem. Soc.* **89**, 6341–6344.

128a. van Bladeren, P. J., Vyas, K. P., Sayer, J. M., Ryan, D. E., Thomas, P. E., Levin, W., and Jerina, D. M. (1984). Stereoselectivity of cytochrome P-450c in the formation of Naphthalene and Anthracene 1,2-oxides. *J. Biol. Chem.* **259**, 8966–8973.

129. Armstrong, R. N., Kedzierski, B., Levin, W., and Jerina, D. M. (1981). Enantioselectivity of microsomal epoxide hydrolase toward arene oxide substrates. *J. Biol. Chem.* **256**, 4726–4733.

130. Watabe, T., Akamatsu, K., and Kiyonaga, K. (1971). Stereoselective hydrolysis of cis- and trans-stilbene oxides by hepatic microsomal epoxide hydrolase. *Biochem. Biophys. Res. Commun.* **44**, 198–204.

131. Dansette, P. M., Ziffer, H., and Jerina, D. M. (1976). Optically active 4-substituted cis-1,2-diphenylethylene oxides and related 1,2-diphenylethane diols. *Tetrahedron* **32**, 2071–2074.

132. Dansette, P. M., Makedonska, V. B., and Jerina, D. M. (1978). Mechanism of catalysis for the hydration of substituted styrene oxides by hepatic epoxide hydrase. *Arch. Biochem. Biophys.* **187**, 290–298.

133. Levin, W., Wood, A. W., Yagi, H., Jerina, D. M., and Conney, A. H. (1976). (±)-trans-7,8-Dihydroxy-7,8-dihydrobenzo[a]pyrene: A potent skin carcinogen when applied topically to mice. *Proc. Natl. Acad. Sci. U.S.A.* **73**, 3867–3871.

134. Kapitulnik, J., Levin, W., Conney, A. H., Yagi, H., and Jerina, D. M. (1977). Benzo[a]pyrene 7,8-dihydrodiol is more carcinogenic than benzo[a]pyrene in newborn mice. *Nature (London)* **266**, 378–380.

135. Levin, W., Wood, A. W., Wislocki, P. G., Kapitulnik, J., Yagi, H., Jerina, D. M., and Conney, A. H. (1977). Carcinogenicity of benzo ring derivatives of benzo[a]pyrene on mouse skin. *Cancer Res.* **37**, 3356–3361.

136. Kapitulnik, J., Wislocki, P. G., Levin, W., Yagi, H., Jerina, D. M., and Conney, A. H. (1978). Tumorigenicity studies with diol epoxides of benzo[a]pyrene which indicate that (±)-trans-7$\beta$,8$\alpha$-dihydroxy-9$\alpha$,10$\alpha$-epoxy-7,8,9,10-tetrahydrobenzo[a]pyrene is an ultimate carcinogen in newborn mice. *Cancer Res.* **38**, 354–358.

137. Slaga, T. J., Bracken, W. M., Viaje, A., Berry, D. L., Fischer, S. M., Miller, D. R., Levin, W., Conney, A. H., Yagi, H., and Jerina, D. M. (1978). Tumor-initiating and -promoting activities of various benzo[a]pyrene metabolites in mouse skin. *In* "Polynuclear Aromatic Hydrocarbons: Second International Symposium on Chemistry and Biology" (P. W. Jones and R. I. Freudenthal, eds.), pp. 371–382. Raven Press, New York.

138. Jerina, D. M., Yagi, H., Thakker, D. R., Karle, J. M., Mah, H. D., Boyd, D. R., Gadaginamath, G., Wood, A. W., Buening, M., Chang, R. L., Levin, W., and Conney, A. H. (1979). Stereoselective metabolic activation of polycyclic aromatic hydrocarbons. *In* "Advances in Pharmacology and Therapeutics" (Y. Cohen, ed.), Vol. 9, pp. 53–62. Pergamon Press, New York.

138a. Yagi, H., and Jerina, D. M. (1982). Absolute configuration of the trans-9,10-dihydrodiol metabolite of the carcinogen benzo[a]pyrene. *J. Am. Chem. Soc.* **104**, 4026–4027.

139. Borgen, A. O., Dover, H., Costagnoli, N., Crucker, T. T., Rasmussen, R. C., and Wong, I. Y. (1973). Metabolic conversion of benzo[a]pyrene by Syrian hamster liver microsomes and binding of metabolites to deoxyribonucleic acid. *J. Med. Chem.* **16**, 502–506.
140. McCaustland, D. J., and Engel, J. F. (1975). Metabolites of polycyclic aromatic hydrocarbons. II. *Tetrahedron Lett.* **30**, 2549–2552.
141. Yagi, H., Hernandez, O., and Jerina, D. M. (1975). Synthesis of (±)-7$\beta$,8$\alpha$-dihydroxy-9$\beta$,10$\beta$-epoxy-7,8,9,10-tetrahydrobenzo[a]pyrene, a potential metabolite of the carcinogen benzo[a]pyrene with stereochemistry related to the antileukemic triptolides. *J. Am. Chem. Soc.* **97**, 6881–6883.
142. Jerina, D. M., Yagi, H., Hernandez, O., Dansette, P. M., Wood, A. W., Levin, W., Chang, R. L., Wislocki, P. G., and Conney, A. H. (1976). Synthesis and biologic activity of potential benzo[a]pyrene metabolites. *In* "Polynuclear Aromatic Hydrocarbons: Chemistry, Metabolism, and Carcinogenesis" (R. I. Freudenthal and P. W. Jones, eds.), pp. 91–113. Raven Press, New York.
143. Jerina, D. M., Yagi, H., and Hernandez, O. (1977). Stereoselective synthesis and reactions of a diol epoxide derived from benzo[a]pyrene. *In* "Biological Reactive Intermediates" (D. J. Jollow, J. J. Kocsis, R. Synder, and H. Vainio, eds.), pp. 371–378. Plenum, New York.
144. Beland, F. A., and Harvey, R. G. (1976). The isomeric 9,10-oxides of *trans*-7,8-dihydroxy-7,8-dihydrobenzo[a]pyrene. *J. Chem. Soc., Chem. Commun.* pp. 84–85.
145. Yagi, H., Thakker, D. R., Hernandez, O., Koreeda, M., and Jerina, D. M. (1977). Synthesis and reactions of the highly mutagenic 7,8-diol 9,10-epoxides of the carcinogen benzo[a]pyrene. *J. Am. Chem. Soc.* **99**, 1604–1611.
146. Wislocki, P. G., Wood, A. W., Chang, R. L., Levin, W., Yagi, H., Hernandez, O., Jerina, D. M., and Conney, A. H. (1976). High mutagenicity and toxicity of a diol epoxide derived from benzo[a]pyrene. *Biochem. Biophys. Res. Commun.* **68**, 1006–1012.
147. Wood, A. W., Wislocki, P. G., Chang, R. L., Levin, W., Lu, A. Y. H., Yagi, H., Hernandez, O., Jerina, D. M., and Conney, A. H. (1976). Mutagenicity and cytotoxicity of benzo[a]pyrene benzo ring epoxides. *Cancer Res.* **36**, 3358–3366.
148. Newbold, R. F., and Brookes, P. (1976). Exceptional mutagenicity of a benzo[a]pyrene diol epoxide in cultured mammalian cells. *Nature (London)* **261**, 52–54.
149. Malaveilla, C., Bartsch, H., Grover, P. L., and Sims, P. (1975). Mutagenicity of non-K region diols and diol epoxides of benz[a]anthracene and benzo[a]pyrene in *S. typhimurium* TA 100. *Biochem. Biophys. Res. Commun.* **66**, 693–700.
150. Thomas, P. E., Korzeniowski, D., Ryan, D., and Levin, W. (1979). Preparation of monospecific antibodies against two forms of rat liver cytochrome *P*-450 and quantitation of these antigens in microsomes. *Arch. Biochem. Biophys.* **192**, 524–528.
151. Thomas, P. E., Reik, L. M., Ryan, D. E., and Levin, W. (1981). Regulation of the forms of cytochrome *P*-450 and epoxide hydrolase in rat liver microsomes: Effect of age, sex and induction. *J. Biol. Chem.* **256**, 1044–1052.
152. Yang, S. K., McCourt, D. W., Roller, P. P., and Gelboin, H. V. (1976). Enzymatic conversion of benzo[a]pyrene leading predominantly to *r*-7, *t*-8-dihydroxy-*t*-8,10-oxy-7,8,9,10-tetrahydrobenzo[a]pyrene through a single enantiomer of *r*-7, *t*-8-dihydroxy-7,8-dihydrobenzo[a]pyrene. *Proc. Natl. Acad. Sci. U.S.A.* **73**, 2594–2598.
153. Deutsch, J., Vatsis, K. P., Coon, M. J., Lentz, J. C., and Gelboin, H. V. (1979). Catalytic activity and stereoselectivity of purified forms of rabbit liver microsomal *P*-450 in the oxygenation of the (−)- and (+)-enantiomers of *trans*-7,8-dihydroxy-7,8-dihydrobenzo[a]pyrene. *Mol. Pharmacol.* **16**, 1011–1018.
154. Cerniglia, C. G., and Gibson, D. T. (1980). Fungal oxidation of benzo[a]pyrene and (±)-*trans*-7,8-dihydroxy-7,8-dihydrobenzo[a]pyrene. *J. Biol. Chem.* **255**, 5159–5163.

155. Marnett, L. J., Reid, G. A., and Dennison, D. J. (1978). Prostaglandin synthetase-dependent activation of 7,8-dihydro-7,8-dihydroxybenzo[a]pyrene to mutagenic derivatives. *Biochem. Biophys. Res. Commun.* **82**, 210–216.
156. Marnett, L. J., Johnson, J. T., and Biankowski, M. J. (1979). Arachidonic acid-dependent metabolism of 7,8-dihydroxy-7,8-dihydrobenzo[a]pyrene by ram seminal vesicles. *FEBS Lett.* **106**, 13–16.
157. Sivarajah, K., Murchtar, M., and Eling, T. (1979). Arachidonic acid-dependent metabolism of (±)-*trans*-7,8-dihydroxy-7,8-dihydrobenzo[a]pyrene (BP 7,8-diol) to 7,10/8,9-tetrols. *FEBS Lett.* **106**, 17–20.
158. Sivarajah, K., Lasker, J., and Eling, T. (1981). Prostaglandin synthetase-dependent cooxidation of (±)-benzo[a]pyrene 7,8-dihydrodiol by human lung and other mammalian tissues. *Cancer Res.* **41**, 1834–1839.
159. Panthananickel, N., and Marnett, L. S. (1981). Arachidonic acid-dependent metabolism of (±)-7,8-dihydroxy-7,8-dihydrobenzo[a]pyrene to polycyclic acid-binding derivatives. *Chem.-Biol. Interact.* **33**, 239–252.
160. Marnett, L. J., Reid, G. A., and Johnson, J. T. (1977). Prostaglandin synthetase-dependent benzo[a]pyrene oxidation: Products of the oxidation and inhibition of their formation by antioxidants. *Biochem. Biophys. Res. Commun.* **79**, 569–576.
161. Marnett, L. J., and Reid, G. A. (1979). Peroxidatic oxidation of benzo[a]pyrene and prostaglandin biosynthesis. *Biochemistry* **18**, 2923–2929.
162. Booth, J., and Sims, P. (1976). Different pathways involved in the metabolism of the 7,8- and 9,10-dihydrodiols of benzo[a]pyrene. *Biochem. Pharmacol.* **25**, 979–980.
163. Cerniglia, C. E., and Gibson, D. T. (1980) Fungal oxidation of (±)-9,10-dihydroxy-9,10-dihydrobenzo[a]pyrene: Formation of diastereomeric benzo[a]pyrene 9,10-diol 7,8-epoxides. *Proc. Natl. Acad. Sci. U.S.A.* **77**, 4554–4558.
164. Lu, A. J. H., Levin, W., Vore, M., Conney, A. H., Thakker, D. R., Holder, G., and Jerina, D. M. (1976). Metabolism of benzo[a]pyrene by purified liver microsomal cytochrome *P*-448 and epoxide hydrolase. *In* "Polynuclear Aromatic Hydrocarbons: Chemistry, Metabolism, and Carcinogenesis" (R. I. Freudenthal and P. W. Jones, eds.), pp. 115–126. Raven Press, New York.
165. Selkirk, J. K. (1980). Comparison of epoxide and free-radical mechanisms for activation of benzo[a]pyrene by Sprague–Dawley rat liver microsomes. *JNCI, J. Natl. Cancer Inst.* **64**, 771–774.
166. Tomaszewski, J. E., Jerina, D. M., and Daly, J. W. (1975). Deuterium isotope effects during formation of phenols by hepatic monooxygenases. Evidence for an alternative to the arene oxide pathway. *Biochemistry* **14**, 2024–2031.
167. King, H. W. S., Thompson, M. H., and Brookes, P. (1975). The role of 9-hydroxybenzo[a]pyrene in the microsome-mediated binding of benzo[a]pyrene to DNA. *Int. J. Cancer* **18**, 339–344.
168. Cohen, G. M., Ashworst, S. W., Selkirk, J. M., and Slaga, T. J. (1980). Hydrocarbon–deoxyribonucleoside adducts *in vivo* and *in vitro* and their relationship to carcinogenicity. *In* "Polynuclear Aromatic Hydrocarbons: Chemistry and Biological Effects" (A. Bjarselth and A. J. Dennis, eds.), pp. 503–521. Battelle Press, Columbus, Ohio.
169. Lorentzen, R. S., and Ts'o, P.O.P. (1977). Benzo[a]pyrenedione/benzo[a]pyrenediol oxidation–reduction couples and the generation of reactive reduced molecular oxygen. *Biochemistry* **16**, 1467–1473.
170. Jenstrom, B., Vadi, H., and Orrenius, S. (1978). Formation of DNA-binding products from benzo[a]pyrene metabolites in rat liver nuclei. *Chem.-Biol. Interact.* **20**, 311–321.
171. Jenstrom, B., Orrenius, S., Underman, O., Graslund, A., and Ehrenberg, A. (1978). Fluorescence study of DNA-binding metabolites of benzo[a]pyrene formed in hepatocytes isolated from 3-methylcholanthrene-treated rats. *Cancer Res.* **38**, 2600–2607.

172. Alexandrov, K., Dansette, P. C., and Frayssinet, C. (1980). Effect of purified epoxide hydrolase on metabolic activation and binding of benzo[a]pyrene to exogenous DNA. Shift of the activation pathway. *Biochem. Biophys. Res. Commun.* **93**, 611–616.
173. Guenthner, T. M., and Oesch, F. (1981). The effect of modulation of microsomal epoxide hydrolase activity on microsome-catalyzed activation of benzo[a]pyrene and its covalent binding to DNA. *Cancer Lett.* **11**, 175–183.
174. Oesch, F., and Guenthner, T. M. (1983). Effects of the modulation of epoxide hydrolase activity on the binding of benzo[a]pyrene metabolites to DNA in the intact nuclei. *Carcinogenesis (N.Y.)* **4**, 57–65.
175. Wood, A. W., Levin, W., Lu, A. Y. H., Yagi, H., Hernandez, O., Jerina, D. M., and Conney, A. H. (1976). Metabolism of benzo[a]pyrene and benzo[a]pyrene derivatives to mutagenic products by highly purified hepatic microsomal enzymes. *J. Biol. Chem.* **251**, 4882–4890.
176. Lubet, R. A., Capdevila, J., and Prough, R. A. (1979). The metabolic activation of benzo[a]pyrene and 9-hydroxybenzo[a]pyrene by liver microsomal fractions. *Int. J. Cancer* **23**, 353–357.
177. Jakoby, W. B., Habig, W. H., Keen, J. H., Kethey, J. N., and Pabst, M. J. (1976). Glutathione S-transferases: Catalytic aspects. *In* "Glutathione, Metabolism and Function" (I. M. Arias and W. B. Jakoby, eds.), pp. 189–211. Raven Press, New York.
178. Jerina, D. M., and Bend, J. R. (1977). Glutathione S-transferases. *In* "Biological Reactive Intermediates" (D. S. Sullow, S. J. Kocsin, R. Synder, and H. Vainio, eds.), pp. 207–236. Plenum, New York.
179. Jerina, D. M., Yagi, H., and Daly, J. W. (1973). Arene oxides–oxepins. *Heterocycles* **1**, 267–326.
180. Nemoto, N., Gelboin, H. V., Habig, W., Ketley, J. N., and Jakoby, W. B. (1975). K-region benzo[a]pyrene 4,5-oxide is conjugated by homogeneous glutathione S-transferase. *Nature (London)* **255**, 512.
181. Bend, J. R., Ben-Zevi, Z., Van Ander, J., Dansette, P., and Jerina, D. M. (1976). Hepatic and extrahepatic glutathione S-transferase activity toward several arene oxides and epoxides in the rat. *In* "Polynuclear Aromatic Hydrocarbons: Chemistry, Metabolism and Carcinogenesis" (R. Freudenthal and P. W. Jones, eds.), pp. 63–79. Raven Press, New York.
182. Hernandez, O., Walker, M., Cox, R. H., Foureman, G. L., Smith, B., and Bend, J. (1980). Regiospecificity and stereospecificity in the enzymatic conjugation of glutathione with (±)-benzo[a]pyrene 4,5-oxide. Biochem. Biophys. Res. Commun. **96**, 1494–1502.
183. Cobb, D., Boehlert, C., Lewis, D., and Armstrong, R. N. (1983). Stereoselectivity of glutathione S-transferase C toward arene and azarene oxides. *Biochemistry* **22**, 805–812.
184. Watabe, T., Ishizuka, T., Isobe, M., and Ozowa, N. (1982). A 7-hydroxymethyl sulfate ester as an active metabolite of 7,12-dimethylbenz[a]anthracene. *Science* **215**, 403–404.
185. Kinoshita, N., and Gelboin, H. V. (1978). $\beta$-Glucuronidase-catalyzed hydrolysis of benzo[a]pyrene-3-glucuronide and binding to DNA. *Science* **199**, 307–309.
186. Daudel, P., Duquesne, M., Vigney, P., Grover, P. L., and Sims, P. (1975). Fluorescence spectral evidence that benzo[a]pyrene–DNA products in mouse skin arise from diol epoxides. *FEBS Lett.* **57**, 250–253.
187. Meehan, T., Straub, K., and Calvin, M. (1976). Elucidation of hydrocarbon structure in an enzyme-catalyzed benzo[a]pyrene–poly[G] covalent complex *Proc. Natl. Acad. Sci. U.S.A.* **73**, 1437–1441.
188. Osborne, M. R., Belend, F. A., Harvey, R. G., and Brookes, P. (1976). Reactions of (±)-7$\alpha$,8$\beta$-dihydroxy-9$\beta$,10$\beta$-epoxy-7,8,9,10-tetrahydrobenzo[a]pyrene with DNA. *Int. J. Cancer* **18**, 362–368.

189. King, H. W. S., Osborne, S. R., Beland, F. A., Harvey, R. G., and Brookes, P. (1976). (±)-7α,8β-Dihydroxy-9β,10β-epoxy-7,8,9,10-tetrahydrobenzo[a]pyrene is an intermediate in the metabolism and binding to DNA of benzo[a]pyrene. *Proc. Natl. Acad. Sci. U.S.A.* **73**, 2679–2681.
190. Weinstein, I. B., Jeffrey, A. M., Jennette, K. W., Blobstein, S. H., Harvey, R. G., Harris, C., Autrup, W., Kocsis, H., and Nakanishi, K. (1976). Benzo[a]pyrene diol epoxides as intermediates in nucleic acid binding *in vitro* and *in vivo*. *Science* **193**, 592–595.
191. Moore, P. D., Koreeda, M., Wislocki, P. G., Levin, W., Conney, A. H., Yagi, H., and Jerina, D. M. (1977). *In vitro* reactions of the diastereomeric 9,10-epoxides of (+)- and (−)-*trans*-7,8-dihydroxy-7,8-dihydrobenzo[a]pyrene with polyguanylic acid and evidence for formation of an enantiomer of each diastereomeric 9,10-epoxide from benzo[a]pyrene in mouse skin. *In* "Drug Metabolism Concepts" (D. M. Jerina, ed.), pp. 127–154. Am. Chem. Soc., Washington, D.C.
192. Nebert, D. W., Boobis, A. R., Yagi, H., Jerina, D. M., and Kouri, R. E. (1977). Genetic differences in benzo[a]pyrene carcinogenic index *in vivo* and in mouse cytochrome *P*-450-mediated benzo[a]pyrene metabolite binding in DNA *in vitro*. *In* "Biological Reactive Intermediates" (D. J. Jollow, J. J. Kocsis, R. Snyder, and H. Vainio, eds.), pp. 125–145. Plenum, New York.
193. Koreeda, M., Moore, P. D., Wislocki, P. G., Levin, W., Cooney, A. H., Yagi, H., and Jerina, D. M. (1978). Binding of benzo[a]pyrene 7,8-diol 9,10-epoxides to DNA, RNA, and protein of mouse skin occurs with high stereoselectivity. *Science* **199**, 778–781.
194. Koreeda, M., Moore, P. D., Yagi, H., Yeh, H. J. C., and Jerina, D. M. (1976). Alkylation of polyguanylic acid at the 2-amino group and phosphate by the potent mutagen (±)-7β,8α-dihydroxy-9β,10β-epoxy-7,8,9,10-tetrahydrobenzo[a]pyrene. *J. Am. Chem. Soc.* **98**, 6720–6722.
195. Jeffrey, A. M., Jennette, K. W., Blobstein, S. H., Weinstein, I. B., Beland, F. A., Harvey, R. G., Kasai, H., Miura, I., and Nakanishi, K. (1976). Structure of guanosine adducts formed by reaction with a tetrahydrodiol epoxide of benzo[a]pyrene. *J. Am. Chem. Soc.* **98**, 5714–5715.
196. Jeffrey, A. M., Weinstein, I. B., Jennette, K. W., Greskociak, K., Nakanishi, K., Harvey, R. G., Autrup, H., and Harris, C. (1977). Structures of benzo[a]pyrene nucleic acid adducts formed in human and bovine bronchial explants. *Nature (London)* **269**, 348–350.
197. Shinohara, K., and Cerutti, P. A. (1977). Excision repair of BP–deoxyguanosine adducts in baby hamster kidney cells and in secondary mouse embryo fibroblasts. *Proc. Natl. Acad. Sci. U.S.A.* **74**, 979–983.
198. Ramsen, J., Jerina, D. M., Yagi, H., and Cerutti, P. (1976). *In vitro* reactions of radioactive 7β,8α-dihydroxy-9α,10α-7,8,9,10-tetrahydrobenzo[a]pyrene and 7β,8α-dihydroxy-9β,10β-epoxy-7,8,9,10-tetrahydrobenzo[a]pyrene with DNA. *Biochem. Biophys. Res. Commun.* **74**, 934–940.
199. Gamper, H., Mechan, T., Straub, K., Tung, A. S.-C., and Kelvin, M. (1978). DNA and RNA. *In* "Polycyclic Hydrocarbon and Cancer" (H. V. Gelboin and P.O.P. Ts'o, eds.), Vol. 2, pp. 51–61. Academic Press, New York.
200. Osborne, M. R., Harvey, R. G., and Brookes, P. (1978). The reaction of *trans*-7,8-Dihydroxy-*anti*-9,10-epoxy-7,8,9,10-tetrahydrobenzo[a]pyrene with DNA involves attack at the N-7 position of guanine moieties. *Chem.-Biol. Interact.* **20**, 123–130.
201. King, H. W. S., Osborne, M. R., and Brookes, P. (1979). The in vitro and in vivo reaction at the N-7 position of guanine of the ultimate carcinogen derived from benzo[a]pyrene. *Chem.-Biol. Interact.* **24**, 345–353.
202. Meehan, T., Straub, K., and Calvin, M. (1977). Benzo[a]pyrene diol epoxide covalently binds to deoxyguanosine and deoxyadenosine in DNA. *Nature (London)* **269**, 725–727.
203. Kakefuda, T., and Yamamoto, H. A. (1978). Modification of DNA by the benzo[a]-

pyrene metabolite diol epoxide r-7,t-8-dihydroxy-t-9,10-oxy-7,8,9,10-tetrahydrobenzo[a]pyrene. *Proc. Natl. Acad. Sci. U.S.A.* **75**, 415–419.
204. Ames, B. N., Durston, W. C., Yamasaki, E., and Lee, F. D. (1973). Carcinogens are mutagens: A simple test system combining liver homogenates for activation and bacteria for detection. *Proc. Natl. Acad. Sci. U.S.A.* **70**, 2281–2285.
205. Ames, B. N., Lee, F. C., and Durston, W. G. (1973). An improved bacterial test system for the detection of mutagens and carcinogens. *Proc. Natl. Acad. Sci. U.S.A.* **70**, 782–786.
206. Chu, E. Y. H. (1971). Induction and analysis of gene mutations in mammalian cell cultures. *In* "Chemical Mutagens: Principles and Methods for Their Detection" (A. Hollander, ed.), pp. 411–444. Plenum Press, New York.
207. Huberman, K., and Sachs, L. (1974). Cell-mediated mutagenesis of mammalian cells with chemical carcinogens. *Int. J. Cancer* **13**, 326–333.
208. McCann, J. O., Chui, E., Yamasaki, E., and Ames, B. N. (1975). Detection of carcinogen as mutagens in the Salmonella/microsome test assay of 300 chemicals. *Proc. Natl. Acad. Sci. U.S.A.* **72**, 5135–5139.
209. Levin, W., Wood, A. W., Wislocki, P. G., Chang, R. L., Kapitulnik, J., Mah, H. D., Yagi, H., Jerina, D. M., and Conney, A. H. (1978). Mutagenicity and carcinogenicity of benzo[a]pyrene and benzo[a]pyrene derivatives. *In* "Polycyclic Hydrocarbons and Cancer" (H. V. Gelboin and P.O.P. Ts'o, eds.), Vol. 1, pp. 189–202. Academic Press, New York.
210. Glatt, H. R., and Oesch, F. (1976). Phenolic benzo[a]pyrene metabolites are mutagens. *Mutat. Res.* **36**, 379–383.
211. Nagao, M., and Sugimura, T. (1978). Mutagenesis: Microbial systems. *In* "Polycyclic Hydrocarbons and Cancer" (H. V. Gelboin and P.O.P. Ts'o, eds.), Vol. 2, pp. 99–121. Academic Press, New York.
212. Wislocki, P. G., Chang, R. L., Wood, A. W., Levin, W., Yagi, H., Hernandez, O., Mah, H. D., Dansette, P. M., Jerina, D. M., and Conney, A. H. (1977). High carcinogenicity of 2-hydroxybenzo[a]pyrene on mouse skin. *Cancer Res.* **37**, 2608–2611.
213. Kapitulnik, J., Levin, W., Yagi, H., Jerina, D. M., and Conney, A. H. (1976). Lack of carcinogenicity of 4-, 5-, 6-, 7-, 8-, 9- and 10-hydroxybenzo[a]pyrene on mouse skin. *Cancer Res.* **36**, 3625–3628.
214. Slaga, T. J., Viage, A., Berry, D. L., Bracken, W., Buty, S. G., and Scribner, J. D. (1976). Skin tumor-initiating activity of benzo[a]pyrene 4,5- and 7,8-diol and 7,8-diol 9,10-epoxides. *Cancer Lett.* **2**, 115–122.
215. Slaga, T. J., Bracken, W. M., Viage, A., Levin, W., Yagi, H., Jerina, D. M., and Conney, A. H. (1977). Comparison of the tumor-initiating activities of benzo[a]pyrene arene oxides and diol epoxides. *Cancer Res.* **37**, 4130–4133.
216. Chouroulinkov, I., Gentil, A., Grover, P. L., and Sims, P. (1976). Tumor initiating activities on mouse skin of dihydrodiols derived from benzo[a]pyrene. *Br. J. Cancer* **34**, 523–532.
217. Slaga, T. J., Bracken, W. J., Gleason, G., Levin, W., Yagi, H., Jerina, D. M., and Conney, A. H. (1979). Marked differences in the skin tumor-initiating activities of the optical enantiomers of the diastereomeric benzo[a]pyrene 7,8-diol 9-10-epoxides. *Cancer Res.* **39**, 67–71.
210. Wislocki, P. G., Kapitulnik, J., Levin, W., Conney, A. H., Yagi, H., and Jerina, D. M. (1978). Tumorigenicity of benzo[a]pyrene 4,5-, 7,8-, 9,10-, and 11,12-oxides in newborn mice. *Cancer Lett.* **5**, 191–197.
219. Chang, R. L., Wislocki, P. G., Kapitulnik, J., Wood, A. W., Levin, W., Yagi, H., Mah, H. D., Jerina, D. M., and Conney, A. H. (1979). Carcinogenicity of 2-hydroxybenzo[a]pyrene and 6-hydroxybenzo[a]pyrene in newborn mice. *Cancer Res.* **39**, 2660–2664.
220. Jerina, D. M., and Daly, J. W. (1976). Oxidation of Carbon. *In* "Drug Metabolism–

from Microbe to Man" (D. V. Parke and R. L. Smith, eds.), pp. 13-32. Taylor & Francis, London.
221. Jerina, D. M., Lehr, R. E., Yagi, H., Hernandez, O., Dansette, P. J., Wislocki, P. G., Wood, A. W., Chang, R. L., Levin, W., and Conney, A. H. (1976). Mutagenicity of benzo[a]pyrene derivatives and the description of a quantum mechanical model which predicts the ease of carbonium ion formation from diol epoxides. *In* "In Vitro Metabolic Activation in Mutagenesis Testing" (F. J. de Serres, J. R. Fouts, J. R. Bend, and R. M. Philpot, eds.), pp. 159-177. Elsevier/North-Holland Biomedical Press, Amsterdam.
222. Stevenson, J. L., and von Haam, E. (1965). Carcinogenicity of benzo[a]anthracene and benzo[c]phenanthrene. *Am. Ind. Hyg. Assoc. J.* **26**, 475-478.
223. Miller, E. C., and Miller, J. A. (1960). The carcinogenicity of fluoro derivatives of 10-methyl-1,2-benzanthracene. I. 3- and 4'-Monofluoro derivatives. *Cancer Res.* **20**, 133-137.
224. Miller, J. A., and Miller, E. C. (1963). The carcinogenicity of fluoro derivatives of 10-methyl-1,2-benzanthracene. II. Substitution of the K region and 3'-, 6-, and 7-positions. *Cancer Res.* **23**, 229-239.
225. Jerina, D. M., and Lehr, R. E. (1977). The bay region theory: A quantum mechanical approach to aromatic hydrocarbon-induced carcinogenicity. *In* "Microsomes and Drug Oxidations" (V. Ullrich, I. Roots, A. G. Hildebrandt, R. W. Estabrook, and A. H. Conney, eds.), pp. 709-720. Pergamon, Oxford.
226. Wood, A. W., Levin, W., Lu, A. Y. H., Ryan, D., West, S. B., Lehr, R. E., Schaefer-Ridder, M., Jerina, D. M., and Conney, A. H. (1976). Mutagenicity of metabolically activated benzo[a]anthracene 3,4-dihydrodiol: Evidence for bay region activation of carcinogenic polycyclic hydrocarbons. *Biochem. Biophys. Res. Commun.* **72**, 680-686.
227. Wood, A. W., Chang, R. L., Levin, W., Lehr, R. E., Schaefer-Riddler, M., Karle, J. M., Jerina, D. M., and Conney, A. H. (1977). Mutagenicity and cytotoxicity of benz[a]anthracene diol epoxides and tetrahydroepoxides: Exceptional activity of the bay region 1,2-epoxides. *Proc. Natl. Acad. Sci. U.S.A.* **74**, 2746-2750.
228. Wood, A. W., Levin, W., Chang, R. L., Lehr, R. E., Schaefer-Ridder, M., Karle, J. M., Jerina, D. M., and Conney, A. H. (1977). Tumorigenicity of five dihydrodiols of benz[a]anthracene on mouse skin: Exceptional activity of benz[a]anthracene 3,4-dihydrodiol. *Proc. Natl. Acad. Sci. U.S.A.* **74**, 3176-3179.
229. Levin, W., Thakker, D. R., Wood, A. W., Chang, R. L., Lehr, R. E., Jerina, D. M., and Conney, A. H. (1978). Evidence that benzo[a]anthracene 3,4-diol 1,2-epoxide is an ultimate carcinogen on mouse skin. *Cancer Res.* **38**, 1705-1710.
230. Wislocki, P. G., Kapitulnik, J., Levin, W., Lehr, R., Schaefer-Ridder, M., Karle, J. M., Jerina, D. M., and Conney, A. H. (1978). Exceptional carcinogenic activity of benzo[a]anthracene 3,4-dihydrodiol in the newborn mouse and the bay region theory. *Cancer Res.* **38**, 693-696.
231. Slaga, T. J., Huberman, E., Selkirk, J. K., Harvey, R. G., and Bracken, W. M. (1978). Carcinogenicity and mutagenicity of benz[a]anthracene diols and diol epoxides. *Cancer Res.* **38**, 1699-1704.
232. Wislocki, P. G., Buening, M. K., Levin, W., Lehr, R. E., Thakker, D. R., Jerina, D. M., and Conney, A. H. (1979). Tumorigenicity of the diastereomeric benz[a]anthracene 3,4-diol 1,2-epoxides and the (+)- and (−)-enantiomers of benz[a]anthracene 3,4-dihydrodiol in newborn mice. *JNCI, J. Natl. Cancer Inst.* **63**, 201-204.
233. Wood, A. W., Levin, W., Thomas, P. E., Ryan, D., Karle, J. M., Yagi, H., Jerina, D. M., and Conney, A. H. (1978). Metabolic activation of dibenzo[a,h]anthracene and its dihydrodiols to bacterial mutagens. *Cancer Res.* **38**, 1967-1973.
234. Buening, M. K., Levin, W., Wood, A. W., Change, R. L., Yagi, H., Karle, J. M., Jerina, D. M., and Conney, A. H. (1979). Tumorigenicity of the dihydrodiols

of dibenzo[a,h]anthracene on mouse skin and in newborn mice. *Cancer Res.* **39,** 1310–1314.
235. Wood, A. W., Levin, W., Ryan, D., Thomas, P. E., Yagi, H., Mah, H. D., Thakker, D. R., Jerina, D. M., and Conney, A. H. (1977). High mutagenicity of metabolically activated chrysene 1,2-dihydrodiol: Evidence for bay region activation of chrysene. *Biochem. Biophys. Res. Commun.* **78,** 847–854.
236. Levin, W., Wood, A. W., Chang, R. L., Yagi, H., Mah, H. D., Jerina, D. M., and Conney, A. H. (1978). Evidence for bay region activation of chrysene 1,2-dihydrodiol to an ultimate carcinogen. *Cancer Res.* **38,** 1831–1834.
237. Wood, A. W., Change, R. L., Levin, W., Ryan, D. E., Thomas, P. E., Mah, H. D., Karle, J. M., Yagi, H., Jerina, D. M., and Conney, A. H. (1979). Mutagenicity and tumorigenicity of phenanthrene and chrysene epoxides and diol epoxides. *Cancer Res.* **39,** 4069–4077.
238. Buening, M. K., Levin, W., Karle, J. M., Yagi, H., Jerina, D. M., and Conney, A. H. (1979). Tumorigenicity of bay region epoxides and other derivatives of chrysene and phenanthrene in newborn mice. *Cancer Res.* **39,** 5063–5068.
239. Wood, A. W., Change, R. L., Levin, W., Yagi, H., Tada, M., Vyas, K. P., Jerina, D. M., and Conney, A. H. (1982). Mutagenicity of the optical isomers of diastereomeric bay region chrysene 1,2-diol 3,4-epoxides in bacterial and mammalian cells. *Cancer Res.* **42,** 2972–2976.
240. Hodgson, R. M., Pal, K., Grover, P. L., and Sims, P. (1982). The metabolic activation of chrysene by hamster embryo cells. *Carcinogenesis (N.Y.)* **3,** 1051–1056.
241. Chang, R. L., Levin, W., Wood, A. W., Yagi, H., Tada, M., Vyas, K. P., Jerina, D. M., and Conney, A. H. (1983). Tumorigenicity of enantiomers of chrysene 1,2-dihydrodiol and of the diastereomeric bay region chrysene 1,3-diol 3,4-epoxides on mouse skin and in newborn mice. *Cancer Res.* **43,** 192–196
242. Wood, A. W., Chang, R. L., Huang, M.-T., Levin, W., Lehr, R. E., Kumar, S., Thakker, D. R., Yagi, H., Jerina, D. M., and Conney, A. H. (1980). Mutagenicity of benzo[*e*]pyrene and triphenylene tetrahydroepoxides and diol epoxides in bacterial and mammalian cells. *Cancer Res.* **40,** 1985–1989.
243. Chang, R. L., Levin, W., Wood, A. W., Lehr, R. E., Kumar, S., Yagi, H., Jerina, D. M., and Conney, A. H. (1981). Tumorigenicity of the diastereomeric bay region benzo[*e*]pyrene 9,10-diol 11,12-epoxides in newborn mice. *Cancer Res.* **41,** 915–918.
244. Wood, A. W., Chang, R. L., Levin, W., Ryan, D. E., Thomas, P. E., Croisy-Delcey, M., Ittah, Y., Yagi, H., Jerina, D. M., and Conney, A. H. (1980). Mutagenicity of the dihydrodiols and bay region diol epoxides of benzo[*c*]phenanthrene in bacterial and mammalian cells. *Cancer Res.* **40,** 2876–2883.
245. Levin, W., Wood, A. W., Chang, R. L., Ittah, Y., Croisy-Delcy, M., Yagi, H., Jerina, D. M., and Conney, A. H. (1980). Exceptionally high tumor-initiating activity of benzo[*c*]phenanthrene bay region diol epoxides on mouse skin. *Cancer Res.* **40,** 3910–3914.
246. Jerina, D. M., Sayer, J. M., Yagi, H., Croisy-Delcey, M., Ittah, Y., Thakker, D. H., Wood, A. W., Chang, R. L., Levin, W., and Conney, A. H. (1982). Highly tumorigenic bay region diol epoxides from the weak carcinogen benzo[*c*]phenanthrene. *In* "Advances in Experimental Medicine and Biology: Biological Reactive Intermediates IIA" (R. Snyder, D. V. Parke, J. J. Kocsis, D. J. Jollow, C. G. Gibson, and C. M. Witmer, eds.), pp. 501–524. Plenum, New York.
247. Wood, A. W., Chang, R. L., Levin, W., Ryan, D. E., Thomas, P. E., Lehr, R. E., Kumar, S., Sardella, D. J., Boger, E., Yagi, H., Sayer, J. M., Jerina, D. M., and Conney, A. H. (1981). Mutagenicity of the bay region diol epoxides and other benzo ring derivatives of dibenzo[a,h]pyrene and dibenzo[a,i]pyrene. *Cancer Res.* **41,** 2589–2597.

248. Chang, R. L., Levin, W., Wood, A. W., Lehr, R. E., Kumar, S., Yagi, H., Jerina, D. M., and Conney, A. H. (1982). Tumorigenicity of bay region diol epoxides and other benzo ring derivatives of dibenzo[a,h]pyrene and dibenzo[a,i]pyrene on mouse skin and in newborn mice. *Cancer Res.* **42,** 25–29.
249. Wood, A. W., Chang, R. L., Levin, W., Thomas, P. E., Ryan, D., Stoming, T. A., Thakker, D. R., Jerina, D. M., and Conney, A. H. (1978). Metabolic activation of 3-methylcholanthrene and its metabolities to products mutagenic to bacterial and mammalian cells. *Cancer Res.* **38,** 3398–3404.
250. Thakker, D. R., Levin, W., Wood, A. W., Conney, A. H., Stoming, T. A., and Jerina, D. M. (1978). Metabolic formation of 1,9,10-trihydroxy-9,10-dihydro-3-methylcholanthrene: A potential proximate carcinogen from 3-methylcholanthrene. *J. Am. Chem. Soc.* **100,** 645–647.
251. Levin, W., Buening, M. K., Wood, A. W., Chang, R. L., Thakker, D. R., Jerina, D. M., and Conney, A. H. (1979). Tumorigenic activity of 3-methylcholanthrene metabolites on mouse skin and in newborn mice. *Cancer Res.* **39,** 3549–3553.
252. King, H. W. S., Osborne, M. R., and Brookes, P. (1976). The metabolism and DNA binding of 3-methylcholanthrene. *Int. J. Cancer* **20,** 564–571.
253. Vigny, P., Duquesne, M., Coulomb, H., Tierney, B., Grover, P. L., and Sims, P. (1977). Fluorescence spectral studies on the metabolic activation of 3-methylcholanthrene and 7,12-dimethylbenz[a]anthracene in mouse skin. *FEBS Lett.* **82,** 278–282.
254. King, H. W. S., Osborne, M. R., and Brooks, P. (1978). The identification of 3-methylcholanthrene-9,10-dihydrodiol as an intermediate in the binding of 3-methylcholanthrene to DNA in cells in culture. *Chem.-Biol. Interact.* **20,** 367–371.
255. Malaveille, C., Bartsch, H., Marquodt, H., Baka, S., Tierney, B., Hewer, A., Grover, P. C., and Sims, P. (1978). Metabolic activation of 3-methylcholanthrene: Mutagenic and transforming activities of the 9,10-dihydrodiol. *Biochem. Biophys. Res. Commun.* **85,** 1568–1574.
256. Chouroulinkov, I., Gentil, A., Tierney, B., Grover, P. L., and Sims, P. (1979). The initiation of tumors on mouse skin by dihydrodiols derived from 7,12-dimethylbenz[a]anthracene and 3-methylcholanthrene. *Int. J. Cancer* **24,** 455–460.
257. Vigny, P., Duquesne, M., Coulomes, H., Lacombe, C., Tierney, B., Grover, P. L., and Sims, P. (1977). Metabolic activation of polycyclic hydrocarbons: Fluorescence and spectral evidence is consistent with metabolism of the 1,2- and 3,4-double bonds of 7-methylbenz[a]anthracene. *FEBS Lett.* **75,** 9–12.
258. Tierney, B., Hewer, A., Walsh, C., Gover, P. L., and Sims, P. (1977). The metabolic activation of 7-methylbenz[a]anthracene in mouse skin. *Chem.-Biol. Interact.* **18,** 179–193.
259. Malaveille, C., Tierney, B., Grover, P. L., Sims, P., and Bartsch, H. (1977). High microsome-mediated mutagenicity of the 3,4-dihydrodiol of 7-methylbenz[a]anthracene in S. typhimurium TA 98. *Biochem. Biophys. Res. Commun.* **75,** 427–433.
260. Marquardt, H., Baker, S., Tierney, B., Grover, P. L., and Sims, P. (1977). The metabolic activation of 7-methylbenz[a]anthracene: The induction of malignant transformation and mutation in mammalian cells by non-K-region dihydrodiols. *Int. J. Cancer* **29,** 828–833.
261. Chouroulinkov, I., Gentil, A., Tierney, B., Grover, P., and Sims, P. (1977). The metabolic activation of 7-methylbenz[a]anthracene in mouse skin: High tumor-initiating activity of the 3,4-dihydrodiol. *Cancer Lett.* **3,** 247–253.
262. Moschel, R. C., Baird, W. M., and Dipple, A. (1977). Metabolic activation of the carcinogen 7,12-dimethylbenz[a]anthracene for DNA binding. *Biochem. Biophys. Res. Commun.* **76,** 1092–1098.

263. Ivanovic, V., Gaecintov, N. E., Jeffrey, A. M., Fu, P. P., Harvey, R. G., and Weinstein, I. B. (1978). Cell- and microsome-mediated binding of 7,12-dimethylbenz[a]anthracene to DNA studied by fluorescence spectroscopy. *Cancer Lett.* **4**, 131-140.
264. Dipple, A., and Nebzydoski, J. A. (1978). Evidence for the involvement of a diol epoxide in the binding of 7,12-dimethylbenz[a]anthracene to DNA in cells in culture. *Chem.-Biol. Interact.* **20**, 17-26.
265. Bigger, C. A. H., Tomaszewski, J. E., and Dipple, A. (1978). Differences between products of binding of 7,12-dimethylbenz[a]anthracene to DNA in mouse skin and in a rat liver microsomal system. *Biochem. Biophys. Res. Commun.* **80**, 229-235.
266. Malaveille, C., Bartsch, H., Tierney, B., Grover, P. L., and Sims, P. (1978). Microsome-mediated mutagenicities of the dihydrodiols of 7,12-dimethylbenz[a]anthracene: High mutagenic activity of the 3,4-dihydrodiol. *Biochem. Biophys. Res. Commun.* **93**, 1468-1473.
267. Marquardt, H., Baker, S., Tierney, B., Grover, P. L., and Sims, P. (1978). Induction of malignant transformation and mutagenesis by dihydrodiols derived from 7,12-dimethylbenz[a]anthracene. *Biochem. Biophys. Res. Commun.* **85**, 357-362.
268. Huberman, E., and Slaga, T. J. (1979). Mutagenicity and tumor-initiating activity of fluorinated derivatives of 7,12-dimethylbenz[a]anthracene. *Cancer Res.* **39**, 411-414.
269. Chou, M. W., and Yang, S. K. (1978). Identification of four *trans*-3,4-dihydrodiol metabolites of 7,12-dimethylbenz[a]anthracene and their *in vitro* DNA-binding activities upon further metabolism. *Proc. Natl. Acad. Sci. U.S.A.* **75**, 5466-5470.
270. Slaga, T. J., Gleason, G. L., DiGiovanni, J., Sukumaran, K. B., and Harvey, R. G. (1979). Potent tumor-initiating activity of the 3,4-dihydrodiol of 7,12-dimethylbenz[a]anthracene in mouse skin. *Cancer Res.* **39**, 1934-1936.
271. Hecht, S. S., Hirota, N., Loy, M., and Hoffmann, D. (1978). Tumor-initiating activity of fluorinated 5-methylchrysenes. *Cancer Res.* **38**, 1694-1698.
272. Hecht, S. S., LaVoie, E., Mazzarese, R., Amin, S., Bedenko, V., and Hoffmann, D. (1978). 1,2-Dihydro-1,2-dihydroxy-5-methylchrysene, a major activated metabolite of the environmental carcinogen 5-methylchrysene. *Cancer Res.* **38**, 2191-2194.
273. Hecht, S. S., Amin, S., Rivenson, A., and Hoffmann, D. (1979). Tumor-initiating activity of 5,11-dimethylchrysene and structural requirements favoring carcinogenicity of methylated polynuclear aromatic hydrocarbons. *Cancer Lett.* **8**, 65-70.
274. Coombs, M. M., Kissonerghis, A. M., Jeffrey, A. A., and Vase, C. (1979). Identification of proximate and ultimate forms of the carcinogen 15,16-dihydro-11-methylcyclopenta[e]phenanthrene-17-one. *Cancer Res.* **39**, 4160-4165.
275. LaVoie, E. J., Amin, S., Hecht, S. S., Furuya, K., and Hoffmann, D. (1982). Tumor-initiating activity of dihydrodiols of benzo[b]fluoranthene benzo[j]fluoranthene, and benzo[k]fluoranthene. *Carcinogenesis (N.Y.)* **1**, 49-52.
276. MacLeod, M. C., Levin, W., Conney, A. H., Lehr, R. E., Mansfield, B. K., Jerina, D. M., and Selkirk, J. K. (1980). Metabolism of benzo[a]pyrene by rat liver microsomal enzymes. *Carcinogenesis (N.Y.)* **1**, 165-171.
277. Nordqvist, M., Thakker, D. R., Levin, W., Yagi, H., Ryan, D. E., Thomas, P. E., Conney, A. H., and Jerina, D. M. (1979). The highly tumorigenic 3,4-dihydrodiol is a principal metabolite formed from dibenzo[a,h]anthracene by liver enzymes. *Mol. Pharmacol.* **16**, 643-655.
278. Ittah, Y., Thakker, D. R., Levin, W., Croisy-Delcey M., Ryan, D. E., Thomas, P. E., Conney, A. H., and Jerina, D. M. (1983). Metabolism of benzo[c]phenanthrene by rat liver microsomes and by a purified monooxygenase system reconstituted with different forms of cytochrome *P*-450. *Chem.-Biol. Interact.* **45**, 15-28.
279. Thakker, D. R., Levin, W., Yagi, H., Ryan, D., Thomas, P. E., Karle, J. M., Lehr, R. E.,

Jerina, D. M., and Conney, A. H. (1979). Metabolism of benzo[a]anthracene to its tumorigenic 3,4-dihydrodiol. *Mol. Pharmacol.* **15,** 138–153.

280. Thakker, D. R., Levin, W., Stoming, T. A., Conney, A. H., and Jerina, D. M. (1978). Metabolism of 3-methylcholanthrene by rat liver microsomes and a highly purified monooxygenase system with and without epoxide hydrase. *In* "Polynuclear Aromatic Hydrocarbons: Second International Symposium on Chemistry and Biology" (P. W. Jones and R. I. Frendenthal, eds.), pp. 253–264. Raven Press, New York.

281. Buhler, D. R., Unlu, F., Thakker, D. R., Slaga, T. J., Conney, A. H., Wood, A. W., Chang, R. L., Levin, W., and Jerina, D. M. (1983). Effect of a 6-fluoro substituent on the metabolism and biological activity of benzo[a]pyrene. *Cancer Res.* **43,** 1541–1549.

281a. Thakker, D. R., Yagi, H., Sayer, J. M., Kapur, U., Levin, W., Chang, R. L., Wood, A. W., Conney, A. H., and Jerina, D. M. (1984). Effects of a 6-fluoro substituent on the metabolism of benzo[a]pyrene 7,8-dihydrodiol to bay-region diol epoxides by rat liver enzymes. *J. Biol. Chem.* **259,** 11249–11256.

282. Ryan, D. E., Thomas, P. E., Reik, L. M., and Levin, W. (1982). Purification, characterization and regulation of five rat hepatic microsomal cytochrome *P*-450 isozymes. *Xenobiotica* **12,** 727–744.

283. Levin, W., Botelho, L. H., Thomas, J. E., and Ryan, D. (1980). Characterization of three forms of rat hepatic cytochromes *P*-450: Evidence for separate gene product. *In* "Microsomes, Drug Oxidations and Chemical Carcinogenesis" (M. J. Coon, A. H. Conney, R. W. Estabrook, H. V. Gelboin, J. R. Gillette, and P. J. O'Brien, eds.), Vol. 1, pp. 47–57. Academic Press, New York.

284. Hecht, S. S., Mazzarese, R., Amin, S., LaVoie, E., and Hoffmann, D. (1979). On the metabolic activation of 5-methylchrysene. *In* "Polynuclear Aromatic Hydrocarbons" (P. W. Jones and P. Leber, eds.), pp. 733–752. Ann Arbor Sci. Publ., Ann Arbor, Michigan.

285. MacNicoll, A. D., Burden, P. M., Ribeiro, O., Helver, A., Grover, P. L., and Sims, P. (1979). The formation of dihydrodiols in the chemical or enzymatic oxidation of 7-hydroxymethyl-12-methylbenz[a]anthracene and the possible role of hydroxymethyl dihydrodiols in the metabolic activation of 7,12-dimethylbenz[a]anthracene. *Chem.-Biol. Interact.* **26,** 121–132.

286. Wislocki, P. G., Gadek, K. M., Chou, M. W., Yang, S. K., and Lu, A. Y. H. (1980). Carcinogenicity and metagenicity of the 3,4-dihydrodiols and other metabolites of 7,12-dimethylbenz[a]anthracene and its hydroxymethyl derivatives. *Cancer Res.* **40,** 3661–3664.

287. Buening, M. K., Levin, W., Wood, A. W., Chang, R. L. Agranat, I., Buhler, D. R., Mah, H. D., Hernandez, O., Jerina, D. M., Conney, A. H., Miller, E. C., and Miller, J. A. (1983). Fluorine substitution as a probe for the role of the 6-position of benzo[a]pyrene in carcinogenesis. *JNCI, J. Natl. Cancer Inst.* **71,** 309–315.

288. Buhler, D. R., Unlu, F., Thakker, D. R., Slaga, T. J., Newman, M. S., Levin, W., Conney, A. H., and Jerina, D. M. (1982). Metabolism and tumorigenicity of 7-, 8-, 9-, and 10-fluorobenzo[a]pyrenes. Cancer Res. **42,** 4779–4783.

289. Pataki, J., and Higgins, C. (1969). Molecular site of substituents of benz[a]anthracene related to carcinogenicity. *Cancer Res.* **29,** 506–509.

290. Van Duuren, B. L., Sivak, A., Goldschmidt, B. M., Katz, C., and Melchionne, S. (1970). Initiating activity of aromatic hydrocarbons in two-stage carcinogenesis. *J. Natl. Cancer Inst. (U.S.).* **44,** 1167–1173.

291. Hecht, S. S., Bondinell, W. E., and Hoffmann, D. (1974). Chrysene and methylchrysenes: Presence in tobacco smoke and carcinogenicity. *J. Natl. Cancer Inst. (U.S.)* **52,** 1121–1133.

292. Coombs, M. M., Bhatt, T. S., and Croft, C. J. (1973). Correlation between carcinogenicity and chemical structure in cyclopentaphenanthrenes. *Cancer Res.* **33**, 832–837.
293. Iyer, R. P., Lyga, J. W., Secrist, J. A., III, Daub, G. H., and Slaga, T. J. (1980). Comparative tumor-initiating activity of methylated benzo[a]pyrene derivatives in mouse skin. Cancer Res. **40**, 1073–1076.
294. LaVoie, E. J., Tulley-Freiler, L., Bedenko, V., and Hoffmann, D. (1981). Mutagenicity, tumor-initiating activity and metabolism of methylphenanthrenes. *Cancer Res.* **41**, 3441–3447.
295. LaVoie, E. J., Bendenko, V., Tulley-Freiler, L., and Hoffmann, D. (1982). Tumor-initiating activity and metabolism of polymethylated phenanthrenes. *Cancer Res.* **42**, 4045–4049.
296. DiGiovanni, J., Diamond, L., Harvey, R. G., and Slaga, T. J. (1983). Enhancement of the skin tumor-initiating activity of polycyclic aromatic hydrocarbons by methyl substitution at nonbenzo bay region positions. Carcinogenesis (N.Y.) **4**, 403–407.
297. Levin, W., Wood, A. W., Chang, R. L., Newman, M. S., Thakker, D. R., Conney, A. H., and Jerina, D. M. (1983). The effect of steric strain in the bay region of polycyclic aromatic hydrocarbons: Tumorigenicity of alkyl-substituted benz[a]anthracene. *Cancer Lett.* **20**, 139–146.
298. Jerina, D. M., Michaud, D. P., Feldmann, R. J., Armstrong, R. N., Vyas, K. P., Thakker, D. R., Yagi, H., Thomas, P. E., Ryan, D. E., and Levin, W. (1982). Stereochemical modeling of the catalytic site of cytochrome P-450c. In "Microsomes, Drug Oxidations, and Drug Toxicity" (R. Sato and R. Kato, eds.), pp. 195–201. Jpn. Sci. Soc. Press Tokyo.
299. Thakker, D. R., Levin, W., Yagi, H., Turujman, S., Kapadia, D., Conney, A. H., and Jerina, D. M. (1979). Absolute stereochemistry of the *trans*-dihydrodiols formed from benzo[a]anthracene by liver microsomes. *Chem.-Biol. Interact.* **27**, 145–161.
300. Yagi, H., Thakker, D. R., Ittah, Y., Croisy-Delcey, M., and Jerina, D. M. (1983). Synthesis and assignment of absolute configuration to the *trans*-3,4-dihydrodiols and 3,4-diol 1,2-epoxides of benzo[c]phenanthrene. *Tetrahedron Lett.* **24**, 1349–1352.
301. van Bladeren, P. J., Armstrong, R. N., Cobb, D., Thakker, D. R., Ryan, D. E., Thomas, P. E., Sharma, N. D., Boyd, D. R., Levin, W., and Jerina, D. M. (1982). Stereoselective formation of benz[a]anthracene (+)-(5S,6R)-oxide and (+)-(8R,9S)-oxide by a highly purified and reconstituted system containing cytochrome P-450c. *Biochem. Biophys. Res. Commun.* **106**, 602–609.
302. Thakker, D. R., Yagi, H., Whalen, D. L., Levin, W., Wood, A. S., Conney, A. H., and Jerina, D. M. (1980). Metabolic formation and reactions of bay region diol epoxides: Ultimate carcinogenic metabolites of polycyclic aromatic hydrocarbons. In "Environmental Health Chemistry—The Chemistry of Environmental Agents as Potential Human Hazards" (J. D. McKinney, ed.), pp. 383–401. Ann Arbor Sci. Publ., Ann Arbor, Michigan.
303. Levin, W., Wood, A. W., Chang, R. L., Slaga, T. J., Yagi, H., Jerina, D. M., and Conney, A. H. (1977). Marked differences in the tumor-initiating activity of optically pure (+)- and (−)-*trans*-7,8-dihydroxy-7,8-dihydrobenzo[a]pyrene on mouse skin. *Cancer Res.* **37**, 2721–2725.
304. Kapitulnik, J., Wislocki, P. G., Levin, W., Yagi, H., Thakker, D. R., Akagi, H., Koreeda, M., Jerina, D. M., and Conney, A. Y. (1978). Marked differences in the carcinogenic activity of optically pure (+)- and (−)-*trans*-7,8-dihydroxy-7,8-dihydrobenzo[a]pyrene in newborn mice. *Cancer Res.* **38**, 2661–2665.
305. Wood, A. W., Chang, R. L., Levin, W., Yagi, H., Thakker, D. R., Jerina, D. M., and Conney, A. H. (1977). Differences in mutagenicity of the optical enantiomers of the

diastereomeric benzo[a]pyrene 7,8-diol 9,10-epoxides. *Biochem. Biophys. Res. Commun.* **77**, 1389–1396.

306. Buening, M. K., Wislocki, P. G., Levin, W., Yagi, H., Thakker, D. R., Akagi, H., Koreeda, J., Jerina, D. M., and Conney, A. H. (1978). Tumorigenicity of the optical enantiomers of the diastereomeric benzo[a]pyrene 7,8-diol 9,10-epoxides in newborn mice: Exceptional activity of (+)-7β,8α-dihydroxy-9α,10α-epoxy-7,8,9,10-tetrahydrobenzo[a]pyrene. *Proc. Natl. Acad. Sci. U.S.A.* **75**, 5358–5361.

Chapter 8

# Furans

LEO T. BURKA AND MICHAEL R. BOYD

*Laboratory of Experimental Therapeutics and Metabolism*
*National Cancer Institute*
*National Institutes of Health*
*Bethesda, Maryland*

|      |                                |     |
|------|--------------------------------|-----|
| I.   | Chemistry                      | 243 |
| II.  | Occurrence                     | 245 |
| III. | Enzymology                     | 246 |
| IV.  | Reactive Intermediates         | 247 |
| V.   | Fate of Reactive Intermediates | 252 |
| VI.  | Summary                        | 253 |
|      | References                     | 254 |

## I. Chemistry

Furans are one of a group of cyclopentadienoid heterocycles including pyrrole (**1**), furan (**2**), and thiophene (**3**) (Fig. 1). The chemistry of these heterocycles has been outlined in several publications including books by Albert,[1] Acheson,[2] and Dunlop and Peters.[3] Although these books are not recent, they provide a good background on the chemistry of furans.

Salient features of the chemistry of the furan ring, especially those points that may have some bearing on metabolism, will be discussed briefly. Furan is aromatic, but it does not derive as much stability from aromaticity as does pyrrole or thiophene and certainly not as much as benzene. The reactivity of furan is somewhere between that of the very reactive enol ether moiety

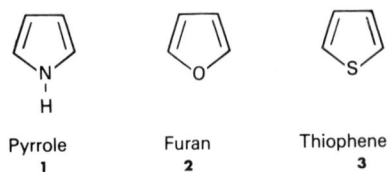

Fig. 1. Cyclopentadienoid heterocycles (1–3).

contained in its structure and an aromatic species. The stability of furans is greatly influenced by substituents; electron-withdrawing substituents tend to make the furan ring more resistant toward oxidation or acid hydrolysis. The effect of substituents on the ease of oxidation may be important in determining what metabolic pathways are preferred for a particular furan compound.

Alkyl-substituted furans tend to be easily oxidized by peracids and other oxidizing agents. Gingerich et al.[4] found that normenthofuran (4) was easily oxidized by m-chloroperoxybenzoic acid (MCPBA), in fact both double bonds of normenthofuran (4) were oxidized to give the enallactone (5) (Fig. 2). We have also observed that simple alkylfurans are easily oxidized. 2-Methylfuran is oxidized by m-chloroperoxybenzoic acid almost instantaneously, even at low (−30°C) temperatures (L. T. Burka, unpublished observation). Substitution of the furan with a carbonyl group imparts considerable stability toward peracid oxidation. Specifically, ethyl 3-furoate and 4-ipomeanol (6) are recovered unchanged from attempted peracid oxidation after 2 to 3 hr at room temperature (L. T. Burka, unpublished observation). Piancatelli et al.[5,6] have observed similar behavior in the pyridinium chlorochromate oxidation of furans. Oxidation of alkylfurans by pyridinium chlorochromate results in ring cleavage to form enediones[5]; furans substituted with the electron-withdrawing nitro group are resistant to oxidation by the reagent.[6]

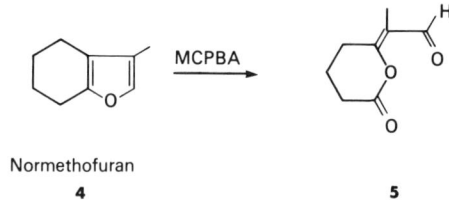

Fig. 2. Oxidation of normenthofuran (4–5).

## II. Occurrence

Furans are ubiquitous in nature. Liberally hundreds of furan-containing natural products have been found in both terrestrial and marine organisms. Many natural furans are terpenoids, but other biogenetic classes are also well represented. A review by Maga[7] lists numerous foods and beverages in which furan compounds have been found; 96 furans (including tetrahydrofurans and furanones) are found in coffee alone.

Some of the naturally occurring furans have been implicated in outbreaks of livestock poisoning (Fig. 3). Examples include ngaione (**7**), which is found in several shrubs and plants in Australia[8] and New Zealand,[9] tetradymol (**8**), from *Tetradymia glabrata* in the western United States,[10] 4-ipomeanol (**6**) and related compounds, from mold-damaged sweet potatoes in the southeastern United States,[11] and perilla ketone (**9**), from the perilla mint plant in many parts of the United States.[12] An additional compound of interest originating from natural sources is 3-methylfuran. This pneumotoxic compound was identified in smog and is believed to arise from photodecomposition of naturally occurring terpenoids.[13]

Several furans are important articles of commerce. Furfural is produced in large quantities from acid hydrolysis and steam distillation of pentose-containing agricultural by-products such as oat hulls and corn cobs. Furfural is the starting point for production of many furan derivatives. Furosemide (**10**) and nitrofurantoin (**11**) are widely used furan-containing therapeutic agents (Fig. 4). The toxicity of the latter two compounds has been the subject of investigation.[14,15]

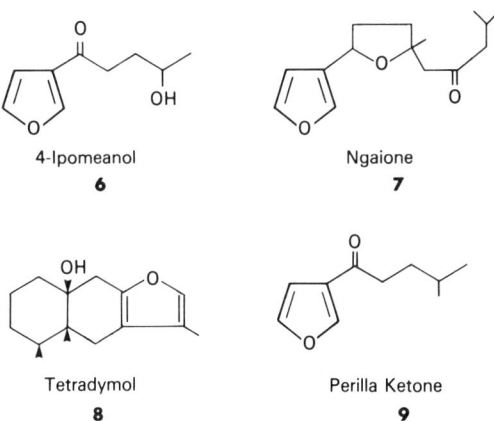

Fig. 3. Some naturally occurring furans (**6–9**).

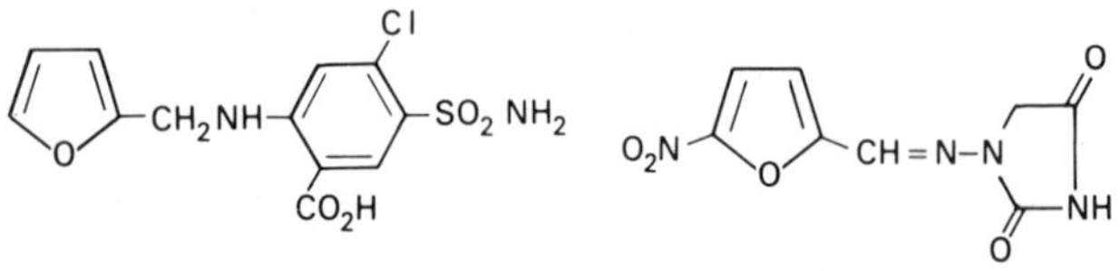

Furosemide
**10**

Nitrofurantoin
**11**

Fig. 4. Therapeutic agents containing a furan ring (**10**, **11**).

## III. Enzymology

Several of the furan compounds studied thus far require bioactivation to cause hepatic, renal, or pulmonary toxicity. In all but one case the bioactivation is catalyzed by the cytochrome $P$-450-containing mixed-function oxidase system. Nitrofurantoin is apparently not activated through a cytochrome $P$-450-catalyzed reaction.[15] The bioactivation of this agent seems to arise by the action of nitroreductase on the nitro group and is thus beyond the scope of this chapter.

Evidence for the involvement of cytochrome $P$-450 in the bioactivation of furans comes from a variety of experimental findings. Seawright and Hrdlicka[16] found the $LD_{50}$ of the hepatotoxic furan, ngaione, to be increased in mice if the animals were treated with the mixed-function oxidase inhibitor SKF-525A. Phenobarbital treatment also increased the $LD_{50}$, but shifted the site of liver necrosis from the midzonal region in untreated mice to the periportal region in phenobarbital-treated mice. Results from *in vitro* investigations of the furan-containing compounds 4-ipomeanol,[17] 3-methylfuran,[18] furosemide,[14] furamide (**12**),[19] and 2-($N$-ethylcarbamoylhydroxymethyl)furan (**13**)[20] (see Fig. 5), and others were consistent with bioactivation by mixed-function oxidases. Thus, in microsomal preparations, covalent binding of the compounds required $O_2$ and NADPH and was inhibited by carbon monoxide and other inhibitors of cytochrome $P$-450 metabolism including piperonyl butoxide, SKF 525A, and cobaltous chloride.

Furamide
**12**

2-($\underline{N}$-Ethylcarbamoyl-hydroxymethyl)furan
**13**

Fig. 5. Additional furans of toxicological interest (**12**, **13**).

At least two compounds, 4-ipomeanol and 2-(N-ethylcarbamoylhydroxymethyl)furan, have been investigated in reconstituted systems containing apparently homogeneous cytochromes $P$-450. Guengerich found the major cytochrome $P$-450 isozyme induced in rabbit liver by phenobarbital to have the highest turnover number for both 4-ipomeanol and carbamate (13) compared to other isozymes isolated.[21] He also showed that a major cytochrome $P$-450 isozyme isolated from rabbit lung efficiently metabolized 4-ipomeanol and carbamate (13).[22] Slaughter et al.[23] and Wolf et al.[24] have investigated in greater detail the metabolism of 4-ipomeanol by microsomes and by purified cytochrome $P$-450 from rabbit lungs.[23] These authors found two isozymes, cytochromes $P$-450$_I$ and $P$-450$_{II}$, which contributed about 49 and 39%, respectively, of the 4-ipomeanol-metabolizing capability of rabbit lung microsomes. However, these two isozymes constitute only a very small percentage of the rabbit hepatic microsomal cytochrome $P$-450.[24]

## IV. Reactive Intermediates

In furan compounds requiring cytochrome $P$-450-catalyzed bioactivation, the furan ring seems to be the functionality acted on by the oxidase enzyme, and covalent binding to tissue macromolecules takes place at the furan functionality. These conclusions are based on experimental data from several compounds. For example, when the covalent bonding of furosemide, labeled either in the furan moiety or the anthranilic acid moiety, was determined, the amount of covalent binding was the same regardless of the part of the molecule labeled.[14] If the protein–furosemide adduct was subjected to mild acid hydrolysis, more label was lost if the label originally was in the anthranilic acid portion of the molecule than in the furan moiety.[25] Control experiments showed that mild acid hydrolysis of furosemide resulted in cleavage of the furanylmethyl carbon–nitrogen bond. Incubation of the [$^{35}$S]tetrahydrofuranyl analog of furosemide with microsomal preparations resulted in no significant amount of covalent binding.[25] Guengerich has demonstrated that hydrolytic conditions that decarbamylate 2-(N-ethylcarbamoylhydroxymethyl)furan do not result in loss of radioactivity from the [$^{14}$C]furan–protein adduct.[20]

The reactive intermediate formed from the metabolic activation of furan compounds is electrophilic in nature, and the observed covalent binding to tissue probably results from the attack of the electrophilic intermediate on proteins, nucleic acids, or other macromolecules. In the compounds investigated thus far, addition of glutathione or other sulfhydryl compounds to *in vitro* incubation mixtures results in decreased covalent binding. Specific examples of this decrease include an ~90% reduction in covalent binding of furosemide in mouse liver microsomes in the presence of 0.2 m$M$ glutathione, cysteine, or cysteamine,[14] a 90% decrease in covalent binding of 4-ipomeanol to rat lung microsomes in the presence of 1 m$M$ glutathione,[17] and a 90% reduction in covalent binding of carbamoylfuran (13) to rat liver microsomes in the presence of 1 m$M$ glutathione.[20]

Manipulation of glutathione concentrations *in vivo* also supports the role of glutathione as a nucleophilic reaction partner for metabolically activated furans. Treatment with diethyl maleate, a compound that lowers tissue glutathione concentrations, decreased the LD$_{50}$ of 4-ipomeanol in rats from 24 to 6.3 mg/kg.[26] The increased toxicity was accompanied by an approximately two- to threefold increase in covalent binding to lung and liver. The effect of diethyl maleate could result from mechanisms other than simple depletion of a nucleophilic species. Diethyl maleate may alter glutathione-dependent or independent metabolism in such a way that increased toxicity would be observed. Statham and Boyd[27] found the relative tissue concentration of unmetabolized 4-ipomeanol and the total amount of excreted ipomeanol 4-glucuronide to be the same after diethyl maleate treatment. However, diethyl maleate treatment resulted in increased covalent binding of 4-ipomeanol metabolites and increased concentrations of nonbound solvent-extractable 4-ipomeanol metabolites. The conclusion reached by Statham and Boyd was that the increased toxicity of 4-ipomeanol after diethyl maleate treatment was due mainly to reduced levels of a soluble nucleophile. In complementary studies[28] it was found that increasing the concentration of soluble nucleophiles by treatment with cysteine or cysteamine led to decreased covalent binding of 4-ipomeanol in rat lung and liver; a corresponding increase in the LD$_{50}$ was observed.

The stability of the electrophilic species from furan metabolism has not been determined directly for any of the compounds discussed. Guengerich estimated the half-life of the reactive intermediate from 2-($N$-ethylcarbamoylhydroxymethyl)furan (13), in the absence of tissue nucleophiles, by quickly filtering the microsomal preparation after a 10-min incubation with carbamate (13) and NADPH. Addition of albumin to samples of the filtrate over a period of time led to the conclusion that the alkylating species has a chemical half-life of 30 to 50 min. The stability of the intermediate was corroborated by Neal *et al.*[29] with a diffusion apparatus in which the activated intermediate must diffuse through a glass fiber membrane before reacting with DNA. The active metabolite of 2-($N$-ethylcarbamoylhydroxymethyl)furan was able to bind covalently to DNA separated by the membrane barrier from the microsomal site of activation. The alkylating species from metabolism of aflatoxin B$_1$ was not stable enough to diffuse through the membrane and alkylate the DNA in the second chamber of the apparatus.

From studies on 4-ipomeanol, it appears that the *in vivo* half-life of the active intermediate is short. For example, autoradiography of lungs of animals treated with tritiated 4-ipomeanol showed that most of the covalently bound radioactivity was in Clara cells,[30] a major site of mixed-function oxidase activity in the lungs.[30,31] From both *in vivo* and *in vitro* studies, it appears that the reactive intermediate is not so reactive that it immediately binds to cytochrome *P*-450 on formation. The protective effect of thiol nucleophiles presumably is brought about by interception of the reactive intermediate when it is free of the cytochrome *P*-450. These nucleophiles not only decrease the amount of covalent binding from furan-containing compounds, as discussed earlier, but also tend to increase the metabolic lifetime of microsomal preparations. On the other hand, the reactive intermediate from 4-ipomeanol is sufficiently reactive that glutathione *S*-transferases are not important in the formation of glutathione conjugates.[32]

The structure of the activated intermediate from mixed-function oxidase metabolism of furan-containing compounds has been the subject of much speculation. However, definitive proof for any intermediate has not been forthcoming. It is, of course, tempting to postulate epoxidation of the furan ring as the activating process in analogy to the major mechanism of metabolism for benzenoid compounds. In addition, aflatoxin $B_1$ is almost certainly activated by epoxidation of the dihydrofuran ring contained in its structure. Glutathione[33] and DNA adducts of aflatoxin $B_1$ have been isolated, and the structure of the major nucleic acid–aflatoxin adduct has been determined.[34,35] The dihydrodiol of aflatoxin is a major metabolite in rat liver microsomes.[36] The structures of these metabolites are consistent with epoxidation of the 2,3-double bond in aflatoxin $B_1$.

If the literature is examined for evidence that a furan epoxide is the intermediate in mixed-function oxidase metabolism of furans, one finds that no furan epoxides have been isolated and that no metabolites resulting from attack of simple nucleophiles (e.g., water, to form dihydrodiols) have been found. Although strong evidence exists for formation of glutathione adducts of 4-ipomeanol,[32] the structure of the adducts have thus far not been determined.

Indirect evidence for epoxidation of furosemide has been reported. Wirth *et al.*[25] found that addition of 1 m$M$ 1,2-epoxy-3,3,3-trichloropropane to mouse liver microsomal incubation mixtures resulted in a nearly twofold increase in covalent binding. The increase was attributed to the inhibition of epoxide hydrolase by the epoxypropane, resulting in higher concentrations of the active metabolite in the microsomes. 1,2-Epoxy-3,3,3-trichloropropane has no significant effect on the covalent binding of either 4-ipomeanol[17] or 2-(*N*-ethylcarbamoylhydroxymethyl)furan.[20] Either an epoxide is not formed in the latter two cases or an epoxide is formed, but it is not a substrate for the hydrolase; alternatively, an epoxide may be formed that is so reactive as to undergo other reactions before reaching the hydrolase. If peracid oxidation mimics oxidation by cytochrome $P$-450, it might be expected that the alkyl-substituted furans (which include furosemide and carbamate, **13**) may undergo epoxidation. Furans substituted with electron-withdrawing groups (4-ipomeanol, perilla ketone) might be activated by alternate mechanisms.

Other hypothetical mechanisms of furan activation can be envisioned. Evidence from a number of laboratories indicates that cytochrome $P$-450-catalyzed oxidations may proceed by discrete, one-electron steps. Oxidation by removal of a hydrogen atom or an electron as an initial step in cytochrome $P$-450-catalyzed metabolism has been postulated for several substrates, for example, vinyl halides,[37,38] cyclopropyl amines,[39,40] organosulfides,[41] and norbornane.[42] The result of such a one-electron oxidation of a furan by the cytochrome $P$-450 perferryl oxide intermediate (**14**) is formation of radical cation **16** in which the radical or the cationic site can be stabilized by the furan oxygen (Fig. 6). Such a species may be stable enough to leave the site of formation, but too reactive to migrate far. Cation radicals could undergo radical–radical reactions or hydrogen atom abstraction, or could react with nucleophiles at the cationic center. Hydrogen atom ab-

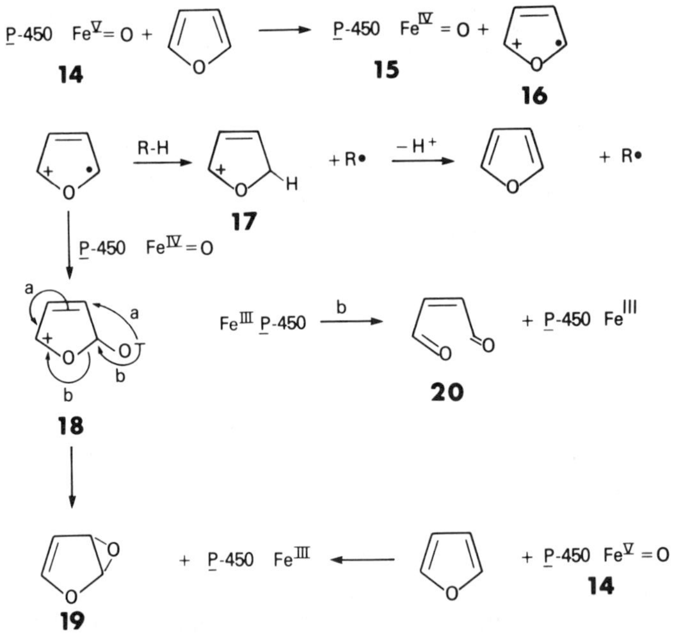

Fig. 6. Furan activation by one-electron oxidation and/or epoxidation (**14–20**).

straction from hydrogen atom donors (e.g., mercaptans) to give **17** followed by loss of a proton would give the parent furan, but a radical chain reaction might ensue with no obvious change in the furan. No direct proof of radical intermediates in furan metabolism have been found, but addition of simple mercaptans such as dithiothreitol or cysteamine to microsomal incubations of 4-ipomeanol decrease covalent binding 80–90%. However, no mercaptan–furan adducts can be found chromatographically (L. T. Burka and M. R. Boyd, unpublished observations). Some phenols, such as propyl gallate and 2,6-di-*tert*-butyl-4-carboxyphenol, are also capable of greatly reducing covalent binding of 4-ipomeanol *in vitro* (L. T. Burka and J. C. Walker, unpublished observations). These results are consistent with radical intermediates being involved in the metabolic activation of 4-ipomeanol. On examining the one-electron oxidation hypothesis further, it seems that a radical–radical reaction between the furan radical cation and the Fe(IV)-O species (**15**, which has radical character) would be facile. Intermediate **18** obtained from the radical reaction could give rise to an epoxide (**19**). A lower energy pathway, however, might be a ring-opening reaction to form enedial (**20**) without going through the epoxide as an intermediate. Enedial (**20**) is quite reactive and could react either via a Michael addition to the double bond or by nucleophilic addition to the aldehyde.

Support for this latter hypothesis has been obtained.[43] Microsome suspensions containing 2- or 3-methylfuran were incubated in the presence of semicarbazone as a trapping agent, and the respective enedial products, acetylacrolein and 2-methylbutenedial, were subsequently isolated as the bissemicarbazones. Addition of semicarbazide to microsomal incubations of radiolabeled 2- and 3-methylfuran markedly decreased covalent binding to microsomal protein, an observation that is consistent with the enedial species being the active intermediate from these two alkylfurans.

Garst and Wilson have proposed an alternate hypothesis for the activation of 3-substituted furans.[44] By plotting modified Hammett $\sigma_{para}$ constants, obtained from a plot of Hammett $\sigma_{para}$ vs the $^{13}C$ NMR chemical shift of carbon-5 in a series of furan compounds, and the log of the $LD_{50}$ in mice of eight toxic furan compounds, the authors obtained a linear ($r = 0.992$) correlation. The slope of this biological Hammett plot was positive (+3). A

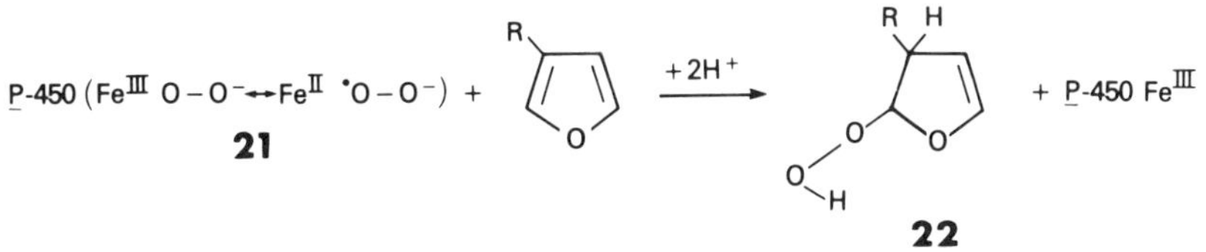

Fig. 7. Furan activation by hydroperoxide formation (**21**, **22**).

positive slope in a Hammett plot is usually taken as evidence for formation of at least a partial negative charge in the rate-determining step of the reaction. Garst and Wilson postulated that the rate-determining step for cytochrome $P$-450 activation of 3-substituted furans involves nucleophilic attack of the one-electron-reduced cytochrome $P$-450 heme–dioxygen intermediate (21), a ferric peroxide–ferrous superoxide species, on the furan (Fig. 7). The subsequent reactions of the resulting furan hydroperoxide (22) is speculated to be the cause of the pulmonary toxicity of 3-substituted furans.

## V. Fate of Reactive Intermediates

Whether the activated intermediates from furan metabolism are epoxides or radicals or other species, it is clear that alkylating species are formed and that alkylation of tissue nucleophiles predominates in the organs affected by the toxin. Amounts of covalent binding of furosemide to liver parallel the severity of liver necrosis.[14] Inhibitors of microsomal metabolism, such as piperonyl butoxide, result in a decrease of furosemide covalent binding and a decrease in severity of liver necrosis.[14] Tissue distribution studies of 4-ipomeanol in the rat reveal that the greatest amount of binding, in terms of nanomoles of 4-ipomeanol bound per milligram of tissue protein, occurs in the lung.[26] In species in which 4-ipomeanol is not a lung toxin, or not exclusively a lung toxin, the covalent binding distribution changes accordingly. For instance, in avian species, where there is no significant covalent binding to the lung, the covalent binding of 4-ipomeanol to the liver is consistent with the hepatotoxicity observed.[45] Renal toxicity is observed for 4-ipomeanol in the male mouse; the preferential covalent binding to the renal cortical tubules corresponds to the predominant site of necrosis within this organ.[46]

The extent of binding of furans to subcellular fractions and to components of these fractions has been investigated. The covalent binding of 4-ipomeanol, given to rats at a dose of 10 mg/kg, was more or less equally distributed (based on nanomoles of 4-ipomeanol bound per milligram of protein) among nuclei, mitochondria, microsomes, and soluble fractions of both lung and liver.[26] Most of the covalent binding was to protein with lesser amounts bound to nucleic acids and very little binding to lipids. Treatment of rats with 40 mg/kg of 2-($N$-ethylcarbamoylhydroxymethyl)furan resulted in an approximately equal distribution (based on percentage of total bound material) of covalent binding among nuclei, mitochondria, microsomes, and cytosol in liver, lungs, and kidney with the exception that covalent binding to liver nuclei was appreciably greater than to the other liver subfractions.[20] In this study, appreciable radioactivity was bound to RNA or DNA, or both, in all liver fractions.

In addition to reaction with nucleophilic sites on tissue macromolecules, reaction with soluble nucleophiles, such as glutathione, may be a major fate of activated furan intermediates. Glutathione has been shown to reduce the amount of covalent binding *in vitro* in a number of studies, and reduction of glutathione concentrations by diethyl maleate *in vivo* leads to increased toxicity as described earlier. The results of decreased covalent binding to protein in the presence of glutathione should lead to glutathione conjugates *in vitro* and mercapturic acids *in vivo*. The best evidence for a glutathione–furan conjugate has been obtained from studies with 4-ipomeanol. High-pressure anion exchange chromatography revealed the presence of two glutathione conjugates from microsomal incubations of 4-ipomeanol in the presence of glutathione.[32] These conjugates no longer contain the furoyl chromaphore, as shown by their lack of UV absorption in the 250-nm region, and were detected by monitoring radioactivity in the column effluent. Two major peaks of radioactivity were detected when either tritiated 4-ipomeanol or [$^3$H]glutathione were included in the incubation mixture. The glutathione-derived metabolites were produced in incubations containing either hepatic or pulmonary microsomes, and the ratio (~ 1 : 2) of the two adducts was about the same from both sources.

The formation of the two glutathione conjugates during metabolism of 4-ipomeanol by purified lung cytochrome *P*-450 isozymes has also been investigated in some detail.[24] Thus, the isozyme designated as $P\text{-}450_I$ gave about five times as much of one adduct as of the other. Cytochrome $P\text{-}450_{II}$, on the other hand, gave about a 1 : 1 ratio of the two conjugates. The structure of neither conjugate is known at this time; the significance of the different ratio of adducts from the two isozymes is also unknown. The two adducts could be stereoisomers or could, perhaps, represent different oxidation states of the 4-ipomeanol portion of the molecule. These details need to be resolved. It should be pointed out that a glutathione adduct of 1,4-ipomeanol (i.e., 4-ipomeanol with the carbonyl group reduced to an alcohol) has been detected in the same high-pressure anion exchange system, and it does not correspond to either of the 4-ipomeanol conjugates.[32]

## VI. Summary

The number of furans of toxicological interest is small, but it is growing. The furan ring seems important in the toxicity of these compounds, because tetrahydrofuranyl or other analogs are not generally as toxic as the parent compound. Among the compounds that have received attention, it is apparent that metabolism converts the furan to an electrophilic species. The subsequent reaction of the electrophilic species with tissue nucleophiles takes place primarily at the site of necrosis. The consensus at this time appears to be that an epoxide is the active intermediate from furan metabolism; however, the evidence on which this consensus is based is not compelling. Substituents, especially those with electron-withdrawing character, alter the chemistry of the furan ring markedly. It seems quite probable that there are multiple pathways for metabolism of furans resulting from these differences in chemistry.

4-Ipomeanol, the furan compound investigated most thoroughly to this point, has proved to be an especially useful toxicological tool. The Clara cell was shown to be a primary site of xenobiotic metabolism in the mammalian lung with this agent. By exploiting interspecies differences in metabolism and toxicity of 4-ipomeanol, it has been possible to demonstrate that covalent binding (i.e., reactive metabolite formation) and tissue damage are closely linked. Good evidence for formation of glutathione adducts of furan metabolites was also obtained with 4-ipomeanol. Thus, a great deal is known about the biological mechanisms of toxicity for this furan. However, the actual chemistry of metabolic activation of this or any other furan is in large part unknown. Our work on simple alkyl-substituted furans indicates that ring opening to an enedial is a route of metabolism for this class of compounds. We are unaware of any other furan metabolite that has been isolated that allows us to infer a specific mechanism of furan ring metabolism. Continued investigation into chemical mechanisms of metabolism of furans by mixed-function oxidase systems is required if the complete story on furan toxicity is to be told.

## References

1. Albert, A. (1968). *In* "Heterocyclic Chemistry," 2nd ed., pp. 256–296. Athlone Press, London.
2. Acheson, R. M. (1967). *In* "An Introduction to the Chemistry of Heterocyclic Compounds," 2nd ed., pp. 93–119. (Wiley Interscience), New York.
3. Dunlop, A. P., and Peters, F. N. (1953). "The Furans." Van Nostrand-Reinhold, Princeton, New Jersey.
4. Gingerich, S. B., Campbell, W. H., Bricca, C. E., Jennings, P. W., and Campana, C. F. (1981). An unexpected product from the peracid oxidation of furan derivatives and a new $\epsilon$ lactone synthesis. *J. Org. Chem.* **46,** 2589.
5. Piancatelli, G., Scettri, A., and D'Auria, M. (1980). Oxidative ring opening of furan derivatives to $\alpha,\beta$-unsaturated $\gamma$-dicarbonyl compounds, useful intermediates for 3-oxycyclopentenes synthesis. *Tetrahedron* **36,** 661.
6. D'Auria, M., Piancatelli, G., and Scettri, A. (1980). A useful preparation of 5-nitro-2-furan derivatives. *Tetrahedron* **36,** 1877.

7. Maga, J. A. (1979). Furans in foods. *CRC Crit. Rev. Food Sci. Nutr.* **11**, 355.
8. Hegarty, B. F., Kelly, J. R., Park, R. J., and Sutherland, M. D. (1970). Terpenoid chemistry XVII. (−)-Ngaione, a toxic constituent of *Myoporum deserti.* The absolute configuration of (−)-ngaione. *Aust. J. Chem.* **23**, 107.
9. McDowall, F. H. (1925). Constituents of *Myoparum lactum,* Forst. (The "Ngaio"). Part I. *J. Chem. Soc.* **127**, 2200.
10. Jennings, P. W., Reeder, S. C., Hurley, J. C., Caughlan, C. N., and Smith, G. D. (1974). Isolation and structure determination of one of the toxic constituents from *Tetradymia glabrata. J. Org. Chem.* **39**, 3392.
11. Boyd, M. R., Burka, L. T., Harris, T. M., and Wilson, B. J. (1974). Lung-toxic furanoterpenoids produced by sweet potatoes *(Ipomoea batatas)* following microbial infection. *Biochim. Biophys. Acta* **337**, 184 (1974).
12. Wilson, B. J., Garst, J. E., Linnabary, R. D., and Channell, R. B. (1977). Perilla ketone: A potent lung toxin from the mint plant, *Perilla frutescens* Britton. *Science* **197**, 573 (1977).
13. Saunders, R., Griffith, J., and Saalfeld, F. (1974). Identification of some organic smog components based on rain water analysis. *Biomed. Mass Spectrum.* **1**, 192.
14. Mitchell, J. R., Nelson, W. L., Potter, W. Z., Sasame, H. A., and Jollow, D. J. (1976). Metabolic activation of furosemide to a chemically reactive, hepatotoxic metabolite. *J. Pharmacol. Exp. Ther.* **199**, 41.
15. Boyd, M. R., Catignani, G. L., Sasame, H. A., Mitchell, J. R., and Stiko, A. W. (1979). Acute pulmonary injury in rats by nitrofurantoin and modification by vitamin E, dietary fat and oxygen. *Am. Rev. Respir. Dis.* **120**, 93.
16. Seawright, A. A., and Hrdlicka, J. (1972). The effect of prior dosing with phenobarbitone and $\beta$-diethylaminoethyl diphenylpropyl acetate (SKF525A) on the toxicity and liver lesion caused by ngaione in the mouse. *Br. J. Exp. Pathol.* **53**, 242.
17. Boyd, M. R., Burka, L. T., Wilson, B. J., and Sasame, H. A. (1978). *In vitro* studies on the metabolic activation of the pulmonary toxin, 4-ipomeanol, by rat lung and liver microsomes. *J. Pharmacol. Exp. Ther.* **207**, 677.
18. Boyd, M. R., Statham, C. N., Franklin, R. B., and Mitchell, J. R. (1978). Pulmonary bronchiolar alkylation and necrosis by 3-methylfuran, a naturally occurring potential atmospheric contaminant. *Nature (London)* **272**, 270.
19. McMurtry, R. J., and Mitchell, J. R. (1977). Renal and hepatic necrosis after metabolic activation of 2-substituted furans and thiophenes, including furosemide and cephaloridine. *Toxicol. Appl. Pharmacol.* **42**, 285.
20. Guengerich, F. P. (1977). Studies on the activation of a model furan compound—toxicity and covalent binding of 2-($N$-ethylcarbamoylhydroxymethyl)furan. *Biochem. Pharmacol.* **26**, 1909.
21. Guengerich, F. P. (1977). Separation and purification of multiple forms of microsomal cytochrome $P$-450. Activities of different forms of cytochrome $P$-450. *J. Biol. Chem.* **252**, 3970.
22. Guengerich, F. P. (1977). Preparation and properties of highly purified cytochrome $P$-450 and NADPH–cytochrome $P$-450 reductase from pulmonary microsomes of untreated rabbits. *Mol. Pharmacol.* **13**, 911.
23. Slaughter, S. R., Statham, C. N., Philpot, R. M., and Boyd, M. R. (1983). Covalent binding of metabolites of 4-ipomeanol to rabbit pulmonary and hepatic microsomal proteins and to the enzymes of the pulmonary cytochrome $P$-450-dependent monooxygenase system. *J. Pharmacol. Exp. Ther.* **224**, 252.
24. Wolf, C. R., Statham, C. N., McMenamin, M. G., Bend, J. F., Boyd, M. R., and Philpot, R. M. (1982). The rabbit pulmonary monooxygenase system. *Mol. Pharmacol.* **22**, 738.
25. Wirth, P. J., Bettis, C. J., and Nelson, W. L. (1975). Microsomal metabolism of furose-

mide: Evidence for the nature of the reactive intermediate involved in covalent binding. *Mol. Pharmacol.* **12**, 759.
26. Boyd, M. R., and Burka, L. T. (1978). *In vivo* studies on the relationship between target organ alkylation and the pulmonary toxicity of a chemically reactive metabolite of 4-ipomeanol. *J. Pharmacol. Exp. Ther.* **207**, 687.
27. Statham, C. N., and Boyd, M. R. (1982). Distribution and metabolism of the pulmonary alkylating agent and cytotoxin, 4-ipomeanol, in control and diethylmaleate-treated rats. *Biochem. Pharmacol.* **31**, 1585.
28. Boyd, M. R., Stiko, A., Statham, C. N., and Jones, R. B. (1982). Protective role of endogenous pulmonary glutathione and other sulfhydryl compounds against lung damage by alkylating agents: Investigations with 4-ipomeanol in the rat. *Biochem. Pharmacol.* **31**, 1579.
29. Neal, G. E., Mattocks, A. R., and Judah, D. J. (1979). The microsomal activation of aflatoxin $B_1$ and 2-(N-ethylcarbamoyloxymethyl)furan *in vitro* using a novel diffusion apparatus. *Biochim. Biophys. Acta* **585**, 134.
30. Boyd, M. R. (1977). Evidence for the Clara cell as a site of cytochrome $P$-450-dependent mixed-function oxidase activity in lung. *Nature (London)* **269**, 713.
31. Serabjit-Singh, C. J., Wolf, C. R., Plopper, G. C., and Philpot R. M. (1980). Cytochrome $P$-450: Localization in rabbit lung. *Science* **207**, 1469.
32. Buckpitt, A. R., and Boyd, M. R. (1980). The *in vitro* formation of glutathione conjugates with the microsomally activated pulmonary bronchiolar alkylating agent and cytotoxin, 4-ipomeanol. *J. Pharmacol. Exp. Ther.* **215**, 97.
33. Degen, G. N., and Neumann, H. (1978). The major metabolite of aflatoxin $B_1$ in the rat is a glutathione conjugate. *Chem.-Biol. Interact.* **22**, 239.
34. Lin, J. K., Miller, J. A., and Miller, E. C. (1977). 2,3-Dihydro-2-(guan-7-yl)-3-hydroxyaflatoxin $B_1$, a major acid hydrolysis product of aflatoxin $B_1$–DNA or–ribosomal RNA adducts formed in hepatic microsome-mediated reactions and in rat liver *in vivo*. *Cancer Res.* **37**, 4430.
35. Eissigman, J. M., Croy, R. G., Nadzan, A. M., Busby, W. F., Reinhold, V. N., Buchi, G., and Wogan, G. N. (1977). Structural identification of the major DNA adduct formed by aflatoxin $B_1$ *in vitro*. *Proc. Natl. Acad. Sci. U.S.A.* **74**, 1870.
36. Neal, G. E., and Colley, P. J. (1979). The formation of 2,3-dihydro-2, 3-dihydroxy aflatoxin $B_1$ by the metabolism of aflatoxin $B_1$ *in vitro* by rat liver microsomes. *FEBS Lett.* **101**, 382.
37. Ortiz de Montellano, P. R., Kunze, K. L., Beilan, H. S., and Wheeler, C. (1982). Destruction of cytochrome $P$-450 by vinyl fluoride, fluroxene and acetylene. Evidence for a radical intermediate in olefin oxidation. *Biochemistry* **21**, 1331.
38. Guengerich, F. P., and Strickland, T. W. (1977). Metabolism of vinyl chloride: Destruction of the heme of highly purified liver microsomal cytochrome $P$-450 by a metabolite. *Mol. Pharmacol.* **13**, 993.
39. Macdonald, T. L., Zirvi, K., Burka, L. T., Peyman, P., and Guengerich, F. P. (1982). Mechanism of cytochrome $P$-450 inhibition by cyclopropylamines. *J. Am. Chem. Soc.* **104**, 2050.
40. Hanzlik, R. P., and Tullman, R. H. (1982). Suicidal inactivation of cytochrome $P$-450 by cyclopropylamines. Evidence for cation-radical intermediates. *J. Am. Chem. Soc.* **104**, 2048.
41. Watanabe, Y., Iyanagi, T., and Oae, S. (1980). Kinetic study on enzymatic S-oxygenation promoted by a reconstituted system with purified cytochrome $P$-450. *Tetrahedron Lett.* **21**, 3685.
42. Groves, J. T., McClusky, G. A., White, R. E., and Coon, M. J. (1973). Aliphatic hydroxy-

# 8. FURANS

lation by highly purified liver microsomal cytochrome $P$-450. Evidence for a carbon radical intermediate. *Biochem. Biophys. Res. Commun.* **81**, 154.
43. Ravindranath, V., Burka, L. T., and Boyd, M. R. (1984). Reactive metabolites from the bioactivation of toxic methylfurans. *Science* **224**, 884.
44. Garst, J. E., and Wilson, B. J. (1981). Preliminary evidence supporting iron peroxide intermediates and nucleophilic addition in the lung cytochrome $P$-450-catalyzed bioactivation and toxicity of 3-substituted furans. *In* "Oxygen and Oxy-Radicals in Chemistry and Biology" (M. A. J. Rodgers and E. L. Powers, eds.), pp. 507–520. Academic Press, New York.
45. Buckpitt, A. R., Statham, C. N., and Boyd, M. R. (1982). *In vivo* studies on the target tissue metabolism, covalent binding, glutathione depletion and toxicity of 4-ipomeanol in birds, species deficient in pulmonary enzymes for metabolic activation. *Toxicol. Appl. Pharmacol.* **65**, 38.
46. Dutcher, J. S., and Boyd, M. R. (1979). Species and strain differences in target organ alkylation and toxicity by 4-ipomeanol. Predictive value of covalent binding in studies of target organ toxicities by reactive metabolites. *Biochem. Pharmacol.* **28**, 3367.

Chapter 9

# Phenols, Catechols, and Quinones

RICHARD D. IRONS
*Chemical Industry Institute of Toxicology*
*Research Triangle Park, North Carolina*

TADASHI SAWAHATA
*Toxicology Laboratory*
*Toray Industries, Inc.*
*Sonoyama, Otsu, Japan*

|   |   |
|---|---|
| I. Introduction | 259 |
| II. Phenols | 261 |
|    A. Chemical Properties | 261 |
|    B. Intrinsic Toxicity and Proposed Reaction Mechanisms | 262 |
|    C. Metabolism and Bioactivation | 263 |
| III. Catechols and Hydroquinones | 266 |
|    A. Chemical Properties | 266 |
|    B. Biosynthesis | 267 |
|    C. Intrinsic Toxicity | 268 |
|    D. Metabolism and Bioactivation | 268 |
| IV. Quinones | 269 |
|    A. Biosynthesis | 269 |
|    B. Chemical Properties | 271 |
|    C. Toxicity and Reaction Mechanisms | 273 |
|    References | 277 |

## I. Introduction

Phenols are hydroxylated products of aromatic hydrocarbons and comprise a vast group of compounds with wide distribution in chemistry and nature. Because they readily undergo oxidation, a discussion of the bioacti-

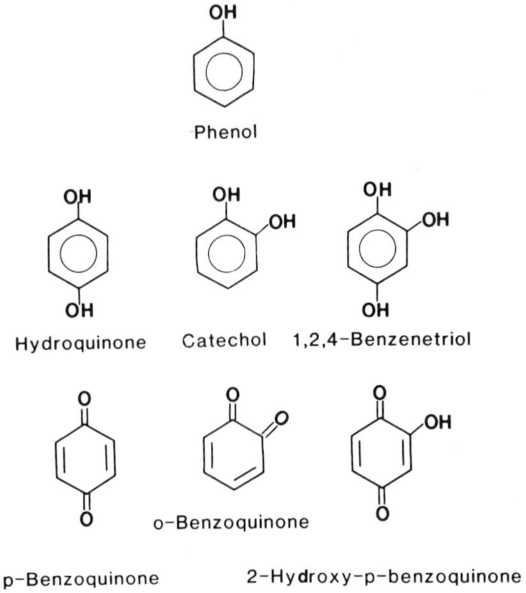

Phenol

Hydroquinone   Catechol   1,2,4-Benzenetriol

o-Benzoquinone

p-Benzoquinone   2-Hydroxy-p-benzoquinone

Fig. 1. Dihydroxy and diketo derivatives of benzene.

vation of phenols must also include a discussion of their oxidation products. Quinones represent the ultimate oxidation products of phenols and similarly make up a large group of compounds; some 200 quinones occur naturally. The chemical and biological reactivity of phenols and quinones is reflected in the large number that are in current use as drugs.

Evidence is accumulating for an important role of secondary metabolites of phenols in the production of toxicity by aromatic hydrocarbons.[1-3] Many aromatic hydrocarbons are bioactivated by cytochrome $P$-450-dependent monooxygenase-catalyzed epoxidation.[4,5] Arene oxides, the primary metabolites of aromatic hydrocarbons, have been identified as precursors of other reactive metabolites, such as epoxydiols,[6,7] epoxyphenols,[8,9] semiquinone radicals, and quinones,[10,11] even though arene oxides themselves may in part be responsible for the toxicity of the parent compounds.

A comprehensive discussion of the bioactivation of phenols, catechols, and quinones is beyond the scope of this chapter. It will therefore be confined to an outline of general characteristics common to phenols, catechols, and quinones and, with certain exceptions, will use the hydroxy and diketo derivatives of benzene as examples (Fig. 1). In addition, the distinction between phenols, catechols, and hydroquinones, vis-à-vis their intrinsic toxicity, is an artificial contrivance because, in the broadest sense, all can be

# 9. PHENOLS, CATECHOLS, AND QUINONES

classified as phenols. Nevertheless, in terms of their propensity to undergo further oxidation to more reactive structures, such distinctions can be made.

## II. Phenols

### A. CHEMICAL PROPERTIES

Phenol is produced commercially by hydrolysis of diazonium salts, the alkali fusion of aromatic sulfonates, or the treatment of halogenated benzenes with alkali, processes without analogy in biology.[12,13] Direct aromatic hydroxylation has little synthetic value because yields are low; however, a number of oxidation systems employing hydrogen peroxide and metal salts have been examined as models for enzymatic hydroxylation.[14] The biosynthesis of monophenols, involving initial aromatic oxidation, is dealt with in Chapter 6. Although epoxide hydrolase catalyzes hydrolysis of arene oxides, uncatalyzed rearrangement of arene oxides to the corresponding phenols appears to predominate both *in vivo* and *in vitro*.[1,15,16]

Phenols are readily susceptible to oxidation, presumably because the aromatic $\pi$-electron structure renders their anionic and radical forms relatively stable (Fig. 2). As a result, the phenolic hydroxyl group behaves less like an alcohol and more like a weak acid,[13] hence the derivation of the term *carbolic acid,* which was used as a synonym for phenol in the classical literature. The phenolic hydroxyl group has a mesomeric (electron-donating) effect and activates aromatic nuclei at the ortho and para positions. Phenols are therefore weakly nucleophilic and participate in a variety of electrophilic substitution reactions of importance in organic chemistry. An example of an enzymatic aromatic substitution reaction involving phenols is the biosynthesis of the thyroid hormones in which tyrosine is progressively iodinated and coupled to form thyroxine.[17,18]

Phenols can be oxidized by a variety of oxidants to catechols and hydroquinones or, alternatively, to a mixture of $o,o'$-, $o,p'$- or $p,p'$-biphenols.

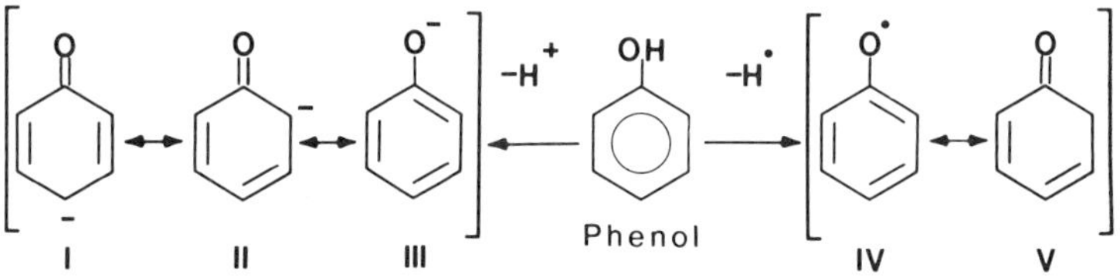

**Fig. 2.** Ionic and radical forms of phenol. **I–III** and **IV–V** represent intermediate forms of the phenolic ion and phenoxy radical, respectively.

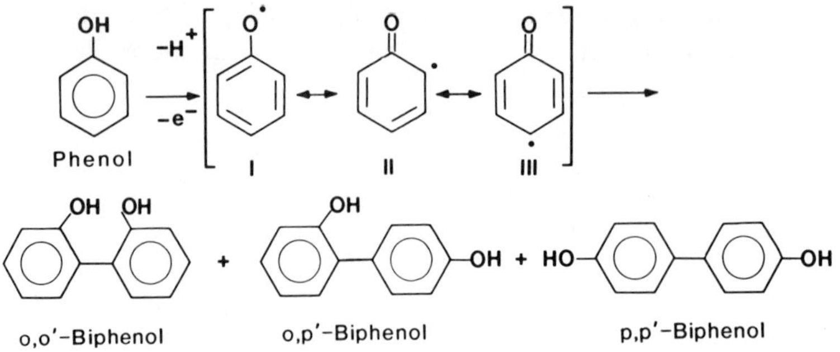

Fig. 3. Products of the potassium ferricyanide-catalyzed oxidation of phenol. I–III represent the intermediate forms of the phenoxy radical.

Whether phenoxide radical reacts with hydroxyl radical to form catechol and hydroquinone or with another phenoxy radical to form biphenols is presumably dependent on the concentration of hydroxyl radical and the rate of phenoxide radical generation. Oxidation of phenol with potassium ferricyanide results in the formation of all isomers of diphenol (Fig. 3) and, as will be discussed later, is of interest from the point of view of comparison with peroxidase-catalyzed oxidation of phenol.[19]

## B. INTRINSIC TOXICITY AND PROPOSED REACTION MECHANISMS

Aside from their behavior as acids, phenols interfere with a variety of processes within cells. Phenols (e.g., 2,4-dinitrophenol) are uncouplers of oxidative phosphorylation. Although the mechanisms of uncoupling are incompletely understood, the activity of 2,4-dinitrophenol can be explained by the fact that it is a lipophilic weak acid, capable of migrating through mitochondrial membranes in both unionized and ionized forms, thus washing out the proton gradient.[20] Phenols are effective inhibitors of a number of FAD- and $NAD^+$-containing oxidases and dehydrogenases via reaction mechanisms that exhibit complex kinetics. Phenols form charge transfer complexes with FAD and $NAD^+$ and, in addition, compete with these coenzymes for binding to enzymes.[21] Because they are easily oxidized by free-radical intermediates, phenolic compounds have long been employed as antioxidants. Hemler and co-workers demonstrated that phenol and 1-naphthol inhibit cyclooxygenase and postulated that the most likely mechanism is the removal of intermediate radicals essential for enzyme activity.[22] Although phenols possess intrinsic biological activity, their toxicity is usually less than that of their oxidation products. For example, the $EC_{50}$ of phenol for suppression of lymphocyte growth in culture is over three orders of magnitude greater than that of hydroquinone or p-benzoquinone.[23]

## C. METABOLISM AND BIOACTIVATION

The major route for the elimination of phenols is via conjugation. Phenols are readily converted to their respective ethereal sulfates or O-glucuronides via pathways that are described elsewhere.[24-26] Although conjugation constitutes the major metabolic route for phenols,[27] their oxidation products are of principal interest from the standpoint of their bioactivation.

There appear to be three major pathways for the bioactivation of phenols: (1) nonenzymatic oxidation, (2) enzymatic one-step oxidation or hydroxylation, and (3) enzymatic two-step oxidation, first to catechols or hydroquinones and ultimately to reactive forms responsible for toxicity. The latter pathway will be discussed in the context of the bioactivation of catechols and hydroquinones.

Some phenols readily undergo autoxidation to the corresponding phenoxy radical. One of the most reactive phenols is 6-hydroxybenzo[a]pyrene, a major metabolite of benzo[a]pyrene, which cannot be detected under normal conditions because of its extreme susceptibility to oxidation by dioxygen.[28] This compound and its oxidation products illustrate the complexity that often characterizes attempts to elucidate mechanisms of bioactivation of phenols. The 6-hydroxybenzo[a]pyrene radical is considered to be responsible for the covalent binding of 6-hydroxybenzo[a]pyrene to cellular macromolecules, such as DNA and proteins, as well as an obligatory intermediate in the formation of diketo derivatives of benzo[a]pyrene that also react with DNA.[29,30] The relative contribution of each of these metabolites to the toxicity of the parent compound remains to be determined.

Phenols that are resistant to autoxidation may undergo activation by serving as substrates for a variety of monooxygenases. A number of different enzyme systems that catalyze the hydroxylation of phenols have been isolated from plant and animal sources. For example, tyrosinase is a copper-containing enzyme that catalyzes the hydroxylation of tyrosine. The mammalian enzyme is active in the biosynthesis of melanin and is found in skin, brain, liver, and adrenal glands.[31] Tyrosinase-catalyzed oxidation of phenols may explain their toxicity to melanocytes,[32] and the enzyme isolated from mushrooms catalyzes the direct hydroxylation of many phenols and the oxidation of the resulting catechols to their respective quinones.[33] The mammalian enzyme exhibits a narrower range of substrate specificity, and it is questionable whether xenobiotic monophenols can be oxidized via this route.[34,35] Although a bacterial NADPH-dependent flavoprotein monoox-

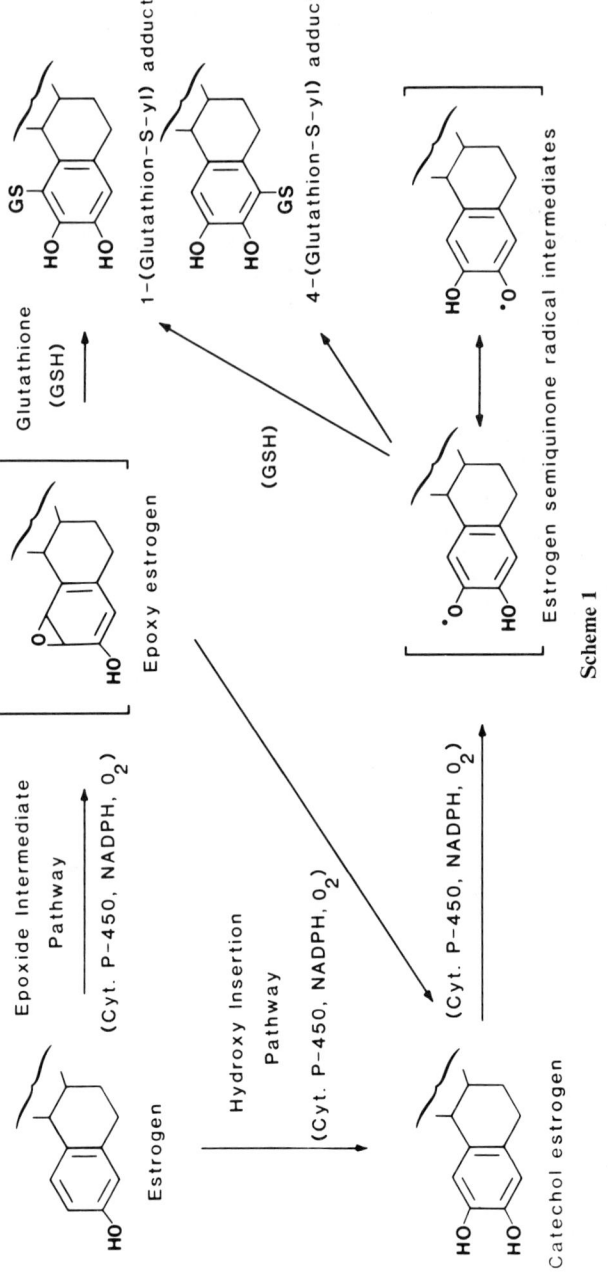

Scheme 1

## 9. PHENOLS, CATECHOLS, AND QUINONES

ygenase has been described that possesses aromatic hydroxylating activity,[36] there is no evidence that the mammalian FAD-containing monooxygenases catalyze a similar reaction.[37]

One of the characteristic cytochrome P-450-catalyzed reactions is the hydroxylation of a variety of aromatic compounds.[4,5] There are two alternative mechanisms for aromatic hydroxylation catalyzed by cytochrome P-450: one via the formation of arene oxide as an obligatory intermediate, the other via a direct hydroxylation. The hydroxylation of phenols by the latter mechanism results in the formation of diphenols, whereas hydroxylation via an arene oxide mechanism results in the formation of hydroxyarene oxides (epoxyphenols) that are more reactive than their respective nonhydroxylated arene oxides.[4] These alternative proposals for phenol hydroxylation have been the subject of controversy, and evidence exists supporting both as mechanisms of phenol bioactivation. For example, estrogens, such as estrone and estradiol, have been postulated to be converted to the corresponding catechols, which, in turn, are oxidized to the corresponding semiquinone or quinone. This has been proposed as a mechanism to account for covalent binding of estrogens to protein.[38-40] Nambara and co-workers, however, suggested that protein adducts of estrogens may be formed by alternative pathways.[41] They postulated that the phenolic moiety of A-ring estrogens is oxidized by a cytochrome P-450-dependent monooxygenase to the corresponding hydroxyarene oxide (Scheme 1).[42] In addition, they provided evidence for the involvement of a species of cytochrome P-450 different from that responsible for formation of the catechol estrogens.[43] Acetaminophen is a phenol with an N-acetylamino substituent for which an epoxy intermediate has been dismissed as the reactive metabolite responsible for the covalent binding of the parent compound to cellular macromolecules.[44] Alternatively, acetaminophen is postulated to undergo oxidation by cytochrome P-450 to form N-acetyl-p-benzoquinonimine[45,46] (Scheme 2).

Acetaminophen has been reported to be oxidized to a radical by horseradish peroxidase.[47] Peroxidases are known to catalyze the oxidation of

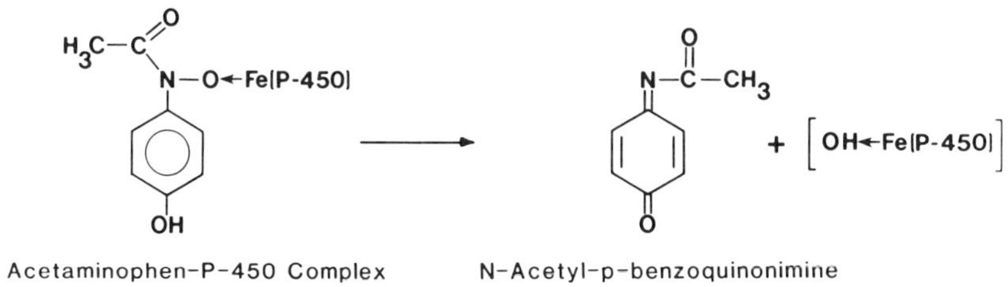

Acetaminophen-P-450 Complex      N-Acetyl-p-benzoquinonimine

Scheme 2

a wide variety of phenols through the formation of phenoxy radical intermediates.[48-50] In the case of acetaminophen, spin-trapping measurements have demonstrated that the radical is centered on a carbon atom of the benzene ring and not on the nitrogen atom, indicating that the peroxidase-catalyzed covalent binding of acetaminophen to protein involves a carbon atom.[47] Prostaglandin endoperoxidase also catalyzes the bioactivation of acetaminophen via an arachidonic acid-dependent hydroperoxidase mechanism.[51] Although not fully understood, this bioactivation mechanism may explain the renal damage that is occasionally observed after massive doses of acetaminophen.

In contrast to the oxidation of phenol by potassium ferricyanide, the reaction catalyzed by horseradish peroxidase gives rise to a single product, $o,o'$-biphenol, indicating that the phenoxy radical remains bound to the enzyme throughout the reaction.[52] Although peroxidase-catalyzed bioactivation of phenols has not received wide attention, studies on the biotransformation of phenol in bone marrow reveal phenol-associated covalent binding and confirm the formation of $o,o'$-biphenol as the sole free product (T. Sawahata, R. D. Irons, and R. A. Neal, unpublished). The significance of biphenol as a metabolite of phenol has not been examined, but is of interest with respect to its possible contribution to benzene-associated myelotoxicity in that phenol is the primary metabolite of benzene.[1,15,16]

## III. Catechols and Hydroquinones

### A. CHEMICAL PROPERTIES

Catechols and hydroquinones are dihydroxy derivatives of aromatic hydrocarbons and are well known for their propensity to undergo further oxidation.[53] In general, conditions that oxidize phenols will also oxidize their catechol and hydroquinone derivatives. The enhanced susceptibility of hydroquinones and catechols to oxidation, relative to the monophenols, may be accounted for by the mesomeric effect of two phenolic groups. For example, both catechol and hydroquinone undergo autoxidation more readily than phenol via semiquinone radical intermediates to form $o$- and $p$-benzoquinone, respectively.[53,54] Under physiological conditions, however, only hydroquinone undergoes appreciable autoxidation.[10,54] The reduced susceptibility of catechol to autoxidize may be accounted for by hydrogen bonding of the two adjacent hydroxyl groups, partly ameliorating their electron-donating effects.

Catechols are oxidized readily to benzoquinones by oxidants such as sodium periodate and silver oxide. The copper(II)-induced oxidation of cate-

# 9. PHENOLS, CATECHOLS, AND QUINONES

chol, which yields o-benzoquinone or *cis,cis*-muconic acid, depending on the reaction conditions,[55] is analogous to the conversion of catechols to o-quinones by tyrosinase, a copper-containing enzyme,[56] as well as to the enzymatic cleavage of catechol by bacterial pyrocatechase.[57] Ring cleavage has been suggested as a route for the bioactivation of benzene, but pyrocatechase activity has not been demonstrated in mammalian tissues. Park and Williams have presented evidence that neither phenol nor catechol are precursors of muconic acid production *in vivo*.[58]

## B. BIOSYNTHESIS

The biosynthesis of catechols in mammalian tissues appears to involve the cytochrome *P*-450-dependent monooxygenase system, which is important because of its broad substrate specificity.[59] As discussed previously, there is considerable controversy over the mechanism of aromatic hydroxylation that results in the formation of catechols. In the case of acetaminophen, a catechol metabolite has been identified (3-hydroxyacetaminophen), but its formation is not altered by the addition of epoxide hydrolase or glutathione indicating that an arene oxide intermediate is not a precursor of the 3-hydroxymetabolite.[60] However, it should be pointed out that both the hydroxylarene oxide and the hydrolysis product of hydroxyarene oxide by epoxide hydrolase would be highly unstable and would undergo rearrangement or dehydration to the corresponding catechol (Scheme 3).

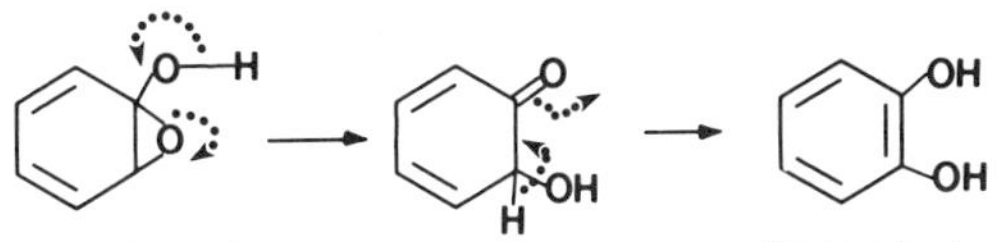

Epoxyphenol → → Catechol

Scheme 3

It has been postulated that catechol can be formed by a sequence of reactions involving epoxidation by cytochrome *P*-450, hydrolysis by epoxide hydrolase, and aromatization by a dehydrogenase.[15] Nevertheless, it is unlikely that this reaction sequence represents a major pathway for the formation of catechol, because the arene oxide is so unstable that it readily undergoes rearrangement to phenol. Two examples suffice. Phenol is the major metabolite of benzene, and benzene dihydrodiol is not detectable *in vitro*.[1] The biotransformation of butamoxane to its 6,7-dihydroxymetabolite has been shown to proceed via two consecutive hydroxylations and not via dihydrodiol or endoperoxide formation.[61]

## C. INTRINSIC TOXICITY

In general, the intrinsic toxicity of catechols and hydroquinones is similar to that of phenols; however, the instability of catechols and hydroquinones often makes it difficult to distinguish their intrinsic effects from the toxicity of their oxidation products. For example, catechol and hydroquinone produce acute effects on the central nervous system involving initial stimulation followed by depression, coma, and death.[62,63] Both compounds, being diphenols, would be expected to participate in the formation of charge transfer complexes, and Borazan and co-workers have described complexes between catechol and both amino acids and nucleotide bases, which may alter nerve conduction.[64,65] Alternatively, catechol-induced seizures may involve a cholinergic pathway. The question about the mechanisms involved and the actual role of catechol or hydroquinone in acute toxicity remains unanswered, in part because quinones, the oxidation products of catechols and hydroquinones, also produce acute effects on the central nervous system, including twitching, excitation, and convulsions.[63]

## D. METABOLISM AND BIOACTIVATION

The most extensive studies of the bioactivation of catechols have been concerned with the catechol estrogens and catecholamines. Three major enzyme systems catalyze the oxidation of catechols and hydroquinones. Two that have been studied in the context of the bioactivation of catechols are the cytochrome $P$-450-dependent monooxygenase system and tyrosinase. In addition, peroxidase oxidizes catechol in the presence of hydrogen peroxide, although the products of this reaction are unknown.[66] The hepatic microsomal $P$-450-dependent monooxygenase system is involved in the biotransformation of catechols, resulting in covalent binding to microsomal protein.[38] Because typical trapping agents for $o$-quinones, such as

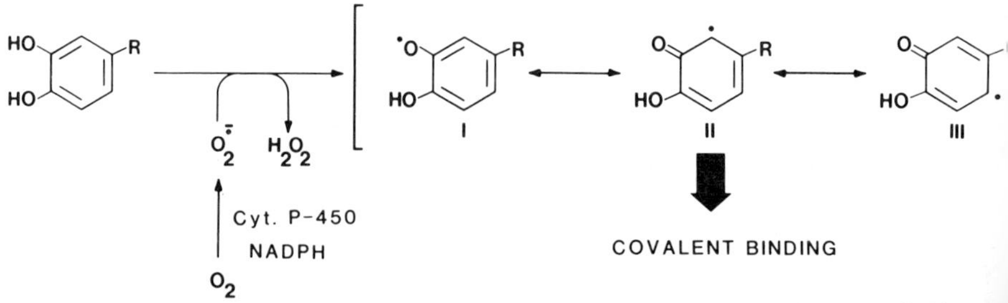

Fig. 4. Proposed scheme for the activation of substituted catechols, such as $\alpha$-methyldopa, by cytochrome $P$-450 via semiquinone intermediates (I–III).

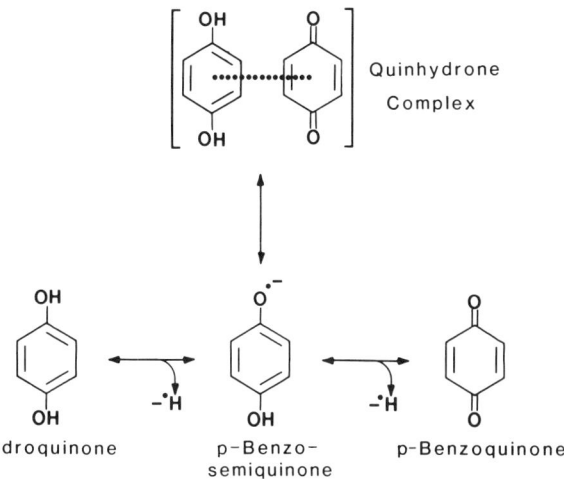

**Fig. 5.** Disproportionation of hydroquinone–benzoquinone and their intermediate forms.

ethylenediamine, fail to trap the reactive metabolites or to inhibit covalent binding, the $o$-quinone has been excluded as the reactive species. The ability of reducing agents, such as glutathione or ascorbic acid to inhibit the covalent binding of quinones has led to the postulate that a semiquinone radical is the reactive metabolite (Fig. 4). In contrast, hydroquinone undergoes oxidation to $p$-benzoquinone via the semiquinone intermediate[10] (Fig. 5), and although the effects of hydroquinone *in vitro* are oxygen dependent, sulfhydryl compounds protect against the effects of hydroquinone *in vitro* or in culture, implicating $p$-benzoquinone as the active species.[23,67]

Tyrosinase catalyzes the oxidation of catechols to quinones. The tyrosinase-catalyzed covalent binding of catechols is inhibited by an $o$-quinone-trapping agent, ethylenediamine, as well as by sulfhydryl compounds such as glutathione and cysteine,[33,68] whereas the cytochrome $P$-450-catalyzed covalent binding of catechols is inhibited by thiols but not by ethylenediamine.

## IV. Quinones

### A. BIOSYNTHESIS

Quinones are diketone derivatives of aromatic hydrocarbons. They may be simple (monocyclic) or complex (polycyclic), substituted or unsubstituted. Quinoid structures are also characteristic of certain amino-substituted aromatic compounds (e.g., $p$-phenylenediamine) that possess many of the same chemical properties as quinones; however, they will not be considered in this chapter. A polycyclic structure may be classified as a quinone whether or not the carbonyl groups are substituents of the same or different rings, but, again, discussion will be limited to quinones in which the carbonyl groups reside on the same ring (Fig. 6).

Quinones are prepared by the oxidation of phenols, catechols, and hydroquinones, although condensation reactions are also employed.[69] A large number of quinones occur naturally in plants and arise from different synthetic pathways[70]; however, because of the inability to synthesize aromatic rings (estrogen biosynthesis excepted), quinone biosynthesis in mammals is limited to pathways utilizing quinone precursors or involving the oxidation of phenols, catechols, or hydroquinones. The biosynthesis of ubiquinone (coenzyme Q), for example, proceeds from phenylalanine, tyrosine, or phenolic acids as starting compounds.[71]

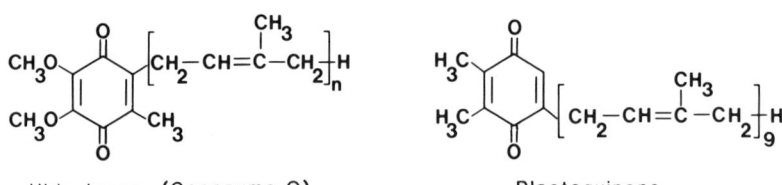

Fig. 6. Representative examples of simple, complex, partially, and completely substituted quinones.

## B. CHEMICAL PROPERTIES

Quinones are highly reactive species that participate in a number of reactions of biological significance. In general, their chemical reactivity is related to their oxidation–reduction potentials and their ability to arylate nucleophilic groups, primarily sulfhydryls. Variations in oxidation–reduction potentials for various quinones appear to be related to differences in resonance energy between the quinone and its corresponding hydroquinone.[72] At neutral pH, $p$-benzoquinone is a potent oxidizing agent ($E = +0.699V$) and is readily reduced to hydroquinone. Because hydroquinone autoxidizes under physiological conditions, a disproportionation reaction occurs between hydroquinone, $p$-benzoquinone, and the 1,4-semiquinone radical intermediate, which, in theory, may also dimerize to form a quinhydrone complex (Fig. 5). An important characteristic of quinone–hydroquinone systems is that one will readily oxidize another with a lower redox potential.

Quinones do not exhibit aromaticity but behave much like $\alpha,\beta$-unsaturated diketones. An unsaturated diketone structure results in a highly polarized carbon–carbon bond, which is subject to nucleophilic attack and accounts for the noted sulfhydryl-arylating capability of quinones.

Covalent interactions of quinones with sulfhydryls are theoretically independent of redox potential and are properly referred to as alkylation, because the immediate product of the reaction possesses no aromatic character (Fig. 7A). However, because this product cannot be isolated and the stable adduct is aromatic, the term arylation will be used. The reaction is often depicted as shown in Fig. 7A; however, it is possible to obtain the tetrathio-derivative,[73] and the primary reaction products are probably described more accurately as the tautomeric isomers (Fig. 7B). Although reactions between quinones and amino acids, amines, and other N-containing structures have been described, sulfhydryl addition occurs much more readily.[63] Moreover, under physiological conditions, the oxidation product of hydroquinone does not react with nucleophilic nitrogen atoms in ATP or GTP.[74]

Variations in the structure of quinones alter their behavior, but the effects of such alterations on oxidation or arylating potential are often difficult to predict. Nevertheless, some generalizations appear applicable. The redox potential of $o$-quinones is usually greater than their corresponding $p$-quinones, as is their susceptibility to nucleophilic attack.[75,76] Although the greater oxidation–reduction potential of $o$-quinones may be explicable in terms of the stability of the reduced forms brought about by hydrogen bonding (e.g., catechol), the greater nucleophilicity cannot be rationalized in this manner. The addition of substituent groups to benzoquinones or naphtho-

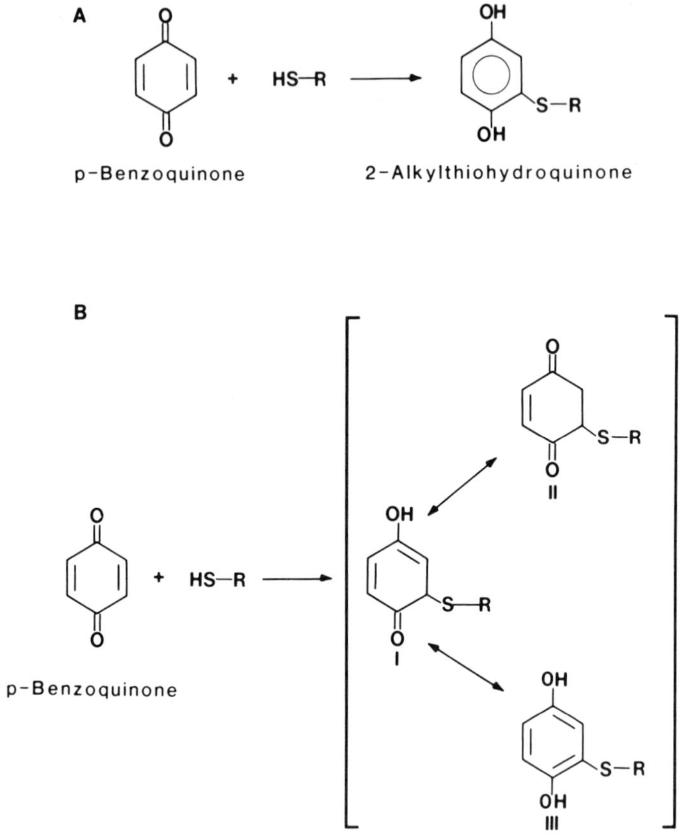

Fig. 7. Schematic representation of the arylation of alkyl sulfhydryls by benzoquinone. The reaction is commonly written as in (A), although 2-alkyl-thiohydroquinone (2-alkylthio-1,4-dihydroxybenzene) is probably more accurately depicted (B) in equilibrium with its tautomeric forms (I–III).

quinones modifies their behavior, though not always predictably. Addition of a hydroxyl group at position 2 reduces both the redox potential of the 1,4-quinones[75] and their sulfhydryl-arylating activity.[32] Similar effects are described for the addition of methyl groups, whereas chlorine addition at the 2-position tends to increase quinone reactivity.[63] These effects are presumed to be due to changes in redox potential, hydroquinone ionization potential, semiquinone stability, or sulfhydryl-arylating activity. The results of substitution experiments with quinones must be interpreted with caution, because they may alter simultaneously more than one property of the molecule. Benzohydroquinone autoxidizes more slowly than 1,2,4-benzenetriol under physiological conditions,[10] but the potential of the triol to inhibit sulfhydryl-dependent reactions is much less than the hydroquinone, presumably because hydroxyl substitution at the 2-position renders the quinone much less reactive toward sulfhydryl groups.[67] Fully substituted quinones (e.g., ubiquinones) possess no arylating potential, but retain redox capability, albeit less than that of unsubstituted quinones.

## C. TOXICITY AND REACTION MECHANISMS

The presence of multiple reactive forms (i.e., quinones, hydroquinones, semiquinone radicals, and quinhydrones) makes it difficult to define the precise mechanisms of quinone toxicity. Nevertheless, quinones are highly reactive species that produce relatively specific effects on cells. Aside from their previously described effects on the central nervous system, quinones produce a broad spectrum of toxicities when administered *in vivo*. Repeated administration of quinones is associated with agranulocytosis, pancytopenia, and suppression of lymphocyte function.[77-79] In addition, quinones produce abnormalities in embryonic development[80] and mitotic arrest in germinal cells in the mouse intestine.[81]

The mechanisms most frequently invoked to explain quinone toxicity are sulfhydryl arylation, interference with electron transport, generation of superoxide or hydroxyl radicals, and arylation by semiquinone intermediates. In addition, two alternative mechanisms have been proposed that deserve brief mention, namely, complex formation and sulfhydryl oxidation. Complex formation with substrates or metal ions has been invoked to explain enzyme inhibition by quinones.[63] Because some hydroquinone must always be present in equilibrium with the quinone, the precise contribution of each to complex formation is often difficult to determine. Quinones are also known to oxidize sulfhydryl groups, a reaction that is critically dependent on pH. Again, it is somewhat difficult to distinguish between the effects of sulfhydryl oxidation and sulfhydryl arylation in a biological system. Simple oxidation of sulfhydryls may result in reversible inhibition of sulfhydryl-dependent enzyme function. In contrast, quinone inhibition of sulfhydryl-dependent enzymes is often irreversible. Most investigators have invoked covalent interaction with sulfhydryls to explain enzyme inhibition by quinones.[63] The controversy over the potential of certain quinones to participate in redox reactions with sulfhydryl groups extends to the question of the role of vitamin E as an antioxidant in mammalian tissues. Vitamin E deficiency has been associated with increased lipid peroxidation and a decline in tissue respiration, which has been attributed to the general ability of quinoid metabolites of vitamin E to maintain intracellular thiols (glutathione) in a reduced state. Alternatively, it has been suggested that the participation of these quinones in a redox system with sulfhydryl groups is unlikely, but that quinoid derivatives of vitamin E may form addition compounds with sulfhydryl groups as a requirement for respiratory enzyme integrity.[82] A similar mechanism has been proposed for plastoquinone, an important cofactor in biochemical reactions utilizing water as a reducing source, for which interaction with sulfhydryls or other reactive groups on enzymes may be an essential feature.[83]

Sulfhydryl arylation appears to be the most important mechanism of toxicity of simple quinones. Quinones are potent inhibitors of a number of sulfhydryl-dependent enzymes including DNA polymerase[32] and catechol $O$-methyltransferase.[76] The $IC_{50}$ values for a variety of quinones in these enzyme systems range from 0.002 to 1.0 m$M$. The inhibitory effects of quinones on cell growth and division occur at even lower concentrations. Lehman first described the ability of quinones to block cell division in mitosis at concentrations that do not produce cell death.[84] $p$-Benzoquinone is 100 times more potent than colchicine in this respect. These findings have been confirmed and extended in a series of experiments in our laboratory, which demonstrate that $p$-benzoquinone inhibits lectin-induced lymphocyte blastogenesis, with an $EC_{50}$ of approximately 0.0007 m$M$.[85] At these concentrations, $p$-benzoquinone does not alter cellular ATP concentrations or membrane permeability.

The ability of $p$-benzoquinone to suppress cell growth in the absence of cytotoxicity suggested that it might act by a mechanism similar to cytochalasin A or colchicine and inhibit cell blastogenesis by interfering with microtubule function. Microtubule integrity is an important early requirement in blastogenesis as well as in spindle formation during metaphase.[83] Furthermore, the effects of $p$-benzoquinone on cell growth are prevented by low molecular weight thiols but not by other nucleophiles, including amines, hydroxyl, or imidazole groups.[67,85,86] The interaction of quinones with microtubule assembly appears to be highly specific, and quinones apparently react with a single sulfhydryl group at or near the GTP-binding site on tubulin.[67,85] The presence of a particularly nucleophilic sulfhydryl group involved in GTP–tubulin binding has been described previously.[87]

Quinones interfere with electron transport either by serving as analogs of endogenous quinones or by inhibiting enzymes active in the electron transport chain. Although these alternatives may be simple to distinguish in theory, the effects of quinones on electron transport are often difficult to interpret. With the exception of unsubstituted benzoquinones, most quinones possess redox potentials between the cytochromes and the flavins and pyridine nucleotides.[63] Quinones can therefore oxidize their corresponding hydroquinones or accept electrons from other molecules such as enzymes involved in electron transport or oxidative phosphorylation. In doing so, the resulting hydroquinones may be reoxidized or transfer electrons to enzymes within or outside the electron transport chain. As such, quinone–hydroquinone systems participate as mobile (not membrane-bound) components in the electron transport chain. A brief description of quinones normally involved in these processes seems warranted.

One of the most extensively studied groups of naturally occurring quinones is the ubiquinones (coenzyme Q), which are completely substituted isoprenoid quinones found in mitochondrial preparations isolated from organisms that use hydrogen as an ultimate source of reducing equivalents.[88] The function of ubiquinone in mammalian systems is the transport of electrons from flavoproteins to cytochrome $c$. Plastoquinones, which contain an unsubstituted position on the quinone ring, retain some sulfhydryl-arylating activity and, as previously mentioned, are found in photosynthetic systems and organisms that use water as a reducing source.

Hydroquinones can donate electrons to an electron transport system, and quinones can divert electrons from the electron transport system. Finally, quinone–hydroquinone systems, acting in concert, can bypass segments of the electron transport chain. For example, NADPH–cytochrome $c$ reductase catalyzes the oxidation of $p$-benzosemiquinone to $p$-benzoquinone, although the semiquinone may directly reduce cytochrome $c$.[66] Some hydroquinones, such as the hydroquinone of 2-methoxy-6-propylbenzoquinone, are not oxidized by the electron transport chain, but can utilize external acceptors such as ferricyanide.[89] The effects of quinone–hydroquinone systems can be quite specific. 1,2-Naphthoquinone will stimulate the oxidation of NADPH, but not the reduction of cytochrome $c$. Alternatively, menadione accelerates the reduction of cytochrome $c$ by NADPH.[63] The number of enzymes known to react with quinone–hydroquinone systems is vast and includes cytochrome $c$ reductase, succinate:quinone reductase, DT-diaphorase, and vitamin K reductase.[63] Because of the numerous enzymes potentially affected as well as the complex behavior of quinones in these systems, it is extremely hazardous to extrapolate results from *in vitro* studies to explain the effects of such agents on respiration in whole cells.

A number of heterocyclic quinones are in current use as chemotherapeutic agents. The generation of oxygen free radicals has been proposed as a primary mechanism for the toxicity of these compounds, although the exact contribution of free radicals to cytotoxicity remains obscure. Adriamycin and daunomycin, both anthracycline derivatives, and mitomycin C, an N-heterocyclic quinone, are examples of quinones used as chemotherapeutic agents. Both groups stimulate NADPH-dependent oxygen consumption by cells and generate superoxide radicals, hydrogen peroxide, or both by a cycling hydroquinone–semiquinone–quinone redox mechanism.[66,90-92] Nevertheless, a number of alternative mechanisms have been postulated to account for the toxicity of these anticancer drugs. Semiquinone radicals have been postulated to account for the action of these agents on DNA. However, free-radical and semiquinone mechanisms cannot operate simultaneously because superoxide generation necessarily destroys the semiquinone, and the stability of the semiquinone metabolites is probably not sufficient to allow macromolecular interactions to compete with electron transfer to oxygen.[66] Alternatively, the anthraquinones and heterocyclic quinones are known to intercalate DNA. Yet for adriamycin, but not mitomycin C,[92] DNA binding prevents superoxide formation.[66] This may reflect differences in DNA-binding affinity for adriamycin and mitomycin C. To add further to the confusion, anthraquinones are known to possess sulfhydryl-arylating ability as well, although this mechanism alone cannot account for the toxicity of anthraquinones on whole cells.[86] Adriamycin[86] and daunomycin[93] interfere with microtubule assembly *in vitro* in a manner analogous to *p*-benzoquinone; nevertheless, at effective concentrations, adriamycin toxicity cannot be attributed solely to sulfhydryl-dependent disruption of the cytoskeleton.[86,93]

Semiquinone radicals have been postulated to be important effectors of the toxicity of catechols, hydroquinones, and quinones. Benzosemiquinone has been postulated to be the species responsible for the toxicity of the analogous polyphenolic metabolites of benzene,[10] although their effects on cell growth response can more likely be attributed to sulfhydryl arylation by benzoquinone.[67,74,85,86] The role of semiquinone formation in anthraquinone toxicity has also been reviewed. The strongest, albeit indirect, demonstration of a role for semiquinones in covalent binding has been previously described for the bioactivation of catechol estrogens by cytochrome *P*-450. The dilemma is related, in part, to the instability of semiquinones as well as to their relatively weak reactivity when compared with other organic free radicals. As a result, a direct covalent interaction involving a free semiquinone radical has yet to be demonstrated directly.

Phenolic and quinone metabolites of aromatic hydrocarbons represent a heterogeneous group of reactive molecules, which possess significant biological reactivity. Our treatment of these compounds is not intended to be encyclopedic, and in fact, we have not discussed a vast array of substituted phenolic and quinone structures for which a significant literature exists. In this chapter, we have attempted to convey an appreciation of the complexity of the chemistry and biological behavior of these compounds. Although the diversity exhibited by these compounds renders elucidation of their bioactivation a challenge, their reactivity provides a powerful tool for furthering our understanding of processes at work in cell and molecular biology.

## References

1. Tunek, A., Platt, K. L., Bently, P., and Oesch, F. (1978). Microsomal metabolism of benzene to species irreversibly binding to microsomal protein and effects of modification of this metabolism. *Mol. Pharmacol.* **14**, 920–929.
2. Hesse, S., and Mezger, M. (1979). Involvement of phenolic metabolites in the irreversible protein-binding of aromatic hydrocarbons: Reactive metabolites of [$^{14}$C]naphthalene and [$^{14}$C]1-naphthol formed by rat liver microsomes. *Mol. Pharmacol.* **16**, 667–675.
3. Hess, S., Wolff, T., and Mezger, M. (1980). Involvement of phenolic metabolites in the irreversible protein binding of [$^{14}$C]bromobenzene catalyzed by rat liver microsomes. *Arch. Toxicol., Suppl.* **4**, 358–362.
4. Daly, J. W., Jerina, D. M., and Witkop, B. (1972). Arene oxides and the NIH shift: The metabolism, toxicity and carcinogenicity of aromatic compounds. *Experientia* **28**, 1129–1149.
5. Jerina, D. M., and Daly, J. W. (1974). Arene oxides: A new aspect of drug metabolism. *Science* **185**, 573–582.
6. Levin, W., Wood, A. W., Wislocki, P. G., Chang, R. L., Kapiyulinik, J., Mah, H. D., Yagi, H., Jerina, D. M., and Conney, A. H. (1978). Mutagenicity and carcinogenicity of benzo[*a*]pyrene and benzo[*a*]pyrene derivatives. *In* "Polycyclic Hydrocarbons and Cancer" (H. V. Gelboin and P. O. P. Ts'o, eds.), Vol. 1, pp. 189–202. Academic Press, New York.
7. Yang, S., Deutsch, J., and Gelboin, H. V. (1978). Benzo[*a*]pyrene metabolism: Activation and detoxication. *In* "Polycyclic Hydrocarbons and Cancer" (H. V. Gelboin and P. O. P. Ts'o, eds.), Vol. 1, pp. 205–213. Academic Press, New York.
8. Jernstrom, B., Vadi, H., and Orrenius, S. (1978). Formation of DNA-binding products from isolated benzo[*a*]pyrene metabolites in rat liver nuclei. *Chem.-Biol. Interact.* **20**, 311–321.
9. Vigry, P., Ginot, Y. M., Kindts, M., Cooper, C. S., Grover, P. L., and Sims, P. (1980). Fluorescence spectral evidence that benzo[*a*]pyrene is activated by metabolism in mouse skin to a diol epoxide and a phenol epoxide. *Carcinogenesis (N.Y.)* **1**, 945–950.
10. Greenlee, W. F., Sun, J. D., and Bus, J. S. (1981). A proposed mechanism of benzene toxicity: Formation of reactive intermediates from polyphenol metabolites. *Toxicol. Appl. Pharmacol.* **59**, 187–195.
11. Tunek, A., Platt, K. L., Przybylski, M., and Oesch, F. (1980). Multistep metabolic activation of benzene. Effect of superoxide dismutase on covalent binding to microsomal macromolecules and identification of glutathione conjugates using high-pressure liquid chromatography and field desorption mass spectrometry. *Chem.-Biol. Interact.* **33**, 1–17.
12. Fieser, L. F., and Fieser, M. (1967–1979). "Reagents for Organic Synthesis," Vols. I–VII. Wiley, New York.
13. Morrison, R. T., and Boyd, R. N. (1973). "Organic Chemistry," 3rd ed., pp. 787–814. Allyn & Bacon, Boston, Massachusetts.
14. Ullrich, V. (1972). Enzymatic hydroxylation with molecular oxygen. *Agnew. Chem., Int. Ed. Engl.* **11**, 701–712.
15. Jerina, D., Daly, J., Witkop, B, Zaltzman-Nirenberg, P., and Udenfriend, S. (1968). Role of the arene oxide–oxepin system in the metabolism of aromatic substrates. I. In vitro conversion of benzene oxide to a premercapturic acid and a dihydrodiol. *Arch. Biochem. Biophys.* **128**, 176–183.
16. Porteous, J. W., and Williams, R. T. (1949). Studies in detoxication 20. The metbolism of benzene. II. The isolation of phenol, catechol, quinol, and hydroxyquinol from the ethereal sulfate fraction of the urine of rabbits receiving benzene orally. *Biochem. J.* **44**, 56–61.

17. Morris, D. R., and Hager, L. P. (1966). Mechanism of the inhibition of enzymatic halogenation by antithyroid agents. *J. Biol. Chem.* **241**, 3582–3589.
18. Maloof, F., and Soodak, M. (1965). The oxidation of thiourea, a new thyroid function. *In* "Current Topics in Thyroid Research" (C. Cassano and M. Andreoli, eds.), pp. 277–290. Academic Press, New York.
19. Hendricson, J. B., Cram, D. J., and Hammond, G. S. (1970), "Organic Chemistry," p. 823. McGraw-Hill, New York.
20. Yang, C. S. (1981). Structure and function of mitochondria. *In* "Advanced Cell Biology" (L. M. Schwartz and M. M. Azar, eds), pp 593–621. Von Nostrand Reinhold, New York.
21. Henneke, C. M., and Wedding, R. T. (1975). NAD–Phenol complex formation, the inhibition of malate dehydrogenase by phenols, and the influence of phenol substituents in inhibitory effectiveness. *Arch. Biochem. Biophys.* **168**, 443–449.
22. Hemmler, M. E., and Lands, W. E. M. (1980). Evidence for a peroxide-initiated free radical mechanism of prostaglandin biosynthesis. *J. Biol. Chem.* **255**, 6253–6261.
23. Pfeifer, R. W., and Irons, R. D. (1981). Inhibition of lectin-stimulated lymphocyte agglutination and mitosis by hydroquinone: Reactivity with intracellular sulfhydryl groups. *Exp. Mol. Pathol.* **35**, 189–198.
24. Goldstein, A., Aranow, L., and Kalman, S. M. (1969). "Principles of Drug Action", pp. 238–239, 245–246. Harper, New York.
25. Kasper, C. B., and Henton, D. (1980). Glucuronidation. *In* "Enzymatic Basis of Detoxication" (W. B. Jacoby, ed.), Vol. 2, pp. 3–36. Academic Press, New York.
26. Jacoby, W. B., Sekura, R. D., Lyon, E. S., Marcus, C. J., and Wang, J.-L. (1980). Sulfotransferases. *In* "Enzymatic Basis of Detoxication" (W. B. Jacoby, ed.), Vol. 2, pp. 199–228. Academic Press, New York.
27. Kao, J., Bridges, J. W., and Faulkner, J. K. (1979). Metabolism of [$^{14}$C]phenol by sheep, pig and rat. *Xenobiotica* **9**, 141–147.
28. Ts'o, P. O. P., Caspary, W. J., and Lorentzen, R. J. (1977). The involvement of free radicals in chemical carcinogenesis. *In* "Free Radicals in Biology" (W. A. Pryor, ed.), Vol. 3, pp. 251–303. Academic Press, New York.
29. Schechtman, L. A., Lesko, S. A., Lorentzen, R. J., and Ts'o, P. O. P. (1974). A cellular system for the study of the chemical reactivity and transforming ability of benzo[a]pyrene and its derivatives. *Proc. Am. Assoc. Cancer Res.* **15**, 66.
30. Ts'o, P. O. P., Caspary, W. J., Leavitt, J., Lesko, S. A., and Lorentzen, R. J. (1976). One-electron oxidation of benzo[a]pyrene in chemical and metabolic processes. *In* "*In Vitro* Metabolic Activation in Mutagenesis Testing" (F. J. DeSerres, J. R. Fouts, J. R. Bend, and R. M. Philpot, eds.), pp. 223–242. Elsevier, Amsterdam.
31. Oser, B. L. (1965). "Hawk's Physiological Chemistry," p. 249. McGraw-Hill, New York.
32. Graham, D. G., Tiffany, S. M., and Vogel, F. S. (1978). The toxicity of melanin precursors. *J. Invest. Dermatol.* **70**, 113–116.
33. Pugh, C. E. M., and Raper, H. S. (1929). The action of tyrosinase on phenols. With some observations on the classification of oxidases. *Biochem. J.* **21**, 1370–1383.
34. Boekelheide, K., Graham, D. G., Mize, P. D., Anderson, C. W., and Jeffs, P. W. (1979). Synthesis of γ-glutaminyl-[3,5,-$^3$H]-hydroxybenzene and the study of reactions catalyzed by the tyrosinase of *Agaricus bisporus*. *J. Biol. Chem.* **254**, 12185–12191.
35. Boekelheide, K., Graham, D. G., Mize, P. D., and Koo, E. H. (1980). Melanocytotoxicity and the mechanism of activation of γ-L-glutaminyl-4-hydroxybenzene. *J. Invest. Dermatol.* **75**, 322–327.
36. Husain, M., Entsch, B., Ballou, D. P., Massey, M., and Chapman, P. (1979). Fluoride elimination from substrates in hydroxylation reactions catalyzed by p-hydroxybenzoate hydroxylase. *J. Biol. Chem.* **255**, 4189–4197.
37. Ziegler, D. M., and Poulsen, L. L. (1978). Hepatic microsomal mixed-function amine

oxidase. *In* "Methods in Enzymology" (S. Fleischer and L. Packer, eds.), Vol. 52, Part C, 142-151. Academic Press, New York.
38. Marks, F., and Hecker, E. (1969). Metabolism and mechansim of action of oestrogens. XII. Structure and mechanism of formation of water-soluble and protein-bound metabolites of oestrone in rat liver microsomes in vitro and *in vivo. Biochim. Biophys. Acta* **187**, 250-265.
39. Jellick, P. H., and Irwin, L. (1963). Interaction of estrogen quinones with ethylenediamine. *Biochim. Biophys. Acta* **78**, 778-780.
40. Hecker, E., Walter, G., and Marks, F. (1965). Zur Bildung wasserlöslicher Metaboliten aus Oestron in Ratteneber-microsomes. *Biochim. Biophys. Acta* **111**, 546-548.
41. Numazawa, M., Tanaka, Y., Monono, Y., and Nambara, T. (1974). Occurrence of the cysteine conjugate of 2-hydroxyestrone in rat bile, with special reference to its formation mechanism. *Chem. Pharm. Bull.* **22**, 663-668.
42. Numazawa, M., and Nambara, T. (1977). A new mechanism of in vitro formation of catechol estrogen glutathione conjugates by rat liver microsomes. *J. Steroid Biochem.* **8**, 835-840.
43. Numazawa, M., Shirao, R., Soeda, N., and Nambara, T. (1978). Properties of enzyme systems involved in the formation of catechol estrogen glutathione conjugates in rat liver microsomes. *Biochem. Pharmacol.* **27**, 1833-1838.
44. Nelson, S. D., Forte, A. J., and Dahlin, D. C. (1980). Lack of evidence for $N$-hydroxyacetaminophen as a reactive metabolite of acetaminophen *in vitro. Biochem. Pharmacol.* **29**, 1617-1620.
45. Hinson, J. A., Pohl, L. R., Monks, T. J., and Gillette, J. R. (1981). Acetaminophen-induced hepatotoxicity. *Life Sci.* **29**, 107-116.
46. DeVries, J. (1981). Hepatotoxic metabolic activation of paracetamol and its derivatives phenacetin and benorilate: Oxygenation or electron transfer? *Biochem. Pharmacol.* **30**, 399-402.
47. Nelson, S. D., Dahlin, D. C., Rauckmen, E. J., and Rosen, G. M. (1981). Peroxidase-mediated formation of reactive metabolites of acetaminophen. *Mol. Pharmacol.* **20**, 195-199.
48. Putter, J. (1974). Peroxidases. *In* "Methods in Enzymtic Analysis" (H. V. Bergmeyer, ed.), Vol. 2, pp. 685-690. Springer-Verlag, Berlin and New York.
49. Yamazaki, I., Mason, H. S., and Pitte, L. (1960). Identification, by electron paramagnetic resonance spectroscopy, of free radicals generated from substrates by peroxidase. *J. Biol. Chem.* **235**, 2444-2449.
50. Yamazaki, I. (1977). Free radicals in enzyme-substrate reactions. *In* "Free Radicals in Biology" (W. A. Pryor, ed.), Vol. 3, pp. 183-218. Academic Press, New York.
51. Mohandas, J., Duggin, G. G., Horvath, J. S., and Tiller, D. J. (1981). Metabolic oxidation of acetaminophen (paracetamol) mediated by cytochrome $P$-450 mixed-function oxidase and prostaglandin endoperoxide synthetase in rabbit kidney. *Toxicol. Appl. Pharmacol.* **61**, 252-259.
52. Danner, D. J., Brignac, P. J., Arceneaux, D., and Patel, V. (1973). The oxidation of phenol and its reaction product by horseradish peroxidase and hydrogen peroxide. *Arch. Biochem. Biophys.* **156**, 759-763.
53. Walling, C. (1957). "Free Radicals in Solution," pp. 403-406. Wiley, New York.
54. Irons, R. D., Greenlee, W. F., Wierda, D., and Bus, J. S. (1982). Relationship between benzene metabolism and toxicity: A proposed mechanism for the formation of reactive intermediates from polyphenol metabolites. *In* "Biological Reactive Intermediates—II" (R. Snyder, D. V. Parke, J. J. Kocsis, and D. J. Jollow, G. G. Gibson, eds.), pp. 229-243. Plenum, New York.
55. Demmin, T. R., Swerdloff, M. D., and Rogic, M. M. (1981). Copper(II)-induced oxida-

tions of aromatic substrates: Catalytic conversion of catechols to $o$-benzoquinones. Copper phenoxides as intermediates in the oxidation of phenol and a single-step conversion of phenol, ammonia, and oxygen to muconic acid mononitrile. *J. Am. Chem. Soc.* **103**, 5795–5804.
56. Nagatsu, T., Levitt, M., and Udenfriend, S. (1964). Tyrosine hydroxylase, the initial step in norepinephrine biosynthesis. *J. Biol. Chem.* **239**, 2910–2917.
57. Que, L., Jr., Lipscomb, J. D., Munck, E., and Wood, J. M. (1977). Protocatechuate 3,4-dioxygenase. Inhibitor studies and mechanistic implications. *Biochim. Biophys. Acta* **485**, 60–74.
58. Parke, D. V., and Williams, R. T. (1953). Studies in detoxication 54. The metabolism of benzene. (a) the formation of phenylglucuronide and phenylsulphuric acid from [$^{14}$C]benzene. (b) the metabolism of [$^{14}$C]phenol. *Biochem. J.* **55**, 337–340.
59. Daly, J., Inscoe, J. K., and Axelrod, J. (1965). The formation of O-methylated catechols by microsomal hydroxylation of phenols and subsequent catechol O-methylation. Substrate specificity. *J. Med. Chem.* **8**, 153–157.
60. Hinson, J. A., Nelson, S. D., Gillette, J. R., and Guengerich, F. P. (1980). 3-Hydroxyacetaminophen: A microsomal metabolite of acetaminophen. Evidence against an epoxide as the reactive metabolite of acetaminophen. *Drug. Metab. Dispos.* **8**, 289–294.
61. Murphy, P. J., Bernstein, J. R., and McMahon, R. E. (1974). The formation of catechols by consecutive hydroxylations: A study of the microsomal hydroxylation of butamoxane. *Mol. Pharmacol.* **10**, 634–639.
62. Wierda, D., and Irons, R. D. (1982). Hydroquinone and catechol reduce the number of progenitor B lymphocytes in mouse spleen and bone marrow. *Immunopharmacology* **4**, 41–54.
63. Webb, J. L. (1966). "Enzymes and Enzyme Inhibitors," Vol. 3, pp. 493–594. Academic Press, New York.
64. Al-Obeidi, F. A., and Borazan, H. N. (1976). Interactions of nucleic acid bases with catechol: UV studies. *J. Pharm. Sci.* **65**, 892–895.
65. Borazan, H. N., and Ajeena, Y. H. (1980). Indole–catechol charge transfer complexes. I. *J. Pharm. Sci.* **69**, 990–991.
66. Mason, R. P. (1982). Free radical intermediates in the metabolism of toxic chemicals. *In* "Free Radicals in Biology" (W. A. Pryor, ed.), Vol. 5, pp. 161–222. Academic Press, New York.
67. Irons, R. D., Neptun, D. A., and Pfeifer, R. W. (1981). Inhibition of lymphocyte transformation and microtubule assembly y quinone metabolites of benzene: Evidence for a common mechanism. *J. Reticuloendothel. Soc.* **30**, 359–372.
68. Bolt, H. M., and Kappus, H. (1974). Irreversible binding of ethynylestradiol metabolites to protein and nucleic acids as catalyzed by rat liver microsomes and mushroom tyrosinase. *J. Steroid Biochem.* **5**, 179–184.
69. Bruce, J. M. (1981). Benzoquinones and related compounds. *In* "Rodd's Chemistry of Carbon Compounds" (M. F. Ansell, ed.), Vol. 3, Part B, p. 19. Elsevier/North-Holland, New York.
70. Bently, R. (1975). Biosynthesis of quinones. *Chem. Soc., Spec. Per. Rep.* **3**, 181–246.
71. Glover, J. (1965). Biosynthesis of biologically active quinones and related compounds. *In* "Biochemistry of Quinones" (R. A. Morton, ed.), pp. 207–255. Academic Press, New York.
72. Berliner, E. (1946). A relation between the oxidation–reduction potentials of quinones and the resonance energies of quinones and of hydroquinones. *J. Am. Chem. Soc.* **68**, 49–51.
73. Schubert, M. (1947). The interaction of thiols and quinones. *J. Am. Chem. Soc.* **69**, 712.

74. Irons, R. D., and Neptun, D. A. (1980). Effects of the principal hydroxy-metabolites of benzene on microtubule polymerization. *Arch. Toxicol.* **45**, 297–305.
75. Morton, R. A. (1965). Introductory account of quinones. *In* "Biochemistry of Quinones" (R. A. Morton, ed.), pp. 1–19. Academic Press, New York.
76. Borchardt, R. T. (1975). Affinity labeling of catechol O-methyltransferase by the oxidation products of 6-hydroxydopamine. *Mol. Pharmacol.* **11**, 436–449.
77. Kracke, R. R., and Parker, F. P. (1934). The etiology of granulocytopenia. With special reference to drugs containing the benzene ring. *Am. J. Clin. Pathol.* **4**, 453–469.
78. Molitor, H., and Robinson, H. J. (1940). Oral and parenteral toxicity of vitamin K; phthiocol and 2-methyl-1,4-naphthoquinone. *Proc. Soc. Exp. Biol. Med.* **43**, 125–128.
79. Wierda, D., and Irons, R. D. (1982). Hydroquinone and catechol reduce the frequency of progenitor B lymphocytes in mouse spleen and bone marrow. Immunopharmacology **4**, 41–54.
80. Bellairs, R. (1954). The effects of tetrasodium 2-methyl 1,4-napthohydroquinone diphosphate on early chick and amphibian embryos. *Br. J. Cancer* **8**, 685–692.
81. Zylberszac, S. (1939). Effect of diphenols on the small intestine. *Acta Brevia Neerl., Physiol., Pharmacol., Microbiol.* **9**, 240.
82. Green. J., and McHale, D. (1965). Quinones related to vitamin E. *In* "Biochemistry of Quinones" (R. A. Morton, ed.), pp. 261–283. Academic Press, New York.
83. Edelman, G. M. (1976). Surface modulation in cell recognition and cell growth. *Science* **192**, 218–226.
84. Lehmann, F. E. (1947). Chemical infuences on cell division. *Experientia* **3**, 223–232.
85. Irons, R. D., Neptun, D. A., and Pfeifer, R. W. (1982). Epigenetic mechanisms of benzene toxicity. *In* "Genotoxic Effects of Airborne Agents" (R. R. Tice, D. L. Costa, and K. Schaich, eds.), pp. 241–256. Plenum, New York.
86. Pfeifer, R. W., and Irons, R. D. (1983). Alteration of lymphocyte function by quinones through a sulfydryl-dependent disruption of microtubule assembly. *Intt. J. Immunopharmacol.* **5**, 463–470.
87. Mann, K., Giesel, M., Fasold, H., and Haase, W. (1978). Isolation of native microtubules from procine brain and characterization of SH groups essential for polymerization at the GTP binding site. *FEBS Lett.* **92**, 45–48.
88. Green, D. E., and Brierley, G. P. (1965). The role of coenzyme Q in electron transfer. *In* "Biochemistry of Quinones" (R. A. Morton, ed.), pp. 405–428. Academic Press, New York.
89. Redfearn, E. R. (1965). Plastoquinone. *In* "Biochemistry of Quinones" (R. A. Morton, ed.), pp. 149–179. Academic Press, New York.
90. Bachur, N. R., Gordon, S. L., Gee, M. V., and Kon, H. (1979). NADPH cytochrome P-450 reductase activation of quinone anticancer agents to free radicals. *Proc. Natl. Acad. Sci. U.S.A.* **76**, 954–957.
91. Tomaz, M. (1976). $H_2O_2$ generation during the redox cycle of mitomycin C ad DNA-bound mitomycin C. *Chem.-Biol. Interact.* **13**, 89–97.
92. Na, C., and Timasheff, S. N. (1977). Physical-chemical study of daunomycin–tubulin interactions. *Arch. Biochem. Biophys.* **182**, 147–154.
93. Freeman, R. W., MacDonald, J. S., Olson, R. D., Boerth, R. C., Oates, J. A., and Harbison, R. D. (1980). Effect of sulfhydryl-containing compounds on the antitumor effects of adriamycin. *Toxicol. Appl. Pharmacol.* **54**, 168–175.

Chapter 10

# Halogenated Alkanes

**M. W. ANDERS**
Department of Pharmacology
School of Medicine and Dentistry
University of Rochester
Rochester, New York

**LANCE R. POHL**
Laboratory of Chemical Pharmacology
National Heart, Lung, and Blood Institute
National Institutes of Health
Bethesda, Maryland

|      |      |
|------|------|
| I. Chemistry of the Halogen–Carbon Bond. | 284 |
| II. Oxidative Dehydrohalogenation Mechanism | 285 |
|    A. Evidence for the Mechanism. | 285 |
|    B. Reactions of Carbonyl Metabolites. | 288 |
|    C. Prediction of Metabolic Reactivity. | 290 |
| III. Oxygenation of Halocarbon Radicals: Reductive-Oxygenation Pathway of Metabolism. | 290 |
|    A. Evidence for the Mechanism. | 290 |
|    B. Prediction of Metabolic Reactivity. | 294 |
|    C. Potential Toxicity of Metabolites Produced by the Reductive-Oxygenation Pathway of Metabolism. | 295 |
| IV. Cytochrome $P$-450-Dependent Reductive Reactions of Halogenated Hydrocarbons | 296 |
| V. Glutathione-Dependent Metabolism of Halogenated Hydrocarbons. | 302 |
|    References | 306 |

BIOACTIVATION OF FOREIGN COMPOUNDS
Copyright © 1985 by Academic Press, Inc.
All rights of reproduction in any form reserved.
ISBN 0-12-059480-3

## I. Chemistry of the Halogen–Carbon Bond

Compounds containing aliphatic carbon–halogen bonds may be enzymatically dehalogenated by several different mechanisms. These include substitution, elimination, reduction, oxidation, and reductive oxygenation. In many cases, reactive electrophilic intermediates are produced, which may react with a variety of tissue molecules.

The diversity of chemical and metabolic reactions that halogen-containing compounds undergo is attributed to the unique physical and chemical properties of the halogens and their respective carbon–halogen bonds. For this reason, a brief description of these properties will be discussed before examining in greater detail the various pathways of metabolic dehalogenation.

Electronegativity is one of these properties. The order of electronegativity of the various halogens, based on a scale devised by Pauling, is F (4.0) > Cl (3.0) > Br (2.8) > I (2.5).[1] Halogens tend to withdraw electrons away from the carbon atom, making it electrophilic in character and more susceptible to nucleophilic attack. The induced positive charge on the carbon atom may also cause a decrease in bond length and an increase of the bond strength between the carbon atom and another substituent, such as a hydrogen atom.[2,3]

Carbon–halogen bond energies, which range from C–F 116, C–Cl 79, C–Br 69, to C–I 50 kcal/mol, are an important factor that can determine the course of a reaction.[4]

The van der Waals radii of covalently bonded halogens differ considerably and decrease in order of C–I (2.15Å) > C–Br (1.95Å) > C–Cl (1.80 Å) > C–F (1.35 Å).[4] Therefore, nonbonding steric interactions would be greatest for iodine and least for fluorine.[4]

Polarizability of carbon–halogen bonds decreases in the order of C–I > C–Br > C–Cl > C–F.[3] This property is a measure of the fluidity of a particular bond. It can lead to induced-dipole–induced-dipole interactions of a halogenated compound with a hydrophobic site of an enzyme. It may also directly affect the reactivity of a compound by allowing it to accommodate more readily to the steric constraints of the transition state of a particular reaction.[5]

The oxidation and reduction potentials of carbon–halogen bonds follow the trend C–I > C–Br > C–Cl > C–F.[6] Polarographic studies show that oxidation of organic halides is almost completely confined to iodo compounds, in that the more electronegative halogens either make the appropriate compounds so difficult to oxidize that they fall outside the accessible range of anodic potentials or cause reactions to take place in other parts of the molecules. In contrast, simple alkyl iodides, bromides, and chlorides can be electrochemically reduced, whereas the isolated C–F bond is resistant. The susceptibility to electrochemical reduction increases as the number of halogens bonded to carbon increases. Moreover, stereochemistry can affect the susceptibility toward reduction. For example, 1,2-dihalides, in which the transcoplanar arrangement of carbon–halogen bonds is conformationally favored or fixed by a rigid structure, are reduced more readily than 1,2-dihalides with other conformations.

An important property of halogen atoms is their ability to stabilize $\alpha$-carbon carbonium ions, free radicals, carbanions, and carbenes. The relative effectiveness of halogens to stabilize $\alpha$-carbon carbonium ions and free radicals in general follows the order Br > Cl > F, although there are exceptions.[2] The effect of iodine does not appear to have been studied thoroughly. Carbanion stabilization follows the order I > Br > Cl > F,[2] whereas carbene stabilization follows the trend F > Cl > Br > I.[7]

All of the properties that have been discussed can influence the way a particular halogenated compound reacts. Unfortunately, it is still not possible to determine which property or combination of properties exerts the most influence on the rates of the various reactions to be discussed. In many cases, however, reasonable rationalizations of the observed experimental results are possible.

## II. Oxidative Dehydrohalogenation Mechanism

### A. EVIDENCE FOR THE MECHANISM

The most common pathway of metabolic dehalogenation is by an oxidative dehydrohalogenation mechanism. Compounds must contain an $sp^3$ carbon atom bonded to at least one hydrogen atom and one halogen atom for this pathway to occur. The first step of the reaction is catalyzed by cytochrome $P$-450-dependent monooxygenases and involves the oxidative cleavage of the carbon–hydrogen bond to produce an alcohol (*gem*-halohydrin) derivative [reaction (1)]. The *gem*-halohydrins are transient interme-

$$X-\underset{|}{\overset{|}{C}}-H \xrightarrow[O_2]{P\text{-}450} \left[ X-\underset{|}{\overset{|}{C}}-O-H \right] \longrightarrow -\overset{O}{\overset{\|}{C}}- + HX \qquad (1)$$

*gem*-Halohydrin

diates that eliminate hydrohalic acids to form carbonyl derivatives. Extensive evidence has accumulated over the past several years to support this reaction mechanism.

Because cytochrome P-450 oxidizes hydrocarbons to alcohols,[8] it seemed reasonable that halogenated compounds would be hydroxylated to *gem*-halohydrins. Although these intermediates have never been identified as products of these reactions, a compelling body of indirect evidence exists for their transient formation. For example, it is known that *gem*-halohydrins are unstable and spontaneously eliminate hydrohalic acids to form carbonyl derivatives.[9,10] The type of carbonyl product produced depends on the number of halogen atoms bonded to the carbon atom, and, as seen in Fig. 1, this may include aldehydes, ketones, acyl halides, or carbonyl halides (Fig. 1a, b, c, and d, respectively). Aldehydes, acyl halides, and carbonyl halides have been identified as metabolites of several halogen-containing compounds (Fig. 2). Liver microsomal cytochromes P-450, particularly those isozymes that are induced by phenobarbital, catalyze the formation of these products. Moreover, in the cases of chloroform,[26] dichloromethane,[14] and dibromomethane,[14] the carbonyl oxygens have been shown unequivocally to be derived from dioxygen by conducting incubations in atmospheres of [$^{18}$O]dioxygen and analyzing the chemically trapped carbonyl products by mass spectrometry. The carbonyl oxygens in the other metabolites listed in Fig. 2 were also likely derived from dioxygen, because, in most cases, the formation of the products is dependent on the presence of dioxygen.

The similarity between the mechanisms of hydroxylation of alkanes and halogenated alkanes is further illustrated by the finding of primary deuter-

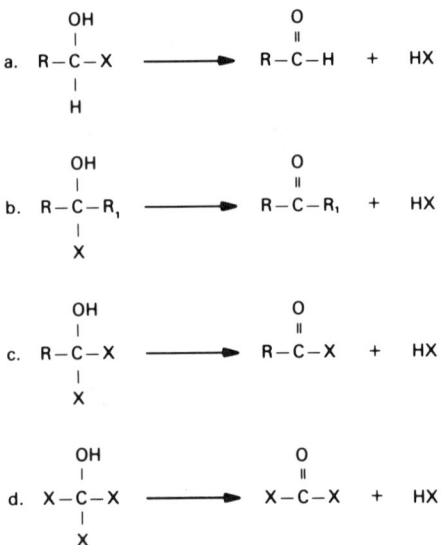

Fig. 1. Decomposition of *gem*-halohydrins to carbonyl derivatives (X = halogen).

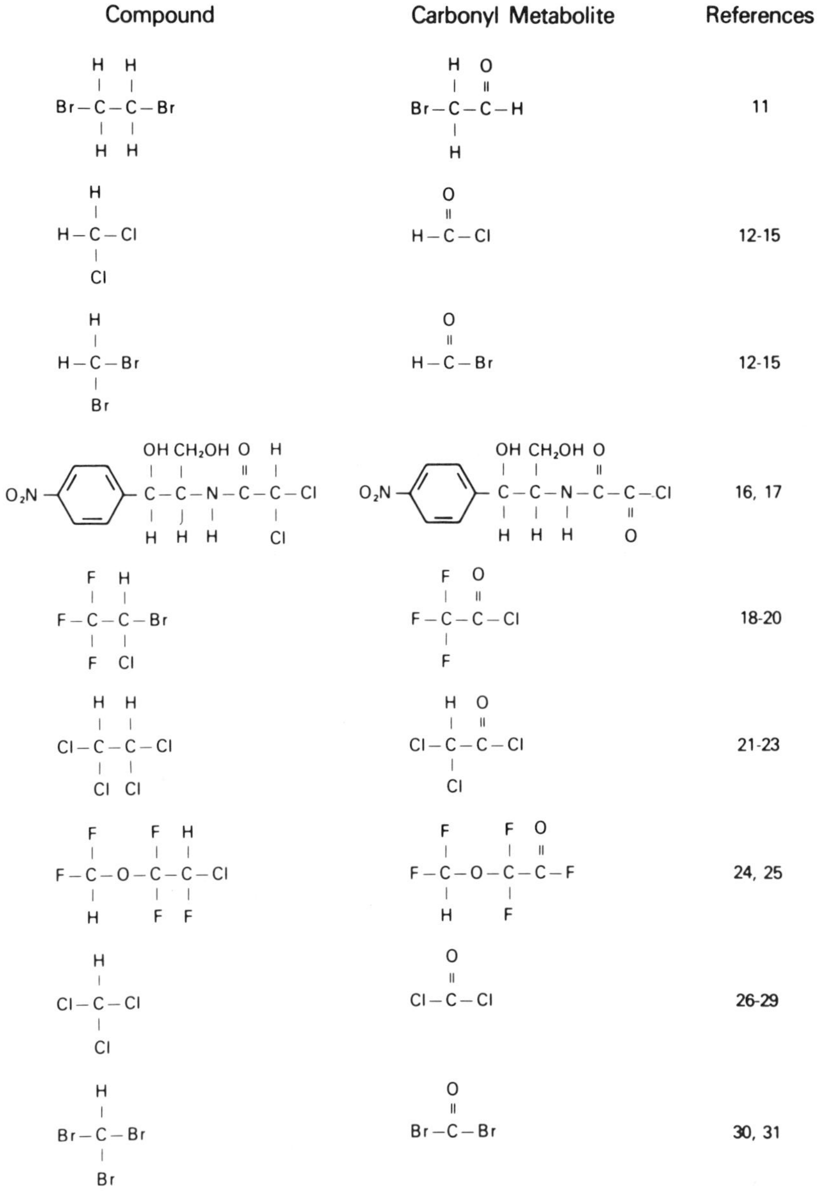

10. HALOGENATED ALKANES

| Compound | Carbonyl Metabolite | References |
|---|---|---|
| $BrCH_2-CH_2Br$ | $BrCH_2-CHO$ | 11 |
| $CH_2Cl_2$ | $HCOCl$ | 12-15 |
| $CH_2BrH$ (i.e., $CH_2HBr$ / $H_2CHBr$) — dihalide $H-CHBr-$ see fig | $HCOBr$ | 12-15 |
| $O_2N-C_6H_4-CH(OH)-CH(CH_2OH)-NH-CO-CH_2Cl$ | $O_2N-C_6H_4-CH(OH)-CH(CH_2OH)-NH-CO-CHO$ | 16, 17 |
| $CF_2H-CHBrCl$ | $CF_2H-COCl$ (shown as F–C(F)–C(=O)–Cl with H) | 18-20 |
| $CHCl_2-CHCl_2$ | $CHCl_2-COCl$ | 21-23 |
| $CF_3-O-CHF-CH_2Cl$ (F–C(F)(H)–O–C(F)(F)–CH_2Cl shown) | $CF_3-O-CF(H)-COF$ (carbonyl fluoride metabolite) | 24, 25 |
| $CHCl_3$ | $COCl_2$ | 26-29 |
| $CHBr_3$ | $COBr_2$ | 30, 31 |

**Fig. 2.** Examples of halogenated compounds metabolized to reactive carbonyl derivatives.

ium isotope effects in the metabolic oxidations of hydrocarbons,[32,33] chloroform[28,34] bromoform,[30,31] dichloromethane,[14] halothane,[20] and enflurane.[24]

## B. REACTIONS OF CARBONYL METABOLITES

The carbonyl metabolites produced during oxidative dehydrohalogenation can react nonenzymatically with a variety of tissue molecules. In several cases, these reactions appear to be responsible, at least in part, for the toxicity produced by some compounds containing carbon–halogen bonds.

### 1. Acyl and Carbonyl Halides

The acyl and carbonyl halide products are very reactive, and they have never been isolated as metabolites. They have, however, been characterized as either chemically trapped products or as decomposition products. For example, phosgene, which is a hepatotoxic metabolite of chloroform, has been identified as its cysteine derivative 2-oxothiazolidine-4-carboxylic acid[26-29] [reaction (2)]. Phosgene has also been shown to react with gluta-

$$\begin{array}{c}\text{H H O}\\\text{H–C—C–C–OH}\\\text{H–S: :N–H}\\\text{H}\\\text{Cl–C–Cl}\\\text{O}\end{array} \xrightarrow{\text{2 HCl}} \begin{array}{c}\text{H H O}\\\text{H–C—C–C–OH}\\\text{S N–H}\\\text{C}\\\text{O}\end{array} \qquad (2)$$

thione (GSH), either *in vitro* in rat liver microsomes or *in vivo*, to form diglutathionyl dithiocarbonate[35] [reaction (3)]. Carbonyl bromide ($COBr_2$)

$$\text{GSH + Cl–C(=O)–Cl} \xrightarrow{\text{HCl}} \text{GS–C(=O)–Cl} \xrightarrow{\text{GSH}} \text{GS–C(=O)–SG + HCl} \qquad (3)$$

produced during the oxidative dehydrobromination of bromoform reacts preferentially with two molecules of glutathione to yield carbon monoxide and glutathione disulfide (GSSG)[30,36-38] [reaction (4)]. Although the reason

$$\text{GSH + Br–C(=O)–Br} \xrightarrow{\text{HBr}} \text{GS–C(=O)–Br} \xrightarrow{\text{GSH}} \text{GS–SG + C + HBr} \qquad (4)$$

# 10. HALOGENATED ALKANES

for the apparent difference in the chemistry of phosgene and carbonyl bromide is not presently understood, it appears that glutathione protects sensitive tissue components from reacting with these compounds. In the case of chloramphenicol ($RNHCOCHCl_2$), the acyl halide metabolite (RNHCOCOCl) acylates selectively the cytochrome $P$-450 molecule that catalyzed its formation.[39-41] A lysine residue has been identified as a major site of acylation of cytochrome $P$-450. Moreover, this activation and covalent interaction appears to occur specifically with a form of cytochrome $P$-450 induced by phenobarbital and to produce an irreversible inactivation of enzyme activity.[16,39-41]

Although attempts have been made to trap the formyl halide (HCOX) metabolites of dihalomethanes, this has not been accomplished in biological systems,[14] perhaps because formyl halides undergo dehydrohalogenation to yield hydrohalic acids and carbon monoxide.[42,43] Carbon monoxide has been identified as an end product of the cytochrome $P$-450-catalyzed oxidative dehydrohalogenation of dihalomethanes.[12-15] The rapid decomposition of formyl halides may explain, at least partially, why dihalomethanes are significantly less hepatotoxic[44] and nephrotoxic[45] than the trihalomethanes. Exposure to dihalomethanes can lead to carboxyhemoglobin concentrations that exceed those allowable from exposure to carbon monoxide itself.[46]

Most acyl and carbonyl halide metabolites react with water to produce carboxylic acid and halide metabolites. In general, the products of this reaction are relatively innocuous except for fluoride ion, which may produce kidney damage.[47,48]

## 2. Aldehydes

Aldehyde metabolites are not nearly as reactive as acyl and carbonyl halides, and, in the case of 1,2-dibromoethane, the aldehyde metabolite bromoacetaldehyde has been identified.[11] Most aldehydes can be either enzymatically reduced to alcohols[49] or oxidized to carboxylic acids.[50] The monohaloacetaldehydes, which are formed from the oxidative dehydrohalogenation of 1,2-dihaloethanes, are electrophilic and appear to react with a variety of tissue nucleophiles. For instance, bromoacetaldehyde reacts with tissue glutathione to form a glutathione conjugate ($GSCH_2CHO$), which is further metabolized to a mercapturic acid.[51] The initial reaction with glutathione may be catalyzed by the glutathione $S$-transferases. The haloacetaldehydes can also react with molecules containing nucleophilic nitrogen atoms, such as cytidine and adenosine residues in ribo- and deoxyribonucleotides[52,53] [reaction (5)]. This interaction may be responsible, at least partially, for the mutagenicity and carcinogenicity of 1,2-dihaloeth-

$$\underset{\mathbf{1}}{\overset{\text{H}_2\text{C}\diagdown\overset{\displaystyle\text{C-H}}{\underset{\text{N}}{\overset{\text{O}}{\|}}}\diagup:\text{NH}_2}{\underset{\text{Cl}}{|}}} \longrightarrow \underset{\mathbf{1}}{\overset{\text{H-O}\diagdown\overset{\displaystyle\text{C}}{}\diagup\text{H}}{\text{H}_2\text{C}\diagdown\underset{\underset{\text{Cl}}{|}}{\text{N-H}}}} \xrightarrow{-\text{HCl}} \underset{\mathbf{1}}{\overset{\text{H-O}\diagdown\overset{\displaystyle\text{C}}{}\diagup\text{H}}{\text{H}_2\text{C}\diagdown\text{N}}} \qquad (5)$$

anes,[54,55] although several studies implicate the glutathione $S$-transferases in the mutagenicity of 1,2-dihaloethanes.[56-58]

## C. PREDICTION OF METABOLIC REACTIVITY

Several investigations on the effect of structure on the rate of metabolic oxidative dehydrohalogenation have been reported.[13,15,25,31,59,60] The results of these studies show that (1) when all other factors are the same, *gem*-dihaloalkanes (R-CHX$_2$) are oxidized more rapidly by cytochrome $P$-450-dependent monooxygenases than are monohaloalkanes (R-CH$_2$X)[59,60]; (2) chlorine substituents are more activating than fluorine substituents[25]; (3) the relative activating influences of I, Br, and Cl are not as clearly understood. For example, the order of halogen activation of the metabolism of dihalomethanes to carbon monoxide was I > Br > Cl,[13,15] whereas in the metabolism of haloforms to carbonyl halides there was no apparent difference in the activating influences between Cl and Br atoms, and iodoform was not metabolized to detectable levels of carbonyl iodide.[31]

Although attempts have been made to explain the results of several of the metabolism studies, they have only been partially successful.[60-63] Clearly, more studies of the effect of structure on the oxidative dehydrohalogenation reaction are needed before the physical and chemical factors determining the course of the reaction are better understood. Similar detailed studies are also needed to define the factors that determine the reactivities with tissue molecules of the various carbonyl products produced during this metabolic reaction.

## III. Oxygenation of Halocarbon Radicals: Reductive-Oxygenation Pathway of Metabolism

### A. EVIDENCE FOR THE MECHANISM

Pulse radiolysis studies indicate that the trichloromethyl radical reacts rapidly with dioxygen to form the trichloromethylperoxyl radical.[64,65] This intermediate is unstable and appears to decompose to phosgene and an

# 10. HALOGENATED ALKANES

unidentified electrophilic form of chlorine.[66-69] Some investigations have suggested that halogenated carbon radicals produced metabolically by reductive dehalogenation may react with dioxygen by a similar pathway [reaction (6)].

$$CX_4 \xrightarrow[1e^-]{P\text{-}450} CX_3{}^{\cdot} + X^-$$
$$\downarrow O_2$$
$$COX_2 + \text{Electrophilic } X \xleftarrow{\quad} CX_3-O-O\cdot \qquad (6)$$

For example, it has been shown that carbon tetrachloride and bromotrichloromethane are metabolized by rat liver microsomes to phosgene[35,70,71] and to an electrophilic form of chlorine.[72] The reactive chlorine metabolite was trapped in the reaction mixtures with 2,6-dimethylphenol to form 4-chloro-2,6-dimethylphenol [reaction (7)]. The reactions were catalyzed by

$$CX_4 \xrightarrow[O_2]{P\text{-}450} \text{Electrophilic } X$$

(with 2,6-dimethylphenol reacting to form 4-halo-2,6-dimethylphenol) (7)

cytochrome $P$-450[71,73-75] and showed an absolute requirement for dioxygen.[71,72,75] The reductive-oxygenation nature of the reaction was uncovered when the rate of metabolism of carbon tetrachloride to phosgene and electrophilic chlorine was found to increase to maximum value as the oxygen concentration of the medium was decreased from 100 to 5%.[75-77] The rates of formation of both products were negligible when dioxygen was absent.

The increase in rates of formation of both phosgene and electrophilic halogen as the dioxygen concentration was decreased from 100 to 5% is consistent with an initial metabolism of carbon tetrachloride to the trichloromethyl radical, because the rates of reductive dehalogenations characteristically increase as the dioxygen concentration is decreased.[78,79] This presumably results from the competition between dioxygen and substrate for electrons donated by cytochrome $P$-450. At very low concentrations of dioxygen, the rates of formation of phosgene and electrophilic halogen may decrease, because the trapping of the trichloromethyl radical by dioxygen becomes rate determining and alternate reactions of the radical become kinetically more important. Thus, at low concentrations of dioxygen, the trichloromethyl radical abstracts a hydrogen atom from the medium to produce chloroform,[70,80] is further reduced by cytochrome $P$-450 to form the dichlorocarbene,[81] or irreversibly reacts with microsomal lipid.[82]

The reductive-oxygenation mechanism is further supported by the findings that bromotrichloromethane is both reductively dehalogenated[83] and metabolized to phosgene[35] and electrophilic chlorine[72] more rapidly than carbon tetrachloride. In addition, chloroform, which is not reductively dechlorinated,[84] is not metabolized to electrophilic chlorine.[72]

Alternate mechanisms for the metabolism of tetrahalomethanes to phosgene and electrophilic halogen have been considered in detail, but they do not appear to be major pathways of metabolism. For example, the possibility that the electrophilic halogen is formed by an oxidation of halide ion, similar to the reactions catalyzed by chloroperoxidase or myeloperoxidase, was excluded by showing that the electrophilic chlorine metabolite of carbon tetrachloride was derived exclusively from carbon tetrachloride.[85] The involvement of superoxide anion radical in the reaction was investigated, because it is known to react rapidly with carbon tetrachloride,[86] can act as a reducing agent, and is produced during reactions catalyzed by cytochrome $P$-450.[87] Superoxide anion radical does not appear to be involved in the metabolism of carbon tetrachloride to electrophilic chlorine and phosgene, however, because superoxide dismutase did not inhibit the formation of these products.[75-77] Catalase also had no effect on the course of the reactions,[75-77] suggesting that hydrogen peroxide was not involved in the metabolism of carbon tetrachloride to phosgene and electrophilic chlorine.

Another pathway of metabolism considered involved the oxygenation of carbon tetrachloride to trichloromethyl hypochlorite by a mechanism analogous to the oxidation of carbon–hydrogen bonds catalyzed by cytochrome $P$-450[72,88] [reaction (8)]. Although trichloromethyl hypochlorite has not

$$\begin{array}{c} Cl \\ | \\ Cl-C-Cl \\ | \\ Cl \end{array} \xrightarrow[O_2]{P\text{-}450} \begin{array}{c} Cl \\ | \\ Cl-C-O-Cl \\ | \\ Cl \end{array} \tag{8}$$

been synthesized, on the basis of the chemistry of analogs, it would be expected to be a chlorinating agent that would release phosgene on reacting with various nucleophiles[72,88] [reaction (9)]. The finding that the addition of cumene hydroperoxide or other oxidizing agents to rat liver microsomes did not support the metabolism of carbon tetrachloride to electrophilic chlorine or phosgene, but did support the oxidations of carbon–hydrogen

## 10. HALOGENATED ALKANES

$$\text{Cl}-\underset{\underset{\text{Cl}}{|}}{\overset{\overset{\text{Cl}}{|}}{\text{C}}}-\text{O}-\text{Cl} + :\text{Nu}- \longrightarrow \underset{\text{Cl}^-}{\downarrow} \longrightarrow \text{COCl}_2 + \text{Cl}-\text{Nu}- \quad (9)$$

:Nu— = Nucleophile

bonds,[75,77] as well as the unique dependence of the reaction on the concentration of dioxygen, indicated that the formation of the hypochlorite did not occur in rat liver microsomes.

Although the mechanism by which trichloromethylperoxyl radical reacts to produce phosgene and electrophilic chlorine is not known, three pathways can be proposed.[77] One pathway involves the dimerization of two trichloromethylperoxyl radicals to form a tetraoxide intermediate. This product would be expected to decompose spontaneously to form phosgene and an electrophilic chlorine radical[89] [reaction (10)]. The remaining two path-

$$2\ \text{Cl}-\underset{\underset{\text{Cl}}{|}}{\overset{\overset{\text{Cl}}{|}}{\text{C}}}-\text{O}-\text{O}\cdot \longrightarrow \text{Cl}-\underset{\underset{\text{Cl}}{|}}{\overset{\overset{\text{Cl}}{|}}{\text{C}}}-\text{O}-\text{O}-\text{O}-\text{O}-\underset{\underset{\text{Cl}}{|}}{\overset{\overset{\text{Cl}}{|}}{\text{C}}}-\text{Cl} \longrightarrow 2\ \text{COCl}_2 + 2\text{Cl}\cdot + \text{O}_2 \quad (10)$$

ways involve an initial complexation of trichloromethylperoxyl radical with ferric or ferrous cytochrome *P*-450, analogous to the reactions of carbon radicals with iron porphyrins to form σ complexes[90] [reaction (11)]. The

$$\underset{+2\text{ or }+3}{\boxed{\text{Fe}}} + \text{Cl}-\underset{\underset{\text{Cl}}{|}}{\overset{\overset{\text{Cl}}{|}}{\text{C}}}-\text{O}-\text{O}\cdot \longrightarrow \underset{\underset{\underset{\underset{\text{Cl}\ \ \ \text{Cl}}{\diagdown\ \diagup}}{\text{C}}}{\underset{\text{O}\diagdown\ \diagup\text{Cl}}{|}}}{\underset{\underset{\text{O}}{|}}{\overset{+3\text{ or }+4}{\boxed{\text{Fe}}}}} \quad (11)$$

heme–trichloromethylperoxyl σ complex may decompose by at least two different mechanisms. One mechanism involves a homolytic cleavage of the peroxyl bond to produce phosgene and an electrophilic chlorine radical. The mechanism of this decomposition is analogous to that ascribed to the decomposition of a cytochrome *P*-450–cumene hydroperoxide complex[91] [reaction (12)]. Alternatively, the heme–trichloromethylperoxyl complex may undergo an intramolecular rearrangement to produce phosgene and an electrophilic chlorine–iron hypochlorite complex. A similar intermediate has been postulated as the electrophilic chlorinating agent of chloroperoxidase[92] [reaction (13)]. The ferric hypochlorite complex may also react with

$$\underset{\substack{|\\O\\|\\O\diagdown\diagup_{C}\diagdown^{Cl}\\Cl\hspace{0.5em}Cl}}{\boxed{Fe}^{+3\text{ or }+4}} \longrightarrow COCl_2 + Cl\cdot + \underset{\substack{|\\O\\|\\\dot{}}}{\boxed{Fe}^{+3\text{ or }+4}} \quad (12)$$

$$\underset{\substack{|\\O\\|\\O\diagdown\diagup_{C}\diagdown^{Cl}\\Cl\hspace{0.5em}Cl}}{\boxed{Fe}^{+3\text{ or }+4}} \longrightarrow \underset{\substack{|\\O\\|\\Cl}}{\boxed{Fe}^{+3\text{ or }+4}} + COCl_2 + Cl^- \quad (13)$$

$$\downarrow H_2O$$

$$\underset{\substack{|\\O\\\diagdown H}}{\boxed{Fe}^{+3\text{ or }+4}} + HOCl$$

water to produce the electrophilic chlorinating agent hypochlorous acid (HOCl).

Therefore, the electrophilic halogen produced during the decomposition of trichloromethylperoxyl radical in rat liver microsomes might be a radical or ionic species, or both may be involved.

## B. PREDICTION OF METABOLIC REACTIVITY

To undergo this pathway of metabolism, a halogenated compound must be susceptible to metabolic reductive dehalogenation. The reduction must also occur in the presence of a sufficient concentration of dioxygen to trap the halogenated carbon radical as a peroxyl derivative. The type of carbonyl product produced would depend on the number of halogen atoms bonded to the carbon atom containing the peroxyl group (Fig. 3).

The reductive-oxygenation pathway of metabolism may not be observed with all compounds that are reductively dehalogenated.[77] One reason for this is that some compounds, such as halothane, are reductively dehalogenated only at very low concentrations of dioxygen, presumably because of their inherently slow rates of reduction.[78] Consequently, insufficient dioxygen may be available to trap the halocarbon radicals. Alternatively, other compounds, such as hexachloroethane, can be dechlorinated in air, but are

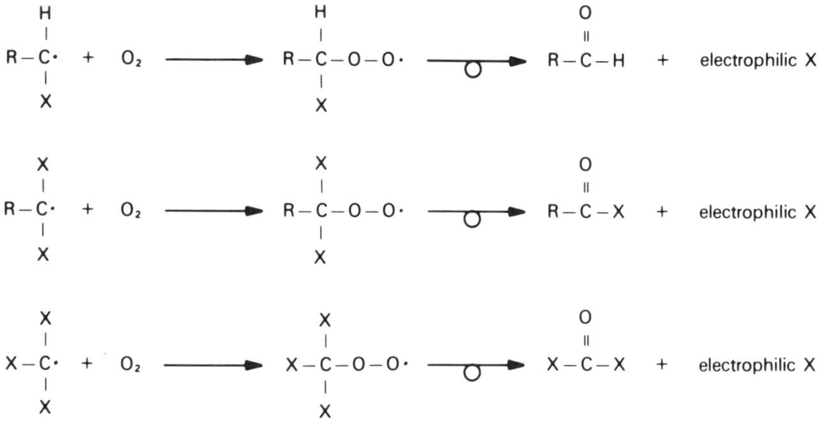

## 10. HALOGENATED ALKANES

$$\begin{array}{c}H\\|\\R-C\cdot\\|\\X\end{array} + O_2 \longrightarrow \begin{array}{c}H\\|\\R-C-O-O\cdot\\|\\X\end{array} \xrightarrow{-O} \begin{array}{c}O\\\|\\R-C-H\end{array} + \text{electrophilic X}$$

$$\begin{array}{c}X\\|\\R-C\cdot\\|\\X\end{array} + O_2 \longrightarrow \begin{array}{c}X\\|\\R-C-O-O\cdot\\|\\X\end{array} \xrightarrow{-O} \begin{array}{c}O\\\|\\R-C-X\end{array} + \text{electrophilic X}$$

$$\begin{array}{c}X\\|\\X-C\cdot\\|\\X\end{array} + O_2 \longrightarrow \begin{array}{c}X\\|\\X-C-O-O\cdot\\|\\X\end{array} \xrightarrow{-O} \begin{array}{c}O\\\|\\X-C-X\end{array} + \text{electrophilic X}$$

**Fig. 3.** Possible reactions of substituted halogenated radicals with dioxygen (X = halogen).

still not metabolized to detectable levels of electrophilic halogen.[77] In the case of this compound, it appears that the intermediate pentachloroethyl radical is rapidly reduced to the pentachloroethyl carbanion by cytochrome $P$-450, and the carbanion undergoes an $\alpha$ elimination to yield tetrachloroethene.[79] The finding that tetrachloroethene and pentachloroethane were formed in a ratio of approximately 200:1 supports this view. Therefore, little pentachloroethyl radical appears to be released from cytochrome $P$-450 either to abstract a hydrogen atom or to react with dioxygen to form pentachloroethane and pentachloroethylperoxyl radical ($Cl_3C$-$CCl_2$-$OO\cdot$), respectively.

### C. POTENTIAL TOXICITY OF METABOLITES PRODUCED BY THE REDUCTIVE-OXYGENATION PATHWAY OF METABOLISM

The lipid peroxidation produced by several halogenated compounds may be due, at least in part, to metabolites formed during reductive oxygenation. For instance, the observation that the order of lipid peroxidation potency[93] and metabolism by reductive oxygenation[72] correlate in the series bromotrichloromethane > carbon tetrachloride > chloroform supports this view. Moreover, it is known that carbon tetrachloride-induced lipid peroxidation[94] and reductive oxygenation[75-77] both occur at maximal rates at a dioxygen concentration of approximately 5%. In this regard, the trichloromethylperoxyl radical has been postulated as being the toxic metabolite of carbon tetrachloride that causes lipid peroxidation.[95] This radical may initiate lipid peroxidation by abstracting a hydrogen atom from a diallylic C–H bond of a phospholipid polyunsaturated fatty acid. The fatty acid carbon radical could then react with dioxygen to produce an unstable lipid hydroperoxide.

Similarly, halogen radicals may initiate lipid peroxidation either by reacting with double bonds of phospholipid polyunsaturated fatty acids or by abstracting a diallylic hydrogen atom from the fatty acid residue. In both cases, fatty acid carbon radicals would be produced, which could react with dioxygen and initiate lipid peroxidation.

If the electrophilic chlorine produced during reductive oxygenation was polar in character, as is hypochlorous acid, it could react with hemoproteins, nucleotides, and thiols.[96] These types of interactions have been implicated in the bactericidal effects of hypochlorous acid.[96]

Carbonyl metabolites would also be expected to react with various tissue components and may be toxic. In this regard, studies indicate that the hepatoxicity of carbon tetrachloride might be partially caused by its metabolism to the hepatotoxic phosgene.[97]

## IV. Cytochrome P-450-Dependent Reductive Reactions of Halogenated Hydrocarbons

The reductive metabolism of carbon tetrachloride is one of the most studied and oldest known biotransformations of halogenated hydrocarbons. In 1961, Butler[98] described the metabolism of carbon tetrachloride to chloroform and suggested the intermediacy of a radical in the reaction. Subsequent studies in several laboratories have confirmed this observation as well as the central role of radicals as intermediates in the reductive metabolism of other polyhalogenated alkanes.

It is now well known that polyhalogenated hydrocarbons undergo both one- and two-electron reductions; a general scheme for this reaction is shown in Fig. 4. The first step in the reaction is the enzyme-catalyzed transfer of a single electron to the antibonding ($\sigma^*$-orbital) or $d$-orbital of the carbon–halogen bond to yield a radical anion as a transient intermediate (Fig. 4, reaction **A**). This intermediate may eliminate halide to form a carbon-centered radical (Fig. 4, reaction **B**), which may abstract a hydrogen atom from lipids to form a reduced polyhalogenated alkane (Fig. 4, reaction **C**). The radical may, while bound to cytochrome P-450, accept a second electron to yield a carbanionic intermediate (Fig. 4, reaction **D**), which may undergo an $\alpha$ elimination to yield a carbene (Fig. 4, reaction **E**) or a $\beta$ elimination to yield an alkene (Fig. 4, reaction **F**).

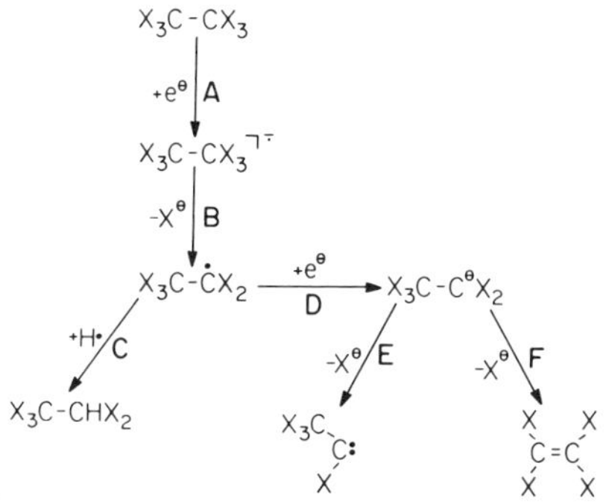

# 10. HALOGENATED ALKANES

[Scheme showing reduction pathway: X₃C-CX₃ → (+e⁻, A) → X₃C-CX₃·⁻ → (−X⁻, B) → X₃C-ĊX₂ → (+H·, C) X₃C-CHX₂; or (+e⁻, D) → X₃C-C:⁻X₂ → (−X⁻, E) X₃C-C:-X; or (−X⁻, F) → X₂C=CX₂]

**Fig. 4.** One- and two-electron reduction of polyhalogenated alkanes (X = halogen, alkane, etc.).

The reduction of carbon tetrachloride will be used as model to discuss the enzymatic reduction of polyhalogenated alkanes, because much of the knowledge concerning this reaction has been derived from studies with carbon tetrachloride. The first step in the reduction of carbon tetrachloride is the donation of one electron to the substrate by cytochrome $P$-450 (Fig. 5, reaction **A**). In effect, the ferrocytochrome $P$-450–carbon tetrachloride complex competes with the ferrocytochrome $P$-450–oxygen complex for electrons; hence, because oxygen is reduced more readily than carbon tetrachloride, little carbon tetrachloride reduction is seen in the presence of dioxygen. The intermediate radical anion rearranges to yield the trichloro-

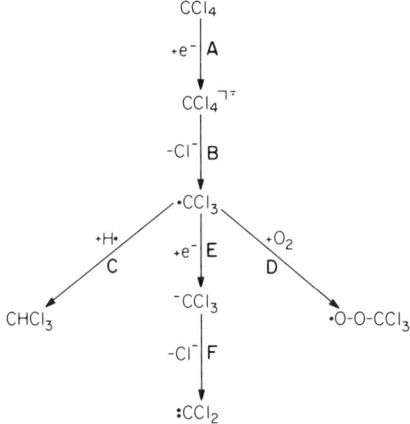

**Fig. 5.** One- and two-electron reductions of carbon tetrachloride.

methyl radical and chloride ion (Fig. 5, reaction **B**). The trichloromethyl radical may abstract a hydrogen radical from the polyunsaturated lipids of the endoplasmic reticulum to yield chloroform (Fig. 5, reaction **C**) as a metabolite and to initiate lipid peroxidative changes, which are characteristic of carbon tetrachloride intoxication. Alternatively, the radical may, depending on the dioxygen concentration of the medium, react with dioxygen to yield the trichloromethylperoxyl radical (Fig. 5, reaction **D**); the formation and fate of this species has been discussed in detail earlier in this chapter.

There is considerable experimental evidence that supports the pathway just described. Although it was suggested earlier that NADPH–cytochrome *c* (cytochrome *P*-450) reductase was involved,[99] subsequent studies have shown conclusively the role of cytochrome *P*-450. The observations that the biotransformation of carbon tetrachloride to chloroform was inhibited by carbon monoxide, a specific inhibitor of cytochrome *P*-450-dependent reactions, and by treatment of rats with cobaltous chloride, which decreases the cytochrome *P*-450 but not NADPH–cytochrome *c* content of hepatic microsomes, showed that cytochrome *P*-450, rather than NADPH–cytochrome *c*, was involved.[100,101] The finding that the metabolism of carbon tetrachloride to chloroform is catalyzed by a purified reconstituted cytochrome *P*-450-containing system clarifies conclusively the enzymology of the reduction.[80]

Considerable evidence is also available supporting the concept that the trichloromethyl radical is formed from carbon tetrachloride. Hexachloroethane, which may arise by the dimerization of trichloromethyl radicals, has been identified in the tissues of rabbits given carbon tetrachloride[102] and in incubation mixtures containing carbon tetrachloride.[103]

Although earlier studies on the direct identification of the trichloromethyl radical by electron spin resonance spectrometry (ESR) have proved controversial and uncertain,[104–108] investigations of carbon tetrachloride metabolism in the presence of spin-trapping agents have supported the formation of the radical.[109,110] The radical identified originally as the spin-trapped trichloromethyl radical may actually have been a spin-trapped carbon-centered dienyl lipid radical.[111] Studies with [$^{13}$C]carbon tetrachloride yielded an ESR spectrum consistent with the formation of a spin-trapped [$^{13}$C]trichloromethyl radical adduct.[112,113]

The identification by mass spectrometry of trichloromethyl radical–lipid adducts formed when human cytochromes *P*-450 incorporated into phospholipid vesicles were incubated in the presence of carbon tetrachloride also supports the existence of the trichloromethyl radical.[114] Similarly, the formation of an adduct between the trichloromethyl radical and cholesterol has been reported, but, apart from the molecular weight as estimated by chemical ionization mass spectrometry, the chemical nature of the adduct has not been determined.[115]

Reaction mechanism studies on the biotransformation of carbon tetrachloride to chloroform also support the formation of the trichloromethyl radical as an intermediate. When the metabolism of carbon tetrachloride to chloroform was studied in the absence of dioxygen but in the presence of deuterium oxide, little deuteriochloroform was formed.[81,116] This finding suggests strongly that the trichloromethyl radical abstracts a hydrogen atom from neighboring lipids to yield chloroform and tends to exclude the formation and protonation of the trichloromethyl carbanion as a mechanism for chloroform formation.

The one-electron reduction mechanism described for carbon tetrachloride can be generalized to other polyhalogenated compounds. For example, the reduction of halothane to 2-chloro-1,1,1-trifluoroethane and 2-chloro-1,1-difluoroethene,[117,118] of hexachloroethane to pentachloroethane and tetrachloroethene,[79] of fluorotrichloromethane to dichlorofluoromethane,[119] of chloramphenicol to dechlorochloramphenicol 1-(4-nitrophenyl)-2-monochloroacetamido-1,3-propanediol,[120] and of DDT [1,1,1-trichloro-2,2-bis(p-chlorophenyl)-ethane] to DDD [1,1-dichloro-2,2-bis(p-chlorophenyl)-ethane][121] has been reported, and all likely involve a radical intermediate.

The reduction of halothane to 2-chloro-1,1,1-trifluoroethane has been studied in detail, and there is considerable evidence for a radical intermediate in this reaction. The reduction of halothane is catalyzed by cytochrome $P$-450-dependent enzymes. When halothane was incubated with human or rabbit cytochrome $P$-450, NADPH–cytochrome $P$-450 reductase, and cytochrome $b$, reconstituted into phospholipid vesicles containing dioleoylphosphatidylcholine and phosphatidylethanolamine, a mixture of 9- and 10-(1-chloro-2,2,2-trifluoroethyl)stearates was formed.[122] Moreover, ESR evidence for a spin-trapped radical, which was suggested to be either the 1-chloro- or the 1-bromo-2,2,2-trifluoroethyl radical, formed *in vivo* has been presented.[123] Additional evidence for the formation of the 2-chloro-1,1,1-trifluoroethyl radical as an intermediate in halothane metabolism has been presented by Ahr *et al.*[78] These workers found that 2-chloro-1,1,1-trifluoroethane formation from halothane was catalyzed by cytochrome $P$-450 with NADPH as an electron donor and with heat-inactivated microsomal fractions with dithionite as the electron donor; thus, this metabolite may be formed by either enzymatic or nonenzymatic processes. In addition, rate of formation of 2-chloro-1,1,1-trifluoroethane was not altered when cytochrome $b_5$ was incorporated into microsomal fractions, although the rate of formation of the two-electron reduction product 2-chloro-1,1-difluoroethene was increased, and antibodies to cytochrome

$b_5$, did not inhibit the metabolism of halothane to 2-chloro-1,1,1-trifluoroethane. The authors conclude that 2-chloro-1,1,1-trifluoroethane is the major one-electron reduction product of halothane, and this is consistent with the observation that the reactive halothane metabolite that binds to lipids contains $^{36}Cl$ or $^{3}H$ when [$^{36}Cl$]- or [$^{3}H$]halothane is the substrate.[124] Finally, it has been demonstrated that 2-chloro-1,1,1-trifluoroethane does not arise by protonation of the 2-chloro-1,1,1-trifluoroethyl carbanion, but is derived from the 2-chloro-1,1,1-trifluoroethyl radical, because, when the reductive reaction is studied in the presence of deuterium oxide, no deuterated alkane is formed.[116]

Additional studies showing that nonenzymatic processes may play a role in the transformation of halothane have been reported by Van Dyke and co-workers,[125] who showed that the release of inorganic fluoride from halothane was catalyzed by heat-denatured hemoglobin and by hemin; fluoride release under reductive conditions was not observed with 2-chloro-1,1,-difluoroethene, 2-bromo-1,1-difluoroethene, 2-chloro-1,1-difluoroethane, or 2-bromo-1,1-difluoroethane when hemin was the catalyst. These findings are similar to the report that iron(II) porphyrins are readily oxidized by a variety of alkyl halides.[126]

The 1-chloro-2,2,2-trifluoroethyl radical formed during the reductive metabolism of halothane may be responsible for the halothane-induced destruction of cytochrome $P$-450 seen *in vivo*[127] and *in vitro*.[128] The loss of cytochrome $P$-450 was inversely proportional to the dioxygen concentration; presumably, halothane is metabolized to trifluoroacetic acid in the presence of adequate dioxygen concentrations.[128] Studies have shown that the loss of cytochrome $P$-450 seen after incubation of halothane with microsomal fractions in the absence of oxygen is associated with the covalent binding of [$^{14}C$]halothane metabolites to the hemoprotein[129]; although the nature of the adduct was not elucidated, evidence against the formation of an N-substituted heme was presented.

Studies on the metabolism of hexachloroethane to pentachloroethane suggest that the reaction mechanism is similar to that reported for halothane,[79] except that, in the case of hexachloroethane, the one-electron reduction product pentachloroethane is a minor metabolite, whereas in the case of halothane, the one-electron reduction product 2-chloro-1,1,1-trifluoroethane is the major product.

The two-electron reduction of polyhalogenated hydrocarbons has also been investigated in detail. In the case of carbon tetrachloride, the initial one-electron reduction product, the trichloromethyl radical, may form a complex with ferric cytochrome $P$-450 [$Fe(III) \cdot CCl_3$], which may accept a second electron to yield a ferrous cytochrome $P$-450–trichloromethyl radical complex [$Fe(II) \cdot CCl_3$] whose mesomeric structure is equivalent to the trichloromethyl carbanion–ferric cytochrome P-450 complex [Fe(III)·CCl$_3$] (Fig. 5, reaction E). The trichloromethyl carbanion–ferric cytochrome P-450 complex may accept an electron and undergo an α-elimination reaction to yield dichlorocarbene (Fig. 5, reaction F), which reacts with ferrous cytochrome P-450 to yield a cytochrome P-450–dichlorocarbene adduct [Fe(II):CCl$_2$].[81,130]

The available evidence supports carbene formation during the reductive biotransformation of polyhalogenated methanes.[131] With carbon tetrachloride, binding of the carbene to cytochrome P-450 yields a characteristic difference spectrum.[132,133] This spectrum has been attributed to the formation of a low-spin ligand complex of cytochrome P-450 and dichlorocarbene.[130] Chemically formed dichlorocarbene–porphyrin complexes have proved to be extraordinarily stable and have been identified by their mass spectra, $^1$H and $^{13}$C NMR spectra,[134] and the crystal and molecular structure of iron(II) porphyrin–dichlorocarbene complexes have been described.[135] Decomposition of the dichlorocarbene formed from carbon tetrachloride yields carbon monoxide as the product.[130,136] Dichlorocarbene has been conclusively identified as a metabolite of carbon tetrachloride when it was trapped in rat liver microsomes with 2,3-dimethyl-2-butene to form 1,1-dichloro-2,2,3,3-tetramethylcyclopropane.[137]

Early studies on the metabolism of halothane suggested that trifluoromethylcarbene might be formed as an intermediate because, when halothane is metabolized under reductive conditions, a characteristic difference spectrum is formed.[136] Moreover, the electronic spectrum of a model cytochrome P-450–trifluoromethylcarbene complex has been described, and the calculated spectrum is consistent with the formation of a ferrous cytochrome P-450–carbene complex when halothane is metabolized in the absence of dioxygen.[138]

Subsequent studies, however, cast doubt on the concept that the peak seen at 470 nm in the difference spectrum when halothane is incubated with cytochrome P-450 in the absence of dioxygen is attributable to carbene formation.[78] The product formed on decomposition of the intermediate complex absorbing at 470 nm was 2-chloro-1,1-difluoroethene; this suggests that the peak seen in the difference spectrum is associated with the formation of the 1-chloro-2,2,2-trifluoroethyl carbanion, which may undergo a β elimination to yield the observed product. Similarly, when hexa- or pentachloroethane is incubated with cytochrome P-450 under reducing conditions, the observed products are tetra- and trichloroethene, respectively[79]; these are the expected products of a β-elimination reaction from a cytochrome P-450-bound carbanion.

Thus, the evidence for the formation of a carbene intermediate in the metabolism of polyhalogenated methanes is persuasive. In contrast, more study is needed to determine the relative roles of carbenic and carbanionic metabolites in the metabolism of polyhalogenated alkanes, but the evidence presently available supports the formation of carbanions as intermediates.

## V. Glutathione-Dependent Metabolism of Halogenated Hydrocarbons

Many halogenated hydrocarbons, particularly monohalogenated or *gem*-dihalogenated alkanes, undergo glutathione $S$-transferase-catalyzed substitution reactions to yield $S$-substituted glutathione derivatives. This initial conjugation reaction proceeds without a change in the oxidation number of the carbon atom at the reaction site, but the final product may be either reduced or oxidized depending on the compound under study and the subsequent processing of intermediates.

The conjugation of monohaloalkanes with glutathione is catalyzed by glutathione $S$-transferases. This is a common route of biotransformation, and $S$-alkylglutathione derivatives are formed as metabolites; for example, iodomethane is metabolized to $S$-methylglutathione,[139] and 1-chloropropane is metabolized to $S$-propylglutathione.[140] Similar results have been obtained for other simple monohalogenated alkanes, and this information has been summarized.[141,142]

These reactions appear, in general, to be examples of $S_N2$ reactions in which glutathione, probably as the thiolate anion, serves as the nucleophile, and the halogen is eliminated as inorganic halide. Indeed, studies on the mechanism of glutathione $S$-transferase activity show that these enzymes catalyze a nucleophilic attack of enzyme-bound glutathione on the electrophilic center of the second substrate.[143] This finding is supported by the observation that the glutathione $S$-transferase A-catalyzed metabolism of $(S)$-1-chloro-1-phenylethane to $S$-$(R)$-1-phenylethylglutathione proceeds with inversion of configuration.[144]

In most instances, the conjugation of halogenated alkanes with glutathione yields conjugates that are processed further to stable $S$-substituted $N$-acetylcysteine derivatives (mercapturic acids), which lack toxicity. There are, however, a number of examples of glutathione conjugation reactions that yield unstable intermediates as products, and these products may be associated with toxic reactions.

*gem*-Dihaloalkanes are metabolized to unstable products by the glutathione $S$-transferases. Dihalomethanes, for example, yield formaldehyde as a metabolite.[145] Reaction mechanism studies show that the initial product of the reaction of a dihalomethane with glutathione is the $S$-halomethylglutathione intermediate (Fig. 6, reaction A); this intermediate undergoes rapid

# 10. HALOGENATED ALKANES

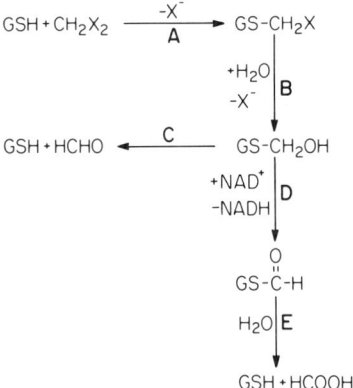

**Fig. 6.** Glutathione-dependent metabolism of dihalomethanes to halide (X), formaldehyde, and formic acid. GSH, glutathione.

hydrolysis to yield the hemimercaptal $S$-hydroxymethylglutathione (Fig. 6, reaction **B**), which gives rise to glutathione and formaldehyde (Fig. 6, reaction **C**).[146] The $S$-hydroxymethylglutathione intermediate can be diverted to yield formic acid as a metabolite by the action of formaldehyde dehydrogenase (Fig. 6, reaction **D**), which converts $S$-hydroxymethylglutathione to $S$-formylglutathione,[147] and by $S$-formylglutathione hydrolase (Fig. 6, reaction **E**), which converts $S$-formylglutathione to formic acid and glutathione.[148]

Chloramphenicol undergoes dechlorination in a glutathione-dependent manner analogous to that seen with dihalomethanes.[149] In this reaction, chloramphenicol is presumably converted to an $\alpha$-halothioether conjugate of glutathione, which undergoes nonenzymatic hydrolysis to the corresponding hemimercaptal; the hemimercaptal eliminates glutathione to yield chloramphenicol aldehyde.

*vic*-Dihaloalkanes also undergo glutathione-dependent metabolism, and reactive electrophilic metabolites may be produced. The glutathione-dependent metabolism of *vic*-dihaloethanes and butanes has been studied in detail. For example, 1,2-dichloro-, 1,2-dibromo-, and 1-bromo-2-chloroethane yield ethene as a metabolite.[150] Stereochemical studies have clarified the mechanism of this reaction.[151] *meso*-2,3-Dibromobutane and *erythro*-2-bromo-3-chlorobutane yield predominantly ($Z$)-2-butene as a product, and racemic 2,3-dibromobutane and *threo*-2-bromo-3-chlorobutane are metabolized to ($E$)-2-butene. These results are consistent with a glutathione $S$-transferase-catalyzed E2 elimination reaction (Fig. 7, pathway **B**). In contrast, *meso*-1,2-dideuterio-1,2-dichloroethane was metabolized exclusively to ($Z$)-1,2-dideuterioethene (Fig. 7, pathway **A**); this result suggests

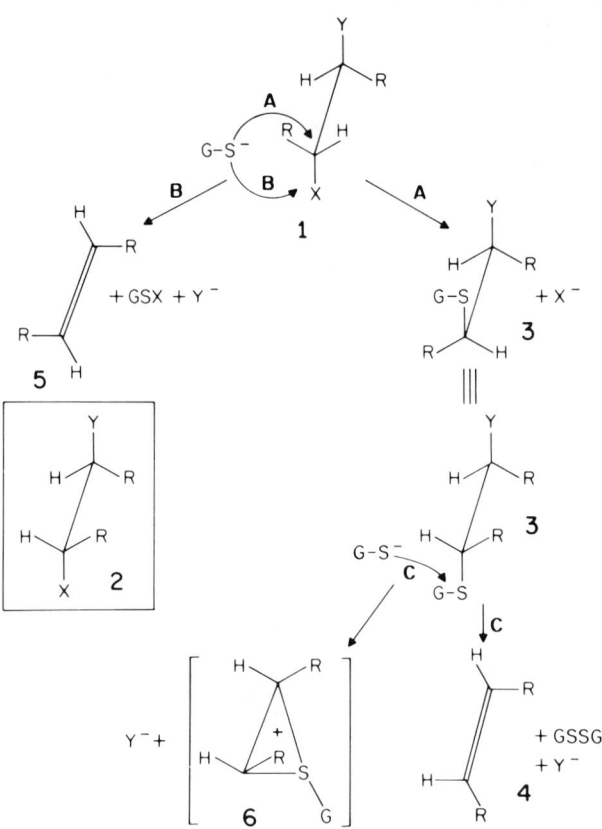

**Fig. 7.** (**1**) *meso*-2,3-Dibromobutane (R = CH$_3$; X = Y = Br), *erythro*-2-bromo-3-chlorobutane (R = CH$_3$; X = Br, Y = Cl), or *meso*-1,2-dideuterio-1,2-dichloroethane (R = $^2$H; X = Y = Cl). Inset (**2**) shows corresponding racemic or threo-isomeric configuration. (**3**) Glutathione conjugate of (**1**); (**4**) (Z)-2-butene (R = CH$_3$) or (Z)-1,2-dideuterioethylene (R = $^2$H); (**5**) (E)-2-butene (R = CH$_3$) or (E)-1,2-dideuterioethylene (R = $^2$H); (**6**) ethylene-S-glutathionylepisulfonium. G–S$^-$, glutathione; GSX, glutathione sulfenyl halide. (Reproduced with permission from ref. 151.)

that *vic*-dihaloethanes undergo an initial glutathione S-transferase-catalyzed substitution reaction to yield a S-(2-haloethyl)glutathione intermediate (Fig. 7, **3**); attack of glutathione on the sulfur atom of the intermediate (Fig. 7, pathway **C**) followed by anti elimination of halide ion yields the (Z)-1,2-dideuterioethene.

The intermediate S-(2-haloethyl)glutathione compound is thought to be responsible for the mutagenicity of 1,2-dihaloethanes, which is associated with glutathione S-transferase activity rather than with monooxygenase activity.[56-58] This has been confirmed in several studies, which showed that both 1,2-dibromoethane and glutathione are bound irreversibly in equimolar amounts to calf thymus DNA when 1,2-dibromoethane and glutathione are incubated with glutathione S-transferase; thus, S-(2-bromoethyl)glutathione or the corresponding episulfonium ion serves as the alkylating intermediate.[152]

The glutathione-dependent reduction of α-haloketones to methylketones has been observed[153,154]; for example, 2,2',4'-trichloroacetophenone is metabolized to 2',4'-dichloroacetophenone. The actual substrate for the enzyme is S-(2,4-dichlorophenacyl)glutathione, which is formed by the nonenzymatic reaction of glutathione and the α-haloketone. Although the reaction is catalyzed by glutathione-dependent cytosolic enzymes, these enzymes are different from the glutathione S-transferases as well as from several other known glutathione-dependent cytosolic enzymes (M. Kitada and M. W. Anders, unpublished data). The reaction mechanism probably involves the attack of glutathione on the sulfur atom of S-(2,4-dichlorophenacyl)glutathione to yield glutathione disulfide and the 2,4-dichlorophenacyl carbanion, which would be stabilized by enolization; protonation of the carbanion would give the observed product.

The enzyme cysteine conjugate β-lyase has been implicated in the bioactivation of several S-substituted cysteines. Cysteine conjugate β-lyase is a pyridoxal-dependent enzyme, which catalyzes the cleavage of the C–S bond of the substrate to yield a thiol, pyruvate, and ammonia. The early interest in this enzyme centered on S-(1,2-dichlorovinyl)cysteine, which was identified as the toxic factor present in trichloroethylene-extracted soybean oil meal that caused aplastic anemia in calves; Anderson and Schultz[155] showed that a lyase cleaves S-(1,2-dichlorovinyl)cysteine to yield pyruvate, ammonia, chloride, and an unidentified sulfur-containing metabolite, which alkylated proteins and glutathione. Later studies have emphasized the role of cysteine conjugate β-lyase in the formation of methylthio-containing metabolites; in this pathway, cysteine conjugate β-lyase converts cysteine conjugates to the corresponding thiols, which are methylated to yield the observed methyl alkyl or aryl sulfides.[156] The purification of the enzyme from rat liver has been reported,[157] and a facile assay for cysteine conjugate β-lyase activity is available.[158]

Currently studies in a number of laboratories implicate cysteine conjugate β-lyase in the nephrotoxicity of halogenated alkenes. For example, S-(1,2-dichlorovinyl)cysteine, a known substrate for cysteine conjugate β-lyase,[157] is nephrotoxic.[159,160] Hexachlorobutadiene is also nephrotoxic,[161,162] and some results suggest that cytochrome P-450-catalyzed reactions are not involved in hexachlorobutadiene-induced nephrotoxicity.[163] Nash et al.[164] have shown that hexachlorobutadiene is metabolized to a glutathione conjugate, S-(1,1,2,3,4-pentachloro-1,3-butadienyl)glutathione, and the corresponding mercapturic acid N-acetyl-S-(1,1,2,3,4-pentachloro-1,3-butadienyl)-L-cysteine; moreover, both compounds are nephrotoxic when given to rats. Nash et al.[164] suggest that hexachlorobutadiene forms a glutathione conjugate, which is converted enzymatically to the corresponding cysteine conjugate; cysteine conjugate β-lyase, in turn, is thought to activate the cysteine conjugate to the ultimate nephrotoxic species. The synthesis of S-(1,1,2,3,4-pentachloro-1,3-butadienyl)cysteine has been reported, and the compound is nephrotoxic.[165] Chlorotrifluoroethene is also nephrotoxic, and the toxicity is not associated with fluoride release.[166,167] Dohn and Anders[168] have shown that chlorotrifluoroethene forms a glutathione conjugate, which was identified as S-(2-chloro-1,1,2-trifluoroethyl)glutathione by $^1$H and $^{19}$F NMR; this glutathione conjugate is also nephrotoxic (D. R. Dohn, A. J. Quebbemann, and M. W. Anders, unpublished data), but a role for cysteine conjugate β-lyase in the nephrotoxicity of S-(2-chloro-1,1,2-trifluoroethyl)glutathione has not been established. Finally, S-(2-chloroethyl)-DL-cysteine, a probable metabolite of 1,2-dichloroethane, is nephrotoxic.[169] This compound is not a substrate for hepatic[157] or renal (A. A. Elfarra and M. W. Anders, unpublished data) cysteine conjugate β-lyase. Present evidence suggests that S-(2-chloroethyl)-DL-cysteine is a direct-acting nephrotoxin, because analogs that cannot form reactive episulfonium ions, such as S-ethyl-L-cysteine, S-(2-hydroxyethyl)-DL-cysteine, and S-(3-chloropropyl)-DL-cysteine, are not nephrotoxic (A. A. Elfarra and M. W. Anders, unpublished data).

The evidence for the hypothesis that glutathione conjugates of halogenated alkanes and alkenes are processed in the kidney to yield the corresponding cysteine conjugates, which may be direct-acting nephrotoxins or may be activated by cysteine conjugate β-lyase, has been reviewed[170].

## References

1. March, M. J. (1977). "Advanced Organic Chemistry. Reactions, Mechanisms, and Structure," pp. 17-20. McGraw-Hill, New York.
2. Modena, G., and Scorrano, G. (1973). Directing, activating and deactivating effects. *In* "The Chemistry of the Carbon-Halogen Bond" (S. Patai, ed.), Part 1, pp. 301-406. Wiley, New York.
3. Wagniere, G. H. (1973). General and theoretical aspects of the carbon-halogen bond. *In* "The Chemistry of the Carbon-Halogen Bond" (S. Patai, ed.), Part 1, pp. 14-22. Wiley, New York.
4. March, M. J. (1977). "Advanced Organic Chemistry. Reactions, Mechanisms, and Structure," pp. 26-28. McGraw-Hill, New York.
5. Hendrickson, J. B., Cram, D. J., and Hammond, G. S. (1970). "Organic Chemistry," pp. 393-398. McGraw-Hill, New York.

10. HALOGENATED ALKANES

6. Casanova, J., and Eberson, L. (1973). Electrochemistry of the carbon–halogen bond. *In* "The Chemistry of the Carbon–Halogen Bond" (S. Patai, ed.), Part 2, pp. 979–1047. Wiley, New York.
7. Hine, J., and Ehrenson, S. J. (1958). The effect of the structure on the relative stability of dihalomethylenes. *J. Am. Chem. Soc.* **80**, 824–830.
8. Van Der Hoeven, T. A., and Coon, M. J. (1974). Preparation and properties of partially purified cytochrome $P$-450 and reduced nicotinamide adenine dinucleotide phosphate–cytochrome $P$-450 reductase from rabbit liver microsomes. *J. Biol. Chem.* **249**, 6302–6310.
9. Henrickson, J. B., Cram, D. J., and Hammond, G. S. (1970). "Organic Chemistry," p. 452. McGraw-Hill, New York.
10. Kloter, G., and Seppelt, K. (1979). Trifluoromethanol ($CF_3OH$) and trifluoromethylamine ($CF_3NH_2$). *J. Am. Chem. Soc.* **101**, 347–349.
11. Hill, D. L., Shih, T. W., Johnston, T. P., and Struck, R. F. (1978). Macromolecular binding and metabolism of the carcinogen 1,2-dibromoethane. *Cancer Res.* **38**, 2438–2442.
12. Kubic, V. L., Anders, M. W., Engel, R. R., Barlow, C. H., and Caughey, W. S. (1974). Metabolism of dihalomethanes to carbon monoxide. I. *In vivo* studies. *Drug Metab. Dispos.* **2**, 53–57.
13. Kubic, V. L., and Anders, M. W. (1975). Metabolism of dihalomethanes to carbon monoxide. II. *In vitro* studies. *Drug Metab. Dispos.* **3**, 104–112.
14. Kubic, V. L., and Anders, M. W. (1978). Metabolism of dihalomethanes to carbon monoxide. III. Studies on the mechanism of the reaction. *Biochem. Pharmacol.* **27**, 2349–2355.
15. Stevens, J. L., Ratnayake, J. H., and Anders, M. W. (1980). Metabolism of dihalomethanes to carbon monoxide. IV. Studies in isolated hepatocytes. *Toxicol. Appl. Pharmacol.* **55**, 484–489.
16. Pohl, L. R., and Krishna, G. (1978). Study of the mechanism of metabolic activation of chloramphenicol by rat liver microsomes. *Biochem. Pharmacol.* **27**, 335–341.
17. Pohl, L. R., Nelson, S. D., and Krishna, G. (1978). Investigation of the mechanism of the metabolic activation of chloramphenicol by rat liver microsomes. Identification of a new metabolite. *Biochem. Pharmacol.* **27**, 491–496.
18. Cohen, E. N., Trudell, J. R., Edmunds, H. N., and Watson, E. (1975). Urinary metabolites of halothane in man. *Anesthesiology* **43**, 392–401.
19. Karashima, D., Hirokata, Y., Shigematsu, A., and Furikawa, T. (1977). The *in vitro* metabolism of halothane (2-bromo-2-chloro-1,1,1-trifluoroethane) by hepatic microsomal cytochrome $P$-450. *J. Pharmacol. Exp. Ther.* **203**, 409–416.
20. Sipes, I. G., Gandolfi, A. J., Pohl, L. R., Krishna, G, and Burnell, B. R. (1980). Comparison of the biotransformation and hepatoxicity of halothane and deuterated halothane. *J. Pharmacol. Exp. Ther.* **214**, 716–720.
21. Halpert, J., and Neal, R. A. (1981). Cytochrome $P$-450-dependent metabolism of 1,1,2,2-tetrachloroethane to dichloroacetic acid *in vitro*. *Biochem. Pharmacol.* **30**, 1366–1368.
22. Ivanetich, K. M., and Van Der Honert, L. H. (1981). Chloroethanes: Their metabolism by hepatic cytochrome $P$-450 *in vitro*. *Carcinogenesis (N.Y.)* **2**, 697–702.
23. Halpert, J. (1982). Cytochrome $P$-450-dependent covalent binding of 1,1,2,2-tetrachloroethane *in vitro*. *Drug Metab. Dispos.* **10**, 465–468.
24. Burke, T. R., Martin, J. L., George, J. W., and Pohl, L. R. (1980). Investigation of the mechanism of defluorination of enflurane in rat liver microsomes with specifically deuterated derivatives. *Biochem. Pharmacol.* **29**, 1623–1626.

25. Burke, T. R., Branchflower, R. V., Lees, D. E. and Pohl, L. R. (1981). Mechanism of defluorination of enflurane. Identification of an organic metabolite in rat and man. *Drug. Metab. Dispos.* **9,** 19–24.
26. Pohl, L. R., Bhooshan, B., Whittaker, N. F., and Krishna, G. (1977). Phosgene: A metabolite of chloroform. *Biochem. Biophys. Res. Commun.* **79,** 684–691.
27. Mansuy, D., Beaune, P., Cresteil, T., Lange, M., and Leroux, J. P. (1977). Evidence for phosgene formation during liver microsomal oxidation of chloroform. *Biochem. Biophys. Res. Commun.* **79,** 513–517.
28. Pohl, L. R., George, J. W., Martin, J. L., and Krishna, G. (1979). Deuterium isotope effect in *in vivo* bioactivation of chloroform to phosgene. *Biochem. Pharmacol.* **28,** 561–563.
29. Pohl, L. R., Martin, J. L., and George, J. W. (1980). Mechanism of metabolic activation of chloroform by rat liver microsomes. *Biochem. Pharmacol.* **29,** 3271–3276.
30. Stevens, J. L., and Anders, M. W. (1979). Metabolism of haloforms to carbon monoxide. III. Studies on the mechanism of the reaction. *Biochem. Pharmacol.* **28,** 3189–3194.
31. Pohl, L. R., Martin, J. L., Taburet, A. M., and George, J. W. (1980). Oxidative bioactivation of haloforms into hepatotoxins. *In* "Microsomes, Drug Oxidations, and Chemical Carcinogenesis" (M. J. Coon, A. H. Conney, R. W. Estabrook, H. V. Gelboin, J. R. Gillette, and P. J. O'Brien, eds.), Vol. 2, pp. 881–884. Academic Press, New York.
32. Hjelmeland, L. M., Aronow, L., and Trudell, J. R. (1977). Intramolecular determination of primary kinetic isotope effects in hydroxylations catalyzed by cytochrome $P$-450. *Biochem. Biophys. Res. Commun.* **76,** 541–549.
33. Groves, J. T., McClusky, G. A., White, R. E., and Coon, M. J. (1978). Aliphatic hydroxylation by highly purified liver microsomal cytochrome $P$-450. Evidence for a carbon radical intermediate. *Biochem. Biophys. Res. Commun.* **81,** 154–160.
34. Pohl, L. R., and Krishna, G. (1978). Deuterium isotope effect in bioactivation and hepatotoxicity of chloroform. *Life Sci.* **23,** 1067–1072.
35. Pohl, L. R., Branchflower, R. M., Highet, R. J., Martin, J. L., Nunn, D. S., Monks, T. J., George, J. W., and Hinson, J. A. (1981). The formation of diglutathionyl dithiocarbonate as a metabolite of chloroform, bromotrichloromethane, and carbon tetrachloride. *Drug Metab. Dispos.* **9,** 334–339.
36. Ahmed, A. E., Kubic, V. L., and Anders, M. W. (1977). Metabolism of haloforms to carbon monoxide. I. *In vitro* studies. *Drug Metab. Dispos.* **5,** 198–204.
37. Anders, M. W., Stevens, J. L., Sprague, R. W., Shaath, Z., and Ahmed, A. E. (1978). Metabolism of haloforms to carbon monoxide. II. *In vivo* studies. *Drug Metab. Dispos.* **6,** 556–560.
38. Stevens, J. L., and Anders, M. W. (1981). Metabolism of haloforms to carbon monoxide. IV. Studies on the reaction mechanism *in vivo*. *Chem.-Biol. Interact.* **37,** 365–374.
39. Halpert, J., and Neal, R. A. (1980). Inactivation of purified rat liver cytochrome $P$-450 by chloramphenicol. *Mol. Pharmacol.* **17,** 427–431.
40. Halpert, J. (1981). Covalent modification of lysine during the suicide inactivation of rat liver cytochrome $P$-450 by chloramphenicol. *Biochem. Pharmacol.* **30,** 875–881.
41. Halpert, J. (1982). Further studies of the suicide inactivation of purified rat liver cytochrome $P$-450 by chloramphenicol. *Mol. Pharmacol.* **21,** 166–172.
42. Krauskopf, K. B., and Rollefson, G. K. (1934). The photochemical reaction between chlorine and formaldehyde. The preparation of formyl chloride. *J. Am. Chem. Soc.* **56,** 2542–2548.
43. Hisatsune, I. C., and Heicklen, J. (1973). Infrared spectrum of formyl chloride. *Can. J. Spectrosc.* **18,** 77–81.

44. Kutob, S. D., and Plaa, G. L. (1962). A procedure for estimating the hepatotoxic potential of certain industrial solvents. *Toxicol. Appl. Pharmacol.* **4,** 354–361.
45. Plaa, G. L., and Larson, R. E. (1965). Relative nephrotoxic properties of chlorinated methane, ethane, and ethylene derivatives in mice. *Toxicol. Appl. Pharmacol.* **7,** 37–44.
46. Stewart, R. D., Fisher, T. N., Hosko, M. J., Peterson, J. E., Baretta, E. D., and Dodd, H. C. (1972). Experimental human exposure to methylene chloride. *Arch. Environ. Health* **25,** 342–348.
47. Mazze, R. I., Calverley, R. K., and Smith, N. T. (1977). Inorganic fluoride nephrotoxicity. Prolonged enflurane and halothane anesthesia in volunteers. *Anesthesiology* **46,** 265–271.
48. Mazze, R. I., Woodruff, R. E., and Heerdt, M. E. (1982). Isoniazid-induced enflurane defluorination in humans. *Anesthesiology* **57,** 5–8.
49. Morris, P. L., Burke, T. R., George, J. W., and Pohl, L. R. (1982). A new pathway for the oxidative metabolism of chloramphenicol by rat liver microsomes. *Drug Metab. Dispos.* **10,** 439–445.
50. Koivula, T., and Koivusalo, M. (1975). Different forms of rat liver aldehyde dehydrogenase and their subcellular distribution. *Biochim. Biophys. Acta* **397,** 9–23.
51. Van Bladeren, P. J., Breimer, D. D., Van Huijgevoort, J. A. T. C. M., Vermeulen, N. P. E., and Van der Gen, A. (1981). The metabolic formation of *N*-acetyl-*S*-2-hydroxyethyl-L-cysteine from tetradeutero-1,2-dibromoethane. Relative importance of oxidation and glutathione conjugation *in vivo*. *Biochem. Pharmacol.* **30,** 2499–2502.
52. Kusmierek, J. T., and Singer, B. (1982). Chloroacetaldehyde-treated ribo- and deoxyribopolynucleotides. 1. Reaction products. *Biochemistry* **21,** 5717–5722.
53. Kusmierek, J. T., and Singer, B. (1982). Chloroacetaldehyde-treated ribo- and deoxyribopolynucleotides. 2. Errors in transcription by different polymerases resulting from ethenocytosine and its hydrated intermediate. *Biochemistry* **21,** 5723–5728.
54. Olson, W. A., Habermann, R. T., Weisburger, E. K., Ward, J. M., and Weisburger, J. H. (1973). Induction of stomach cancer in rats and mice by halogenated aliphatic fumigants. *J. Natl. Cancer Inst. (U.S.)* **51,** 1993–1995.
55. Fishbein, L. (1976). Industrial mutagens and potential mutagens. I. Halogenated aliphatic derivatives. *Mutat. Res.* **32,** 267–308.
56. Rannug, U., Sundvall, A., and Ramel, C. (1978). The mutagenic effect of 1,2-dichloroethane on *Salmonella typhimurium*. I. Activation through conjugation with glutathione *in vitro*. *Chem.-Biol. Interact.* **20,** 1–16.
57. Shih, T. W., and Hill, D. L. (1981). Metabolic activation of 1,2-dibromoethane by glutathione transferase and by microsomal mixed-function oxidase: Further evidence for formation of two reactive metabolites. *Res. Commun. Chem. Pathol. Pharmacol.* **33,** 449–461.
58. Sundheimer, D. W., White, R. D., Brendel, K., and Sipes, I. G. (1982). The bioactivation of 1,2-dibromoethane in rat hepatocytes: Covalent binding to nucleic acids. *Carcinogenesis (N.Y.)* **3,** 1129–1133.
59. Van Dyke, R. A., and Wineman, C. G. (1971). Enzymatic dechlorination. Dechlorination of chloroethanes and propanes *in vitro*. *Biochem. Pharmacol.* **20,** 463–470.
60. Salmon, A. G., Jones, R. B., and Mackrodt, W. C. (1981). Microsomal dechlorination of chloroethanes: Structure–reactivity relationships. *Xenobiotica* **11,** 723–734.
61. Loew, G., Trudell, J., and Motulsky, H. (1973). Quantum chemical studies of the metabolism of a series of chlorinated ethane anesthetics. *Mol. Pharmacol.* **9,** 152–162.
62. Loew, G., Motulsky, H., Trudell, J., Cohen, E., and Hjelmeland, L. (1974). Quantum chemical studies of the metabolism of the inhalation anesthetics methoxyflurane, enflurane, and isoflurane. *Mol. Pharmacol.* **10,** 406–418.

63. Loew, G. H., and Rebagliati, M., and Poulsen, M. (1984). Metabolism and relative carcinogenic potency of chloroethanes: A quantum chemical structure–activity study. *Cancer Biochem. Biophys.* **7**, 109–132.
64. Hesse, C., Leray, N., and Roncin, J. (1971). Etude par résonance paramagnétique électronique de la structure du radical $CCl_3$. *Mol. Phys.* **22**, 137–145.
65. Packer, J. E., Willson, R. L., Bahnemann, D., and Asmus, K. D. (1980). Electron transfer reactions of halogenated aliphatic peroxyl radicals: Measurement of absolute rate constants by pulse radiolysis. *J. Chem. Soc.* pp. 296–299.
66. Chapman, A. T. (1935). The peroxidation of chloroform. *J. Am. Chem. Soc.* **57**, 419–422.
67. Kwai, S. (1966). Discussion on decomposition of chloroform. *Yakugaku Zasshi* **86**, 1125–1132.
68. Hautecloque, S. (1980). On the photoxidation of gaseous $HCCl_3$ and ClO radical formation. *J. Photochem.* **14**, 157–165.
69. Lyons, E. H., and Dickinson, R. G. (1935). The photooxidation of liquid carbon tetrachloride. *J. Am. Chem. Soc.* **57**, 443–446.
70. Shah, H., Hartman, S. P., and Weinhouse, S. (1979). Formation of carbonyl chloride in carbon tetrachloride metabolism by rat liver *in vitro. Cancer Res.* **39**, 3942–3947.
71. Kubic, V. L., and Anders, M. W. (1980). Metabolism of carbon tetrachloride to phosgene. *Life Sci.* **26**, 2151–2155.
72. Mico, B. A., Branchflower, R. V., Pohl, L. R., Pudzianowski, A. T., and Loew, G. H. (1982). Oxidation of carbon tetrachloride, bromotrichloromethane and carbon tetrabromide by rat liver microsomes to electrophilic halogens. *Life Sci.* **30**, 131–137.
73. Mico, B. A., and Pohl, L. R. (1982). Metabolism of carbon tetrachloride to electrophilic chlorine by rat liver cytochrome *P*-450. Exclusion of a chloroperoxidase mechanism. *Fed. Proc. Fed. Am. Soc. Exp. Biol.* **41**, 1222.
74. Mico, B. A., Branchflower, R. V., and Pohl, L. R. (1983). Formation of electrophilic chlorine from carbon tetrachloride: Involvement of cytochrome *P*-450. *Biochem. Pharmacol.* **32**, 2357–2359.
75. Pohl, L. R., Morris, P. L., Schulick, R. D., Highet, R. J., and George, J. W. (1984). Reductive oxygenation mechanism of metabolism of carbon tetrachloride to phosgene by rat liver cytochrome *P*-450. *Mol. Pharmacol.* **25**, 318–321.
76. Mico, B. A., and Pohl, L. R. (1982). Reductive oxygenation: A possible mechanism of electrophilic chlorine formation during the microsomal metabolism of carbon tetrachloride. *Pharmacologist* **24**, 135.
77. Mico, B. A., and Pohl, L. R. (1983). Reductive oxygenation of carbon tetrachloride: Trichloromethylperoxyl radical as a possible intermediate in the conversion of carbon tetrachloride to electrophilic chlorine. *Arch. Biochem. Biophys.* **225**, 596–609.
78. Ahr, H. J., King, L. J., Nastainczyk, W., and Ullrich, V. (1982). The mechanism of reductive dehalogenation of halothane by liver cytochrome *P*-450. *Biochem. Pharmacol.* **31**, 383–390.
79. Nastainczyk, W., Ahr, H. J., and Ullrich, V. (1982). The reductive metabolism of halogenated alkanes by liver microsomal cytochrome *P*-450. *Biochem. Pharmacol.* **31**, 391–396.
80. Wolf, C. R., Harrelson, W. G., Nastainczyk, W. M., Philpot, R. M., Kalyanaraman, B., and Mason, R. P. (1980). Metabolism of carbon tetrachloride in heptatic microsomes and reconstituted monooxygenase systems and its relationship to lipid peroxidation. *Mol. Pharmacol.* **18**, 553–558.
81. Ahr, H. J., King, L. J., Nastainczyk, W., and Ullrich, V. (1980). The mechanism of

chloroform and carbon monoxide formation from carbon tetrachloride by microsomal cytochrome $P$-450. *Biochem. Pharmacol.* **29,** 2855–2861.
82. Kieczka, H., Remmer, H., and Kappus, H. (1981). Influence of oxygen on the inhibition of liver microsomal activation of carbon tetrachloride by the catechol 2-hydroxyestradiol-17$\beta$. *Biochem. Pharmacol.* **30,** 319–324.
83. Bini, A., Vecchi, G., Vivol, G., Vannini, V., and Cessi, C. (1975). Detection of early metabolites in rat liver after administration of $CCl_4$ and $CBrCl_3$. *Pharmacol. Res. Commun.* **7,** 143–149.
84. Fry, B. J., Taylor, T., and Hathway, D. E. (1972). Pulmonary elimination of chloroform and its metabolite in man. *Arch. Int. Pharmacodyn. Ther.* **196,** 98–111.
85. Mico, B. A., and Pohl, L. R. (1982). Metabolism of carbon tetrachloride to electrophilic chlorine by liver microsomes. Exclusion of cytochrome $P$-450-catalyzed chloroperoxidase reaction. *Biochem. Biophys. Res. Commun.* **107,** 27–31.
86. Roberts, J. L., and Sawyer, D. T. (1981). Facile degradation by superoxide ion of carbon tetrachloride, chloroform, methylene chloride, and $p,p'$-DDT in aprotic media. *J. Am. Chem. Soc.* **103,** 714–715.
87. Kuthan, H., and Ullrich, V. (1982). Oxidase and oxygenase function of the microsomal cytochrome $P$-450 monooxygenase system. *Eur. J. Biochem.* **126,** 583–588.
88. Pudzianowski, A. T., Loew, G. H., Mico, B. A., Branchflower, R. V., and Pohl, L. R. (1983). A molecular orbital study of model cytochrome $P$-450 oxidation of $CCl_4$ and $CHCl_3$. *J. Am. Chem. Soc.* **105,** 3434–3438.
89. Howard, J. A., Bennett, J. E., and Brunton, G. (1981). Absolute rate constants for hydrocarbon autoxidation. 30. On the self-reaction of the $\alpha$-cumylperoxy radical in solution. *Can. J. Chem.* **59,** 2253–2260.
90. Brault, D., and Neta, P. (1981). Reactions of iron porphyrins with methyl radicals. *J. Am. Chem. Soc.* **103,** 2705–2710.
91. White, R. E., and Coon, M. J. (1980). Oxygen activation by cytochrome $P$-450. *Annu. Rev. Biochem.* **49,** 315–356.
92. Libby, R. D., Thomas, J. A., Kaiser, L. W., and Hager, L. P. (1982). Chloroperoxidase halogenation reactions. Chemical versus enzymic halogenating intermediate. *J. Biol. Chem.* **257,** 5030–5037.
93. Moody, D. E., James, J. R., Clawson, G. A., and Smuckler, E. A. (1981). Correlations among the changes in hepatic microsomal components after intoxication with alkyl halides and other hepatotoxins. *Mol. Pharmacol.* **20,** 685–693.
94. Kieczka, H., and Kappus, H. (1980). Oxygen dependence of $CCl_4$-induced lipid peroxidation *in vitro* and *in vivo*. *Toxicol. Lett.* **5,** 191–196.
95. Albano, E., Lott, K. A. K., Slater, T. F., Stier, A., Symons, M. C. R., and Tomasi, A. (1982). Spin-trapping studies on the free-radical products formed by metabolic activation of carbon tetrachloride in rat liver microsomal fractions, isolated hepatocytes and *in vivo* in the rat. *Biochem. J.* **204,** 593–603.
96. Albrich, J. M., McCarthy, C. A., and Hurst, J. K. (1981). Biological reactivity of hypochlorous acid: Implications for microbicidal mechanisms of leucocyte myeloperoxidase. *Proc. Natl. Acad. Sci. U.S.A.* **78,** 210–214.
97. Harris, R. N., and Anders, M. W. (1981). 2-Propanol treatment induces selectively the metabolism of carbon tetrachloride to phosgene. Implications for carbon tetrachloride hepatotoxicity. *Drug Metab. Dispos.* **9,** 551–556.
98. Butler, T. C. (1961). Reduction of carbon tetrachloride *in vivo* and reduction of carbon tetrachloride and chloroform *in vitro* by tissues and tissue constituents. *J. Pharmacol. Exp. Ther.* **134,** 311–319.

99. Slater, T. F., and Sawyer, B. C. (1971). The stimulatory effect of carbon tetrachloride on peroxidative reactions in rat liver fractions *in vitro*. Interaction sites in the endoplasmic reticulum. *Biochem. J.* **123,** 815–821.
100. Reiner, O., Athanassopoulos, S., Hellmer, K. H., Murray, R. E., and Uehleke, H. (1972). Bildung von Chloroform aus Tetrachlorkohlenstoff in Lebermikrosomen, Lipidperoxidation und Zerstörung von Cytochrom *P*-450. *Arch. Toxikol.* **29,** 219–233.
101. Suarez, K. A., and Bhonsle, P. (1976). The relationship of cobaltous chloride-induced alterations of hepatic microsomal enzymes to altered carbon tetrachloride hepatotoxicity. *Toxicol. Appl. Pharmacol.* **37,** 23–37.
102. Fowler, J. S. L. (1969). Carbon tetrachoride metabolism in the rabbit. *Br. J. Pharmacol.* **37,** 733–737.
103. Uehleke, H., Hellmer, K. H., and Tabarelli, S. (1973). Binding of [$^{14}$C]carbon tetrachloride to microsomal proteins *in vitro* and formation of $CHCl_3$ by reduced liver microsomes. *Xenobiotica* **3,** 1–11.
104. Burdino, E., Gravela, E., Ugazio, G., Vannini, V., and Calligaro, A. (1973). Initiation of free-radical reactions and hepatotoxicity in rats poisoned with carbon tetrachloride or bromotrichloromethane. *Agents Actions* **4,** 244–253.
105. Calligaro, A., and Vannini, V. (1975). Electron spin resonance study of homolytic cleavage of carbon tetrachloride in rat liver: Trichloromethyl free radicals. *Pharmacol. Res. Commun.* **7,** 323–329.
106. Sancier, K. M. (1976). Reply to "Electron spin resonance study of homolytic cleavage of carbon tetrachloride in rat liver: Trichloromethyl free radicals." *Pharmacol. Res. Commun.* **8,** 429–430.
107. Calligaro, A., and Vannini, V. (1976). Reply to the comments of K. M. Sancier concerning the communication: "Electron spin resonance study of homolytic cleavage of carbon tetrachloride in rat liver: Trichloromethyl free radicals." *Pharmacol. Res. Commun.* **8,** 431–434.
108. Mason, R. P. (1979). Free-radical metabolites of foreign compounds and their toxicological significance. *Rev. Biochem. Toxicol.* **1,** 151–200.
109. Lai, E. K., McCay, P. B., Noguchi, R., and Fong, K. L. (1979). *In vivo* spin-trapping of trichloromethyl radicals formed from $CCl_4$. *Biochem. Pharmacol.* **28,** 2231–2235.
110. Poyer, F. L., Floyd, R. A., McCay, P. B., Janzen, E. G., and Davis, E. R. (1978). Spin-trapping of the trichloromethyl radical produced during enzymic NADPH oxidation in the presence of carbon tetrachloride or bromotrichloromethane. *Biochim. Biophys. Acta* **539,** 402–409.
111. Kalyanaraman, B., Mason, R. P., Perez-Reyes, E., Chignell, C. F., Wolf, C. R., and Philpot, R. M. (1979). Characterization of the free radical formed in aerobic microsomal incubations containing carbon tetrachloride and NADPH. *Biochem. Biophys. Res. Commun.* **89,** 1065–1072.
112. Poyer, J. L., McCay, P. B., Lai, E. K., Janzen, E. G., and Davis, E. R. (1980). Confirmation of assignment of the trichloromethyl radical spin adduct detected by spin-trapping during [$^{13}$C]carbon tetrachloride metabolism *in vitro* and *in vivo*. *Biochem. Biophys. Res. Commun.* **94,** 1154–1160.
113. Tomasi, A., Albano, E., Lott, K. A. K., and Slater, T. F. (1980). Spin-trapping of free radical of $CCl_4$ activation using pulse radiolysis and high-energy radiation procedures. *FEBS Lett.* **133,** 303–306.
114. Trudell, J. R., Bösterling, B., and Trevor, A. J. (1982). Reductive metabolism of carbon tetrachloride by human cytochromes *P*-450 reconstituted in phospholipid vesicles; Mass spectral identification of trichloromethyl radical bound to dioleoyl phosphatidylcholine. *Proc. Natl. Acad. Sci. U.S.A.* **79,** 2678–2682.

115. Ansari, G. A. S., Moslen, M. T., and Reynolds, E. S. (1982). Evidence for *in vivo* covalent binding of •CCl$_3$ derived from CCl$_4$ to cholesterol of rat liver. *Biochem. Pharmacol.* **31**, 3509–3510.
116. Kubic, V. L., and Anders, M. W. (1981). Mechanism of the microsomal reduction of carbon tetrachloride and halothane. *Chem.-Biol. Interact.* **34**, 201–207.
117. Mukai, A., Morio, M., Fujii, K., and Hanaki, C. (1977). Volatile metabolites of halothane in the rabbit. *Anesthesiology* **47**, 248–251.
118. Sharp, J. H., Trudell, J. R., and Cohen, E. N. (1979). Volatile metabolites and decomposition products of halothane in man. *Anesthesiology* **50**, 2–8.
119. Wolf, C. R., King, L. J., and Parke, D. V. (1978). The anaerobic dechlorination of trichlorofluoromethane by rat liver preparations *in vitro*. *Chem.-Biol. Interact.* **21**, 277–288.
120. Morris, P. L., Burke, T. R., Jr., and Pohl, L. R. (1983). Reductive dechlorination of chloramphenicol by rat liver microsomes. *Drug Metab. Dispos.* **11**, 126–130.
121. Walker, C. H. (1969). Reductive dechlorination of $p,p'$-DDT by pigeon liver microsomes. *Life Sci.* **8**, 1111–1115.
122. Trudell, J. R., Bösterling, B., and Trevor, A. J. (1982). Reductive metabolism of halothane by human and rabbit cytochrome $P$-450. Binding of 1-chloro-2,2,2-trifluoroethyl radical to phospholipids. *Mol. Pharmacol.* **21**, 710–717.
123. Poyer, J. L., McCay, P. B., Weddle, C. C., and Downs, P. E. (1981). *In vivo* spin-trapping of radicals formed during halothane metabolism. *Biochem. Pharmacol.* **30**, 1517–1519.
124. Wood, C. L., Gandolfi, A. J., and Van Dyke, R. A. (1976). Lipid binding of a halothane metabolite. *Drug Metab. Dispos.* **4**, 305–313.
125. Baker, M. T., Nelson, R. M., and Van Dyke, R. A. (1983). The release of inorganic fluoride from halothane and halothane metabolites by cytochrome $P$-450, hemin, and hemoglobin. *Drug Metab. Dispos.* **11**, 308–311.
126. Wade, R. S., and Castro, C. E. (1973). Oxidation of iron(II) porphyrins by alkyl halides. *J. Am. Chem. Soc.* **95**, 226–230.
127. Reynolds, E. S., and Moslen, M. T. (1974). Liver injury following halothane anesthesia in phenobarbital-pretreated rats. *Biochem. Pharmacol.* **23**, 189–195.
128. de Groot, H., Harnisch, U., and Noll, T. (1983). Suicidal inactivation of microsomal cytochrome $P$-450 by halothane under hypoxic conditions. *Biochem. Biophys. Res. Commun.* **107**, 885–891.
129. Kreiter, P. A., and Van Dyke, R. A. (1983). Cytochrome $P$-450 and halothane metabolism. Decrease in rat liver microsomal $P$-450 *in vitro*. *Chem.-Biol. Interact.* **44**, 219–235.
130. Wolf, C. R., Mansuy, D., Nastainczyk, W., Deutschmann, G., and Ullrich, V. (1977). The reduction of polyhalogenated methanes by liver microsomal cytochrome $P$-450. *Mol. Pharmacol.* **13**, 698–705.
131. Mansuy, D. (1980). New iron–porphyrin complexes with metal–carbon bond—Biological implications. *Pure Appl. Chem.* **52**, 681–690.
132. Reiner, O., and Uehleke, H. (1971). Bindung von Tetrachlorkohlenstoff an reduziertes mikrosomales Cytochrom $P$-450 und an Häm. *Hoppe-Seyler's Z. Physiol. Chem.* **352**, 1048–1052.
33. Ullrich, V., and Schnable, K. H. (1971). Formation and binding of carbanions by cytochrome $P$-450 of liver microsomes. *Drug Metab. Dispos.* **1**, 176–182.
34. Mansuy, D., Lange, M., Chottard, J. C., and Guérin, P. (1977). Reaction of carbon tetrachoride with 5,10,15,20-tetraphenyl-porphinatoiron(II) [(TPP)Fe$^{II}$]: Evidence for the formation of the carbene complex [(TPP)Fe$^{II}$(CCl$_2$)]. *J. Chem. Soc., Chem. Commun.* pp. 648–649.

135. Mansuy, D., Lange, M., Chottard, J. C., Bartoli, J. F., Chevrier, B., and Weiss, R. (1978). Dichlorocarbene complexes of iron(II)–porphyrins—crystal and molecular structure of Fe(TPP)(CCl$_2$) (H$_2$O). *Angew. Chem., Int. Ed. Engl.* **17**, 781–782.
136. Mansuy, D., Nastainczyk, W., and Ullrich, V. (1974). The mechanism of halothane binding to microsomal cytochrome *P*-450. *Naunyn-Schmiedeberg's Arch. Pharmacol.* **285**, 315–324.
137. Pohl, L. R., and George, J. W. (1983). Trapping of dichlorocarbene as a metabolite of carbon tetrachloride in rat liver microsomes. *Biochem. Biophys. Res. Commun.* **117**, 367–371.
138. Loew, G., and Goldblum, A. (1980). Electronic spectrum of model cytochrome *P*-450 complex with postulated carbene metabolite of halothane. *J. Am. Chem. Soc.* **102**, 3657–3659.
139. Johnson, M. K. (1966). Metabolism of iodomethane in the rat. *Biochem. J.* **98**, 38–43.
140. Tachizawa, H., MacDonald, T. L., and Neal, R. A. (1982). Rat liver microsomal metabolism of propyl halides. *Mol. Pharmacol.* **22**, 745–751.
141. Chasseaud, L. F. (1973). The nature and distribution of enzymes catalyzing the conjugation of glutathione with foreign compounds. *Drug Metab. Rev.* **2**, 185–220.
142. Chasseaud, L. F. (1979). The role of glutathione and glutathione *S*-transferases in the metabolism of chemical carcinogens and other electrophilic agents. *Adv. Cancer Res.* **29**, 175–274.
143. Keen, J. H., Habig, W. H., and Jakoby, W. B. (1976). Mechanism for the several activities of the glutathione *S*-transferases. *J. Biol. Chem.* **251**, 6183–6188.
144. Mangold, J. B., and Abdel-Monem, M. M. (1983). Stereochemical aspects of conjugation reactions catalyzed by rat liver glutathione *S*-transferase isozymes. *J. Med. Chem.* **26**, 66–71.
145. Ahmed, A. E., and Anders, M. W. (1976). Metabolism of dihalomethanes to formaldehyde and inorganic halide. I. *In vitro* studies. *Drug Metab. Dispos.* **4**, 357–361.
146. Ahmed, A. E., and Anders, M. W. (1978). Metabolism of dihalomethanes to formaldehyde and inorganic halide. II. Studies on the mechanism of the reaction. *Biochem. Pharmacol.* **27**, 2021–2025.
147. Uotila, L., and Koivusalo, M. (1974). Formaldehyde dehydrogenase from human liver. *J. Biol. Chem.* **249**, 7653–7663.
148. Uotila, L., and Koivusalo, M. (1974). Purification and properties of *S*-formylglutathione hydrolase from human liver. *J. Biol. Chem.* **249**, 7664–7672.
149. Martin, J. L., George, J. W., and Pohl, L. R. (1980). Glutathione-dependent dechlorination of chloramphenicol by cytosol of rat liver. *Drug Metab. Dispos.* **8**, 93–97.
150. Livesey, J. C., and Anders, M. W. (1979). *In vitro* metabolism of 1,2-dihaloethanes to ethylene. *Drug Metab. Dispos.* **7**, 199–203.
151. Livesey, J. C., Anders, M. W., Langvardt, P. W., Putzig, C. L., and Reitz, R. H. (1982). Stereochemistry of the glutathione-dependent biotransformation of *vicinal*-dihaloalkanes to alkenes. *Drug Metab. Dispos.* **10**, 201–204.
152. Ozawa, N., and Guengerich, F. P. (1983). Evidence for formation of an *S*-[2-($N^7$-guanyl)ethyl]glutathione adduct in glutathione-mediated binding of the carcinogen 1,2-dibromoethane to DNA. *Proc. Natl. Acad. Sci. U.S.A.* **80**, 5266–5270.
153. Hutson, D. H., Holmes, D. S., and Crawford, M. J. (1976). The involvement of glutathione in the reductive dechlorination of a phenacylhalide. *Chemosphere* **5**, 79–84.
154. Brundin, A., Ratnayake, J. H., Sunram, J. M., and Anders, M. W. (1982). Glutathione dependent reductive dehalogenation of 2,2′,4-trichloroacetophenone to 2′,4′-dichloroacetophenone. *Biochem. Pharmacol.* **31**, 3885–3890.

155. Anderson, P. M., and Schultze, M. O. (1965). Cleavage of S-(1,2-dichlorovinyl)-L-cysteine by an enzyme of bovine origin. *Arch. Biochem. Biophys.* **111**, 593–602.
156. Tateishi, M., Suzuki, S., and Shimizu, H. (1978). Cysteine conjugate β-lyase in rat liver. *J. Biol. Chem.* **253**, 8854–8859.
157. Stevens, J., and Jakoby, W. B. (1983). Cysteine conjugate β-lyase. *Mol. Pharmacol.* **23**, 761–765.
158. Dohn, D. R., and Anders, M. W. (1982). Assay of cysteine conjugate β-lyase activity with S-(2-benzothiazolyl)cysteine as the substrate. *Anal. Biochem.* **120**, 379–386.
159. Gandolfi, A. J., Nagle, R. B., Soltis, J. J., and Plescia, F. H. (1981). Nephrotoxicity of halogenated vinyl cysteine compounds. *Res. Commun. Chem. Pathol. Pharmacol.* **33**, 249–261.
160. Hassall, C. D., Gandolfi, A. J., and Brendel, K. (1983). Effect of halogenated vinyl cysteine conjugates on renal tubular active transport. *Toxicology* **26**, 285–294.
161. Lock, E. A., and Ishmael, J. (1979). The acute toxic effects of hexachloro-1:3-butadiene on the rat kidney. *Arch. Toxicol.* **43**, 47–57.
162. Berndt, W. O., and Mehendale, H. M. (1979). Effects of hexachlorobutadiene (HCBD) on renal function and renal organic ion transport in the rat. *Toxicology* **14**, 55–65.
163. Hook, J. B., Rose, M. S., and Lock, E. A. (1982). The nephrotoxicity of hexachloro-1:3-butadiene in the rat: Studies of organic anion and cation transport in renal slices and the effect of monooxygenase inducers. *Toxicol. Appl. Pharmacol.* **65**, 373–382.
164. Nash, J. A., King, L. J., Lock, E. A., and Green, T. (1984). The metabolism and disposition of hexachloro-1:3-butadiene in the rat and its relevance to nephrotoxicity. *Toxicol. Appl. Pharmacol.* **73**, 124–137.
165. Jaffe, D. R., Hassall, C. D., Brendel, K., and Gandolfi, A. J. (1983). *In vivo* and *in vitro* nephrotoxicity of the cysteine conjugate of hexachlorobutadiene. *J. Toxicol. Environ. Health* **11**, 857–867.
166. Potter, C. L., Gandolfi, A. J., Nagle, R., and Clayton, J. W. (1981). Effects of inhaled chlorotrifluoroethylene and hexafluoropropene on the rat kidney. *Toxicol. Appl. Pharmacol.* **59**, 431–440.
167. Buckley, L. A., Clayton, J. W., Nagle, R. B., and Gandolfi, A. J. (1982). Chlorotrifluoroethylene nephrotoxicity in rats: A subacute study. *Fundam. Appl. Toxicol.* **2**, 181–186.
168. Dohn, D. R., and Anders, M. W. (1982). The enzymatic reaction of chlorotrifluoroethylene with glutathione. *Biochem. Biophys. Res. Commun.* **109**, 1339–1345.
169. Elfarra, A. A., Baggs, R. B., and Anders, M. W. (1983). Nephrotoxicity of S-(2-chloroethyl)-DL-cysteine: A possible metabolic product of 1,2-dichloroethane in the rat. *Pharmacologist* **25**, 104.
170. Elfarra, A. A., and Anders, M. W. (1984). Renal processing of glutathione conjugates: Role in nephrotoxicity. *Biochem. Pharmacol.* (in press).

## Chapter 11

# Halogenated Alkenes and Alkynes

DIETRICH HENSCHLER

*Institut für Pharmakologie und Toxikologie*
*Universität Würzburg*
*Würzburg, Federal Republic of Germany*

I. Chemical Reactivity as a Basis for Predicting Biotransformation Pathways and Rates . . . . . . . . . . . . . . . . . . 317
II. Halogenated Ethylenes . . . . . . . . . . . . . . . . . 318
   A. General Reaction Schemes . . . . . . . . . . . . . 318
   B. Vinyl Chloride, Vinyl Bromide, and Vinyl Fluoride . . . 320
   C. Vinylidene Chloride and Vinylidene Fluoride . . . . . . 323
   D. *cis*- and *trans*- 1,2-Dichloroethylenes . . . . . . . . . 327
   E. Trichloroethylene . . . . . . . . . . . . . . . . . . 328
   F. Tetrachloroethylene . . . . . . . . . . . . . . . . . 333
III. Halogenated Allyl Compounds . . . . . . . . . . . . . 335
IV. Halogenated Alkynes . . . . . . . . . . . . . . . . . 339
   References . . . . . . . . . . . . . . . . . . . . . . 341

## I. Chemical Reactivity as a Basis for Predicting Biotransformation Pathways and Rates

Halogenated aliphatic compounds differ widely in their chemical reactivities as a result of the interplay of variations in the electron densities of the halogen and carbon atoms involved in the respective aliphatic system. In general, the electron-withdrawing effect of a halogen atom interferes with the mesomeric donor effect on the first and second carbon atoms, resulting in an electron deprivation in the adjacent C—C bond system. The consequences for the chemical reactivity are completely different in alkanes, alkenes, and alkynes: in alkanes and alkynes, a destabilization in the C ··· C system is found, whereas in olefins the decrease in the electron density results in a stabilization of the adjacent double bond.

From these basic considerations, some tentative predictions can be made as to the biotransformation mechanisms and the reactivities of the metabolites produced enzymatically from certain types of halogenated aliphatic compounds by enzymes providing activated oxygen (oxidases, peroxidases, mixed-function oxygenases). The electronegativity of activated oxygen will induce, in halogenated alkanes, the formation of alkanols and of free radicals, whereas in olefins the formation of oxiranes will be favored [reaction (1)].

$$\begin{array}{c}\diagdown\\ \diagup\end{array}\!\!C\!=\!C\!\!\begin{array}{c}Cl\\ \diagdown\end{array} \rightarrow \begin{array}{c}\diagdown\\ \diagup\end{array}\!\!C\!\!\overset{O}{\underset{}{\text{---}}}\!\!C\!\!\begin{array}{c}Cl\\ \diagdown\end{array} \qquad (1)$$

For halogenated acetylenes, the formation of free radicals with very high reactivity is expected.

Allyl halides constitute a special class of compounds. Besides epoxidation and radical formation, halide ion and allyl cation formation may occur [reaction (2)].

$$CH_2=CH-CH_2Cl \rightarrow [CH_2 \doteq CH \doteq CH_2]^+Cl^- \qquad (2)$$

The allyl cation is stabilized by resonance. Because this is a nonenzymatic reaction, it may compete with or even exclude the enzyme-catalyzed oxidative reactions.

In addition, another effect of halogen substitution may come into play, at least with chlorine, bromine, and iodine: steric protection from electrophilic attacks by the bulky halogen atoms. From this, one would assume an antagonism of the destabilizing negative induction effect ($-$I effect) with increasing numbers of halogen substitutions in alkanes, and a synergistic increase in stability in olefins, in parallel with a decrease in metabolization rates.[1]

## II. Halogenated Ethylenes

### A. GENERAL REACTION SCHEMES

The first report in the literature on the metabolic formation of an oxirane from a chlorinated ethylene dates back to 1945, when Powell postulated that 2,2,3-trichlorooxirane was an intermediate in trichloroethylene metabolism.[2] Oxirane formation seems to be a common pathway for all chlorin ated ethylenes as the first step in oxidative biotransformation, as judged from the final metabolites produced in mammalian systems.[3] The rate of metabolite formation, in the isolated perfused rat liver preparation, parallels the reaction rates of ethylene:vinyl chloride:trichloroethylene:tetrachloroethylene with ozone, which are 2500:1180:3.6:1,[4] thus confirming the stabilizing effect of chlorine substitution for the oxidative metabolic conversion.[3]

Chlorinated oxiranes are electrophilic compounds, which, in biological systems, may undergo a variety of secondary reactions (see Fig. 1): pathway (1) alkylation of nucleophilic sites of cellular macromolecules (e.g., DNA or RNA bases); pathway (2) conjugation with soluble low molecular weight nucleophiles [e.g., glutathione (GSH), cysteine] both spontaneously and enzymatically (glutathione transferases); pathway (3) hydrolysis to $vic$-diols, again nonenzymatically or possibly by epoxide hydrolases. These diols are, at least in the case of more than one halogen substituent, very unstable compounds and may undergo a variety of decomposition reactions; pathway (4) intramolecular rearrangement, with C—O heterolysis and sometimes intramolecular chlorine migration, to halogenated aldehydes or acyl halides, or both.

Pathways (2)–(4) in Fig. 1 constitute, in chemical and biological terms, deactivation mechanisms. Thus, the toxic effects of halogenated ethylenes

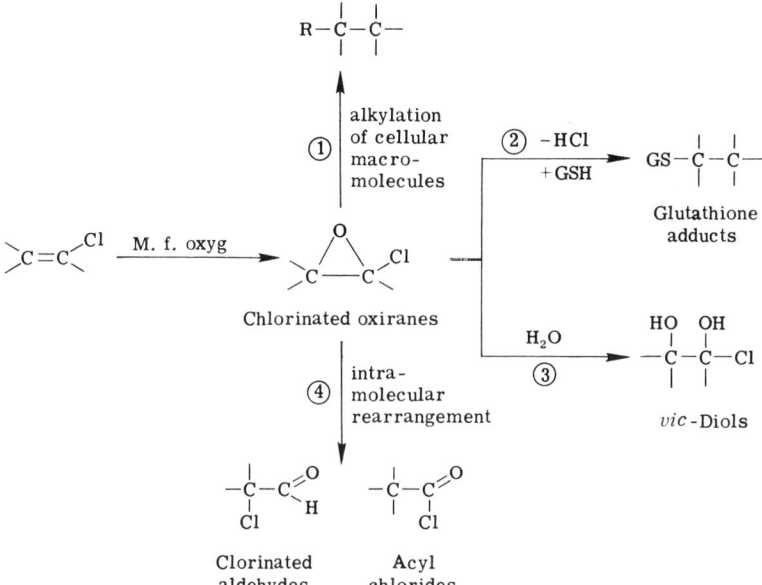

**Fig. 1.** Formation of oxiranes by mixed-function oxygenases (M. f. oxyg.) from haloethylenes, and possible secondary reactions. (See text for explanation.)

will depend on pharmacokinetic parameters (e.g., solubility, vapor pressure), on the relative rates of the formation of the oxiranes, and on ensuing pathways (1) vs (2)–(4). The mechanistic evaluation must consider the rates of all steps of biotransformation and the biological significance of all reaction products.

## B. VINYL CHLORIDE, VINYL BROMIDE, AND VINYL FLUORIDE

Vinyl chloride (**1**) metabolism has been studied extensively since the discovery of its mutagenic[5] and carcinogenic[6] potential. There is overwhelming evidence[1,7-9] that vinyl chloride is activated, by cytochrome $P$-450-catalyzed monooxygenases, to monochlorooxirane (**2**), which rapidly rearranges to chloroacetaldehyde (**3**) [reaction (3)].

$$\underset{1}{\underset{H}{\overset{H}{>}}C=C\underset{Cl}{\overset{H}{<}}} \xrightarrow[\text{P-450}]{\underset{\text{NADPH}}{O_2}} \underset{2}{\underset{H}{\overset{H}{>}}C\overset{O}{-}C\underset{Cl}{\overset{H}{<}}} \rightarrow \underset{3}{H-\underset{Cl}{\overset{H}{\underset{|}{C}}}-C\overset{O}{\underset{H}{<}}} \quad (3)$$

Chlorooxirane (**2**) may be attacked by epoxide hydrolase to form the *vic*-diol monochloroglycol (**4**), which rapidly decomposes by dehydrochlorination of the *gem*-chlorohydrin (**4**) to glycol aldehyde (**5**) [reaction (4)].[10]

$$\underset{2}{\underset{H}{\overset{H}{>}}C\overset{O}{-}C\underset{H}{\overset{Cl}{<}}} \xrightarrow[\text{epoxide hydrolase}]{H_2O} \underset{4}{H-\underset{H}{\overset{OH}{\underset{|}{C}}}-\underset{Cl}{\overset{OH}{\underset{|}{C}}}-H} \xrightarrow{-HCl} \underset{5}{H-\underset{H}{\overset{OH}{\underset{|}{C}}}-C\overset{O}{\underset{H}{<}}} \quad (4)$$

Chloroacetaldehyde (**3**) can be reduced by alcohol dehydrogenase (ADH) to 2-chloroethanol (**6**)[10] or oxidized by aldehyde oxidase (Aldox) to chloroacetic acid (**7**) [reaction (5)].

$$\underset{6}{ClCH_2-CH_2-OH} \xleftarrow[\text{ADH}]{\text{NADPH}} \underset{3}{ClCH_2-C\overset{O}{\underset{H}{<}}} \xrightarrow[\text{Aldox}]{\text{NADH}} \underset{7}{ClCH_2-C\overset{O}{\underset{OH}{<}}} \quad (5)$$

The major pathway of metabolic deactivation of chlorooxirane (**2**) is however, conjugation of both the oxirane and its rearrangement product chloroacetaldehyde (**3**) with glutathione (GSH) (**8**). The glutathione adduct is subsequently transformed by well-known pathways of mercapturic acid formation to *S*-(2-hydroxyethyl)cysteine (**9**), its *N*-acetyl product (**10**), and thiodiglycolic acid (**11**) [reaction (6)].

$$\begin{array}{c} H_2C \overset{O}{-} CHCl \\ \mathbf{2} \end{array} \quad \begin{array}{c} ClCH_2-CHO \\ \mathbf{3} \end{array} \to GS-CH_2-CHO \to \to HO-CH_2-CH_2-S-\underset{NH_2}{\overset{}{CH}}-C\overset{O}{\underset{OH}{\diagdown}} \quad (6)$$

(structures: **2** chlorooxirane; **3** chloroacetaldehyde; **7** ClCH₂—COOH; intermediates GS—CH₂—CHO and GS—CH₂—COOH; **9** HO—CH₂—CH₂—S—CH(NH₂)—COOH; **10** HO—CH₂—CH₂—S—CH(NHCOCH₃)—COOH; **11** HOOC—CH₂—S—CH₂—COOH)

Chloroacetic acid (**7**), if administered to rodents, may also be conjugated with glutathione (**8**) to form thiodiglycolic acid (**11**) and carboxymethylcysteine (**12**)[11] [see reaction (12)] and by this way, in the case of vinyl chloride metabolism, add to the mercapturic acid pool.

Vinyl chloride metabolites bind covalently to macromolecules. Another binding site is the heme moiety of cytochrome *P*-450, the activity of which is markedly decreased by giving vinyl chloride (**1**).[12] Chlorooxirane (**2**) and chloroacetaldehyde (**3**) can form identical alkylation products with DNA bases and proteins. However, under the influence of various enzymes in the

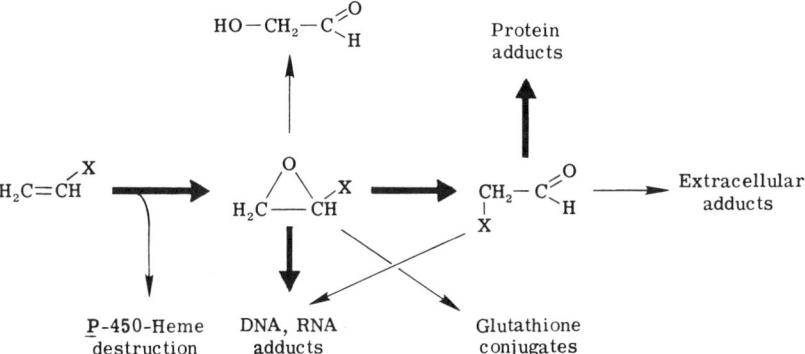

**Fig. 2.** Metabolic fate of monohalo ethylenes (X = halogen), and major and minor pathways of secondary reactions. (Modified from Ref. 14.)

microsomal and cytosolic multienzyme complexes, preferential binding patterns will result: the more reactive oxirane (2) binds primarily to DNA bases, whereas chloroacetaldehyde (3) binds preferentially to proteins.[13] A semiquantitative and sequential scheme of vinyl halide metabolism and binding has been proposed recently (Fig. 2).[14] Covalent binding to the thiol group of coenzyme A has also been reported.[15]

DNA and RNA adducts of vinyl chloride (1) and vinyl bromide (13) metabolites have been described (see formulas for compound's 14–17). All are

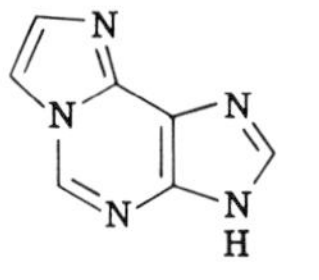

1-$N^6$-Ethenoadenine
**14**

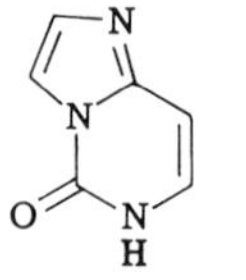

3-$N^6$-Ethenocylosine
**15**

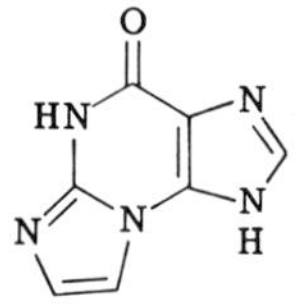

$N^6$-3-Ethenoguanine
**16**

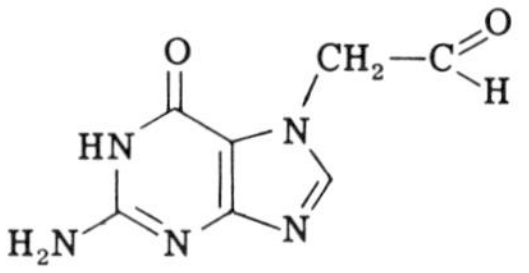

7,(2-Oxoethyl)guanine
**17**

formed by cyclization reactions of halooxiranes or the respective acetaldehydes with an exocyclic amino group and an endocyclic nitrogen atom [reaction (7)].[16-19] $N^7$-Alkylation of guanine leads to 7-oxoethylguanine

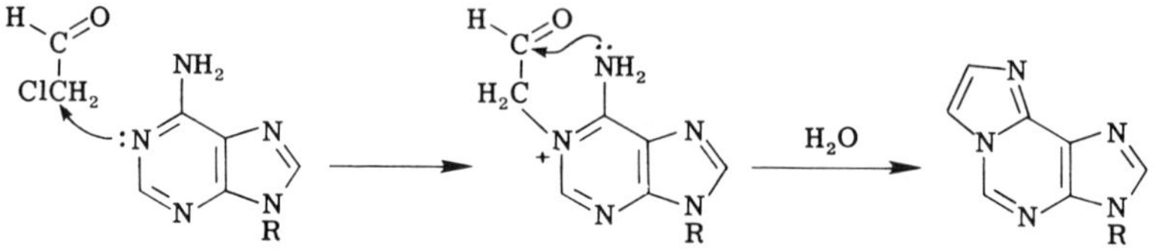

$$(7)$$

(**17**),[20] which seems to be the major, if not the only, DNA alkylation product formed *in vivo*.[21] It may be in equilibrium with the hemiacetal $O^6$-7-(1-hydroxyethano)deoxyguanine (**18**) [reaction (8)].[21,22] Protein binding sites for

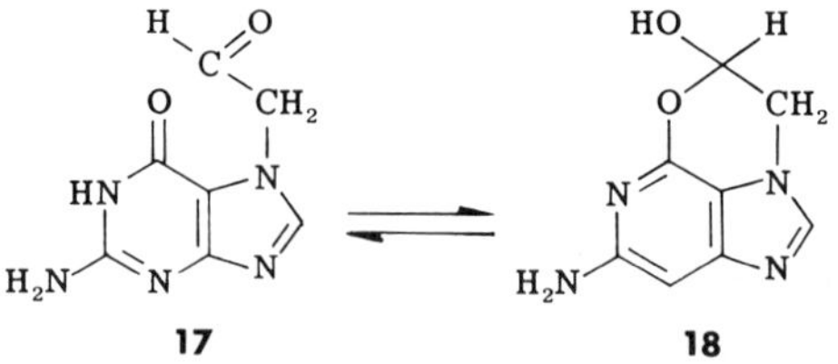

$$(8)$$

**17**      **18**

activated vinyl chloride metabolites have been identified as the thiol groups of cysteine[23] and the $N^{-1}$ and $N^{-3}$ positions of histidine.[20] The coplanar position of the imidazole ring formed by ethenoderivatization indicates that interference by miscoding with translation and transcription mechanisms of DNA and RNA bases may occur.[24]

Although there is no experimental proof for the occurrence of chlorooxirane as a metabolic intermediate, its formation can be postulated by indirect evidence, such as trapping with model nucleophiles in microsomal incubation of vinyl chloride.[25,26] Chlorooxirane (**2**) is mutagenic in *in vitro* systems[27,28] and is carcinogenic in animals[29]; chloroacetaldehyde (**3**) is much less mutagenic[27] and carcinogenic.[29,30]

Vinyl bromide (**13**) and vinyl fluoride (**19**) have not been investigated systematically, and detailed metabolism studies have not been carried out.

$$\begin{array}{ccc} \text{H} \quad \text{Br} & \text{H} \quad \text{F} & \text{H} \quad \text{O} \quad \text{F} \\ \diagdown \diagup & \diagdown \diagup & \diagdown \diagup \diagdown \diagup \\ \text{C}=\text{C} & \text{C}=\text{C} & \text{C}-\text{C} \\ \diagup \diagdown & \diagup \diagdown & \diagup \diagdown \\ \text{H} \quad \text{H} & \text{H} \quad \text{H} & \text{H} \quad \text{H} \\ \textbf{13} & \textbf{19} & \textbf{20} \end{array}$$

However, occasional comparative covalent binding investigations strongly indicate that both analogs are also metabolized through oxiranes: vinyl bromide (**13**) binds similarly to microsomal proteins[31] and produces the same DNA and RNA adducts as vinyl chloride (**1**).[32] The compound is mutagenic after metabolic activation[33] and is carcinogenic in rats after inhalation, producing the same type of tumor as vinyl chloride (**1**).[34] The carcinogenic potential of vinyl fluoride (**19**) is suggested by the induction of preneoplastic lesions in rat liver after inhalation in newborn rats[35]; from this, the oncogenic effect of the three haloethylenes has been suggested to be of the order: vinyl chloride > vinyl fluoride > vinyl bromide.[36] The metabolic rate in inhalation experiments in intact rats is, however, in the order (micromoles per hour per kilogram body weight): vinyl chloride:vinyl bromide:vinyl fluoride = 111 : 40 : 7.[37] Fluorooxirane (**20**) is, according to molecular orbital studies, under greater ring-strain than chlorooxirane, and this is less stable.[38]

## C. VINYLIDENE CHLORIDE AND VINYLIDENE FLUORIDE

Vinylidene chloride (**21**) represents a special case in the series of chlorinated ethylenes. One would expect it to be the most reactive member of the series because of its most uneven electron distribution, resulting from the asymmetric chlorine substitution. Although the structural analogy to vinyl chloride (**1**) led most investigators to presume formation of an oxirane in the metabolic activation mechanism,[39-41] the oxirane (22) would be expected to be the most polar and unstable member of the series [reaction (9)].

$$\underset{21}{\underset{H}{\overset{H}{>}}C=C\underset{Cl}{\overset{Cl}{<}}} \rightarrow \underset{22}{\underset{H}{\overset{H}{>}}C\underset{\overline{\phantom{xx}}}{\overset{O}{\overset{\diagup\diagdown}{\phantom{x}}}}C\underset{Cl}{\overset{Cl}{<}}} \quad (9)$$

Indeed, the oxirane (22) cannot be synthesized by conventional methods because of immediate rearrangement[3]; the only isolable product from the oxidation of vinylidene chloride (21) with m-chloroperbenzoic acid is chloroacetyl chloride.[3] Therefore, the evidence for its metabolic formation is less well established, because it is not available, as are the oxiranes of the other chlorinated ethylenes, for comparative model experiments.

Theoretically, there are two possibilities for rearrangement of this oxirane [reactions (10) and (11)].

$$\underset{22}{\underset{H}{\overset{H}{>}}C\underset{\phantom{x}}{\overset{O}{\overset{\diagup\diagdown}{\phantom{x}}}}C\underset{Cl}{\overset{Cl}{<}}} \rightarrow \underset{23}{ClCH_2-C\underset{Cl}{\overset{\diagdown O}{<}}} \quad (10)$$

$$\underset{22}{\underset{H}{\overset{H}{>}}C\underset{\phantom{x}}{\overset{O}{\overset{\diagup\diagdown}{\phantom{x}}}}C\underset{Cl}{\overset{Cl}{<}}} \rightarrow \underset{24}{\underset{H}{\overset{O}{>}}C-CHCl_2} \quad (11)$$

The formation of monochloroacetyl chloride (23), or its secondary metabolic products, is well established in whole-animal studies.[3,40-42] The occurrence of dichloroacetaldehyde (24) has, however, also been postulated[39] and demonstrated in experiments with isolated liver microsomes.[43] No dichloro compounds have been detected in intact animal studies.[40-42] This raises the question of a specific biological (enzymatic?) influence on the rearrangement mechanism similar to that seen with trichloroethylene,[1,44] which might occur only in intact (liver) cells. A careful reevaluation of the situation under *in vitro* and *in vivo* conditions seems mandatory, because it may add decisively to the understanding of acute toxic effects, as well as to the mutagenicity and carcinogenicity, of vinylidene chloride.

The epoxide rearrangement product chloroacetyl chloride (23) or its hydrolysis product chloroacetic acid (7) may react with glutathione (8) to yield, after subsequent degradation of the glutathione adduct, carboxymethylcysteine (12), N-acetylcarboxymethylcysteine (25), thiodiglycolic acid (11), thioglycolic acid (26), and dithioglycolic acid (27) [reaction (12)].[40-42]

Chloroacetic acid (7) and carbon dioxide ($CO_2$) (28) have also been found. Quantitative differences in these conjugation products are seen between mice and rats.[40]

A concomitant, but different, pathway has been derived from the urinary metabolite methylthioacetylaminoethanol (29), which might result from the reaction of 1,1-dichlorooxirane (22) or chloroacetyl chloride (23) with phosphatidylethanolamine (30), followed by hydrolysis and S-methylation [reaction (13)].

This reaction could explain, through the attack on the main phospholipid in lipid membranes, the extraordinary hepatotoxicity of vinylidene chloride.[42]

The covalent binding of [$^{14}$C]vinylidene chloride (**21**) to proteins[45] and to liver and kidney DNA[46] has been reported, but no alkylation products of macromolecules have been identified. They may differ from those described for vinyl chloride (**1**) in that the etheno-cyclization reaction with adenine and cytosine will carry an oxygen into the adduct (**31**) [e.g., reaction (14)]. This might have specific implications for translation and transcription.[47]

Vinylidene chloride (**21**) is mutagenic after metabolic activation.[48,49] Positive[50,51] and negative carcinogenicity experiments have been reported.[46] The conflicting results may be due to interference of the acute toxicity of vinylidene chloride with tumor promotion.

With vinylidene fluoride, $F_2C=CH_2$ (**32**), no studies on metabolite formation have been reported. The compound is metabolized 100 times more slowly than vinyl chloride (**1**)[37] and produces preneoplastic liver foci in rats when given by inhalation shortly after birth[52] indicating a very low carcinogenic potential.

## D. CIS- AND TRANS-1,2-DICHLOROETHYLENES

The oxiranes (**35,36**) of cis- (**33**) and trans- (**34**) 1,2-dichloroethylenes have been synthesized and were found to rearrange exclusively to dichloroacetaldehyde (**24**) [reaction (15)].[3,53]

$$
\begin{array}{c}
\text{Cl}\diagdown\!\!\!\!\!\diagup\text{Cl} \\
\text{C}=\text{C} \\
\text{H}\diagup\!\!\!\!\!\diagdown\text{H} \\
\mathbf{33}
\end{array}
\longrightarrow
\begin{array}{c}
\text{Cl}\;\;\text{O}\;\;\text{Cl} \\
\text{C}-\text{C} \\
\text{H}\;\;\;\;\text{H} \\
\mathbf{35}
\end{array}
\searrow
$$

$$\text{Cl}_2\text{CH}-\text{C}(=\text{O})\text{H} \quad \mathbf{24} \tag{15}$$

$$
\begin{array}{c}
\text{H}\diagdown\!\!\!\!\!\diagup\text{Cl} \\
\text{C}=\text{C} \\
\text{Cl}\diagup\!\!\!\!\!\diagdown\text{H} \\
\mathbf{34}
\end{array}
\longrightarrow
\begin{array}{c}
\text{H}\;\;\text{O}\;\;\text{Cl} \\
\text{C}-\text{C} \\
\text{Cl}\;\;\;\;\text{H} \\
\mathbf{36}
\end{array}
\nearrow
$$

Metabolism studies in the isolated perfused rat liver revealed, as metabolites, dichloroacetic acid and dichloroethanol.[3] Microsomal preparations converted both isomers to dichloroacetaldehyde (**24**) and dichloroethanol (**37**),[39,54] whereas dichloroacetic acid (**38**) was found only with the trans isomer.[54] No monochloroacetic acid (**7**) was detected. The mechanism of conversion[3] is shown in reaction (16).

$$
\begin{array}{c}
\text{(Cl)}\;\;\text{O}\;\;\text{Cl} \\
\text{C}-\text{C} \\
\text{H}\;\;\;\;\text{H} \\
\mathbf{35}
\end{array}
\searrow
\quad
\begin{array}{c}
\text{(Cl)}\;\;\text{O}\;\;\text{H} \\
\text{C}-\text{C} \\
\text{H}\;\;\;\;\text{Cl} \\
\mathbf{36}
\end{array}
\nearrow
\quad \text{Cl}_2\text{CH}-\text{C}(=\text{O})\text{H} \quad \mathbf{24}
\quad \longrightarrow
\begin{array}{l}
\text{Cl}_2\text{CH}-\text{CH}_2\text{OH} \quad \mathbf{37} \\
\\
\text{Cl}_2\text{CH}-\text{C}(=\text{O})\text{OH} \quad \mathbf{38}
\end{array}
\tag{16}
$$

This is consistent with the expectation that no simple hydride shift occurs in the course of the intramolecular rearrangement of the oxiranes, which would result in the formation of chloroacetyl chloride; indeed, this type of shift has not been found in any of the chlorooxiranes investigated thus far and has low probability on theoretical grounds.[55]

Overwhelming evidence indicates cytochrome *P*-450-dependent monooxygenases as enzymes involved in 2,3-dichlorooxirane formation.[1,39,54] The metabolic rate of the cis isomer is higher than of the trans isomer[3,54,56]; this is probably a result of the steric opening of the cis molecule. Cytochrome *P*-450 is inhibited by covalently bound metabolites produced during bioactivation of both compounds.[54,57] No specific information on the enzymes converting dichloroacetaldehyde (**24**) to dichloroethanol (**37**) and dichloroacetic acid (**38**) is available.

Both 1,2-dichloroethylenes have been found nonmutagenic in *E. coli* K12 with and without addition of metabolizing enzymes.[58]

## E. TRICHLOROETHYLENE

Because of the widespread use of trichloroethylene (**39**) as an organic solvent and general anesthetic in surgery and gynecology, extensive studies on the metabolism of this compound were carried out long before interest in vinyl and vinylidene chlorides arose. Oxirane formation was proposed[2] as the first step in the metabolism of trichloroethylene more than a decade before information on the nature of the metabolizing enzymes became available. The basis for suggesting oxirane formation in the metabolic conversion of trichloroethylene (**39**) was the identification of 1,1,1-trichloro compounds as urinary metabolites—that is, trichloroacetic acid (**40**) and trichloroethanol (**41**) or its glucuronide (**42**)—which can only be explained by a chlorine migration in an unstable intermediate [reaction (17)].

$$Cl_2C=CHCl \;(\mathbf{39}) \rightarrow \text{oxirane} \;(\mathbf{43}) \rightarrow Cl_3C-CHO \;(\mathbf{44}) \rightarrow \begin{cases} Cl_3C-CH_2OH \;(\mathbf{41}) \rightarrow Cl_3C-CH_2-O-\text{glu} \;(\mathbf{42}) \\ Cl_3C-COOH \;(\mathbf{40}) \end{cases} \quad (17)$$

These pathways of metabolism have been confirmed with *in vitro* liver cell fractions as the source of metabolizing enzymes.[59-61] 2,2,3-Trichlorooxirane (**43**) is the only representative of the entire series of chlorinated oxiranes for which direct evidence of its metabolic formation has been obtained. The specific binding spectrum with cytochrome *P*-450 is identical after addition of trichloroethylene (**39**) or the oxirane itself (**43**) to the metabolizing system.[62] The formation of the oxirane (**43**) from trichloroethylene (**39**) by a microsomal preparation, and its chromatographic identification as an adduct to 4-nitrobenzylpyridine has been described[62a]; the structure of this adduct, however, has not been determined.

Subsequent work has demonstrated, however, that the conditions of the formation, reactivity, rearrangement, and ensuing reactions are much more complicated than previously assumed. The thermal rearrangement of the synthetically obtained pure oxirane in nonpolar solvents has been carefully studied[1,3,63,64] and found not to yield chloral (**44**) but predominantly dichloroacetyl chloride (**45**) [reaction (18)].

$$\begin{array}{c} \text{Cl}_2\text{CH—C(=O)Cl} \\ \textbf{45} \\ \textbf{43} \xrightarrow{\text{thermal}} \\ \xrightarrow{\text{in vivo}} \\ \text{Cl}_3\text{C—C(=O)H} \\ \textbf{44} \end{array} \quad (18)$$

A theoretical explanation for the preference of thermal rearrangment to dichloroacetyl chloride (**45**) has been given[1,65]; the rearrangement process is most probably initiated by oxirane ring opening through C—O heterolysis yielding α-ketocarbonium ions.[46,47]. The opening can take place at two positions, as shown in reactions (19) and (20).

$$\textbf{43} \rightarrow \textbf{46} \rightarrow \text{Cl}_2\text{CH—C(=O)Cl} \quad \textbf{45} \quad (19)$$

$$\textbf{43} \rightarrow \textbf{47} \rightarrow \text{C(=O)—CCl}_3 \quad \textbf{44} \quad (20)$$

C—O Heterolysis at the singly substituted carbon atom in **43** results in a carbonium ion (**46**), which is bonded to one chlorine atom [reaction (19)]; this intermediate is more stable than the geminally substituted carbonium ion (**47**) in reaction (20), which makes this intermediate (**47**) much less stable. Consequently, rearrangement according to reaction (19) is favored, and dichloroacetyl chloride is found as the thermal rearrangement product.

Furthermore, an explanation has been offered,[1,65] and later supported,[66] for the controversial situation *in vivo* where only 1,1,1-trichlorometabolites[3] are found. The rearrangement of 2,2,3-trichlorooxirane (**43**) to chloral (**44**) can be effected by Lewis acids such as AlCl$_3$,[44] or FeCl$_3$,[1] where the attack of the electron acceptor molecule can best be assumed at the site of the "bay region" of the oxirane, with an interaction at the nearest chlorine atom or at the oxirane oxygen, or at both [reaction (21)]. Such a Lewis acid-like inter-

$$\text{Lewis acid} \longleftarrow \underset{\mathbf{43}}{\begin{array}{c}H\\ \diagup O \diagdown \\ C - C \\ | \quad | \\ Cl \quad Cl \\ \text{Cl}\end{array}} \longrightarrow \underset{\mathbf{44}}{\overset{O}{\underset{H}{\diagdown}}C - CCl_3} \qquad (21)$$

action is also conceivable in the biological environment of metabolic oxirane formation.[3] Cytochrome $P$-450 activates oxygen through its heme iron and forms $Fe^{3+}$ after transferring the oxygen to the substrates; in this state, the heme iron could act as a Lewis acid.[44] Support for this hypothesis is found in experiments in which model heme complexes with a fifth ligand to Fe have been shown to effect, like $FeCl_3$, 2,2,3-trichlorooxirane (**43**) rearrangement to chloral (**44**).[67]

A different explanation of chloral (**44**) formation also has been proposed; it postulates the formation of trichloroethylene (**39**) adducts to cytochrome $P$-450 heme iron, rather than oxirane (**43**) formation.[62a] The evidence for this alternative is rather indirect [comparison of kinetics of rearrangement and hydrolysis products of the oxirane (**43**)].

The failure of 2,2,3-trichlorooxirane (**43**) to rearrange to dichloroacetyl chloride (**45**) constitutes a protective mechanism, because chloral (**44**) is not chemically reactive, whereas dichloroacetyl chloride (**45**) is. Because under *in vivo* normal conditions no dichloroacetic acid (**38**) is produced through trichloroethylene metabolism, the probability prevails that the enzymatically formed oxirane is completely deactivated by rearrangement to chloral before leaving the catalytic site of the enzyme. This assumption is strongly supported by additional experimental data. 2,2,3-Trichlorooxirane (**43**) decomposes rapidly in aqueous environments, undergoing a variety of reactions,[44] which are summarized in Fig. 3. The detection of one-carbon products such as formic acid (**48**) and carbon monoxide (**49**), as well as of glyoxylic acid (**50**), implicates hydrolysis to the *vic*-diol (**51**), subsequent dehydrochlorination, and a carbon–carbon fission. Because neither formic acid nor carbon monoxide has been identified as a metabolite in intact animals, the escape of the intact oxirane from the hydrophobic site of the enzyme seems unlikely.

When given large oral doses of trichloroethylene (**39**), mice excrete, in addition to 1,1,1-trichloro compounds, small amounts of dichloroacetic acid (**38**). It has been hypothesized from this that the higher cytochrome $P$-450 activity of this species, compared to rats, might overpower the protective rearrangement mechanism to chloral (**44**) and thus render mice more susceptible to an assumed carcinogenic activity of trichloroethylene.[66] However, under similar exposure conditions, rats excrete even more dichloroacetic acid (**38**)[68] and are resistant to trichloroethylene-induced carci-

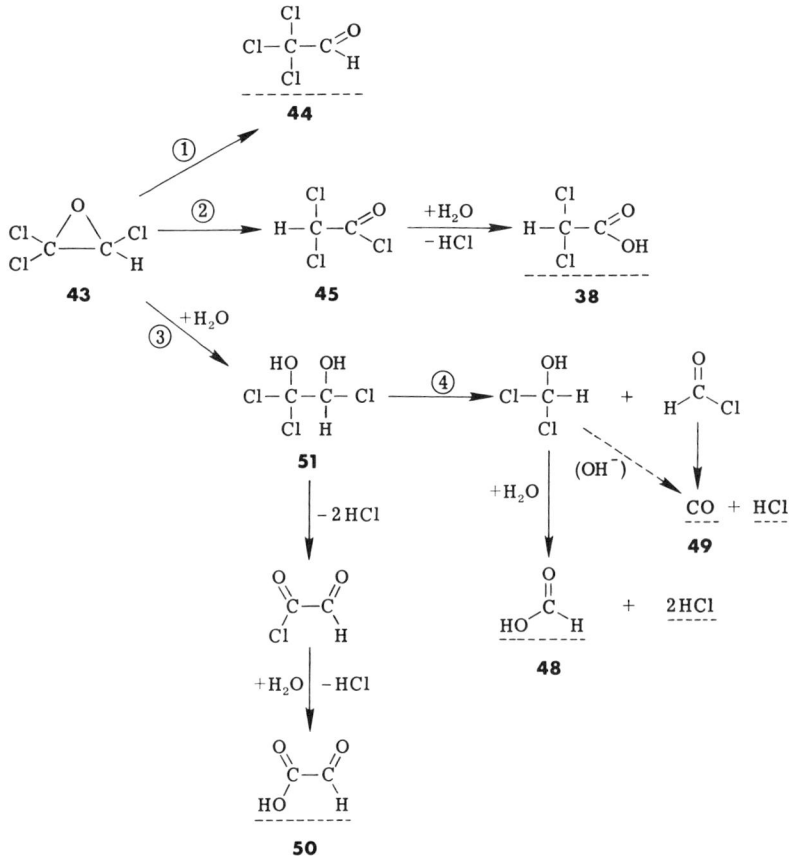

**Fig. 3.** Reactions of 2,2,3-trichlorooxirane in aqueous systems.[66] Identified products have a dashed line beneath them. Reaction pathways (1) and (2), intramolecular rearrangements; Pathway (3), hydrolysis with subsequent dehydrochlorination and hydrolytic dechlorination; pathway (4), C···C fission with subsequent homolysis or dehydrochlorination and hydrolysis, respectively.

nogenicity. In addition, several metabolites were identified that had previously been overlooked: oxalic acid (**52**), carbon dioxide (**28**), and the ethanolamide of glycolic acid, HO—$CH_2$—CO—NH—$CH_2$—$CH_2$—OH (**53**).[68] The oxalic acid (**52**) originates from the *vic*-diol (**51**), which is dechlorinated to yield glyoxylic acid (**50**) (see Figs. 3 and 4). Carbon dioxide may arise from the decarboxylation of these acids. The ethanolamide may be formed from a reaction of the oxirane with phosphatidylethanolamine,[68] as outlined in reaction (22).

$$\underset{43}{\underset{Cl}{\overset{H}{>}}C\underset{Cl}{\overset{O}{-}}C{<}} \rightarrow \rightarrow \rightarrow HO-CH_2-\underset{\underset{53}{NH-CH_2-CH_2-OH}}{\overset{O}{C}} \qquad (22)$$

No glutathione adducts or mercapturic acids have been detected as trichloroethylene metabolites,[68] which again supports the hypothesis that the oxirane rapidly rearranges or reacts with structures around the site of its formation before gaining access to the cytosol. [$^{14}$C]Trichloroethylene (**39**) binds covalently to liver microsomal protein, *in vitro*[69,70] as well as *in vivo*.[69,71] The amount of binding is consistent with species-related cytochrome *P*-450 activities[71] and is highest in microsomes and lowest in cytosolic fractions.[71] Binding to microsomal structures is paralleled by a decrease in cytochrome *P*-450 activity.[72]

[$^{14}$C]Trichloroethylene metabolites bind to DNA[70] and RNA,[73] when trichloroethylene (**39**) is added to *in vitro* preparations, and 2,2,3-trichlorooxirane (**43**) undergoes, in nonpolar solvents, a variety of reactions with low molecular weight model nucleophiles.[74] The adducts formed with RNA are not identical with binding patterns found with vinyl chloride (**1**).[73]

Attempts to demonstrate unequivocally DNA and RNA binding *in vivo*, however, have been unsuccessful. The extremely low amount of binding in mice treated orally with high doses of [$^{14}$C]trichloroethylene (**39**) could not be associated with typical DNA bases.[75] Small amounts of $^{14}$C activity in liver DNA or orally treated rats and mice, in connection with evidence of considerable $^{14}CO_2$ formation,[76,77] were correlated with metabolic incorporation of one-carbon fission products (see earlier) into DNA and RNA[78]; again, no specific DNA adducts as reaction products of trichloroethylene metabolites with nucleophilic sites of DNA bases were found.[78]

In the light of these findings and the activation and deactivation mechanisms just described, it is not surprising that reports on the mutagenicity and carcinogenicity of trichloroethylene are negative or, at best, controversial. The situation is complicated by the use of technical-grade or otherwise impure samples of trichloroethylene, which contained, as stabilizers, epoxides such as epichlorohydrin (**54**) or 1,2-epoxybutane (**55**), or both, the mutagenicity

$$\underset{54}{H_2C\overset{O}{-\!\!-\!\!-}CH-CH_2Cl} \qquad \underset{55}{H_2C\overset{O}{-\!\!-\!\!-}CH-CH_2-CH_3}$$

or carcinogenicity of which has been demonstrated.[79] Besides many slightly positive or borderline results with *in vitro* mutagenicity tests, there are reports of unambiguously negative tests in microbial systems when pure trichloroethylene was used.[33,79] Also, the oxirane (**43**) itself is negative in *Salmonella typhimurium* and *E. coli* tester strains.[80]

One positive carcinogenicity finding with trichloroethylene in mice, but not in rats, at maximum tolerated oral doses has been reported.[81] Tumor formation was restricted to a dose-related increase in the high spontaneous rate of liver cell carcinomas specific to the strain of test mouse (B6C3F$_1$). A technical grade of trichloroethylene, which contained epoxide stabilizers, was used in this experiment. On the other hand, an inhalation experiment with pure trichloroethylene (100 and 500 ppm, 18 months) in three animal species (rat, mouse, hamster) was negative except for a slight increase in spontaneous lymphoma rate in female mice at the high dose level, which is not regarded as proof of carcinogenicity.[82] Also, a gavage study with large doses of highly purified trichloroethylene in Swiss mice was negative, whereas addition of epichlorohydrin (**54**) and 1,2-epoxybutane (**55**) produced forestomach cancers.[83] Short-term tumor induction models in mice were also negative with trichloroethylene.[30]

Interpretation of the available results of mutagenicity and carcinogenicity testing in context with the mechanistic evaluation of formation, reactivity, and deactivation of electrophilic intermediates, indicates that trichloroethylene has no genotoxic potential, but may exhibit borderline activity under extreme exposure conditions where the normally operational detoxication reactions are overpowered. This may be true for some *in vitro* systems, such as microsomal preparations, thereby explaining positive mutagenicity results, or for some hypersensitive species, given enzyme inducers. Further research should therefore scrutinize the model systems used and should focus on intact organisms, humans included, under realistic exposure conditions.

## F. TETRACHLOROETHYLENE

By analogy to other chlorinated ethylenes, tetrachloroethylene (**56**) has also been suggested to be oxidized by cytochrome *P*-450-catalyzed monooxygenases to an oxirane (**57**), which undergoes an intramolecular rearrangement, with migration of one chlorine atom as the only possible mechanism [reaction (23)].

$$Cl_2C=CCl_2 \rightarrow \underset{57}{\text{oxirane}} \rightarrow Cl_3C-COCl \rightarrow Cl_3C-COOH \quad (23)$$

**56**  **57**  **58**  **59** (**40**)

The resulting trichloroacetyl chloride (**58**) may be hydrolyzed to yield trichloroacetic acid (**59**). Support for this pathway is provided by several findings. No chlorinated metabolite other than trichloroacetic acid (**40**) is found.[3,84,85] Only one $^{36}$Cl atom is eliminated per metabolized molecule of [$^{36}$Cl]tetrachloroethylene (**56**) in the rat.[84] The thermal rearrangement of

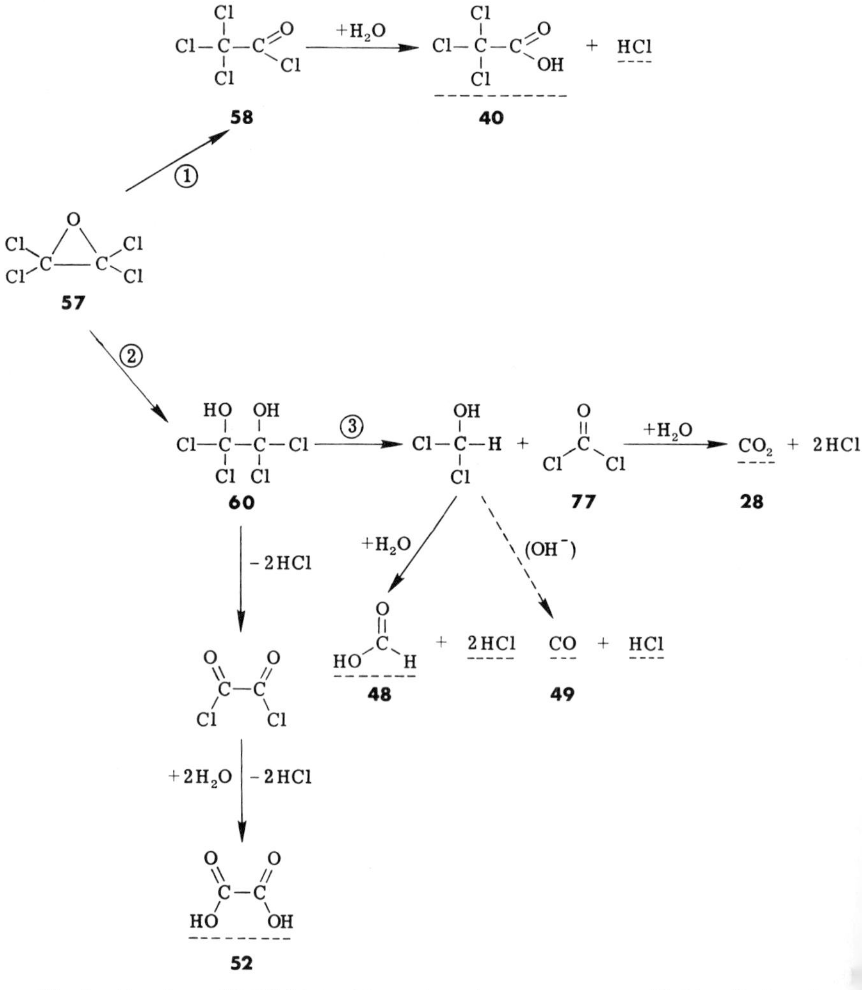

Fig. 4. Reactions of 2,2,3,3-tetrachlorooxirane in aqueous systems.[87] Identified products have a dashed line beneath them. Reaction pathway (1), intramolecular rearrangement with subsequent hydrolytic dechlorination; reaction pathway (2), hydrolytic diol formation with subsequent dehydrochlorination and hydrolytic dechlorination; reaction pathway (3), C···C fission with subsequent dehydrochlorination and hydrolytic dechlorination; carbon monoxide formation occurs only under alkaline conditions.

the synthetic 2,2,3,3-tetrachlorooxirane (**57**) to trichloroacetyl chloride (**58**) has been reported.[3,86] However, as with trichloroethylene (**39**), there is some additional dechlorination, which probably stems from the *vic*-diol (**60**) of 2,2,3,3-tetrachlorooxirane after hydrolysis. The hydrolytic decomposition of this oxirane in aqueous systems has also been investigated[87] and found to yield a variety of products (Fig. 4). From whole-animal studies, no evidence is available for the formation of one-carbon fragments ($CO$, $CO_2$, formic acid) from carbon–carbon bond fission, but oxalic acid (**52**) has been identified as an *in vivo* metabolite.[85,88] Additional investigations are needed to further our understanding of all mechanisms involved in tetrachloroethylene metabolism.

The covalent binding of [$^{14}C$]tetrachloroethylene (**56**) after metabolic activation has been described by several authors.[89-91] Part of this binding may be due to acylation from trichloroacetylchloride (**58**), because acid hydrolysis of tissue after metabolic transformation in the isolated rat liver preparation yields free trichloroacetic acid (**40**).[3] The significance of this type of binding for toxic reactions is not clear. No covalent binding to DNA could be demonstrated,[91] indicating a similar protective mechanism through forced rearrangement before the formed oxirane can gain access to essential macromolecules.

This hypothesis is in line with findings from mutagenicity and carcinogenicity testing. 2,2,3,3-Tetrachlorooxirane (**57**) is mutagenic in *S. typhimurium*,[80] whereas tetrachloroethylene (**56**), under the influence of activting enzyme systems, is not.[33,58] One carcinogenicity study with oral maxium-tolerated doses of a technical tetrachloroethylene sample revealed an increase in hepatocellular carcinomas in $B6C3F_1$ mice,[92] whereas an inhalation study in rats was negative.[93] Also, tetrachloroethylene did not induce preneoplastic liver foci in rats, but vinyl halides did.[89] Negative results were also reported from four short-term tumor induction models in mice.[30] In conclusion, there is at present overwhelming evidence that tetrachloroethylene is noncarcinogenic.

### III. Halogenated Allyl Compounds

As noted at the outset, allyl halides, although similar in their basic structure to vinyl halides, differ fundamentally in their reactivities and bioactivation mechanisms. The assumption that the epoxidation of the double bond and the oxirane thereby formed may be responsible for the alkylating and genotoxic potencies[94] of alkyl halides can no longer be accepted, because typical allyl halides may act as direct alkylating agents by $S_N1$ reactivity as in reaction (24), where X stands for the leaving group.

$$CH_2=CH-CH_2-X \xrightarrow{-X^-} [\overset{\frac{1}{2}+}{CH_2}=CH=\overset{\frac{1}{2}+}{CH_2}] \quad (24)$$

The formed allyl cation is stabilized by resonance; by $S_N2$ reactivity, and a particular form [see reaction (25)], $S_N2'$ reactivity in combination with a nucleophile (Nu);

$$Nu^- + {}^\frown CH_2=CH-CH_2-X \rightarrow [Nu-CH_2-CH=CH_2]^+X^- \quad (25)$$

or by radical mechanisms, as in reaction (26)

$$CH_2=CH-CH_2X \xrightarrow{-X\cdot} [CH_2=CH-CH_2 \rightleftharpoons \cdot CH_2-CH=CH_2] \quad (26)$$

where the radical is stabilized by resonance.

This direct alkylating property of allyl halides has long been overlooked. Present evidence from comparative experiments on the direct and indirect mutagenic activity in microbial systems and the direct alkylating potential indicates that direct alkylation plays a predominant if not exclusive role in the mutagenicity and carcinogenicity displayed by these compounds.[95-98] Systematic metabolism studies provide an opportunity to discriminate between the roles of direct alkylation and alkylation via epoxidation, because the reaction products with macromolecules will differ in their chemical structure.

The nature of the halogen leaving group should determine the strength of the alkylating effect. Indeed, the alkylating as well as the direct mutagenic potencies in *S. typhimurium* TA100 increase in the order: Cl < Br < I, reflecting the decreasing stability of the carbon–halogen bond.[95] A special case is seen in benzyl chloride (**61**), a well-known alkylating agent,[99] which can be taken as an allylic compound with a greatly enhanced stability due to resonance of the cation,[98] as shown in reaction (27). As expected from this,

(27)

## 11. HALOGENATED ALKENES AND ALKYNES

benzyl chloride is 60 times more mutgenic than allyl chloride (**62**),[96] whereas 1-chloro-2-cyclohexene (**63**) equals allyl chloride in alkylating as well as in direct mutagenic activity,[96] as their structural formulas show.

$$\overset{3}{C}H_2=\overset{2}{C}H-\overset{1}{C}H_2Cl \qquad \text{[cyclohexene]}-Cl$$

$$\textbf{62} \qquad \textbf{63}$$

Additional chlorine substitution at carbon atoms 1 and 3 greatly enhances alkylating potency, due to the dominance of the +M effect over the −I effect

$$:\ddot{C}l-CH=CH-CH_2-X \rightarrow \begin{bmatrix} Cl-CH=CH-CH_2^+ \\ \updownarrow \\ Cl-\overset{+}{C}H-CH=CH_2 \\ \updownarrow \\ \overset{+}{C}l=CH-CH=CH_2 \end{bmatrix} X^- \qquad (28)$$

[reaction (28)], whereas substitution in carbon atom 2 (−I effect only), as in 2,3-dichloro-1-propene (**64**), is much less active. *cis*-1,3-Dichloropropene (**65**) is the most active alkylating compound in the series tested, and exceeds the trans isomer (**66**) in activity because of steric hindrance and neighboring effects of the chlorine atoms, which favor cation (**67**) stabilization [see reaction (29)].[96]

$$\underset{\textbf{65}}{\overset{H}{\underset{:\ddot{C}l:}{\diagdown}}\!\!\!C\!=\!C\!\!\!\overset{H}{\underset{\overset{|}{:\ddot{C}l:}}{\diagup}}} \rightarrow \begin{bmatrix} H\diagdown\overset{H}{\underset{Cl}{C}}\diagup H \\ C\overset{\oplus}{}C \\ \diagup \quad \diagdown \\ Cl \quad H \end{bmatrix} \underset{\textbf{67}}{} \quad Cl^- \not\!\!\!\!\leftarrow \underset{H}{\overset{Cl}{\diagdown}}C\!=\!C\!\!\!\overset{H}{\underset{\overset{|}{Cl}}{\diagup}} \underset{\textbf{66}}{} \qquad (29)$$

Alkyl substitution also increases, through the agency of +I and +M effects, reactivity by increasing instability of the carbon–halogen bond and cation by stabilizing through resonance, as exemplified in

$$\begin{matrix} CH_3-CH=CH-CH_2Cl \\ \\ CH_2=CH-CH-CH_3 \\ | \\ Cl \end{matrix} \Bigg\rangle \Big[ \overset{\frac{1}{2}+}{C}H_2 = CH = \overset{\frac{1}{2}+}{C}H - CH_3 \Big] Cl^- \qquad (30)$$

reaction (30). Therefore, the monochlorobutenes, which are of great practical significance, display extraordinary alkylating and direct mutagenic potencies. Molecules with halogen substitution in nonallylic positions, such as 1-chloro-1-propene (ClCH=CH—CH$_3$) **(68)** or 2-chloro-2-butene (CH$_3$—CH=CCl—CH$_3$) **(69)**,[98] lack any alkylating and direct genotoxic potential.

In most allyl halides, good quantitative correlation between alkylating and direct mutagenic potencies is found (Fig. 5). Little, if any, metabolic activation through epoxidation of the double bond occurs in mammalian organisms. In the rat, most of the radioactivity from $^{14}$C-labeled allyl bromide is found in the DNA of different organs in the form of allyl adducts, such as $O^6$-allylguanine, 7-allylguanine, $N^2$-allylguanine, 3-allyladenine, and $N^6$-allyladenine; no binding has been detected in the form of adducts from the epoxides.[100,101] These findings necessitate a revision of former concepts,[102]

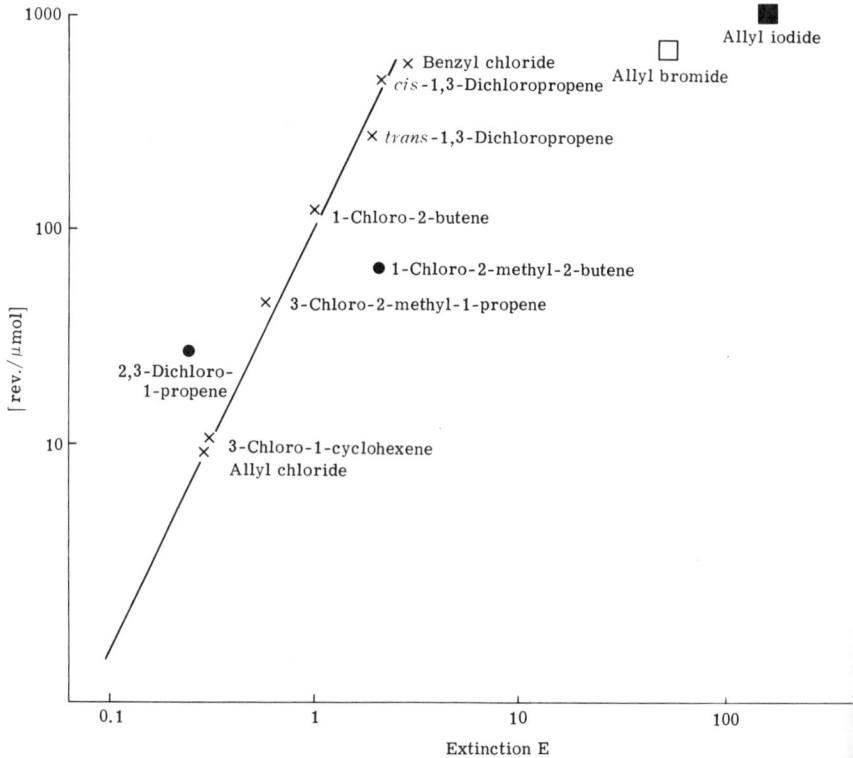

**Fig. 5.** Correleation between alkylating [abscissa: extinction at 560 nm after reaction with 4-($p$-nitrobenzyl)pyridine] and mutagenic potency (*S. typhimurium* TA100, without metabolizing enzyme systems) in allyl halides. Aberrations with allyl bromide and allyl iodide are due to instability or nonspecific reaction of the compounds with constituents of the microbial testing media, or both.[97]

## 11. HALOGENATED ALKENES AND ALKYNES

wherein epoxidation was the only or the predominant activation mechanism for halogenated alkenes. 2,3-Dichloro-1-propene (**70**) is an exception. This compound is more mutagenic in *S. typhimurium* test systems in the presence of an S-9 activating liver enzyme fraction than in the absence of the S-9 fraction. This can be explained by the formation of an epoxide (**71**), which rearranges to the extremely reactive molecule 1,3-di-chloroacetone (**72**) or, after hydrolysis, to the similarly reactive 1-chloro-3-hydroxyacetone (**73**)[98]. This sequence is shown in reaction (31).

$$
\begin{array}{c}
\text{CH}_2=\text{C}\begin{array}{c}\text{CH}_2\text{Cl}\\\text{Cl}\end{array} \rightarrow \text{H}_2\text{C}\underset{:\text{Cl}:}{\overset{\text{O}}{\triangle}}\text{C}\begin{array}{c}\text{CH}_2\text{Cl}\end{array} \xrightarrow[\text{hydrolysis}]{\text{rearrangement}} \begin{array}{c}\text{Cl}-\text{CH}_2-\overset{\text{O}}{\underset{\|}{\text{C}}}-\text{CH}_2\text{Cl}\\\textbf{72}\\\\\text{HO}-\text{CH}_2-\overset{\text{OH}}{\underset{\text{Cl}}{\text{C}}}-\text{CH}_2\text{Cl}\\\downarrow -\text{HCl}\\\text{HO}-\text{CH}_2-\overset{\text{O}}{\underset{\|}{\text{C}}}-\text{CH}_2\text{Cl}\\\textbf{73}\end{array}
\end{array} \quad (31)
$$

**70**      **71**

## IV. Halogenated Alkynes

Halogenated ethynes (ClC≡CCl, **74**; BrC≡CBr, **75**; IC≡CI, **76**) are common agents in organic chemistry for the introduction of the ethyne moiety into organic molecules. The underlying radical mechanisms are well understood by the electron deprivation in the triple-bond system, which destabilizes the carbon–halogen bonds. These extremely unstable compounds decompose rapidly in air through the electrophilic attack of oxygen [reaction (32)].

$$\text{Cl}-\text{C}\equiv\text{C}-\text{Cl} \xrightarrow{O_2} \text{CO} + \text{COCl}^2 \qquad (32)$$

**74**      **77**

They may be stabilized, however, by the presence of suitable concentrations of acetylene, trichloroethylene, or tetrachloroethylene, and thus may cause intoxications.

The toxicological significance of halogenated ethynes arises mainly from their formation as by-products in the synthesis of haloolefins or, more important, through alkaline decomposition of haloalkanes or haloalkenes [reactions (33) and (34)].

$$\begin{array}{c}\text{Cl}\diagdown\!\!\!\!\!\diagup\text{Cl}\\ \text{C}\!=\!\text{C}\\ \text{H}\diagup\!\!\!\!\!\diagdown\text{Cl}\\ \textbf{39}\end{array} \xrightarrow{-HCl} \text{Cl}-\text{C}\!\equiv\!\text{C}-\text{Cl} \quad (33)$$

$$\begin{array}{c}\text{Cl  Cl}\\ |\ \ |\\ \text{H}-\text{C}-\text{C}-\text{H}\\ |\ \ |\\ \text{Cl  Cl}\end{array} \xrightarrow{-2HCl} \textbf{74}$$

$$\begin{array}{c}\text{Cl  Cl}\\ |\ \ |\\ \text{Cl}-\text{C}-\text{C}-\text{H}\\ |\ \ |\\ \text{H  H}\end{array} \xrightarrow{-2HCl} \text{Cl}-\text{C}\!\equiv\!\text{C}-\text{H} \quad (34)$$

$$\begin{array}{c}\text{Cl  H}\\ |\ \ |\\ \text{Cl}-\text{C}-\text{C}-\text{H}\\ |\ \ |\\ \text{Cl  H}\end{array} \xrightarrow{-2HCl} \textbf{75}$$

No metabolism studies have been carried out so far with halogenated alkynes. From toxicological evidence, however, bioactivation processes seem likely. The highly reactive dichloroacetylene does not damage lung tissue after inhalation, but does produce kidney, liver, and central and peripheral nervous system injury. Cancer may be a further consequence.

The findings of Reichert and co-workers on the chemical behavior of dichloroacetylene (**74**) shed some light on the possible bioactivation mechanisms. After controlled reaction with oxygen, a variety of decomposition products could be identified, the formation of which can only be explained by radical mechanisms (Fig. 6).[102a] Among these, phosgene (**77**), trichloroacetyl chloride (**58**), hexachlorobutadiene (**78**), and trichloroacryloyl chloride (**79**) are compounds of high toxic potential. One common toxicological feature is seen with dichloroacetylene, hexachlorobutadiene, and chloroform (**80**): they share the capability of inducing renal tumors in rats and mice.[103-105] In addition, phosgene (**77**) has been identified as a metabolic intermediate of chloroform (**80**).[106] The similarity of the toxicological pattern of these compounds suggests common bioactivation pathways; these, however, have not been evaluated.

In humans, dichloroacetylene (**74**) induces irreversible degeneration of central nerves, predominantly the trigeminal nerve[107]; in mice and rats, it causes nephrotoxic and hepatotoxic effects.[108,109] It is also highly carcinogenic in mice and rats.[103] The compound displays direct mutagenic activity in *S. typhimurium* TA100.[110]

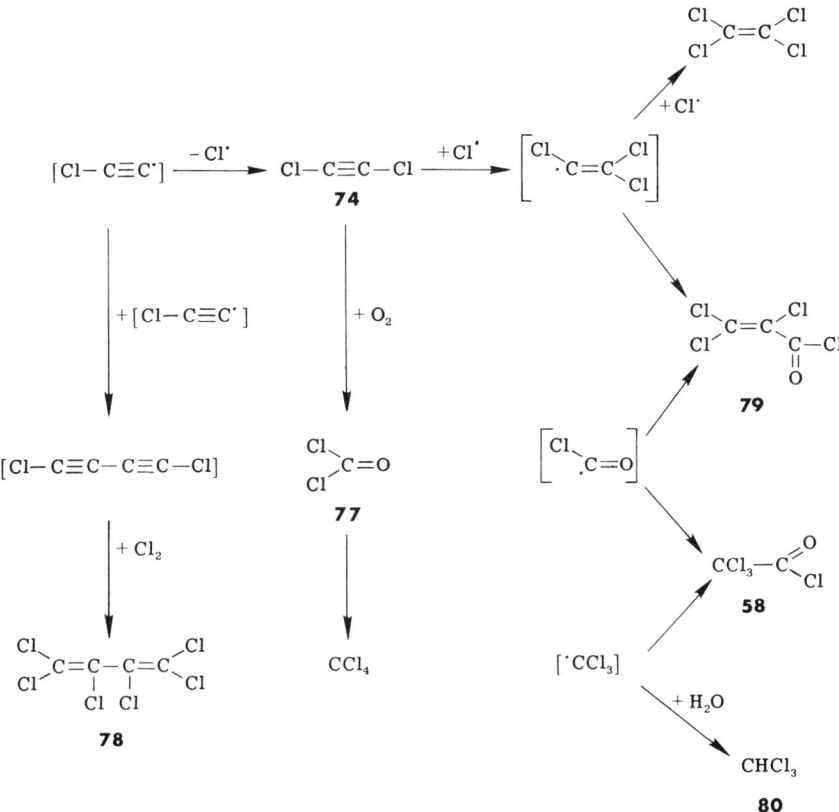

**Fig. 6.** Scheme of decomposition of dichloroacetylene (74) in the presence of oxygen. Hypothetical intermediates are shown in brackets.[102a]

## Acknowledgments

I am grateful to Mrs. V. Schneider for careful editorial work, Mrs. B. Hasenmüller for technical assistance, and Drs. Dekant, Eder, and Reichert for helpful discussions of scientific aspects.

## References

1. Bonse, G., and Henschler, D. (1976). Chemical reactivity, biotransformation, and toxicity of polychlorinated aliphatic compounds. *CRC Crit. Rev. Toxicol.* **5**, 395.
2. Powell, J. F. (1945). Trichloroethylene: Absorption, elimination and metabolism. *Br. J. Ind. Med.* **2**, 142.

3. Bonse, G., Urban, T., Reichert, D., and Henschler, D. (1975). Chemical reactivity metabolic oxirane formation and biological reactivity of chlorinated ethylene in the isolated perfused liver preparation. *Biochem. Pharmacol.* **24,** 1829.
4. Williamson, D. G., and Cvetanovic, R. J. (1968). Rates of reactions of ozone with chlorinated and conjugated olefins. *J. Am. Chem. Soc.* **90,** 4248.
5. Rannug, U., Johansson, A., Ramel, C., and Wachtmeister, C. A. (1974). The mutagenicity of vinyl chloride after metabolic activation. *Ambio* **3,** 194.
6. Maltoni, C., and Lefemine, G. (1974). Carcinogenicity bioassays of vinyl chloride. I. Research plan and early results. *Environ. Res.* **7,** 387.
7. Green, T., and Hathway, D. E. (1975). The biological fate in rats of vinyl chloride in relation to its oncogenicity. *Chem.-Biol. Interact.* **11,** 545.
8. Hefner, R. E., Jr., Watanabe, P. G., and Gehring, P. J. (1975). Preliminary studies of the fate of inhaled vinyl chloride monomer (VCM) in rats. *Ann. N.Y. Acad. Sci.* **246,** 135.
9. Bolt, H. M., Kappus, H., Buchter, A., and Bolt, W. (1976). Disposition of 1,2-[$^{14}$C]vinyl chloride in the rat. *Arch. Toxicol.* **35,** 153.
10. Guengerich, F. P., Crawford, W. M., Jr., and Watanabe, P. G. (1979). Activation of vinyl chloride to covalently bound metabolites: Roles of 2-chloroethylene oxide and 2-chloroacetaldehyde. *Biochemistry* **18,** 5177.
11. Yllner, S. (1971). Metabolism of chloroacetate-1-[$^{14}$C] in the mouse. *Acta Pharmacol. Toxicol.* **30,** 69.
12. Guengerich, F. P., and Strickland, T. W. (1977). Metabolism of vinyl chloride: Destruction of the heme of highly purified liver microsomal cytochrome P-450 by a metabolite. *Mol. Pharmacol.* **13,** 993.
13. Guengerich, F. P., Mason, P. S., Stott, W. S., Fox, T. R., and Watanabe, P. G. (1981). Irreversible binding of metabolites of vinyl halides to protein and DNA: Roles of 2-haloethylene oxides and 2-haloacetaldehydes derived from vinyl bromide and vinyl chloride. *Cancer Res.* **41,** 4391–4398.
14. Guengerich, F. P. (1982). Metabolism of vinyl halides: In vitro studies on roles of potential activated metabolites. *Adv. Exp. Med. Biol.* **136A–136B,** 685.
15. Bolt, H. M., Filser, J. G., and Laib, R. J. (1982). Covalent binding of haloethylenes. *Adv. Exp. Med. Biol.* **136A–136B,** 667.
16. Green, T., and Hathway, D. E. (1978). Interactions of vinyl chloride with rat liver DNA in vivo. *Chem.-Biol. Interact.* **22,** 211.
17. Laib, R. J., and Bolt, H. M. (1977). Alkylation of RNA by vinyl chloride metabolites in vitro and in vivo: Formation of 1,$N^6$-ethenoadenosine. *Toxicology* **8,** 185.
18. Laib, R. J., and Bolt, H. M. (1978). Formation of 3,$N^4$-ethenocytidine moieties in RNA by vinyl chloride metabolites in vitro and in vivo. *Arch. Toxicol.* **39,** 235.
19. Doerjer, G., and Oesch, F. (1981). Isolation of guanine derivatives in the reaction of chloroacetaldehyde with DNA in vitro. *J. Cancer Res. Clin. Oncol.* **99,** AZ4.
20. Osterman-Golkar, S., Hultmark, D., Segerbäck, D., Calleman, C. J., Göthe, R., and Ehrenberg, C. A. (1977). Alkylation of DNA and proteins in mice exposed in vinyl chloride. *Biochem. Biophys. Res. Commun.* **63,** 259.
21. Laib, R. J., Gwinner, L. M., and Bolt, H. M. (1981). DNA alkylation by vinyl chloride metabolites: Etheno derivatives or 7-alkylation of guanine. *Chem.-Biol. Interact.* **37,** 219.
22. Scherer, E., van der Laken, C. J., Gwinner, L. M., Laib, R. J., and Emmelot, P. (1981). Modification of deoxyguanosine by chloroethylene oxide. *Carcinogenesis (N.Y.)* **2,** 671.
23. Kappus, H., Bolt, H. M., Buchter, A., and Bolt, W. (1975). Rat liver microsomes catalyze covalent binding of [$^{14}$C]vinyl chloride to macromolecules. *Nature (London)* **257,** 134.

24. Hathway, D. E. (1981). Mechanisms of vinyl chloride carcinogenicity/mutagenicity. *Br. J. Cancer* **44**, 597.
25. Göthe, R., Calleman, C. J., Ehrenberg, L., and Wachtmeister, C. A. (1974). Trapping with 3,4-dichlorobenzenethiol of reactive metabolites formed in vitro from the carcinogen vinyl chloride. *Ambio* **3**, 234.
26. Barbin, A., Brésil, H., Croisy, A., Jacquignon, P., Malaveille, C., Montesano, R., and Bartsch, H. (1975). Liver microsome-mediated formation of alkylating agents from vinyl bromide and vinyl chloride. *Biochem. Biophys. Res. Commun.* **67**, 596.
27. Rannug, U., Göthe, R., and Wachtmeister, C. A. (1976). The mutagenicity of chloroethylene oxide, chloroacetaldehyde, 2-chloroethanol and chloroacetic acid, conceivable metabolites of vinyl chloride. *Chem.-Biol. Interact.* **12**, 251.
28. Huberman, E., Bartsch, H., and Sachs, L. (1975). Mutation induction in Chinese hamster V79 cells by two vinyl chloride metabolites, chloroethylene oxide and chloroacetaldehyde. *Int. J. Cancer* **16**, 539.
29. Zajdela, F., Croisy, A., Barbin, A., Malaveille, C., Tomatis, L., and Bartsch, H. (1980). Carcinogenicity of chloroethylene oxide, an ultimate reactive metabolite of vinyl chloride, and bis(chloromethyl)ether after subcutaneous administration and in initiation–promotion experiments in mice. *Cancer Res.* **40**, 352.
30. van Duuren, B. L., Goldschmidt, B. M., Loewengart, G., Smith, A. C., Melchionne, S., Seidman, I., and Roth, D. (1979). Carcinogenicity of halogenated olefinic and aliphatic hydrocarbons in mice. *JNCI, J. Natl. Cancer Inst.* **63**, 1433.
31. Bolt, H. M., Filser, J. G., and Hinderer, R. K. (1978). Rat liver microsomal uptake and irreversible protein binding of 1,2-[$^{14}$C]vinyl bromide. *Toxicol. Appl. Pharmacol.* **44**, 481.
32. Ottenwälder, H., Laib, R. J., Filser, J. G., and Bolt, H. M. (1979). Disposition and metabolic activation of vinyl bromide in rats. *Adv. Pharmacol. Ther., Proc. Int. Congr. Pharmacol., 7th, 1978* Abstract 719.
33. Bartsch, H., Malaveille, C., Barbin, A., and Planche, G. (1979). Mutagenic and alkylating metabolites of halo-ethylenes, chlorobutadienes and dichlorobutenes produced by rodent or human liver tissues. *Arch. Toxicol.* **41**, 249.
34. Huntingdon Research Center (1978). "18-Month Sacrifice Report (Vinyl Bromide)," HRC Proj. 7511-253. Huntingdon Res. Cent., New York.
35. Bolt, H. M., Laib, R. J., and Klein, K. P. (1981). Formation of preneoplastic hepatocellular foci by vinyl fluoride in newborn rats. *Arch. Toxicol.* **47**, 71.
36. Bolt, H. M., Laib, R. J., and Filser, J. G. (1982). Reactive metabolites and carcinogenicity of halogenated ethylenes. *Biochem. Pharmacol.* **31**, 1.
37. Filser, J. G., and Bolt, H. M. (1981). Inhalation pharmacokinetics based on gas uptake studies. I. Improvement of kinetic models. *Arch. Toxicol.* **47**, 279.
38. Furman, E. G., and Meleshevich, A. P. (1977). Study of the effect of the nature of a substituent on the electron state of an epoxy ring by the CNDO/2 method. *Teor. Eksp. Khim.* **13**, 328; quoted in *Chem. Abstr.* **87**, 167379.
39. Leibman, K. C., and Ortiz, E. (1977). Metabolism of halogenated ethylenes. *Environ. Health Perspect.* **21**, 91.
40. Jones, B. K., and Hathway, D. E. (1978). The biological fate of vinylidene chloride in rats. *Chem.-Biol. Interact.* **20**, 27.
41. McKenna, M. I., Zempel, J. A., Madrid, E. O., and Braun, W. H. (1978). Metabolism and pharmacokinetic profile of vinylidene chloride in rats following oral administration. *Toxicol. Appl. Pharmacol.* **45**, 821.
42. Reichert, D., Werner, H. W., Metzler, M., and Henschler, D. (1979). Molecular mechanism of 1,1-dichloroethylene toxicity: Excreted metabolites reveal different pathways of reactive intermediates. *Arch. Toxicol.* **42**, 159.

43. Costa, A. K., and Ivanetich, K. M. (1982). Vinylidene chloride: Its metabolism by hepatic microsomal cytochrome $P$-450 in vitro. *Biochem. Pharmacol.* **31**, 2083.
44. Henschler, D., Hoos, W. R., Fetz, H., Dallmeier, E., and Metzler, M. (1979). Reactions of trichloroethylene epoxide in aqueous systems. *Biochem. Pharmacol.* **28**, 543.
45. McKenna, M. I., Watanabe, P. G., and Gehring, P. J. (1977). Pharmacokinetics of vinylidene chloride in the rat. *Environ. Health Perspect.* **21**, 99.
46. Reitz, R. H., Watanabe, P. G., McKenna, M. J., Quast, J. F., and Gehring, P. J. (1980). Effects of vinylidene chloride on DNA synthesis and DNA repair in the rat and mouse: A comparative study with dimethylnitrosamine. *Toxicol. Appl. Pharmacol.* **52**, 357.
47. Hathway, D. E. (1977). Comparative mammalian metabolism of vinyl chloride and vinylidene chloride in relation to oncogenic potential. *Environ. Health Perspect.* **21**, 55.
48. Henschler, D., Bonse, G., and Greim, H. (1976). Carcinogenic potential of chlorinated ethylenes—tentative molecular rules. *Colloq.—Inst. Natl. Sante Rech. Med.* **52**, 171.
49. Bartsch, H., Malaveille, C., Montesano, R., and Tomatis, L. (1975). Tissue mediated mutagenicity of vinylidene chloride and 2-chloroprene in Salmonella typhimurium. *Nature (London)* **255**, 641.
50. Lee, C. C., Bhanderi, J. C., Winston, J. M., and House, W. B. (1978). Carcinogenicity of vinyl chloride and vinylidene chloride. *J. Toxicol. Environ. Health* **4**, 15.
51. Maltoni, C., and Tovoli, D. (1979). First experimental evidence of the carcinogenic effects of vinylidene chloride. *Med. Lav.* **70**, 363.
52. Stöckle, G., Laib, R. J., Filser, J. G., and Bolt, H. M. (1979). Vinylidene fluoride: Metabolism and induction of preneoplastic hepatic foci in relation to vinyl chloride. *Toxicol. Lett.* **3**, 337.
53. Griesbaum, G., Kibar, R., and Pfeffer, B. (1975). Synthese und Stabilität von 2,3-Dichloroxiranen. *Liebigs Ann. Chem.* No. 2, p. 214.
54. Costa, A. K., and Ivanetich, K. M. (1982). The 1,2-dichloroethylenes: Their metabolism by hepatic cytochrome $P$-450 in vitro. *Biochem. Pharmacol.* **31**, 2093.
55. McDonald, R. N., and Schwab, P. A. (1963). Molecular rearrangements. II. Chlorine migration in the epoxide–carbonyl rearrangement. *J. Am. Chem. Soc.* **85**, 4004.
56. Filser, J. G., and Bolt, H. M. (1979). Pharmacokinetics of halogenated ethylenes in rats. *Arch. Toxicol.* **42**, 123.
57. Freundt, K. J., and Macholz, J. (1978). Inhibition of mixed-function oxidases in rat liver by *trans*- and *cis*-1,2-dichloroethylene. *Toxicology* **10**, 131.
58. Greim, H., Bonse, G., Radwan, Z., Reichert, D., and Henschler, D. (1975). Mutagenicity in vitro and potential carcinogenicity of chlorinated ethylenes as a function of metabolic oxirane formation. *Biochem. Pharmacol.* **24**, 2013.
59. Leibman, K. C. (1965). Metabolism of trichloroethylene in liver microsomes. I. Characteristics of the reaction. *Mol. Pharmacol.* **1**, 239.
60. Byington, K. H., and Leibman, K. C. (1965). Metabolism of trichloroethylene (Tri) in liver microsomes. II. Identification of the reaction products as chloral hydrate. *Mol. Pharmacol.* **1**, 247.
61. Leibman, K. C., and Allister, W. J. (1967). Metabolism of trichloroethylene in liver microsomes. III. Induction of the enzymatic activity and its effect on excretion of metabolites. *J. Pharmacol. Exp. Ther.* **157**, 574.
62. Uehleke, H., Tabarelli-Poplawski, S., Bonse, G., and Henschler, D. (1977). Spectral evidence for 2,2,3-trichlorooxirane formation during microsomal trichloroethylene oxidation. *Arch. Toxicol.* **37**, 95.
62a. Miller, R. E., and Guengerich, F. P. (1983). Metabolism of trichloroethylene in isolated hepatocytes, microsomes, and reconstituted enzyme systems containing cytochrome $P$-450. *Cancer Res.* **43**, 1145–1152.

63. Poluektov, V. A., and Mekhryushev, Y. Y. (1973). Kinetics of the thermal isomerisation of trichloroepoxyethane. *Russ. J. Phys. Chem. (Engl. Transl.)* **47**, 959.
64. Oesterreicher, F. (1967). Ph.D. Thesis, Wien.
65. Henschler, D. (1977). Metabolism and mutagenicity of halogenated olefins — a comparison of structure and activity. *Environ. Health Perspect.* **21**, 61.
66. Hathway, D. E. (1980). Consideration of the evidence for mechanisms of 1,1,2-trichloroethylene metabolism, including new identification of its dichloroacetic acid and trichloroacetic metabolites in mice. *Cancer Lett.* **8**, 263.
67. Henschler, D., Bonse, G., and Dekant, W. (1982). Mechanisms of formation and reactions of electrophilic intermediates of halogenated olefins. (1983). *Proc. Int. Cancer Congr., 13th,* **1**, 175–183.
68. Dekant, W., and Henschler, D. (1982). Dechlorination reactions in course of the metabolism of trichloroethylene. *Naunyn-Schmiedebergs Arch. Pharmacol.* **319**, Suppl. R64.
69. Bolt, H. M., and Filser, J. G. (1977). Irreversible binding of chlorinated ethylenes to macromolecules. *Environ. Health Perspect.* **21**, 167.
70. Banerjee, S., and van Duuren, B. L. (1978). Covalent binding of the carcinogen trichloroethylene to hepatic microsomal proteins and to exogenous DNA in vitro. *Cancer Res.* **38**, 776.
71. Uehleke, H., and Poplawski-Tabarelli, S. (1977). Irreversible binding of [14]C-labelled trichloroethylene to mice liver constituents in vivo and in vitro. *Arch. Toxicol.* **37**, 289.
72. Pessayre, D., Allemand, H., Wandscheer, J. C., Descatoire, V. Artigon, J.-Y., and Benhamou, J.-P. (1979). Inhibition, activation, destruction, and induction of drug-metabolizing enzymes by trichloroethylene. *Toxicol. Appl. Pharmacol.* **49**, 355.
73. Laib, R. J., Stöckle, G., Bolt, H. M., and Kunz, W. (1979). Vinyl chloride and trichloroethylene: Comparison of alkylating effects of metabolites and induction of preneoplastic enzyme deficiencies in rat liver. *J. Cancer Res. Clin. Oncol.* **94**, 139.
74. Kline, S. A., and van Duuren, B. L. (1977). Reactions of epoxy-1,1,2-trichloroethane with nucleophiles. *J. Heterocycl. Chem.* **14**, 455.
75. Stott, W. T., Quast, J. F., and Watanabe, P. G. (1982). The pharmacokinetics and molecular interactions of trichloroethylene in mice and rats. *Toxicol. Appl. Pharmacol.* **62**, 137.
76. Parchman, L. G., and Magee, P. (1980). Production of [14]CO$_2$ from trichloroethylene in rats and mice and a possible interaction of a trichloroethylene metabolite with DNA. *19th Society of Toxicology Meet. 1980* Abstracts, No. 153.
77. Magee, P. N., Chu, C. K., Gombar, C. T., Jensen, D. E., and Parchman, L. G. (1982). Interaction of reaction intermediates with DNA. *Adv. Exp. Med. Biol.* **136A–136B**, 1335.
78. Bergman, K. (1983). Interactions of trichloroethylene with DNA in vitro and with RNA and DNA of various mouse tissues in vivo. *Arch. Toxicol.* **54**, 181–194.
79. Henschler, D., Eder, E., Neudecker, T., and Metzler, M. (1977). Carcinogenicity of trichloroethylene: Fact or artifact? *Arch. Toxicol.* **37**, 233.
80. Kline, S. A., McCoy, E. C., Rosenkranz, H. S., and van Duuren, B. L. (1982). Mutagenicity of chloroalkene epoxides in bacterial systems. *Mutat. Res.* **101**, 115.
81. U.S. Dept. of Health, Education and Welfare (1976). "NCI Carcinogenesis Bioassay of Trichloroethylene, Cas No. 79-01-6, NCI Tech. Rep. Ser. No. 2, DHEW Publ. No. 76-802. USDHEW, Washington, D.C.
82. Henschler, D., Romen, W., Elsässer, H. M., Reichert, D., and Radwan, Z. (1980). Carcinogenicity study of trichloroethylene by longterm inhalation in three animal species. *Arch. Toxicol.* **43**, 237.
83. Henschler, D., Elsässer, H. M., Romen, W., and Eder, E. (1984). Carcinogenicity study

of trichloroethylene, with and without epoxide stabilisers, in mice. *J. Cancer Res. Clin. Oncol.* **107**, 149–156.
84. Daniel, J. W. (1963). The metabolism of $^{36}$Cl-labelled trichloroethylene and tetrachloroethylene in the rat. *Biochem. Pharmacol.* **12**, 795.
85. Yllner, S. (1961). Urinary metabolites of [$^{14}$C]tetrachloroethylene in mice. *Nature (London)* **191**, 820.
86. Frankel, D. M., Johnson, C. E., and Pitt, H. M. (1957). Preparation and properties of tetrachloroethylene oxide. *J. Org. Chem.* **22**, 1119.
87. Henschler, D., and Hoos, W. R. (1982). Metabolic activation of di-, tri- and tetrachloroethylenes. *Adv. Exp. Med. Biol.* **136A–136B**, 659–666.
88. Pegg, D. G., Zempel, J. A., Braun, W. H., and Watanabe, P. G. (1979). Disposition of tetrachloro [$^{14}$C] ethylene following oral and inhalation exposure in rats. *Toxicol. Appl. Pharmacol.* **51**, 465.
89. Bolt, H. M., and Link, B. (1980). Zur Toxikologie von Perchloräthylen. *Verh. Dtsch. Ges. Arbeitsmed.* **20**, 463.
90. Costa, A. K., and Ivanetich, K. M. (1980). Tetrachloroethylene metabolism by the hepatic microsomal cytochrome *P*-450 system. *Biochem. Pharmacol.* **29**, 2863.
91. Schumann, A. M., Quast, J. F., and Watanabe, P. G. (1980). The pharmacokinetics and macromolecular interactions of perchloroethylene in mice and rats as related to oncogenicity. *Toxicol. Appl. Pharmacol.* **55**, 207.
92. U.S. Dept. of Health, Education and Welfare (1977). "NCI Bioassay of Tetrachloroethylene for Possible Carcinogenicity," NCI-GG-TR-13, DHEW Publ. NIH-77-813. USDHEW, Washington, D.C.
93. Rampy, L. W., Quast, J. F., Leong, B. K. J., and Gehring, P. J. (1978). Results of longterm inhalation toxicity studies on rats of 1,1,1-trichloroethane and perchloroethylene formulations. *Proc. Int. Congr. Toxicol., 1st, 1977* Abstract, p. 27.
94. van Duuren, B. L., Goldschmidt, B. M., and Seidman, J. (1975). Carcinogenic activity of di- and trifunctional α-chloroethers and of 1,4-dichlorobutene-2 in ICR/ha Swiss mice. *Cancer Res.* **35**, 2553.
95. Eder, E., Neudecker, T., Lutz, D., and Henschler, D. (1980). Mutagenic potential of allyl and allylic compounds. Structure–activity relationship as determined by alkylating and direct in vitro mutagenic properties. *Biochem. Pharmacol.* **29**, 993.
96. Neudecker, T., Lutz, D., Eder, E., and Henschler, D. (1980). Structure–activity relationship in halogen and alkyl-substituted allyl and allylic compounds: Correlation of alkylating and mutagenic properties. *Biochem. Pharmacol.* **29**, 2611.
97. Eder, E., Neudecker, T., Lutz, D., and Henschler, D. (1982). Correlation of alkylating and mutagenic activities of allyl and allylic compounds: Standard alkylation test vs. kinetic investigation. *Chem.-Biol. Interact.* **38**, 303.
98. Eder, E., Henschler, D., and Neudecker, T. (1982). Mutagenic proprties of allylic and α,β-unsaturated compounds: Consideration of alkylating mechanisms. *Xenobiotica* **12**, 831–848.
99. Preussmann, R., Schneider, H., and Epple, F. (1969). Untersuchungen zum Nachweis alkylierender Agentien. *Arzneim.-Forsch.* **19**, 1059.
100. Eder, E., Lutz, D., and Henschler, D. (1981). In vitro alkylation of mutagenic allyl compounds. *Naunyn-Schmiedeberg's Arch. Pharmacol.* **316**, Suppl., 51.
101. Eder, E., and Sebeikat, D. (1982). In vivo alkylation of DNA by allyl bromide. *Naunyn-Schmiedeberg's Arch. Pharmacol.* **319**, Suppl., 80.
102. van Duuren, B. L. (1977). Chemical structure, reactivity, and carcinogenicity of halocarbons. *Environ. Health Perspect.* **21**, 17.
102a. Reichert, D., Metzler, M., and Henschler, D. (1980). Decomposition of the neuro- an

nephrotoxic compound dichloroacetylene in the presence of oxygen: Separation and identification of novel products. *J. Environ. Pathol. Toxicol.* **4,** 525–532.
103. Reichert, D., Spengler, U., Romen, W., and Henschler, D. (1980). Carcinogenic potential of dichloroacetylene. *In* "Mechanisms of Toxicity and Hazard Evaluation" (B. Holmstedt *et al.,* eds.), p. 269. Elsevier North-Holland Biomedical Press, Amsterdam.
104. Kociba, R. J., Keyes, D. G., Jersey, G. C., Ballard, J. J., Dittenber, D. A., Quast, J. F., Wade, C. E., Humiston, C. G., and Schwetz, B. A. (1977). Results of a two year chronic toxicity study with hexachlorobutadiene in rats. *Am. Ind. Hgy. Assoc. J.* **38,** 589.
105. International Agency for Research on Cancer (1979). NCI bioassay program, report on carcinogenicity of chloroform, Bethesda, Maryland. *IARC Monogr. Eval. Carcinog. Risk Chem. Hum.* **20,** 401–427.
106. Cresteil, T., Beane, P., Levaux, J. P., and Mansuy, D. (1979). Biotransformation of chloroform by rat and human liver microsomes; in vitro effect on some enzyme activities and mechanism of irreversible binding to macromolecules. *Chem.-Biol. Interact.* **24,** 153.
107. Henschler, D., Broser, F., and Hopf, H. C. (1970). "Polyneuritis cranialis" durch Vergiftung mit chlorierten Acetylenen beim Umgang mit Vinylidenchlorid-Copolymeren. *Arch. Toxicol.* **26,** 62.
108. Reichert, D., Henschler, D., and Bannasch, P. (1978). Nephrotoxic and hepatotoxic effects of dichloroacetylene. *Food Cosmet. Toxicol.* **16,** 227.
109. Reichert, D., Liebaldt, G., and Henschler, D. (1976). Neurotoxic effects of dichloroacetylene. *Arch. Toxicol.* **37,** 23.
110. Reichert, D., Neudecker, T., Spengler, U., and Henschler, D. (1983). The mutagenicity of dichloroacetylene and its degradation products 3-chloroacetyl chloride, 3-chloroacryloyl chloride and hexachlorobutadiene. *Mutat. Res.* **117,** 21–29.

Chapter 12

# Arylamines and Arylamides: Oxidation Mechanisms

SIDNEY D. NELSON
*Department of Medicinal Chemistry*
*University of Washington*
*Seattle, Washington*

|  |  |
|---|---|
| I. Introduction. | 349 |
| II. Chemical Properties of Arylamines and Arylamides. | 350 |
| III. Reaction Mechanisms of N-Oxidation for Arylamines and Arylamides. | 352 |
|     A. Oxidation with N–O Bond Formation | 352 |
|     B. Oxidation without N–O Bond Formation | 357 |
| IV. Fate of Arylamine and Arylamide Oxidation Products. | 363 |
|     A. Further Oxidation. | 363 |
|     B. Reduction | 364 |
|     C. Rearrangement, Condensation, and Hydrolysis | 364 |
| V. Summary. | 366 |
|     References | 366 |

## I. Introduction

Arylamines and their amide derivatives form a chemical class of compounds that are extensively used as drugs, cosmetics, dyestuffs, pesticides, and synthetic intermediates. Particular attention is paid to reaction mechanisms involved in the bioactivation of this class of compounds inasmuch as several aromatic amines and amides are known or suspected human mutagens, carcinogens, or directly acting cytotoxins.[1-3]

Reports of a high incidence of bladder cancer in the dye industry in the late nineteenth century led to studies by Hueper and associates[4] that implicated 2-naphthylamine as a causative agent. Soon thereafter, the acetyl amide derived from 2-fluorenamine (2-FA), 2-fluorenylacetamide (2-FAA), was found to induce tumors in rats, but not at their site of application.[5] This report prompted a multitude of studies, many of which are still ongoing, on the mechanisms by which 2-FAA causes tumor development. A key discovery was the finding that N-hydroxylation is the initial metabolic reaction required for tumorigenesis.[6,7]

In developments that chronologically paralleled those on arylamine- and arylamide-induced carcinogenesis, it was found that aniline[8] and acetanilide[9] caused methemoglobinemia and hemolysis. Heubner[10] proposed that these toxic effects on red blood cells might be caused by $N$-hydroxy metabolites, and later investigations by Kiese[11] showed that methemoglobin formation was related to the formation of arylhydroxylamines.

Subsequent studies have implicated N-oxidized metabolites of the anilide analgesics acetaminophen and phenacetin as causative agents in the toxic effects associated with the abuse of these two drugs.[12-16] Large doses of acetaminophen can cause liver cell necrosis and kidney tubular necrosis, whereas the abuse of phenacetin has been associated with methemoglobinemia and hemolysis, renal papillary necrosis, and renal pelvic carcinoma. Although $N$-hydroxyphenacetin is a known metabolite of phenacetin[17] and is carcinogenic in rats,[18,19] $N$-hydroxyacetaminophen is apparently not a metabolite of acetaminophen.[20-22]

This chapter will explore several N-oxidation mechanisms, including $N$-oxide formation and hydroxylation, as well as oxidation without N–O bond formation. The fate of various N-oxidized products will also be examined with special emphasis on reactive intermediates formed from N-oxidation products of arylamines and arylamides.

## II. Chemical Properties of Arylamines and Arylamides

The properties of arylamines and arylamides that best relate to their chemistry in biological systems are basicity, nucleophilicity, and ionization potential. Underlying the chemical properties of all nitrogen compounds is the arrangement of atomic orbitals in the nitrogen atom, whose configuration ($1s^2 2s^2 2px 2py 2px$) allows the formation of three covalent bonds plus a lone pair of electrons. It is this lone pair of electrons and the hybridization state of the molecular orbitals into which they are distributed that account

## 12. ARYLAMINES AND ARYLAMIDES: OXIDATION MECHANISMS

for much of the chemical reactivity and structural versatility of nitrogen compounds.[23,24]

For example, aliphatic amines are considered to be strong organic bases ($pK_a$ of the protonated form $\simeq 9$), because the lone pair of electrons in the rather diffuse $sp^3$ orbital can serve in an electron-donating capacity to form new covalent bonds. This is in contrast to arylamines ($pK_a$ protonated form $\simeq 4.5$), which are approximately as basic as $sp^2$-hybridized imines. In part, the decreased basicity of arylamines is caused by resonance effects of the aryl group so that the lone pair of electrons on nitrogen is delocalized onto the aromatic structure through $\pi$-molecular orbital overlap. Additionally, field electron-withdrawing effects and differential solvation of protonated and unprotonated forms play an important role in the relative basicities of alkyl- and arylamines. (For a discussion, see March.[25])

Resonance delocalization of the lone pair of electrons on arylamide nitrogen occurs onto both the aromatic ring and the carbonyl oxygen. Because the electrons are so highly delocalized, arylamides do not act as bases in aqueous media; that is, their conjugate acids ($pK_a$ of protonated form of amides $\simeq -1$) are stronger than the hydronium ion. In fact, protonation of amides occurs on the carbonyl oxygen and not on nitrogen.[26]

$$R-\overset{H}{\underset{..}{N}}-\overset{O}{\overset{\|}{C}}-R' \quad \xrightarrow{H^+} \quad R-\overset{H}{\underset{+}{N}}-\overset{\overset{+}{O}-H}{\overset{\|}{C}}-R' \quad \longleftrightarrow \quad R-\overset{H}{\underset{+}{N}}=\overset{O-H}{\overset{\|}{C}}-R'$$

In a very broad sense, the relative basicities of arylamines and arylamides influence the way in which they interact with the enzymes involved in their oxidation, inasmuch as the unprotonated forms apparently are the substrates that bind.[27,28] Many primary alkyl- and arylamines, as well as $sp^2$-hybridized nitrogen-containing compounds, form ferrihemochrome complexes with cytochrome P-450,[29] which reflects the nucleophilic character of their polarizable valence electrons. Arylamides, on the other hand, do not form ferrihemichrome complexes with cytochrome P-450, although many of them perturb heme ligand binding in an unusual way and induce reverse type I spectral changes that are indicative of weak ligand formation to the heme iron of cytochrome P-450.[29] Amides are known to coordinate chemically to metal complexes through their carbonyl oxygen atoms.[26] Differences in resonance delocalization of the nitrogen valence electrons in arylamines and arylamides also markedly affect ionization potential.[23,24] Arylamines have significantly lower gas phase one-electron ionization potentials than arylamides ($\sim 5$ vs $\sim 10$ eV), and ionization is calculated to occur preferentially from carbonyl oxygen valence electrons rather than from nitrogen in the amide.

Clearly, several factors other than basicity, nucleophilicity, and ionization potential are important in the interactions of arylamines and arylamides with their oxidizing enzymes, and attempts to correlate such chemical properties of arylamines[30] with their enzymatic oxidation rates have not been successful. Part of the problem lies in the fact that the microsomal oxygenases that are responsible for oxidizing arylamines and arylamides comprise a number of enzymes and isozymes with various substrate selectivities and product distributions. These enzymes will be discussed briefly as the various oxidation processes of arylamines and arylamides are described. Another part of the problem is the multitude of enzymatic and nonenzymatic products that can be formed by oxidative metabolism of arylamines and arylamides.[31,32] These are composed of products of one-electron oxidation (radicals and cation radicals), two-electron oxidation (carbinolamines, hydroxylamines, hydroxamic acids, $N$-oxides), and their further oxidation and decomposition products (iminium ions, nitrenium ions, nitrones, dealkylation products, nitroxyl radicals, nitroso compounds, and nitro compounds). Chemical and biochemical mechanisms for the formation of these various arylamine and arylamide derivatives will be discussed in succeeding sections of this chapter.

## III. Reaction Mechanisms of N-Oxidation for Arylamines and Arylamides

Because detailed knowledge of the catalytic mechanisms of the various enzymes that are involved in oxidations of arylamines and arylamides is lacking, the description of various reaction mechanisms will proceed from a knowledge of the end products of the reaction. Thus, from product analysis, two types of N-oxidation can occur: with or without observable N—O bond formation. Within each type of N-oxidation, multiple mechanisms that require one- or two-electron transfer equivalents, or both, can occur. Liberal use will be made of what little is known about enzyme and substrate intermediates to describe rational pathways from substrates to products.

### A. OXIDATION WITH N-O BOND FORMATION

Tertiary arylamines and secondary alkyl arylamines are oxidized by both a microsomal flavin-containing monooxygenase[28] and cytochrome $P$-450[31] to tertiary amine $N$-oxides and hydroxylamines, respectively. The mechanism for the reaction catalyzed by the flavin-containing monooxygenase has been rather rigorously defined by the combined use of enzyme kinetics[33] and

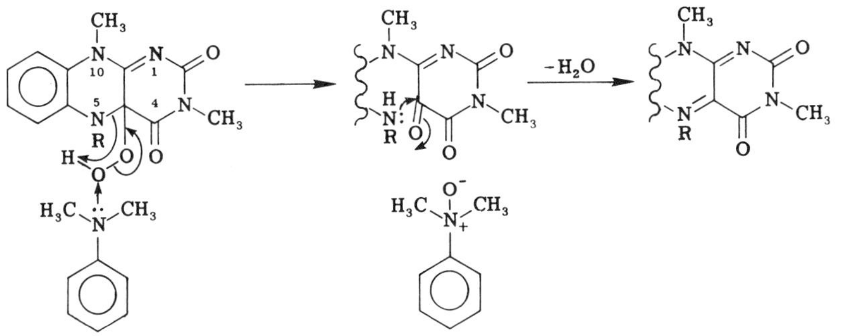

Fig. 1. Reaction of $N,N$-dimethylaniline with a 4a-(hydroperoxy)flavin.

model peroxyflavins.[34] Kinetic studies are consistent with the formation of a peroxyflavin followed by the addition of oxidizable organic substrate. Spectral characteristics of the intermediate NADPH–$O_2$–FAD intermediate closely resemble model 4a-(hydroperoxy)flavins, which catalyze the reaction shown in Fig. 1. The N–O bond of the peroxyflavin is inductively polarized by $N^1$, $N^5$, $N^{10}$, and $C^4$—O of the isoalloxazine ring and essentially leaves the 4a-carbon atom in the oxidation state of a peroxy acid carbonyl group. (For a discussion of oxidation states in organic compounds, see Hendrickson et al.[35]) This electronegativity of the 4a-position yields a reactive peroxidizing agent; the reactions that it undergoes are similar to those of peroxy acids, although reactions with 4a-(hydroperoxy)flavins are more stereoselective.[36]

There is no evidence for radical intermediates in the N-oxidations catalyzed by model peroxyflavins, and rates of catalysis increase with increasing nucleophilicity of the amine substrates.[34,36] This information coupled with the ordered Ter-Bi kinetic mechanism for catalysis,[33] tend to rule out the mechanisms of flavoprotein-catalyzed N-oxidation proposed by Arrhenius[37] and by Beckett and Bélanger[38] (Fig. 2). The intermediacy of a 4a-(hydroperoxy)flavin can account for all other detectable products of amine oxidations (hydroxylamines and nitrones) formed by the microsomal flavin-con-

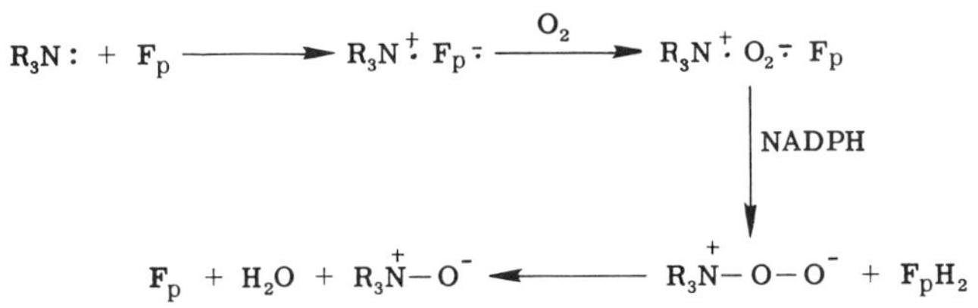

Fig. 2. Mechanism for flavoprotein (Fp)—catalyzed N-oxidations that involves cation radical formation.

taining monooxygenase without invoking the formation of radical cations and N-hydroperoxy intermediates.[28,34]

A feature of the flavin monooxygenase that cannot be chemically rationalized is substrate selectivity. Although the lack of oxygenation of amides and carbamates can be attributed to weak nucleophilicity, lack of oxygenation of most primary arylamines and pyridines cannot be explained.[28]

Cytochrome P-450 apparently can catalyze the array of N-oxidation reactions involving N–O bond formation.[31,39] The mechanism of oxidation for this monooxygenase reaction has not been well characterized. Inasmuch as cytochrome P-450 is a mixture of isozymes with differing substrate selectivities,[40] it is not certain that the same oxidation mechanism is responsible for the various reactions catalyzed by cytochrome P-450. However, most of the evidence supports a common mechanism for oxygen activation.[41,42]

Several schemes can be described for N-oxidation of arylamines by using hypothetical intermediates that are consistent with cytochrome P-450 oxygenations as postulated by White and Coon.[41] In an oxenoid pathway, either a thiolate–ferric peroxide complex (**I**) or a thiolate–ferricperoxy acid complex (**II**) could react by two-electron oxidation of arylamines in a manner paralleling their chemical reactions with peroxides and peracids (Fig. 3A). However, if decomposition of peroxy intermediates occurs before substrate oxidation, two basic reactions can be envisaged (Fig. 3B, C). Oxidation by a perferryl form of cytochrome P-450 (**III**) is equivalent to reaction of the amine with singlet oxene (Fig. 3B). Alternatively, reaction with a thiolate–ferryl oxylide complex (**IV**) is equivalent to reaction of the amine with triplet oxene and requires a series of one-electron transfers (Fig.

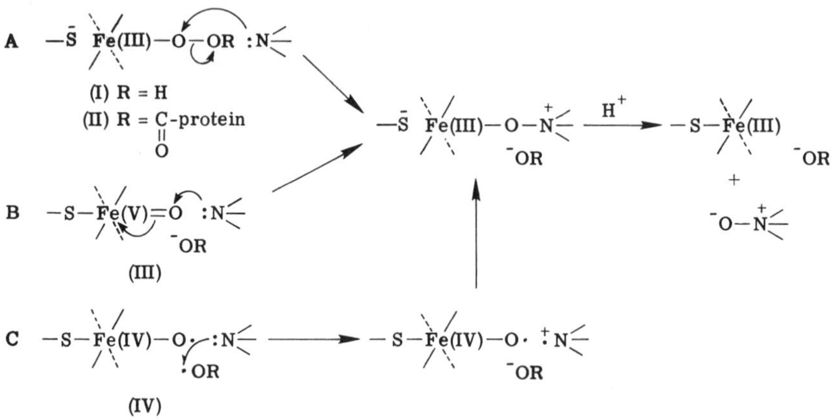

**Fig. 3.** Possible mechanisms for N-oxygenation, catalyzed by cytochrome P-450. The use of thiolate–ferryloxy resonance forms for the oxygenating complex is arbitrary and can be described in several other ways.[41]

3C).  Evidence has been accumulating for one-electron transfer pathways in cytochrome $P$-450-catalyzed oxidations of alkanes[43,44] and sulfoxides,[45] although mechanisms other than the oxenoid pathway may be involved.[46] The formation of amine cation radicals in the process of amine oxidation is also consistent with mechanisms proposed for oxidative N-dealkylation,[47-49] a subject that will be discussed later in this chapter.

Two additional mechanisms for the formation of N-oxygenated products of arylamines are reaction of the arylamine with hydrogen peroxide or hydroxyl radical, which are apparently generated from superoxide released in an autoxidation reaction of cytochrome $P$-450.[41,42]  Ring hydroxylation of aniline by hydroxyl radical, in both cytochrome $P$-450 and hemoglobin-dependent reactions, has been demonstrated.[50]  However, no attempt was made to assay for phenylhydroxylamine, and therefore, it is not known if N-oxygenation can result from by-products of the autoxidation of cytochrome $P$-450.

The hydroxylation of arylamines and arylamides to hydroxylamines and hydroxamic acids, respectively, has been the subject of an extensive review.[51]  The reactions are cytochrome $P$-450 catalyzed,[52,53] although the source of oxygen atom has not been reported for any arylamide hydroxylation and only for two arylamine oxidations.[54]  Possible oxidation mechanisms would be similar to those already proposed (Fig. 3) with the proper hydrogen transfers.  In addition, three alternative mechanisms for hydroxylation of arylamides will be considered (Fig. 4).

As in the reactions shown in Fig. 3B, the reaction of a singlet-type oxene with amides may occur with the iminol tautomer to generate an oxazirane (V), which may exist in tautomeric equilibrium with the hydroxamic acid (Fig. 4A).  The few synthetic methods available for the preparation of hydroxamic acids (albeit in low yields) probably proceed through such intermediates, although an oxazirane has been isolated in only one case.[55,56] Furthermore, gas-chromatographic and mass spectral evidence has been published for the existence of the oxazirane tautomer of $N$-hydroxy-2-fluorenylacetamide.[57]

Resonance delocalization of the nitrogen valence electrons in the amide structure apparently reverses the normal order of ionization potentials so that one-electron oxidation of amides is calculated to occur from the carbonyl oxygen nonbonding lone pair of electrons[23] (see Section II).  If this order of orbital energies holds at the active site of cytochrome $P$-450, the oxidation mechanism may proceed by one-electron transfers (Fig. 4B). Removal of an electron from the nonbonding carbonyl lone pair by a thiolate–ferryl oxylide form of cytochrome $P$-450 would produce a radical cation.  Proton transfer and rehybridization of the amide radical cation would give the resonance-stabilized amidyl radical (VI), which could react

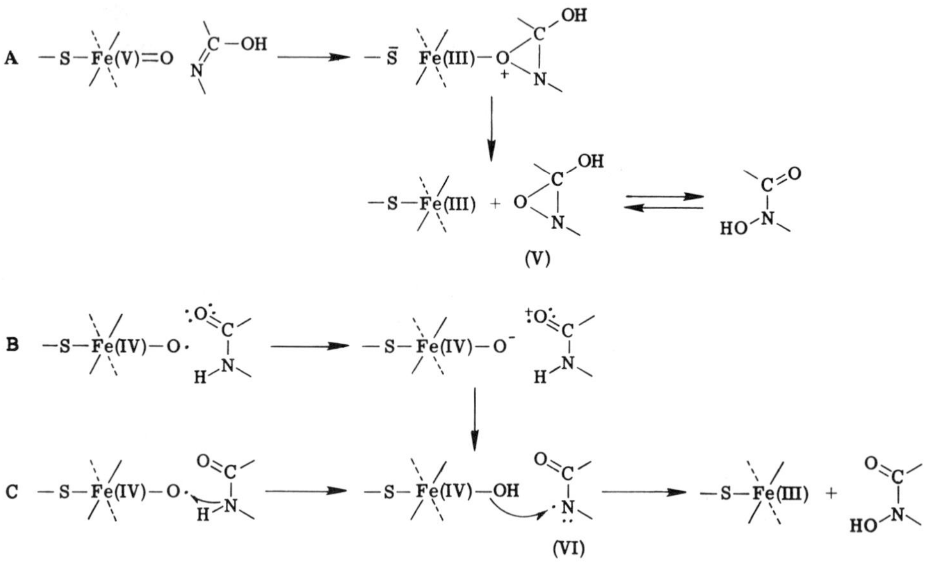

Fig. 4. Possible mechanisms for amide N-oxygenation catalyzed by cytochrome $P$-450.

by oxygen rebound to generate the hydroxamic acid. This mechanism would be difficult to distinguish from one involving direct hydrogen atom transfer to form **VI** (Fig. 4C).

The reactions shown in Figs. 3 and 4 are highly speculative, in that no studies have specifically addressed mechansims of N–O bond formation by cytochrome $P$-450. Support for one-electron transfer mechanisms comes from studies of N-oxidation in which N–O bond formation is not observed (see Section III,B). Few quantitative structure–activity relationship studies have examined relative rates of N-hydroxylation and arylamine or arylamide structure. Hanna and co-workers[58,59] found that both electronegative and electropositive groups in the 4′-position of 4-acetamidostilbenes decreased rates of N-hydroxylation by hamster liver microsomes. Although the Michaelis constants were different for some substrates, most had similar $K_m$ values but either were metabolized at slower rates or followed alternate pathways of metabolism. In a limited series of phenacetin analogs, Kapetanovic et al.[60] found that N-hydroxylation rates increased as lipophilicity increased. Several other laboratories have investigated the effects of arylamine and arylamide structure on mutagenicity and carcinogenicity but failed to detect quantitative structure–activity correlations.[61–65] It would be interesting to examine rates of N-hydroxylation in a series of substituted anilines and acetanilides with purified isozymes of cytochrome $P$-450 that

## B. OXIDATION WITHOUT N–O BOND FORMATION

Oxidation without oxygenation requires the removal of electrons from a molecule and is the basis for electrochemical reactions. Arylamines were among the first compounds to be oxidized chemically and electrochemically to stable free radicals. (For a discussion, see Forrester *et al.*[66]) Oxidation of *p*-phenylenediamines in acetic acid with bromine produced deeply colored Würster salts, which were shown to be cation radicals, analogous to semiquinone radicals. Ceruloplasmin, a copper-containing enzyme found in plasma, oxidizes phenylenediamine (**VII**) by single-electron oxidation to a Würster cation radical (**VIII**). Interestingly, the $K_m$ of ceruloplasmin for various substrates correlates with the energy of the highest-occupied molecular orbital.[67]

One of the earliest biochemical demonstrations of a free radical by electron spin resonance (ESR) techniques was of the amino radical of *p*-aminobenzoic acid that was formed by a peroxidase.[68] Since that time several cation free radicals of aromatic or other resonance-stabilized amines have been detected as products of peroxidase-catalyzed reactions.[69] The red radical cation (**IX**) of chlorpromazine has been detected as a product of both horseradish peroxidase- and catalase-catalyzed oxidations.[70] This radical has been suggested to be a cause of the phototoxic effects of chlorpromazine,[71] although other radicals may also be involved.

The relative nephrotoxicity of a series of aminophenols correlates well with the ease of oxidation of the compounds, because substituents that decreased the oxidation potential ($E_o$) of the arylamines increased nephrotoxicity.[72] Studies with various metabolic inhibitors and inducers strongly indicated that cytochrome P-450 was not the major enzyme responsible for oxidation of the aminophenols to nephrotoxic substances, although kidney microsomal enzymes were apparently involved.[73] Because aminophenols are rapidly oxidized by compound I of peroxidase,[74] it is tempting to speculate that a peroxidative mechanism is involved in the formation of the reactive metabolites.

Work by Zenser and associates[75,76] has shown that benzidine, a nephrotoxin and bladder carcinogen, is metabolized to reactive intermediates by prostaglandin synthetase, which is present in microsomes of the renal inner medulla. Thus, arylamines with suitable redox potentials may be cooxidized in reactions catalyzed by the hydroperoxidase of prostaglandin synthetase (Fig. 5). The nature of the oxidant generated in the hydroperoxidase reaction is unknown; but it produces an arylamine radical cation, as detected by ESR in the horseradish peroxidase-catalyzed oxidation of p-aminophenol.[77] Such radicals further oxidize to electrophilic quinone imines.

Some phenolic arylamides, notably acetaminophen (X), are also metabolized by prostaglandin synthetase in a hydroperoxidase-catalyzed cooxidation reaction that apparently produces the same arylating metabolite of acetaminophen as that formed by cytochrome P-450 catalyzed oxidation.[78] The characteristics of the reaction of acetaminophen with prostaglandin synthetase were similar to those of acetaminophen with horseradish peroxidase–$H_2O_2$, a reaction that produces a semiquinoneimine radical of

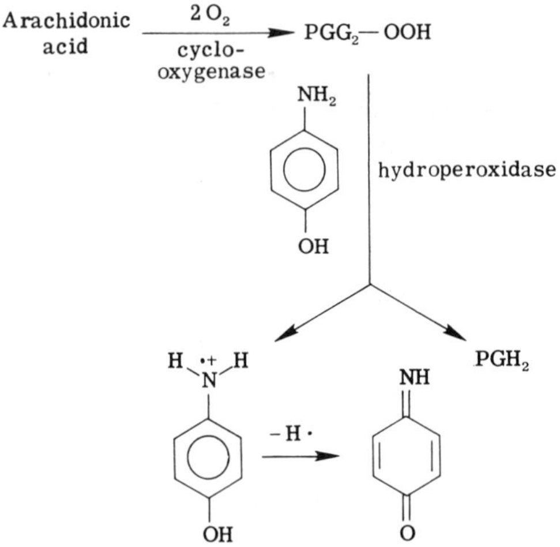

Fig. 5. Hypothetical scheme for prostaglandin synthetase-catalyzed oxidation of p-aminophenol. $PGG_2$ and $PGH_2$ are normal prostaglandin products of the reaction.

## 12. ARYLAMINES AND ARYLAMIDES: OXIDATION MECHANISMS

acetaminophen.[79] The similar nature of the reactive metabolites of acetaminophen that are formed from peroxidases and cytochrome $P$-450 led us to postulate that cytochrome $P$-450 may catalyze two one-electron oxidations of acetaminophen (**X**) to $N$-acetyl-$p$-benzoquinoneimine (NAPQI, **XI**), as depicted in Fig. 6A. Although the acetaminophen radical has not been detected or spin-trapped in cytochrome $P$-450-catalyzed oxidations, it may rapidly be further oxidized to NAPQI in a radical cage reaction. This is consistent with the instability of $N$-acetyl-$p$-benzosemiquinoneimine as determined both chemically and biochemically.[80] Inasmuch as $N$-methylacetaminophen forms arylating metabolites at a much slower rate than acetaminophen,[79] $N$-oxidation, and not just phenolic radical formation, is apparently involved.

Alternatively, NAPQI could be formed in cytochrome $P$-450-catalyzed reactions without the formation of semiquinone-like intermediates (Fig. 6B). The initial step of the oxidation with the perferryl form of cytochrome $P$-450 is the same as that shown in Fig. 3B. The decomposition of the

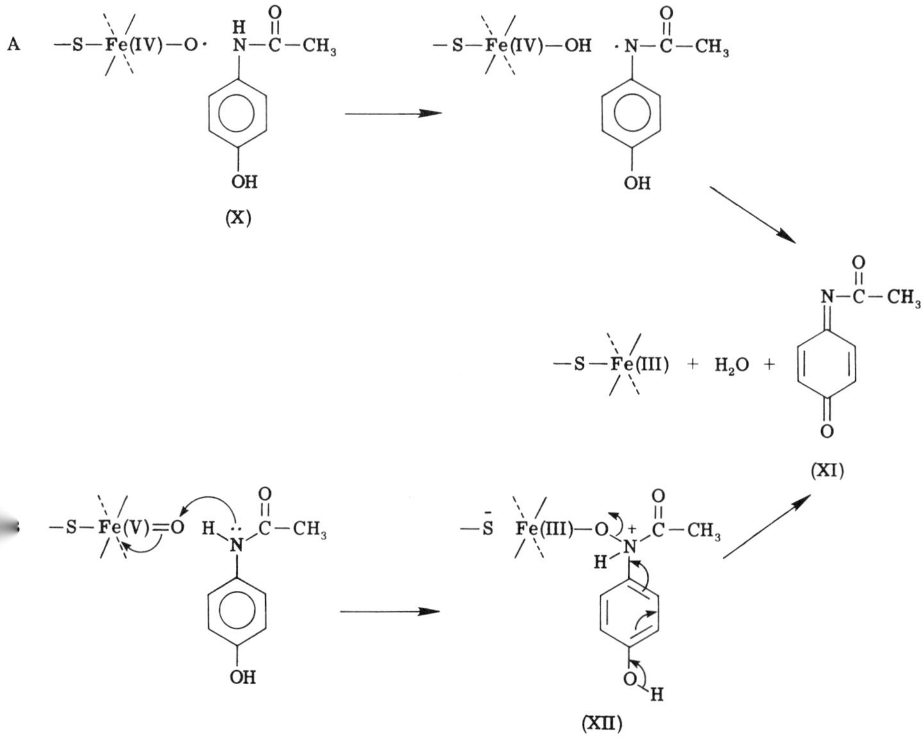

**Fig. 6.** Possible mechanisms for oxidation of acetaminophen by cytochrome $P$-450.

ferric-oxyamide complex (**XII**) is essentially an active site-catalyzed dehydration reaction, which can only occur with amides that have protons at resonance-stabilized anionic sites, such as the *p*-hydroxy group. Thus, phenacetin and other O-alkylated analogs of acetaminophen would yield hydroxamic acids in the same oxidation reaction.[17,60]

NAPQI has been prepared from acetaminophen by chemical oxidation with both lead tetraacetate[16] and silver oxide,[81,82] and by electrochemical oxidation,[83] and indirect evidence for its formation in enzyme-catalyzed oxidations is accumulating.[81-85] Recently, NAPQI has been detected as an oxidation product of acetaminophen in a reaction mediated by purified cytochrome *P*-450 and cumene hydroperoxide, and indirect evidence for its formation in NADPH-mediated reactions was compelling even though NAPQI is rapidly reduced back to acetaminophen by NADPH and NADPH-cytochrome *P*-450 reductase.[86]

N-Dealkylation of arylamines and arylamides may also proceed by N-oxidation without N-oxygenation and with or without accompanying C-oxygenation. McMahon[87] proposed, in 1966, that microsomal oxidative N-dealkylation proceeds via radical intermediates; tangible evidence has since been presented to support this hypothesis.[47-49] Although it is not always easy to distinguish between α-carbon hydrogen atom abstraction and electron transfer, the latter process appears to precede dealkylation of *N, N*-dimethylaniline,[88-90] and aminopyrine[91-94] by various heme-peroxide and hemoprotein-peroxide systems where radical cation formation has been detected by ESR techniques.

Therefore, a mechanism for oxidative N-dealkylation has been proposed (Fig. 7) that involves radical abstraction to form an aminium cation radi-

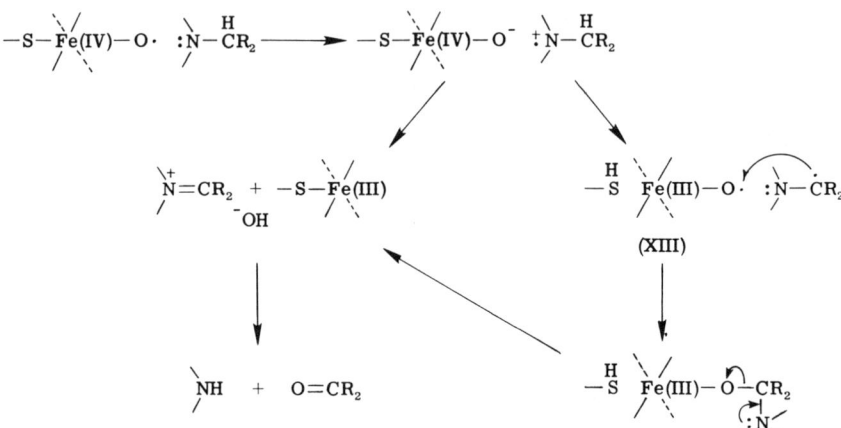

Fig. 7. Mechanism for oxidative N-deaklylation that involve iminium ion formation before carbinolamine formation.

cal.[90,91] Formation of an iminium ion is postulated to occur either by hydrogen radical abstraction or by loss of the $\alpha$-carbon hydrogen as a proton followed by addition to a thiolate–ferric oxylide complex (**XIII**) and subsequent elimination. Hydrolysis of the iminium ion produces the dealkylated amine and aldehyde.

Both abstraction and $\alpha$-hydrogen abstraction have been observed in amine dealkylations by chemical oxidants.[95,96] $\alpha$-Hydrogen abstraction was found to compete with electron abstraction only in systems where its dissociation energy was considerably lowered, such as in methylene groups adjacent to aromatic systems. Therefore, for aromatic amines, where the amine lone pair is adjacent to the aromatic system, one would expect that electron abstraction would be predominant. Electrochemical oxidation of amines also produces predominantly aminium cation radicals.[97-99]

Many characteristics of oxidative amine dealkylation reactions by chemical oxidants, anodic oxidation, and heme–peroxide and hemoprotein–peroxide systems are similar to cytochrome *P*-450 catalyzed oxidative N-dealkylation reactions. Although aminium cation radicals have not been detected in cytochrome *P*-450-dependent systems that utilize oxygen, subsequent indirect evidence suggests their formation.[47-49] Both *N*-benzylcyclopropylamine (**XIV**) and its methyl analog (**XV**) are suicide inactivators of cytochrome *P*-450; thus it is unlikely that a carbinolamine is involved in the observed heme destruction (Fig. 8). Formation of an aminium cation radical can account for the observed results, although the structures of the bound adducts need to be determined to clarify the reaction of heme with activated metabolite.

More direct evidence for one-electron oxidation of amines to aminium cation radicals has come from studies with another suicide inactivator of cytochrome *P*-450, 4-ethyl-3,5-bis(ethoxycarbonyl)-2,6-dimethyl-1,4-dihydropyridine (**XVI**).[100] This compound is oxidized by cytochrome *P*-450 to the pyridine (**XVIII**), with loss of the ethyl radical, which was spin-trapped during the reaction. Additional studies with *N*-alkyl-substituted analogs of **XVI** strongly support the conclusions that the intermediate in the oxidation is the cation radical, **XVII**.

Low intermolecular and intramolecular isotope effects ($k_H/k_D \sim 1.5$) for oxidative amine dealkylations by microsomal oxygenases[49,101-104] are also consistent with rate-determining electron transfer from substrate to enzyme

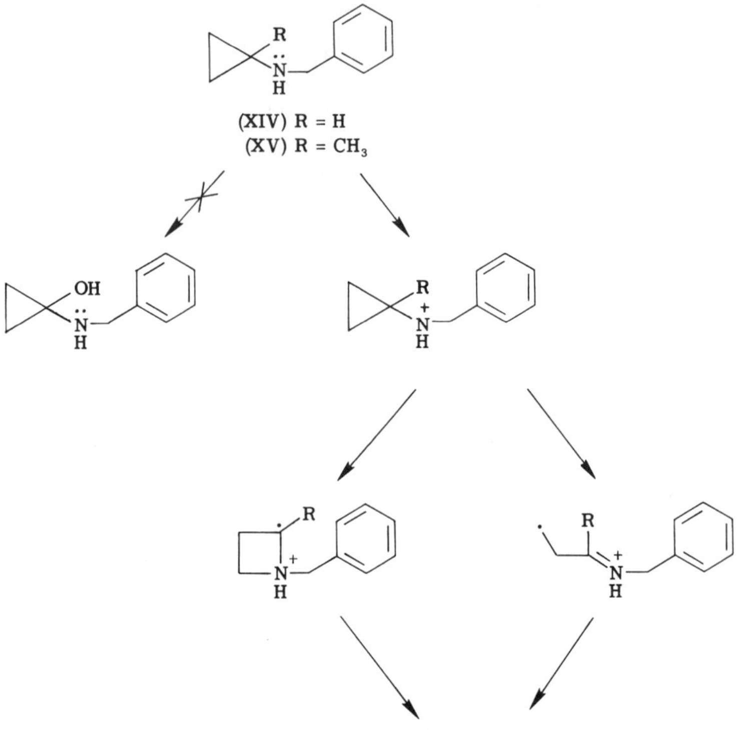

**Fig. 8.** Mechanisms that have been proposed for cytochrome P-450 catalyzed oxidation of N-benzylcyclopropylamine (**XIV**) and its methyl analog (**XV**), and subsequent reactions.[48,100]

followed by $\alpha$-C–H bond cleavage. Similar secondary isotope effects are observed in both chemical[95,96] and electrochemical[49,97-99] oxidations of amines with subsequent dealkylation. The only experimental observation that is not entirely consistent with the mechanism illustrated in Fig. 7 is that oxygen from molecular oxygen is incorporated, to a significant extent, into the carbinolamine[105] or aldehyde[106] products in microsomal NADPH-dependent oxygenase-catalyzed reactions. Modification of the mechanism, as shown in Fig. 9, is consistent with the observation. Pathway A, involving both carbinolamine and iminium ion formation, appears to be most likely because carbinolamines have been isolated as microsomal metabolites of some amines,[107-109] and iminium ions have been trapped with cyanide as intermediates in microsomal oxygenase-catalyzed amine dealkylations.[110-113]

Thus, free-radical mechanisms are most consistent with oxidation of arylamides and arylamines where N–O bond formation is not observed.

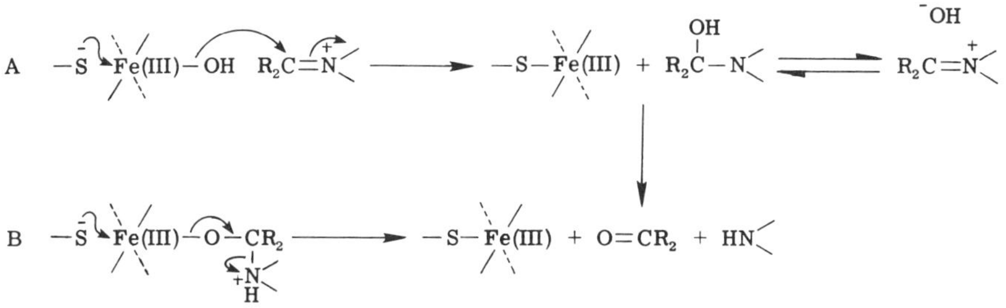

12. ARYLAMINES AND ARYLAMIDES: OXIDATION MECHANISMS  363

A $\quad -S-Fe(III)-OH \quad R_2C=\overset{+}{N}\overset{\diagdown}{\diagup} \quad \longrightarrow \quad -S-Fe(III) \; + \; R_2\overset{OH}{\underset{|}{C}}-N\overset{\diagdown}{\diagup} \quad \rightleftharpoons \quad R_2C=\overset{+}{N}\overset{\diagdown}{\diagup}\overset{^-OH}{}$

B $\quad -S-Fe(III)-O-CR_2 \quad \longrightarrow \quad -S-Fe(III) \; + \; O=CR_2 \; + \; HN\overset{\diagdown}{\diagup}$
$\qquad\qquad\qquad \underset{H}{\overset{|}{\underset{+N}{}}}\overset{\diagdown}{\diagup}$

**Fig. 9.** Mechanisms for oxidative N-dealkylation that involve carbinolamine formation before iminium ion formation.

## IV. Fate of Arylamine and Arylamide Oxidation Products

Certain aspects of the fate of arylamine and arylamide oxidation products will be discussed in this section, including further oxidations, reductions, rearrangements, and binding to tissue constituents. Conjugation reactions and the reaction of conjugated products are discussed in Chapter 13.

### A. FURTHER OXIDATION

Arylhydroxylamines undergo rapid oxidation by oxyhemoglobin to arylnitroso compounds with concomitant formation of hydrogen peroxide and methemoglobin.[11] Although the mechanism for this oxidation has been variously described,[11,32,42] the most rational mechanism[42] recognizes the fact that superoxide anion is a strong base but a weak oxidant. Thus, ferrous oxyhemoglobin (diagrammatically shown as a ferric superoxide complex, Fig. 10) is reduced to methemoglobin and the perhydroxyl radical by a hydroxylamine. The resulting nitroxyl radical can be further oxidized to the nitroso compound. Aliphatic nitroso compounds have also been shown to be oxidized further to nitro compounds by superoxide anion generated from cytochrome $P$-450.[114] In the initial step of this reaction, superoxide anion may act as a nucleophile with attack on the nitroso nitrogen atom. However, it has not been determined if arylnitroso compounds can be oxidized to arylnitro compounds by the same mechanism.

$$Fe^{3+}O_2^- \; + \; HON\underset{H}{R} \quad \longrightarrow \quad Fe^{3+} \; + \; \cdot O-N\underset{H}{R} \; + \; \cdot O_2H \quad \longrightarrow \quad Fe^{3+} \; + \; O=NR \; + \; H_2O_2$$

**Fig. 10.** Possible mechanism for oxyhemoglobin-catalyzed oxidation of a hydroxylamine to a nitroso compound.

For some secondary arylhydroxylamines, it has been shown that further oxidation can be catalyzed by microsomal flavin-containing monooxygenases to yield nitrones, which can subsequently be hydrolyzed to dealkylated products.[115] Thus, more than one mechanism may lead to arylamine N-dealkylation.

Further oxidation of hydroxylated arylamides, such as N-hydroxy-2-FAA, proceeds through nitroxyl free radicals to yield arylnitroso compounds and N-acetoxyarylamides by dismutation.[116,117] The initial one-electron oxidation product can be metabolized by a number of peroxidative enzymes and chemical systems.[118]

## B. REDUCTION

Aromatic amine N-oxides, arylhydroxylamines, and arylhydroxamic acids are reduced to their corresponding amines or amides by several chemical reducing agents, the mildest of which appears to be titanium(III) chloride.[119] Reduction of several hydroxylamines and N-oxides by reductases, including cytochrome P-450 acting as a reductase, has been observed, although the mechanisms are not clear (for a review, see ref. 32). The proposed major toxic oxidative metabolite of the arylamide acetaminophen, NAPQI, is readily reduced to acetaminophen by ascorbate, thiol-containing compounds, reduced pyridine nucleotides, and NADPH–cytochrome P-450 reductase.[80-86] Thus, several amine metabolic products exist in quasi-equilibrium in biological systems.

## C. REARRANGEMENT, CONDENSATION, AND HYDROLYSIS

Aromatic amine N-oxides undergo a variety of chemical rearrangements that are both thermally and metal-catalyzed, including N-dealkylation, Cope elimination, Meisenheimer rearrangement, and the Polonovski reaction (for a review, see ref. 32). There is controversy about whether or not these reactions are enzymatically catalyzed.[120] The same ambiguity exists with regard to the condensation of arylnitrosamines and arylhydroxylamines to produce coupled azo and azoxy compounds.

In contrast, it has been shown that Bamberger-type rearrangements of arylhydroxylamines and arylhydroxamic acids can be catalyzed by unidentified enzymes in rat liver.[121,122] The first step in this reaction is thought to be formation of a reactive nitrenium ion (Fig. 11). Molecular orbital studies have calculated that arylnitrenium ions are resonance-stabilized singlet ions.[123] and that carcinogenic and mutagenic amines are characterized by

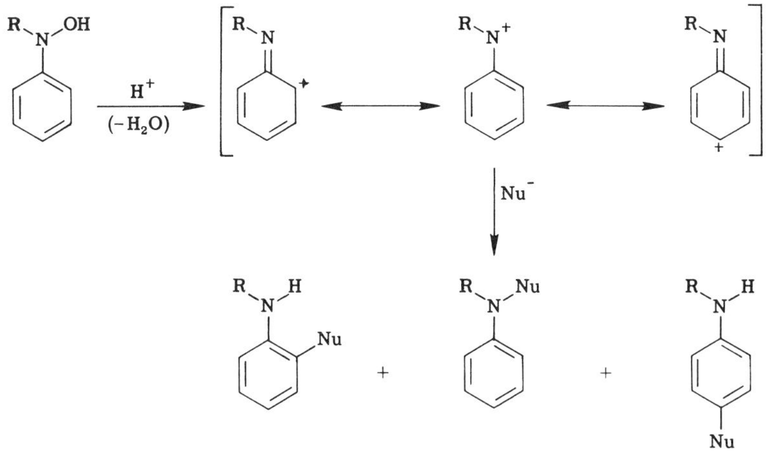

## 12. ARYLAMINES AND ARYLAMIDES: OXIDATION MECHANISMS

Fig. 11. Bamberger-like rearrangement of arylhydroxylamines or arylhydroxamic acids with incipient nitrenium ion formation and subsequent reaction with tissue nucleophiles (Nu⁻).

their greater singlet-state stability.[124] Tissue nucleophiles, such as glutathione, and nucleophilic groups on proteins and DNA apparently form adducts by reacting with the resonance-stabilized ion. It is thought that these adducts alter cellular biochemistry, leading to the array of toxicological responses produced by arylamines and arylamides.[7,125-128] These adducts appear to form biologically from acetyl, sulfate, or glucuronide derivatives of the hydroxylated arylamines or arylamides, which are much more reactive than the N-hydroxy metabolite. A discussion of the formation and fate of these secondary metabolites will be presented in Chapter 13.

Although short-lived radicals and nitrenium ions have been regarded as important intermediates in toxic reactions, subsequent studies with a model system of phenylhydroxylamine and bisulfite suggest that nucleophilic addition–elimination reactions might provide alternative mechanisms for reactions of arylhydroxylamines and arylhydroxyamides.[129] Further tests of this model with tissue macromolecules are required to substantiate further the significance of this mechanism.

Carbinolamines of the arylamine antitumor agent hexamethylmelamine[130] and of the carcinogen N,N-dimethyl-4-aminoazobenzene[131] can apparently react with tissue nucleophiles to give substitution products at the methylol carbon atom. It is not known if such reactions are important in the therapeutic or toxic properties of these agents, but such reactions point to a possible physiological significance for the formation of relatively stable carbinolamines, that is, carbinolamines with relatively nonbasic amine functions.

Hydrolysis of the *N*-hydroxyamide metabolites of 2-fluorenylacetamide (2-FAA) and phenacetin yields the corresponding arylhydroxylamines. This reaction is carried out by a microsomal amidase and is required for activation of *N*-hydroxy-2-FAA and *N*-hydroxyphenacetin to mutagenic metabolites.[132,133] The nature of the interaction of these metabolites with DNA has not been fully elucidated. However, indirect evidence suggests that a radical product of *N*-hydroxy-FA is involved in the mutagenic process, in that ascorbic acid promotes the mutagenesis but decreases the binding to protein. Unfortunately, the effect on binding to DNA was not determined. This would be important, because the extent of FA–DNA adduct formation, particularly *N*-(deoxyguanosin-8-yl)-2-FA, appears to correlate with mutagenicity and carcinogenicity caused by *N*-hydroxy-2-FAA.[134]

Finally, hydrolysis of a reactive metabolite of acetaminophen has been found to produce acetamide and presumably benzoquinone.[85] Benzoquinone and acetamide are also hydrolytic products of NAPQI,[82,83] and this provides strong support for the hypothesis that NAPQI is formed by cytochrome *P*-450-catalyzed oxidation of acetaminophen. The fate of the hydrolytic products is unknown, although benzoquinone presumably binds to proteins in a conjugate addition reaction.

## V. Summary

Arylamines and arylamides are a widely used class of chemicals whose oxidation products are often toxic. Therefore, an understanding of the mechanisms of their oxidation and of the nature of the enzymes involved is important to a better assessment of the toxicological properties, of arylamines and arylamides. This chapter has explored oxidation mechanisms of arylamines and arylamides that involve both one- and two-electron transfers. The lone pair of electrons on nitrogen appears to play a crucial role in the oxidation reactions. Some products of these reactions are formed with N—O bonds, and others are not. However, in all cases N-oxidation has been proposed. A significant amount of work remains to be carried out to determine if the mechanisms proposed are, in fact, correct.

## References

1. Weisburger, E. K. (1981). N-substituted aryl compounds in carcinogenesis and mutagenesis. *Natl. Cancer Inst. Monogr.* **58**, 1–10.
2. Beutler, E. (1969). Drug-induced hemolytic anemia. *Pharmacol. Rev.* **21**, 73–102.
3. Gillette, J. R., Nelson, S. D., Mulder, G. J., Jollow, D. J., Mitchell, J. R., Pohl, L. R., and

Hinson, J. A. (1982). Formation of chemically reactive metabolites of phenacetin and acetaminophen. *In* "Biological Reactive Intermediates—II" (R. Snyder, D. V. Park, J. J. Kocsis, D. J. Jollow, C. G. Gibson, and C. M. Witner, eds.), pp. 931–950. Plenum, New York.
4. Hueper, W. C., Wiley, F. H., and Wolfe, H. D. (1938). Experimental production of bladder tumors in dogs by administration of $\beta$-naphthylamine. *J. Ind. Hyg. Toxicol.* **20**, 46–84.
5. Wilson, R. H., Deeds, F., and Cox, A. J., Jr. (1941). The toxicity and carcinogenic activity of 2-acetylaminofluorene. *Cancer Res.* **1**, 595–608.
6. Cramer, J. W., Miller, J. A., and Miller, E. C. (1960). A new metabolic reaction observed in the rat with the carcinogen 2-acetylaminofluorene. *J. Biol. Chem* **235**, 885–888.
7. Miller, J. A. (1970). Carcinogenesis by chemicals: An overview. *Cancer Res.* **30**, 559–576.
8. Leymann, M. (1910). Erkrankungsverhältnisse in einer Anilinfarbenfabrik. *Concordia* **17**, 355–356.
9. Hénocque, A. (1887). Mode de l'action de l'acétanilide (antifibrine) sur le sange et sur l'activité de la réduction de l' oxyhémoglobine. *C. R. Seances Soc. Biol. Ses. Fil.* **4**, 498–500.
10. Heubner, W. (1913). Studien über Methämoglobinbildung. *Arch. Exp. Pathol. Pharmakol.* **72**, 241–281.
11. Kiese, M. (1959). Oxydation von Anilin zu Nitrosobenzol im Hunde. *Naunyn-Schmiedebergs Arch. Exp. Pathol. Pharmakol.* **235**, 354–359.
12. Mitchell, J. R., and Jollow, D. J. (1975). Metabolic activation of drugs to toxic substances. *Gastroenterology* **68**, 392–410.
13. Nelson, S. D., Boyd, M. R., and Mitchell, J. R. (1977). Role of metabolic activation in chemical-induced tissue injury. *In* "Drug Metabolism Concepts" (D. M. Jerina, ed.), pp. 155–185. Am. Chem. Soc., Washington, D.C.
14. Hinson, J. A. (1981). Biochemical toxicology of acetaminophen. *Rev. Biochem. Toxicol.* **2**, 103–130.
15. Nery, R. (1971). The possible role of N-hydroxylation in the biological effects of phenacetin. *Xenobiotica* **1**, 27–31.
16. Calder, I. C., Creek, M. J., Williams, P. J., Funder, C. C., Green, C. R., Ham, K. N., and Tange, J. D. (1973). N-Hydroxylation of *p*-acetophenetidide as a factor in nephrotoxicity. *J. Med. Chem.* **16**, 499–502.
17. Hinson, J. A., and Mitchell, J. R. (1976). N-Hydroxylation of phenacetin by hamster liver microsomes. *Drug Metab. Dispos.* **4**, 430–435.
18. Calder, I. C., Goss, D. E., Williams, P. J., Funder, C. C., Green, C. R., Ham, K. N., and Tange, J. D. (1976). Neoplasia in the rat induced by *N*-hydroxyphenacetin, a metabolite of phenacetin. *Pathology* **8**, 1–6.
19. Johansson, S. L. (1981). Carcinogenicity of analgesics: Long-term treatment of Sprague–Dawley rats with phenacetin, phenazone, caffeine, and paracetamol (acetaminophen). *Int. J. Cancer* **27**, 521–529.
20. Hinson, J. A., Pohl, L. R., and Gillette, J. R. (1979). *N*-Hydroxyacetaminophen: a microsomal metabolite of *N*-hydroxyphenacetin but apparently not of acetaminophen. *Life Sci.* **24**, 2133–2138.
21. Nelson, S. D., Forte, A. J., and Dahlin, D. C. (1980). Lack of evidence for *N*-hydroxyacetaminophen as a reactive metabolite of acetaminophen *in vitro*. *Biochem. Pharmacol.* **29**, 1617–1620.
22. Calder, I. C., Hart, S. J., Healey, K., and Ham, K. (1981). *N*-Hydroxyacetaminophen: A postulated toxic metabolite of acetaminophen. *J. Med. Chem.* **24**, 988–993.

23. Daudel, R. (1968). General and theoretical. *In* "The Chemistry of the Amino Group" (S. Patai, ed.), pp. 1–35. Wiley (Interscience), New York.
24. Robin, M. R., Bovey, F. A., and Basch, H. (1970). Molecular and electronic structure of the amide group. *In* "The Chemistry of Amides" (J. Zabicky, ed.), pp. 1–72. Wiley (Interscience), New York.
25. March, J. (1977). "Advanced Organic Chemistry," pp. 243–244. McGraw-Hill, New York.
26. Homer, R. B., and Johnson, C. D. (1970). Acid–base and complexing properties of amides. *In* "The Chemistry of the Amino Group" (S. Patai, ed.), pp. 187–243. Wiley (Interscience), New York.
27. White, R. E., Oprian, D. P., and Coon, M. J. (1980). Resolution of multiple equilibria in binding of small molecules to cytochrome $P$-450$_{LM}$. *In* "Microsomes, Drug Oxidations, and Chemical Carcinogenesis" (M. J. Coon, A. H. Cooney, R. W. Estabrook, H. V. Gelboin, J. R. Gillette, and P. J. O'Brien, eds.), pp. 243–250. Academic Press, New York.
28. Ziegler, D. M. (1980). Microsomal flavin-containing monooxygenase: Oxygenation of nucleophilic nitrogen and sulfur compounds. *In* "Enzymatic Basis of Detoxification" (W. B. Jakoby, ed.), Vol. I, pp. 201–230. Academic Press, New York.
29. Schenkman, J. B., Sligar, S. G., and Cinti, D. L. (1981). Substrate interaction with cytochrome $P$-450. *Pharmacol. Ther.* **12,** 43–71.
30. Gorrod, J. W. (1978). The current status of the p$K_a$ concept in the differentiation of enzymic N-oxidation. *In* "Biological Oxidation of Nitrogen" (J. W. Gorrod, ed.), pp. 201–210. Elsevier/North-Holland, Amsterdam.
31. Hlavica, P. (1982). Biological oxidation of nitrogen in organic compounds and disposition of N-oxidized products. *CRC Crit. Rev. Biochem.* **12,** 39–101.
32. Lindeke, B. (1982). The non- and postenzymatic chemistry of N-oxygenated molecules. *Drug Metab. Rev.* **13,** 71–121.
33. Poulsen, L. L., and Ziegler, D. M. (1979). The liver microsomal FAD-containing monooxygenase—Spectral characterization and kinetic studies. *J. Biol. Chem.* **254,** 6449–6455.
34. Ball, S., and Bruice, T. C., (1980). Oxidation of amines by a 4$a$-hydroperoxyflavin. *J. Am. Chem. Soc.* **102,** 6498–6503.
35. Hendrickson, J. B., Cram, D. J., and Hammond, G. S. (1970). "Organic Chemistry," 3rd ed., pp. 739–746. McGraw-Hill, New York.
36. Oae, S., Asada, K., and Yoshimura, T. (1983). The mechanistic mode of oxidation of substituted $N,N$-dimethylanilines, thionaisoles, and methyl phenyl sulfoxides by 4$a$-FlEtOOH. *Tetrahedron Lett.* **24,** 1265–1268.
37. Arrhenius, E. (1969). Correlation of N- and C-oxygenation of aromatic amines with conditions which increase the carcinogenicity of these compounds. *FEBS Symp.* **16,** 209–225.
38. Beckett, A. H., and Bélanger, P. M. (1976). Metabolic N-oxidation of secondary and primary aromatic amines as a route to ring hydroxylation, to various N-oxygenated products, and to dealkylation of secondary amines. *Biochem. Pharmacol.* **25,** 211–214.
39. Gorrod, J. W., and Damani, L. A. (1979). The effect of various potential inhibitors, activators and inducers on the N-oxidation of 3-substituted pyridines *in vitro*. *Xenobiotica* **9,** 219–226.
40. Lu, A. Y. H., and West, S. B. (1980). Multiplicity of mammalian microsomal cytochromes $P$-450. *Pharmacol. Rev.* **31,** 277–295.
41. White, R. E., and Coon. M. J. (1980). Oxygen activation by cytochrome $P$-450. *Annu Rev. Biochem.* **49,** 315–356.

42. Trager, W. F. (1982). The postenzymatic chemistry of activated oxygen. *Drug Metab. Rev.* **13**, 51–69.
43. Groves, J. T., McClusky, G. A., White, R. E., and Coon, M. J. (1978). Aliphatic hydroxylation by highly purified liver microsomal cytochrome *P*-450. Evidence for a carbon radical intermediate. *Biochem. Biophys. Commun.* **81**, 154–160.
44. Gelb, M. H., Heimbrook, D. C., Malkonen, P., and Sligar, S. G. (1982). Stereochemistry and deuterium isotope effects in camphor hydroxylation by the cytochrome $P\text{-}450_{cam}$ monooxygenase system. *Biochemistry* **21**, 370–377.
45. Watanabe, Y., Iyanagi, T., and Oae, S. (1982). One-electron transfer mechanism in the enzymatic oxygenation of sulfoxide to sulfone promoted by a reconsituted system with purified cytochrome *P*-450. *Tetrahedron Lett.* **23**, 533–536.
46. White, R. E., Sligar, S. G., and Coon, M. J. (1980). Evidence for a homolytic mechanism of peroxide oxygen–oxygen bond cleavage during substrate hydroxylation by cytochrome *P*-450. *J. Biol. Chem.* **255**, 11108–11111.
47. Hanzlik, R. P., and Tullman, R. H. (1982). Suicidal inactivation of cytochrome *P*-450 by cyclopropylamines. Evidence for cation-radical intermediates. *J. Am. Chem. Soc.* **104**, 2048–2050.
48. MacDonald, T. L., Zirvi, K., Burka, L. T., Peyman, P., and Guengerich, F. P. (1982). Mechanism of cytochrome *P*-450 inhibition by cyclopropylamines. *J. Am. Chem. Soc.* **104**, 2050–2052.
49. Shono, T., Toda, T., and Oshino, N. (1982). Electron transfer from nitrogen in microsomal oxidation of amine and amide. Simulation of microsomal oxidation by anodic oxidation. *J. Am. Chem. Soc.* **104**, 2639–2641.
50. Ingleman-Sundberg, M., and Ekström, G. (1982). Aniline is hydroxylated by the cytochrome *P*-450-dependent hydroxyl radical-mediated oxygenation mechanism. *Biochem. Biophys. Res. Commun.* **106**, 625–631.
51. Weisburger, J. H., and Weisburger, E. K. (1973). Biochemical formation and pharmacological, toxicological, and pathological properties of hydroxylamines and hydroxamic acids. *Pharmacol. Rev.* **25**, 1–66.
52. Thorgeirsson, S. S., Jollow, D. J., Sasame, H. A., Green, I., and Mitchell, J. R. (1973). The role of cytochrome *P*-450 in N-hydroxylation of 2-acetylaminofluorene. *Mol. Pharmacol.* **9**, 398–404.
53. Lotlikar, P. D., and Hong, Y. S. (1981). Microsomal N- and C- oxidations of carcinogenic aromatic amines and amides. *Natl. Cancer Inst. Monogr.* **58**, 101–107.
54. Schmidt, H. L., Kexel, H., and Weber, N. (1972). Mikrosomale oxydationen am stickstoffatom aromatischer Amine. *Biochem. Pharmacol.* **21**, 1641–1648.
55. Matlin, S. A., Sammes, P. G., and Upton, R. M. (1979). The oxidation of trimethylsilylated amides to hydroxamic acids. *J. Chem. Soc., Perkins Trans. I*, 2481–2487.
56. Black, D. St. C., Brown, R. F. C., and Wade, A. M. (1971). Peracid oxidation of *O*-alkylimino esters of hexanolactam. *Tetrahedron Lett.* pp. 4519–4520.
57. Lhoest, G., Razzouk, C., and Mercier, M. (1976). Biological implications of the reaction possibilities of the proximate carcinogenic compound, *N*-hydroxy-2-fluorenylacetanide. *Biomed. Mass. Spectrom.* **3**, 21–27.
58. Gammans, R. E., Sehon, R. D., Anders, M. W., and Hanna, P. E. (1977). Microsomal N-hydroxylation of *trans*-4′-halo-4-acetamidostilbenes. *Drug. Metab. Dispos.* **5**, 310–316.
59. Hanna, P. E., Gammans, R. E., Sehon, R. D., and Lee, M. K. (1980). Metabolic N-hydroxylation. Use of substituent variation to modulate the in vitro bioactivation of 4-acetamidostilbenes. *J. Med. Chem.* **23**, 1038–1044.
60. Kapetanovic, I. M., Strong, J. M., and Mieyal, J. J. (1979). Metabolic structure–activity

relationship for a homologous series of phenacetin analogs. *J. Pharmacol. Exp. Ther.* **209**, 20–24.
61. Miller, J. A., Sandin, R. B., Miller, E. C., and Rusch, H. P. (1955). The carcinogenicity of compounds related to 2-acetylaminofluorene. II. Variations in the bridges and the 2-substituents. *Cancer Res.* **15**, 188–199.
62. Morris. H. P., Velat, L. A., Wager, B. P., Dahlgard, N., and Roy, F. E. (1960). Studies of carcinogenicity in the rate of derivatives of aromatic amines related to *N*-2-fluorenylacetamide. *J. Natl. Cancer Inst. (U.S.)* **24**, 149–180.
63. Nelson, W. L., and Thorgeirsson, S. S. (1976). Structural requirements for mutagenic activity of 2-acylaminofluorenes in the *Salmonella* test system. *Biochem. Biophys. Res. Commun.* **71**, 1201–1206.
64. Radomski, J. L. (1979). The primary aromatic amines: Their biological properties and structure–activity relationships. *Annu. Rev. Pharmacol. Toxicol.* **19**, 129–157.
65. Soderlund, E. J., Dybing, E., Nordenson, S., and Tjelta, E. (1980). The role of ethyl and fluorine substitution in the 4′-position for *N*,*N*-diethyl-4-aminoazobenzene mutagenicity and azo reduction. *Acta Pharmacol. Toxicol.* **47**, 175–182.
66. Forrester, A. R., Hay, J. M., and Thompson, R. H. (1968). "Organic Chemistry of Stable Free Radicals," pp. 247–280. Academic Press, New York.
67. Pettersson, G. (1970). Electronic characteristics of substrates for ceruloplasmin. *Acta Chem. Scand.* **24**, 1838–1839.
68. Chance, B. (1952). The kinetics and stoichiometry of the transition from the primary to the secondary peroxidase peroxide complexes. *Arch. Biochem. Biophys.* **41**, 416–424.
69. Mason, R. P. (1979). Free-radical metabolites of foreign compounds and their toxicological significance. *Rev. Biochem. Toxicol.* **1**, 160–164.
70. Piette, L. H., Bulow, G., and Yamazaki, I. (1964). Electron paramagnetic resonance studies of the chlorpromazine free radical formed during enzymic oxidation by peroxidase–hydrogen peroxide. *Biochim. Biophys. Acta* **88**, 120–129.
71. Blois, M. S. Jr, (1965). On chlorpromazine binding *in vivo*. *J. Invest. Dermatol.* **45**, 475–481.
72. Calder, I. C., Williams, P. J., Woods, R. A., Funder, C. C., Green, C. R., Ham, K. N., and Tange, J. D. (1975). Nephrotoxicity and molecular structure. *Xenobiotica* **5**, 303–307.
73. Calder, I. C., Yong, A. C., Woods, R. A., Crowe, C. A., Ham, K. N., and Tange, J. D. (1979). The nephrotoxicity of *p*-aminophenol. II. The effect of metabolic inhibitors and inducers. *Chem.-Biol. Interact.* **27**, 245–254.
74. Job, D., and Danford, H. B. (1976). Substituent effect on the oxidation of phenols and aromatic amines by horseradish peroxidase compound I. *Eur. J. Biochem.* **66**, 607–614.
75. Zenser, T. V., Mattamal, M. B., Armbrecht, H. J., and Davis, B. B. (1980). Benzidine binding to nucleic acids mediated by the peroxidative activity of prostaglandin endoperoxide synthetase. *Cancer Res.* **40**, 2839–2845.
76. Rapp, N. S., Zenser, T. V., Brown, W. W., and Davis, B. B. (1980). Metabolism of benzidine by a prostaglandin–mediated process in renal inner medullary slices. *J. Pharmacol. Exp. Ther.* **215**, 401–406.
77. Josephy, P. D., Eling, T., and Mason, R. P. (1983). Oxidation of *p*-aminophenol catalyzed by horseradish peroxidase and prostaglandin synthase. *Mol. Pharmacol.* **23**, 461–466.
78. Moldeus, P., Andersson, B., Rahimtula, A., and Berggren, M. (1982). Prostaglandin synthetase-catalyzed activation of paracetamol. *Biochem. Pharmacol.* **31**, 1363–1368.
79. Nelson, S. D., Dahlin, D. C., Rauckman, E. J., and Rosen, G. M. (1981). Peroxidase-mediated formation of reactive metabolites of acetaminophen. *Mol. Pharmacol.* **20**, 195–199.

80. Powis, G., Svingen, B. A., Dahlin, D. C., and Nelson, S. D. (1984). Enzymatic and non-enzymatic reduction of N-acetyl-p-benzoquinoneimine and some properties of the N-acetyl-p-benzosemiquinoneimine radical. *Biochem. Pharmacol.* **33,** 2367–2370.
81. Blair, I. A., Boobis, A. R., Davies, D. S., and Cresp, T. M. (1980). Paracetamol oxidation: Synthesis and reactivity of N-acetyl-p-benzoquinoneimine. *Tetrahedron Lett.* **21,** 4947–4950.
82. Dahlin, D. C., and Neslon, S. D. (1982). Synthesis, decomposition kinetics, and preliminary toxicological studies of pure N-acetyl-p-benzoquinoneimine, a proposed toxic metabolite of acetaminophen. *J. Med. Chem.* **25,** 885–886.
83. Miner, D. J., and Kissinger, P. T. (1979). Evidence for the involvement of N-acetyl-p-benzoquinoneimine in acetaminophen metabolism. *Biochem. Pharmacol.* **28,** 3285–3290.
84. Corcoran, G. B., Mitchell, J. R., Vaishnav, Y. N., and Horning, E. C. (1980). Evidence that acetaminophen and N-hydroxyacetaminophen form a common arylating intermediate, N-acetyl-p-benzoquinoneimine. *Mol. Pharmacol.* **18,** 536–542.
85. Hinson, J. A., Pohl, L. R., Monks, T. J., and Gillette, J. R. (1981). Minireview: Acetaminophen-induced toxicity. *Life Sci.* **29,** 107–116.
86. Dahlin, D. C., Miwa, G. T., Lu, A. H., and Nelson, S. D. (1984). N-acetyl-p-benzoquinoneimine: A cytochrome P-450-mediated oxidation product of acetaminophen. *Proc. Natl. Acad. Sci. USA* **81,** 1327–1331.
87. McMahon, R. E. (1966). Microsomal dealkylation of drugs. *J. Pharm. Sci.* **55,** 457–465.
88. Griffin, B. W. (1978). Evidence for a free-radical mechanism of N-demethylation of N,N-dimethylaniline and an analog by heme protein–$H_2O_2$ systems. *Arch. Biochem. Biophys.* **190,** 850–853.
89. Kedderis, G. L., Koop, D. R., and Hollengerg, P. F. (1980). N-Demethylation reactions catalyzed by chloroperoxidase. *J. Biol. Chem.* **255,** 10174–10182.
90. Miyata, N., Kiuchi, H., and Hirobe, M. (1981). Oxidative dealkylation of tertiary amines by iron(III) porphyrin–iodosoxylene system as a model of cytochrome P-450. *Chem. Pharm. Bull.* **29,** 1489–1492.
91. Griffin, B. W., and Ting, P. L. (1978). Mechanism of N-demethylation of aminopyrine by hydrogen peroxide catalyzed by horseradish peroxidase, methemoglobin, and protohemin. *Biochemistry* **17,** 2206–2211.
92. Sayo, H., and Hosokawa, M. (1980). Electron spin resonance studies on the cumene hydroperoxide-supported oxidation of aminopyrine by catalase. *Chem. Pharm. Bull.* **28,** 2077–2082.
93. Griffin, B. W., Marth, C., Yasukochi, Y., and Masters, B. S. S. (1980). Radical mechanism of aminopyrine oxidation by cumene hydroperoxide catalyzed by purified liver microsomal cytochrome P-450. *Arch. Biochem. Biophys.* **205,** 543–553.
94. Lasker, J. M., Sivarajah, K., Mason, R. P., Kalyanaraman, B., Abou-Donia, M. B., and Eling, T. E. (1981). A free-radical mechanism of prostaglandin synthase-dependent aminopyrine demethylation. *J. Biol. Chem.* **256,** 7764–7767.
95. Hull, L. A., Davis, G. T., Rosenblatt, D. H., Williams, H. K. R., and Weglein, R. C. (1967). Oxidation of amines. III. Duality of mechanism in the reaction of amines with chlorine dioxide. *J. Am. Chem. Soc.* **89,** 1163–1170.
96. Rosenblatt, D. H., Davis, G. T., Hull, L. A., and Forberg, G. D. (1968). Oxidation of amines. V. Duality of mechanism in the reaction of aliphatic amines with permanganate. *J. Org. Chem.* **33,** 1649–1651.
97. Seo, E. T., Nelson. R. F., Fritsch, J. M., Marcoux, L. S., Leedy, D. W., and Adams, R. N. (1966). Anodic oxidation pathways of aromatic amines. Electrochemical and electron paramagnetic resonance studies. *J. Am. Chem. Soc.* **88,** 3498–3503.

98. Sayo, H., and Masui, M. (1973). Anodic oxidation of amines. Part III. Cyclic voltammetry and controlled potential electrolysis of 4-dimethylaminoantipyrine in acetonitrile. *J. Chem. Soc., Perkins, Trans. 2* pp. 1640–1645.
99. Lindsay-Smith, J. R., and Masheder, O. (1976). Amine oxidation. Part IX. electrochemical oxidation of some tertiary amines: The effect of structure on reactivity. *J. Chem. Soc., Perkin. Trans. 2* pp. 47–51.
100. Augusto, O., Beilan, H. S., and Ortiz de Montellano, P. R. (1982). The catalytic mechanism of cytochrome *P*-450. *J. Biol. Chem.* **257,** 11288–11295.
101. Thompson, J. A., and Holtzman, J. L. (1974). Deuterium and tritium isotope effects on the microsomal N-demethylation of ethylmorphine. *Drug. Metab. Dispos.* **2,** 577–582.
102. Abdel-Monem, M. M. (1975). Isotope effects in enzymatic N-demethylation of tertiary amines. *J. Med. Chem.* **18,** 427–430.
103. Nelson, S. D., Pohl, L. R., and Trager, W. F. (1975). Primary and β-secondary deuterium isotope effects in N-deethylation reactions. *J. Med. Chem.* **18,** 1062–1065.
104. Miwa, G. T., Garland, W. A., Hodshon, B. J., Lu, A. Y. H., and Northrup, D. B. (1980). Kinetic isotope effects in cytochrome *P*-450-catalyzed oxidation reactions. *J. Biol. Chem.* **255,** 6049–6054.
105. Shea, J. P., Valentine, G. L., and Nelson, S. D. (1982). Source of oxygen in cytochrome *P*-450-catalyzed carbinolamine formation. *Biochem. Biophys. Res. Commun.* **109,** 231–235.
106. McMahon, R. E., Culp, H. W., Craig, J. C., and Ekwuribe, N. (1979). Mechanism of the dealkylation of tertiary amines by hepatic oxygenases: Stable isotope studies with 1-benzyl-4-cyano-4-phenylpiperidine. *J. Med. Chem.* **22,** 1100–1103.
107. Gorrod, J. W., and Temple, D. J. (1976). The formation of an *N*-hydroxymethyl intermediate in the N-demethylation of *N*-methylcarbazole *in vivo* and *in vitro*. *Xenobiotica* **6,** 265–274.
108. Novak, R. F., Koop, D. R., and Hollenberg, P. F. (1980). Liver microsomal metabolism of *N*-methylcarbazole: Structural identification of the four major metabolites of *N*methylcarbazole using H'-Fourier transform NMR spectroscopy. *Mol. Pharmacol.* **17,** 128–136.
109. Gescher, A., Hickman, J. A., and Stevens, M. F. (1979). Oxidative metabolism of some *N*-methyl-containing xenobiotics can lead to stable progenitors of formaldehyde. *Biochem. Pharmacol.* **28,** 3235–3238.
110. Murphy, P. J. (1973). Enzymatic oxidation of nicotine to nicotine $\Delta^{1'(5')}$-immonium ion. *J. Biol. Chem.* **248,** 2796–2800.
111. Nguyen, T. L., Gruenke, L. D., and Castagnoli, N., Jr. (1979). Metabolic oxidation of nictone to chemically reactive intermediates. *J. Med. Chem.* **22,** 259–263.
112. Ho, B., and Castagnoli, N., Jr. (1980). Trapping of metabolically generated electrophilic species with cyanide ion: Metabolism of 1-benzylpyrrolidine. *J. Med. Chem.* **23,** 133–139.
113. Ziegler, R., Ho, B., and Castagnoli, N., (1981). Trapping of metabolically generated electrophilic species with cyanide ion: Metabolism of methapyrilene. *J. Med. Chem.* **24,** 1133–1138.
114. Maynard, M. S., and Cho, A. K. (1981). Oxidation of *N*-hydroxyphentermine to 2-methyl-2-nitro-1-phenylpropane by liver microsomes. *Biochem. Pharmacol.* **30,** 1115–1119.
115. Prough, R. A., and Ziegler, D. M. (1977). The relative participation of liver microsomal amine oxidase and cytochrome *P*-450 in N-demethylation reactions. *Arch. Biochem. Biophys.* **180,** 363–373.

116. Bartsch, J., and Hecker, E. (1971). On the metabolic activation of the carcinogen N-hydroxy-N-2-acetylaminofluorene. III. Oxidation with horseradish peroxidase to yield 2-nitrosofluorene and N-acetylaminofluorene. *Biochim. Biophys. Acta.* **237**, 567–578.
117. Floyd, R. A., and Soong, L. M. (1977). Obligatory free-radical intermediate in the oxidative activation of the carcinogen N-hydroxy-2-acetylaminofluorene. *Biochim. Biophys. Acta.* **298**, 244–249.
118. Floyd, R. A. (1981). Free radicals in arylamine carcinogenesis. *Natl. Cancer Inst. Monogr.* **58**, 123–131.
119. McMurry, J. E. (1974). Organic chemistry of low-valent titanium. *Acc. Chem. Res.* **7**, 281–286.
120. Lindeke, B., and Cho, A. K. (1982). N-Dealkylation and deamination. In "Metabolic Basis of Detoxication" (W. B. Jakoby, J. R. Bend, and J. Caldwell, eds.), pp. 105–126. Academic Press, New York.
121. Gutmann, H. R., and Erickson, R. R. (1972). The conversion of the carcinogen N-hydroxy-2-fluorenylacetamide to o-amidophenols by rat liver *in vitro*. *J. Biol. Chem.* **247**, 660–666.
122. Sternson, L. A., and Gammans, R. E. (1975). A mechanistic study of aromatic hydroxylamine rearrangement in the rat. *Bioorg. Chem.* **4**, 58–63.
123. Ford, G. P., and Scribner, J. D. (1981). MNDO molecular orbital study of nitrenium ions derived from carcinogenic aromatic amines and amides. *J. Am. Chem. Soc.* **103**, 4281–4291.
124. Hartman, G. D., and Schlegel, H. B. (1981). The relationship of the carcinogenic/mutagenic potential of arylamines to their singlet–triplet nitrenium ion energies. *Chem.-Biol. Interact.* **36**, 319–330.
125. Kriek, E., and Westra, J. G. (1979). Metabolic activation of aromatic amines and amides and interactions with nucleic acids. In "Chemical Carcinogens and DNA (P. L. Grover, ed.), pp. 1–28. CRC Press, Boca Raton, Florida.
126. Kadlubar, F. F., Unruh, L. E., Beland, F. A., Straub, K. M., and Evans, F. E. (1981). Formation of DNA adducts by the carcinogen N-hydroxy-2-naphthylamine. *Natl. Cancer Inst. Monogr.* **58**, 1–10.
127. Mulder, G. J., Unruh, E., Evans, R. E., Ketterer, B., and Kadlubar, F. F. (1982). Formation and identification of glutathione conjugates from 2-nitrosofluorene and N-hydroxy-2-aminofluorene. *Chem.-Biol. Interact.* **39**, 111–127.
128. Meerman, J. H. N., Beland, F. A., Ketterer, B., Srai, S. K. S., Bruins, A. P., and Mulder, G. J. (1982). Identification of glutathione conjugates formed from N-hydroxy-2-acetylaminofluorene in the rat. *Chem.-Biol. Interact.* **39**, 149–168.
129. Sternson, L. A., Dixit, A. S., and Becker, A. R. (1983). Reaction of phenylhydroxylamine with bisulfite. A possible model for amine-mediated carcinogenesis. *J. Org. Chem.* **48**, 57–60.
130. Ames, M. M, Sanders, M. E., and Tiede, W. S. (1983). Role of N-methylolpentamethylmelamine in the metabolic activation of hexamethylmelamine. *Cancer Res.* **43**, 500–504.
131. Coles, B., Srai, S. K. S., Ketterer, B., Waynforth, B., and Kadlubar, F. F. (1983). Identification of 4'-sulfonyloxy-N-(glutathion-S-methylene)-4-aminoazobenzene, a compound conjugated with both sulphate and glutathione, which is a major biliary metabolite of N,N-dimethyl-4-aminoazobenzene. *Chem. Biol. Interact.* **43**, 123–129.
132. Andrews, L. S., Hinson, J. A., and Gillette, J. R. (1978). Studies on the mutagenicity of N-hydroxy-2-acetylaminofluorene in the Ames *Salmonella* mutagenesis test system. *Biochem. Pharmacol.* **27**, 2399–2408.

133. Wirth, P. J., Dybing, E., von Bahr, C., and Thorgeirsson, S. S. (1980). Mechanism of N-hydroxyacetylarylamine mutagenicity in the *Salmonella* test system: Metabolic activation of N-hydroxyphenacetin by liver and kidney fractions from rat, mouse, hamster, and man. *Mol. Pharmacol.* **18,** 117–127.
134. Beranek, D. T., White, G. L., Helflich, R. H., and Beland, F. A. (1982). Aminofluorene–DNA adduct formation in *Salmonella typhimurium* exposed to the carcinogen N-hydroxy-2-acetylaminofluorene. *Proc. Natl. Acad. Sci. U.S.A.* **79,** 5175–5178.

Chapter 13

# Arylhydroxylamines and Arylhydroxamic Acids: Conjugation Reactions

PATRICK E. HANNA

*Departments of Medicinal Chemistry and Pharmacology*
*University of Minnesota*
*Minneapolis, Minnesota*

R. BRUCE BANKS

*Department of Chemistry*
*University of North Carolina*
*Greensboro, North Carolina*

| | |
|---|---|
| I. Introduction . . . . . . . . . . . . . . . . . . . . | 376 |
| II. Bioactivation of Arylhydroxylamines and Arylhydroxamic Acids by Conjugation Reactions . . . . . . . . . . . . . . . | 376 |
|    A. Bioactivation of Arylhydroxamic Acids by O-Sulfonation . | 377 |
|    B. Bioactivation of Arylhydroxylamines by O-Sulfonation . . | 382 |
|    C. Bioactivation of Heterocyclic N-Hydroxy Compounds By O-Sulfonation . . . . . . . . . . . . . . . . . | 382 |
|    D. Bioactivation of Arylhydroxamic Acids by Glucuronidation. | 383 |
|    E. Bioactivation of Arylhydroxylamines by Glucuronidation . | 385 |
|    F. Bioactivation of Arylhydroxamic Acids by N-Arylhydroxamic Acid $N,O$-Acyltransferase . . . . . . . . . . . . . | 386 |
| III. Fate of Reactive Intermediates Generated by Conjugation of Arylhydroxylamines and Arylhydroxamic Acids . . . . . . | 392 |
| References. . . . . . . . . . . . . . . . . . . . | 395 |

## I. Introduction

Certain members of the arylhydroxylamine (**1**) and arylhydroxamic acid (**2**) classes of compounds have long been associated with the production of untoward biological effects, ranging from methemoglobinemia to tissue necrosis and tumorigenesis.[1-4]

$$\text{ArNHOH} \quad \text{ArNOHCOR}$$
$$\mathbf{1} \qquad\qquad \mathbf{2}$$

Although environmental exposure of mammals to either **1** or **2** is probably not a common occurrence, it is well established that compounds belonging to these chemical classes may accumulate in mammalian tissues as a result of the metabolic N-oxidation of arylamines (**3**) and arylamides (**4**).[1] Indeed, the presence of arylamides in mammalian tissues often is due to metabolic N-acetylation of arylamines (**3**) rather than to exposure to arylamides (**4**) themselves.[5] Similarly, arylhydroxylamines **1** may arise from the action of deacylases on arylhydroxamic acids (**2**)[6,7] or from the effects of reductases on arylnitro compounds.[1] Corbett and co-workers have demonstrated the potential for arylhydroxamic acid production from the interaction of arylnitroso compounds with certain thiamin-dependent enzymes.[8]

$$\text{ArNH}_2 \quad \text{ArNHCOR}$$
$$\mathbf{3} \qquad\qquad \mathbf{4}$$

There are a number of differences in the physical properties and chemical behavior of **1** and **2**. The conversion of an arylamine (**3**) to an arylhydroxylamine (**1**) results in a slight decrease in the basicity of the compound, whereas the conversion of an arylamide (**4**) to an arylhydroxamic acid (**2**) involves the formation of an acidic product from a functional group that, under normal circumstances, is nonionizable (neutral). Most arylhydroxylamines (**1**) are susceptible to chemical or enzymatic oxidation processes that convert them to arylnitroso and arylnitro compounds, whereas most arylhydroxamic acids (**2**) are less readily oxidized. Similarly, the chemical and enzymatic reduction of arylhydroxylamines (**1**) to arylamines (**3**) is a more facile process than is the reduction of arylhydroxamic acids (**2**) to arylamides (**4**).

## II. Bioactivation of Arylhydroxylamines and Arylhydroxamic Acids by Conjugation Reactions

The pioneering research efforts of James and Elizabeth Miller and others have led to wide acceptance of the concept that the ultimate carcinogenic forms of nearly all organic chemical carcinogens, including arylhydroxylamines (1) and arylhydroxamic acids (2), are electrophilic (i.e., electron-deficient) intermediates that react nonenzymatically with the nucleophilic functional groups of cellular constituents.[9,10] Covalent bond formation between the carcinogen and critical nucleophilic target sites is believed to be a key factor in the initiation of neoplasia. The reactive-electrophile concept not only accounts for much of the data relating to the mechanism of action of chemical carcinogens and mutagens, but also affords at least a partial explanation of the mechanisms of toxicity of a variety of drugs and chemicals.[11] Unfortunately, the critical nucleophilic functional group target sites for most carcinogens, mutagens, and toxic organic chemicals have not been determined with certainty.

Although it has been demonstrated that the arylhydroxylamine carcinogen $N$-hydroxy-2-aminofluorene can become covalently bound to DNA *in vitro* (pH 7.5) by a nonenzymatic process,[12] most arylhydroxylamines (1) and arylhydroxamic acids (2) do not exhibit an inherent high degree of electrophilic reactivity. Therefore, either chemical or enzymatic activation is required before most compounds of these classes form significant amounts of covalent adducts with biological target sites. The major focus of research in this area has been on bioactivation processes involving the sulfotransferases, the UDPglucuronosyltransferases, and the $N,O$-acyltransferases. Although this chapter will emphasize studies with these three enzyme systems, it is important to note that a growing body of literature indicates potentially important roles for other processes in the bioactivation of 1 and 2. Bartsch and Hecker,[13] Floyd,[14] and Forrester and co-workers[15] have demonstrated that arylhydroxamic acids can be activated chemically and enzymatically through a free-radical mechanism that results in the formation of arylnitroso derivatives and $N$-acetoxyarylamides, both of which are more reactive than the arylhydroxamic acids from which they are formed.

The carcinogenic and mutagenic tryptophan pyrolysate 3-amino-1-methyl-5$H$-pyrido[4,3-$b$]indole is metabolized to its $N$-hydroxy derivative, which becomes covalently bound to DNA either without further activation[16] or on incubation with prolyl–tRNA synthetase present in rat liver cytosol.[17] The mechanism of this latter bioactivation process is presumed to involve conjugation of the arylhydroxylamine with the amino acid proline. Yeast seryl–tRNA synthetase has been shown to be capable of catalyzing the covalent binding of both 4-hydroxyaminoquinoline 1-oxide and several $N$-hydroxy-4-aminoazobenzene dyes to nucleic acids.[18,19]

## A. BIOACTIVATION OF ARYLHYDROXAMIC ACIDS BY O-SULFONATION

By 1970 it was well established that the arylhydroxamic acid carcinogen $N$-hydroxy-2-acetamidofluorene (5) (Fig. 1) undergoes both O-glucuronida-

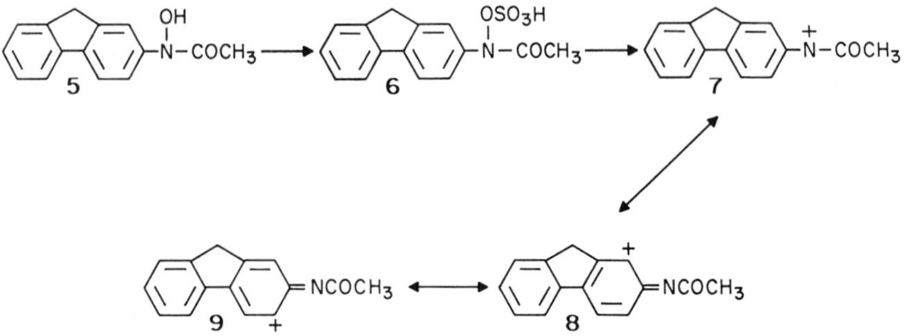

Fig. 1. Bioactivation of N-hydroxy-2-acetamidofluorene (5) by sulfate conjugation.

tion and O-sulfonation *in vitro* and *in vivo,* and that both types of conjugates display sufficient chemical reactivity to form covalent bonds with nucleophilic functional groups. The early studies in this area have been reviewed by Irving.[20,21]

The earliest reports implicating the sulfotransferase system in the bioactivation of the carcinogen N-hydroxy-2-acetamidofluorene (5) were those of DeBaun et al.[22,23] and of King and Phillips.[24,25] The recognition that the *in vitro* reaction of various synthetic esters of 5, including the sulfate (6) (Fig. 1), with methionine and other nucleophiles produced the same adducts obtained when 5 was administered to rats and the observation that hepatic sulfotransferase activity levels correlated roughly with sex and species susceptibility to hepatocarcinogenesis by 5 led to the suggestion that 6 is one of the ultimate reactive and carcinogenic metabolites of 5 (Fig. 1). It was proposed that the N–O bond of 6 undergoes heterolytic cleavage to generate the electrophilic N-acetyl-N-arylnitrenium ion (7), as well as the canonical forms 8 and 9. The participation of intermediates 7, 8, and 9 accounts for the products formed when 6 and related synthetic esters of 5 are allowed to react with nucleosides, nucleotides, and nucleic acids under physiologically relevant conditions.[4]

The sulfotransferases are cytosolic enzymes that catalyze the transfer of the sulfate residue to hydroxyl and amino groups of organic compounds, a process that results in the formation of sulfates and sulfamates, respectively. The sulfate donor is 3'-phosphoadenosine 5'-phospholsulfate (PAPS) ("activated sulfate"), which is formed from adenosine 5'-phosphosulfate and ATP in a reaction catalyzed by adensosine 5'-phosphosulfate kinase. The biochemical, pharmacological, and toxicological properties of the sulfo transferases have been reviewed.[26-28] Mammalian tissues contain several sulfotransferases that exhibit different substrate specificities. The rat live sulfotransferase that is responsible for the bioactivation of 5 (Fig. 1) has bee purified 2000-fold by Wu and Straub.[29]

A number of studies were undertaken in an attempt to define the relationship between arylhydroxamic acid bioactivation by sulfate conjugation and adduct formation, toxicity, and carcinogenicity in various species and tissues. DeBaun *et al.* found that the amount of 1- and 3-methylmercapto-2-acetamidofluorene obtained from the hepatic protein of male rats was proportional to the dose of **5** and paralleled sulfotransferase activity and the susceptibility of male rat liver to carcinogenesis by **5**. Female rats and male mice, hamsters, rabbits, and guinea pigs were much less susceptible to hepatic carcinogenesis by **5**, formed less hepatic protein-bound 1- and 3-methylmercapto-2-acetamidofluorene, and exhibited less hepatic sulfotransferase activity.[22] Gutmann *et al.* also concluded that the susceptibility of male rats to hepatocarcinogenesis by **5** is related to hepatic sulfotransferase activity,[30] and DeBaun *et al.*[22] and Lotlikar[31] demonstrated that administration of testosterone to female rats enhanced hepatic sulfotransferase activity; no significant sex differences in hepatic sulfotransferase activity were observed in mice, hamsters, and guinea pigs.[31] When the $LD_{50}$ of the hydroxamic acid **5** was determined in several strains of rats of both sexes, it was found that the toxic dose was lower in animals with high levels of sulfotransferase activity.[32]

The use of inhibitors of hepatic sulfotransferase activity, such as 4-nitrophenol and pentachlorophenol, has provided compelling evidence for a causal relationship between rat hepatic sulfotransferase activity and both the hepatotoxicity of **5** and the formation of covalent adducts of **5** that retain the $N$-acetyl group.[33-37] The results of earlier studies, in which acetanilide and 4-hydroxyacetanilide were found to inhibit the hepatocarcinogenicity of **5**, were interpreted in part on the basis of the presumed abilities of acetanilide and 4-hydroxyacetanilide to deplete endogenous sulfate.[38-40] The data from the studies with acetanilide and 4-hydroxyacetanilide, however, are less definitive than those with 4-nitrophenol and pentachlorophenol, because 4-hydroxyacetanilide has effects on the consumption and metabolism of **5**[41] and because both acetanilide and its 4-hydroxy analog can inhibit enzyme systems other than the sulfotransferases, which may be involved in the bioactivation of **5**.[42]

## Structure–Activity Relationships

Although a number of arylhydroxamic acids are known to undergo conversion to sulfate derivatives by rat hepatic sulfotransferases, the rates at which an arylhydroxamic acid is converted to a sulfate conjugate often is not measured readily because of the lability of the product. Therefore, the rate at which the electrophiles generated by sulfate conjugation form adducts with nucleophiles has been used as an indirect indication of the rate of sulfate conjugation. The activities of the N-hydroxy compounds in these *in vitro* assays have been measured by the rate of PAPS-dependent production of o-methylmercapto adducts by the use of methionine as a nucleophilic trapping agent[23,31,43,44] and by formation of adducts with nucleosides, nucleic acids,[25,43,45-47] or protein.[48,49] Such assays, however, may not accurately reflect the relative conjugation rates of arylhydroxamic acids, because the rate of adduct formation depends on both the stability of the sulfate conjugates themselves and the reactivity of the electrophiles generated by the conjugates. Another factor that must be considered is the ability of the nucleophiles used in the assays to react with the electrophilic species produced from the sulfate conjugates. Irving[50,51] used the rate of PAPS-dependent formation of water-soluble metabolites as an indication of sulfotransferase activity and concluded that N-hydroxy-2-acetamidophenanthrene is a good substrate, even though it was poorly active in the methionine-trapping assay.[23] The activity of N-arylhydroxamic acids as sulfotransferase substrates has also been estimated by the rate of p-nitrophenol formation when p-nitrophenyl sulfate and 3′,5′-adenosine diphosphate were used to generate PAPS.[48,49]

DeBaun *et al.*[23] measured the o-methylmercapto adducts obtained after incubation of N-arylhydroxamic acids with male rat liver cytosol fortified with PAPS. N-Hydroxy-2-acetamidofluorene (**5**) yielded the largest amount of adduct, followed by N-hydroxy-4-acetamidobiphenyl, N-hydroxy-4-acetamidoazobenzene, and N-hydroxy-4-acetamidostilbene. N-Hydroxy-2-acetamidophenanthrene formed a very small amount of adduct, and both N-hydroxy-1-acetamidonaphthalene and N-hydroxy-2-acetamidonaphthalene were inactive in the assay. Lotlikar reported a similar order of activity for the fluorene, biphenyl, stilbene, and phenanthrene compounds in the methionine-trapping assay.[31]

The results just described indicate that the structure of the aryl portion of N-arylhydroxamic acids is a determining factor in the generation of reactive intermediates in PAPS-dependent processes. However, the data do not permit conclusions with regard to the particular role of the aryl group in determining the rate of sulfate conjugation, the stability of the conjugates, or the reactivity of the electrophiles with methionine. Somewhat more insight into the role of the aryl group in sulfotransferase-catalyzed activation processes is obtained by examination of the results of Mulder *et al.*,[48] who used the p-nitrophenyl sulfate–3′,5′-adenosine diphosphate system to demonstrate that N-hydroxyphenacetin (**10**), N-hydroxyacetanilide, N-hydroxy-p-chloroacetanilide, N-hydroxy-2-acetamidofluorene (**5**), and N-hydroxy-2-acetamidonaphthalene undergo sulfate conjugation at similar rates in the presence of the postmicrosomal supernatant obtained from male rat liver. In contrast, only **10** and **5** exhibited extensive covalent binding to protein

$$H_5C_2O-C_6H_4-N(OH)-COCH_3$$

**10**

under these conditions. Because the measurement of covalent binding in these experiments depended on the retention of the radiolabeled acetyl group in each of the covalently bound residues, any conclusions regarding structure–activity relationships require that no loss of the $N$-acetyl group occurred under the experimental conditions. With this qualification, it is reasonable to conclude that the structure of the aryl group was not a significant factor in the rate of sulfate conjugation, but that the $p$-ethoxy group of **10** greatly enhanced the reactivity of the sulfate conjugate compared to the effect of a $p$-chloro group or $p$-hydrogen. Similarly, the 2-substituted fluorene ring system exhibited greater reactivity than the 2-substituted naphthalene ring.

The position of the aromatic ring to which the $N$-hydroxyacetamido group is attached would be expected to influence the rate of bioactivation of arylhydroxamic acids by sulfate conjugation. Unfortunately, little information is available on this aspect of structure–activity relationships. Zieve and Gutmann found that incubation of radiolabeled $N$-hydroxy-2-acetamidofluorene (**5**) with soluble proteins of rat liver and the cofactor required for sulfate conjugation resulted in covalent binding of radioactivity to tRNA.[43] However, the treatment of the positional isomer, $N$-hydroxy-3-acetamidofluorene, under the same conditions did not result in the formation of adducts with tRNA. Although it is possible that the 3-substituted compound is not a sulfotransferase substrate, it seems more likely that the lack of adduct formation is due to low reactivity of the sulfate conjugate. It was demonstrated that the synthetic ester $N$-acetoxy-3-acetamidofluorene is unreactive, whereas the corresponding derivative of **5** reacts readily with methionine to form methylmercapto adducts.[43] $N$-Hydroxy-3-acetamidofluorene may be an exception to the general observation that the esters of carcinogenic arylhydroxic acids readily form electrophilic intermediates.

Weeks *et al.* studied the sulfotransferase activity of the $N$-formyl, $N$-propionyl, and $N$-chloroacetyl analogs of **5** by quantification of the formation of adducts between tRNA and the radiolabeled arylhydroxamic acids in the presence of rat liver supernatant and a PAPS-generating system. None of the compounds was more active than **5**, and there was less than a fourfold range of activity for the four compounds.[45] Although the structure variation in this set of compounds is limited, the results indicate that sulfotransferase-catalyzed activation of arylhydroxamic acids is not highly sensitive to the structure of the acyl group.

## B. BIOACTIVATION OF ARYLHYDROXYLAMINES BY O-SULFONATION

Incubation of the hepatocarcinogenic dye N-hydroxy-N-methyl-4-aminoazobenzene (11) with rat hepatic cytosol in the presence of PAPS resulted in the formation of a reactive intermediate, presumably the sulfate conjugate, which formed adducts with methionine, guanosine, rRNA, and glutathione.[52,53] The properties of the sulfotransferases that activate the hydroxylamine derivatives of the aminoazo dyes appeared to be different from those that activate the arylhydroxamic acid (5). Female rat liver had approximately 25% of the activity of male rat liver, and male rat kidney had 20% of the activity of liver. Under optimal incubation conditions for sulfate formation, considerably less adduct was formed between nucleophiles and the electrophiles produced from 11 than from the PAPS-dependent activa-

$$\langle\bigcirc\rangle-N=N-\langle\bigcirc\rangle-\underset{\underset{OH}{|}}{N}-CH_3$$

11

tion of 5.[52] Rabbit, guinea pig, and mouse hepatic cytosols had approximately 25% of the activity of male rat liver, but very little activity was found in hamster liver cytosol.

### Structure-Activity Relationships

The PAPS-dependent system that was found to be optimal for the bioactivation of 11 was used to study the sulfotransferase-catalyzed activation of several other N-arylhydroxylamines.[52] The activities of N-hydroxy-N-ethyl-4-aminoazobenzene, N-hydroxy-4-aminoazobenzene, N-hydroxy-4-aminobiphenyl, and N-hydroxy-1- and 2-naphthylamine were 20–50% of that of 11 as measured by the PAPS-dependent disappearance of substrate or the formation of methylmercaptoarylamine adducts. Those arylhydroxylamines found to be inactive under these conditions were trans-N-hydroxy-4-aminostilbene, N-hydroxy-2-aminofluorene, N-hydroxyaniline, and N-hydroxy-N-methyl-N-benzylamine. The reported inactivities cannot be attributed to a lack of reactivity of the sulfates or to rearrangement of the sulfates to undetected products, because, in each case, there was an absence of PAPS-dependent substrate loss.[52]

## C. BIOACTIVATION OF HETEROCYCLIC N-HYDROXY COMPOUNDS BY O-SULFONATION

Several oxidized purines are carcinogenic in rats and are substrates for rat hepatic sulfotransferases.[54] 3-Hydroxyxanthine (12) is activated by sulfo

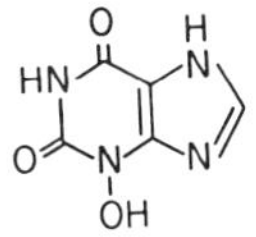

**12**

transferases, and possibly by phosphotransferases, to an electrophilic intermediate that reacts with methionine to form 8-methylmercaptoxanthine.[55] Although a number of N-hydroxypurines are converted to electrophiles by liver and fibroblast sulfotransferases, the relationship between O-sulfonation and carcinogenicity is not well established for this class of compounds, because some that are carcinogens do not appear to be subject to conjugation by sulfotransferases.[54] Similarly, the high degree of reactivity of esters of certain N-hydroxypurines may preclude carcinogenicity, because the esters may react with noncritical nucleophiles before they reach target nucleophiles that are relevant to tumor induction.[56]

## D. BIOACTIVATION OF ARYLHYDROXAMIC ACIDS BY GLUCURONIDATION

The search for endogenous reactive derivatives of N-arylhydroxamic acids that could account for the covalently bound adducts formed *in vivo* has included a number of investigations of O-glucuronide conjugates. Most N-arylhydroxamic acids are readily conjugated with glucuronic acid by glucuronyltransferases that are found in the endoplasmic reticulum and nuclear envelope of the cell.[57] The glucuronyltransferases catalyze the transposition of glucuronic acid from UDPglucuronic acid to an appropriate nucleophilic acceptor group as illustrated for the glucuronidation of **5** in Fig. 2.

The glucuronide (**13**) (Fig. 2) undergoes reaction *in vitro* at pH 7 to form adducts with methionine and guanosine that retain the acetyl group and are identical with the products formed from the synthetic N-acetoxy ester of **5**.[58] Thus, the glucuronide appears to undergo heterolytic decomposition to form the same electrophilic intermediates (**7–9**) produced by the sulfate conjugate (Fig. 1). However, the glucuronide (**13**) is much less reactive than

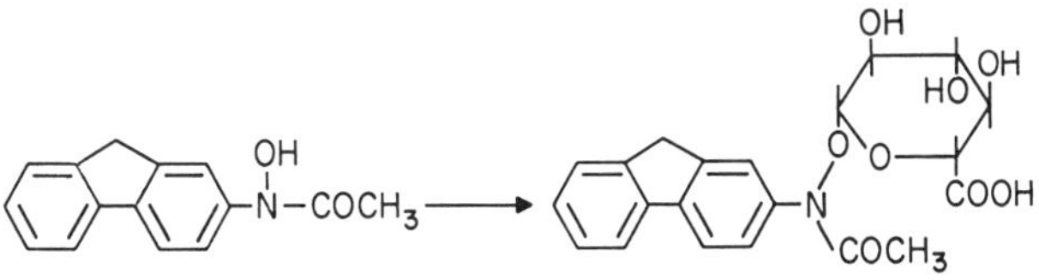

Fig. 2. Glucuronide conjugation of N-hydroxy-2-acetamidofluorene (**5**).

the $N$-acetoxy ester of **5**, and an increase in the pH of the incubation medium to 7.5 results in the formation of a substantial portion of adducts that do not retain the $N$-acetyl group.[58,59] An increase in pH also caused an increase in the total amount of adduct formed from **13**, but a smaller increase in adduct formation was observed with the $O$-glucuronide of $N$-hydroxy-4-acetamidobiphenyl, which was much less reactive thn **13** and which also formed mainly deacetylated adducts.[60]

The possibility that the $O$-glucuronide of $N$-hydroxy-2-aminofluorene might be an intermediate in a glucuronyltransferase-catalyzed bioactivation sequence leading to the formation of 2-aminofluorene adducts with biological nucleophiles prompted Irving and Russell to synthesize this compound, which proved to be much more reactive *in vitro* with nucleic acids than **13**.[61] Although no evidence has appeared in support of the glucuronyltransferase-catalyzed formation of the $O$-glucuronide of $N$-hydroxy-2-aminofluorene, Cardona and King found that **13** undergoes deacetylation in the presence of guinea pig hepatic microsomes to produce a reactive intermediate, presumably the $O$-glucuronide, that forms aminofluorene adducts with tRNA.[62]

$O$-Glucuronides constitute prominent proportions of the metabolites of most $N$-arylhydroxamic acids, but it is not known whether they play a significant role in the production of carcinogenic effects. Only **13** has been tested for carcinogenicity, and its effects were relatively weak.[50] It has been suggested that glucuronides of N-hydroxylated amines and amides serve as transport forms of the carcinogens, and this may play an important role in the distribution of the carcinogens to various tissues (see Section II,E).

### Structure–Activity Relationships

An investigation of the reactivity *in vitro* of the $O$-glucuronides of several carcinogenic $N$-arylacetohydroxamic acids revealed that **13** was the most reactive, as measured by the formation of adducts with nucleic acids.[63] The order of reactivity of the $O$-glucuronides with nucleic acids was **13** > glucuronide of $N$-hydroxy-*trans*-4-acetamidostilbene > glucuronide of $N$-hydroxy-4-acetamidophenyl > glucuronide of $N$-hydroxy-2-acetamidophenanthrene.

Phenacetin is an analgesic drug that is metabolically converted to several reactive intermediates and to the relatively unreactive $N$-hydroxy derivative (**10**).[48,64] Compound **10** is of considerable toxicological interest, because it is carcinogenic and because it has been identified as a proximate mutagenic metabolite of phenacetin.[65,66] The glucuronide conjugate of **10** is less reactive than the corresponding sulfate conjugate (Section II,A), but it is much more reactive than the glucuronide (**13**), as determined by the extent of covalent binding to protein *in vitro*. The glucuronide of $N$-hydroxy-4 chloroacetanilide formed at a faster rate than the glucuronide of **10**, but the conjugate of the 4-chloro compound did not undergo covalent binding to protein.[48] This indicates that the electron-releasing properties of the 4-substituent are an important factor in determining the reactivity of the conjugates of N-hydroxyacetanilides. Studies of the decomposition and reaction with nucleophiles of the glucuronide of **10** indicate that the ethyl group is lost in the process of covalent binding to protein and that the mechanisms involved are complex.[49,67]

## E. BIOACTIVATION OF ARYLHYDROXYLAMINES BY GLUCURONIDATION

O-Glucuronides of N-arylhydroxylamines have not been directly detected or isolated from biological systems, presumably because of their extreme instability. As mentioned previously, Irving and Russell synthesized the O-glucuronide of N-hydroxy-2-aminofluorene, which reacted readily with guanosine 5'-monophosphate,[61] and Radomski et al. later described the synthesis of the O-glucuronide of N-hydroxy-4-aminobiphenyl, which underwent rapid decomposition.[68]

Because UDPglucuronyltransferase activity is present in most tissues[69] the possibility must be considered that O-glucuronidation may be responsible for bioactivation of some fraction of the N-arylhydroxylamine to which a tissue is exposed. It should be noted, however, that the UDPglucuronate-dependent conjugation of N-hydroxy-2-naphthylamine in the presence of DNA and rRNA resulted in less than 0.02% conversion of the hydroxylamine to nucleic acid-bound adducts.[70] Thus, if the O-glucuronide was formed under these conditions, it can be assumed to have undergone either intramolecular rearrangement or reaction with nucelophiles other than those present on nucleic acids.

Neither N-glucuronide nor sulfamate conjugates of N-arylhydroxylamines are considered to be compounds that readily lead to the generation of reactive electrophiles. Although sulfamates, such as that of N-hydroxy-2-naphthylamine (**14**), tend to undergo rearrangement to o-sulfate esters and

**14**

o-hydroxy derivatives,[52] the N-glucuronide (**15**) is stable under physiological conditions and appears to serve as a transport form of the carcinogen.[68,70-76] Compound **15**, although it is stable at normal physiological pH, is hydro-

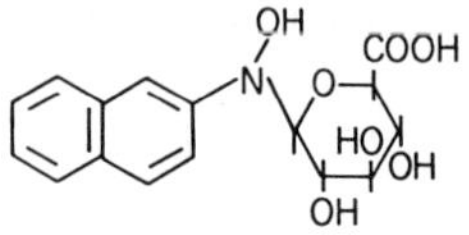

                                15

lyzed at pH 5 to the N-hydroxyarylamine, which is converted to the highly reactive arylnitrenium ion.[70] These observations led to the development of the concept whereby 15 serves as a proximate carcinogen, and the N-hydroxy derivative is proposed to be the ultimate carcinogenic form of the bladder carcinogen 2-naphthylamine.[70,72–76]

## F. BIOACTIVATION OF ARYLHYDROXAMIC ACIDS BY N-ARYLHYDROXAMIC ACID N,O-ACYLTRANSFERASE

The occurrence of aminoarene adducts as the predominant molecular lesions of DNA after *in vivo* administration of N-hydroxy-2-acetamidofluorene[77–79] stimulated interest in enzymes that activate the carcinogen to electrophiles with simultaneous loss of the acyl group. A cytosolic acyltransferase has been implicated in the covalent binding of aminoarene moieties to DNA and in the production of hepatic and extrahepatic tumors by arylhydroxamic acids. The available evidence indicates that the enzyme catalyzes the rearrangement of arylhydroxamic acids (2) to N-acyloxyarylamines[80], and that the N-acyloxyarylamines (16) resulting from this process are extremely reactive electrophiles that react with a variety of biological nucleophiles, including amino acids,[80] RNA[42] DNA[81] and proteins[25] (Fig. 3). An N,O-acyl transfer mechanism has also been proposed to

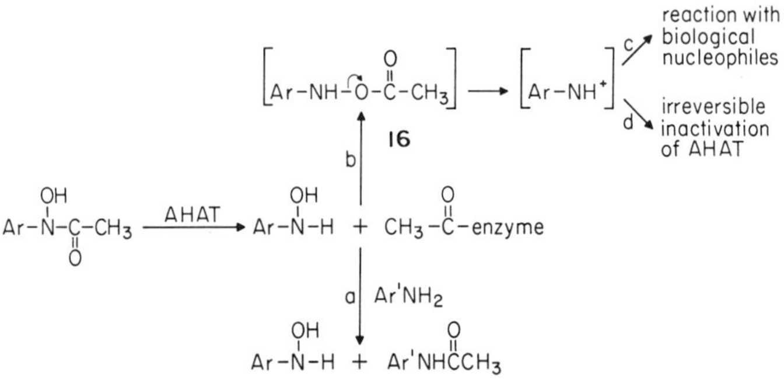

**Fig. 3.** Bioactivation of N-arylhydroxamic acids by N-arylhydroxamic acid N,O-acyltransferase (AHAT).

be involved in the bioactivation of arylhydroxamic acids by liver microsomal deacylases.[82]

In addition to promoting the conversion of arylhydroxamic acids to electrophiles, $N,O$-acyltransferase catalyzes the transfer of acyl groups from arylhydroxamic acids to arylamines, such as 4-aminoazobenzene, 2-aminofluorene, 4-aminobiphenyl, and benzidine (Fig. 3, pathway a). This transacylation activity, which was first reported by Booth,[83] has been shown to copurify with $N,O$-acyltransferase.[80] Rabbit liver $N,O$-acyltransferase is identical with an acetyl Coenzyme A: arylamine $N$-acetyltransferase (EC 2.3.1.5) that catalyzes the acetylation of carcinogens, such as 2-aminofluorene and benzidine, and drugs, such as sulfamethazine and isoniazid.[84] $N,O$-Acyltransferase is widely distributed in mammalian tissues, including those of humans.[80,85,86] In most species, the highest levels of the enzyme occur in the liver; rabbits, hamsters and rats have particularly high levels of the hepatic $N,O$-acyltransferase.[80,86] Several tissues contain multiple forms of the enzyme that are distinguishable by their immunochemical properties, chromatographic behavior, molecular weight, and substrate specificity,[81,87-89], and purification studies have been conducted with $N,O$-acyltransferase from rat and rabbit liver and rat mammary gland.[42,84,90,91] The general properties, purification methods, and assays for cytosolic arylhydroxamic acid $N,O$-acyltransferase have been reviewed.[88,92,93]

## 1. Mechanism of Arylhydroxamic Acid Activation by N,O-Acyltransferase

The mechanism of bioactivation of arylhydroxamic acids by $N,O$-acyltransferase appears to be more complex than the processes involved in sulfotransferase- or glucuronyltransferase-catalyzed bioactivation (Figs. 1 and 2). Studies of the action of $N,O$-acyltransferase support a two-step mechanism for the transformation of arylhydroxamic acids to electrophiles and for the transacylation of aromatic amines by arylhydroxamic acids. The involvement of the $N$-acetoxy intermediate (16) was originally proposed by Bartsch and co-workers[80] and was incorporated into the model proposed by King.[42] The initial step in the mechanism is enzymatic deacylation of the arylhydroxamic acid to liberate an arylhydroxylamine (1); this step results in the formation of an acyl–enzyme complex, which can subsequently acylate either the oxygen of the arylhydroxylamine to form an electrophilic $N$-acyloxyarylamine (16) or the nitrogen of an aromtic amine to form an amide (Fig. 3). The ability of $N,O$-acyltransferase to accept acetyl-CoA as a substrate is easily accommodated by this model if it is assumed that both arylhydroxamic acids and acetyl-CoA can function as acyl donors to form an acyl–enzyme complex.

Several experiments support the proposed mechanism and reveal details about individual enzymatic steps. The ability of rat liver cytosol to catalyze the incorporation of aminofluorene residues from N-hydroxy-2-acetamidofluorene (5) into RNA was demonstrated by King and Phillips in 1968.[24] Experiments with purified enzyme preparations have confirmed that loss of the acyl group of arylhydroxamic acids is associated with N,O-acyltransferase activity.[42,84,91]

Bartsch and co-workers showed that coincubation of an arylhydroxamic acid, an arylhydroxylamine, and rat liver cytosol resulted in the production of electrophiles that could be trapped as methylthio adducts with the nucleophile N-acetylmethionine.[80] No adducts were obtained when arylhydroxylamines were incubated with enzyme in the absence of arylhydroxamic acid. Incubations of N-hydroxy-2-acetamidofluorene (5) with N-hydroxy-4-aminobiphenyl produced methylthio adducts containing the biphenyl ring of the arylhydroxylamine, as well as adducts containing the fluorene ring of the arylhydroxamic acid. Adducts between arylhydroxamic acids and N-acetylmethionine were also formed in the absence of arylhydroxylamine, but longer incubation times were required. These results are consistent with the initial formation and release of arylhydroxylamines during deacylation of arylhydroxamic acids by N,O-acyltransferase and with the participation of these arylhydroxylamines in the formation of electrophiles. Booth has reported the production of arylhydroxylamines during incubation of arylhydroxamic acids with 4-aminoazobenzene and rat liver cytosol.[83] The observations that arylhydroxylamines exert a stimulatory effect on adduct formation and that mixed adducts are formed during incubations of arylhydroxamic acids and arylhydroxylamines containing different ring systems provide strong evidence that enzymatic N,O-acyl transfer is an intermolecular, rather than strictly intramolecular, process.

Treatment of arylhydroxylamines with acetic anhydride in the presence of N-acetylmethionine was shown to produce the same methylthio adducts as formed during incubations of the corresponding arylhydroxamic acids.[80] This result is consistent with the formation of reactive N-acetoxyarylamines (16) (Fig. 3) by acetylation of the hydroxylamine oxygen. The N-acetoxyarylamines are expected to be reactive electrophiles by analogy with N-acetoxy-N-arylacetamides and N-benzoyloxy-N-methyl-4-aminoazobenzene, which react nonenzymatically with nucleophiles subsequent to loss of the acyloxy group.[94,95]

In addition to acting as substrates, certain carcinogenic arylhydroxamic acids irreversibly inhibit N,O-acyltransferase *in vivo* and act *in vitro* as mechanism-based irreversible inhibitors (suicide substrates).[96,97] Incubation of partially purified rat or hamster hepatic N,O-acyltransferase with N-hydroxy-2-acetamidofluorene results in an irreversible time-dependent loss of the enzyme's ability to catalyze subsequent arylhydroxamic acid activation or 4-aminoazobenzene transacylation. Cysteine and other nucleophiles retard, but do not prevent, the inactivation.[97] The experimental data for suicide inactivation of $N,O$-acyltransferase are consistent with the activation of arylhydroxamic acids to electrophiles that can be trapped immediately by nucleophilic functional groups at or near the active site of the enzyme or that can escape the enzyme to form covalent adducts with nucleic acids and other biological nucleophiles, including those present on $N,O$-acyltransferase itself (Fig. 3). The partial protection of the enzyme by cysteine and other small nucleophiles has been attributed to prevention of the inactivation of $N,O$-acyltransferase by electrophiles that are released from the enzyme.[97]

Hanna and co-workers have shown that acetyl-CoA-dependent $N$-acetyltransferase activities are also irreversibly inactivated during incubation of partially purified hamster hepatic $N,O$-acyltransferase preparations with $N$-hydroxy-2-acetamidofluorene (**5**.)[97] These results support the previous conclusion that $N,O$-acyltransferase and $N$-acetyltransferase activities are associated with the same enzyme.[84] However, $p$-aminobenzoic acid $N$-acetyltransferase activity, in contrast with sulfamethazine $N$-acetyltransferase and $N,O$-acyltransferase activities, is protected against inactivation when cysteine is included in the incubation mixtures. This indicates that, although $N,O$-acyltransferase activity may be associated with sulfamethazine $N$-acetyltransferase in hamster liver, $p$-aminobenzoic acid $N$-acetyltransferase is a different enzyme. The successful separation from rapid acetylator hamster liver of two $N$-acetyltransferases, which have different substrate selectivities for sulfamethazine and $p$-aminobenzoic acid[98] and which differ in their abilities to bioactivate arylhydroxamic acids,[99] supports the conclusions drawn from the suicide inactivation experiments with regard to $N$-acetyltransferase mutiplicity.

## 2. Structure–Activity Relationships

Structural effects on $N,O$-acyltransferase-catalyzed reactions operate at several levels of the proposed enzymatic mechanism. For example, substrate structure affects the individual rates of deacylation and subsequent acyl transfer reactions, and, likewise, the chemical fate of enzymatically generated $N$-acyloxyarylamines (**16**, Fig. 3) are subject to structural effects on their reactivity (e.g., hydrolysis, rearrangement, reaction with nucleophiles).

A free $N$-hydroxyl group is required for the conversion of arylhydroxamic acids to reactive electrophiles by $N,O$-acyltransferase. $N$-Methoxy-2-acetamidofluorene did not form adducts with tRNA under conditions that supported transacetylation of an arylamine in the presence of rat liver cytosol.[42] Movement of the hydroxy group from the nitrogen to the ortho positions of the aromatic ring or to a carbon atom of the acyl group dramatically diminishes the ability of potential substrates to support enzymatic transacylation.[100] The presence of an *N*-hydroxyl group, however, is not strictly required for compounds to act as acyl donors to *N,O*-acyltransferase; for example, *N*-2-naphthoylhydroxylamine *O*-acetate is an efficient acetyl donor for 4-aminoazobenzene in the presence of rat liver cytosol,[101] and 3-hydroxy-4-acetamidobiphenyl and *N*-methoxy-2-acetamidofluorene also support transacylation, although at much lower rates than the corresponding *N*-arylhydroxamic acids.[42,100]

Replacement of the *N*-aryl group of *N*-arylhydroxamic acids with cycloalkyl or alkyl groups drastically reduced the ability of hydroxamic acids to function as acyl donors in *N,O*-acyltransferase-catalyzed reactions. These types of compounds are ineffective as acetyl donors to *N*-hydroxy-4-aminobiphenyl or 4-aminoazobenzene, indicating that the presence of an *N*-aryl group enhances the deacylation of hydroxamic acids by *N,O*-acyltransferase to form the initial acyl–enzyme complex (Fig. 3).[102]

The ability of *N*-hydroxy-*N*-phenylacetamides (**17**) to serve as *N,O*-acyltransferase substrates is strongly dependent on the type of para substituent that is present on the ring. Mangold and Hanna found that such compounds become more effective substrates for rat or hamster hepatic *N,O*-acyltransferase-catalyzed arylamine transacetylation as the para substituent of **17** increases in length from hydrogen to *n*-pentyl. The presence of unsat-

$$R\text{–}C_6H_4\text{–}N(OH)\text{–}COCH_3$$

17

uration in the para substituent further enhances substrate activity.[103] 4-Cyclohexyl-*N*-hydroxy-*N*-phenylacetamide (**17**, R = cyclohexyl) was activated by hamster hepatic *N,O*-acyltransferase to an electrophile, which could be trapped with 2-mercaptoethanol.[103] Vaught and co-workers have shown that *N*-hydroxyphenacetin (**10**), a hepatocarcinogen in the rat, is a substrate for purified rat and rabbit hepatic *N,O*-acyltransferase.[46]

A large number of *N*-arylhydroxamic acids containing polycyclic aromatic systems act as substrates for *N,O*-acyltransferase; these include arylhydroxamic acids derived from 2-acetamidofluorene, 4-acetamidobiphenyl, 4-acetamidoazobenzene, diacetylbenzidine, 4-acetamidostilbene, 2-acetamidonaphthalene, and 2-acetamidophenanthrene.[80,83,104] *N*-Hydroxy-4-acetamidobiphenyl has been reported to form more *N*-acetylmethionine adduct than *N*-hydroxy-2-acetamidofluorene (**5**) when these compounds are incubated with rat hepatic or mammary cytosols.[80,105] However, as measured by the RNA-binding assay, 5 was a considerably better substrate than N-hydroxy-4-acetamidobiphenyl for human and rat hepatic and rat mammary cytosolic N,O-acyltransferases.[85,91,92]

The position of substitution of the aromatic ring by the N-hydroxyamide moiety is an important determinant of N,O-acyltransferase activity. For example, N-hydroxy-1-acetamidonaphthalene and N-hydroxy-3-acetamidofluorene were ineffective as acetyl donors to N-hydroxy-2-aminofluorene with rat liver N,O-acyltransferase, while their isomeric 2-substituted arylhydroxamic acids were excellent substrates. In contrast, N-hydroxy-3-acetamidofluorene was an efficient substrate for the transacylation of 4-aminoazobenzene.[80,88]

Substituents in the 7-position of the fluorene ring markedly affect the acyl donor activity of N-hydroxy-2-acetamidofluorenes in the presence of hamster hepatic N,O-acyltransferase; however, the effect of the substituents (halogens, cyano, acetyl, alkoxy) differed for 4-aminoazobenzene transacetylation and for conversion of the arylhydroxamic acids to reactive electrophiles.[106]

Substrates for N,O-acyltransferase are subject to strict stereoelectronic requirements for the acyl group. King and co-workers examined the ability of several N-acyl derivatives of N-hydroxy-2-aminofluorene to serve as acyl donors for partially purified rat liver N,O-acyltransferase.[45,107] The order of enzymatic activity, as determined in the tRNA-binding assay, was acetyl > propionyl > formyl ≫ monochloroacetyl = trifluoro derivative. Beland and co-workers found a similar order, acetyl > propionyl > formyl, for liver N,O-acyltransferases partially purified from rat, guinea pig, monkey, and baboon; however, a different rank order, formyl ≫ acetyl > propionyl, was observed with human and pig liver enzymes.[81] Acetyl-specific and formyl-specific N,O-acyltransferases were partially resolved by ammonium sulfate fractionation of rat and baboon liver cytosols,[81] and, subsequently, separation of enzymes of similar substrate specificity from rat mammary gland cytosol has been achieved.[89,108] Yeh and Hanna investigated the ability of 12 N-acyl derivatives of N-hydroxy-2-aminofluorene to serve as substrates for hamster hepatic N,O-acyltransferase.[109] Although N-acetyl, N-propionyl, and N-methoxyacetyl derivatives exhibited relatively high levels of activity, as measured by 4-aminoazobenzene acylation or formation of N-acetylmethionine adducts, compounds containing larger acyl groups were only marginally active. N-Hydroxyurea and N-hydroxycarbamate derivatives of N-hydroxy-2-aminofluorene did not show significant N,O-acyltransferase activity even when the steric bulk of the analogs was similar to that of the more effective substrates; this result confirms previous conclusions that electronic factors, as well as steric factors, are important determinants of acyl donor ability.[45]

## III. Fate of Reactive Intermediates Generated by Conjugation of Arylhydroxylamines and Arylhydroxamic Acids

Sulfate conjugation and, perhaps, glucuronidation of $N$-arylhydroxamic acids ultimately leads to the formation of charge-delocalized arylnitrenium ions (Fig. 1). These positively charged electrophiles are expected to have both nitrenium ion and carbenium ion character and, therefore, are expected to capture nucleophiles either at nitrogen or at aromatic ring positions. Bioactivation by $N,O$-acyltransferase results in the formation of resonance-stabilized electrophiles that are analogous to those shown in Fig. 1 but do not retain the acyl group (Fig. 3). Although it has been suggested that triplet radical cation intermediates may be responsible for the formation of adducts with nucleic acids and proteins,[110-112] the types of adducts formed can be readily accounted for by the involvement of singlet-state arylnitrenium ions. Additionally, theoretical considerations led to the conclusion that aryl- and $N$-acetyl-$N$-arylnitreniun ions are preferentially stabilized as ground-state singlets,[113] whereas aliphatic nitrenium ions are known to undergo ready spin inversion to the triplet state.[114] The results of another theoretical study suggested that noncarcinogenic, nonmutagenic arylamines produce nitrenium ions whose singlet states are much less stable than the triplet, whereas carcinogenic, mutagenic arylamines are characterized by nitrenium ions whose singlet states are of similar or greater stability than the triplet.[115] These and other semiempirical molecular orbital calculations support extensive delocalization of charge in arylnitrenium ions and provide other interesting predictions, but they do not correctly predict the predominant reaction of nucleic acids with the nitrogen atom of the arylnitrenium ions.[116,117]

The administration of various carcinogenic arylamines, arylamides, arylhydroxylamines, and arylhydroxamic acids to animals results in the formation of covalently bound adducts with proteins and nucleic acids. The relationship of the protein-bound derivatives of carcinogens, such as $N$-hydroxy-2-acetamidofluorene (5) and $N$-methyl-4-aminoazobenzene, to their toxicity or carcinogenicity is unknown. Degradation of the protein-bound adducts formed *in vivo* from these compounds yields $o$-methylmercapto derivatives, the same products that are obtained by the *in vitro* reaction of methionine with synthetic esters of either $N$-hydroxy-$N$-methyl-4-aminoazobenzene (11) or 5. (For a review of the early work in this area, see ref. 9.)

The administration of 5 to rats results in the biliary excretion of the glutathione conjugates 1- and 3-(glutathion-S-yl)-$N$-acetyl-2-aminofluorene.[118] When $N$-acetoxy-2-acetylaminofluorene, an analog of the reactive sulfate conjugate (6), was allowed to react *in vitro* with glutathione, adducts were formed at the 4- and 7-positions of the fluorene ring as well as at the 1- and 3-positions.[118] The only previous report of reaction of a nucleophile at the 7-position of an activated 2-amidofluorene involved the formation of the 7-phosphate adduct.[49] These results indicate that delocalization of the positive charge of the arylnitrenium ion (7) (Fig. 1) includes the 7-position, although no products resulting from reaction of methionine or nucleic acids at the 7-position have been reported. Glutathione also forms adducts with $N$-methyl-4-aminoazobenzene *in vivo* and with the reactive ester $N$-benzoyloxy-$N$-methyl-4-aminoazobenzene *in vitro*.[53,119] Although glutathione may play a role in the detoxication of bioactivated arylamine and arylamide carcinogens, it is not highly effective as an inhibitor of covalent binding of a number of such compounds to nucleic acids and proteins *in vitro*.[53,64]

Because of the general belief that the initiation stage of chemical carcinogenesis often involves the formation of covalently bound adducts between nucleic acids and electrophilic reactants, considerable effort has been directed toward characterization of the products formed as a result of the reaction of DNA, RNA, and various nucleosides with the electrophiles generated by conjugative activation of arylhydroxylamines and arylhydroxamic acids. Much of this research has been reviewed by Kriek and Westra.[4] The administration of 2-acetamidofluorene or its N-hydroxylated derivative (5) to rats results in the formation of covalent adducts with hepatic DNA and RNA. The major RNA adduct is $N$-(guanosin-8-yl)-2-acetamidofluorene, whereas most of the material bound to DNA does not retain the acetyl group.[4] The predominant acetylated and nonacetylated DNA adducts are 18 and 19, respectively. A minor DNA adduct that retains the acetyl group

18  R = COCH$_3$
    R' = deoxyribose

19  R = H
    R' = deoxyribose

and that has a significantly longer *in vivo* half-life than 18 is the product of the reaction at the N-2 position of guanosine (20).

20  R = deoxyribose

Although the relative half-lives of DNA adducts may be an important factor in chemical carcinogenesis, Beland and co-workers have presented data indicating that the persistence of such adducts is not itself sufficient for tumorigenesis; other factors appear to be involved.[120] Furthermore, although the carcinogenicity of arylhydroxamic acids has long been associated with the sulfotransferase-catalyzed formation of acetylated adducts such as **18** and **20**, Scribner et al.[121] and Beranke et al.[122] have proposed that nonacetylated 8-arylaminoguanine adducts in DNA are more relevant to carcinogenesis and to mutagenesis. The work of Stout et al., who concluded that N-hydroxy-2-aminofluorene is the principal mutagenic metabolite of **5** in the *Salmonella typhimurium* system when soluble liver enzymes are used as an activation system, is also supportive of the importance of nonacetylated adducts to mutagenesis.[123] Kriek, on the basis of results of *in vivo* binding studies with N-hydroxy-2-acetamidofluorene (**5**) and with 4'-fluoro-4-acetamidobiphenyl, also suggested that deacetylated reaction products with DNA may be more important for the initiation of the carcinogenic process than acetylated derivatives.[4]

Despite the results just mentioned, which suggest the importance of nonacetylated DNA adducts in the carcinogenic and mutagenic activities of **5** and related compounds, much remains to be determined regarding the structure and biological significance of the various nucleic acid adducts formed subsequent to conjugative bioactivation of arylhydroxamic acids and arylhydroxylamines. For example, Tarpley et al. reported that reaction of N-acetoxy-2-acetamidofluorene and N-benzoyloxy-N-methyl-4-aminozobenzene with DNA appeared to result in the formation of N-7 adducts with guanine, resulting in depurination. The structures of these guanine derivatives have not been elucidated.[124] The structures of the nucleic acid adducts formed *in vivo* by the carcinogen 2-acetamidophenanthrene have also not been determined, although Scribner and Koponen found that they were not identical to those adducts isolated after the treatment of DNA with either the sulfate or acetate ester of N-hydroxy-2-acetamidophenanthrene *in vitro*.[125] In contrast, identical adducts are formed *in vivo* and *in vitro* from the activated forms of a number of arylamines and arylamides including 2-naphthylamine,[126] benzidine and N-acetylbenzidine,[127] N-methyl-4-aminoazobenzene,[128] and 3-amino-1-methyl-5H-pyrido[4,3-b-]indole (TRP-2).[129,130] The relationships between the formation of the various adducts and the toxicity, mutagenicity, and carcinogenicity of the parent compounds, the roles of conjugative enzyme systems in the formation of reactive intermediates, and the chemical mechanisms involved in the reaction of the intermediates with biological nucleophiles are problems that continue to stimulate vigorous research.

## References

1. Weisburger, J. H., and Weisburger, E. K. (1973). Biochemical formation and pharmacological, toxicological and pathological properties of hydroxylamines and hydroxamic acids. *Pharmacol. Rev.* **25**, 1–66.
2. Kiese, M. (1966). The biochemical production of ferrihemoglobin-forming derivatives from aromatic amines, and mechanisms of ferrihemoglobin formation. *Pharmacol. Rev.* **18**, 1091–1161.
3. Miller, E. C., and Miller, J. A. (1969). Studies on the mechanism of activation of aromatic amine and amide carcinogens to ultimate carcinogenic electrophilic reactants. *Ann. N. Y. Acad. Sci.* **163**, 731–750.
4. Kriek, E., and Westra, J. G. (1979). Metabolic activation of aromatic amines and amides and interactions with nucleic acids. *In* "Chemical Carcinogens and DNA" (P. L. Grover. ed.), Vol. II, pp. 1–28. CRC Press, Boca Raton, Florida.
5. Weber, W. W., and Glowinski, I. B. (1980). Acetylation. *In* "Enzymatic Basis of Detoxication" (W. B. Jakoby, ed.), Vol. 2, pp. 169–186. Academic Press, Inc., New York.
6. Irving, C. C. (1966). Enzymatic deacetylation of $N$-hydroxy-2-acetylaminofluorene by liver microsomes. *Cancer Res.* **26**, 1390–1396.
7. Weber, W. W., Radtke, H. E., and Tannen, R. H. (1980). Extrahepatic $N$-acetyltransferases and $N$-deacetylases. *In* "Extrahepatic Metabolism of Drugs and Other Foreign Compounds" (T. E. Gram, ed.), pp. 493–531. SP Medical and Scientific Books, New York.
8. Corbett, M. D., Corbett, B. R., and Doerge, D. R. (1982). Hydroxamic acid production and active site-induced Bamberger rearrangement from the action of $\alpha$-ketoglutarate dehydrogenase on 4-chloronitrosobenzene. *J. Chem. Soc., Perkin Trans. 1* pp. 345–350.
9. Miller, E. C., and Miller, J. A. (1981). Searches for ultimate chemical carcinogens and their reactions with cellular macromolecules. *Cancer* **47**, 2327–2345.
10. Miller, J. A., and Miller, E. C. (1977). Ultimate chemical carcinogens as reactive mutagenic electrophiles. *Cold Spring Harbor Conf. Cell Proliferation* **4**, Book B, 605–627.
11. Nelson, S. D. (1982). Metabolic activation and drug toxicity. *J. Med. Chem.* **25**, 753–765.
12. Frederick, C. B., Mays, J. B., Ziegler, D. M., Guengerich, F. P., and Kadlubar, F. F. (1982). Cytochrome $P$-450 and flavin-containing monooxygenase-catalyzed formation of the carcinogen $N$-hydroxy-2-aminofluorene and its covalent binding to nuclear DNA. *Cancer Res.* **42**, 2671–2677.
13. Bartsch, H., and Hecker, E. (1971). On the metabolic activity of the carcinogen $N$-hydroxy-$N$-2-acetylaminofluorene. III. Oxidation with horseradish peroxidase to yield 2-nitrosofluorene and $N$-acetoxy-$N$-2-acetylaminofluorene. *Biochim. Biophys. Acta* **237**, 567–578.
14. Floyd, R. A. (1981). Free-radical events in chemical and biochemical reactions involving carcinogenic arylamines. *Radiat. Res.* **86**, 243–263.
15. Forrester, A. R., Ogiloy, M. M., and Thomson, R. H. (1970). Mode of action of carcinogenic amines. Part I. Oxidation of $N$-arylhydroxamic acids. *J. Chem. Soc. C* pp. 1081–1083.
16. Mita, S., Ishii, K., Yamazoe, X., Kamataki, T., Kato, R., and Sugimura, T. (1981). Evidence for the involvement of N-hydroxylation of 3-amino-1-methyl-5$H$-pyriod[4,3-$b$]indole by cytochrome $P$-450 in the covalent binding to DNA. *Cancer Res.* **41**, 3610–3614.

17. Yamazoe, Y., Shimada, M., Kamataki, T., and Kato, R. (1982). Covalent binding of N-hydroxy-TRP-2 to DNA by a cytosolic proline-dependent system. *Biochem. Biophys. Res. Commun.* **107,** 165–172.
18. Tada, M., and Tada, M. (1975). Seryl-tRNA synthetase and activation of the carcinogen 4-nitroquinoline 1-oxide. *Nature (London)* **255,** 510–512.
19. Hashimoto, Y., Degawa, M., Watanabe, H. K., and Tada, M. (1981). Amino acid conjugation of N-hydroxy-4-aminoazobenzene dyes: A possible activation of the ultimate mutagenic or carcinogenic metabolites. *Gann* **72,** 937–943.
20. Irving, C. C. (1970). Conjugates of N-hydroxy compounds. *In* "Metabolic Conjugation and Metabolic Hydrolysis" (W. H. Fishman, ed.), Vol. 1, pp. 53–119. Academic Press, New York.
21. Irving, C. C. (1971). Metabolic activation of N-hydroxy compounds by conjugation. *Xenobiotica* **1,** 387–398.
22. DeBaun, J. R., Rowley, J. Y., Miller, E. C., and Miller, J. A. (1968). Sulfotransferase activation of N-hydroxy-2-acetylaminofluorene in rodent livers susceptible and resistant to this carcinogen. *Proc. Soc. Exp. Biol. Med.* **129,** 268–273.
23. DeBaun, J. R., Miller, E. C., and Miller, J. A. (1970). N-Hydroxy-2-acetylaminofluorene sulfotransferase: Its probable role in carcinogenesis and in protein–(methion-S-yl) binding in rat liver. *Cancer Res.* **30,** 577–595.
24. King, C. M., and Phillips, B. (1968). Enzyme-catalyzed reactions of the carcinogen N-hydroxy-2-fluorenylacetamide with nucleic acid. *Science* **159,** 1351–1353.
25. King, C. M., and Phillips, B. (1969). N-Hydroxy-2-fluorenylacetamide. Reaction of the carcinogen with guanosine, ribonucleic acid, deoxyribonucleic acid, and protein following enzymatic deacetylation or esterification. *J. Biol. Chem.* **244,** 6209–6216.
26. Bock, K. W. (1977). Dual role of glucuronyl and sulfotransferases converting xenobiotics into reactive or biologically inactive and easily excretable compounds. *Arch. Toxicol.* **39,** 77–85.
27. Reichert, D. (1981). Toxication of foreign substances by conjugation reactions. *Angew. Chem., Int. Ed. Engl.* **20,** 135–142.
28. Jakoby, W. B., Skura, R. D., Lyon, E. S., Marcus, C. J., and Wang, J. L. (1980). Sulfotransferases. *In* "Enzymatic Basis of Detoxication" (W. B. Jakoby, ed.), Vol. 2, pp. 199–208. Academic Press, New York.
29. Wu, S.-C. G., and Straub, K. D. (1976). Purification and characterization of N-hydroxy-2-acetylaminofluorene sulfotransferase from rat liver. *J. Biol. Chem.* **251,** 6529–6536.
30. Gutmann, H. R., Malejka-Giganti, D., Barry, E. J., and Rydell, R. E. (1972). On the correlation between hepatocarcinogenicity of the carcinogen, N-2-fluorenylacetamide, and its metabolic activation by the rat. *Cancer Res.* **32,** 1554–1561.
31. Lotlikar, P. D. (1970). Effects of sex hormones on enzymic esterification of 2-(N-hydroxyacetamido)fluorene by rat liver cytosol. *Biochem. J.* **120,** 409–416.
32. Irving, C. C. (1975). Comparative toxicity of N-hydroxy-2-acetylaminofluorene in several strains of rats. *Cancer Res.* **35,** 2959–2961.
33. Dybing, E., Soderlund, E. J., and Thorgeirsson, S. S. (1978). Oxidation and conjugation of 2-acetylaminofluorene in isolated liver cells. *In* "Conjugation Reactions in Drug Biotransformation" (A. Aitio, ed.), pp. 283–292. Elsevier/North-Holland and Biomedical Press, Amsterdam.
34. Dybing, E., Soderlund, E., Haug, L. T., and Thorgeirsson, S. S. (1979). Metabolism and activation of 2-acetylaminofluorene in isolated rat hepatocytes. *Cancer Res.* **39,** 3268–3275.
35. Meerman, J. H. N., van Doorn, A. B. D., and Mulder, G. J. (1980). Inhibition of sulfate

conjugation of N-hydroxy-2-acetylaminofluorene in isolated perfused rat liver and in the rat in vivo by pentachlorophenol and low sulfate. *Cancer Res.* **40,** 3772-3779.
36. Merrman, J. H. N., Beland, F. A., and Mulder, G. J. (1981). Role of sulfation in the formation of DNA adducts from N-hydroxy-2-acetylaminofluorene in rat liver in vivo. Inhibition of N-acetylated aminofluorene adduct formation by pentachlorophenol. *Carcinogenesis (N.Y.)* **2,** 413-416.
37. Meerman, J. H. N., and Mulder, G. J. (1981). Prevention of the hepatotoxic action of N-hydroxy-2-acetylaminofluorene in the rat by inhibition of N-O sulfation by pentachlorophenol. *Life Sci.* **28,** 2361-2365.
38. Weisburger, J. H., Yamamoto, R. S., Williams, G. M., Grantham, P. H., Matsushima, T., and Weisburger, E. K. (1972). On the sulfate ester of N-hydroxy-N-2-fluorenylacetamide as a key ultimate hepatocarcinogen in the rat. *Cancer Res.* **32,** 491-500.
39. Yamamoto, R. S., Williams, G. M., Richardson, H. L., Weisburger, E. K., and Weisburger, J. H. (1973). Effect of p-hydroxyacetanilide on liver cancer induction by N-hydroxy-N-2-fluorenylacetamide. *Cancer Res.* **33,** 454-457.
40. Weisburger, J. H., Weisburger, E. K., Madison, R. M., Wenk, M. L., and Klein, D. S. (1973). Effect of acetanilide and p-hydroxyacetanilide on the carcinogenicity of N-2-fluorenylacetamide and N-hydroxy-N-2-fluorenylacetamide and N-hydroxy-N-2-fluorenylacetamide in mice, hamsters, and female rats. *J. Natl. Cancer Inst. (U.S.)* **51,** 235-240.
41. Mohan, L. C., Grantham, P. H., Weisburger, E. K., Weisburger, J. H., and Idoine, J. B. (1976). Mechanisms of the inhibitory action of p-hydroxyacetanilide on carcinogenesis by N-2-fluorenylacetamide or N-hydroxy-N-2-fluorenylacetamide. *J. Natl. Cancer Inst. (U.S.)* **56,** 763-768.
42. King, C. M. (1974). Mechanism of reaction, tissue distribution and inhibition of arylhydroxamic acid acyltransferase. *Cancer Res.* **34,** 1503-1575.
43. Zieve, F. J., and Gutmann, H. R., (1971). Reactivities of the carcinogens, N-hydroxy-2-fluorenylacetamide and N-hydroxy-3-fluorenylacetamide, with tissue nucleophiles. *Cancer Res.* **31,** 471-476.
44. Morton, K. C., Beland, F. A., Evans, F. E., Fullerton, N. F., and Kadlubar, F. F. (1980). Metabolic activation of N-hydroxy-N,N-diacetylbenzidine by hepatic sulfotransferase. *Cancer Res.* **40,** 751-757.
45. Weeks, C. E., Allaben, W. T., Tresp, N. M., Louis, S. C., Lazear, E. J., and King, C. M. (1980). Effects of structure of N-acyl-N-fluorenylhydroxylamines on arylhydroxamic acid acyltransferase, sulfotransferase, and deacylase activities, and on mutation in *Salmonella typhimurium,* TA 1538. *Cancer Res.* **40,** 1204-1211.
46. Vaught, J. B., McGarvey, P. B., Lee, M.-S., Garner, C. D., Wang, C. Y., Linsmaier-Bednar, E. M., and King, C. M. (1981). Activation of N-hydroxyphenacetin to mutagenic and nucleic acid-binding metabolites by acyl transfer, deacylation and sulfate conjugation. *Cancer Res.* **41,** 3424-3429.
47. Wirth, P. J., and Thorgeirsson, S. S. (1981). Mechanism of N-hydroxy-2-acetylaminofluorene mutagenicity in the *Salmonella* test system. Role of N,O-acyltransferase and sulfotransferase from rat liver. *Mol. Pharmacol.* **19,** 337-344.
48. Mulder, G. J., Hinson, J. A., and Gillette, J. R. (1977). Generation of reactive metabolites of N-hydroxyphenacetin by glucuronidation and sulfation. *Biochem. Pharmacol.* **26,** 189-196.
49. Andrews, L. S., Hinson, J. A., and Gillette, J. R. (1978). Studies on the mutagenicity of N-hydroxy-2-acetylminofluorene in the Ames *Salmonella* mutagenesis test system. *Biochem. Pharmacol.* **27,** 2399-2408.

50. Irving, C. C. (1978). Reactivity of conjugates of N-hydroxylated arylamines and arylamides. *In* "Biological Oxidation of Nitrogen" (J. W. Gorrod, ed.), pp. 325–334. Elsevier/North-Holland and Biomedical Press, Amsterdam.
51. Irving, C. C. (1979). Species and tissue variation in the metabolic activation of aromatic amines. *In* "Carcinogens: Identification and Mechanisms of Action" (A. C. Griffin and C. R. Shaw, eds.), pp. 211–227. Raven Press, New York.
52. Kadlubar, F. F., Miller, J. A., and Miller, E. C. (1976). Hepatic metabolism of N-hydroxy-N-methyl-4-aminoazobenzene and other N-hydroxyarylamines to reactive sulfuric acid esters. *Cancer Res.* **36**, 2350–2359.
53. Kadlubar, F. F., Ketterer, B., Flammang, T. J., and Christodoulides, L. (1980). Formation of 3-(glutathion-S-yl)-N-methyl-4-aminoazobenzene and inhibition of aminoazo dye–nucleic acid binding *in vitro* by reaction of glutathione with metabolically generated N-methyl-4-aminoazobenzene N-sulfate. *Chem.-Biol. Interact.* **31**, 265–278.
54. McDonald, J. J., Stohrer, G., and Brown, G. B. (1973). Oncogenic purine N-oxide derivatives as substrates for sulfotransferase. *Cancer Res.* **33**, 3319–3323.
55. Stohrer, G., Corbin, E., and Brown, G. B. (1972). Enzymatic activation of the oncogen 3-hydroxyxanthine. *Cancer Res.* **32**, 637–642.
56. Lee, T.-C., Teller, M. W., Budinger, J. M., Klotzer, W. K., and Brown, G. B. (1979). Chemical reactivities and oncogenicities of a series of N-hydroxy heterocycles. *Chem.-Biol. Interact.* **25**, 369–372.
57. Kasper, C. B., and Henton, D. (1980). Glucuronidation. *In* "Enzymatic Basis of Detoxication" (W. B. Jakoby, ed.), Vol. 2, pp. 3–36. Academic Press, New York.
58. Miller, E. C., Lotlikar, P. D., Miller, J. A., Butler, B. W., Irving, C. C., and Hill, J. T. (1968). Reactions *in vitro* of some tissue nucleophiles with the glucuronide of the carcinogen N-hydroxy-2-acetylaminofluorene. *Mol. Pharmacol.* **4**, 147–154.
59. Irving, C. C., Veazey, R. A., and Russell, L. T. (1969/70). Possible role of the glucuronide conjugate in the biochemical mechanism of binding of carcinogen N-hydroxy-2-acetylaminofluorene to rat liver deoxyribonucleic acid *in vivo*. *Chem.-Biol. Interact.* **1**, 19–26.
60. Irving, C. C., Russell, L. T, and Kriek, E. (1972). Biosynthesis and reactivity of the glucuronide of N-hydroxy-4-acetylaminobiphenyl. *Chem.-Biol. Interact.* **5**, 37–46.
61. Irving, C. C., and Russell, L. T. (1970). Synthesis of the O-glucuronide of N-2-fluorenylhydroxylamine. Reaction with nucleic acids and with guanosine 5′-monophosphate. *Biochemistry* **9**, 2471–2476.
62. Cardona, R. A., and King, C. M. (1976). Activation of the O-glucuronide of the carcinogen N-hydroxy-N-2-fluorenylacetamide by enzymatic deacetylation *in vitro*: Formation of fluorenylamine–tRNA adducts. *Biochem. Pharmacol.* **25**, 1051–1056.
63. Irving, C. C. (1977). Influence of the aryl group on the reactions of glucuronides of N-arylacetohydroxamic acids with polynucleotides. *Cancer Res.* **37**, 524–528.
64. Mulder, G. J., Hinson, J. A., and Gillette, J. R. (1978). Conversion of the N-O-glucuronide and N-O-sulfate conjugates of N-hydroxyphenacetin to reactive intermediates. *Biochem. Pharmacol.* **27**, 1641–1649.
65. Calder, I. C., Goss, D. E., Williams, P. J., Funder, C. C., Green, C. R., Ham, K. N., and Tange, J. D. (1976). Neoplasia in the rat induced by N-hydroxyphenacetin, a metabolite of phenacetin. *Pathology* **8**, 1–6.
66. Camus, A.-M., Friesen, M., Croisy, A., and Bartsch, H. (1982). Species-specific activation of phenacetin into bacterial mutagens by hamster liver enzymes and identification o N-hydroxyphenacetin O-glucuronide as a promutagen in the urine. *Cancer Res.* **42** 3201–3208.
67. Hinson, J. A., Andrews, L. S., and Gillette, J. R. (1979). Kinetic evidence for multipl

chemically reactive intermediates in the breakdown of phenacetin N-O-glucuronide. *Pharmacology* **19**, 237–248.
68. Radomski, J. L., Hearn, W. L., Radomski, T., Moreno, H., and Scott, W. E. (1977). Isolation of the glucuronic acid conjugate of N-hydroxy-4-aminobiphenyl from dog urine and its mutagenic activity. *Cancer Res.* **37**, 1757–1762.
69. Aitio, A., and Marniemi, J. (1980). Extrahepatic glucuronide conjugation. In "Extrahepatic Metabolism of Drugs and Other Compounds" (T. E. Gram. ed.), pp. 365–387. S P Medical and Scientific Books, New York.
70. Kadlubar, F. F., Miller, J. A., and Miller, E. C. (1977). Hepatic microsomal N-glucuronidation and nucleic acid binding of N-hydroxyarylamines in relation to urinary bladder carcinogenesis. *Cancer Res.* **37**, 805–814.
71. Radomski, J. L., Rey, A. A., and Brill, E. (1977). Evidence for a glucuronic acid conjugate of N-hydroxy-4-aminobiphenyl in the urine of dogs given 4-aminobiphenyl. *Cancer Res.* **33**, 1284–1289.
72. Poupko, J. M., Hearn, W. L., and Radomski, J. L. (1979). N-glucuronidation of N-hydroxy aromatic amines: A mechanism for their transport and bladder-specific carcinogenicity. *Toxicol. Appl. Pharmacol.* **50**, 479–484.
73. Kadlubar, F. F., Unruh, L. E., Flammang, T. J., Sparks, D., Mitchum,. R. K., and Mulder, G. J. (1981). Alteration of urinary levels of the carcinogen, N-hydroxy-2-naphthylamine, and its N-glucuronide in the rat by control of urinary pH, inhibition of metabolic sulfation, and changes in biliary excretion. *Chem.-Biol. Interact.* **33**, 129–147.
74. Moreno, H. R., and Radomski, J. L. (1978). Synthesis of the urinary glucuronic acid conjugate of N-hydroxy-4-aminobiphenyl. *Cancer Lett.* **4**, 85–88.
75. Oglesby, L. A., Flammang, T. J., Tullis, D. L., and Kadlubar, F. F. (1981). Rapid absorption, distribution, and excretion of carcinogenic N-hydroxyarylamines after direct urethral instillation into the rat urinary bladder. *Carcinogenesis (N.Y.)* **2**, 15–20.
76. Kadlubar, F., Flammang, T., and Unruh, L. (1978). The role of N-hydroxyarylamine N-glucuronides in arylamine-induced urinary bladder carcinogenesis: Metabolite profiles in acidic, neutral and alkaline urines of 2-naphthylamine- and 2-nitronaphthalene-treated rats. In "Conjugation Reactions in Drug Biotransformation" (A. Aitio, eds.), pp. 443–454. Elsevier/North-Holland Biomedical Press, Amsterdam.
77. Irving, C. C., and Veazey, R. (1969). Persistent binding of 2-acetylaminofluorene to rat liver DNA *in vivo* and consideration of the mechanisms of binding of N-hydroxy-2-acetylaminofluorene to rat liver nucleic acids. *Cancer Res.* **24**, 1799–1804.
78. Kriek, E. (1969). On the mechanism of action of carcinogenic aromatic amines. I. Binding of 2-acetylaminofluorene and N-hydroxy-2-acetylaminofluorene to rat liver nucleic acids *in vivo*. *Chem.-Biol. Interact.* **1**, 3–17.
79. Kriek, E. (1972). Persistent binding of a new reaction product of the carcinogen N-hydroxy-2-acetylaminofluorene with guanine in rat liver DNA *in vivo*. *Cancer Res.* **32**, 2042–2048.
80. Bartsch, H., Dworkin, M., Miller, J. A., and Miller, E. C. (1972). Electrophilic N-acetoxyaminoarenes derived from carcinogenic N-hydroxy-N-acetylaminoarenes by enzymatic deacetylation and transacetylation in liver. *Biochim. Biophys. Acta* **286**, 272–298.
81. Beland, F. A., Allaben, W. T., and Evans, F. E. (1980). Acyltransferase-mediated binding of N-hydroxyarylamides to nucleic acids. *Cancer Res.* **40**, 834–840.
82. Glowinski, I. B., Savage, L., Lee, M.-S., and King, C. M. (1983). Relationship between nucleic acid adduct formation and deacylation of arylhydroxamic acids. *Carcinogenesis (N.Y.)* **4**, 67–75.
83. Booth, J. (1966). Acetyl transfer in arylamine metabolism. *Biochem. J.* **100**, 745–752.
84. Glowinski, I. B., Weber, W. W., Fysh, J. M., Vaught, J. B., and King, C. M. (1980).

Evidence that arylhydroxamic acid $N,O$-acyltranferase and the genetically polymorphic $N$-acetyltransferase are properties of the same enzyme in rabbit liver. *J. Biol. Chem.* **255**, 7883–7890.
85. King, C. M., Olive, C. W., and Cardona, R. A. (1975). Activation of carcinogenic arylhydroxamic acids by acyltransferase of human tissues. *J. Natl. Cancer Inst. (U.S.)* **55**, 285–287.
86. King, C. M., and Olive, C. W. (1975). Comparative effects of strain, species and sex on the acyltransferase- and sulfotransferase-catalyzed activations of $N$-hydroxy-$N$-2-fluorenylacetamide. *Cancer Res.* **35**, 906–912.
87. Olive, C. W., and King, C. M. (1975). Evidence for a second arylhydroxamic acid acyltransferase species in the small intestine of the rat. *Chem.-Biol. Interact.* **11**, 599–604.
88. King, C. M., and Allaben, W. T. (1978). The role of arylhydroxamic acid $N,O$-acyltransferase in the carcinogenicity of aromatic amines. *In* "Conjugation Reactions in Drug Biotransformation" (A. Aito, ed.), pp. 431–441. Elsevier/North-Holland Biomedical Press, Amsterdam.
89. Allaben, W. T., Weeks, C. E., Weis, C. C., Burger, G. T., and King, C. M. (1982). Rat mammary gland carcinogenesis after local injection of $N$-hydroxy-$N$-acyl-2-aminofluorenes: Relationship to metabolic activation. *Carcinogenesis (N.Y.)* **3**, 233–240.
90. Allaben, W. T., and King, C. M. (1977). Purification and characterization of arylhydroxamic acid acyltransferase. *Fed. Proc., Fed. Am. Soc. Exp. Biol.* **36**, 349.
91. King, C. M., Traub, N. R., Lortz, Z. M., and Thissen, M. R. (1979). Metabolic activation of arylhydroxamic acids by $N,O$-acyltransferase of rat mammary gland. *Cancer Res.* **39**, 3369–3372.
92. King, C. M., and Allaben, W. T. (1980). Arylhydroxamic acid acyltransferase. *In* "Enzymatic Basis of Detoxication" (W. B. Jakoby, ed.), Vol. 2, pp. 187–197. Academic Press, New York.
93. Weber, W. W., and King, C. M. (1981). $N$-Acetyltransferase and arylhydroxamic acid acyltransferase. *In* "Methods in Enzymology" (W. B. Jakoby, ed.), Vol. 77, pp. 272–280. Academic Press, New York.
94. Scribner, J. D., Miller, J. A., and Miller, E. C. (1970). Nucleophilic substitution on carcinogenic $N$-acetoxy-$N$-arylacetamides. *Cancer Res.* **30**, 1570–1579.
95. Poirier, L. A., Miller, J. A., Miller, E. C., and Sato, K. (1967). $N$-Benzoyloxy-$N$-methyl-4-aminoazobenzene: Its carcinogenic activity in the rat and its reactions with proteins and nucleic acids and their constituents *in vitro*. *Cancer Res.* **27**, 1600–1613.
96. Banks, R. B., and Hanna, P. E. (1979). Arylhydroxamic acid $N,O$-acyltransferase. Apparent suicide inactivation by carcinogenic $N$-arylhydroxamic acids. *Biochem. Biophys. Res. Commun.* **91**, 1423–1429.
97. Hanna, P. E., Banks, R. B., and Marhevka, V. C. (1982). Suicide inactivation of hamster hepatic arylhydroxamic acid $N,O$-acyltransferase. A selective probe of $N$-acetyltransferase multiplicity. *Mol. Pharmacol.* **21**, 159–165.
98. Hein, W. D. (1982). The biochemical basis of the N-acetylation polymorphism and its toxicological consequences. Ph.D. Thesis, University of Michigan, Ann Arbor.
99. Smith, T. J., and Hanna, P. E. (1983). Inactivation of hamster hepatic $N$-acetyltransferase by $N$-hydroxyphenacetin. *Fed. Proc., Fed. Am. Soc. Exp. Biol.* **42**, 889.
100. Banks, R. B., Smith, T. J., and Hanna, P. E. (1982). $N$-Arylhydroxamic acid $N,O$-acyl transferase. Positional requirements for the substrate hydroxyl group. *J. Med. Chem.* **25**, 842–846.
101. Wang, C. Y., Linsmaier, Bednar, E. M., and Lee, M.-S. (1981). Mutagenicity of the O-esters of $N$-acylhydroxylamines for *Salmonella*. *Chem.-Biol. Interact.* **34**, 267–278.

102. Elfarra, A. A., Yeh, H.-M., and Hanna, P. E. (1982). Synthesis and evaluation of $N$-(alkylphenyl)acetohydroxamic acids as potential substrates for $N$-arylhydroxamic acid $N,O$-acyltransferase. *J. Med. Chem* **25**, 1189–1192.
103. Mangold, B. L. K., and Hanna, P. E. (1982). Arylhydroxamic acid $N,O$-acyltransferase substrates. Acetyl transfer and electrophile-generating activity of 4-alkyl-, 4-alkenyl-, and 4-cyclohexyl-$N$-hydroxy-$N$-phenylacetamides. *J. Med. Chem.* **25**, 630–638.
104. Morton, K. C., King, C. M., and Baetcke, K. P. (1979). Metabolism of benzidine to $N$-hydroxy-$N,N'$-diacetylbenzidine and subsequent nucleic acid binding and mutagenicity. *Cancer Res.* **39**, 3107–3113.
105. Bartsch, H., Dworkin, C., Miller, E. C., and Miller, J. A. (1973). Formation of electrophilic $N$-acetoxyarylamines in cytosols from rat mammary gland and other tissues by transacetylation from the carcinogen $N$-hydroxy-4-acetamidobiphenyl. *Biochim. Biophys. Acta* **304**, 42–55.
106. Hanna, P. E., Marhevka, V. C., and Elfarra, A. A. (1984). In preparation.
107. King, C. M., Allaben, W. T., Lazear, E. J., Louie, S. C., and Weeks, C. E. (1978). Influence of the acyl group on arylhydroxamic acid $N,O$-acyltransferase-catalyzed mutagenicity and metabolic activation of $N$-acyl-2-fluorenylhydroxylamines. *In* "Biological Oxidation at Nitrogen" (J. E. Gorrod, ed.), pp. 335–340. Elsevier/North-Holland and Biomedical Press, Amsterdam.
108. Shirai, T., Fysh, J. M., Lee, M. S., Vaught, J. T., and King, C. M., (1981). Relationship of metabolic activation of $N$-hydroxy-$N$-acylarylamines to biological response in the liver and mammary gland of the female CD rat. *Cancer Res.* **41**, 4346–4353.
109. Yeh, H.-M., and Hanna, P. E. (1982). Arylhydroxamic acid bioactivation via acyl group transfer. Structural requirements for transacylating and electrophile-generating activity of $N$-(2-fluorenyl)hydroxamic acids and related compounds. *J. Med. Chem.* **25**, 842–846.
110. Scribner, J. D., and Naimy, N. K. (1973). Reactions of esters of $N$-hydroxy-2-acetamidophenanthrene with cellular nucleophiles and the formation of free radicals upon decomposition of $N$-acetoxy-$N$-arylacetamides. *Cancer Res.* **32**, 1159–1164.
111. Scribner, J. D., and Naimy, N. K. (1975). Destruction of triplet nitrenium ion by ascorbic acid. *Experientia* **31**, 470–471.
112. Bobst, A. M., Wang, T. V., and Cerutti, P. A. (1981). Mechanism of reaction of carcinogen $N$-acetoxy-2-acetylaminofluorene with DNA. *Experientia* **37**, 597–598.
113. Ford, G. P., and Scribner, J. D. (1981). MNDO molecular orbital study of nitrenium ions derived from carcinogenic aromatic amines and amides. *J. Am. Chem. Soc.* **103**, 4281–4291.
114. Gassman, P. G. (1970). *Acc. Chem. Res.* **3**, 26–33.
115. Hartman, G. D., and Schlegel, H. B. (1981). The relationship of the carcinogenic/mutagenic potential of arylamines to their singlet–triplet nitrenium ion energies. *Chem.-Biol. Interact.* **36**, 319–330.
116. Loew, G. H., Phillips, J., and Pack, G. (1979). Quantum chemical studies of the metabolism of polycylic aromatic amines and the stabilities and electrophilcities of their acrylnitrenium ions in relation to their mutagenic/carcinogenic potencies. *Cancer Biochem. Biophys.* **3**, 101–110.
117. Lowe, G., Sudhindra, B. S., Burt, S., Pack, G. R., and MacElroy, R. (1979). Aromatic amine carcinogenesis: Activation of interaction with nucleic acid bases. *Int. J. Quantum Chem., Quantum Biol. Symp.* **6**, 259–281.
118. Meerman, J. H. M., Beland, F. A., Ketterer, B., Srai, S. K. S., Bruins, A. P., and Mulder, G. J. (1982). Identification of glutathione conjugates formed from $N$-hydroxy-2-acetylaminofluorene in the rat. *Chem.-Biol. Interact.* **39**, 149–168.

119. Ketterer, B., Kadlubar, F., Flammang, T., Carne, T., and Enderby, G. (1979). Glutathione adducts of N-methyl-4-aminoazobenzene formed *in vivo* and by reaction of N-benzoyloxy-N-methyl-4-aminoazobenzene with glutathione. *Chem.-Biol. Interact.* **25**, 7–21.
120. Beland, F. A., Dooley, K. L., and Jackson, C. D. (1982). Persistence of DNA adducts in rat liver and kidney after multiple doses of the carcinogen N-hydroxy-2-acetylaminofluorene. *Cancer Res.* **42**, 1348–1354.
121. Scribner, J. D., Scribner, N. K., and Koponen, G. (1982). Metabolism and nucleic acid binding of 7-fluoro-2-acetamidofluorene in rats: Oxidative defluorination and apparent dissociation from hepatocarcinogenesis of 8-(N-arylamide)guanine adducts on DNA. *Chem.-Biol. Interact.* **40**, 27–43.
122. Beranek, D. T., White, G. L., Helflich, R. H., and Beland, F. A. (1982). Aminofluorene–DNA adduct formation in *Salmonella typhimurium* exposed to the carcinogen N-hydroxy-2-acetylaminofluorene. *Proc. Natl. Acad. Sci. U.S.A.* **79**, 5175–5178.
123. Stout, D. L., Baptist, J. N., Matney, T. S., and Shaw, C. R. (1976). N-Hydroxy-2-aminofluorene: The principal mutagen produced from N-hydroxy-2-acetylaminofluorene by a mammalian supernatant enzyme preparation. *Cancer Lett.* **1**, 269–274.
124. Tarpley, W. G., Miller, J. A., and Miller, E. C. (1982). Rapid release of carcinogen–guanine adducts from DNA after reaction with N-acetoxy-2-acetylaminofluorene or N-benzoyloxy-N-methyl-4-aminoazobenzene. *Carcinogenesis (N.Y.)* **3**, 81–88.
125. Scribner, J. D., and Koponen, G. (1979). Binding of the carcinogen 2-acetamidophenanthrene to rat liver nucleic acids: Lack of correlation with carcinogenic activity, and failure of the hydroxamic acid ester model for *in vivo* activation. *Chem. Biol. Interact.* **15**, 201–209.
126. Kadlubar, F. F., Anson, J. F., Dooley, K. L., and Beland, F. A. (1981). Formation of urothelial and heaptic DNA adducts from the carcinogen 2- naphthylamine. *Carcinogenesis (N.Y.)* **2**, 467–470.
127. Martin, C. N., Beland, F. A., Roth, R. W., and Kadlubar, F. F. (1982). Covalent binding of benzidine and N-acetylbenzidine to DNA at the C-8 atom of deoxyguanosine *in vivo* and *in vitro*. *Cancer Res.* **42**, 2678–2686.
128. Beland, F. A., Tullis, D. L., Kadlubar, F. F., Straub, K. M., and Evans, F. E. (1980). Characterization of DNA adducts of the carcinogen N-methyl-4-aminoazobenzene *in vitro* and *in vivo*. *Chem.-Biol. Interact.* **31**, 1–17.
129. Hashimoto, Y., Shudo, K., and Okamoto, T. (1979). Structural identification of a modified base in DNA covalently bound with mutagenic 3-amino-1-methyl-5H-pyrido[4,3-b]indole. *Chem. Pharm. Bull.* **27**, 1058–1060.
130. Hashimoto, Y., Shudo, K., and Okamoto, T. (1982). Modification of nucleic acids with muta-carcinogenic heteroaromatic amines in vivo. Identification of modified bases in DNA extracted from rats injected with 3-amino-1-methyl-5H-pyrido[4,3-b]indole and 2-amino-6-methyldipyrido[1,2-a:3′,2′-d]imidazole. *Mutat. Res.* **105**, 9–13.

Chapter 14

# Nitrosamines

MICHAEL C. ARCHER AND GEORGE E. LABUC

Department of Medical Biophysics
University of Toronto
Ontario Cancer Institute
Toronto, Ontario, Canada

I. Introduction . . . . . . . . . . . . . . . . . . . 403
II. Chemical Properties of Nitrosamines . . . . . . . . . . 404
III. Mechanisms in the Bioactivation of Nitrosamines . . . . . 405
IV. Enzymology of the Bioactivation of Nitrosamines . . . . . 417
V. Fate of Reactive Intermediates from Nitrosamines . . . . . 419
References . . . . . . . . . . . . . . . . . . . . 420

## I. Introduction

$N$-Nitrosamines have been known since the nineteenth century. However, intensive research on the chemical and particularly the biological properties of nitrosamines stems in large part from the observations of Magee and Barnes in 1956[1] on the carcinogenicity of nitrosodimethylamine. Since this first report, the majority of over a hundred other nitrosamines have been shown to be carcinogenic in test animals.[2] Nitrosamines can induce tumors in a large number of species including members of the mammals, birds, amphibia, and fish.[3] An interesting feature of nitrosamine carcinogenesis is the diversity of organs in which tumors can be induced.[2] The tissue specificity of a nitrosamine is not usually affected by its route of administration; the susceptibility of a tissue depends to a large extent on the chemical structure of the compound. Major sites for carcinogenic action of nitrosamines include liver, esophagus, respiratory tract, kidney, and bladder. Another important property of nitrosamine carcinogenesis is that a single administration of at least some compounds can result in subsequent tumor formation in high incidence (e.g., Refs. 4, 5). Tumors can also be produced by continuous exposure to extremely low doses.[6]

Although nitrosamines are an interesting and valuable series of compounds with which to study mechanisms of the chemical induction of cancer, additional interest derives from the occurrence of a number of nitrosamines in the environment and the potential for their actual formation in the body.[7,8] Nitrosamines, therefore, are prime suspects in the search for causes of human cancer.[9]

This chapter will focus on mechanisms involved in the bioactivation of nitrosamines to reactive intermediates, which are believed to be the proximate or ultimate carcinogenic forms of the compounds. In order to provide background information necessary to understand the biological activity of nitrosamines, some of their chemical properties will first be reviewed.

## II. Chemical Properties of Nitrosamines

Polar resonance forms contribute as much as 48% to the ground-state electronic configuration of nitrosamines (Fig. 1).[10,11] The resonance leads to a partial double-bond character of the N-N bond, which results in a planar configuration and a barrier to free rotation of the nitroso group—23 kcal/mol for nitrosodimethylamine.[12] The two configurational isomers can generally be distinguished by nuclear magnetic resonance spectroscopy (e.g., Refs. 12–15, and references therein), and in some cases they can be separated chromatographically.[14–17] An important consequence of the dipolar nature of nitrosamines is the acidity of the $\alpha$-hydrogen atoms, which readily undergo base-catalyzed exchange in deuterium oxide.[18] This acidity has been utilized by Seebach and Enders[19] in a major synthetic reaction for nucleophilic $\alpha$-(secondary amino)-alkylation via $\alpha$-metalated nitrosamines.

Nitrosamines can readily be converted to their corresponding secondary amines (denitrosation) by reaction with a number of nucleophilic species [e.g., $Cl^-$, $Br^-$, $I^-$, $SCN^-$, $SC(NH_2)_2$] in acid solution.[20] This process is normally reversible, with the equilibrium lying well over on the side of the nitrosamine. Another reaction of nitrosamines in dilute acidic solution that

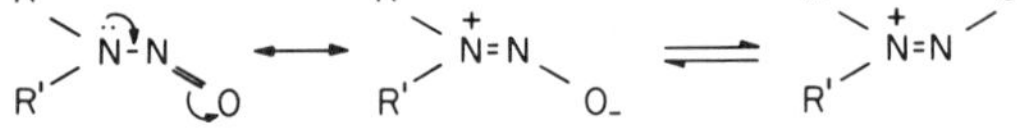

Fig. 1. Polar resonance forms of nitrosamines.

is catalyzed by nucleophilic species is the transfer of the nitroso group directly to other amines without the intermediacy of nitrous acid. Under appropriate conditions, aromatic, aliphatic, and alicyclic nitrosamines all serve as transnitrosating agents.[21-24] Since the early work of Fischer and Hepp,[25] aromatic N-nitrosamines have been known to rearrange in acid solution to give the corresponding p-nitroso isomers. This reaction, which is commonly used as a method for preparing aromatic C-nitroso compounds, proceeds by a mechanism in which rearrangement and denitrosation occur concurrently by two separate reactions of the protonated nitrosamine.[26]

In the presence of dilute acid, nitrosamines undergo various photoreactions that are essentially the chemistry of aminium radicals in the presence of NO. These reactions include photoelimination, photoreduction, and photoaddition.[27]

Nitrosamines may be reduced to 1,1-disubstituted hydrazines by reagents such as zinc dust in acid.[28] Reductive denitrosation to secondary amines takes place with metal halide–sodium borohydride or hydrogen–Raney nickel.[29,30] Nitrosamines are oxidized to the corresponding nitramine with peroxytrifluoroacetic acid.[31]

In two interesting reactions the nitroso group has been shown to act as an internal nucleophile. N-Nitroso-α-amino acids, on treatment with acetic anhydride, cyclize to yield the mesoionic sydnones,[32] while the N-nitroso group displays an exceptionally powerful neighboring group effect in the solvolysis of derivatives of β-hydroxynitrosodialkylamines.[33,34]

## III. Mechanisms in the Bioactivation of Nitrosamines

Most of the initial evidence for the mechanism of activation of nitrosamines was accumulated from experiments with the simplest nitrosamine, nitrosodimethylamine. Studies on the concentration of nitrosodimethylamine in blood and organs of rats, rabbits, and mice led Magee[35] to the conclusion that the compound is rapidly distributed throughout body water with no selective concentration in its target organ, the liver. This result has been confirmed in subsequent studies.[36,37] The rapid fall in the concentration of nitrosodimethylamine in the whole animal and the low recovery of unchanged nitrosamine in excreta led Magee to the additional conclusion that it is rapidly metabolized.[35] Hepatectomized rats gave increased recoveries of unchanged nitrosodimethylamine in excretion products, which indicated that the liver is the major organ that metabolizes the nitrosamine.[35] Dutton and Heath[38,39] showed that the principal metabolic product following nitrosodimethylamine administration to rats and mice is $CO_2$, from which they concluded that demethylation takes place. Dutton and Heath also concluded that the biological effects of nitrosodimethylamine were due to a metabolite rather than to the nitrosamine itself. Studies of nitrosodimethylamine metabolism in subcellular fractions of the liver showed a requirement for $O_2$ and NADPH and localization of the activity in the 9000-$g$ supernatant fraction.[40-42] Brouwers and Emmelot[43] showed that formaldehyde was a major product of nitrosodimethylamine metabolism *in vitro* with hepatic microsomes.

Next, labeled cellular constituents were detected following administration of [$^{14}$C]nitrosodimethylamine to rats. Labeled products are to be expected, of course, because formaldehyde is a metabolite that can enter the normal one-carbon pool of the cell. However, unusual cellular products, such as 1- and 3-methylhistidine, were detected in protein hydrolsates.[44,45] Magee and Farber[46] made the important observation that DNA and RNA from rat liver was highly labeled, and they identified 7-methylguanine as the principal product.

As a result of these observations, the reaction sequence shown in Fig. 2 was formulated to describe the metabolic activation of nitrosamines.[44,47,48] Enzymatic hydroxylation at the carbon atom $\alpha$ to the $N$-nitroso group is the critical initial step in the biotransformation. Spontaneous cleavage of the carbon–nitrogen bond in the $\alpha$-hydroxynitrosamine (**2**) leads to the production of an aldehyde and the alkyldiazohydroxide (**3**). The diazohydroxide may then produce either the diazoalkane (**4**) or cationic products (**5** and **6**), which may finally be trapped by water as alcohols or react at a nucleophilic site on a biomolecule such as DNA.

Evidence has subsequently been provided for the production of methanol during both the *in vivo* and the *in vitro* metabolism of nitrosodimethylamine in the rat[49] and for methylation of calf thymus DNA by nitrosodimethylamine in the presence of rat liver microsomes and an NADPH-regenerating system.[50] The rate of DNA methylation *in vitro* correlates well with the rate of formaldehyde production as predicted by the reaction sequence shown in Fig. 2.[51]

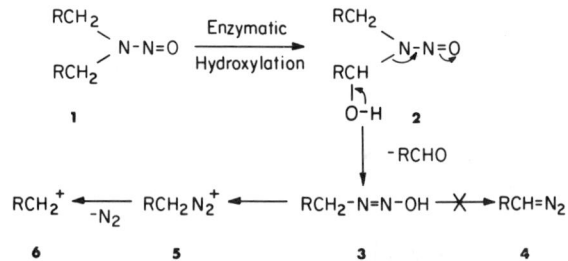

**Fig. 2.** Metabolic activation of nitrosamines by $\alpha$ hydroxylation.

Generation of the diazoalkane (4) as the ultimate alkylating agent has been ruled out by the demonstration that, when nitrosodimethyl-$d_6$-amine was administered to rats, the resulting hepatic nucleic acids contained methyl groups that retained all three deuterium atoms.[52] A similar result as obtained with nitrosodiethyl-$d_{10}$-amine.[53] Indirect evidence, based on an association of decreasing selectivity toward nucleophilic sites in DNA with increasing tendency for unimolecular reactivity, has been presented for participation of carbonium ion intermediates in the alkylation of DNA by nitrosamines.[54] Results in which propionaldehyde and both *n*-propanol and isopropanol were detected[55] following incubation of nitrosodi-*n*-propylamine with rat liver fractions also supported the hypothesis that a reaction sequence leading to carbonium ions is initiated by microsomal oxidation of nitrosamines (the first-formed primary propyl cation has a tendency to rearrange to the more stable secondary ion before reaction with water to form the alcohol). However, Park *et al.*[56] have shown that administration of nitrosodi-*n*-propylamine to rats leads to the formation of 7-*n*-propylguanine, but not 7-isopropylguanine, in hepatic DNA. This result suggests that some intermediate in the reaction sequence is intercepted by nucleophilic sites in DNA before a carbonium ion can be formed. Park *et al.*[56] suggested that the ultimate carcinogenic form is probably one of the earlier electrophilic intermediates, such as the alkyldiazohydroxide, the alkyldiazonium ion, or even the first-formed α-hydroxynitrosamine, which could participate in a concerted or nearly concerted $S_N2$ reaction with a nucleophilic site in the nucleic acids, possibly via a partially stabilized hydrogen-bounded complex. This conclusion is supported by the theoretical study of Andreozzi *et al.*[57] Other data suggest that, whereas the N-7 position of guanine reacts with activated nitrosodi-*n*-propylamine to form only the *n*-propyl adduct, the O-6 position, which is less nucleophilic, reacts to form mainly the isopropyl adduct.[58] Scribner and Ford[58] postulate that the guanine moiety reacts directly with the *n*-propyldiazonium ion with rearrangement occurring before, or concomitant with, the loss of $N_2$. The propyl moiety would then be more carbenium ionlike in the transition state for reaction at the O-6 position than for reaction at the 7-position. It is clear that more work is required in this area before the precise alkylation mechanism can be defined.

A number of nitrosamines have now been studied with respect to formation of the aldehyde (and in some cases the alcohol) that results from *in vitro* microsomal α oxidation, and such evidence for an α-oxidation pathway has been obtained for nitrosodiethylamine,[59-61] nitrosodipropylamine, and its β-oxidized analogs,[55,62,63] nitrosodibutylamine,[64] and nitrosodiallylamine.[65] β-Fluorination has been shown to influence the microsomal metabolism of nitrosodiethylamine at nonfluorinated positions.[66,67] Thus, fluorine substitution at the β position reduced the susceptibility of the molecule to α oxidation, possibly via electronic effects. Mutagenic and carcinogenic activity of the β-fluorinated compounds correlated with the extent of metabolism at the α position.[66,67] Unsymmetrical nitrosodialkylamines may, of course, be oxidized at either of the two α-carbon atoms to yield two aldehydes. Nitrosomethylethylamine,[61,68] nitrosomethylbutylamine,[68] nitrosomethyl(2-phenylethyl)amine[68] and nitrosomethylbenzylamine[69] appear to be oxidized at a slower rate at the methyl moiety than at the other group. For the first three of these compounds, the rate of metabolism ($V_{max}$) at the methyl group is about half that at the other group with rat liver microsomes. For nitrosomethylbenzylamine, oxidation at the methyl carbon is 10 times slower than oxidation at the benzylic carbon with hepatic microsomes, whereas this differential is 100-fold with microsomes from rat esophagus, the target organ for this compound.[69]

There is evidence that cyclic nitrosamines are also activated by α hydroxylation. For these compounds, metabolism followed by reaction of the unstable intermediate with water yields a product containing both alcohol and aldehyde groups. These reactions are illustrated in Fig. 3 for nitrosopiperidine. In this case, the expected product, 5-hydroxypentanal, exists predominantly as the tautomeric cyclic hemiacetal. This metabolite was detected after metabolism of nitrosopiperidine by rat liver microsomes.[70] Similarly, the cyclic hemiacetal of 5-hydroxybutanal, 2-hydroxytetrahydrofuran, was identified as the metabolite from nitrosopyrrolidine.[71-73] Other cyclic nitrosamines for which reaction products resulting from in vitro α hydroxylation have been identified include nitrosomorpholine,[74] nitrosohexamethyleneimine,[75,76] and the tobacco-specific nitrosamines nitrosonornicotine[77] and 4-(N-methyl-N-nitrosamino)-1-(3-pyridyl)-1-butanone.[78] No evidence has yet been obtained with cyclic nitrosamines for rearrangement of electrophilic intermediates formed before their reaction with water as the nucleophile.

Several additional lines of evidence suggest that α hydroxylation is the key activation pathway for nitrosamines. Early structure–activity studies[48]

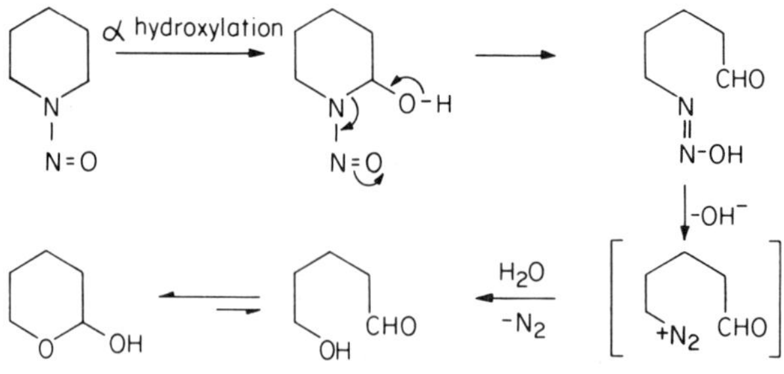

**Fig. 3.** Metabolic activation of nitrosopiperidine by α hydroxylation.

showed that compounds such as nitrosodiisopropylamine and nitrosodicyclohexylamine, in which enzymatic $\alpha$ hydroxylation is possibly impeded by steric hindrance, were only weakly carcinogenic or inactive. Compounds such as nitrosodibenzylamine and nitrosoproline ethyl ester, however, were also not carcinogenic.[48] Interpretation of such structure–reactivity relationships is difficult, because substitution at the $\alpha$-carbon atom changes more than one physicochemical characteristic of the molecule. When factors such as water–hexane partition coefficients and electronic parameters are taken into account in a quantitative manner, however, reactivity at the $\alpha$-carbon atom may be implicated in the carcinogenic activity of a number of nitrosamines.[79]

An intriguing observation that supports the $\alpha$-hydroxylation mechanism was made by Keefer et al. in 1973,[80] who found that the incidence of liver tumors was 26% in rats receiving 5 ppm nitrosodimethylamine in their drinking water for 30 weeks. A significantly different tumor incidence of 3% was observed in a second group of rats receiving nitrosodimethyl-$d_6$-amine under identical conditions. Dagani and Archer[81,82] found that deuterated nitrosodimethylamine is metabolized more slowly than nitrosodimethylamine itself ($V_{max}^H / V_{max}^D = 2.8$). The isotope effect indicates that the rate-limiting step in the enzymatic decomposition of the nitrosamine, as measured by formaldehyde production, is breaking of the C—H bond. This result, together with the biological observation of Keefer et al.,[80] suggests that the rate of formation of a proximate or ultimate carcinogen may be important in determining the subsequent tumorigenic potency of nitrodimethylamine. A similar deuterium isotope effect was found for both the carcinogenicity[83] and mutagenicity[84,85] of nitrosomorpholine and nitrosomorpholine-$\alpha$-$d_4$, again implicating the importance of the $\alpha$ position in the formation of active metabolites. Subsequent studies with unsymmetrical nitrosamines labeled with deuterium at various positions have provided a more complex picture and suggest that there seems to be a balance between several pathways of metabolism, the result of which determines the carcinogenic potency of a particular compound.[86-88] For example, with nitrosomethylethylamine, deuterium labeling at the methyl group reduced its carcinogenic potency, while deuterium labeling at the $\alpha$-methylene group greatly increased its carcinogenicity.[86] In addition, deuterium substitution was shown to influence the organ specificity of nitrosomethylethylamine. For unknown reasons, compounds with deuterium in the terminal methyl of the ethyl group tended to induce esophageal as well as liver tumors.[86] Singer and Lijinsky[89] have shown that compounds that give pronounced biological isotope effects tend to undergo exchange to a small extent, but that no biological isotope effect is found for compounds that undergo extensive exchange. Thus, extremely labile $\alpha$ protons might be expected to equilibrate faster, even at physiological pH, than metabolic processes leading to carcinogenesis.

Studies of the chemical and biological properties of acetoxy derivatives of α-hydroxynitrosamines have provided further support for α hydroxylation as the activation pathway. Nitrosomethyl(α-acetoxymethyl)amine was first prepared by Roller et al. in 1975,[90] who also showed that its solvolysis led to formation of an equimolar mixture of acetic acid, formaldehyde, and methanol as predicted from the activation mechanism of nitrosodimethylamine (Fig. 2). The acetoxy compound was hydrolyzed by hog liver esterase, but the α-hydroxy compound so formed was too short lived to be isolated. The reactive alkylating agent produced by the action of esterases was shown to produce a spectrum of alkylated products in DNA identical to those found when nitrosodimethylamine is administered to rats.[91] Tumor induction by nitrosomethyl(α-acetoxymethyl)amine takes place in organs at or near their site of application, or in those first exposed to the compound by systemic circulation.[91-97] The correlation of the tissue susceptibility with the extent of DNA alkylation indicates that the α-acetoxy compound is rapidly taken up by cells and is activated by esterases that are present in most tissues.[91]

A number of other nitroso(α-acetoxyalkyl)alkylamines have been prepared and have been shown to be generally potent bacterial mutagens that do not require microsomal activation.[98-102] These results are again in accord with the hypothesis that α-acetoxynitrosamines are hydrolyzed either chemically or by esterases to yield the α-hydroxynitrosamine, which is the proximate mutagen. In one study, the mutagenicity of a series of acetoxynitrosamines was inversely related to their half-lives in aqueous solution.[99] Nitroso(α-acetoxymethyl)-*tert*-butylamine was not mutagenic, a finding that was attributed to the low reactivity of the *tert*-butyl carbonium ion.[99] In support of this conclusion, both nitrosomethyl-*tert*-butylamine and nitroso-(α-acetoxymethyl)-*tert*-butylamine have been shown not to be carcinogenic in the Syrian hamster.[103] *tert*-Butyldiazotic acid was shown to produce *tert*-butanol in 63% yield, but evidently cannot alkylate the critical cellular site for steric reasons.[103]

In a study of the mutagenicity of nitrosomethylbenzylamine, the two α-acetoxy isomers were prepared.[104] Nitrosomethyl(α-acetoxybenzyl)amine, a methylating agent, was strongly mutagenic for *Salmonella typhimurium,* whereas nitroso(α-acetoxymethyl)benzylamine, a benzylating agent, was inactive.[104] The biological activity of these two acetoxy derivatives toward *S. typhimurium* is in accord with additional data that show that nitrosomethylbenzylamine acts exclusively as a methylating agent in vivo.[105] A different result, however, might have been predicted from the chemical properties of the acetoxy compounds. Although the acetoxyben zyl derivative is hydrolyzed via the highly resonance-stabilized benzylnitrosimminium ion more rapidly than the acetoxymethyl derivative,[106] the favored alkylation reaction *in vitro* with 2-aminopyridine as nucleophile is benzylation.[107] The latter result is interpreted as a reflection of the relative reactivities of the two electrophilic alkylating intermediates. Lack of benzylation of DNA *in vivo* by nitrosomethylbenzylamine is likely to be due, at least in part, to the specificity of the mixed-function oxidase, which favors oxidation at the benzylic carbon to produce a methylating agent.[69]

Perhaps the best evidence for the $\alpha$-hydroxylation pathway of nitrosamine activation is the report of the actual isolation and characterization of nitrosomethyl($\alpha$-hydroxymethyl)amine and nitrosobutyl($\alpha$-hydroxymethyl)amine in carefully dried aprotic solvents.[108] These compounds were prepared from the corresponding nitrosohydroperoxymethylamines by deoxygenation. The hydroperoxy compounds were prepared either by substitution of nitroso($\alpha$-acetoxymethyl)alkylamines with hydrogen peroxide in acetic acid[109] or by oxygenation of the $\alpha$-lithiated nitrosamines with oxygen.[110] Almost quantitative deoxygenation of the hydroperoxy compounds was achieved with triphenylphosphine in chloroform or in somewhat lower yield with aqueous sodium bisulfite followed by extraction into chloroform. Both hydroxymethyl compounds were unstable in aqueous solution. Products of decomposition were identified as formaldehyde, 1-butanol, and 2-butanol for the $\alpha$-hydroxybutyl compound, and formaldehyde and methanol for the $\alpha$-hydroxymethyl compound. In the presence of the nucleophile thiophenol, butylation (1.2% yield) or methylation (26% yield) was observed. From pH 1 to 5, decomposition rates of the $\alpha$-hydroxynitrosamines in phosphate solution were almost constant, with half-lives of about 5 min. Above pH 6 the decomposition rates began to increase so that at pH 7, both compounds had half-lives of only about 10 sec, and at pH 8, 1 sec. These values are likely to be lower in the presence of cellular nucleophiles, but nevertheless still allow ample time for intra- or even intercellular diffusion of the activated nitrosamine, as has been observed for nitrosodimethylamine.[111]

All of the experiments just described point to the importance of $\alpha$ hydroxylation as the activation pathway for carcinogenesis by nitrosamines. In a carefully performed experiment, evolution of $^{15}$N-labeled molecular nitrogen was used to gauge the extent of $\alpha$ hydroxylation of doubly $^{15}$N-labeled nitrosodimethylamine and nitroso-*N*-methylaniline by rat liver homogenate.[112] It was estimated that 34% of nitrosodimethylamine and 19% of nitrosomethylaniline are metabolized by $\alpha$ hydroxylation. A number of detoxication pathways have been elucidated that include denitrosation[113] and reduction to unsymmetrical hydrazines.[114] These reactions will not be considered here. Two other metabolic reactions, $\beta$ oxidation and $\omega$ oxidation, are involved in the activation of a number of nitrosamines and will now be considered.

A somewhat surprising feature of the biological reactivity of at least two simple nitrosodialkylamines, nitrosodipropylamine and nitrosodibutylamine, was discovered by Krüger in 1971.[115] He found that, in addition to direct transfer of an intact propyl or butyl group to DNA and RNA, the two nitrosamines also act as methylating agents.[115] When these nitrosamines are administered to rats, 7-methylguanine is the major alkylation product in hepatic nucleic acids. To explain these results, Krüger suggested that the nitrosodialkylamines are metabolically degraded in a manner similar to fatty acids, by two consecutive $\beta$-oxidation reactions followed by cleavage of the acyl fragment from the $\beta$-ketonitrosamine (possibly by participation of coenzyme A) to yield the nitrosomethylalkylamine.[115] The latter compound finally acts as a methylating agent by hydroxylation at the $\alpha$-carbon of the alkyl group (Fig. 4). In accordance with this mechanism, nitrosodipropylamine and nitrosodibutylamine labeled with $^{14}C$ at the $\alpha$-carbon atoms both yielded radioactive 7-methylguanine in rat liver.[115] No labeled 7-methylguanine, however, was observed when nitrosodipropylamine labeled with $^{14}C$ at the $\beta$-carbon atom was used. Additional support for this mechanism was provided by administration of nitroso-2-hydroxypropylpropylamine or nitroso-2-oxopropylpropylamine to rats, after which the levels of 7-methylguanine in hepatic nucleic acids were considerably higher than after administration of nitrosodipropylamine.[116,117]

Investigation of the metabolism of nitrosodipropylamine *in vitro* by hepatic microsomes showed that, in addition to oxidation at the $\alpha$-carbon atom, $\beta$ oxidation also takes place.[62] Nitroso-2-hydroxypropylpropylamine was isolated and characterized as a product when nitrosodipropylamine was

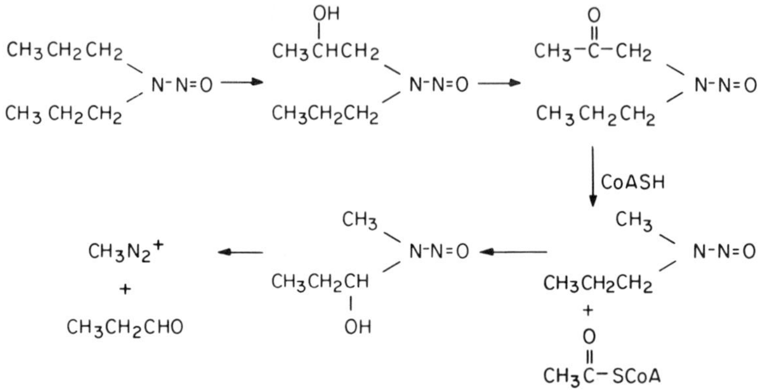

Fig. 4. Hypothetical pathway of Krüger[115] for the metabolic degradation of nitrosodipropylamine.

substrate; furthermore, nitroso-2-oxopropylpropylamine was isolated as a product when nitroso-2-hydroxypropylpropylamine was the substrate. These results therefore provide direct evidence for the first two steps of Krüger's hypothetical reaction sequence.[115]

In experiments that confirm and extend Krüger's observations on the methylating properties of nitrosodipropylamine and its $\beta$-oxidized derivatives, Leung et al.[118] have shown that administration of nitroso-2-oxopropylpropylamine to rats leads not only to formation of 7-methylguanine, but also to formation of $O^6$-methylguanine in hepatic DNA. The $O^6$- to 7-methylguanine ratio was 0.07, which is similar to the ratio of these two methylated guanines obtained after nitrosodimethylamine administration at comparable dose levels and times.[119] Lawley[120] has shown that the $O^6$- to 7-methylguanine ratio in DNA is indicative of the reactivity of the methylating agent. The results of Leung et al.[118] therefore support the hypothesis that nitrosomethylpropylamine may be the methylating agent formed from nitrosodipropylamine.

In subsequent experiments, however, formation of nitrosomethylpropylamine from nitroso-2-oxopropylpropylamine has not been detected. Nitroso-2-oxopropylpropylamine is converted into nitrosomethylpropylamine in a base-catalyzed, nonenzymatic reaction, but this takes place only at high pH; there was no detectable reaction at physiological pH even after 16 hr.[118] In an extensive search, no rat liver fraction capable of catalyzing this conversion has been found.[118,121]

Alternative mechanisms to account for the methylating properties of $\beta$-oxidized nitrosamines have been proposed. Lawson et al.[122] have suggested that $\beta$-ketonitrosamines may undergo $\omega$ oxidation to the keto acid followed by decarboxylation. A further $\omega$-oxidation step, followed again by decarboxylation, would yield the nitrosomethylalkylamine. There is no evidence, however, that such reactions take place. Michejda and co-workers[33,34,123] have suggested that conjugation of $\beta$-hydroxynitrosamines with sulfate could result in formation of very effective alkylating agents (oxadiazolium ions) via participation of the nitroso group as an internal nucleophile (Fig. 5). This mechanism, of course, requires prior formation of the nitrosomethyl-2-hydroxyalkylamine (R = H, in Fig. 5) to account for methylation reactions. Leung and Archer[124] have isolated the $\beta$-glucuronide conjugate of nitroso-2-hydroxypropylpropylamine in high yield from the urine of rats administered nitrosodipropylamine or its $\beta$-hydroxy or $\beta$-oxo derivatives. The glucuronyl moiety is a poor leaving group, however, and there is no evidence for formation of sulfate conjugates.

Leung and Archer[125] have investigated, with the aid of model compounds, the possibility that nitroso-2-oxopropylpropylamine may act as a methylating agent via formation of 2-oxopropyldiazotate. This intermediate is

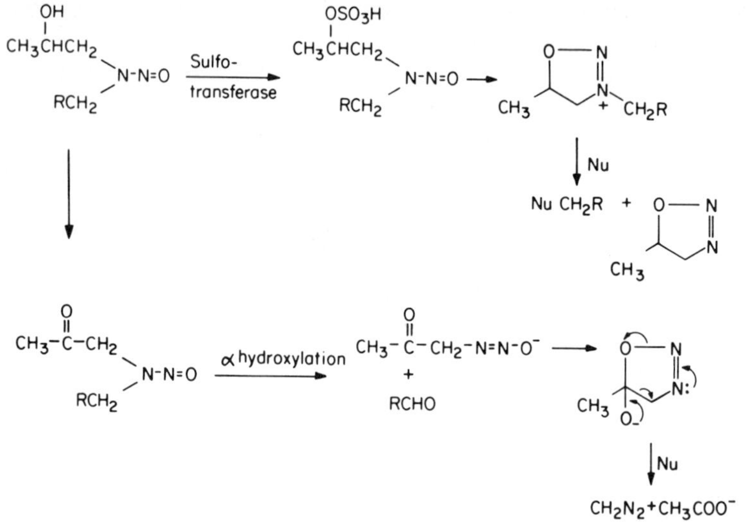

Fig. 5. Mechanisms proposed by Michejda[123] and Leung and Archer[125] to account for the methylating property of β-oxidized nitrosamines.

formed after metabolic α hydroxylation of this nitrosamine on the propyl side chain with subsequent loss of propionaldehyde from the α-hydroxynitrosamine (Fig. 5). Two reactions that form the same putative intermediate, 2-oxopropyldiazotate, are base-catalyzed decomposition of N-(2-oxopropyl)-N-nitrosourea and enzymatic hydrolysis of N-nitroso-N-acetoxymethyl-N-2-oxopropylamine. Characterization of the products of these reactions, as well as the ability of the compounds to methylate nucleophiles including DNA, showed that 2-oxopropyldiazotate undergoes an internal cyclization reaction to yield an oxadiazoline (Fig. 5). This intermediate spontaneously decomposes to yield acetic acid and the methylating agent diazomethane. Although diazomethane was detected as a product of the decomposition of the oxadiazoline *in vitro*, it is possible that the ultimate methylating agent *in vivo* is not diazomethane itself. The oxadiazoline intermediate may react concertedly with nucleophilic sites on DNA to form methylated bases without release of diazomethane. Leung and Archer have also provided evidence[121] that nitroso-2-oxopropylpropylamine does indeed undergo α hydroxylation on the n-propyl group in a reaction catalyzed by rat hepatic microsomal cytochrome P-450 to yield propionaldehyde and, presumably, 2-oxopropyldiazotate.

Okada and co-workers (reviewed in Ref. 126) and Blattmann and Preussmann[127-132] have provided evidence for a chain-shortening mechanism for dialkylnitrosamines via metabolic hydroxylation at the terminal (ω) carbon atom. For example, nitrosobutyl-4-hydroxybutylamine, a potent bladder carcinogen, yields nitrosobutyl-3-carboxypropylamine as the major urinary metabolite[126] (Fig. 6). Several nitrosamines were characterized as minor metabolites, including the glucuronic acid conjugates of nitrosobutyl-4-hydroxybutylamine and nitrosobutyl-3-carboxypropylamine and compounds produced from the 3-carboxy compound by subsequent β oxidation, including nitrosobutyl-3-carboxy-2-hydroxypropylamine, nitrosobutylcarboxymethylamine, and nitrosobutyl-2-oxopropylamine (Fig. 6). The successive removal of two carbon fragments from these nitrosamines is analogous to the Kroop mechanism of fatty acid metabolism. Nitrosodibutylamine underwent ω, ω-1, and ω-2 oxidations in the rat, and the metabolites so produced also underwent conjugation with glucuronic acid or further oxidation to the corresponding oxo compounds[126] (Fig. 6). Preussmann and Blattmann examined the *in vivo* metabolism of a homologous series of nitrosodialkylamines from diethyl to dipentyl and showed that ω oxidation occurred in every case.[127]

Nitrosobutyl-3-carboxypropylamine, the principal urinary metabolite of both nitrosobutyl-4-hydroxybutylamine and nitrosodibutylamine, was shown to be a proximate form of these nitrosamines in their action as bladder carcinogens in the rat.[127] The butyl group of nitrosobutyl-4-hydroxybutylamine can be replaced by other alkyl groups without altering the

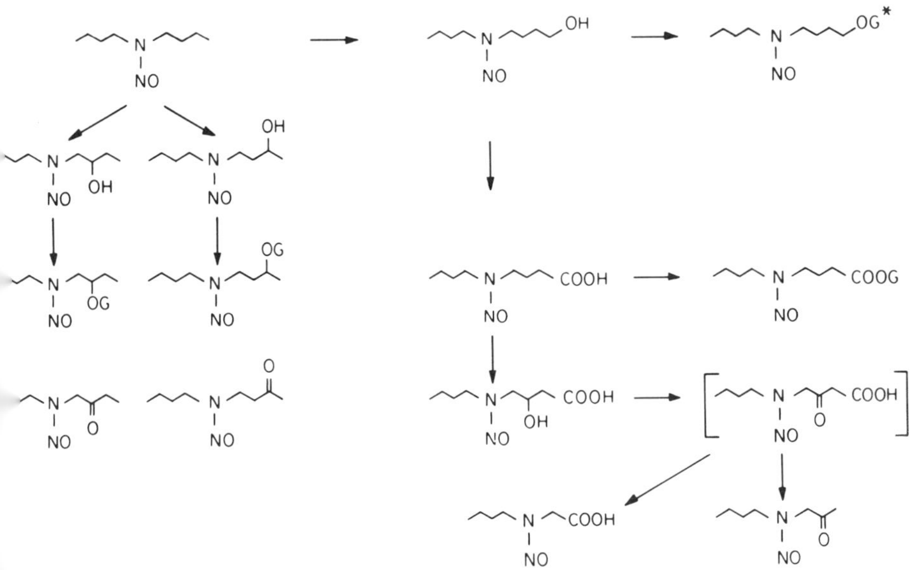

Fig. 6. Urinary metabolites of nitrosodibutylamine and nitrosobutyl-4-hydroxybutylamine in the rat.[126] *G, glucuronic acid conjugate.

pattern of urinary metabolites or the carcinogenicity of the compounds.[133,134] However, the number of carbon atoms between the nitrogen atom and the terminal hydroxyl or carboxyl groups markedly affects the carcinogenic activity of the nitrosamines. For example, nitrosoalkyl-2-carboxyethylamines do not produce bladder tumors.[126] In view of these results, Mochizuki et al.[135] have suggested that nitrosoalkyl-3-carboxypropylamines are proximate carcinogens that undergo further metabolism via α hydroxylation to compounds that can readily form stable γ-lactones (Fig. 7). The corresponding α- or β-lactones would be less likely to form. Mochizuki et al.[135] synthesized 4-(N-butylnitrosamino)-4-hydroxybutyric acid lactone and showed that it is a potent mutagen and gene-damaging agent without activation. They suggested that nitrosoalkyl-4-hydroxybutylamines are transformed in the liver into nitrosoalkyl-3-carboxypropylamines, which are then excreted into the urine. The epithelial cells of the bladder finally activate the 3-carboxypropyl compounds by α oxidation, and the resulting ultimate carcinogenic metabolite is stabilized as the γ-lactone.

According to this mechanism, only nitrosomethylalkylamines with even-numbered alkyl chains should give rise to bladder tumors. This hypothesis was supported by the finding that nitrosomethyldodecylamine induced bladder tumors,[136] but liver tumors were produced by the undecyl homolog.[137] In another study,[138] bladder tumors were induced by the n-octyl, n-dodecyl, and n-tetradecyl homologs, but not the homologous compounds with odd numbers of carbon atoms. The principal urinary metabolite of the compounds with odd-numbered chains was nitrosomethyl-2-carboxyethylamine.[139] The compounds with even-numbered chains yielded nitrososarcosine and nitrosomethyl-3-carboxypropylamine as major urinary metabolites.

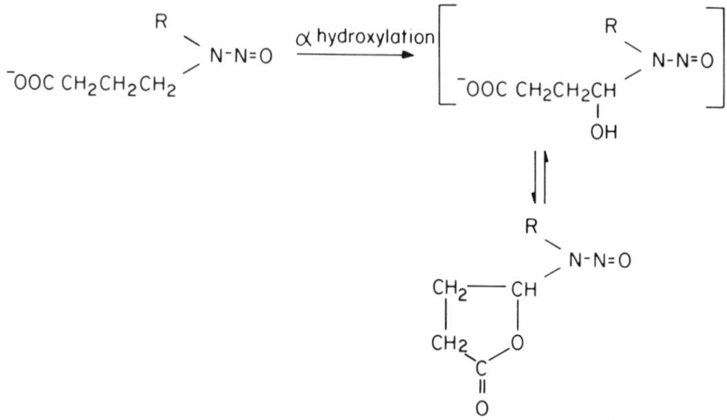

Fig. 7. Formation of γ-lactone from nitrosoalkyl-3-carboxypropylamine.

## IV. Enzymology of the Bioactivation of Nitrosamines

The majority of studies on the enzymology of nitrosamine activation have centered on the nitrosodimethylamine demethylase activity of rodent liver. Subcellular fractionation of rat liver demonstrated that demethylase activity is primarily located in the microsomal fraction.[40-43,140] The demethylase requires NADPH and oxygen,[40-43] is inhibited by carbon monoxide,[141,142] and is repressed by pretreatment with cobaltous chloride.[142] Reconstituted systems containing hepatic cytochrome $P$-450 and NADPH–cytochrome reductase activity possess demethylase activity.[143,144] Other evidence that points to an involvement of cytochrome $P$-450 includes inhibition by SKF-525A[140,145] and optical difference and electron paramagnetic resonance spectroscopy, which indicate that nitrosodimethylamine and other nitrosamines interact with microsomes and bind to cytochrome $P$-450 both as substrate and as ligand.[146] There has been much confusion over reports that cytochrome $P$-450 inducers both increase and repress nitrosodimethylamine demethylase activity. Kinetic studies from a number of laboratories have now implicated the existence of at least two forms of microsomal demethylase, a high-affinity demethylase I and a low-affinity demethylase II.[140,147-150] These two enzyme forms respond very differently after treatment of rats with mixed-function oxidase inducers. Whereas the activity of demethylase II is normally increased, the activity of demethylase II is repressed by such treatments.[147-159] Only demethylase I is expected to play a physiological role in nitrosodimethylamine activation by the liver, because nitrosamine concentrations obtained *in vivo* are generally several orders of magnitude lower than those required for significant metabolism by demethylase II.[160,161]

Lake and co-workers have provided evidence that a cytochrome $P$-450-independent enzyme may be involved, at least in part, in the demethylation of nitrosodimethylamine.[140,162,163] They have found a greater stability of the demethylase compared to other hepatic mixed-function oxidase activities and the inhibition of demethylation by compounds that do not alter other cytochrome $P$-450-catalyzed reactions. Furthermore, they failed to obtain evidence for an interaction of nitrosodimethylamine and cytochrome $P$-450 by optical-difference spectroscopy. Lake *et al.* have suggested that at least one component of nitrosodimethylamine demethylase is an amine oxidase, possibly monoamine oxidase, because demethylation is inhibited by substrates and inhibitors of the mitochondrial momoamine oxidase, and microsomes have been reported to contain monoamine oxidase activity.[162,164] This view is not supported by Lai *et al.*,[165] who could find no evidence for an involvement of an amine oxidase in nitrosodimethylamine demethylation or for microsomal monoamine oxidase activity. Identification of the enzymes involved in the microsomal demethylation of nitrosodimethylamine and their assignment to the different kinetic forms clearly require further study.

In addition to the microsomal demethylase, Kroeger-Koepke and Michejda[150] have reported the existence of a nitrosodimethylamine demethylase in rat liver cytosol. This cytosolic enzyme may explain the three different $K_m$ values obtained for nitrosodimethylamine with rat liver postmitochondrial supernatant compared to the two $K_m$ values obtained with microsomes.[140] There are contradictory reports of a nitrosodimethylamine demethylase activity in the hepatic nucleus. Grandjean et al.[166] reported the activation of nitrosodimethylamine and several other nitrosamines into alkylating agents by rat liver nuclei. In addition, although not apparently linked to nitrosodimethylamine demethylation, Floyd[167] reported formation of nitrosamine free radicals when nitrosodimethylamine and several other nitrosamines were incubated with rat liver nuclei. In contrast, Lai et al.[165] could not detect metabolism of nitrosodimethylamine to formaldehyde or to a methylating agent using either rat liver nuclei or purified nuclear membranes.

Reasons for the discrepancies in the literature concerning the properties of nitrosodimethylamine demethylase are not known. Argus and Arcos[168] have pointed out that high concentrations of nitrosodimethylamine could cause membrane perturbations and protein denaturation. Discrepancies could also occur from the use of different methods to measure enzyme activity, such as measurement of formaldehyde production by colorimetric or radiometric methods or incorporation of radioactivity from the nitrosamine into tissue macromolecules. Production of formaldehyde and a methylating agent exhibit similar responses to various treatments, in accordance with their stoichiometric formation from nitrosodimethylamine. Alterations in this general pattern, however, have been reported,[169,170] but it is not known whether these alterations may be attributed to other activation pathways or to effects of the treatments on the stability of the methylating agent. Another complication may be caused by contamination of commercially available [$^{14}$C]nitrosodimethylamine by [$^{14}$C]formaldehyde.[69,171] Finally, freezing tissue fractions before assay results in a gradual loss of activity for conversion of nitrosodimethylamine to formaldehyde and to a mutagen by inducer-pretreated tissues, but not by control tissues.[152,172] It is clear that technical difficulties may account for the various discrepancies reported in the literature in this area.

Distribution of nitrosodimethylamine demethylase activity has not been studied among the various cell types of the liver, all of which are susceptible to tumor induction by the nitrosamine. The demonstration that the methylating agent formed from nitrosodimethylamine has sufficient stability to diffuse between cells, however, suggests that all target cells within the liver need not possess activating capacity.[111,173] Metabolism of nitrosodimethylamine has been demonstrated in a number of extrahepatic tissues, including kidney, lung, and small intestine,[174,175] but the nature of the enzymes is not yet known.

As described earlier, a number of other nitrosamines, including cyclic compounds, undergo metabolism by $\alpha$ oxidation. Although poorly studied with respect to enzymology, at least in some cases the $\alpha$ oxidation appears to be catalyzed by cytochrome $P$-450, as demonstrated by use of inducers[65,69,176-178] and inhibitors,[62,65,69] and by spectroscopic methods.[146] In contrast, preliminary evidence has been obtained for activation of nitrosopiperidine[179] and nitrosopyrrolidine[180] by a pathway that does not involve cytochrome $P$-450. Also described earlier, oxidation of nitrosamines at $\beta$- and $\omega$-carbon atoms may in some cases be involved in their activation. $\beta$ Oxidation of nitrosodipropylamine occurs in the microsomal fraction of rat liver and is inhibited by SKF-525A.[62] Oxidation of nitroso-2-hydroxypropylpropylamine to the 2-oxo compound also takes place with rat liver microsomes.[62] The enzymology of $\omega$ oxidation of dialkylnitrosamines has not been studied. Hydroxylation of several cyclic nitrosamines at various ring positions appears to be microsomal.[181-184]

## V. Fate of Reactive Intermediates from Nitrosamines

As discussed earlier, the exact chemical identity of the ultimate alkylating agent from nitrosamines has not been determined. Once formed, however, the intermediate reacts rapidly with cellular nucleophiles. Interaction with water results in the formation of the corresponding alcohol; alkylated products are formed after reaction with other nucleophiles, notably nucleic acids and proteins. Because DNA is generally considered to be the critical cellular target for carcinogens during tumor initiation, most attention has been focused on DNA alkylation by nitrosamines.

Although 7-alkylguanine is the most abundant modified base in DNA produced by dialkylnitrosamines, reaction of electrophilic intermediates with a wide variety of other nucleophilic sites can also occur.[120,185,186] These sites include the 1-, 3-, and 7-positions of adenine, the 3- and O-6 positions of guanine, the 3- and O-2 positions of cytosine, the 3-, O-2, and O-4 positions of thymine, and the phosphate groups. There is evidence, however, that not all of these products have the same biological importance. While the extent of alkylation at the 7-position of guanine shows no correlation with carcinogenic activity, a striking correlation has been obtained between tissue susceptibility to tumor induction by nitroso compounds and the initial extent of formation and subsequent persistence of $O^6$-alkylguanine residues (reviewed in Refs. 187–189). A number of exceptions to this correlation have been reported, however, in both liver and extrahepatic tissues.[187,190,191]

The rate and extent of reaction *in vivo* of the alkylating intermediates formed from nitrosamines with sites on DNA depend on the nature of both electrophile and nucleophile. For example, as mentioned earlier, the reactive intermediate from nitrosomethyl-*tert*-butylamine containing the *tert*-butyl group, reacts readily with water, but, presumably for steric reasons, not with DNA.[103] A number of studies have demonstrated that alkylation of DNA in chromatin by nitrosodimethylamine and other nitroso compounds is nonrandom (e.g., Refs. 192–195). These studies suggest that chromatin proteins or other microenvironmental factors, such as pH or ionic strength, may impair access of the alkylating intermediate to the DNA itself.

Little is known concerning the interaction of electrophilic intermediates from heterocyclic nitrosamines with DNA. Early work by Lee and Lijinsky[196] suggested that nitrosopyrrolidine, nitrosomorpholine, and nitrosopiperidine form 7-methylguanine in hepatic RNA, but this observation has not been confirmed. Three cyclic nitrosamines were shown to cause single-strand breaks in hepatic DNA.[197] Ross and Mirvish[198] reported that metabolism of nitrosohexamethyleneimine gave 1,6-hexanediol bound to rat liver nucleic acids. Hunt and Shank[199] have provided evidence for an unidentified fluorescent adduct in hepatic DNA from rats given nitrosopyrrolidine. Very recently, Chung and Hecht[200] have shown that $\alpha$-acetoxynitrosopyrrolidine, upon treatment with porcine liver esterase, reacts with deoxyguanosine to yield a $1,N^2$-substituted tricyclic derivative. There is no evidence that such an adduct occurs in the DNA of nitrosopyrrolidine-treated animals.

## Acknowledgment

The authors acknowledge support from the Ontario Cancer Treatment and Research Foundation, Grant MT-7025 from the Medical Research Council of Canada, and PHS Grant CA 26651 awarded by the National Cancer Institute, DHHS.

## References

1. Magee, P. N., and Barnes, J. M. (1956). The production of malignant primary hepatic tumors in the rat by feeding dimethylnitrosamine. *Br. J. Cancer* **10**, 114–122.
2. Shank, P. C., and Magee, P. N. (1981). Toxicity and carcinogenicity of nitroso compounds. *In* "Mycotoxins and *N*-Nitroso Compounds: Environmental Risks" (R. C. Shank, ed.), Vol. 1, pp. 185–217. CRC Press, Boca Raton, Florida.

3. Schmähl, D., Habs, M., and Ivankovic, S. (1978). Carcinogenesis of $N$-nitrosodiethylamine (DENA) in chickens and domestic cats. *Int. J. Cancer* **22**, 552–557.
4. Magee, P. N., and Barnes, J. M. (1959). The experimental production of tumors in the rat by dimethylnitrosamine ($N$-nitrosodimethylamine). *Acta Unio Int. Cancrum* **15**, 187–190.
5. Pour, P., Salmasi, S. Z., and Runge, P. C. (1978). Selective induction of pancreatic ductular tumors by single doses of $N$-nitroso-bis(2-oxopropyl)amine in Syrian golden hamsters. *Cancer Lett.* **4**, 317–323.
6. Anderson, L. M., Priest, L. J., and Budinger, J. M. (1979). Lung tumorigenesis in mice after chronic exposure in early life to a low dose of dimethylnitrosamine. *JNCI, J. Natl. Cancer Inst.* **62**, 1553–1555.
7. Preussmann, R., Eisenbrand, G., and Spiegelhalder, B. (1979). Occurrence and formation of $N$-nitroso compounds in the environment and *in vivo*. *In* "Environmental Carcinogenesis" (P. Emmelot and E. Kriek, eds.), pp. 51–71. Elsevier/North-Holland Biochemical Press, Amsterdam.
8. Archer, M. C. (1982). Hazards of nitrate, nitrite and nitrosamines in human nutrition. *In* "Nutritional Toxicology" (J. N. Hathcock, ed.), Vol. 1, pp. 327–381. Academic Press, New York.
9. Magee, P. N., ed. (1982). "The Possible Role of Nitrosamines in Human Cancer," Banbury Rep. 12. Cold Spring Harbor Lab., Cold Spring Harbor, New York.
10. Haszeldine, R. N., and Jander, J. (1955). Further remarks on the spectra of nitrites and nitrosamines. *J. Chem. Phys.* **23**, 979–980.
11. Tanaka, J. (1957). III. Electronic structure and electronic spectra of a nitroso group and related compounds. *J. Chem. Soc. Jpn.* **78**, 1647–1650.
12. Looney, C. E., Phillips, W. D., and Reilly, E. L. (1957). Nuclear magnetic resonance and infrared study of hindered rotation in nitrosamines. *J. Am. Chem. Soc.* **79**, 6136–6142.
13. Harris, R. K., Pryce-Jones, T., and Swinbourne, F. J. (1980). Nuclear magnetic resonance studies of $N$-nitrosamines. Part 4. Barriers to rotation about the N-N bond for some cyclic compounds. *J. Chem. Soc., Perkin Trans. 2* pp. 476–482.
14. Suzuki, E., Iiyoshi, M., and Okada, M. (1980). Nuclear magnetic resonance spectra of $N$-alkyl-$N$-(hydroxy- and oxoalkyl) nitrosamines and chromatographic separation of their (Z)- and (E)-conformers. *Chem. Pharm. Bull.* **28**, 979–983.
15. Suzuki, E., Iiyoshi, M., and Okada, M. (1980). Nuclear magnetic resonance spectra of $N$-alkyl-$N$-($\omega$-carboxyalkyl)-nitrosamines and their esters, and chromatographic separation of the (Z)- and (E)-conformers of the esters. *Chem. Pharm. Bull.* **28**, 1612–1618.
16. Iwaoka, W. T., Hanse, T., Hsieh, S.-T., and Archer, M. C. (1975). Chromatographic separation of conformers of substituted asymmetric nitrosamines. *J. Chromatogr.* **103**, 349–354.
17. Iwaoka, W., and Tannenbaum, S. R. (1976). Liquid chromatography of $N$-nitrosoamino acids and their syn and anti conformers. *J. Chromatogr.* **124**, 105–110.
18. Keefer, L. K., and Fodor, C. H. (1970). Facile hydrogen isotope exchange as evidence for an $\alpha$-nitrosamino carbanion. *J. Am. Chem. Soc.* **92**, 5747–5748.
19. Seebach, D., and Enders, D. (1975). Umpolung of amine reactivity. Nucleophilic $\alpha$-(secondary amino)-alkylation via metalated nitrosamines. *Angew. Chem., Int. Ed. Engl.* **14**, 15–32.
20. Hallett, G., and Williams, D. L. (1979). The reactivity of thiourea, alkylthioureas, cysteine, glutathione, $S$-methylcysteine, and methionine towards $N$-methyl-$N$-nitrosoaniline in acid solution. *J. Chem. Soc., Perkin Trans. 2* pp. 624–627.
21. Challis, B. C., and Osborne, M. R. (1973). The chemistry of nitroso compounds. Part VI. Direct and indirect transnitrosation reactions of $N$-nitrosodiphenylamine. *J. Chem. Soc., Perkin Trans. 2* pp. 1526–1533.

22. Challis, B. C., and Osborne, M. R. (1972). Chemistry of nitroso compounds. The reaction of N-nitrosodiphenylamine with N-methylaniline—a direct transnitrosation. J. Chem. Soc., Chem. Commun. pp. 518–519.
23. Singer, S. S. (1978). Kinetics and mechanism of aliphatic transnitrosation. J. Org. Chem. 43, 4612–4616.
24. Singer, S. S., Singer, G. M., and Cole, B. B. (1980). Alicyclic nitrosamines and nitrosamino acids as transnitrosating agents. J. Org. Chem. 45, 4931–4935.
25. Fischer, O., and Hepp, E. (1886). Zur Kenntniss der Nitrosamine (und Nitrobasen). Ber. Dtsch. Chem. Ges. 19, 2991–2995.
26. Williams, D. L. H. (1975). The mechanism of the Fischer–Hepp rearrangement of aromatic N-nitroso-amines. Tetrahedron 31, 1343–1349.
27. Chow, Y. L. (1973). Nitrosamine photochemistry: Reactions of aminium radicals. Acc. Chem. Res. 6, 354–360.
28. Fischer, C. (1875). Über die Hydrazinverbindungen der Fettreihe. Ber. Dtsch. Chem. Ges. 8, 1587–1590.
29. Enders, D., Hassel, T., Pieter, R., Renger, B., and Seebach, D. (1976). Reductive denitrosation of nitrosamines to secondary amines with hydrogen/Raney nickel. Synthesis 8, 548–550.
30. Kano, S., Tanaka, Y., Sugino, S., Shibuya, S., and Hibino, S. (1980). Reductive denitrosation of nitrosamines to secondary amines with metal halide/sodium borohydride. Synthesis 9, 741–742.
31. Emmons, W. D. (1954). Peroxytrifluoroacetic acid. I. The oxidation of nitrosamines to nitramines. J. Am. Chem. Soc. 76, 3468–3470.
32. Stewart, F. H. C. (1964). The chemistry of the sydnones. Chem. Rev. 64, 129–147.
33. Michejda, C. J., and Koepke, S. R. (1978). Powerful anchimeric effect of the N-nitroso group. J. Am. Chem. Soc. 100, 1959–1960.
34. Koepke, S. R., Kupper, R., and Michejda, C. J. (1979). Unusually facile solvolysis of primary tosylates. A case, for participation by the N-nitroso group. J. Org. Chem. 44, 2718–2722.
35. Magee, P. N. (1956). Toxic liver injury. The metabolism of dimethylnitrosamine. Biochem. J. 64, 676–682.
36. Wishnok, J. S., Rogers, A. E., Sanchez, O., and Archer, M. C. (1978). Dietary effects on the pharmacokinetics of three carcinogenic nitrosamines. Toxicol. Appl. Pharmacol. 43, 391–398.
37. Johansson, E. B., and Tjälve, H. (1978). The distribution of [$^{14}$C]dimethylnitrosamine in mice: Autoradiographic studies in mice with inhibited and non-inhibited dimethylnitrosamine metabolism and a comparison with the distribution of [$^{14}$C]formaldehyde. Toxicol. Appl. Pharmacol. 45, 565–575.
38. Dutton, A. H., and Heath, D. F. (1956). Demethylation of dimethylnitrosamine in rats and mice. Nature (London) 178, 644.
39. Heath, D. F., and Dutton, A. (1958). The detection of metabolic products from dimethylnitrosamine in rats and mice. Biochem. J. 70, 619–626.
40. Magee, P. N., and Vandekar, M. (1958). Toxic liver injury. The metabolism of dimethylnitrosamine in vitro. Biochem. J. 70, 600–605.
41. Mizrahi, I. J., and Emmelot, D. (1962). The effect of cysteine on the metabolic changes produced by two carcinogenic N-nitrosodialkylamines in rat liver. Cancer Res. 22, 339–351.
42. Mizrahi, I. J., and Emmelot, P. (1963). Counteraction by sulphydryl compounds of enzymatic conversion of and the metabolic lesions produced by two carcinogenic N-nitrosodialkylamines in rat liver. Biochem. Pharmacol. 12, 55–63.
43. Brouwers, J. A., and Emmelot, P. (1960). Microsomal N-demethylation and the effect of

the hepatic carcinogen dimethylnitrosamine on amino acid incorporation into the proteins of rat livers and hepatomas. *Exp. Cell Res.* **19,** 467–474.
44. Magee, P. N., and Barnes, J. M. (1967). Carcinogenic nitroso compounds. *Adv. Cancer Res.* **10,** 163–246.
45. Craddock, V. M. (1965). Reaction of the carcinogen dimethylnitrosamine with proteins and with thiol compounds in the intact animal. *Biochem. J.* **94,** 323–330.
46. Magee, P. N., and Farber, E. (1962). Toxic liver injury and carcinogensis. Methylation of rat liver nucleic acids by dimethylnitrosamine *in vivo*. *Biochem. J.* **83,** 114–124.
47. Heath, D. F. (1962). The decomposition and toxicity of dialkylnitrosamines in rats. *Biochem. J.* **85,** 72–91.
48. Druckrey, H., Preussmann, R., Ivankovic, S., and Schmähl, D. (1967). Organotrope carcinogene Wirkungen bei 65 verschiedenen N-Nitroso-Verbindungen an BD-Ratten. *Z. Krebsforsch.* **69,** 103–201.
49. Lake, B. G., Minski, M. J., Phillips, J. C., Gangolli, S. D., and Lloyd, A. G. (1975). Investigations into the hepatic metabolism of dimethylnitrosamine in the rat. *Life Sci.* **17,** 1599–1606.
50. Chin, A. E., and Bosmann, H. B. (1980). Hepatic microsomal metabolism of N-nitrosodimethylamine to its methylating agent. *Toxicol. Appl. Pharmacol.* **54,** 76–89.
51. Jensen, D. E., Lotlikar, P. D., and Magee, P. N. (1981). The in vitro methylation of DNA by microsomally activated dimethylnitrosamine and its correlation with formaldehyde production. *Carcinogenesis (N.Y.)* **2,** 349–354.
52. Lijinsky, W., Loo, J., and Ross, A. E. (1968). Mechanism of alkylation of nucleic acids by nitrosodimethylamine. *Nature (London)* **218,** 1174–1176.
53. Ross, A. E., Keefer, L., and Lijinsky, W. (1971). Alkylation of nucleic acids of rat liver and lung by deuterated N-nitrosodiethylamine in vivo. *J. Natl. Cancer Inst. (U.S.)* **47,** 789–795.
54. Lawley, P. D. (1976). Carcinogenesis by alkylating agents. *ACS Monogr.* **173,** 83–245.
55. Park, K. K., Wishnok, J. S., and Archer, M. C. (1977). Mechanism of alkylation by N-nitroso compounds: Detection of rearranged alcohol in the microsomal metabolism of N-nitrosodi-n-propylamine and base-catalyzed decomposition of N-n-propyl-N-nitrosourea. *Chem.-Biol. Interact.* **18,** 349–354.
56. Park, K. K., Archer, M. C., and Wishnok, J. S. (1980). Alkylation of nucleic acids by N-nitrosodi-n-propylamine: Evidence that carbonium ions are not significantly involved. *Chem.-Biol. Interact.* **29,** 139–144.
57. Andreozzi, P., Klopman, G., and Hopfinger, A. J. (1980). Theoretical study of N-nitrosamines and their presumed proximate carcinogens. *Cancer Biochem. Biophys.* **4,** 209–220.
58. Scribner, J. D., and Ford, G. P. (1982). n-Propyldiazonium ion alkylates $O^6$ of guanine with rearrangement, but alkylates N-7 without rearrangement. *Cancer Lett.* **16,** 51–56.
59. Magour, S., and Nievel, J. G. (1971). Effect of inducers of drug-metabolizing enzymes on diethylnitrosamine metabolism and toxicity. *Biochem. J.* **123,** 89–98.
60. Arcos, J. C., Bryant, G. M., Pastor, K. M., and Argus, M. F. (1976). Structural limits of specificity of methylcholanthrene-repressible nitrosamine N-dealkylases. Inhibition by analog substrates. *Z. Krebsforsch.* **86,** 171–183.
61. Chau, I. Y., Dagani, D., and Archer, M. C. (1978). Kinetic studies on the hepatic microsomal metabolism of dimethylnitrosamine, diethylnitrosamine and methylethylnitrosamine in the rat. *JNCI, J. Natl. Cancer Inst. (U.S.)* **61,** 517–521.
62. Park, K. K., and Archer, M. C. (1978). Microsomal metabolism of N-nitrosodi-n-propylamine: Formation of products resulting from α- and β-oxidation. *Chem.-Biol. Interact.* **22,** 83–90.
63. Park, K. K., and Archer, M. C. (1978). Metabolism of N-nitroso-2-oxopropylpropyl-

amine by rat liver: Formation of products resulting from both oxidation and reduction. *Cancer Biochem. Biophys.* **3**, 37-40.
64. Blattmann, L., and Preussmann, R. (1977). Oxidative biotransformation of di-*n*-butylnitrosamine. Formation in vitro of aldehydes in the presence of rat liver microsomes. *Z. Krebsforsch.* **88**, 311-314.
65. Grandjean, C. J., Knepper, S., and Morris, N. (1980). Microsomal metabolism of diallylnitrosamine. *In* "Microsomes, Drug Oxidations and Chemical Carcinogenesis" (M. J. Coon, ed.), Vol. 2, pp. 1145-1148. Academic Press, New York.
66. Preussmann, R., Habs, M., Pool, B., Stummeyer, D., Lijinsky, W., and Reuber, M. D. (1981). Fluoro-substituted *N*-nitrosamines. 1. Inactivity of *N*-nitrosobis(2,2,2-trifluoroethyl)amine in carcinogenicity and mutagenicity tests. *Carcinogenesis (N.Y.)* **2**, 753-756.
67. Janzowski, C., Pool, B. L., Preussmann, R., and Eisenbrand, G. (1982). Fluorosubstituted *N*-nitrosamines. 2. Metabolism of *N*-nitrosodiethylamine and of fluorinate analogs in liver microsomal fractions. *Carcinogenesis (N.Y.)* **3**, 155-159.
68. Farrelly, J. G., Stewart, M. L., Saavedra, J. E., and Lijinsky, W. (1982). Relationship between carcinogenicity and in vitro metabolism of nitrosomethylethylamine, nitrosomethyl-*N*-butylamine and nitrosomethyl-(2-phenylethyl)amine labeled with deuterium in the methyl and α-methylene positions. *Cancer Res.* **42**, 2105-2109.
69. Labuc, G. E., and Archer, M. C. (1982). Esophageal and hepatic microsomal metabolism of *N*-nitrosomethylbenzylamine and *N*-nitrosodimethylamine in the rat. *Cancer Res.* **42**, 3181-3186.
70. Leung, K. H., Park, K. K., and Archer, M. C. (1978). α-Hydroxylation in the metabolism of *N*-nitrosopiperidine by rat liver microsomes: Formation of 5-hydroxypentanal. *Res. Commun. Chem. Pathol. Pharmacol.* **19**, 201-211.
71. Hecht, S. S., Chen, C. B., and Hoffmann, D. (1978). Evidence for metabolic α-hydroxylation of *N*-nitrosopyrrolidine. *Cancer Res.* **38**, 215-218.
72. Chen, C. B., McCoy, G. D., Hecht, S. S., Hoffmann, D., and Wynder, E. L. (1978). High-pressure liquid chromatographic assay for α-hydroxylation of *N*-nitrosopyrrolidine by isolated rat liver microsomes. *Cancer Res.* **38**, 3812-3816.
73. Hecker, L. I., Farrelly, J. G., Smith, J. H., Saavedra, J. E., and Lyon, P. A. (1979). Metabolism of the liver carcinogen, *N*-nitrosopyrrolidine by rat liver microsomes. *Cancer Res.* **39**, 2679-2686.
74. Hecht, S. S., and Young, R. (1981). Metabolic α-hydroxylation of *N*-nitrosomorpholine and 3,3,5,5-tetradeutero-*N*-nitrosomorpholine in the F344 rat. *Cancer Res.* **41**, 5039-5043.
75. Grandjean, C. J. (1976). Metabolism of *N*-nitrosohexamethyleneimine. *J. Natl. Cancer Inst. (U.S.)* **57**, 181-185.
76. Hecker, L. I., and McClusky, G. A. (1982). Comparison of the in vitro metabolism of *N*-nitrosohexamethyleneimine by rat liver and lung microsomal fractions. *Cancer Res.* **42**, 59-64.
77. Chen, C. B., Hecht, S. S., and Hoffmann, D. (1978). Metabolic α-hydroxylation of the tobacco-specific carcinogen *N'*-nitrosonornicotine. *Cancer Res.* **38**, 3639-3645.
78. Hecht, S. S., Young, R., and Chen, C. B. (1980). Metabolism in the F344 rat of 4-(*N*-methyl-*N*-nitrosamino)-1-(3-pyridyl)-1-butanone, a tobacco-specific carcinogen. *Cancer Res.* **40**, 4144-4150.
79. Wishnok, J. S., Archer, M. C., Edelman, A. S., and Rand, W. M. (1978). Nitrosamine carcinogenicity: A quantitative Hansch-Taft structure-activity relationship. *Chem.-Biol. Interact.* **20**, 43-54.
80. Keefer, L. K., Lijinsky, W., and Garcia, H. (1973). Deuterium isotope effect on the

carcinogenicity of dimethylnitrosamine in the liver. *J. Natl. Cancer Inst. (U.S.)* **51**, 299–302.
81. Dagani, D., and Archer, M. C. (1976). Deuterium isotope effect in the microsomal metabolism of dimethylnitrosamine. *J. Natl. Cancer Inst. (U.S.)* **57**, 955–957.
82. Dagani, D., and Archer, M. C. (1978). Deuterium isotope effect in the microsomal metabolism of dimethylnitrosamine—correction. *JNCI, J. Natl. Cancer Inst.* **61**, 619.
83. Lijinsky, W., Taylor, H. W., and Keefer, L. (1976). Reduction of rat liver carcinogenicity of nitrosomorpholine by α-deuterium substitution. *J. Natl. Cancer Inst. (U.S.)* **57**, 1311–1315.
84. Charnley, G., and Archer, M. C. (1977). Deuterium isotope effect in the activation of nitrosomorpholine into a bacterial mutagen. *Mutat. Res.* **46**, 265–268.
85. Elespuru, R. K. (1976). Deuterium isotope effects in mutagenesis by nitroso compounds. *Mutat. Res.* **38**, 377–378.
86. Lijinsky, W., and Reuber, M. D. (1980). Carcinogenicity in rats of nitrosomethylethylamines labeled with deuterium in several positions. *Cancer Res.* **40**, 19–21.
87. Lijinsky, W., Reuber, M. D., Saavedra, J. E., and Blackwell, B. N. (1980). The effect of deuterium on the carcinogenicity of nitrosomethyl-*n*-butylamine. *Carcinogenesis* **1**, 157–160.
88. Lijinsky, W., and Reuber, M. D. (1980). Carcinogenicity of deuterium-labeled *N*-nitroso-*N*-methylcyclohexylamine in rats. *JNCI, J. Natl. Cancer Inst. (U.S.)* **64**, 1535–1536.
89. Singer, G. M., and Lijinsky, W. (1979). Relative extents of hydrogen–deuterium exchange of nitrosamines: Relevance to biological isotope effect studies. *Cancer Lett.* **8**, 29–34.
90. Roller, P. P., Shimp, D. R., and Keefer, L. K. (1975). Synthesis and solvolysis of methyl(acetoxymethyl)nitrosamine. Solution chemistry of the presumed carcinogenic metabolite of dimethylnitrosamine. *Tetrahedron Lett.* **25**, 2065–2068.
91. Kleihues, P., Doerjer, G., Keefer, L. K., Rice, J. M., Roller, P. P., and Hodgson, R. M. (1979). Correlation of DNA methylation by methyl(acetoxymethyl)nitrosamine with organ-specific carcinogenicity in rats. *Cancer Res.* **39**, 5136–5140.
92. Berman, J. J., Rice, J. M., Wenk, M. L., and Roller, P. P. (1979). Intestinal tumors induced by a single intraperitoneal injection of methyl(acetoxymethyl)nitrosamine in three strains of rats. *Cancer Res.* **39**, 1462–1466.
93. Berman, J. J., Rice, J. M., Wenk, M. L., and Roller, P. P. (1979). Dependence of tumor spectrum on route of administration in Sprague–Dawley rats as a result of single or multiple injections of methyl(acetoxymethyl)nitrosamine, *JNCI, J. Natl. Cancer Inst. (U.S.)* **63**, 93–100.
94. Joshi, S. R., Rice, J. M., Wenk, M. L., Roller, P. P., and Keefer, L. K. (1977). Selective induction of intestinal tumors in rats by methyl(acetoxymethyl)nitrosamine, an ester of the presumed reactive metabolite. *J. Natl. Cancer Inst. (U.S.)* **58**, 1531–1535.
95. Ward, J. M., Roller, P. P., and Wenk, M. L. (1977). Natural history of intestinal neoplasms induced in rats by a single injection of methyl(acetoxymethyl)nitrosamine. *Cancer Res.* **37**, 3046–3052.
96. Habs, M., Schmähl, D., and Wiessler, M. (1978). Carcinogenicity of acetoxymethyl-methyl-nitrosamine after subcutaneous, intravenous and intrarectal applications in rats. *Z. Krebsforsch.* **91**, 217–221.
97. Wiessler, M., and Schmähl, D. (1976). Zur carcinogen Wirkung von *N*-Nitroso-Verbindungen. 5. Acetoxmethyl-methyl-nitrosamine. *Z. Krebsforsch.* **85**, 47–49.
98. Baldwin, J. E., Branz, S. E., Gomez, R. F., Kraft, P. L., Sinskey, A. J., and Tannenbaum,

S. R. (1976). Chemical activation of nitrosamines into mutagenic agents. *Tetrahedron Lett.* **5**, 333–336.
99. Camus, A. M., Wiessler, M., Malaveille, C., and Bartsch, H. (1978). High mutagenicity of *N*-(α-acyloxy)alkyl-*N*-alkylnitrosamines in *S. typhimurium:* Model compounds for metabolically activated *N,N*-dialkylnitrosamines. *Mutat. Res.* **49**, 187–194.
100. Mochizuki, M., Suzuki, E., Anjo, T., Wakabayashi, Y., and Okada, M. (1979). Mutagenic and DNA-damaging effects of *N*-alkyl-*N*-(α-acetoxyalkyl)nitrosamines, models for metabolically activated *N,N*-dialkylnitrosamines. *Gann* **70**, 663–670.
101. Mochizuki, M., Suzuki, E., Anjo, T., Wakabayashi, Y., and Okada, M. (1980). Mutagenic and DNA-damaging effects of *N*-(ω-acetoxyalkyl and ω-methoxy-carbonylalkyl)-*N*-(α-acetoxyalkyl)nitrosamines, models for metabolically activated *N,N*-dialkylnitrosamines with an ω-functional group. *Gann* **71**, 124–130.
102. Pool, B. L., and Wiessler, M. (1981). Investigations on the mutagenicity of primary and secondary α-acetoxynitrosamines with *Salmonella typhimurium:* Activation and deactivation of structurally related compounds by S-9. *Carcinogenesis (N.Y.)* **2**, 991–997.
103. Gold, B., Salmasi, S., Linder, W., and Althoff, J. (1981). Biological and chemical studies involving methyl-*t*-butylnitrosamine, a noncarcinogenic nitrosamine. *Carcinogenesis (N.Y.)* **2**, 529–532.
104. Tannanbaum, S. R., Kraft, P., Baldwin, J., and Branz, S. (1977). The mutagenicity of methylbenzylnitrosamine and its α-acetoxy derivatives. *Cancer Lett.* **2**, 305–310.
105. Hodgson, R. M., Schweinsberg, F., Wiessler, M., and Kleihues, P. (1982). Mechanism of esophageal tumor induction in rats by *N*-nitrosomethylbenzylamine and its ring-methylated analog *N*-nitrosomethyl(4-methylbenzyl)amine. *Cancer Res.* **42**, 2836–2840.
106. Baldwin, J. E., Scott, A., Branz, S. E., Tannenbaum, S. R., and Green, L. (1978). Chemical studies on carcinogenic nitrosamines. 1. Hydrolysis of α-acetoxynitrosamines. *J. Org. Chem.* **43**, 2427–2431.
107. Skipper, P. L., Tannenbaum, S. R., Baldwin, J. E., and Scott, A. (1977). Alkylation by α-acetoxy-*N*-nitrosamines: Models for *N*-nitrosamine metabolites. *Tetrahedron Lett.* **49**, 4269–4272.
108. Mochizuki, M., Anjo, T., and Okada, M. (1980). Isolation and characterization of *N*-alkyl-*N*-(hydroxymethyl)nitrosamines from *N*-alkyl-*N*-(hydroperoxymethyl)nitrosamines by deoxygenation. *Tetrahedron Lett.* **21**, 3693–3696.
109. Mochizuki, M., Anjo, T., Wakabayashi, Y., Sone, T., and Okada, M. (1980). Formation of *N*-alkyl-*N*-(1-hydroperoxyalkyl)nitrosamines from *N*-alkyl-*N*-(1-acetoxyalkyl)nitrosamines. *Tetrahedron Lett.* **21**, 1761–1764.
110. Mochizuki, M., Sone, T., Anjo, T., and Okada, M. (1980). Synthesis of *N*-alkyl-*N*-(1-hydroperoxyalkyl)nitrosamines by oxygenation of lithiated dialkylnitrosamines. *Tetrahedron Lett.* **21**, 1765–1766.
111. Umbenhauer, D. R., and Pegg, A. E. (1981). Alkylation of intracellular and extracellular DNA by dimethylnitrosamine following activation by isolated rat hepatocytes. *Cancer Res.* **41**, 3471–3474.
112. Kroeger-Koepke, M. B., Koepke, S., McCusky, G. A., Magee, P. N., and Michejda, C. J. (1981). α-Hydroxylation pathway in the in vitro metabolism of carcinogenic nitrosamines: *N*-nitrosodimethylamine and *N*-nitroso-*N*-methylaniline. *Proc. Natl. Acad. Sci. U.S.A.* **78**, 6489–6493.
113. Appel, K. E., and Graf, H. (1982). Metabolic nitrite formation from *N*-nitrosamines: Evidence for a cytochrome *P*-450-dependent reaction. *Carcinogenesis (N.Y.)* **3**, 293–296.
114. Grilli, S., and Prodi, G. (1975). Identification of dimethylnitrosamine metabolites in vitro. *Gann* **66**, 473–480.
115. Krüger, F. W. (1971). Metabolismus von Nitrosaminen in vivo. I. Über die β-Oxida-

tion Aliphatischer Di-*n*-alkylnitrosamine: Die Bildung von 7-Methylguanine neben 7-Propylbzw. 7-Butylguanin nach Applikation von Di-*n*-propyl-oder Di-*n*-butylnitrosamin. *Z. Krebsforsch.* **76**, 145–154.
116. Krüger, F. W. (1973). Metabolism of nitrosamines in vivo. II. On the methylation of nucleic acids by aliphatic di-*n*-alkyl-nitrosamines in vivo, caused by $\beta$-oxidation: The increased formation of 7-methylguanine after application of $\beta$-hydroxypropylpropyl-nitrosamine compared to that after application of di-*n*-nitrosamine. *Z. Krebsforsch.* **79**, 90–97.
117. Krüger, F. W., and Bertram, B. (1973). Metabolism of nitrosamines in vivo. III. On the methylation of nucleic acids by aliphatic di-*n*-alkylnitrosamines in vivo resulting from $\beta$-oxidation: The formation of 7-methylguanine after application of 2-oxo-propylpropylnitrosamine and methylpropyl-nitrosamine. *Z. Krebsforsch.* **80**, 189–196.
118. Leung, K. M., Park, K. K., and Archer, M. C. (1980). Methylation of DNA by *N*-nitroso-2-oxo-propylpropylamine: Formation of $O^6$- and 7-methylguanine and studies on the methylation mechanism. *Toxicol. Appl. Pharmacol.* **53**, 29–34.
119. Nicoll, J. W., Swann, P. F., and Pegg, A. E. (1975). Effect of dimethylnitrosamine on persistence of methylated guanines in rat liver and kidney DNA. *Nature (London)* **254**, 201–262.
120. Lawley, P. D. (1976). Carcinogenesis by alkylating agents. *ACS Monogr.* **173**, 83–244.
121. Leung, K.-H., and Archer, M. C. (1985). Mechanism of DNA methylation by *N*-nitroso-2-oxopropylpropylamine. *Carcinogenesis* (in press).
122. Lawson, T. A., Helgeson, A. S., Grandjean, C. J., Wallcave, L., and Nagel, D. (1981). The formation of *N*-nitrosomethyl(2-oxopropyl)amine from *N*-nitroso-bis(2-oxopropyl)amine in vivo. *Carcinogenesis (N.Y.)* **2**, 845–849.
123. Michejda, C. J., Kroeger-Koepke, M. B., Koepke, S. R., and Kupper, R. J. (1979). Oxidative activation of *N*-nitrosamines: Model compounds. *ACS Symp. Ser.* **101**, 77–89.
124. Leung, K.-H., and Archer, M. C. (1981). Urinary metabolites of *N*-nitrosodipropylamine, *N*-nitroso-2-hydroxypropylpropylamine and *N*-nitroso-2-oxopropylpropylamine in the rat. *Carcinogenesis (N.Y.)* **2**, 859–862.
125. Leung, K.-H., and Archer, M. C. (1984). Studies on the metabolic activation of $\beta$-ketonitrosamines: Mechanism of DNA methylation by *N*-(2-oxopropyl)-*N*-nitrosourea and *N*-nitroso-*N*-acetoxymethyl-*N*-2-oxopropylamine. *Chem.-Biol. Interact.* **48**, 169–180.
126. Okada, M., and Ishidate, M. (1977). Metabolic fate of *N*-*n*-butyl-*N*-(4-hydroxybutyl)nitrosamine and its analogues. Selective induction of urinary bladder tumors in the rat. *Xenobiotica* **7**, 11–24.
127. Blattmann, L., and Preussmann, R. (1973). Struktur von Metaboliten carcinogenen Dialkylnitrosamin in Rattenurin. *Z. Krebsforsch.* **79**, 3–5.
128. Blattmann, L., Joswig, N., and Preussmann, R. (1974). Struktur von Metaboliten des carcinogenen Methyl-*n*-butyl-nitrosamin in Rattenurin. *Z. Krebsforsch.* **81**, 71–73.
129. Blattmann, L., and Preussmann, R. (1974). Biotransformation von carcinogenen Dialkylnitrosaminen Weitere Urinmetaboliten von Di-*n*-butyl-und Di-*n*-pentyl-nitrosamin. *Z. Krebsforsch.* **81**, 75–78.
130. Blattmann, L., and Preussmann, R. (1975). Metaboliten von (2-hydroxybutyl)-*n*-butylnitrosamin in Rattenurin. *Z. Krebsforsch.* **83**, 125–127.
131. Blattmann, L., and Preussmann, R. (1977). Oxidative biotransformation of di-*n*-butylnitrosamine. *Z. Krebsforsch.* **88**, 311–314.
132. Blattmann, L. (1977). Direct alkyl chain cleavage after C-hydroxylation of dialkylnitrosamines in rats. A new pathway of oxidative biotransformation. *Z. Krebsforsch.* **88**, 315–322.
133. Okada, M., and Hashimoto, Y. (1974). Carcinogenic effect of *N*-nitrosamines related to

butyl(4-hydroxybutyl)nitrosamine in ACI/N rats, with special reference to induction of urinary bladder tumors. *Gann* **65**, 13-19.
134. Okada, M., Suzuki, E., and Hashimoto, Y. (1976). Carcinogenicity of *N*-nitrosamines related to *N*-butyl-*N*-(4-hydroxybutyl)nitrosamine and *N,N*-dibutylnitrosamine in ACI/N rats. *Gann* **67**, 825-834.
135. Mochizuki, M., Irving, C. C., Anjo, T., Wakabayashi, Y., Suzuki, E., and Okada, M. (1980). Synthesis and mutagenicity of 4-(*N*-butylnitrosamino)-4-hydroxybutyric acid lactone, a possible activated metabolite of the proximate bladder carcinogen *N*-butyl-*N*-(3-carboxypropyl)nitrosamine. *Cancer Res.* **40**, 162-165.
136. Lijinsky, W., and Taylor, H. W. (1975). Induction of urinary bladder tumors in rats by administration of nitrosomethyldodecylamine. *Cancer Res.* **35**, 958-961.
137. Lijinsky, W., Taylor, H. W., Mangino, M., and Singer, G. M. (1978). Carcinogenesis of nirosomethylundecylamine in Fisher rats. *Cancer Lett.* **5**, 209-213.
138. Lijinsky, W., Saavedra, J. E., and Reuber, M. D. (1981). Carcinogenesis in Fischer rats by methylalkylnitrosamines. *Cancer Res.* **41**, 1288-1292.
139. Singer, G. M., Lijinsky, W., Buettner, L., and McClusky, G. A. (1981). Relationships of rat urinary metabolites of *N*-nitrosomethyl-*N*-alkylamine to bladder carcinogenesis. *Cancer Res.* **41**, 4942-4946.
140. Lake, B. G., Phillips, J. C., Heading, C. E., and Gangolli, S. D. (1976). Studies on the in vitro metabolism of dimethylnitrosamine by rat liver. *Toxicology* **5**, 297-309.
141. Czygan, P., Greim, H., Garro, A. J., Hutterer, F., Schaffner, F., Popper, H., Rosenthal, O., and Cooper, D. Y. (1973). Microsomal metabolism of dimethylnitrosamine and the cytochrome *P*-450 dependency of its activation to a mutagen. *Cancer Res.* **33**, 2983-2986.
142. Argus, M. F., Arcos, J. C., Pastor, K. M., Wu, B. C., and Venkatesan, N. (1976). Dimethylnitrosamine demethylase: Absence of increased enzyme catabolism and multiplicity of effector sites in repression. Hemoprotein involvement. *Chem.-Biol. Interact.* **13**, 127-140.
143. Lotlikar, P. D., Baldy, W. J., and Dwyer, E. N. (1975). Dimethylnitrosamine demethylation by reconstituted liver microsomal cytochrome *P*-450 enzyme system. *Biochem. J.* **152**, 705-708.
144. Guengerich, F. P. (1977). Separation and purification of multiple forms of microsomal cytochrome *P*-450. *J. Biol. Chem.* **252**, 3970-3979.
145. Godoy, H. M., Diaz Gomez, M. I., and Castro, J. A. (1978). Mechanism of dimethylnitrosamine metabolism and activation in rats. *JNCI, J. Natl. Cancer Inst.* **61**, 1285-1289.
146. Appel, K. E., Ruf, H. H., Mahr, B., Schwarz, M., Rickart, R., and Kunz, W. (1979). Binding of *N*-nitrosamines to cyt. *P*-450 of liver microsomes. *Chem.-Biol. Interact.* **28**, 17-33.
147. Arcos, J. C., Davies, D. L., Brown, C. E. L., and Argus, M. F. (1977). Repressible and inducible enzymic forms of DMN-demethylase. *Z. Krebsforsch.* **89**, 181-199.
148. Lotlikar, P. P., Hang, Y. S., and Baldy, W. J. (1978). Effect of dimethylnitrosamine concentration on its demethylation by liver microsomes from control and 3-methylcholanthrene-pretreated rats, hamsters and guinea pigs. *Cancer Lett.* **4**, 355-361.
149. Sipes, I. G., Slocumb, M. L., and Holtzman, G. (1978). Stimulation of microsomal dimethylnitrosamine-*N*-demethylase by pretreatment of mice with acetone. *Chem.-Biol. Interact.* **21**, 155-166.
150. Kroeger-Koepke, M. B., and Michejda, C. J. (1979). Evidence for several demethylase enzymes in the oxidation of dimethylnitrosamine and phenylmethylnitrosamine by rat liver fractions. *Cancer Res.* **39**, 1587-1591.

151. Arcos, J. C., Bryant, G. M., Venkatesan, N., and Argus, M. F. (1975). Repression of dimethylnitrosamine demethylase by typical inducers of microsomal mixed-function oxidases. *Biochem. Pharmacol.* **24**, 1544-1547.
152. Guttenplan, J. B., and Garro, A. J. (1977). Factors affecting the induction of dimethylnitrosamine demethylase by Aroclor 1254. *Cancer Res.* **37**, 329-330.
153. Guttenplan, J. B., Hutterer, F., and Garro, A. J. (1976). Effects of cytochrome $P$-448 and $P$-450 inducers on microsomal dimethylnitrosamine demethylase activity and the capacity of isolated microsomes to activate dimethylnitrosamine to a mutagen. *Mutat. Res.* **35**, 415-422.
154. Lake, B. G., Heading, C. E., Phillips, J. C., Gangolli, S. D., and Lloyd, A. C. (1974). Studies on the effects of phenobarbitone and 20-methylcholanthrene pretreatments on the metabolism and toxicity of dimethylnitrosamine in the rat. *Biochem. Soc. Trans.* **2**, 882-885.
155. Appel, K. E., Rickart, R., Schwarz, M., and Kunz, H. W. (1979). Influence of drugs on activation and inactivation of hepatocarcinogenic nitrosamines. *Arch. Toxicol., Suppl.* **2**, 471-477.
156. Appel, K. E., Schwarz, M., Rickart, R., and Kunz, H. W. (1979). Influences of inducers and inhibitors of the microsomal monooxygenase system on the alkylating intensity of dimethylnitrosamine in mice. *J. Cancer Res. Clin. Oncol.* **94**, 47-61.
157. Haag, S. M., and Sipes, I. G. (1980). Differential effects of acetone on Aroclor 1254 pretreatment on the microsomal activation of dimethylnitrosamine to a mutagen. *Mutat. Res.* **74**, 431-438.
158. Anderson, L. M., and Angel, M. (1980). Induction of dimethylnitrosamine demethylase activity in mouse liver by polychlorinated biphenyls and 3-methylcholanthrene. *Biochem. Pharmacol.* **29**, 1375-1383.
159. Mostafa, M. H., Ruchirawat, M., and Weisburger, E. K. (1981). Comparative studies on the effects of microsomal enzyme inducers on the $N$-demethylation of dimethylnitrosamine. *Biochem. Pharmacol.* **30**, 2007-2011.
160. Kunz, W., Appel, K. E., Rickart, R., Schwartz, M., and Stöckle, G. (1978). Enhancement and inhibition of carcinogenic effectiveness of nitrosamines. *In* "Primary Liver Tumors" (H. Remmer, H. M. Bolt, P. Bannasch, and H. Popper, eds.), pp. 261-283. University Park Press, Baltimore, Maryland.
161. Magee, P. N. (1980). Metabolism of nitrosamines: An overview. *In* "Microsomes, Drug Oxidations and Chemical Carcinogenesis" (M. J. Coon, A. H. Conney, R. W. Estabrook, H. V. Gelboin, J. R. Gillette, and P. J. O'Brien, eds.), pp. 1081-1092. Academic Press, New York.
162. Lake, B. G., Phillips, J C., Cottrell, R. C., and Gangolli, S. D. (1978). The possible involvement of a microsomal amine oxidase enzyme in hepatic dimethylnitrosamine degradation in vitro. *In* "Biological Oxidation of Nitrogen" (J. W. Gorrod, ed.), pp. 131-135. Elsevier/North-Holland, Amsterdam.
163. Phillips, J. C., Lake, B. G., Gangolli, S. D., Grasso, P., and Lloyd, A. G. (1977). Effects of pyrazole and 3-amino-1,2,4-triazole on the metabolism and toxicity of dimethylnitrosamine in the rat. *J. Natl. Cancer Inst. (U.S.)* **58**, 629-633.
164. Rowland, I. R., Lake, B. G., Phillips, J. C., and Gangolli, S. D. (1980). Substrates and inhibitors of hepatic amine oxidase inhibit dimethylnitrosamine-induced mutagenesis in *Salmonella typhimurium*. *Mutat. Res.* **72**, 63-72.
165. Lai, D. Y., Myers, S. C., Woo, Y. T., Greene, E. J., Friedman, M. A., Argus, M. F., and Arcos, J. C. (1979). Role of dimethylnitrosamine demethylase in the metabolic activation of demethylnitrosamine. *Chem.-Biol. Interact.* **28**, 107-126.
166. Grandjean, C. J., Gold, B. I., Knepper, S., and Morris, N. (1978). Activation of $N$-ni-

trosamines to alkylating agents by rat liver nuclei and microsomes: Binding to endogenous and exogenous DNA. *Proc. Am. Assoc. Cancer Res.* **19**, 185.

167. Floyd, R. A. (1978). Spin-trapping demonstration of hydroxyl free radical and nitrosamine free radicals produced upon interaction of nitrosamine carcinogens with rat liver microsomes. *Proc. Am. Assoc. Cancer Res.* **19**, 166.
168. Argus, M. F., and Arcos, J. C. (1978). Use of high concentrations of dimethylnitrosamine in bacterial lethality, mutagenesis, and enzymological studies. *Cancer Res.* **38**, 226–228.
169. Godoy, H. M., Diaz Gomez, M. I., and Castro, J. A. (1980). Relationship between dimethylnitrosamine metabolism or activation and its ability to induce liver necrosis in rats. *JNCI, J. Natl. Cancer Inst. (U.S.)* **64**, 533–538.
170. Diaz Gomez, M. I., Godoy, H. M., and Castro, J. A. (1981). Further studies on dimethylnitrosamine metabolism, activation, and its ability to cause liver injury. *Arch. Toxicol.* **47**, 159–168.
171. Den Engelse, L., Gebbink, M., and Phillips, P. E. (1975). [$^{14}$C]Formaldehyde: A possible contaminant of [$^{14}$C]dimethylnitrosamine. *Chem.-Biol. Interact.* **11**, 133–137.
172. Yoshikawa, K., Nohmi, T., Naghara, A., Inokawa, Y., and Ishidate, M. (1980). Stability of S9 during frozen storage in *Salmonella*/S9 assays. *Mutat. Res.* **74**, 389–391.
173. Kim, S., Paik, W. K., Choi, J., Lotlikar, P. D., and Magee, P. N. (1981). Microsome-dependent methylation of erythrocyte proteins by dimethylnitrosamine. *Carcinogenesis (N.Y.)* **2**, 179–182.
174. Montesano, R., and Magee, P. N. (1974). Comparative metabolism in vitro of nitrosamines in various animal species including man. *IARC Sci. Publ.* **10**, 39–56.
175. Johansson, E. B., and Tjälve, H. (1978). The distribution of [$^{14}$C]dimethylnitrosamine in mice. Autoradiographic studies in mice with inhibited and non-inhibited dimethylnitrosamine metabolism and a comparison with the distribution of [$^{14}$C]formaldehyde. *Toxicol. Appl. Pharmacol.* **45**, 565–575.
176. Chen, C. B., Fung, P. T., and Hecht, S. S. (1979). Assay for microsomal $\alpha$-hydroxylation of $N'$-nitrosonornicotine and determination of the deuterium isotope effect for $\alpha$-hydroxylation. *Cancer Res.* **39**, 5057–5062.
177. McCoy, G. D., Chen, C. B., and Hecht, S. S. (1981). Influence of mixed-function oxidase inducers on the in vitro metabolism of $N'$-nitrosonornicotine by rat and hamster liver microsomes. *Drug. Metab. Dispos.* **9**, 168–169.
178. Kato, R., Shoji, H., and Takanaka, A. (1967). Metabolism of carcinogenic compounds. I. Effect of phenobarbital and methylcholanthrene on the activities of N-demethylation of carcinogenic compounds by liver microsomes of male and female rats. *Gann* **58**, 467–469.
179. Lai, D. Y., Arcos, J. C., and Argus, M. F. (1979). Factors influencing the microsome- and mitochondria-catalyzed *in vitro* binding of diethylnitrosamine and *N*-nitrosopiperidine to deoxyribonucleic acid. *Biochem. Pharmacol.* **28**, 3545–3550.
180. Cottrell, R. C., Young, P. J., Walters, D. G., Phillips, J. C., Lake, B. G., and Gangolli, S. D. (1979). Studies of the metabolism of *N*-nitrosopyrrolidine in the rat. *Toxicol. Appl. Pharmacol.* **51**, 101–106.
181. Rayman, M. P., Challis, B. C., Cox, P. J., and Jarman, M. (1975). Oxidation of *N*-nitrosopiperidine in the Udenfriend model system and its metabolism by rat-liver microsomes. *Biochem. Pharmacol.* **24**, 621–626.
182. Hecht, S. S., Chen, C. B., and Hoffmann, D. (1980). Metabolic $\beta$-hydroxylation and N-oxidation of *N*-nitrosonornicotine. *J. Med. Chem.* **23**, 1175–1178.
183. Hecker, L. I., and Saavedra, J. E. (1980). In vitro formation and properties of $\beta$, and $\gamma$-hydroxy-*N*-nitrosohexamethyleneimine. *Carcinogenesis (N.Y.)* **1**, 1017–1025.
184. Hecker, L. I., and McClusky, G. A. (1982). Comparison of the in vitro metabolism o

N-nitrosohexamethyleneimine by rat liver and lung microsomal fractions. *Cancer Res.* **42**, 59–64.
185. Singer, B. (1979). N-Nitroso alkylating agents: Formation and persistence of alkyl derivatives in mamalian nucleic acids as contributing factors in carcinogenesis. *JNCI, J. Natl. Cancer Inst.* **62**, 1329–1339.
186. Magee, P. N., Montesano, R., and Preussmann, R. (1976). N-Nitroso compounds and related carcinogens. *ACS Monogr.* **173**, 491–625.
187. Pegg, A. E. (1980). Formation and subsequent repair of alkylation lesions in tissues of rodents treated with nitrosamines. *Arch. Toxicol., Suppl.* **3**, 55–68.
188. Monesano, R., Pegg, A. E., and Margison, G. P. (1980). Alkylation of DNA and carcinogenicity of N-nitroso compounds. *J. Toxicol. Environ. Health* **6**, 1001–1008.
189. Pegg, A. E. (1977). Formation and metabolism of alkylated nucleosides: Possible role in carcinogenesis by nitroso compounds and alkylating agents. *Adv. Cancer Res.* **25**, 195–269.
190. Kleihues, P., Bamborschke, S., and Doerjer, G. (1980). Persistence of alkylated DNA bases in the Mongolian gerbil *(Meriones unguiculatus)* following a single dose of methylnitrosourea. *Carcinogenesis (N.Y.)* **1**, 111–113.
191. Kleihues, P., Veit, C., Wiessler, M., and Hodgson, R. M. (1981). DNA methylation by N-nitrosomethylbenzylamine in target and non-target tissues of NMR1 mice. *Carcinogenesis (N.Y.)* **2**, 897–899.
192. Cooper, H. K., Margison, G. P., O'Connor, P. J., and Itzhaki, R. F. (1975). Heterogeneous distribution of DNA alkylation products in rat liver chromatin after in vivo administration of $N,N$-di[$^{14}$C]methylnitrosamine. *Chem.-Biol. Interact.* **11**, 483–492.
193. Ramanathan, R., Rajalaskshmi, S., Sarma, D. S. R., and Farber, E. (1976). Nonrandom nature of in vivo methylation by dimethylnitrosamine and the subsequent removal of methylated products from rat liver chromatin DNA. *Cancer Res.* **36**, 2073–2079.
194. Faustman, E. M., and Goodman, J. I. (1981). Alkylation of DNA in specific hepatic chromatin fractions following exposure to methylnitrosourea or dimethylnitrosamine. *Toxicol. Appl. Pharmacol.* **58**, 379–388.
195. Kyrtopoulos, S. A., and Swann, P. F. (1980). The use of radioimmunoassay to study the formation and disappearance of $O^6$-methylguanine in mouse liver satellite and mainband DNA following dimethylnitrosamine administration. *J. Cancer Res. Clin. Oncol.* **98**, 127–138.
196. Lee, K. Y., and Lijinsky, W. (1966). Alkylation of rat liver RNA by cyclic nitrosamines in vivo. *J. Natl. Cancer Inst. (U.S.)* **37**, 401–407.
197. Stewart, B. W., and Farber, E. (1973). Strand breakage in rat liver DNA and its repair following administration of cyclic nitrosamines. *Cancer Res.* **33**, 3209–3215.
198. Ross, A. E., and Mirvish, S. S. (1977). Metabolism of N-nitrosohexamethyleneimine to give 1,6-hexanediol bound to rat liver nucleic acids. *J. Natl. Cancer Inst. (U.S.)* **58**, 651–655.
199. Hunt, E. J., and Shank, R. C. (1982). Evidence for DNA adducts in rat liver after administration of N-nitrosopyrrolidine. *Biochem. Biophys. Res. Commun.* **104**, 1343–1348.
200. Chung, F. L., and Hecht, S. S. (1983). Formation of cyclic $1,N^2$-adducts by reaction of deoxyguanosine with $\alpha$-acetoxy-N-nitrosopyrrolidine, 4-(carbethoxynitrosamine)butanal, or crotonaldehyde. *Cancer Res.* **43**, 1230–1235.

Chapter 15

# Hydrazines

R. A. PROUGH AND S. J. MOLONEY
Department of Biochemistry
The University of Texas Health Science Center
Dallas, Texas

|     |                                                  |     |
| --- | ------------------------------------------------ | --- |
| I.  | Introduction                                     | 433 |
| II. | Chemical Properties of Hydrazines                | 434 |
|     | A. Hydrazide and Hydrazone Formation             | 434 |
|     | B. Oxidation–Reduction Reactions                 | 435 |
| III.| Bioactivation of Hydrazines                      | 436 |
|     | A. Hydrazone Formation                           | 436 |
|     | B. Hydrazide Formation                           | 437 |
|     | C. N-Oxidation Reactions                         | 437 |
| IV. | Fates of Reactive Intermediates of Hydrazines    | 441 |
|     | A. Flavin-Dependent Enzymes                      | 441 |
|     | B. Heme and Hemoproteins                         | 442 |
|     | C. Thiols                                        | 444 |
|     | D. Proteins and Nucleic Acids                    | 445 |
| V.  | Conclusion                                       | 445 |
|     | References                                       | 446 |

## I. Introduction

Over a decade ago, two review articles involving the metabolism of hydrazine derivatives were published by L. B. Colvin[1] and M. R. Juchau and A. Horita.[2] Colvin addressed the biological reactions that linked hydrazine derivatives to small molecules; the conjugation of hydrazines to form hydrazides or hydrazones was suggested to be an important step to aid in the disposition of these toxic compounds. Juchau and Horita were concerned with those hydrazine derivatives that are used in drug therapy and that may have toxic properties.[2] On the basis of reports that cytochrome *P*-450 catalyzed the oxidation of mono- and disubstituted hydrazines,[3-5] Juchau and Horita stressed that the metabolism of hydrazines may involved oxidative transformation of the hydrazine functional group. In the last decade, it has been shown that nearly all hydrazine derivatives are carcinogenic[6] and toxic.[2,7,8] In fact, considerable evidence has been provided to establish that the oxidative metabolism of these compounds can lead to various intermediates that alter normal cellular metabolism. The purpose of this chapter is to describe the reaction mechanism of the formation and decomposition of the chemical species obtained during hydrazine metabolism and to provide a framework for future questions to be addressed with regard to hydrazine carcinogenesis and toxicity.

## II. Chemical Properties of Hydrazines

### A. HYDRAZIDE AND HYDRAZONE FORMATION

Hydrazines can occur with varying degrees of substitution similar to ammonia and its derivatives.[9] Mono- and disubstituted hydrazines can react with acylating agents, such as acid chlorides or acid anhydrides, to form hydrazides (Fig. 1). These reactions are similar to the conversion of amines to amides. In most cases, the highly polar hydrazines are converted into the less polar *N*-acyl compounds. Hydrazine itself can undergo acylation to yield the diacyl derivative.[9]

The reaction of acyl-, monosubstituted, or 1,1-disubstituted hydrazines with unhindered aldehydes and ketones proceeds in an exothermic and stoichiometric manner. This process occurs in two steps: nucleophilic addition across the carbon–oxygen unsaturated bond and dehydration of the

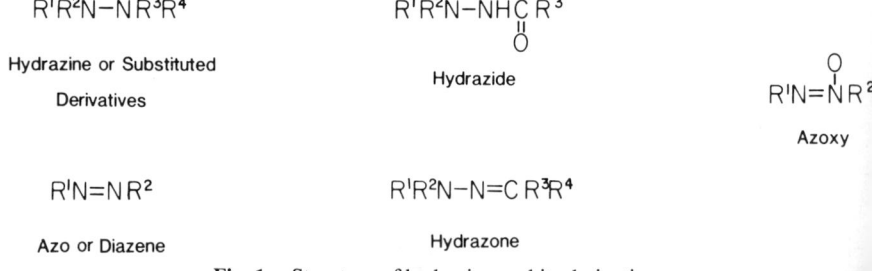

Fig. 1. Structure of hydrazine and its derivatives.

tetrahedral intermediate formed in the first step.[9,10] In fact, the hydrazines are more nucleophilic than their amine counterparts in these reactions, because the hydrazine functional group possesses two adjacent heteroatoms with free electron pairs.[11] Hydrazine and the monosubstituted hydrazines can also condense with two molecules of an unhindered aldehyde to form an aldazine derivative.[9] However, 1,2-di-, tri-, and tetrasubstituted hydrazines do not easily form hydrazides or hydrazones. The hydrazone derivatives are much less polar than the hydrazine precursor and can be extracted from aqueous solutions by nonpolar organic solvents such as ethyl acetate.

## B. OXIDATION-REDUCTION REACTIONS

Hydrazine and its derivatives possess electrochemical properties that allow them to serve as either oxidants or reductants. In the presence of a strong reducing agent, such as Raney nickel or Zn and HCl, they can be reduced to the respective amines and ammonia.[9]

The oxidation of hydrazines is much more complex (Fig. 2). Tetra- and trisubstituted hydrazines form stable radical cations on one-electron oxidatons.[12] These cations can be subsequently dealkylated on a second one-electron oxidation step. Hydrazine and its mono- and disubstituted derivatives generally undergo two- and four-electron oxidation reactions.[9] For 1,2-disubstituted hydrazines, the two-electron oxidation product is a relatively stable azo intermediate. This product is nonpolar and can be easily

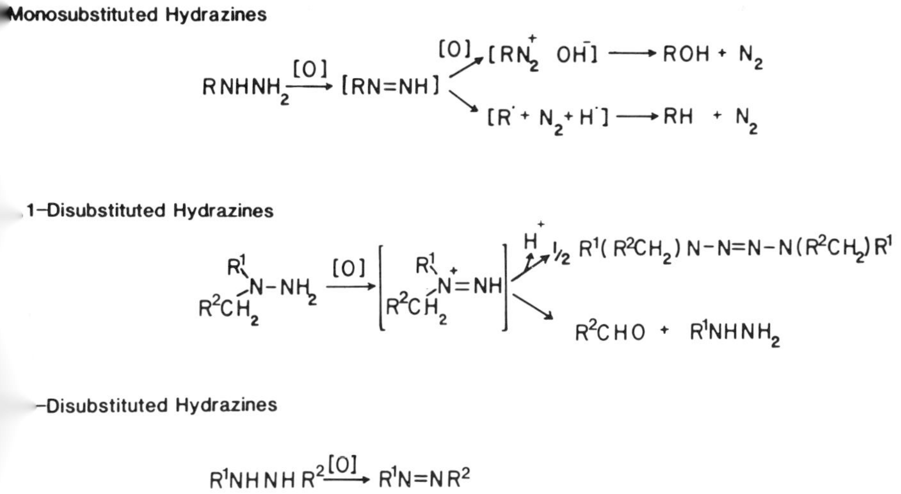

Monosubstituted Hydrazines

$$RNHNH_2 \xrightarrow{[O]} [RN=NH] \underset{\searrow}{\overset{\nearrow}{}} \begin{array}{l} [O], [RN_2^+ \; OH^-] \longrightarrow ROH + N_2 \\ [R^{\cdot} + N_2 + H^{\cdot}] \longrightarrow RH + N_2 \end{array}$$

1-Disubstituted Hydrazines

$$\underset{R^2CH_2}{\overset{R^1}{>}}N-NH_2 \xrightarrow{[O]} \left[ \underset{R^2CH_2}{\overset{R^1}{>}}N^+=NH \right] \underset{\searrow}{\overset{\nearrow H}{}} \begin{array}{l} \tfrac{1}{2} R^1(R^2CH_2)N-N=N-N(R^2CH_2)R^1 \\ R^2CHO + R^1NHNH_2 \end{array}$$

-Disubstituted Hydrazines

$$R^1NHNHR^2 \xrightarrow{[O]} R^1N=NR^2$$

Fig. 2. The oxidation products of substituted hydrazines.

partitioned from an aqueous phase into an organic phase. Unless the azo derivative contains a methylene carbon next to the azo functional group, the derivatives do not readily tautomerize to the corresponding hydrazone.

The mono- and 1,1-disubstituted hydrazines form unstable diazene intermediates on two-electron oxidation. The resultant monosubstituted diazenes undergo one of three reactions: tautomerization to a hydrazone, decomposition to the corresponding hydrocarbon via a free-radical intermediate, or further two-electron oxidation to form a diazonium derivative. The chemistry of diazene intermediates and their reactivity has been characterized by Kosower.[13] Certain of the diazenes, such as phenyldiazene, are good reductants and are easily oxidized in the presence of molecular oxygen. Unlike monosubstituted diazenes, 1,1-disubstituted diazenes tend either to dimerize, forming tetrazenes, or to tautomerize to hydrazones.[14] No chemically reactive intermediates have been noted during the decomposition of 1,1-disubstituted diazenes.

## III. Bioactivation of Hydrazines

A number of reactions have been described that utilize the chemical reactivity of hydrazines to alter enzymatic processes, including enzyme inhibition, sequestration of key biochemical intermediates, and protein modification. This section will provide details regarding the enzymatic nature of these reactions, and Section IV will delineate the fate of the intermediates formed from the hydrazines.

### A. HYDRAZONE FORMATION

Several biochemical entities contain reactive carbonyl groups, which can participate in hydrazone formation. For example, most enzymes that utilize vitamin $B_6$, pyridoxal phosphate, readily form the hydrazone of pyridoxal phosphate. This usually results in inhibition of enzyme function, because the hydrazones are thermodynamically stable and are not easily displaced by the amine substrates. As was seen with the plasma diamine oxidases,[15] kynureninase,[16] and the aminotransferases,[17,18] hydrazone formation with pyridoxal phosphate has pronounced affects on intermediary metabolism. Hucko-Haas and Reed demonstrated that the plasma diamine oxidase is transiently inhibited by stoichiometric quantities of hydrazines such as benzyl- and 1,1-dimethylhydrazine.[15] Their results suggest that the copper-prosthetic group of the enzyme slowly oxidizes the pyridoxal phosphate hydrazone until inhibition is reversed because of depletion of th hydrazine substrate from the reaction mixture. Consideration of these oxidative reactions will be provided later.

One pronounced effect of hydrazine and its derivatives on hepatic metabolism is the inability of liver to synthesize glucose from pyruvate and amino acids such as alanine or aspartate[17-19] in the presence of hydrazine. It is well known that the cytosolic and mitochondrial aminotransferases are potently inhibited by hydrazines. These enzyme steps are required to form the α-ketoacids required for the function of the tricarboxylic acid cycle and for gluconeogenesis in the liver. In addition, hydralazine,[20] phenelzine,[21] and phenylhydrazine[1] are excreted as the α-ketoacid hydrazones. This chemical reaction may serve to deplete the α-ketoacids, which are required for pyruvate oxidation by the tricarboxylic acid cycle. The hydrazone of phenelzine and pyruvate inhibited gluconeogenesis in perfused guinea pig liver when lactate was used as a precursor.[21]

## B. HYDRAZIDE FORMATION

In the presence of N-acetyltransferase, acetyl-CoA is utilized to N-acetylate hydrazines with a free amino functional group. In fact, hydrazine is excreted as the diacetyl derivative.[1] Acetylation of hydrazine derivatives in humans shows a polymorphic distribution of rapid and slow acetylators.[22] This acetylator status has been linked to several deleterious problems including hydralazine-induced systemic lupus erythematosus in slow acetylators[23] and decreased efficacy of phenelzine in depressed patients who are rapid acetylators.[24]

Although acetylation has largely been considered to be a detoxication step, some evidence exists that acetylation of isoniazid leads to enhanced hepatotoxicity of this drug.[25] Hydrolysis of N-acetylisoniazid by an acylamidase to form acetylhydrazine has been proposed to lead to the bioactivation of N-acetylisoniazid. The administration *in vivo* of an acylamidase inhibitor, bis-*p*-nitrophenyl phosphate, protected animals from liver necrosis after treatment with isoniazid and N-acetylisoniazid, but not against N-acetylhydrazine-induced necrosis.[26]

## C. N-OXIDATION REACTIONS

Although evidence exists that certain hydrazines may be intermediates in the reduction of azo compounds to the parent amines,[27] the most relevant biological redox reactions are the oxidation of hydrazines to azo, azoxy, and diazene intermediates. The discussion of the bioactivation and enzymology involved will be presented based on the structure of the hydrazine compounds. Because of the widespread use of trivial names in the literature, we have attempted to use the nomenclature of P. A. S. Smith,[9] except where specific drugs are cited.

## 1. Monosubstituted Hydrazines

As a result of their interest in the mechanisms of the antitumor agent procarbazine [$N$-isopropyl-$\alpha$-(2-methylhydrazino)-$p$-toluamide hydrochloride], Prough, Wittkop, and Reed[3-5] studied the oxidative metabolism of a number of monosubstituted hydrazines. Two principal reactions were detected: alkylhydrazine oxidation, which converted the hydrazines to their parent hydrocarbons, and alkylhydrazine dealkylation, which gave the corresponding aldehyde and dealkylated hydrazine. These reactions required NADPH, oxygen, and liver microsomal protein. The inhibitor studies performed suggested that more than one enzyme system might be involved, but that cytochrome $P$-450 appeared to be involved, at least partially, in both the demethylase and oxidase reactions.

The formation of the corresponding hydrocarbons suggested that the alkyldiazene intermediate must exist and can undergo decomposition via a free radical mechanism to give the alkane and molecular nitrogen. Augusto *et al.* have established that during metabolism of ethylhydrazine there exists an ethyl free radical intermediate, which can be detected with a free radical-trapping agent.[28] The identity of the ethyl radical was demonstrated by both electron spin resonance spectroscopy and mass spectral analysis of the radical adduct of the trapping agent. Similar results were obtained with phenelzine, which demonstrates that a 2-phenylethyl radical is formed during metabolism of this hydrazine.[29] The mechanism of diazene formation is not known, in that either N-oxygenation to an $N$-hydroxyhydrazine (which would rapidly dehydrate) or direct dehydrogenation would lead to the diazene intermediate.

Prough[30] demonstrated that the purified microsomal FAD-containing monooxygenase can catalyze the oxidation of alkylhydrazines, but subsequent work on the substrate affinity of the amine oxidase[31] for hydrazines suggests that cytochrome $P$-450 is the major microsomal enzyme capable of oxidizing the monoalkylhydrazines. Further studies with hydrazine and isopropylhydrazine substantiate this observation.[7,8,25,26] In addition, Coomes and Prough have suggested that the mitochondrial monoamine oxidase can convert alkylhydrazines to alkanes but may only account for a small portion of the total formation of hydrocarbon in rat liver.[32] Although the unstable diazenes are known to undergo air oxidation,[13] no direct evidence exists to document whether the diazenes are further oxidized to dia zoalkane derivatives in biological systems. The hydrocarbon products formed suggest that the alkyl free radical may be the ultimate reactive species of monoalkylhydrazine metabolism. Indeed, monosubstituted diazenes are very unstable and decompose to form carbon-centered radicals, even in the absence of molecular oxygen.[33]

### 2. Hydrazides and Hydrazones

It has been suggested that the hydrazides $N$-acetylisoniazid and iproniazid are hydrolyzed *in vitro* to form acetylhydrazine and isopropylhydrazine.[8,9,25,26,34] As shown earlier, isopropylhydrazine is thought to be oxidized by microsomal cytochrome $P$-450 to give either propane or covalent binding; the formation of either product is consistent with a free-radical intermediate. Studies with [$^{14}$C]acetylhydrazine demonstrated that, in the presence of liver microsomes, NADPH, and oxygen, the hydrazide was metabolized to a species capable of forming covalent bonds to microsomal protein. These results support the contention that hydrazides, like monosubstituted hydrazines, can be oxidized to the reactive diazene intermediates (Fig. 2).

To date, only the oxidative metabolism of the hydrazone analog of procarbazine has been studied. Like monoalkylhydrazines and hydrazides, this hydrazone is rapidly metabolized to give hydrocarbons as a product.[35] Since methylhydrazine was shown to be a poor substrate for cytochrome $P$-450 relative to the hydrazone, it is presumed that certain hydrazones can be oxidized enzymatically to form diazene intermediates, which can decompose via free radical intermediates.

### 3. 1,1-Disubstituted Hydrazines

In liver, the oxidative metabolism of 1,1-disubstituted hydrazines is catalyzed largely by the FAD-containing monooxygenase.[31] In part, this is because the oxidation products, 1,1-disubstituted diazenium ions, can form a complex with the other microsomal monooxygenase, cytochrome $P$-450, and terminate its function.[36] The flavoprotein monooxygenase catalyzes the stoichiometric conversion of several hydrazines to intermediates, which decompose to give unreactive intermediates.[31] For example, 1 mol each of NADPH and oxygen was required to convert 1 mol of 1,1-dimethylhydrazine to a mole of formaldehyde and methylhydrazine. In addition, $N$-aminopiperidine was converted to the dipiperidyltetrazene in a 5–10% yield. These products indicate that the likely intermediate formed in the reaction was a 1,1-disubstituted diazenium ion, which can either tautomerize to give a hydrazone or dimerize to give the corresponding tetrazene.[14] The hydrazone or its hydrolysis products may subsequently be bioactivated.

### 4. 1,2-Disubstituted Hydrazines

As expected from the chemistry of these compounds, one would expect the formation of a stable azo derivative during microsomal metabolism. Indeed, both 1,2-dimethylhydrazine and procarbazine have been shown to be rapidly converted to azomethane[37-39] and azoprocarbazine [$N$-isopropyl-$\alpha$-(2-methylazo)-$p$-toluamide],[40,41] respectively. In the rat hepatocytes, two enzyme systems participate in the reaction to form stable azo derivatives: the mitochondrial monoamine oxidase[32] and cytochrome $P$-450.[32,41]

### 5. Azo and Azoxy Derivatives

The stable azo derivatives of 1,2-dimethylhydrazine and procarbazine are further oxidized by cytochrome $P$-450 to yield azoxy derivatives.[37-42] For procarbazine, there are two position isomers (methyl-NNO and methyl-ONN), and the ratio of these two products depends on the form of cytochrome $P$-450 that catalyzes the reaction.[41,42]

Based on the pathological experiments of Druckrey, it was postulated that azoxymethane is C-hydroxylated to give methylazoxymethanol.[37] Fiala *et al.* have measured urinary metabolites *in vivo*[43,44] and, by using specific inhibitors, demonstrated that the metabolism of azoxymethane most likely results in the generation of a reactive methylating species.[44-46] This methyl group also was shown to enter the "$C_1$" pool and eventually can form $CO_2$. Although methylazoxymethanol is itself unstable,[45] it has been suggested that cytosolic alcohol dehydrogenase enhances decomposition of the C-hydroxy compound into a reactive agent similar, if not identical, to diazomethane.[46]

The metabolism of either [*benzyl*-$^{14}$C]- or [*methyl*-$^{14}$C]azoprocarbazine resulted in covalent binding of microsomal protein preferentially by the methyl portion of the molecule.[42] However, with either of the azoxyprocarbazine derivatives, Prough *et al.* did not detect covalent binding of the $N$-methyl group of either azoxyprocarbazine isomer to microsomal protein in the presence of NADPH and oxygen.[47] In addition, methane is a minor metabolite of procarbazine metabolism. Only azoprocarbazine and its hydrazone tautomer were metabolized by microsomal proteins to methane at a rate sufficient to account for the methane formed.[35,47] In contrast, the azoxy derivatives were not converted to methane and were shown not to be metabolized appreciably by liver microsomal proteins in the presence of NADPH and oxygen. Another enzyme system may exist that can metabolize the azoxy derivatives.

## IV. Fates of Reactive Intermediates of Hydrazines

Based on the two major reactions of hydrazines, hydrazone formation and oxidation to azo–diazene intermediates, it would appear that many of the toxic or carcinogenic effects of hydrazines could occur as a result of these two reactions. Alteration of normal cellular function by hydrazone formation with vital carbonyl compounds, such as pyridoxal phosphate, probably accounts for the effects of hydrazines on central nervous system function. This area will not be covered further in this chapter. However, the question regarding the significance of the oxidation of hydrazines in cancer and toxicology will be addressed in the following section.

### A. FLAVIN-DEPENDENT ENZYMES

To date, two flavin-dependent enzyme systems have been shown to be irreversibly altered by reaction with an oxidized species of hydrazine derivatives: the mammalian monoamine oxidase localized in the outer mitochondrial membrane[48,49] and trimethylamine dehydrogenase from bacteria.[50] Even though Zeller and co-workers[48] had established that hydrazines, in general, are potent inhibitors of monoamine oxidase (MAO), the mechanism of action of the hydrazines in the therapeutic regulation of hypertension and central nervous system depression was not well understood. Patek and Hellerman were able to purify the enzyme from bovine kidney cortex,[49] and the enzyme was shown to have covalently bound flavin attached through a cysteine residue.[51]

When substrates like benzylamine were added to the enzyme, the flavin absorbance maxima at approximately 450 nm was decreased, suggesting flavin reduction.[49] This bleaching phenomenon was reversed on exposure to molecular oxygen, and the products, ammonia and benzaldehyde, could be isolated. These results indicated that the flavin oxidized the amine to form a Schiff's base and could be reoxidized by molecular oxygen. On addition of various hydrazines, an irreversible bleaching of the flavin absorbance at 450 nm with half-times in minutes was observed. If [$^{14}$C]phenylhydrazine was utilized, covalent modification of the flavoprotein occurred with approximately 1.4 mol of inhibitor bound per mole equivalent of enzyme.[49] If phenyldiazene was generated chemically in the presence of the flavoprotein, the inhibition and flavin bleaching also occurred. These results were interpreted to demonstrate that the enzyme oxidized phenylhydrazine to phenyldiazene, which subsequently alkylated the enzyme and resulted in loss of enzyme activity.

Trimethylamine dehydrogenase, isolated from bacteria grown on trimethylamine, has been shown by Colby and Zatman[50] to be potently inhibited by hydrazines. Because this enzyme and MAO are both inhibited by hydrazines and contain a cysteinyl flavin as a prosthetic group, Nagy et al.[52] attempted to demonstrate that phenylhydrazine alkylated the flavin-prosthetic group of the dehydrogenase. After removal of the flavin by proteolysis, they used chemical methods to demonstrate that the flavin moiety was arylated at the C-4A position.

These results show that hydrazines have pronounced effects on both types of amine oxidases (flavin- and pyridoxal phosphate/copper-containing oxidases). As a result, certain of these compounds can be successfully utilized as therapeutic agents, and others may cause major toxicological problems. For example, one can imagine that hydrazine administration might lead to hypertensive crisis. A similar phenomenon has been observed when foods containing high amounts of biogenic amines are ingested[53]; it was found that tyramine, a normal constituent of cheese, could lead to such hypertensive episodes.

## B. HEME AND HEMOPROTEINS

In 1885, Hoppe-Seyler noted the formation of green pigments in the tissues of animals treated with phenylhydrazine,[54] and Heinz reported the formation of a new body in red cells treated with phenylhydrazine.[55] It had been assumed that the changes induced by many hydrazines led to the alteration of red cell function; that is, a decrease in the effective transport of oxygen to the tissues. Until recently, little was known about the chemistry of the changes required to cause Heinz body formation and green pigment production.

Associated with these reactions is the formation of a diverse number of products, including methemoglobin, superoxide anion, and hydrogen peroxide.[56,57] In addition, the reaction of hemoglobin with various hydrazines is known to cause the formation of a ferrihemochrome possessing unique electron paramagnetic resonance and visible absorption properties.[58,59] In addition to its physical properties, this species was shown to contain 1 mol of phenylhydrazine, or an oxidation product, bound per mole of heme.[60] In addition, arylation of the protein portion of hemoglobin could also be demonstrated.

The altered heme of phenylhydrazine-treated hemoglobin has been isolated and characterized by mass spectral and proton nuclear magnetic resonance techniques.[61,62] The arylated heme moieties isolated were shown to be $\beta$-mesophenylbiliverdin IX$\alpha$ and $N$-phenylprotoporphyrin IX. Ortiz de Montellano et al.[63] have described the formation of a globin-stabilized iron phenylheme complex on reaction of phenylhydrazine with hemoglobin. The existence of such an iron-substituted heme species was supported by the work of Lange and Mansuy,[64] who noted the migration of an iron-bound alkyl group to the nitrogen of the porphyrin ring. The iron–carbon bond appeared to be formed by interaction of the radical species derived from phenylhydrazine oxidation; coordination of these intermediates at the heme 6-coordination position is a likely mechanism for generation of the iron–carbon bonded heme species. The reaction with hemoglobin to form $N$-phenylheme is not a stoichiometric reaction, because at least five molecules of benzene (formed by a nondestructive radical decomposition pathway) were formed for every $N$-phenylheme isolated.[63] The mysterious green pigment of Hoppe-Seyler is perhaps related to the phenylated heme-prosthetic group. However, how this relates to erythrocyte lysis is not clear. Infusion of superoxide anion or hydrogen peroxide alone cannot entirely account for membrane destruction, and the phenyl radical itself may initiate some process resulting in cell lysis.

Other heme proteins have been reported to be alkylated by hydrazine derivatives. The apoprotein of horseradish peroxidase is alkylated by the isopropyl group of iproniazid, resulting in inactivation of the protein's function.[65] Similarly, catalase is also alkylated, with concomitant loss in catalytic activity.[65,66] Of particular interest is the ability of metabolites of various hydrazines to form abortive complexes with cytochrome $P$-450.[36,67] Depending on the structure of the hydrazine, the formation of a metabolite complex can be demonstrated either to be reversible or to lead to heme destruction. For example, 1,1-disubstituted hydrazines form spectral intermediates that absorb maximally at 438 nm but that decay with a half-time of 15 to 20 min on removal of the hydrazine or the source of reducing equivalents.[36] The monosubstituted hydrazines, including phenylhydrazine, cause much less distinct absorbance maxima to form, but lead to heme destruction.[67] All of the monosubstituted hydrazines bind to the ferric, substrate-free form of the enzyme and convert it to an intermediate, which does not display the characteristic electron spin resonance of the cytochrome. Like hemoglobin, there are measurable amounts of the corresponding hydrocarbon formed during these reactions. 1,2-Disubstituted hydrazines do not appear to form these complexes on metabolism, but hydrazones and hydrazides appear to lead to heme alteration (Moloney and Prough, unpublished results).

These results suggest that the biological oxidation of hydrazine derivatives by cytochrome $P$-450 leads to formation of diazene intermediates. The contention that a diazene can bind to the 6-coordination position of cytochrome $P$-450[36,67] is supported by crystallographic studies demonstrating the ability of a 1,1-disubstituted diazene to coordinate to hemin.[68] Battioni *et al.*[69] have demonstrated that arylhydrazines and aryldiazenes form stab: ferrous cytochrome *P*-450–diazene complexes, as well as σ-iron(III)–ary complexes (iron-bound aryl groups). The alkylhydrazines and diazenes d not form the σ-iron(III)–alkyl complexes in the presence of oxygen. Th chemistry presumably explains why arylhydrazines are more potent in alky lating heme than alkylhydrazines, in that the ferrous–diazene complexes c the alkyl derivatives decompose directly to hydrocarbons in the presence c molecular oxygen. Either the common iron–diazene intermediate or th σ-iron–carbon bonded intermediate could account for hydrocarbon pro duction and heme *N*-alkylation. These studies provide a chemical rationa for how hydrazines alter hemoprotein function, but further studies are re quired to relate these reactions to the toxicological problems of red cell lys and the other hemolytic problems observed on hydrazine exposure.

### C. THIOLS

Several examples of reactions of the oxidation products of hydrazines wit thiols can be documented. Glutathione and cysteine adducts have bee demonstrated on incubation of isopropylhydrazine or acetylhydrazine, live microsomal protein, NADPH, oxygen, and the corresponding thiol con pound.[26,34] *S*-Acetylglutathione and *N*-acetylcysteine have been identifiec The chemical mechanism, based on sophisticated mass spectral analysi supports the assumption that most monosubstituted hydrazines are oxida tively metabolized to diazenes, which decompose to carbon-centered fre radicals. These radicals are probably involved in the alkylation of thiols

Studies in this laboratory have indicated that another reaction of thiol which may occur during oxidative metabolism of alkylhydrazines and hy drazides, is the enhanced formation of alkanes (S. J. Moloney, P. Wiebkin S. W. Cummings, and R. A. Prough, unpublished results). Addition c thiols to incubation mixtures containing liver microsomes, NADPH, $O_2$ and azoprocarbazine resulted in stimulation of alkane formation, presum ably because of preferential abstraction of a hydrogen from the thiol by th free radical instead of reaction with other electrophiles in the microsoma protein fraction. In addition, a reduction in covalent binding to protein ha been observed concomitant with the stimulation of alkane formation b thiols. This reaction of thiols with free radicals would yield a thiyl radica which ultimately leads to disulfide formation.[70]

A third reaction of thiols with hydrazine oxidation products is the nuclec philic attack of thiols on azo linkages substituted with electron-withdrawin groups on one or both nitrogens. Kosower[13,70-72] demonstrated that az derivatives, such as methyl phenyldiazenecarboxylate or 1,1-azobis(*N,N*-d methylformamide), served as oxidants of intracellular glutathione. In this reaction, two molecules of glutathione are oxidized to glutathione disulfide, and the azo compound is apparently reduced to the parent hydrazine. The antifungal activity of certain hydrazine compounds may result from thiol depletion after hydrazine oxidation by chemical or enzymatic processes.[73] In fact, the use of acetylphenylhydrazine to deplete erythrocyte glutathione levels required molecular oxygen[74] and suggested that formation of an azo derivative was required to yield a species capable of depleting intracellular glutathione.

### D. PROTEINS AND NUCLEIC ACIDS

The studies on hemoglobin, horseradish peroxidase, catalase, and microsomal protein have demonstrated that proteins can be alkylated during hydrazine metabolism. The nature of these protein adducts are not known. However, it is known that, on metabolism, the methyl group of 1,2-dimethylhydrazine (1,2-DMH) is transferred to nucleic acids.[75,76] The specific base adducts isolated were $N^7$- and $O^6$-methylguanine, and the detailed study of DNA adducts formed by 1,2-DMH have been performed in rats.[76,77] The ratio of $O^6$- to $N^7$-methylguanine adducts was highest in the target tissues, colon and liver, and DNA repair in colon removed the $O^6$-methylguanine at a lower rate than in liver. This report has been interpreted to indicate that the persistence of the $O^6$-methylguanine may be of significance to 1,2-DMH carcinogenesis in the lower tract. However, unlike the previous sections, the metabolism of 1,2-DMH is thought to require C-hydroxylation of azoxymethane and the degradation of methylazoxymethanol to give a reactive species similar to diazomethane.[37]

## V. Conclusion

The current state of knowledge related to the oxidative bioactivation of hydrazines strongly suggests that two possible reactive intermediates may exist: formation of unstable diazenes and C-hydroxylation of methylazoxy compounds. The diazenes are easily air oxidized, and future studies must address the possibility that alkyldiazenes may be oxidized to diazoalkanes. These diazoalkanes may also serve as reactive intermediates responsible for the deleterious effects of hydrazines in biological systems. The balance between carcinogenesis and toxicity may be explained, in part, by the existence of these two pathways.

## Acknowledgments

Portions of the work reported were supported by American Cancer Society Grant BC-336 and Robert A. Welch Grant I-616. S. J. M. was a Chilton Foundation postdoctoral fellow.

## References

1. Colvin, L. B. (1969). Metabolic fate of hydrazines and hydrazides. *J. Pharm. Sci.* **58**, 1433–1443.
2. Juchau, M. R., and Horita, A. (1972). Metabolisms of hydrazine derivatives of pharmacologic interest. *Drug Metab. Rev.* **1**, 71–100.
3. Prough, R. A., Wittkop, J. A., and Reed, D. J. (1969). Evidence for the hepatic metabolism of some monoalkylhydrazines. *Arch. Biochem. Biophys.* **131**, 369–373.
4. Wittkop, J. A., Prough, R. A., and Reed, D. J. (1969). Oxidative demethylation of N-methylhydrazines by rat liver microsomes. *Arch. Biochem. Biophys.* **134**, 308–315.
5. Prough, R. A., Wittkop, J. A., and Reed, D. J. (1969). Further evidence on the nature of microsomal metabolism of procarbazine and related alkylhydrazines. *Arch. Biochem. Biophys.* **140**, 450–458.
6. Toth, B. (1975). Synthetic and naturally occurring hydrazines as possible cancer causative agents. *Cancer Res.* **35**, 3693–3697.
7. Nelson, S. D., Mitchell, J. R., Timbrell, J. A., Snodgrass, W. R., and Corcoran, G. B., III (1976). Isoniazid and iproniazid: Activation of metabolites to toxic intermediates in man and rat. *Science* **193**, 901–903.
8. Timbrell, J. A. (1979). The role of metabolism in the hepatotoxicity of isoniazid and iproniazid. *Drug Metab. Rev.* **10**, 125–147.
9. Smith, P. A. S. (1966). "The Chemistry of Open-chain Organic Nitrogen Compounds," Vol. II. Benjamin, New York.
10. Cordes, E. H., and Jencks, W. P. (1962). Nucleophilic catalysis of semicarbazine formation by anilines. *J. Am. Chem. Soc.* **84**, 826–831.
11. Edwards, J. O., and Pearson, R. G. (1962). The factors determining nucleophilic reactivities. *J. Am. Chem. Soc.* **84**, 16–24.
12. Nelson, S. F. (1981). One-electron oxidation of tetraalkylhydrazines. *Acc. Chem. Res.* **14**, 131–138.
13. Kosower, E. M. (1971). Monosubstituted diazenes (diimides). Surprising intermediates. *Acc. Chem. Res.* **4**, 193–198.
14. Lemal, D. M. (1970). Aminonitrenes (1,1-diazenes). *In* "Nitrenes" (W. Lwowski, ed.), pp. 345–403. Wiley (Interscience), New York.
15. Hucko-Haas, J. E., and Reed, D. J. (1970). Hydrazines as substrates for bovine plasma amine oxidase (PAO). *Biochem. Biophys. Res. Commun.* **39**, 396–400.
16. Bender, D. A., and Russell-Jones, R. (1979). Isoniazid-induced pellagra despite vitamin $B_6$ supplementation. *Lancet* **2**, 1125–1126.
17. McKennis, H., Jr., and Weatherby, J. H. (1956). Blood ammonia following administration of various hydrazine compounds. *Fed. Proc., Fed. Am. Soc. Exp. Biol.* **15**, 458–459.
18. Simonsen, D. G., and Roberts, E. (1967). Influence of hydrazine on the distribution of free amino acids in mouse liver. *Proc. Soc. Exp. Biol. Med.* **124**, 806–811.
19. Smith, T. E. (1965). Brain carbohydrate metabolism during hydrazine toxicity. *Biochem. Pharmacol.* **14**, 979–988.
20. Haegele, K. O., McLean, A. J., duSouich, P., Barron, K., Laquer, J., McNay, J. L., and

Carrier, O. (1978). Identification of hydralazine and hydralazine hydrazone metabolites in human body fluids and quantitative *in vitro* comparisons of their smooth muscle relaxant activity. *Br. J. Clin. Pharmacol.* **5,** 489–494.
21. Kleineke, J., Peters, H., and Soling, H. D. (1979). Inhibition of hepatic gluconeogenesis by phenylethylhydrazine (phenelzine). *Biochem. Pharmacol.* **28,** 1379–1389.
22. Price Evans, D. A. (1968). Genetic variation in the acetylation of isoniazid and other drugs. *Ann. N. Y. Acad. Sci.* **151,** 723–733.
23. Lunde, P. K. M., Frislid, K., and Hansteen, V. (1977). Disease and acetylation polymorphism. *Clin. Pharmacokinet.* **2,** 182–197.
24. Johnstone, E. C., and Marsh, W. (1973). Acetylator status and response to phenelzine in depressed patients. *Lancet* **1,** 567–570.
25. Timbrell, J. A., Mitchell, J. R., Snodgrass, W. R., and Nelson, S. D. (1980). Isoniazid hepatotoxicity: The relationship between covalent binding and metabolism *in vivo*. *J. Pharmacol. Exp. Ther.* **213,** 364–369.
26. Nelson, S. D., Mitchell, J. R., Snodgrass, W. R., and Timbrell, J. A. (1978). Hepatotoxicity and metabolism of iproniazid and hydrazine. *J. Pharmacol. Exp. Ther.* **206,** 574–585.
27. Hernandez, P. H., Gillette, J. R., and Mazel, P. (1967). Studies on the mechanism of action of mammalian hepatic azoreductase. I. Azoreductase activity of reduced nicotinamide adenine dinucleotide phosphate–cytochrome *c* reductase. *Biochem. Pharmacol.* **16,** 1859–1875.
28. Augusto, O., Ortiz de Montellano, P. R., and Quintanilha, A. (1981). Spin-trapping of free radicals formed during microsomal metabolism of ethylhydrazine and acetylhydrazine. *Biochem. Biophys. Res. Commun.* **101,** 1324–1330.
29. Ortiz de Montellano, P. R., Augusto, O., Viola, F., and Kunze, K. L. (1983). Carbon radicals in the metabolism of alkylhydrazines. *J. Biol. Chem.* **258,** 8623–8629.
30. Prough, R. A. (1973). The N-oxidations of alkylhydrazines catalyzed by the microsomal mixed-function amine oxidase. *Arch. Biochem. Biophys.* **158,** 442–444.
31. Prough, R. A., Freeman, P. C., and Hines, R. N. (1981). The oxidation of hydrazine derivatives catalyzed by the purified liver microsomal FAD-containing monooxygenase. *J. Biol. Chem.* **256,** 4178–4184.
32. Coomes, M. W., and Prough, R. A. (1983). The mitochondrial metabolism of 1,2-disubstituted hydrazines, procarbazine and 1,2-dimethylhydrazine. *Drug Metab. Dispos.* **11,** 1106–1112.
33. Huang, P. C., and Kosower, E. M. (1967). Diazenes. III. Properties of phenyldiazene. *J. Am. Chem. Soc.* **90,** 2367–2376.
34. Nelson, S. D., Hinson, J. A., and Mitchell, J. A. (1976). Application of chemical ionization mass spectrometry and the twin-ion technique to better define a mechanism in acetylhydrazine toxicity. *Biochem. Biophys. Res. Commun.* **69,** 900–907.
35. Moloney, S. J., and Prough, R. A. (1982). Studies on the pathway of methane formation from procarbazine, a 2-methylbenzylhydrazine derivative, by rat liver microsomes. *Arch. Biochem. Biophys.* **221,** 577–584.
36. Hines, R. N., and Prough, R. A. (1980). The characterization of an inhibitory complex formed with cytochrome P-450 and a metabolite of 1,1-disubstituted hydrazines. *J. Pharmacol. Exp. Ther.* **214,** 80–86.
37. Druckrey, H. (1970). Production of colonic carcinomas by 1,2-dialkylhydrazines and azoxyalkanes. *In* "Carcinoma of the Colon and Antecedent Epithelium" (W. J. Burdette, ed.), pp. 267–279. Thomas, Springfield, Illinois.
38. Fiala, E. S., Kulakis, C., Bobotas, G., and Weisburger, J. H. (1976). Detection and estimation of azomethane in expired air of 1,2-dimethylhydrazine-treated rats. *J. Natl. Cancer Inst. (U.S.)* **56,** 1271–1273.

39. Fiala, E. S. (1977). Investigations into the metabolism and mode of action of the colon carcinogen, 1,2-dimethylhydrazine and azomethane. *Cancer* **40**, 2436–2445.
40. Weinkam, R. J., and Shiba, D. A. (1978). Metabolic activation of procarbazine. *Life Sci.* **22**, 937–946.
41. Dunn, D. L., Lubet, R. A., and Prough, R. A. (1979). Oxidative metabolism of *N*-isopropyl-α-(2-methylhydrazine)-*p*-toluamide hydrochloride (procarbazine) by rat liver microsomes. *Cancer Res.* **39**, 4555–4563.
42. Wiebkin, P., and Prough, R. A. (1980). Oxidative metabolism of *N*-isopropyl-α-(2-methylazo)-*p*-toluamide (azoprocarbazine) by rodent liver microsomes. *Cancer Res.* **40**, 3524–3529.
43. Fiala, E. S., Bobotas, G., Kulakis, C., and Weisburger, J. H. (1977). Inhibition of 1,2-dimethylhydrazine metabolism by disulfiram. *Xenobiotica* **7**, 5–9.
44. Fiala, E. S., Kulakis, C., Christiansen, G., and Weisburger, J. H. (1978). Inhibition of the metabolism of the colon carcinogen, azoxymethane, by pyrazole. *Cancer Res.* **38**, 4515–4521.
45. Benn, M. H., and Kazmaier, P. (1972). Methylation with (*Z*)-methyl-*ONN*-azoxymethanol. The nature of the reactive species. *J. Chem. Soc., Chem. Commun.* pp. 887–888.
46. Feinberg, A., and Zedeck, M. S. (1980). Production of a highly reactive alkylating agent from the organospecific carcinogen methylazoxymethanol by alcohol dehydrogenase. *Cancer Res.* **40**, 4446–4450.
47. Prough, R. A., Coomes, M. W., Cummings, S. W., and Wiebkin, P. (1981). Metabolism of procarbazine [*N*-isopropyl-α-(2-methylhydrazino)-*p*-toluamide HCl. *Adv. Exp. Med. Biol.* **136B**, 983–996.
48. Zeller, E. A., Barsky, J., and Berman, E. R. (1955). Amine oxidases. XI. Inhibition of monoamine oxidase by 1-nicotinyl-2-isopropylhydrazine. *J. Biol. Chem.* **214**, 267–274.
49. Patek, D. R., and Hellerman, L. (1974). Mitochondrial monoamine oxidase. Mechanism of inhibition of phenylhydrazine and by aralkylhydrazines. Role of enzymatic oxidation. *J. Biol. Chem.* **249**, 2373–2380.
50. Colby, J., and Zatman, L. J. (1974). Purification and properties of the trimethylamine dehydrogenase of bacterium 4B6. *Biochem. J.* **143**, 555–567.
51. Kearney, E. B., Salach, J. I., Walker, W. H., Seng, R. L., Kenney, W., Zeszotek, E., and Singer, T. P. (1971). The covalently bound flavin of hepatic monoamine oxidase. I. Isolation and sequence of a flavin peptide and evidence for binding the 8α-position. *Eur. J. Biochem.* **24**, 321–327.
52. Nagy, J., Kenney, W. C., and Singer, T. P. (1979). The reaction of phenylhydrazine with trimethylamine dehydrogenase and with free flavins. *J. Biol. Chem.* **254**, 2684–2688.
53. Blackwell, B. (1967). Hypertensive crises due to monamine oxidase inhibitors. *Lancet* **2**, 733–734.
54. Hoppe-Seyler, G. (1885). Über die Wirkung des Phenylhydrazines auf den Organismus. *Z. Physiol. Chem.* **9**, 34–39.
55. Heinz, R. (1890). 2. Morphologische Veränderungen der roten Blutkörperchen durch Gifte. *Arch. Pathol. Anat. Physiol. Klin. Med.* **122**, 112–116.
56. Cohen, G., and Hochstein, P. (1964). Generation of hydrogen peroxide in erythrocytes by hemolytic agents. *Biochemistry* **3**, 895–900.
57. Goldberg, B., and Stern, A. (1977). The mechanism of oxidative hemolysis produced by phenylhydrazine. *Mol. Pharmacol.* **13**, 832–839.
58. Itano, H. A., Hirota, K., and Vedvick, T. S. (1977). Ligands and oxidants in ferrihemochrome formation. *Proc. Natl. Acad. Sci. U.S.A.* **74**, 2556–2560.
59. Peisach, J., Blumberg, W. E., and Rachmilewitz, E. A. (1975). The demonstration of ferrihemochrome intermediates in Heinz body formation following the reduction of oxyhemoglobin A by phenylhydrazine. *Biochim. Biophys. Acta* **393**, 404–418.

60. Itano, H. A., and Mannen, S. (1976). Reaction of phenyldiazene and ring-substituted phenyldiazenes with ferrihemoglobin. *Biochim. Biophys. Acta* **421**, 87–96.
61. Saito, S., and Itano, H. A. (1981). β-*meso*-Phenylbiliverdin IXα and *N*-phenylprotoporphyrin IX, products of the reaction of phenylhydrazine with oxyhemoproteins. *Proc. Natl. Acad. Sci. U.S.A.* **78**, 5508–5512.
62. Ortiz de Montellano, P. R., and Kunze, K. L. (1981). Formation of *N*-phenylheme in the hemolytic reaction phenylhydrazine with hemoglobin. *J. Am. Chem. Soc.* **103**, 6534–6536.
63. Augusto, O., Kunze, K. L., and Ortiz de Montellano, P. R. (1982). *N*-Phenylprotoporphyrin IX formation in the hemoglobin phenylhydrazine reaction. Evidence for a protein-stabilized iron–phenyl intermediate. *J. Biol. Chem.* **256**, 6231–6241.
64. Lange, M., and Mansuy, O. (1981). N-Substituted porphyrins formation from carbene iron–porphyrin complexes: A possible path for cytochrome *P*-450 heme destruction. *Tetrahedron Lett.* **22**, 2561–2564.
65. Hidaka, H., and Udenfriend, S. (1970). Evidence of a hydrazine-reactive group at the active site of the nonheme portion of horseradish peroxidase. *Arch. Biochem. Biophys.* **140**, 174–180.
66. Ortiz de Montellano, P. R., and Kerr, D. E. (1983). Inactivation of catalase by phenylhydrazine. Formation of a stable aryl–iron heme complex. *J. Biol. Chem.* **258**, 10558–10563.
67. Jonen, H. G., Werringloer, J., Prough, R. A., and Estabrook, R. W. (1982). The reaction of phenylhydrazine with microsomal cytochrome *P*-450. Catalysis of heme modification. *J. Biol. Chem.* **257**, 4404–4411.
68. Mansuy, D., Battioni, P., and Mahy, J. P. (1982). Isolation of an iron–nitrene complex from the dioxygen and iron porphyrin oxidation of a hydrazine. *J. Am. Chem. Soc.* **104**, 4487–4488.
69. Battioni, P., Mahy, J., Delforge, M., and Mansuy, D. (1983). Reaction of monosubstituted hydrazines and diazenes with rat-liver cytochrome *P*-450. Formation of ferrous-diazene and Ferric σ-alkyl complexes. *Eur. J. Biochem.* **134**, 241–248.
70. Kosower, N. S., Song, K. R., and Kosower, E. M. (1969). Glutathione. I. The methyl phenyldiazenecarboxylate (azo ester) procedure for intracellular oxidation. *Biochim. Biophys. Acta* **192**, 1–7.
71. Kosower, N. S., Song, K. R., Kosower, E. M., and Correa, W. S. (1969). Glutathione. II. Chemical aspects of azoester procedure for oxidation to disulfide. *Biochim. Biophys. Acta* **192**, 8–14.
72. Kosower, N. S., Kosower, E. M., Wertheim, B., and Correa, W. S. (1969). *N*-Diamide, a new reagent for the intracellular oxidation of glutathione to disulfide. *Biochem. Biophys. Res. Commun.* **37**, 593–596.
73. Miadera, T. (1975). Biological formation and reactions of hydrazo, azo, and azoxy groups. *In* "The Chemistry of the Hydrazo, Azo, and Azoxy Groups" (S. Patai, ed.), Part 1, pp. 495–539. Wiley, London.
74. Beutler, E., Robson, M., and Buttenwieser, E. (1957). The mechanism of glutathione destruction and protection in drug-sensitive and -nonsensitive erythrocytes. *In vitro* studies. *J. Clin. Invest.* **36**, 617–628.
75. Nagata, Y., and Matsumoto, H. (1969). Studies in methylazoxymethanol: Methylation of nucleic acids in the fetal rat brain. *Proc. Soc. Exp. Biol. Med.* **132**, 383–385.
76. Hawks, A., and Magee, P. N. (1974). The alkylations of nucleic acids of rat and mouse *in vivo* by the carcinogen, 1,2-dimethylhydrazine. *Br. J. Cancer* **30**, 440–446.
77. Rogers, K. J., and Pegg, A. E. (1977). Formation of $O^6$-methylguanine by alkylation of rat liver, colon, and kidney DNA following administration of 1,2-dimethylhydrazine. *Cancer Res.* **37**, 4082–4087.

Chapter 16

# Nitroimidazoles

**P. DAVID JOSEPHY**

*Laboratory of Pulmonary Function and Toxicology*
*National Institute of Environmental Health Sciences*
*Research Triangle Park, North Carolina*

**RONALD P. MASON**

*Laboratory of Environmental Biophysics*
*National Institute of Environmental Health Sciences*
*Research Triangle Park, North Carolina*

| | |
|---|---|
| I. Introduction | 451 |
| II. Preparation of Aminoimidazoles | 454 |
| III. Zinc Reduction of Nitroimidazoles | 455 |
| IV. Enzymatic Reduction of Nitroimidazoles: Diamagnetic Products | 456 |
| V. Free-Radical Intermediates in the Enzymatic Reduction of Nitro Compounds | 458 |
| VI. Catalytic, Radiolytic, and Electrochemical Reduction of Nitroimidazoles | 461 |
| VII. Coreduction of Nitroimidazoles with Macromolecules | 462 |
| VIII. Nonenzymatic Reactions of the Nitro Anion Free Radical | 462 |
| IX. Other Free-Radical Metabolites of Nitro Compounds | 469 |
| X. Bacterial Metabolism of Nitroimidazoles | 472 |
| XI. Metabolism of Misonidazole by Mammalian Cells *in Vitro* | 473 |
| XII. Summary | 475 |
| References | 476 |

## I. Introduction

The nitroaromatic and nitroheterocyclic compounds constitute one of the most significant classes of xenobiotics to which humans are exposed. No single review can provide comprehensive coverage of the vast literature on the chemistry and toxicology of nitro compounds, and we will deal primarily with the nitroimidazole drugs; this is a reflection of the current importance of this group in medicine. On occasion, we will refer to work on other classes of nitro compounds to fill lacunae in the nitroimidazole literature. We have tried to minimize overlap with previous reviews and begin by drawing the reader's attention to some of these. Wardman's review[1] on the reduction of nitroaromatic compounds provides an excellent background, although most of the work on nitroimidazoles has appeared since its publication. Nitrofurans are widely used as antibacterial agents, and the nitroreduction of these compounds has been reviewed.[2,3] Several reviews of free-radical processes in drug metabolism have been published, and all of these contain material on nitro group metabolism.[4,8] General reviews of the metabolism of nitrogen functional groups devote some space to nitro compounds.[9,10] The synthesis and some aspects of the metabolism of nitroimidazoles are covered in several reviews.[10a,10b] Finally, we mention the proceedings of the conference "Toxicity of Nitroaromatic Compounds," Raleigh, North Carolina, January 1982.[11]

There has been a resurgence of interest in nitroheterocyclic drugs. The prototype of this class, 2-nitroimidazole (azomycin), was characterized by Japanese investigators more than 25 years ago.[12] Independently, researchers at the French pharmaceutical firm of Rhone-Poulenc showed that 2-nitroimidazole was responsible for the antiprotozoal activity of extracts of certain *Streptomyces* fungi. A variety of nitroimidazoles was synthesized, culminating in the development of metronidazole (1,2'-hydroxyethyl-2-methyl-5-nitroimidazole), also known as Flagyl. Metronidazole was introduced into clinical practice for the treatment of trichomonal vaginitis and is considered to be the drug of choice[13]; it has proved to be a safe and effective chemotherapeutic agent.

In the early 1970s, researchers in Canada and England discovered that nitrofurans[14] and nitroimidazoles[15] are radiosensitizers of hypoxic (oxygen-deficient) cells. Cellular response to ionizing radiation is strongly influenced by oxygen concentration; hypoxic cells are much more resistant to radiation damage than are normally oxygenated cells. The sensitizing effect of molecular oxygen is probably a consequence of the rapid chemical reactions between oxygen and the short-lived free radicals generated by radiation absorption. Radioresistant hypoxic cells may be present in rapidly growing tumors, limiting the effectiveness of radiation therapy. Adams and Dewey[16] suggested that electron-affinic (or easily reduced) chemicals might mimic the sensitizing action of oxygen and thereby overcome the radioresistance of hypoxic cells. This possibility was borne out by studies of nitroaromatic compounds.[17] The discovery of the radiosensitizing properties of nitroheterocyclics[14,15] led to the current intense interest in the medical application of radiosensitizers. Clinical trials have been conducted with metronidazole[18] and with the newer drug misonidazole [1-(2-nitro-1-imidazolyl)-3-methoxypropanol; RO 07-0582]. Misonidazole was found to produce severe neurotoxicity after multiple high doses,[19] and this has limited subsequent trials to suboptimal drug levels. Although preliminary evidence suggests that some clinical benefit may be obtained with misonidazole, the problem of neurotoxicity must be overcome before significant progress can be achieved.[20] Several reviews of the development of radiosensitizers have been published.[21-24]

Investigators studying the action of nitroheterocyclic radiosensitizers found that these compounds were selectively toxic to hypoxic mammalian cells *in vitro,* even in the absence of radiation.[25-27] This phenomenon of increased toxicity to hypoxic cells may be related to the neuropathic side effects of nitro compounds. If toxicity to normal tissues can be minimized, these drugs might be exploited as chemotherapeutic agents targeted against hypoxic tumor cells.[28] The metabolic reduction of the nitro group is probably of central importance in the toxicity of nitro compounds. The most direct evidence for this hypothesis is the isolation of bacterial mutants resistant to nitro compound toxicity and mutagenicity; such mutants are always characterized by defective or absent nitroreductase enzymes.[2,3] The oxygen sensitivity of the nitro anion free-radical intermediate formed during enzymatic nitroreduction[29] accounts for the selective action of nitro compounds against anaerobic cells; we will consider this point in more detail later.

In the simplest case, reduction of the nitro group yields an amine via two possible diamagnetic (nonradical) intermediates:

| | | |
|---|---|---|
| 0 | Ar—NO$_2$ | Nitro |
| 2 | Ar—N=O | Nitroso |
| 4 | Ar—NHOH | Hydroxylamino |
| 6 | Ar—NH$_2$ | Amino |

The number indicates the reducing equivalents necessary to form each species starting from the nitro compound. A second class of derivatives can arise from the dimerization of the reduced intermediates, yielding species with N–N bonds:

| | | |
|---|---|---|
| 6 | Ar—N=N—Ar (with O on N) | Azoxy |
| 8 | Ar—N=N—Ar | Azo |
| 10 | Ar—NH—NH—Ar | Hydrazo |

Here, the number refers to total reducing equivalents and should be halved to obtain the number of electrons per molecule of nitro compound. In the nitrobenzene series, each of these derivatives has been synthesized and studied.

## II. Preparation of Aminoimidazoles

The study of the reduction of nitroimidazoles has a remarkably long history, which predates the application of these compounds in medicine. Much of this early work remains of interest today but is rarely cited. The first attempts to synthesize 4-aminoimidazole by reduction of 4-nitroimidazole were unsuccessful. (These compounds have also been designated as 5-substituted imidazoles: in the absence of a substituent on an imidazole N atom, the 4- and 5-positions are equivalent.) For example, reduction of the nitro group with tin and hydrochloric acid yielded glycine, ammonia, and other products resulting from cleavage of the imidazole ring.[30] Preparation of 4-aminoimidazole was first accomplished with sodium amalgam in the absence of oxygen and water.[31] In a later study, 4-aminoimidazole and its carboxylic acid derivative were synthesized by catalytic nitroreduction.[32] In both reports, the amines were reported to be very unstable, particularly in aqueous solution or if exposed to air. Several synthetic routes for the preparation of 4-aminoimidazole derivatives are now available, and this has been reviewed.[33] Most of these studies were undertaken out of an interest in purine biosynthesis or for the preparation of triazenoimidazoles, which have been used as antineoplastics. Although reduction of 4-nitroimidazoles usually results in an amine or ring cleavage product, the isolation of a 4-hydroxylaminoimidazole has been reported in the case of 1-methyl-4-nitroimidazole-5-carbonitrile.[34]

The 2-aminoimidazoles may be prepared by reduction of a benzeneazoimidazole[35] or by the condensation of aminoaldehydes with cyanamide.[36] In contrast to the isomeric 4-aminoimidazoles, the 2-aminoimidazoles are quite stable. Diazotization was once thought to be impossible, but was later accomplished and used for the synthesis of 2-nitroimidazoles.[37]

The structural basis for the marked differences between the chemistry of 2- and 5-substituted imidazoles may be the ability of the former, but not the latter, to undergo tautomeric rearrangements.[30,35] 2-Aminoimidazole behaves as an imine and is resonance stabilized by the contributions of structures with positive charge localized on the imidazole ring (Scheme 1). In contrast, 5-aminoimidazole behaves as a "true" amine. Such tautomerism may also be possible for 2-hydroxylaminoimidazoles, although direct evidence is unavailable. The importance of amine–imine tautomerism was demonstrated by a study of the reactions of 2- and 5-aminoimidazole with

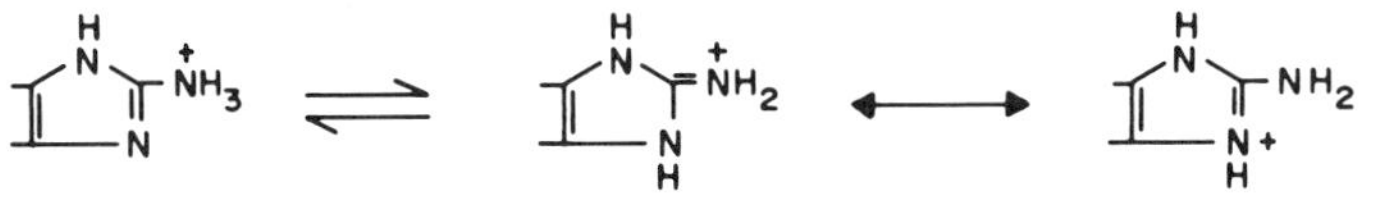

**Scheme 1.** Structure of 2-aminoimidazole. The amine and imine cations are tautomers; the third structure, in which positive charge is transferred to the ring, is a resonance hybrid form of the imine.

picryl fluoride.[38] The 2-isomer yielded 1,3-dipicryl-2-imino-1-imidazole, whereas the 5-isomer yielded 1-picryl-5-picrylaminoimidazole.

## III. Zinc Reduction of Nitroimidazoles

The first attempts to isolate the reduced derivatives of 4-nitroimidazole were unsuccessful, as noted earlier. Reduction of metronidazole with zinc and hydrochloric acid also produced negative results.[39] (A summary of the procedures described in this section is given in Table I.) No products were characterized, and the authors concluded that "5-aminoimidazoles are apparently unstable and rapidly destroyed."[39] Reduction of misonidazole with zinc dust was reported in 1976[40]; the method was based on an earlier procedure for the reductive synthesis of aminofluorene.[41] Although misonidazole reduction was achieved (as measured by loss of UV absorption), the products were not characterized. Examination by thin-layer chromatography (TLC) of the products obtained with this procedure revealed the presence of many colored and fluorescent products.[42] Reduction of misonidazole with zinc dust at room temperature yielded a bright yellow solution, which contained both azomisonidazole and azoxymisonidazole[43]; these were separated by preparative reversed-phase column chromatography and characterized by UV-visible spectroscopy, nuclear magnetic resonance (NMR) spectroscopy, and, after acetylation of the hydroxyl groups, mass spectrometry. Both compounds are stable at room temperature. The biological effects of these derivatives have been examined.[42] There is, however, no clear evidence for the formation of bimolecular derivatives of nitroimidazoles in biological systems,[42] despite earlier suggestions to the contrary.[39] The nature of the yellow or red pigments often observed in the urine of patients receiving metronidazole or misonidazole remains unknown.

The presence of a chloride salt is apparently required for zinc reduction of nitroaromatics, although the reason is not known.[44] Misonidazole reduction with zinc in the presence of ammonium chloride (a method used for the reduction of nitrobenzene to phenylhydroxylamine[44]) yielded a colorless product.[45] Mass spectroscopy suggested the presence of the hydroxylamine

Table I

Zinc Reduction of Nitroimidazoles

| Solvent | Temperature (°C) | Concentration (mg/ml) | | | | References |
|---|---|---|---|---|---|---|
| | | [Misonidazole] | [Zn] | [CaCl$_2$] | [NH$_4$Cl] | |
| EtOH (78%) | Reflux | 30.0$^a$ | 300 | 10 | — | 41 |
| EtOH (80%) | Reflux | 0.5 | 0.4 | 0.25 | — | 40 |
| Aq. | Room | 10.0$^b$ | 17 | — | 7.0 | 48 |
| Aq. | Room | 10.0 | 20 | 10 | | 43 |
| Aq. | 65 | 1.0 | 20 | — | 2.0 | 45 |
| Aq. | 37 | 1-20 | ns$^c$ | — | 0.5-10 | 46 |
| Aq. | 55 | 0.1-20 | 2-400 | — | 2.5-500 | 47 |

$^a$ Nitrofluorene.
$^b$ Metronidazole.
$^c$ ns, not stated.

($m/z = 187$). The initial product obtained after addition of zinc was yellow; a colorless product was obtained later. Optical spectra of the initial yellow product[46] showed a peak near 400 nm, presumably resulting from formation of azo and azoxy derivatives. Further reduction to the colorless product generated at least three products, which were suggested to be the amine, hydroxylamine, and hydrazo compounds.[46] Each of these products was identified by mass spectrometry.[47]

Koch and Goldman[48] reduced metronidazole with zinc and ammonium chloride. No yellow product was observed, and the only product characterized was N-(2-hydroxyethyl)oxamic acid, apparently in low yield. This suggests that a reduced intermediate of metronidazole decomposes with cleavage of the imidazole ring.

## IV. Enzymatic Reduction of Nitroimidazoles: Diamagnetic Products

The xanthine–xanthine oxidase system has been studied by many investigators, because it provides a convenient model for enzymatic nitroreduction. Xanthine oxidase is a soluble enzyme, easily purified from milk; it contains molybdenum, iron–sulfur groups, and FAD.[49] In the presence of oxygen, xanthine oxidase oxidizes both hypoxanthine to xanthine and xanthine to uric acid, and reduces oxygen to superoxide anion radical and hydrogen peroxide.

In the absence of oxygen, may xenobiotics are reduced by the enzyme. Tatsumi et al.[50] demonstrated the reduction of nitrobenzene derivatives to the corresponding hydroxylamines by the xanthine–xanthine oxidase system in the absence of oxygen. Clarke and colleagues[51,51a] studied the reduction of nitroimidazoles by xanthine oxidase and by reduced flavin mononucleotide ($FMNH_2$). Metronidazole, misonidazole, and other nitroimidazoles were reduced by $FMNH_2$, as measured by the formation of FMN after addition of the nitro compound. The stoichiometry of the reaction was consistent with reduction to the hydroxylamine.

Chrystal, Koch, and Goldman[52] characterized a number of products formed by the reduction of [$^{14}$C] metronidazole by the hypoxanthine–xanthine oxidase system. The product mixture was separated by ion exchange column chromatography, and the fractions were analyzed by high pressure liquid chromatography (HPLC) and TLC. Several metabolites were identified by comparison with authentic standards: *N*-(2-hydroxyethyl)oxamic acid, *N*-glycoylethanolamine, *N*-acetylethanolamine, ethanolamine, acetate, acetamide, and glycine. These small molecules arise from the cleavage of the imidazole ring. A mechanism for the nonenzymatic breakdown of reduced metronidazole can be drawn that incorporates the proposal of Koch and Goldman[48] and the known chemistry of enamines (Scheme 2).[53] Tautomerism of 5-aminoimidazole derivatives would result in an unstable imine. Hydrolysis followed by deamination yields a 5-ketoimidazole, and spontaneous cleavage of this species leads to the observed aliphatic products. A similar mechanism was proposed to explain the formation of open-chain nitriles from the nitrofurans nifuradene[54] and nitrofurazone.[55] The validity of this mechanism must be doubted, however, in light of the report that the amine derivative of metronidazole is an isolatable, stable compound (ref. 55a, see later). Possibly an analogous mecha-

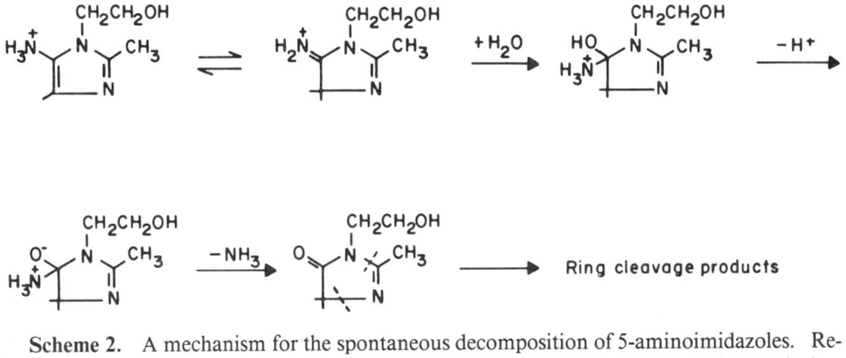

Scheme 2. A mechanism for the spontaneous decomposition of 5-aminoimidazoles. Reduction of the nitroimidazole yields an amine, and tautomeric rearrangement gives an imine. Hydrolysis, deprotonation, and deamination yields an unstable ketoimidazole, which cleaves in more than one position to give aliphatic fragments.

nism occurs, but involves the decomposition of the hydroxylamine rather than the amine.

Reduction of misonidazole by the xanthine–xanthine oxidase system yielded a single major product, as analyzed by TLC and HPLC.[56] The product was colorless and very polar and was formed with four-electron stoichiometry (determined by measurement of conversion of xanthine to uric acid). Mass spectrometry gave a parent peak at $m/z = 187$, consistent with hydroxylaminomisonidazole, but NMR spectroscopy failed to confirm the structure.[42]

One group of investigators[56a] suggests that the misonidazole reduction product isolatable at neutral pH is a ring-opened or otherwise rearranged isomer of the hydroxylamine. This product may undergo further transformation; the dialdehyde glyoxal has been identified as a product of xanthine oxidase-catalyzed reduction of misonidazole.[57,57a]

## V. Free-Radical Intermediates in the Enzymatic Reduction of Nitro Compounds

There are three potential free-radical intermediates of enzymatic nitro-reduction.

| | | | |
|---|---|---|---|
| 0 | Ar—$NO_2$ | | Nitro |
| 1 | Ar—$\dot{NO}_2^-$ | | Nitro anion radical |
| 3 | Ar—$\underset{H}{NO}$ | | Hydronitroxide radical |
| 5 | Ar—$\dot{NH}_2^+$ | | Amino cation radical |
| 6 | Ar—$NH_2$ | | Amine |

Again, the number denotes the reducing equilavents needed to form each product. The results of electron spin resonance (ESR) studies as well as indirect chemical and biochemical evidence suggest the formation of the nitro anion and the hydronitroxide free-radical intermediates in the reduction of nitro compounds by the oxygen-sensitive nitroreductases, which do not form diamagnetic products in the presence of air.[4,5] No evidence for formation of the amino cation free radical in nitroreductase incubations is known to the authors.

The first intermediate in this scheme is the nitro anion free radical. The incubation of misonidazole with xanthine oxidase and hypoxanthine at physiological pH and in the absence of oxygen leads to a multiple-line ESR spectrum (Fig. 1A). Radical formation is dependent on hypoxanthine, which is the ultimate source of the extra electron (Fig. 1B). No signal is

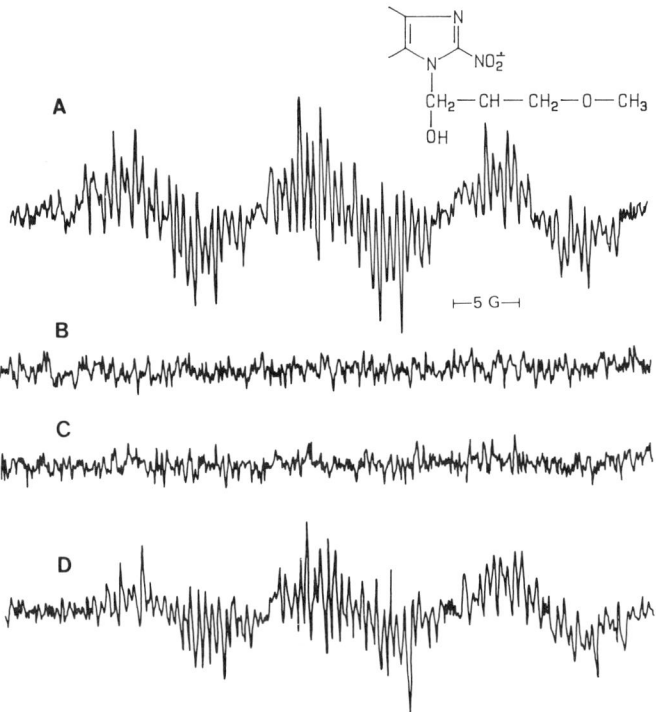

Fig. 1. Reduction of misonidazole by the hypoxanthine–xanthine oxidase system, neutral pH. (A) The incubation contained misonidazole (32 mM), hypoxanthine (32 mM), and xanthine oxidase (0.86 units/ml) under nitrogen, in phosphate buffer, pH 7.8. (B) No hypoxanthine. (C) Heat-denatured enzyme. (D) Initially oxygenated sample (oxygen is consumed by the oxidation of hypoxanthine). Instrumental conditions: modulation amplitude, 0.33 G; scan time, 16 min; time constant, 1 sec; nominal microwave power, 20 mW.

detected with heat-denatured enzyme (Fig. 1C). Although a lack of oxygen is necessary to obtain detectable levels of the nitro anion free radical, the ESR flat cell is a closed system, and the oxidase activity of xanthine oxidase will deoxygenate the incubation in seconds (Fig. 1D). Xanthine oxidase has good activity at basic pH values, and the decay of nitro anion free radicals is acid catalyzed[58,59]; therefore, much better spectra are obtained at pH 10 (Fig. 2A). A computer simulation of the 324-line spectrum identifies the atomic structure of the free radical (Fig. 2B) and confirms that the metabolite is simply misonidazole plus an extra electron. Results of electron spin resonance studies of the NADPH-supported microsomal reduction of a wide variety of nitro compounds and drugs, including nitrobenzene,[29,60] chloramphenicol,[4,5,61] p-nitrobenzoic acid,[29,62] nitrofurantoin,[63,64] nitrofu-

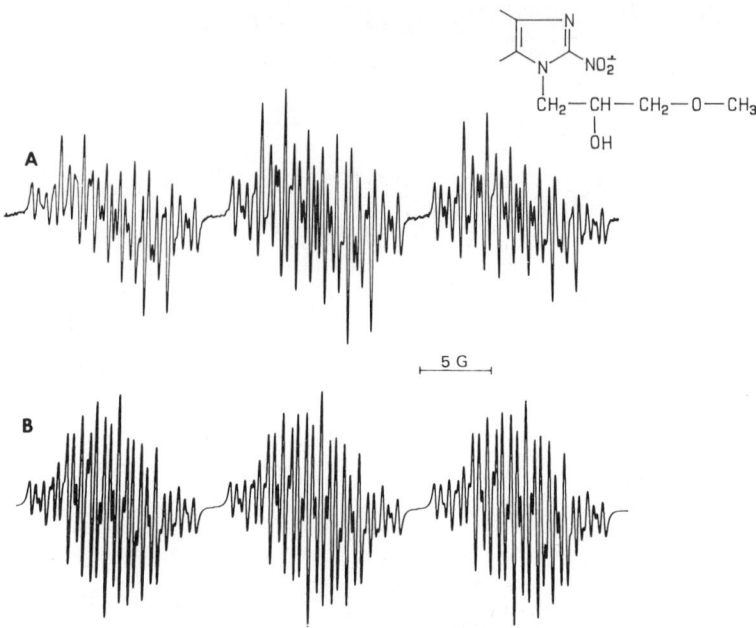

**Fig. 2.** Reduction of misonidazole by the hypoxanthine–xanthine oxidase system, pH 10. (A) Experimental spectrum. The system contained misonidazole (67 m$M$), hypoxanthine (3.3 m$M$), and xanthine oxidase (0.12 units/ml) in carbonate buffer, pH 10. Instrumental conditions: modulation amplitude, 0.13 G; scan time, 30 min; time constant, 0.5 sec; nominal microwave power, 20 mW. (B) Computer simulation of spectrum of misonidazole anion free radical. Hyperfine splitting constants were as follows: $a_{NO_2}^N = 14.22$ G; $a_{ring}^N = 1.65, 0.98$ G; $a_{ring}^H = 0.66, 0.42$ G; $a_{CH_2}^H = 2.65$ G.

razone,[55,64] AF-2,[65,66] nifurtimox,[67,68] FANFT,[60] metronidazole,[4,60,69] and ronidazole,[69] have demonstrated unambiguously the presence of the anion free-radical metabolite. If the nitro anion radical is an intermediate in the nitroreductase pathway, the rate of formation of the radical should be equal to or greater than the rate of formation of the amine product, as has been found with $p$-nitrobenzoate.[29]

After establishing that the anion radical was an obligate metabolite of the microsomal nitroreductase pathway, the nature of the source of the single electrons that are transferred to the nitro compounds was investigated. A number of components of the microsomal preparation could transport the single electron necessary to produce the nitro anion radical; the main candidates are cytochrome $P$-450 and its flavin-containing reductase NADPH–cytochrome $P$-450 reductase. Carbon monoxide does not significantly affect the steady-state concentration of the metronidazole nitro anion radical[69] or the other nitro anion radical.[29,60,68] This result suggests that the nitro compounds are reduced by the NADPH-cytochrome $P$-450($c$) reductase.[29,70-72] The anion radical is not detected in the presence of oxygen.

## VI. Catalytic, Radiolytic, and Electrochemical Reduction of Nitroimidazoles

Reduction of misonidazole by catalytic hydrogenation has been used by several groups. Flockhart et al.[73] used Adams' catalyst in methanol. The amine derivative was isolated by TLC and characterized by mass spectrometry and (as the HCl salt) NMR spectroscopy. A variety of derivatization techniques has been used to study the amine derivative of misonidazole, which has no useful optical absorption band. These methods include reactions with 2,4-dinitro-1-fluorobenzene,[73] trifluoroacetylation,[74] and dansylation.[75] Catalytic hydrogenation of metronidazole with palladium on charcoal gave the amine derivative in good yield.[55a] The product was characterizied by NMR and mass spectrometry; surprisingly, the amine appeared to be stable over a period of hours in aqueous solution.

Radiation reduction has proved to be a valuable technique for the study of nitroimidazole reduction chemistry. Irradiation of misonidazole by $^{60}$Co $\gamma$ or 25-MeV electrons in deoxygenated neutral aqueous solutions containing formate (which scavenges oxidizing radiolysis products and yields a purely reducing system) produced a colorless product[76] identical to that obtained enzymatically.[56] In unbuffered solutions, a rise in pH occurred during radiolysis, and azo and azoxy dimers were formed.[76] Radiolytic reduction of 2-nitroimidazole consumed six reducing equivalents: 2-aminoimidazole was isolated and identified by NMR spectroscopy.[77] In contrast, reduction of 1-methyl-2-nitroimidazole consumed only four equivalents.[77] Thus, N-1 substitution alters the chemistry of 2-nitroimidazole derivatives markedly; the 1-methyl-substituted derivative is also much more mutagenic than 2-nitroimidazole itself in the *Salmonella* assay.[77] One study[56a] investigated the radiolytic reduction of misonidazole at pH 4 and pH 7. At acid pH, the reduction consumes four reducing equivalents and apparently yields the hydroxylamine: NMR spectroscopy confirmed the presence of the imidazole ring protons. The product has a half-life of about 15 hr at pH 4 but decomposes rapidly at neutral pH, with loss of the imidazole ring proton NMR signals.[56a] This suggests that the product ($m/z = 187$) obtained in earlier studies[56,76] at neutral pH is an isomeric aliphatic rearrangement product of the hydroxylamine. The mechanism of base-catalyzed rearrangment is unclear.

Electrochemical reduction of misonidazole gave results comparable to those obtained with enzymatic and radiolytic systems.[78] Electrochemical reduction of metronidazole has been reported[78a]; the nitro group was found to be cleaved from the imidazole ring and was recovered as inorganic nitrite.

## VII. Coreduction of Nitroimidazoles with Macromolecules

The difficulty of isolating reduced products may be circumvented by coreduction techniques, in which the intermediates are "trapped" by reaction with macromolecules. Reduction of metronidazole with dithionite gave four-electron stoichiometry.[51,79] Coreduction of [$^{14}$C] metronidazole by dithionite caused binding to DNA, which was apparently specific to guanine and cytosine residues.[79] Electrochemical reduction of metronidazole or misonidazole in the presence of DNA caused strand breakage.[80,81] Zinc reduction of misonidazole resulted in DNA and protein binding.[46] Electrochemical reduction of misonidazole in the presence of DNA caused release of thymine, thymidine, and thymidine phosphates, but other bases were unaffected.[82]

McManus and colleagues[83] studied the activation of misonidazole to a species capable of binding to macromolecules (mainly protein) in microsomal preparations from rat liver. The binding was NADPH dependent, carbon monoxide insensitive, and strongly inhibited by glutathione. Activation by purified NADPH–cytochrome $P$-450($c$) reductase generates metabolites that bind to bovine serum albumin. These results confirm the role of NADPH–cytochrome $P$-450($c$) reductase noted earlier.[29]

## VIII. Nonenzymatic Reactions of the Nitro Anion Free Radical

The enzymatic formation of reactive nitro anion free-radical metabolites may lead to toxic effects. Three chemical reactions of nitro free radicals could be of importance in the toxicity of these compounds (Scheme 3). The first is the reaction of the free radical with itself, that is, disproportionation. This might by a detoxification reaction, because the free radical destroys itself. Alternatively, disproportionation could be on the pathway to a more toxic species, such as the nitroso product of disproportionation.

A second possible important reaction is the presumed covalent binding of the free-radical metabolite to tissue macromolecules. A highly reactive transient reductive metabolite of nitro drugs, such as metronidazole, nitrofurazone, and nitrofurantoin, binds covalently to macromolecules, espe-

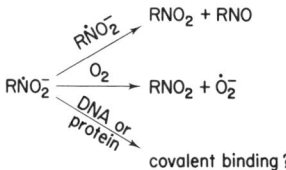

**Scheme 3.** Reactions of the nitro anion free radical. Upper, disproportionation; middle, reduction of oxygen; and lower, covalent binding.

cially proteins.[5,45] The anion free radical has received consideration as the reductive metabolite that binds covalently.[4,5,8,84] In addition, many nitro compounds are mutagens and carcinogens; the role of the nitro anion radical in these processes is not usually considered.

The third reaction, air oxidation of nitro anion free radicals, is the dominant pathway in the presence of oxygen. With the exception of superoxide, free radicals are not usually substrates for enzymes. Most react spontaneously, so that these three chemical reactions may be important not only in the context of the nitroreductase pathway but also in the mechanisms of toxicity of the nitro xenobiotics.

It is generally accepted that mammalian nitroreductases are inhibited by oxygen. Indeed, the results of numerous studies have shown that the rates of disappearance of nitro compounds and appearance of hydroxylamine or amine products are markedly lowered, if not reduced to zero, in the presence of air. The rates of disappearance of nitro compounds and drugs as diverse as $p$-nitrobenzoate, nitrofurazone, and niridazole have all been reported to be inhibited by oxygen.[4,5] The effect of oxygen results from its reaction with the free-radical product and is not related to enzyme inhibition.[29,62] A mechanism for this interaction is depicted in Scheme 4. The rapid air oxidation of the nitroaromatic anion radical is the pivotal event; there is no net reduction of the nitro compound, because the parent compound is regenerated. Because the absence of net reduction of the nitro compound in the presence of oxygen is due to a chemical reversal of the enzymatic reduction, and not the result of a direct action of oxygen on the enzyme itself, we will use the term "oxygen-sensitive" nitroreductase in quotation marks.

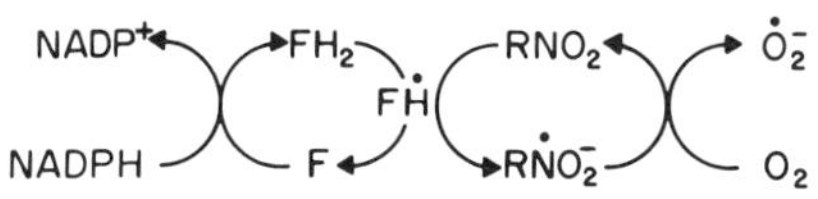

**Scheme 4.** The futile cycle of oxygen reduction, NADPH oxidation, and superoxide formation catalyzed by nitroaromatics. The exact mechanism of the one-electron transfer from flavoproteins to the nitro substrates is unknown; the species F, FH, and $FH_2$ are used figuratively. (From ref. 69 with permission.)

Nitro compounds are therefore catalysts of superoxide production. Superoxide anion free radical is toxic to microbes and may be toxic to mammals, including humans.[4,5,85] For example, paraquat poisoning and the resulting pulmonary fibrosis is probably an oxidative stress resulting from paraquat-catalyzed superoxide formation or the concomitant NADPH depletion, or both,[4,5] although some evidence, such as paraquat induction of lipid peroxidation[86] and the effect of superoxide dismutase *in vivo*,[87] is still controversial. Nitrofurantoin-catalyzed reduction of oxygen to superoxide and hydrogen peroxide may be responsible for some of the toxic effects that occur during nitrofurantoin therapy.[62] The occasional cases of pulmonary edema and fibrosis caused by nitrofurantoin are similar to the effects of paraquat poisoning, apparently because both compounds catalyze superoxide formation by flavoproteins.[88-90]

Nitro anion radicals cannot be detected in nitroreductase incubations containing oxygen, consistent with air oxidation of the nitro anion radicals; alternatively, oxygen might inhibit the enzymatic formation of the radical, possibly as a competitive electron acceptor. Considerable evidence supports the oxidation of the anion radical by oxygen.[4,5] Spin-trapping[64,69] has been used to overcome the fact that superoxide anion does not have a detectable ESR spectrum at room-temperature. The nitrone spin-trap DMPO (dimethylpyrroline *N*-oxide) reacts with superoxide, the hydroxyl radical, and other free radicals to form nitroxide spin adducts.[91] The ESR spectrum of an aerobic rat liver microsomal incubation containing ronidazole, DMPO, and NADPH is shown in Fig. 3. The DMPO-superoxide radical adduct will always decompose into the DMPO-hydroxyl radical adduct[92]; the ESR spectrum shown in Fig. 3 is a composite of both species. The outer two lines are due to the DMPO-hydroxyl radical adduct. Most of the remaining lines (the dotted lines) are due to the 12-line spectrum of DMPO-superoxide, which is spectroscopically distinct and can be suppressed totally by superoxide dismutase. The DMPO-superoxide adduct was observed only until the oxygen was consumed, as evidenced by the appearance of the metronidazole nitro anion free radical and the disappearance of the DMPO-superoxide spectrum.[69] To obtain this result, the magnetic field was fixed at the position indicated by the arrow in Fig. 4A (bottom, left-hand side). At this magnetic field, the DMPO-superoxide spectrum (the dotted lines) had a negative amplitude, and the metronidazole free-radical spectrum had a positive amplitude. When the reaction was initiated by the addition of NADPH (Fig. B), the signal amplitude decreased as a result of the formation of the DMPO-superoxide radical adduct. After reaching a maximum, the concentration of the DMPO-superoxide radical adduct slowly decreased for 13 min, when it was rapidly replaced by the metronidazole anion radical as shown by the positive deflection. This result support

# 16. NITROIMIDAZOLES

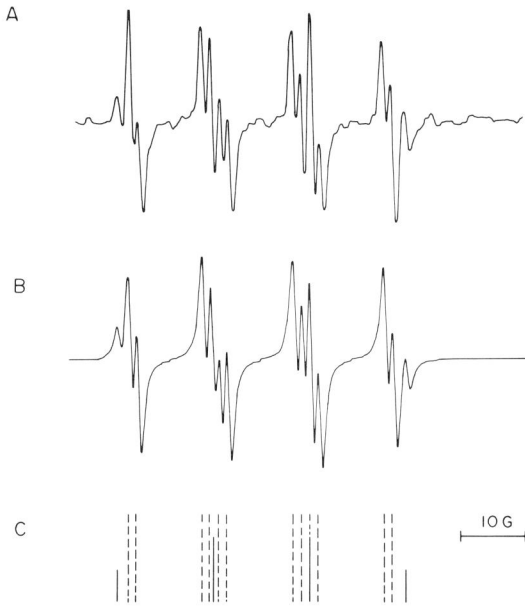

**Fig. 3.** (A) The ESR spectrum of the DMPO–superoxide spin adduct. The incubation contained 2 mg/ml microsomal protein, 10 m$M$ ronidazole, 80 m$M$ DMPO, and the NADPH-generating system. The nominal power was 20 mW, and the modulation amplitude was 0.33 G. (B) A computer simulation of the ESR spectrum shown in (A) with a composite of the DMPO–superoxide spin adduct and the DMPO–hydroxyl spin adduct in the ratio of 5.4:1. (C) The stick diagram demonstrates that at the field position used for kinetic experiments, the spectrum of the DMPO–hydroxyl spin adduct (solid lines) does not contribute to the spectrum of the DMPO–superoxide spin adduct (dotted lines). (From ref. 69 with permission.)

the major prediction of the mechanism of the oxygen "inhibition" of nitro reductases that superoxide anion will be generated in aerobic nitroreductase incubation.

Nitro anion radical formation in the presence of oxygen will not result in the corresponding formation of the ultimate products of nitroreduction, the amine and hydroxylamine compounds. The absence of these reductive metabolites *in vitro* or *in vivo* does not indicate that nitro reduction to the anion radical has not occurred. *In vitro* and whole-animal studies that show no net formation of products may be misleading in ascertaining the importance of free-radical intermediates, because futile metabolism is characteristic of many classes of "phantom" free radicals.[4-6]

The reduction of nitrofurazone by either the "oxygen-sensitive" *Escherichia coli* reductase or the many "oxygen-sensitive" mammalian nitrore-

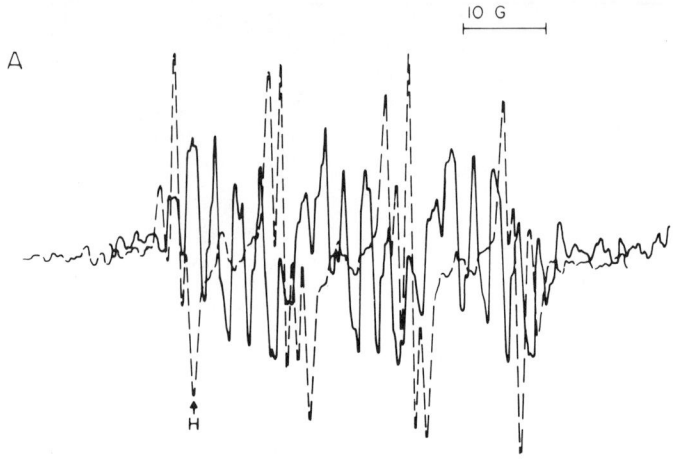

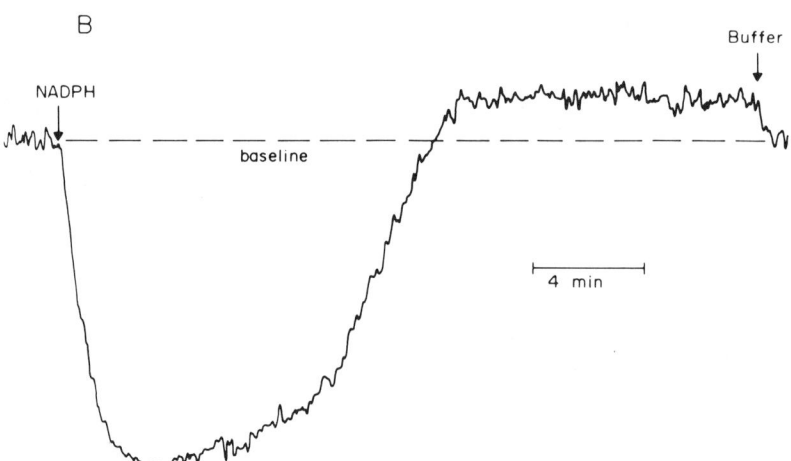

Fig. 4. (A) The superimposition of the ESR spectra of the metroniazole anion radical upon the DMPO-superoxide spin adduct. At the magnetic field position indicated by the arrow, it can be seen that the metronidazole radical has a positive deflection and that the DMPO-superoxide spin adduct has a negative deflection. The DMPO-hydroxyl spin adduct does not absorb at this field position. (B) The time course of the changes in the ESR signal amplitudes at the field position indicated in (A). At first, the DMPO-superoxide spin adduct is formed causing a negative amplitude. The concentration of the DMPO-superoxide spin adduct begins to decrease as the oxygen in the sample is consumed. The concentration of the metronidazole anion radical begins to increase eventually as shown by the positive deflection. (From ref. 69 with permission.)

ductases is shown in Scheme 5. Nitroaromatic anion radicals disproportionate spontaneously in aqueous solutions.[58,59] The kinetics of the radical decay also follow a second-order process in oxygen-free nitroreductase incubations. If the rate of free-radical formation is linear with the enzyme concentration, the steady-state radical concentration will be proportional to the square root of the enzyme concentration.[4,5] This square root relationship has been observed with a variety of nitroreductase and nitro substrates.[55,62,64,68] Because the derivation of this relationship requires that the nitro anion radicals decay by a second-order termination reaction and is consistent with spontaneous disproportionation of nitro radical anions to form the corresponding nitroso compound, these data are indirect evidence of nitroso compound formation. On the other hand second-order kinetics can also result from dimerization, and nitrovinyl compounds form dimers in nitroreductase incubations.[93]

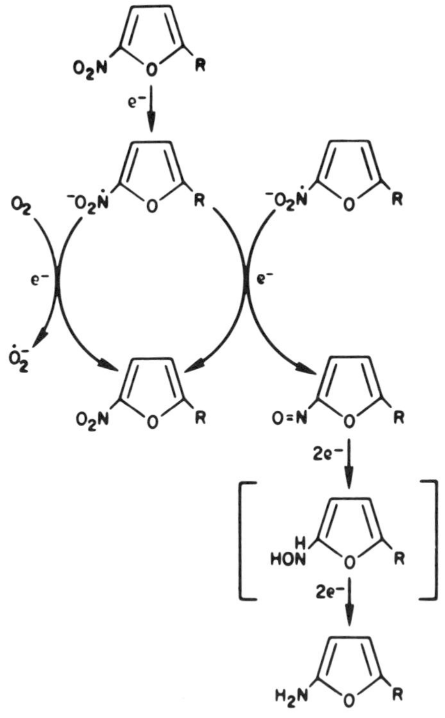

Scheme 5. The anion radical is the first intermediate formed in the reduction of nitrofurans and other nitro compounds by the oxygen-sensitive nitroreductases. This free radical either disproportionates or is oxidized by oxygen. Disproportionation forms the corresponding nitroso compound, which is usually reduced to the corresponding hydroxylamine or amine. (From ref. 55 with permission.)

The nitro anion radical is unlikely to behave as an electrophilic reactant that binds to tissue macromolecules. The steady-state concentrations of the nitrofuran (C. F. Polnaszek, R. P. Mason, F. J. Peterson, and J. L. Holtzman, unpublished) or misonidazole (P. D. Josephy and R. P. Mason, unpublished observations) anion radicals formed in enzymatic systems are unaffected by 100 m$M$ glutathione. The reaction of nitro anion radicals with cysteine is too slow for measurement by pulse radiolysis techniques (second-order rate constant less than $10^4$ $M^{-1}$ sec$^{-1}$; P. O'Neill and E. M. Fielden, personal communication). In the presence of iron salts,[94] sulfhydryl compounds reduce metronidazole and misonidazole to unidentified products, but this reaction may not involve a nitro anion free-radical intermediate. Although the oxygen-insensitive nitroreductase in *E. coli* activates nitrofurazone to an alkylating mutagenic species,[2,3] this enzyme does not form the nitro anion free radical (Fig. 5B). Under identical conditions, the "oxygen-sensitive" *E. coli* nitroreductase forms the radical (Fig. 5A). These incubations have

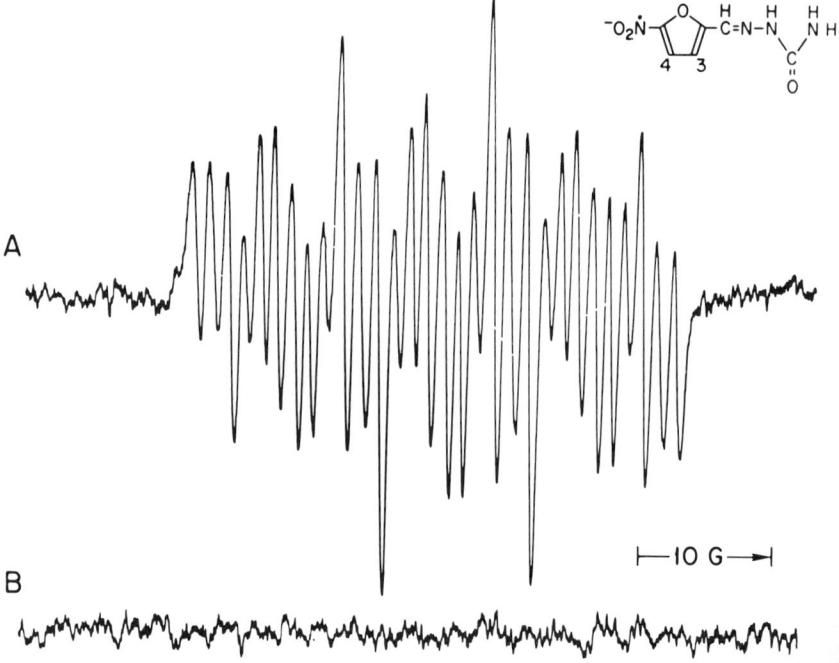

Fig. 5. The ESR spectrum of the nitrofurazone anion free radical (A) observed in an oxygen-free incubation of 1 m$M$ nitrofurantoin with an NADPH-generating system and 1 mg/ml of the oxygen-sensitive *E. coli* reductase. With identical conditions and nitroreductase activity, incubations of the oxygen-insensitive reductase (B) do not give a detectable ESR signal. (From ref. 61 with permission.)

equal nitrofurazone disappearance rates (i.e., equal nitroreductase activity); therefore, the rates of formation of the radical should have been equal, if the radical pathway was operative in both incubations. Apparently, the nitro anion radical is neither the primary alkylating nor the primary mutagenic species formed by nitro reduction; nevertheless, the nitro anion radical is an obligatory intermediate in the formation of these species by all of the "oxygen-sensitive" nitroreductases.

## IX. Other Free-Radical Metabolites of Nitro Compounds

The three-electron reduction product of a nitro compound is the hydronitroxide. (The prefix hydro refers to the $\alpha$-hydrogen attached to nitrogen.) Although nitroxides are a relatively stable class of free radicals, the hydronitroxide has not been observed in nitroreductase incubations. However, nitroso compounds (the two-electron reduction products) can be reduced to the corresponding hydronitroxide by mild reducing agents. The phenylhydronitroxide is formed by the reduction of nitrosobenzene by ascorbate, epinephrine,[95] or isoproterenol.[96]

The reduction of nitroso compounds in microsomal systems has been reported. The reduction of *tert*-nitrosobutane in the presence of oxygen with rat liver microsomes and NADPH results in the formation of *tert*-butyl hydronitroxide.[97] The concentration of this free radical was not in steady state (as is the case with nitro anion free radicals) but increased for more than 30 min. Superoxide dismutase completely inhibited the formation of this free radical, indicating that, in this case, superoxide anion is the one-electron donor.[97] The enzymatic formation of hydronitroxides from nitroso compounds by flavoproteins has not been reported, but enzymatic nitroso reduction is known,[98,99] and enzymatic hydronitroxide formation is to be expected. The biological activities of hydronitroxides are unknown, but in practice they would be difficult to distinguish from the biological activities of hydroxylamines, to which the hydronitroxides can be converted by reducing agents, such as ascorbate.

Ten years ago, a nitroxide metabolite of 4-nitroquinoline *N*-oxide was found in chloroform–methanol extracts of hepatic microsomal incubations.[100] Subsequently, this nitroxide has been shown not to be the hydronitroxide metabolite, but rather a secondary free radical.[101-105] These secondary free radicals probably form via an ene reaction of a nitroso compound with the unsaturated fatty acids of microsomal phospholipids (Scheme 6). An alternative mechanism is the spin-trapping of the primary hydronitroxide by a variety of compounds containing carbon–carbon double bonds.[103]

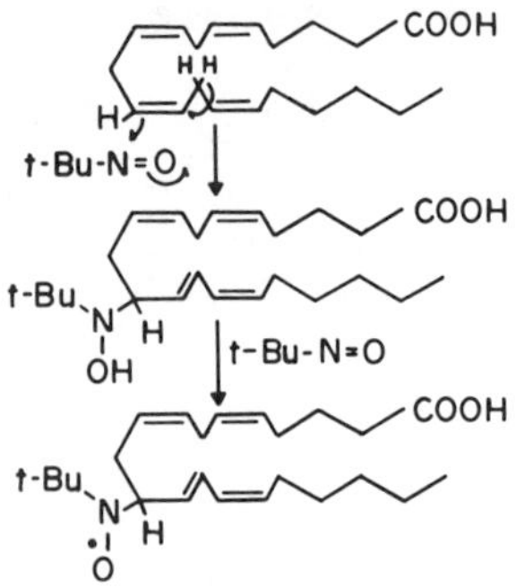

**Scheme 6.** The addition of *tert*-nitrosobutane to the C-11 double bond of arachidonic acid by an ene mechanism. The resulting hydroxylamine can be oxidized under nitrogen by *tert*-nitrosobutane to form the fatty acid nitroxide. (From ref. 97 with permission.)

The observed correlation between secondary nitroxide formation and carcinogenicity of the aromatic amines suggests that these radicals may be involved in tumor initiation or promotion.[103] The basis of this correlation is unclear, because the nitroxide formed from methyl oleate and nitrosofluorene is much less mutagenic than is nitrosofluorene itself.[106]

The nonenzymatic formation of nitroxide radicals in biological systems was also noted in studies of the mechanism of prostaglandin biosynthesis from arachidonic acid. The nonenzymatic reaction of *tert*-nitrosobutane with arachidonic acid will form a fatty acid-containing nitroxide at room temperature under either a nitrogen or an air atmosphere (Fig. 6). Because the addition reaction might take place across any of the four double bonds of arachidonic acid, as many as eight radicals are possible where both $\beta$ substituents are $C_4$ or longer. Either one radical predominates or all of the radicals have very similar ESR spectra, as is consistent with the spectra of related compounds.[104] The $\beta$-hydrogen responsible for the small hydrogen–hyperfine interaction in the secondary nitroxide spectrum does not exchange with the hydrogens of water; therefore, this spectrum will be identical in deuterium oxide buffers. In contrast, the hydronitroxide (the primary nitroxide metabolite) has a larger hydrogen–hyperfine interaction, which is due to an exchangeable hydrogen. In deuterium oxide buffer, the $\alpha$-hydrogen will exchange with deuterium to give a different spectrum, which is related to the first spectrum by the ratio of the deuterium to hydrogen nuclear moments and their respective nuclear spins. Thus, the spectrum of the primary hydronitroxide metabolite is dramatically different in deuterium oxide buffer (Fig. 7), whereas the secondary nitroxides will have ESR spectra that are unchanged in deuterium oxide buffer. This effect should be of general utility in determining the structure of nitroxides in enzymatic incubations, especially those containing unsaturated fatty acids.

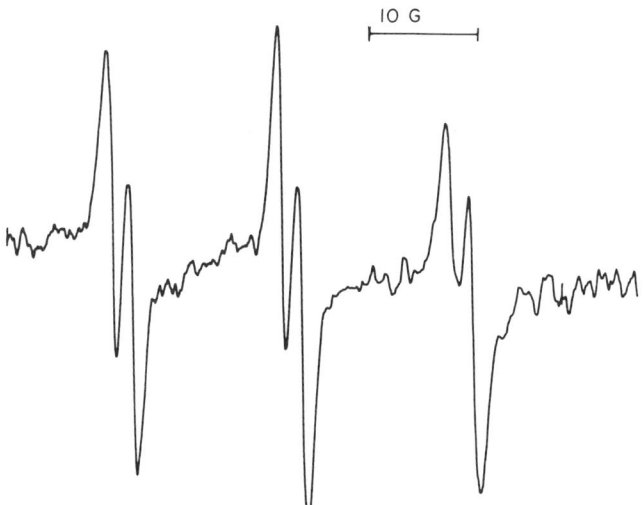

Fig. 6. The ESR spectrum of the nitroxides formed by the reaction of equal volumes of arachidonic acid (50 mg/ml) in absolute ethanol with *tert*-nitrosobutane (1 mg/ml) in Tris buffer (150 m$M$, pH 9.0) under nitrogen at 22°C. (From ref. 55 with permission.)

The five-electron reduction product, the amino cation free radical, is known as the one-electron oxidation product of aromatic amines, but has not been observed in nitroreductase incubations. Both horseradish peroxidase[107] and prostaglandin hydroperoxidase[108] oxidize aromatic amines to amino cation free radicals.

$$R-NH_2 \xrightarrow{-e^-} R-\overset{\cdot}{N}\overset{+}{H_2}$$

These free radicals might be formed by the enzymatic reduction of hydroxylamine compounds or of closely related species, such as the protonated hydroxylamine or its dehydration product, the nitrenium ion. Cytochrome P-450 is an oxygen-sensitive reductase for both phenylhydroxylamine[72] and N-hydroxy-2-acetylaminofluorene.[109] An NADH-dependent, oxygen-insensitive N-hydroxylamine reductase has been isolated from pig liver microsomes.[110] Ascorbate does not reduce N-hydroxy-2-acetylaminofluorene to 2-acetylaminofluorene but does reduce the corresponding nitrenium ion to a free radical,[111,111a] which is the conjugate base of the amino cation radical.

$$R-\overset{+}{N}-Ac + AH_2 \rightarrow R-\overset{\cdot}{N}-Ac + AH^- + H^+$$

Furthermore, Andrews *et al.* found that ascorbate diminished covalent binding to protein[112] and DNA,[113] but increased mutagenesis 12-fold,[112,113] which suggests that the free radical, not the nitrenium ion, may be the ultimate carcinogen.

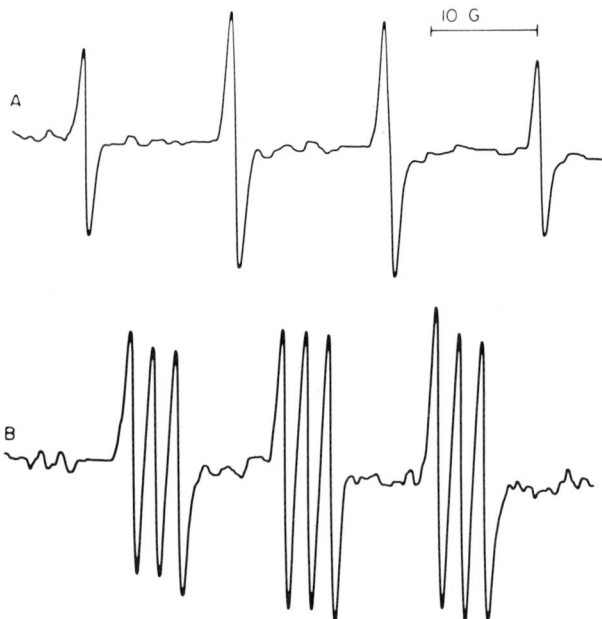

**Fig. 7.** The ESR spectrum of *tert*-butyl hydronitroxide (A) obtained on reduction of *tert*-nitrosobutane with sodium borohydride in water. The ESR spectrum of *tert*-butyl deuteronitroxide (B) obtained on reduction of *tert*-nitrosobutane with sodium borohydride in deuterium oxide. (From ref. 97 with permission.)

## X. Bacterial Metabolism of Nitroimidazoles

Searle and Willson[114] observed reduction of metronidazole by bacteria from rat cecal contents. Reduction was "oxygen sensitive" and was inhibited by *N*-ethylmaleimide and by iron chelators; it was suggested that an iron–sulfur enzyme is responsible for drug reduction. Goldstein and co-workers[115] studied the metabolism of [1-methyl-$^{14}$C]-2-nitro-5-vinylimidazole by *E. coli*. The parent drug (assayed spectrophotometrically) was removed almost completely over a period of 90 min. Some of this material became associated with the trichloroacetic acid (TCA)-precipitable (macromolecular) fraction, increasing to a maximum of about 3% of the total activity. The supernatant fraction remaining after centrifugation of the cells was lyophilized and analyzed by TLC. In a system of chloroform–methanol (9:1), loss of the parent drug was observed; all the metabolites remained in a single peak at the origin. Rechromatography of this material in a more polar solvent system (methanol–acetic acid, 95:1) showed that a least three components were present. These were not identified, but the authors suggested that they were reduced derivatives of the drug. The reduction of metronidazole by rat cecal contents in the absence of oxygen was confirmed by Goldman and colleagues, and several metabolites were separated by HPLC. One of the compounds was identified as N-(2-hydroxyethyl)oxamic acid,[48] which was also obtained by zinc reduction of the drug. This metabolite was recovered from the urine of rats, given [$^{14}$C]metronidazole (about 1% of total radioactivity). Because none of this metabolite was detected in the urine of similarly treated germ-free rats, the bacterial flora appear to be responsible for its formation. Acetamide was identified as another product of metronidazole reduction by either rat cecal contents or by *Clostridium perfringens* cultures.[74] Again, the metabolite was present in the urine of conventional, but not germ-free rats. Both acetamide and N-(2-hydroxyethyl)oxamic acid were present in the urine of patients receiving metronidazole.[116] Recently, the protozoa *Entamoeba histolytica* and *Trichomonas vaginalis* were shown to reduce metronidazole to acetamide.[117] Misonidazole is reduced by rat cecal contents in the absence of oxygen.[118] Aminomisonidazole was recovered from the urine of conventional, but not germ-free, rats given [$^{14}$C]misonidazole (less that 5% of dose). The amine is futher metabolized to carbon dioxide, apparently via imidazole ring cleavage and urea formation.[118]

## XI. Metabolism of Misonidazole by Mammalian Cells *In Vitro*

The first study of misonidazole metabolism by mammalian cells was reported in 1976.[40] Suspensions of CHO (hamster) or KHT (mouse) cells were incubated with [$^{14}$C]misonidazole under hypoxic conditions or in the presence of oxygen. Several unidentified metabolites were resolved by paper chromatographic analysis of extracts of both aerobic and hypoxic cells; however, the hypoxic cells showed considerably higher levels of at least one metabolite. Taylor and Rauth[119] performed similar experiments with CHO and HeLa (human) cells. A steady accumulation of radioactivity within the cells occurred under hypoxia, but no increase over zero-time levels was observed in cells incubated in air. HeLa cells accumulated radioactivity more rapidly than did CHO cells and were also more susceptible to misonidazole toxicity. Several metabolites were detected by paper chromatographic analysis of cell extracts from hypoxic incubations, but there was virtually no product formation in the presence of oxygen. These results were extended in a second report[120] in which the effects of ascorbate and sulfhydryl agents on *in vitro* metabolism of misonidazole were studied.

Again, metabolism was observed only under hypoxic conditions; even when cells were incubated with very high levels of drug (75 m$M$) in the presence of oxygen such that aerobic toxicity was demonstrable, no misonidazole metabolites could be detected. Under hypoxia, the presence of ascorbate resulted in significantly greater accumulation of at least one misonidazole metabolite. This effect may be related to the enhancement of misonidazole toxicity by ascorbate reported earlier.[121,122] Josephy et al.[123] studied the metabolism of misonidazole in CHO cells with labeled drug of higher specific activity and more extensive separation and analysis of metabolites. Very dense cell suspensions were used to maximize metabolism of the drug. Cell samples were disrupted, lyophilized, and extracted with ethyl acetate–methanol. The organic-soluble fraction was studied by TLC and HPLC; the organic-insoluble fraction was separated into TCA-soluble and -insoluble material, and each fraction was counted. Over a period of 3 hr, misonidazole was metabolized completely to very polar products (low $R_f$ on TLC). Activity was lost from the organic-soluble fraction and accumulated primarily in the organic-insoluble, acid-soluble fraction. This material might represent very polar conjugates or ring cleavage products. Radioactivity also became bound to the acid-insoluble pellet. HPLC analysis of the organic-soluble fraction showed the presence of a large number of metabolites, with total radioactivity increasing as a function of time. However, these organic-soluble products represent only a small fraction of the total radioactivity; possibly, they are short-lived and rapidly converted to more polar (organic-insoluble) material. By the use of a dual-labeling technique, it was shown that one of the organic-soluble products cochromatographed with the product of misonidazole reduction by the xanthine–xanthine oxidase system (probably an isomeric rearrangement product of the hydroxylamino derivative of misonidazole).[56a]

A study of misonidazole metabolism in both bacterial and mammalian (Chinese hamster lung fibroblast) cells has been reported.[124] $^{14}$C-Labeled drug was incubated with a confluent monolayer of fibroblasts for 24 hr under hypoxic conditions. Even after this length of time, only about half of the parent drug was metabolized. Several metabolites were characterized by cation exchange chromatography, thin-layer chromatography on cellulose plates, and reversed-phase HPLC. Aminomisonidazole accounted for 6% of metabolites formed by cecal bacteria and for 13% of metabolites in fibroblasts. Both systems yielded ring cleavage products. Urea accounted for approximately 20% of the radioactivity in bacteria and fibroblasts. The major metabolite in the cecal bacteria incubations was (2-hydroxy-3-methoxypropyl)guanidine, which was identified by mass spectrometry and NMR; this product results from a ring cleavage different from that which produces urea. The ratio of the yield of the guanidine derivative to the yield of urea was about 3:1 in bacteria and 1:1 in the mammalian cells. Appaently, the pattern of ring cleavages of a reduced misonidazole derivative is different in the two systems, perhaps suggesting that the process is enzyme catalyzed rather than spontaneous.

Although the study of the reductive metabolism of misonidazole has been hampered by the low rates of metabolism and the highly polar and unstable characteristics of the products, significant progress is now being made. The accelerating rate of publication in this field suggests that a detailed understanding of the chemical mechanisms of nitroheterocyclic toxicity is a reasonable goal.

## XII. Summary

The reductive metabolism of nitro compounds, once regarded as little more than a curiosity, is now recognized as the crucial process in the biotransformation of an important class of xenobiotics. Although attention was focused first on the nitrophenyls, other classes of nitrocompounds are now studied intensively, particularly nitroimidazoles, nitrofurans, and nitro polycyclics. One-electron reduction of nitro compounds to the nitro anion radical is catalyzed by oxygen-sensitive mammalian and bacterial enzymes.

Bacteria possess oxygen-insensitive nitroreductases, which do not form the anion radical and are crucial to the antimicrobial properties of nitro drugs. The diamagnetic products of nitroimidazole metabolism have proved elusive because of their instability, but much progress has now been made. Reductive activation of nitro compounds results in DNA strand breakage and adduct formation, protein binding, cytotoxicity, and mutagenesis. Elucidation of the chemical mechanisms responsible for these interactions should lead to the design of better chemotherapeutic agents and to the control of the potential hazards resulting from the widespread use of nitroaromatics in medicine and industry.

Note that a number of papers and abstracts from the conference on "Radiation and Cytotoxic Drugs," Key Biscayne, Florida, September 1981, may be of interest to the reader. [*Int. J. Radiat. Oncol.: Biol. Phys.* **8,** Nos. 3–4, March–April (1982).] Additional studies have appeared that may also be of interest to the reader.[125-132]

## Acknowledgments

We wish to thank Dr. A. M. Rauth, Dr. J. A. Raleigh, and Dr. D. R. McCalla for helpful discussions and access to unpublished results. We also thank Peggy Ellis for her patient assistance in the preparation of this manuscript. P. D. J. is a Research Fellow of the National Cancer Institute of Canada.

## References

1. Wardman, P. (1977). The use of nitroaromatic compounds as hypoxic cell radiosensitizers. *Curr. Top. Radiat. Res. Q.* **11**, 347–398.
2. McCalla, D. R. (1979). Nitrofurans. *Antibiotics (N.Y.)* **5**, 176–213.
3. McCalla, D. R. (1981). Metabolic activation of nitroheterocyclic compounds in bacteria and mammalian cells. *In* "Short-term Tests for Chemical Carcinogens" (H. F. Stich and R. H. C. San, eds.), pp. 36–47. Springer-Verlag, Berlin and New York.
4. Mason, R. P. (1979). Free radical metabolites of foreign compounds and their toxicological significance. *Rev. Biochem. Toxicol.* **1**, 151–200.
5. Mason, R. P. (1982). Free-radical intermediates in the metabolism of toxic chemicals. *In* "Free Radicals in Biology," (W. A. Pryor, ed.), Vol. 5, pp. 161–222. Academic Press, New York.
6. Kappus, H., and Sies, H. (1981). Toxic drug effects associated with oxygen metabolism: Redox cycling and lipid peroxidation. *Experientia* **37**, 1233–1241.
7. Greenstock, C. L. (1981). Redox processes in radiation biology and cancer. *Radiat. Res.* **86**, 196–211.
8. Biaglow, J. E. (1981). Cellular electron transfer and radical mechanisms for drug metabolism. *Radiat. Res.* **86**, 212–242.
9. Parris, G. E. (1980). Environmental and metabolic transformations of primary aromatic amines and related compounds. *Residue Rev.* **76**, 1–30.
10. Hlavica, P. (1982). Biological oxidation of nitrogen in organic compounds and disposition of N-oxidized products. *CRC Crit. Rev. Biochem.* **12**, 39–101.
10a. Monney, H., Parrick, J., and Wallace, R. G. (1981). Nitromidazole radiosensitizers: Approaches to their chemical synthesis. *Pharmacol. Ther.* **14**, 197–216.
10b. Smithen, C. E., and Hardy, C. R. (1982). The chemistry of nitroimidazole hypoxic cell radiosensitizers. *In* "Advanced Topics on Radiosensitizers of Hypoxic Cells" (A. Breccia, C. Rimondi, and G. E. Adams, eds.), pp. 1–47. Plenum, New York.
11. Rickert, D., ed. (1984). "Toxicity of Nitroaromatic Compounds." Hemisphere Publ., New York.
12. Maeda, K., Osato, T., and Umezawa, H. (1953). A new antibiotic, azomycin. *J. Antibiot., Ser. A* **6**, 182.
13. Goldman, P. (1980). Metronidazole. *N. Engl. J. Med.* **303**, 1212–1218.
14. Reuvers, A. P., Chapman, J. D., and Borsa, J. (1972). Potential use of nitrofurans in radiotherapy. *Nature (London)* **237**, 402–403.
15. Foster, J. L., and Willson, R. L. (1973). Radiosensitization of anoxic cells by metronidazole. *Br. J. Radiol.* **46**, 234–235.
16. Adams, G. E., and Dewey, D. L. (1963). Hydrated electrons and radiobiological sensitization. *Biochem. Biophys. Res. Commun.* **12**, 473–477.
17. Adams, G. E., Asquith, J. C., Dewey, D. I., Foster, J. L., Michael, B. D., and Willson, R. L. (1971). Electron-affinic sensitization. Part II. *para*-Nitroacetophenone: A radiosensitizer for anoxic bacterial and mammalian cells. *Int. J. Radiat. Biol.* **19**, 575–585.
18. Urtasun, R. C., Band, P., Chapman, J. D., Feldstein, M. L., Mielke, B., and Fryer, C. (1976). Radiation and high-dose metronidazole in supratentorial glioblastomas. *N. Engl. J. Med.* **294**, 1364–1367.
19. Urtasun, R. C., Band, P., and Chapman, J. D. (1977). Phase I study of the nitroimidazole Ro 07-0582. A specific radiosensitizer of hypoxic tumor cells. *Radiat. Res.* **70**, 70 (abstr).
20. Wasserman, T. H., Stetz, J., and Phillips, T. C. (1981). Radiation therapy oncology group clinical trials with misonidazole. *Cancer* **47**, 2382–2390.

21. Chapman, J. D. (1979). Hypoxic sensitizers—implications for radiation therapy. *N. Engl. J. Med.* **301**, 1429–1432.
22. Adams, G. E. (1981). Hypoxia-mediated drugs for radiation and chemotherapy. *Cancer* **48**, 696–707.
23. Chapman, J. D., Ngan-Lee, J., and Meeker, B. E. (1981). Mechanistic and pharmacological considerations in the design and use of hypoxic cell radiosensitizers. *In* "Molecular Actions and Targets for Cancer Chemotherapeutic Agents" (A. C. Sartorelli, J. S. Lazo, and J. R. Bertino, eds.), pp. 419–430. Academic Press, New York.
24. Adams, G. E., and Stratford, I. J. (1981). Hypoxia-dependent radiation sensitizers and chemotherapeutic agents. *In* "Molecular Actions and Targets for Cancer Chemotherapeutic Agents" (A. C. Sartorelli, J. S. Lazo, and J. R. Bertino, eds.), pp. 401–418. Academic Press, New York.
25. Sutherland, R. M. (1974). Selective chemotherapy of noncycling cells in an *in vitro* tumour model. *Cancer Res.* **34**, 3501–3503.
26. Moore, B. A., Palcic, B., and Skarsgard, L. D. (1976). Radiosensitizing and toxic effects of the 2-nitroimidazole Ro 07-0582 in hypoxic mammalian cells. *Radiat. Res.* **67**, 459–473.
27. Hall, E. J., and Roizin-Towle, L. (1975). Hypoxic sensitizers: Radiobiological studies at the cellular level. *Radiology* **117**, 453–457.
28. Kennedy, K. A., Teicher, B. A., Rockwell, S., and Sartorelli, A. C. (1980). The hypoxic tumor cell: A target for selective cancer chemotherapy. *Biochem. Pharmacol.* **29**, 1–8.
29. Mason, R. P., and Holtzman, J. L. (1975). The mechanism of microsomal and mitochondrial nitroreductase. Electron spin resonance evidence for nitroaromatic free radical intermediates. *Biochemistry* **14**, 1626–1632.
30. Fargher, R. G. (1920). Orientation of the nitro- and arylazoglyoxalines. Fission of the glyoxalone nucleus. *J. Chem. Soc.* **117**, 668–680.
31. Hunter, G., and Nelson, J. A. (1941). 4- (or 5-)aminoglyoxaline (iminazole). *Can. J. Res., Sect. B* **19**, 296–304.
32. Rabinowitz, J. C. (1956). Purine fermentation by *Clostridium cylindrosporum*. III. 4-Amino-5-imidazole-carboxylic acid and 4-aminoimidazole. *J. Biol. Chem.* **218**, 175–187.
33. Shealy, Y. F. (1970). Syntheses and biological activity of 5-aminoimidazoles and 5-triazenoimidazoles. *J. Pharm. Sci.* **59**, 1533–1588.
34. Taylor, E. C., and Loeffler, P. K. (1959). Studies in purine chemistry. V. 7-Methyladenine-3-oxide. *J. Org. Chem.* **24**, 2035–2036.
35. Burtles, R., and Pyman, F. L. (1925). 2-Amino-4:5-dimethylglyoxaline. *J. Chem. Soc.* **127**, 2012–2018.
36. Lancini, G. C., Arioli, V., Lazzari, E., and Bellani, P. (1969). Synthesis and relationship between structure and activity of 2-nitroimidazole derivatives. *J. Med. Chem.* **12**, 775–780.
37. Beaman, A. G., Tautz, W., and Duschinsky, R. (1967). Studies in the nitroimidazole series. III. 2-Nitroimidazole derivatives, substituted in the 1-position. *Antimicrob. Agents Chemother.* **7**, 520–530.
38. Coburn, M. D., and Neuman, P. N. (1970). The condensation of amino- and nitroimidazoles with picryl halides (1). *J. Heterocycl. Chem.* **7**, 1391–1393.
39. Stambaugh, J. E., Feo, L. G., and Manthei, R. W. (1968). The isolation and identification of the urinary oxidative metabolites of metronidazole in man. *J. Pharmacol. Exp. Ther.* **161**, 373–381.
40. Varghese, A. J., Gulyas, S., and Mohindra, J. K. (1976). Hypoxia-dependent reduction of 1-(2-nitro-1-imidazolyl)-3-methoxy-2-propanol by Chinese hamster ovary cells and KHT tumor cells *in vitro* and *in vivo*. *Cancer Res.* **36**, 3761–3765.

41. Kuhn, W. E. (1933). 2-Nitrofluorene and 2-aminofluorene. *Org. Synth.* **13**, 447–448.
42. Josephy, P. D. (1982). Chemical and biological studies of the radiosensitizer misonidazole. Dissertation Abstracts International, **42**(7), 2669B.
43. Josephy, P. D., Palcic, B., and Skarsgard, L. D. (1980). Synthesis and properties of reduced derivatives of misonidazole. *In* "Radiation Sensitizers and Protectors" (L. W. Brady, ed.), pp. 61–64. Masson, New York.
44. Fieser, L. F., and Fieser, M. (1967). "Reagents for Organic Synthesis," p. 1280. Wiley, New York.
45. Varghese, A. J., and Whitmore, G. F. (1980). Binding of nitroreduction products of misonidazole to nucleic acids and protein. *Cancer Clin. Trials* **3**, 43–46.
46. Varghese, A. J., and Whitmore, G. F. (1980). Binding to cellular macromolecules as a possible mechanism for the cytotoxicity of misonidazole. *Cancer Res.* **40**, 2165–2169.
47. Varghese, A. J., and Whitmore, G. F. (1981). Cellular and chemical reduction products of misonidazole. *Chem.-Biol. Interact.* **36**, 141–151.
48. Koch, R. L., and Goldman, P. (1979). The anaerobic metabolism of metronidazole forms $N$-(2-hydroxyethyl)oxamic acid. *J. Pharmacol. Exp. Ther.* **208**, 406–410.
49. Bray, R. C. (1975). Molybdenum iron–sulfur flavin hydroxylases and related enzymes. *In* "The Enzymes" (P. D. Boyer, ed.), 3rd ed., Vol. 12, pp. 300–414. Academic Press, New York.
50. Tatsumi, K., Kitamura, S., Yoshimura, H., and Kawazoe, Y. (1978). Susceptibility of aromatic nitro compounds to xanthine oxidase catalyzed reduction. *Chem. Pharm. Bull.* **26**, 1713–1717.
51. Clarke, E. D., Wardman, P., and Goulding, K. H. (1980). Anaerobic reduction of nitroimidazoles by reduced flavin mononucleotide and by xanthine oxidase. *Biochem. Pharmacol.* **29**, 2684–2687.
51a. Clarke, E. D., Goulding, K. H., and Wardman, P. (1982). Nitroimidazoles as anaerobic electron acceptors for xanthine oxidase. *Biochem. Pharmacol.* **31**, 3237–3242.
52. Chrystal, E. J. T., Koch, R. L., and Goldman, P. (1980). Metabolites from the reduction of metronidazole by xanthine oxidase. *Mol. Pharmacol.* **18**, 105–111.
53. Stamhuis, E. J. (1969). Hydrolysis of enamines. *In* "Enamines: Synthesis, Structure, and Reactions" (A. G. Cook, ed.), pp. 101–114. Dekker, New York.
54. Gavin, J. J., Ebetino, F. F., Freeman, R., and Waterbury, W. E. (1966). The aerobic degradation of 1-(5-nitrofurfuryl-ideneamino)-2-imidazolidinone (NF-246) by *Escherichia coli*. *Arch. Biochem. Biophys.* **113**, 399–404.
55. Peterson, F. J., Mason, R. P., Hovsepian, J., and Holtzman, J. L. (1979). Oxygen-sensitive and -insensitive nitroreduction by *Escherichia coli* and rat hepatic microsomes. *J. Biol. Chem.* **254**, 4009–4014.
55a. Sullivan, C. E., Tally, F. P., Goldin, B. R., and Vouros, P. (1982). Synthesis of 1-(2-hydroxyethyl)-2-methyl-5-aminoimidazole: A ring-intact reduction product of metronidazole. *Biochem. Pharmacol.* **31**, 2689–2691.
56. Josephy, P. D., Palcic, B., and Skarsgard, L. D. (1981). Reduction of misonidazole and its derivatives by xanthine oxidase. *Biochem. Pharmacol.* **30**, 849–853.
56a. McClelland, R. A., Fuller, J. R., Seaman, N. E., Rauth, A. M., and Battistella, R. (1984). 2-Hydroxylaminoimidazoles: Unstable intermediates in the reduction of 2-nitroimidazoles. *Biochem. Pharmacol.* **33**, 303–309.
57. Liū, S. F., and Raleigh, J. A. (1982). Reductive fragmentation of misonidazole in the presence of xanthine oxidase. Glyoxal formation. *Radiat. Res.* **91**, 376 (abstr).
57a. Raleigh, J. A., and Liu, S. F. (1983). Reductive fragmentation of 2-nitroimidazoles in the presence of nitroreductases. Glyoxal formation from misonidazole. *Biochem. Pharmacol.* **32**, 1444–1446.

58. Corvaja, C., Farnia, G., and Vianello, E. (1966). Kinetics of decay of nitrophenol radical anions and reduction mechanism of nitrophenols in aqueous alkaline media. *Electrochim. Acta* **11**, 919–929.
59. Neta, P., Simic, M. G., and Hoffman, M. Z. (1976). Pulse radiolysis and electron spin resonance studies of nitroaromatic radical anions. Optical absorption spectra, kinetics, and one-electron redox potentials. *J. Phys. Chem.* **80**, 2018–2023.
60. Mason, R. P. (1974). High-resolution ESR of the nitroaromatic anion radical—the first reduced species formed by microsomal nitroreduction. *Fed. Proc., Fed. Am. Soc. Exp. Biol.* **33**, 587.
61. Mason, R. P., and Holtzman, J. L. (1974). ESR spectra of free radicals formed from nitroaromatic drugs by microsomal nitroreductase. *Pharmacologist* **16**, 277.
62. Mason, R. P., and Holtzman, J. L. (1975). The kinetics of nitroreductase anion radical intermediates. *Fed. Proc., Fed. Am. Soc. Exp. Biol.* **34**, 665.
63. Mason, R. P., and Holtzman, J. L. (1975). The role of catalytic superoxide formation in the $O_2$ inhibition of nitroreductase. *Biochem. Biophys. Res. Commun.* **67**, 1267–1274.
64. Sealy, R. C., Swartz, H. M., and Olive, P. L. (1978). Electron spin resonance–spin trapping. Detection of superoxide formation during aerobic microsomal reduction of nitro compounds. *Biochem. Biophys. Res. Commun.* **82**, 680–684.
65. Kalyanaraman, B., Perez-Reyes, E., Mason, R. P., Peterson, F. J., and Holtzman, J. L. (1979). Electron spin resonance evidence for a free radical intermediate in the cis-trans isomerization of furylfuramide by oxygen-sensitive nitroreductases. *Mol. Pharmacol.* **16**, 1059–1064.
66. Kalyanaraman, B., Mason, R. P., Rowlett, R., and Kispert, L. D. (1981). An electron spin resonance investigation and molecular orbital calculation of the anion radical intermediate in the enzymatic cis-trans isomerization of furylfuramide, a nitrofuran derivative of ethylene. *Biochim. Biophys. Acta* **660**, 102–109.
67. Docampo, R., Moreno, S. N. J., Stoppani, A. O. M., Leon, W., Cruz, F. S., Villalta, F., and Muniz, R. F. A. (1981). Mechanism of nifurtimox toxicity in different forms of *Trypanosoma cruzi*. *Biochem. Pharmacol.* **30**, 1947–1951.
68. Docampo, R., Mason, R. P., Mottley, C., and Muniz, R. F. A. (1981). Generation of free radicals induced by nifurtimox in mammalian tissues. *J. Biol. Chem.* **256**, 10930–10933.
69. Perez-Reyes, E., Kalyanaraman, B., and Mason, R. P. (1980). The reductive metabolism of metronidazole and ronidazole by aerobic liver microsomes. *Mol. Pharmacol.* **17**, 239–244.
70. Feller, D. R., Morita, M., and Gillette, J. R. (1971). Enzymatic reduction of niridazole by rat liver microsomes. *Biochem. Pharmacol.* **20**, 203–215.
71. Wang, C. Y., Behrens, B. C., Ichikawa, M., and Bryan, G. T. (1974). Nitroreduction of 5-nitrofuran derivatives by rat liver xanthine oxidase and reduced nicotinamide adenine dinucleotide phosphate–cytochrome *c* reductase. *Biochem. Pharmacol.* **23**, 3395–3404.
72. Harada, N., and Omura, T. (1980). Participation of cytochrome *P*-450 in the reduction of nitro compounds by rat liver microsomes. *J. Biochem. (Tokyo)* **87**, 1539–1554.
73. Flockhart, I. R., Large, P., Troup, D., Malcolm, S. L., and Marten, T. R. (1978). Pharmacokinetic and metabolic studies of the hypoxic cell radiosensitizer misonidazole. *Xenobiotica* **8**, 97–105.
74. Koch, R. L., Chrystal, E. J. T., Beaulieu, B. B., Jr., and Goldman, P. (1979). Acetamide—a metabolite of metronidazole formed by the intestinal flora. *Biochem. Pharmacol.* **28**, 3611–3615.
75. Varghese, A. J. (1981). Detection of the amine derivative of misonidazole in human urine by high-pressure liquid chromatography. *Anal. Biochem.* **110**, 197–200.

76. Whillans, D. W., and Whitmore, G. F. (1981). The radiation reduction of misonidazole. *Radiat. Res.* **86,** 311-324.
77. Rauth, A. M., Battistella, R., McClelland, R. A., Fuller, J. R., and Seaman, E. (1982). Possible role of N-1 substitution on the mutagenicity of 2-nitroimidazoles. *Radiat. Res.* **91,** 376 (abstr.).
78. Middlestadt, M. V., and Rauth, A. M. (1982). Characterization and reactivity of misonidazole reduction products. *Radiat. Res.* **91,** 381 (abstr.).
78a. Gattavecchia, E., Tonelli, D., Breccia, A., and Roffia, S. (1982). The production of nitrite from radiolytic, photolytic and electrolytic degradation of metronidazole. *Int. J. Radiat. Biol.* **42,** 105-109.
79. La Russo, N. F., Tomasz, M., Muller, M., and Lipman, R. (1977). Interaction of metronidazole with nucleic acids *in vitro*. *Mol. Pharmacol.* **13,** 872-882.
80. Knight, R. C., Skolimowski, I. M., and Edwards, D. I. (1978). The interaction of reduced metronidazole with DNA. *Biochem. Pharmacol.* **27,** 2089-2093.
81. Rowley, D. A., Knight, R. C., Skolimowski, I. M., and Edwards, D. I. (1979). The effect of nitroheterocyclic drugs on DNA: An *in vitro* model of cytotoxicity. *Biochem. Pharmacol.* **28,** 3009-3013.
82. Knox, R. J., Knight, R. C., and Edwards, D. I. (1981). Misonidazole-induced thymidine release from DNA. *Biochem. Pharmacol.* **30,** 1925-1929.
83. McManus, M. E., Lang, M. A., Stuart, K., and Strong, J. (1982). Activation of misonidazole by rat liver microsomes and purified NADPH-cytochrome *c* reductase. *Biochem. Pharmacol.* **31,** 547-552.
84. Schroy, C. B., and Biaglow, J. E. (1981). Use of an oxidase electrode to determine factors affecting the *in vitro* production of hydrogen peroxide by Ehrlich cells and 1-chloro-2,4-dinitrobenzene. *Biochem. Pharmacol.* **30,** 3201-3207.
85. Fridovich, I. (1981). Role and toxicity of superoxide in cellular systems. *In* "Oxygen and Oxy-radicals in Chemistry and Biology" (M. A. J. Rodgers and E. L. Powers, eds.), pp. 197-204. Academic Press, New York.
86. Brigelius, R., Haslem, A., and Lengfelder, E. (1981). Paraquat-induced alterations of phospholipids and GSSG-release in the isolated perfused rat liver, and the effect of SOD-active complexes. *Biochem. Pharmacol.* **30,** 349-354.
87. Patterson, C. E., and Rhodes, M. L. (1982). The effect of superoxide dismutase on paraquat mortality in mice and rats. *Toxicol. Appl. Pharmacol.* **62,** 65-72.
88. Boyd, M. R., Catignani, G. L., Sasame, H. A., Mitchell, J. R., and Stiko, A. W. (1979). Acute pulmonary injury in rats by nitrofurantoin and modification by vitamin E, dietary fat, and oxygen. *Am. Rev. Respir. Dis.* **120,** 93-99.
89. Peterson, F. J., Mason, R. P., Holtzman, J. L., and Combs, G. F., Jr. (1980). The effect of selenium and vitamin E deficiency on the toxicity of nitrofurantoin in the chick. *In* "Microsomes, Drug Oxidation, and Chemical Carcinogenesis" (M. J. Coon, A. H. Conney, R. W. Estabrook, H. V. Gelboin, J. R. Gillette, and P. J. O'Brien, eds.), Vol. 2, pp. 873-876. Academic Press, New York.
90. Peterson, F. J., Mason, R. P., Holtzman, J. L., and Combs, G. F., Jr. (1982). The effect of selenium and vitamin E deficiency on the toxicity of nitrofurantoin. *J. Nutr.*
91. Janzen, E. G. (1980). A critical review of spin trapping in biological systems. *In* "Free Radicals in Biology" (W. A. Pryor, ed.), Vol. 4, pp. 115-154. Academic Press, New York.
92. Finkelstein, E., Rosen G. M., and Rauckman, E. F. (1979). Spin trapping of superoxide. *Mol. Pharmacol.* **16,** 676-685.
93. Tatsumi, K., Yamada, H., Yoshimura, H., and Kawazoe, Y. (1982). Studies on enzy

matic reduction of aliphatic nitro compounds: Reductive dimerization of beta-nitrostyrene and 1-nitro-4-phenylbutadiene. *Arch. Biochem. Biophys.* **213,** 689–694.
94. Bahnemann, D., Basaga, H., Dunlop, J. R., Searle, A. J. F., and Willson, R. L. (1978). Metronidazole (Flagyl), Misonidazole (Ro 07-0582), iron, zinc, and sulphur compounds in cancer therapy. *Br. J. Cancer* **37,** Suppl. III, 16–19.
95. Mottley, C., Kalyanaraman, B., and Mason, R. P. (1981). Spin trapping artifacts due to the reduction of nitroso spin traps. *FEBS Lett.* **130,** 12–14.
96. Sridhar, R. (1981). Accelerated oxygen consumption by catecholamines in the presence of aromatic nitro and nitroso compounds. Implications for neurotoxicity of nitro compounds. *In* "Oxygen and Oxy-Radicals in Chemistry and Biology" (M. A. J. Rodgers and E. L. Powers, eds.), pp. 363–365. Academic Press, New York.
97. Kalyanaraman, B., Perez-Reyes, E., and Mason, R. P. (1979). The reduction of nitroso-spin traps in chemical and biological systems. A cautionary note. *Tetrahedron Lett.* pp. 4809–4812.
98. Becker, A. R., and Sternson, L. A. (1980). Nonenzymatic reduction of nitrosobenzene to phenylhydroxylamine by NAD(P)H. *Bioorg. Chem.* **9,** 305–312.
99. Horie, S., Watanabe, T., and Ogura, Y. (1980). Studies on the enzymatic reduction of C–nitroso compounds. *J. Biochem. (Tokyo)* **88,** 847–857.
100. Stier, A., Reitz, I., and Sackmann, E. (1972). Radical accumulation in liver microsomal membranes during biotransformation of aromatic amines and nitro compounds. *Naunyn-Schmiedeberg's Arch. Pharmacol.* **274,** 189–191.
101. Floyd, R. A., Soong, L. M., Stuart, M. A., and Reigh, D. L. (1978). Free radicals and carcinogenesis. Some properties of the nitroxyl free radicals produced by covalent binding of 2-nitrosofluorene to unsaturated lipids of membranes. *Arch. Biochem. Biophys.* **185,** 450–457.
102. Schenk, C., and de Boer, T. J. (1979). C-nitroso compounds. XXXI. The addition of α-chloronitroso compounds to olefins containing allylic hydrogen. *Tetrahedron* **35,** 147–153.
103. Stier, A., Clauss, R., Lucke, A., and Reitz, I. (1980). Redox cycle of stable mixed nitroxides formed from carcinogenic aromatic amines. *Xenobiotica* **10,** 661–673.
104. Mason, R. P., Kalyanaraman, B., Tainer, B. E., and Eling, T. E. (1980). A carbon-centered free radical intermediate in the prostaglandin synthetase oxidation of arachidonic acid. *J. Biol. Chem.* **255,** 5019–5022.
105. Raleigh, J. A., Shum, F. Y., and Liu, S. F. (1981). Nitroreductase-induced binding of nitroaromatic radiosensitizers to unsaturated lipids. *Biochem. Pharmacol.* **30,** 2921–2925.
106. Sridhar, R., Hampton, M. J., Steward, J. E., and Floyd, R. A. (1980). Studies on the mutagenicity and electron spin resonance spectra of nitrosofluorene–lipid adducts. *Appl. Spectrosc.* **34,** 289–293.
107. Griffin, B. W., Davis, D. K., and Bruno, G. V. (1981). Electron paramagnetic resonance study of the oxidation of N-methylsubstituted aromatic amines catalyzed by hemeproteins. *Bioorg. Chem.* **10,** 342–355.
108. Lasker, J. M., Sivarajah, K., Mason, R. P., Kalyanaraman, B., Abou-Donia, M. B., and Eling, T. E. (1981). A free radical mechanism of prostaglandin synthase-dependent aminopyrine demethylation. *J. Biol. Chem.* **256,** 7764–7767.
109. Yamazoe, Y., Ishii, K., Yamaguchi, N., Kamataki, T., and Kato, R. (1980). Reduction of *N*-hydroxy-2-acetylaminofluorene by liver microsomes. *Biochem. Pharmacol.* **29,** 2183–2188.
110. Kadlubar, F. F., and Ziegler, D. M. (1974). Properties of a NADH-dependent *N*-hy-

droxyamine reductase isolated from pig liver microsomes. *Arch. Biochem. Biophys.* **162**, 83–92.
111. Scribner, J. D., and Naimy, N. K. (1975). Destruction of triplet nitrenium ion by ascorbate. *Experientia* **31**, 470–471.
111a. Ford, G. P., and Scribner, J. D. (1981). MNDO molecular orbital study of nitrenium ions derived from carcinogenic aromatic amines and amides. *J. Am. Chem. Soc.* **103**, 4281–4291.
112. Andrews, L. S., Hinson, J. A., and Gillette, J. R. (1978). Studies on the mutagenicity of N-hydroxy-2-acetylaminofluorene in the Ames *Salmonella* mutagenesis test system. *Biochem. Pharmacol.* **27**, 2399–2408.
113. Andrews, L. S., Fysh, J. M., Hinson, J. A., and Gillette, J. R. (1979). Ascorbic acid inhibits covalent binding of enzymatically generated 2-acetylaminofluorene-N-sulfate to DNA under conditions in which it increases mutagenesis in *Salmonella* TA-1538. *Life Sci.* **24**, 59–63.
114. Searle, A. J. F., and Wilson, R. L. (1976). Metronidazole (Flagyl): Degradation by the intestinal flora. *Xenobiotica* **6**, 457–464.
115. Goldstein, B. P., Vidal-Plana, R. R., Cavalleri, B., Zerilli, I., Carniti, G., and Silvestri, L. G. (1977). The mechanism of action of nitro-heterocyclic antimicrobial drugs. Metabolic activation by microorganisms. *J. Gen. Microbiol.* **100**, 283–298.
116. Koch, R. L., Beaulieu, B. B., Jr., Chrystal, E. J. T., and Goldman, P. (1981). A metronidazole metabolite in human urine and its risk. *Science* **211**, 398–400.
117. Beaulieu, B. B., Jr., McLafferty, M. A., Koch, R. L., and Goldman, P. (1981). Metronidazole metabolism in cultures of *Entamoeba histolytica* and *Trichomonas vaginalis*. *Antimicrob. Agents Chemother.* pp. 410–414.
118. Koch, R. L., Beaulieu, B. B., Jr., and Goldman, R. (1980). Role of the intestinal flora in the metabolism of misonidazole. *Biochem. Pharmacol.* **29**, 3281–3284.
119. Taylor, Y. C., and Rauth, A. M. (1978). Differences in the toxicity and metabolism of the 2-nitroimidazole misonidazole (Ro 07-0582) in HeLa and Chinese hamster ovary cells. *Cancer Res.* **38**, 2745–2752.
120. Taylor, Y. C., and Rauth, A. M. (1980). Sulphydryls, ascorbate, and oxygen as modifiers of the toxicity and metabolism of misonidazole *in vitro*. *Br. J. Cancer* **41**, 892–900.
121. Josephy, P. D., Palcic, B., and Skarsgard, L. D. (1978). Ascorbate-enhanced cytotoxicity of misonidazole. *Nature (London)* **271**, 370–372.
122. Koch, C. J., Howell, R. L., and Biaglow, J. E. (1979). Ascorbate anion potentiates cytotoxicity of nitro-aromatic compounds under hypoxic and anoxic conditions. *Br. J. Cancer* **39**, 321–329.
123. Josephy, P. D., Palcic, B., and Skarsgard, L. D. (1981). *In vitro* metabolism of misonidazole. *Br. J. Cancer* **43**, 443–450.
124. Koch, R. L., Rose, C., Rich, T. A., and Goldman, P. (1982). Comparative misonidazole metabolism in anaerobic bacteria and hypoxic Chinese hamster lung fibroblast (V79-473) cells. *Biochem. Pharmacol.* **31**, 411–414.
125. Kagiya, T., Ide, H., Nishimoto, S., and Wada, T. (1983). Radiation-induced reduction of nitroimidazole derivatives in aqueous solution. *Int. J. Radiat. Biol.* **44**, 505–517.
126. Docampo, R., and Moreno, S. N. J. (1984). Free radical metabolites in the mode of action of chemotherapeutic agents and phagocytic cells on *Trypanosoma cruzi*. *Rev. Infect. Dis.* **6**, 223–238.
127. Declerck, P. J., De Ranter, C. J., and Volckaert, G. (1983). Base specific interaction of reductively activated nitroimidazoles with DNA. *FEBS Lett.* **164**, 145–148.
128. Smith, B. R., Born, J. L., and Garcia, D. J. (1983). Influence of hypoxia on the metabo-

lism and excretion of misonidazole by the isolated perfused rat liver — a model system. *Biochem. Pharmacol.* **32,** 1609–1612.
129. Smith, B. R. (1984). Hypoxia-enhanced reduction and covalent binding of [2-$^3$H] misonidazole in the perfused rat liver. *Bochem. Pharmacol.* **33,** 1379–1381.
130. Stratford, I. J., Hoe, S., Adams, G. E., Hardy, C., and Williamson, C. (1983). Abnormal radiosensitizing and cytotoxic properties of ortho-substituted nitroimidazoles. *Int. J. Radiat. Biol.* **43,** 31–43.
131. Varghese, A. J. (1983). Glutathione conjugates of misonidazole. *Biochem. Biophys. Res. Commun.* **112,** 1013–1020.
132. Chapman, J. D., Baer, K., and Lee, J. (1983). Characteristics of the metabolism-induced binding of misonidazole to hypoxic mammalian cells. *Cancer Res.* **43,** 1523–1528.

Chapter 17

# Nitriles

AHMED E. AHMED, MOHAMMED Y. H. FAROOQUI, AND NORMAN M. TRIEFF

*Departments of Pathology, Pharmacology, and Toxicology and
Preventive Medicine and Community Health
The University of Texas Medical Branch
Galveston, Texas*

|      |                                       |     |
|------|---------------------------------------|-----|
| I.   | Introduction                          | 485 |
| II.  | Inorganic Cyanides                    | 486 |
| III. | Nitriles                              | 490 |
| IV.  | Saturated Aliphatic Nitriles          | 491 |
|      | A. Unsubstituted Nitriles             | 491 |
|      | B. Substituted Nitriles               | 493 |
| V.   | Unsaturated Aliphatic Nitriles        | 499 |
|      | A. Acrylonitrile                      | 499 |
|      | B. Crotononitrile                     | 506 |
| VI.  | Alkylaryl Nitriles                    | 506 |
|      | A. Amygdalin (Laetrile)               | 507 |
|      | B. 2-Chlorobenzylidene Malononitrile  | 507 |
|      | C. Bromobenzyl Cyanide                | 508 |
|      | D. Cinnamonitrile                     | 508 |
| VII. | Aryl Nitriles                         | 508 |
| VIII.| Summary                               | 509 |
|      | References                            | 510 |

## I. Introduction

Nitriles or organic cyanides are compounds of the general formula R-CN, where R may be any saturated or unsaturated aliphatic or aromatic moiety.

Nitriles have found valuable industrial applications in the manufacture of synthetic fibers, resins, and vitamins. They also have been used as plasticizers, solvents, elastomers, agricultural insecticides, and high-pressure lubricants.[1] According to a NIOSH[2] estimate, several thousand workers are potentially exposed each day to various aliphatic nitriles. Numerous reports have described fatalities and signs of poisoning in humans exposed to acetonitrile, acrylonitrile, acetone cyanohydrin, malononitrile, and succinonitrile. The toxicology of aliphatic nitriles has been extensively reviewed.[2-4]

Cyanide is a known metabolite of many nitriles. The role of cyanide, liberated by biotransformation of organic nitriles *in vivo*, in their acute toxicity has been emphasized by several investigators.[5-10] To discuss fully the metabolic fate of nitriles, it will be essential to discuss the biological fate of the liberated cyanide moiety. After a discussion of inorganic cyanide ($CN^-$), the bioactivation of other nitriles and the mechanism and extent of cyanide liberation from each individual nitrile will be discussed. Mechanism of other possible toxic interactions of the nitriles will also be considered.

This chapter covers primarily the metabolism and bioactivation of nitriles with less emphasis on toxicology and environmental impact, which is covered elsewhere.[11] Examples of various aliphatic mono- and dinitriles, alkylaryl cyanides, and aromatic nitriles are included.

## II. Inorganic Cyanides

Hydrogen cyanide (HCN) is a colorless or bluish white liquid with the faint, characteristic odor of bitter almonds perceptible only to some people. The chemical and physical properties of HCN have been reviewed.[11,12]

Hydrogen cyanide is toxic by virtue of its ionization to cyanide ion in biological systems. The latter exerts its toxicity by producing cytotoxic hypoxia. Cyanide is a powerful metabolic inhibitor and arrests cellular respiration by inactivating metalloenzymes that are fundamental in the respiratory process.[13] Although it is the cyanide that is the ultimate toxic moiety, it must enter the blood as hydrogen cyanide by either the respiratory or the gastric route.

Either inhaled hydrogen cyanide or ingested cyanide, which is converted to hydrogen cyanide in the stomach, enters the blood and then ionizes to cyanide, which readily enters the cell. All tissues take up cyanide, with the lowest content found in the muscle and the highest in spleen. The blood cyanide level at equilibrium is 3–10 times greater than the tissue concentration in humans.[14,15]

There is much evidence to demonstrate that in most animals, the major portion of sublethal doses of hydrogen cyanide is ultimately excreted in the urine as thiocyanate (SCN⁻). The biotransformation of cyanide to thiocyanate was first demonstrated by Lang as early as 1894.[16] In rabbits, 80% of doses of 1 to 2 mg of hydrogen cyanide was excreted as thiocyanate in 24–48 h.[17-19] This process also occurs in dogs, but the excretion of thiocyanate is slower than in rabbits.

Boxer and Rickards[20] demonstrated an equilibrium between cyanide and thiocyanate *in vivo* according to Eq. (1).

$$CN^- + S \underset{\text{thiocyanate oxidase}}{\overset{\text{rhodanese}}{\rightleftharpoons}} SCN^- \tag{1}$$

With larger doses of cyanide, this equilibrium, on the basis of the laws of mass action, must shift to favor thiocyanate formation. Rats treated with potassium thiocyanate largely excreted unchanged thiocyanate, and only 1–4.5% of the thiocyanate sulfur was excreted as sulfate.[21] The biotransformation of thiocyanate to cyanide has been observed in humans and in dogs by Goldstein and Rieders.[22] It is believed that the conversion of thiocyanate to cyanide might explain some of the toxic effects of thiocyanate, which are similar to those of subacute hydrogen cyanide poisoning. The *in vivo* conversion of thiocyanate to cyanide is not due to a reverse catalysis by rhodanese but to an enzyme found only in erythrocytes and called thiocyanate oxidase.[23] This enzyme has been detected in humans, dogs, rabbits, and rats.

Rhodanese catalyzes the formation of thiocyanate from cyanide very rapidly in the presence of sodium thiosulfate or colloidal sulfur, but cystine, cysteine, glutathione (GSH), and thioethanolamine are poor sulfur donors.[24] The enzyme is specific for free cyanide, having no effect on organically bound cyano groups, such as those in aliphatic or aromatic nitriles.[25] These properties of rhodanese suggest that the excretion of thiocyanate after administration of organic nitriles results not from a direct action of the nitrile itself, but from cyanide, which was released metabolically from the nitrile. Thiocyanate is also formed when sodium sulfide is the source of sulfur, but only in the presence of oxygen.[25]

The catalyst rhodanese is plentiful, so the rate-limiting factor is the availability of sulfur. Lang[24] proposed that thiosulfate, which is a common antidote for hydrogen cyanide, is the source of sulfur; the reaction catalyzed by the enzyme is shown in Fig. 1. Sörbo[26-28] has suggested that rhodanese contains an active disulfide group, which reacts with thiosulfate and cyanide as shown in Fig. 1. This raises the question of the occurrence and origin of thiosulfate in the body. Although thiosulfate has been shown to be a metabolite in higher animals, its source is obscure.[29,30] Another source of sulfur could be sulfur-containing amino acids. Wood and Cooley[31] have demonstrated the production of [³⁵S]thiocyanate after administration of [³⁵S]cys-

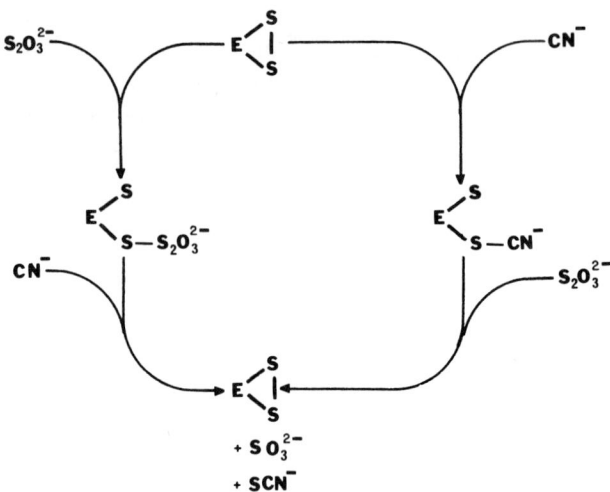

**Fig. 1.** Formal mechanism of thiocyanate formation and substrate interactions with enzyme rhodanese.

tine and cyanide. However, free cysteine or cystine did not act as a substrate for rhodanese *in vitro*.[32] β-Mercaptopyruvic acid (BMV) can also provide sulfur as rapidly as thiosulfate for cyanide detoxication and yields pyruvic acid and colloidal sulfur at pH 7.5–8.5.[33] A possible metabolic pathway of cyanide detoxication by cysteine is shown in Fig. 2.

On excessive intake of hydrogen cyanide, the rhodanese mechanism, because of limited availability of thiosulfate, is incapable of detoxifying all of the cyanide present. Under such cirsumstances, cyanide may attain toxic concentrations in various tissues. However, other minor pathways for cyanide metabolism may take place. These include (a) the oxidation of cyanide to carbon dioxide by the formation of cyanate (HCNO), followed by further oxidation to formic acid. The latter may be excreted or utilized by the cell in

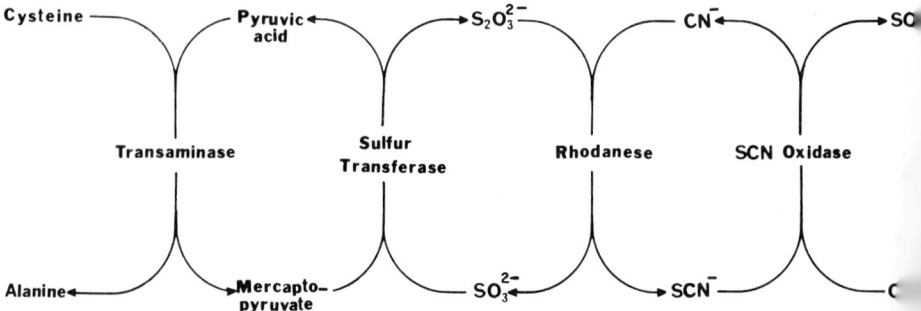

**Fig. 2.** Scheme showing sources of sulfur in cyanide metabolism.

the normal one-carbon metabolic pathway, such as the folate cycle.[20] (b) The reaction of cyanide with cobalamin, yielding vitamin $B_{12}$ (cyanocobalamin), which is an essential hematopoietic vitamin. The role of exogenous cyanide in cyanocobalamin formation is not clear. However, this represents a minor, but significant, route of cyanide disposition *in vivo*. (c) The conversion of 15% of administered hydrogen cyanide to 2-iminothiazolidine-4-carboxylic acid, which is excreted in the urine. The latter was postulated by Wood and Cooley[34] to be a product of the nonenzymatic reaction of cyanide with cystine, as shown in Eq. (2).

$$NH_2CHCH_2-S-S-CH_2CHNH_2 + CN^- \rightarrow NH_2CHCH_2SCN + NH_2CHCH_2S^- \quad (2)$$
$$\underset{COOH}{|} \quad \underset{COOH}{|} \quad \underset{COOH}{|} \quad \underset{COOH}{|}$$

2-Iminothiazolidine-4-carboxylic acid

Figure 3 summarizes the metabolism of cyanide.

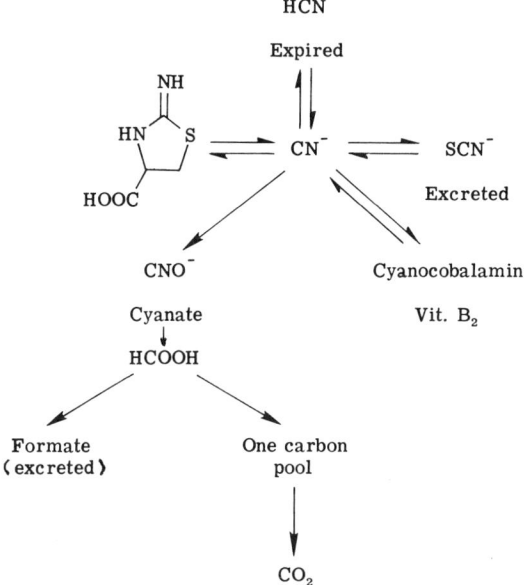

Fig. 3. Scheme showing basic processes involved in cyanide metabolism. (Derived from Williams.[25])

## III. Nitriles

Since the early work of Lang,[16] the mechanism of toxicity of nitriles has been primarily attributed to *in vivo* liberation of cyanide. However, there is no definite chemical reaction known so far that describes completely the direct displacement of cyanide from an organic molecule. Chemically, the cyanide moiety is a strong electron-withdrawing group with polarizable charges on the nitrogen and carbon atoms:

$$R-CH_2-\overset{\delta+}{C}\equiv\overset{\delta-}{N}$$

The reactivity of the nitrile group is thus illustrated by its rapid hydrolysis as shown in Eq. (3).

$$R-CH_2-C\equiv N + HO^- \rightarrow R-CH_2-\underset{OH}{\overset{|}{C}}=NH \rightarrow R-CH_2-\underset{O}{\overset{\|}{C}}-NH_2 \quad (3)$$

In addition, the electron deficiency created by the nitrile group makes the hydrogen atoms on the α-carbon atom acidic and renders them vulnerable to attack by nucleophiles and bases. Hence, the reactivity of the nitriles is localized in the highly acidic α-hydrogen atoms. Similar reactivity was postulated to be essential[24] in the bioactivation of nitriles before cyanide liberation, as shown in Eq. (4).

$$R-CH_2-CN \xrightarrow{[O]} [R-\underset{}{\overset{OH}{\overset{|}{C}H}}-CN] \underset{HCN \xrightarrow{rhodanese} CNS^-}{\overset{R-CHO \rightarrow R-COOH}{\diagup}} \quad (4)$$

aliphatic nitrile    cyanohydrin

Equation (4) represents the only known metabolic change for aliphatic nitriles. It involves oxygen insertion at the α-methylene groups by the microsomal monooxygenase system. The unstable cyanohydrins thus produced decompose to liberate cyanide, the latter being detoxified by rhodanese.[35,36] Metabolic degradation to cyanide explains the toxic effects of some aliphatic nitriles, although some uncertainty exists in the interpretation of the toxicity of acetonitrile and acrylonitrile, both of which release moderate amounts of cyanide in biological systems. As the length of the aliphatic group increases, the acidity of the α-carbon atom also increases, which results in higher susceptibility to nucleophilic attack and thus to α oxidation.

In *in vitro* studies, verification of such a reaction mechanism was accomplished with benzyl cyanide, which was metabolized *in vivo* to cyanide. When incubated with a hepatic microsomal fraction in the presence of NADPH and $O_2$ at pH 7.4, benzyl cyanide was converted to mandelonitrile, a moderately stable cyanohydrin, according to the reaction shown in Eq. (5).

17. NITRILES

$$\underset{\text{Benzyl cyanide}}{\text{H}-\underset{|}{\overset{\text{H}}{\text{C}}}-\text{CN}} \xrightarrow[\text{NADPH, O}_2]{\text{microsomes}} \underset{\text{Mandelonitrile}}{\text{H}-\underset{|}{\overset{\text{OH}}{\text{C}}}-\text{CN}} \qquad (5)$$

In addition, a deuterium isotope effect was observed when [α,α-$^2$H2]benzylcyanide was the substrate; the rate of formation of deuteriomandelonitrile was one-half that of mandelonitrile ($V_H/V_D = 2$). This confirmed the lability of the α-hydrogen in formation of the α-hydroxy derivatives.

## IV. Saturated Aliphatic Nitriles

### A. UNSUBSTITUTED NITRILES

#### 1. Acetonitrile

Acetonitrile ($CH_3CN$) is a colorless liquid, very soluble in water, alcohol, ether, and acetone. It is used as a solvent both in industry and in the laboratory, as a rodenticide, and in the denaturation of alcohol. Current information on human toxicity of acetonitrile comes from well-documented occupational accidents.[37-44]

Acetonitrile metabolism in dogs was demonstrated by Lang,[16] who reported that about 20% of the nitrile administered was converted to thiocyanate in the urine; guinea pigs metabolized acetonitrile to a greater extent (50% of dose excreted as thiocyanate). When the animals were treated with ethanol, acetonitrile metabolism was induced (87% of dose excreted as thiocyanate). Baumann et al.[17] found that rabbits given acetonitrile excreted 27–35% of the dose as thiocyanate. In thyroidectomized rabbits, thiocyanate excretion decreased significantly to 3 to 5% of the dose, and thiocyanate excretion increased after feeding desiccated thyroid. Hunt[45] found that powdered sheep thyroid protected mice against acetonitrile toxicity. However, the role played by thyroid in the detoxication of cyanide to thiocyanate is unclear. Baumann et al.[17] suggested that thyroid may have a greater effect on the cleavage of cyanide from acetonitrile than on the sulfation of cyanide to thiocyanate.

The metabolism of acetonitrile to cyanide was first proposed by Lang[16] as shown in Eq. (6).

$$CH_3CN \xrightarrow{O_2} CH_2\!\!\begin{array}{c}\nearrow\text{OH}\\ \searrow\text{CN}\end{array} \rightarrow CH_2O + HCN \xrightarrow{\text{rhodanese}} SCN^- \qquad (6)$$

$$HCOOH \rightarrow CO_2$$

Frimin and Gray[46] studied the fate of acetonitrile in *Pseudomonas* species and found that [$^{14}$C]acetonitrile is metabolized to citrate, succinate, fumarate, malate, glutamate, pyrrolidonecarboxylic acid, and aspartate. They reported that this species metabolized acetonitrile by direct hydrolysis of the cyanide moiety to acetamide. When the bacteria were supplied with nonradioactive acetamide or acetate, the synthesis of $^{14}$C-labeled acids was diminished after addition of [$^{14}$C]acetonitrile, implying an equilibrium. They found that [1-$^{14}$C]acetate and [2-$^{14}$C]acetonitrile were metabolized to identical products at a similar rate. Fluoroacetate and arsenite, tricarboxylic acid (TCA) cycle inhibitors, greatly inhibited the transfer of label from [2-$^{14}$C]acetonitrile to all products of the cycle. These investigators proposed the pathway shown in Eq. (7) (for the bacterial biotransformation of acetonitrile.

$$^{14}CH_3CN \rightarrow {^{14}CH_3CONH_2} \rightarrow {^{14}CH_3COO^-} \rightarrow \text{tricarboxylic acid cycle intermediates} \quad (7)$$

Whether or not these reactions and the incorporation of the acetonitrile backbone into the energy-generating cycles occur in mammalian systems needs to be carefully investigated. Furthermore, the correlation between such reactions and the toxicity of acetonitrile has to be examined.

## 2. Propionitrile

Propionitrile ($CH_3CH_2CN$, ethyl cyanide) is commercially important as a selective solvent in petroleum refining, as a raw material in the manufacture of certain drugs, and as a setting agent for various resins. Inhaled propionitrile is about three to four times more toxic than acetonitrile. In rats, the primary effect of propionitrile is the formation of duodenal ulcers, which is seen after subcutaneous administration.[47-51]

*In vivo* studies on propionitrile metabolism have been confined to determining cyanide and thiocyanate concentrations in blood and urine of treated animals.[52]

In *in vitro* studies, propionitrile has been found to be activated by the hepatic mixed-function oxidase system to produce its α-hydroxy analog acetaldehyde cyanohydrin, which was in equilibrium with the breakdown products propionaldehyde and hydrogen cyanide. [Eq. (8)].

$$H_3C-\underset{CN}{\overset{OH}{CH}} \quad\quad H_2C-\underset{H}{\overset{O}{C}} + HCN \quad\quad (8)$$

## 3. Butyronitrile

$n$-Butyronitrile [$CH_3CH_2CH_2CN$] and isobutyronitrile [$(CH_3)_2CHCN$] are about 2.4 times as toxic as acetonitrile in rats. Isobutyronitrile has been implicated in several cases of industrial poisoning.[53,54]

Studies on $n$-butyronitrile and isobutyronitrile include those of Haguenoer et al.,[55] Szabo and Reynolds,[56] and Tsurumi and Kawada.[57] Haguenoer et al.[55] determined concentrations of $n$-butyronitrile and cyanide in tissues from various organs of male rats. With a single dose of 1440 mg/kg of $n$-butyronitrile, rats became comatose and cyanotic, and died within 90 min. All organs were found to contain $n$-butyronitrile and cyanide. The highest concentration of $n$-butyronitrile was found in the lungs, and the highest cyanide concentrations were present in the heart and brain. It was concluded that the slow urinary excretion parallels the relatively low water solubility of $n$-butyronitrile.

## B. SUBSTITUTED NITRILES

### 1. Fluoronitriles

Pattison et al.,[58] synthesized and studied the properties and metabolic breakdown of several $\omega$-fluoronitriles, including fluoroacetonitrile, 3-fluoropropionitrile, and 4-fluorohexano-nitrile. Their data showed that the nitriles with an odd number of carbon atoms were toxic. The hydrolysis of fluoronitriles to acids ($RCN \rightarrow RCOOH$) or to amides ($RCN \rightarrow RCONH_2$) was not related to the toxicity, because these products were nontoxic. The high toxicity of these fluoronitriles can be explained by an $\alpha$-oxidation mechanism, which yields the toxic fluoroacids and cyanide.

Such a reaction with 3-fluoropropionitrile may proceed as shown in Eq. (9).

$$FCH_2CH_2CN \rightarrow FCH_2\overset{OH}{\underset{|}{C}}HCN \rightarrow FCH_2\overset{O}{\underset{H}{C}} + HCN \quad (9)$$
$$\text{3-fluoropropionitrile} \qquad \qquad \downarrow$$
$$FCH_2COOH$$
$$\text{fluoroacetic acid}$$

The fluoroacetic acid thus produced is a known inhibitor of the tricarboxylic acid cycle.

### 2. Hydroxynitriles

The introduction of a hydroxyl group in the $\alpha$ position relative to the cyanide moiety enhances the toxicity of nitriles, whereas $\beta$ substitution abolishes the toxicity. Accordingly, β-hydroxypropionitrile (HOCH$_2$CH$_2$CN) lacks animal toxicity[59] in contrast to lactonitrile [CH$_3$CH(OH)CN], which is extremely hazardous.[60] This reflects the absence in the former, and the presence in the latter, of cyanide-forming capability.[61]

### a. β-HYDROXYNITRILES (CYANOHYDRINS)

The metabolic breakdown of cyanohydrins is illustrated by Eqs. (10)–(12).

$$\underset{\substack{\text{glyconitrile}\\ \text{(formaldehyde}\\ \text{cyanohydrin)}}}{\overset{H}{\underset{H}{>}}C\overset{OH}{\underset{CN}{<}}} \rightleftarrows \underset{\text{formaldehyde + hydrogen cyanide}}{\overset{H}{\underset{H}{>}}C=O + HCN} \quad (10)$$

$$\underset{\substack{\alpha\text{-hydroxyisobutyronitrile}\\ \text{(acetone cyanohydrin)}}}{\overset{H_3C}{\underset{H_3C}{>}}C\overset{OH}{\underset{CN}{<}}} \rightleftarrows \underset{\text{acetone + hydrogen cyanide}}{\overset{H_3C}{\underset{H_3C}{>}}C=O + HCN} \quad (11)$$

$$\underset{\text{lactonitrile}}{\overset{H_3C}{\underset{H}{>}}C\overset{OH}{\underset{CN}{<}}} \rightleftarrows \underset{\text{acetaldehyde + hydrogen cyanide}}{\overset{H_3C}{\underset{H}{>}}C=O + HCN} \quad (12)$$

The toxicities of lactonitrile,[60] glyconitrile,[62] and acetone cyanohydrin[63] in animals are associated with release of cyanide.

The cyanide antidotes nitrites and sodium thiosulfate, given separately or in combination, were effective for treating acute acetone cyanohydrin intoxication.[56] However, the chronic toxicity of cyanohydrins was not identical to their acute toxicity, which indicates a possible involvement of metabolites other than cyanide.

### b. β-HYDROXYPROPIONITRILE

Acrylamide polymers are used as flocculants for purifying drinking water. Acrylamide monomer contains β-hydroxypropionitrile (HO—CH$_2$CH$_2$—C≡N, β-HPN) at concentrations of as high as several thousand ppm. β-Hydroxypropionitrile is present in the finished polymer and may be extracted and enter the water that is being flocculated.[64]

Sauerhoff et al.[64] have studied the pharmacokinetics and metabolism of

$\beta$-[1-$^{14}$C]HPN. $\beta$-Hydroxypropionitrile was extensively absorbed from the GI tract, and peak plasma concentrations of $^{14}$C occurred 4 hr after administration. Excretion of $^{14}$C activity amounted to 53.2% in the urine (including thiocyanate), 7.39% in the feces with 0.44% as hydrogen cyanide, and 25.6% as carbon dioxide in the expired air. The rate constant for elimination of [$^{14}$C]cyanide in urine was $0.113/\text{hr} \pm 0.022$ ($t_{1/2} = 6.13$ hr). The net elimination of [$^{14}$C]thiocyanate in urine reached peak values 8–16 hr after administration.

Cyanide production amounted to 80 $\mu$g/48 hr after giving 20 mg/kg of $\beta$-HPN. Thiocyanate production exceeded that of cyanide, and the increase in cyanide was not toxicologically significant.[64] The urinary metabolite other than thiocyanate was the conjugate of $\beta$-HPN. Some $\beta$-HPN was also excreted in the urine.

The biphasic plasma clearance of $^{14}$C activity from $\beta$-HPN-treated animals suggested that the $\beta$ phase might be related to clearance of a metabolite or of the nitrile from a compartment other than plasma. Alternatively, it might be related to the incorporation of $^{14}$C into the one-carbon metabolic pool,[20] as indicated by slow and steady elimination of [$^{14}$C]carbon dioxide during the experiment. Hence, the $\beta$-phase elimination of $^{14}$C may be due to incorporation, metabolism, and excretion of one-carbon fragments.

Because no significant toxic manifestations characteristic of cyanide poisoning were observed in rats after ingestion of 270 mg/kg/day of $\beta$-HPN, it is considered that $\beta$-HPN has an intrinsically low toxicity. This is reasonable in the light of the structure of $\beta$-HPN, in which hydrogen bond formation can be postulated:

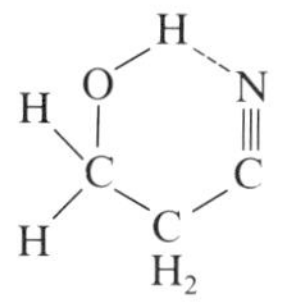

This internal hydrogen bond may minimize dissociation of cyanide. In addition, $\beta$-HPN does not yield cyanide even at moderate temperatures in water, which supports the stability of the molecule. In contrast, acetone cyanohydrin, a closely related molecule, is quite toxic because of the equilibrium reaction [Eq. (11)].

### 3. Aminonitriles

Lathyrism is an ancient disease, which was recognized in ancient India and cited by Hippocrates in 46 B.C. as a neurological disorder secondary to the ingestion of *Lathyrus* seeds (wild peas).[65,66] $\beta$-Aminopropionitrile

($NH_2CH_2CH_2C\equiv N$, β-APN), is the toxic factor in *L. odoratus,* which caused osteolathyrism, including bone lesions and aneurysms. The mechanism of the lathyrogenic effects of β-APN is not known. It has not yet been clarified whether such adverse effects are caused by β-APN or its metabolites. The extent of metabolic conversion of β-APN to the deaminated, oxidized derivative cyanoacetic acid (CAA) by rat liver homogenates was small[67] [Eq. (13)].

$$H_2NCH_2CH_2CN \xrightleftharpoons[\text{homogenate}]{\text{rat liver}} HOOCCH_2CN \quad (13)$$

β-aminopropionitrile      cyanoacetic acid
(β-APN)      (CAA)

Fleisher et al.[68] have reported that in rats significant amounts of unchanged β-APN were excreted within 24 hr. Radioactivity recovered in brain and liver 24 hr after β- [$^{14}C$]APN administration was exclusively found in [$^{14}C$]CAA. Rabbit serum monoamine oxidase (MAO) did not convert β-APN to CAA.[68] However, pargyline inhibited the conversion of β-APN to CAA by rabbit liver homogenates, which indicated that liver is the major site of β-APN metabolism.

In humans, less than half of the administered β-APN was recovered unchanged in the urine, although it disappeared rapidly from plasma.[69] McEwen and Cohen[70] have hypothesized that the disappearance of β-APN from plasma in humans could involve enzymatic degradation by MAO. Alternatively, β-APN may be metabolized to CAA in human tissues.[68,71]

Fleisher et al.[72] reported that human serum did not convert β-APN to the nonlathyrogenic metabolite CAA. Urinary CAA appeared more slowly than β-APN and increased gradually to approximately three times that of urinary β-APN. After stopping β-APN administration, the urinary excretion of CAA was prolonged. These findings, along with earlier findings in animals, suggest that β-APN is sequestered in tissues, where it is metabolized to CAA before being slowly released.

No studies on cyanide release from β-aminopropionitrile have been reported. Perhaps this compound is not metabolized to cyanide, or cyanide release has not been examined in biological systems. However, the presence of β-amino group indicates that the molecule can be stabilized in a manner analogous to β-hydroxypropionitrile:

# 17. NITRILES

The history, toxicology, and chemistry of other aminonitriles have been reviewed.[11]

### 4. Dinitriles

Dinitriles of the formula $CN(CH_2)_nCN$ are highly toxic, because they liberate cyanide *in vivo*. This, along with subsequent thiocyanate formation, is the likely metabolic route in the body. When both the cyanide groups are $\alpha$ to each other, as in malononitrile,

$$H_2C\begin{matrix}\diagup C\equiv N\\ \diagdown C\equiv N\end{matrix}$$

the molecule becomes very reactive chemically because of the highly acidic methylene hydrogens. Hence, exposure to atmospheric oxygen and light causes oxidation and cyanide release. Dinitriles with longer carbon chains where the cyanide groups are not $\alpha$ to each other are stable and require biological activation for the liberation of cyanide. It should be emphasized that removal of cyanide from dinitriles results in the formation of dialdehydes. Dialdehydes may harm cells by interacting with tissue macromolecules and thus producing conformationally fixed products. Indeed, dialdehydes are used currently in tissue fixation.

#### a. MALONONITRILE

Malononitrile (methylene dicyanide) is used primarily as an intermediate in synthesis of drugs and vitamins, in the manufacture of photosensitizers, acrylic fibers, and dyestuffs, and as an oil-soluble, but polar, additive in lubricating oil.

Malononitrile has been used in the treatment of various forms of mental illnesses.[73-75] However, its use as a therapeutic agent has been discontinued because of cyanide release and the concomitant toxicity.

*In vitro* metabolic studies on malononitrile were described by Stern *et al.*[76] In the presence of thiosulfate, brain, liver, and kidney slices metabolized malononitrile to thiocyanate. The formation of thiocyanate from malononitrile and thiosulfate was highest in the presence of liver tissues, lowest with brain, and intermediate with kidney. The liver enzyme system was saturated at 0.0033 $M$ malononitrile; the pH optimum was 7.0. This enzyme system was inhibited by cysteine and glutathione and was inactivated by heat. Stern *et al.*[76] also indicated that the total amount of cyanide and thiocyanate formed from malononitrile by tissue preparation was higher in the presence than in the absence of thiosulfate.

### b. SUCCINONITRILE

Succinonitrile ($N\equiv CCH_2CH_2C\equiv N$) is the dinitrile most widely used for therapeutic purposes. It is employed in Europe as an antidepressant agent. However, Marigo and Pappalardo[77] reported five cases of death in the course of therapeutic treatment with succinonitrile; cyanide was present in the viscera of these patients. Contessa and Santi[78] studied the biological fate and cyanide release from succinonitrile *in vivo* and *in vitro* in rats and rabbits. They reported a sixfold increase in urinary thiocyanate after intravenous administration of succinonitrile, compared to untreated animals, and the thiocyanate content of urine returned to normal 48–120 hr after treatment.

Succinonitrile is metabolized by the liver,[78-82] and the cyanide liberated is thought to be one of the metabolites responsible for its toxicity. About 70 to 90% of the succinonitrile administered to animals is eliminated as thiocyanate. The toxicity of succinonitrile can be altered with mixed-function oxidase modulators. Carbon tetrachloride blocks cyanide release from succinonitrile, whereas ethanol, on the other hand, halves its $LD_{50}$.[81] This effect of ethanol was suggested to be due to both enhanced cyanide liberation from succinonitrile and a decreased sulfur availability for thiocyanate formation.[82] In an *in vitro* metabolic study, Floreani *et al.*[82] concluded that cyanide liberation from succinonitrile is a multistep process in which the mitochondrial membrane and the endoplasmic reticulum are involved. Despite extensive studies on succinonitrile metabolism, no metabolites other than cyanide have been reported. It is important to determine the fate of the remainder of the molecule to verify the role of cyanide and other metabolites in succinonitrile toxicity.

### c. TETRAMETHYLSUCCINONITRILE

Animals treated with tetramethylsuccinonitrile

$$NC-\underset{\underset{CH_3}{|}}{\overset{\overset{CH_3}{|}}{C}}-\underset{\underset{CH_3}{|}}{\overset{\overset{CH_3}{|}}{C}}-CN$$

develop violent convulsions; asphyxia followed by death occurs within 1 min to 5 hr after the convulsion.[83] Sodium thiosulfate and sodium nitrite, the cyanide antidotes, had no effect on tetramethylsuccinonitrile toxicity. However, quick-acting barbiturates and phenobarbital protected against its toxicity. Whether this compound is highly toxic because of cyanide release or via the formation of other highly toxic metabolites is not known; however, if all nitriles are postulated to be metabolized via an $\alpha$ oxidation to release cyanide, tetramethylsuccinonitrile, because it does not contain an α-hydrogen, is not susceptible to oxidation. Therefore, it seems reasonable that no cyanide will be released, and, hence, the observed lack of effectiveness of cyanide antidotes against tetramethylsuccinonitrile toxicity is expected.

d. ADIPONITRILE

Ghiringhelli[84] described human toxicity of adiponitrile ($NCCH_2CH_2CH_2CH_2CN$) as including deep anesthesia, headache, vertigo, vomiting, cyanosis of the skin and mucosa, tachypnea, tachycardia, hypotension, mydriasis, and clonic convulsions of the limbs. Such symptoms were reversed completely by treatment of the patients with glucose and sodium thiosulfate.

Animal studies indicated that the concentrations of thiocyanate in the blood and urine of guinea pigs injected with adiponitrile are proportional to the doses administered. Of the dose of adiponitrile administered, 79% was eliminated as thiocyanate in the urine. Of the cyanide antidotes, thiosulfate was most effective in protecting against adiponitrile poisoning, and nitrite was less effective. However, on the basis of the ratio between administered adiponitrile dose and quantity of cyanide detected, Ghiringhelli[60] concluded that a greater part of the dose is metabolized to cyanide.

## V. Unsaturated Aliphatic Nitriles

α,β-Unsaturated nitriles are highly reactive compounds from a metabolic point of view. Acrylonitrile or vinyl cyanide ($CH_2=CH-CN$) bears a structural resemblance to vinyl chloride, a well-known carcinogen. On the other hand α,β-unsaturated nitriles can undergo Michael additions and act as direct alkylating agents.

A. ACRYLONITRILE

Acrylonitrile ($CH_2=CH-C\equiv N$, VCN) is a monomer widely used in the manufacture of synthetic fibers, rubber, plastics, plasticizers, and adhesives, and as intermediate for various products. It is a demonstrated carcinogen,[85,86] teratogen,[87,88] and potent central nervous system toxin.[5,93-95] Biochemically, VCN has been found to influence tissue thiol content, pyruvate metabolism, red cell metabolism.[9,94,96] and blood coagulation, and to cause adrenal apoplexy in laboratory animals.[97]

Figure 4 illustrates the structure of VCN in terms of (a) its simple formula, (b) the orbitals of the π-bond electrons in the double and triple bonds, (c) the

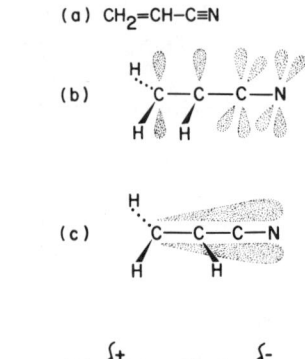

**Fig. 4.** Structure and electronic configuration of acrylonitrile.

electron delocalization of this conjugated molecule, and (d) the effective charge polarization due to the electron delocalization.

Because of the electron deficiency of the $\beta$-carbon, VCN readily adds to nucleophiles (RẌH) by cyanoethylation [Eq. (14)].

$$RẌH + CH_2CHCN \rightarrow RXCH_2CH_2CN \qquad (14)$$

This Michael addition reaction occurs almost quantitatively with alcohols, phenols, sulfhydryls, and amines with or without a catalyst.[98] The double carbon–carbon bond with the partial positive charge on the $\beta$-carbon is susceptible to oxidation reactions.[99] The triple nitrile bond is susceptible to acid- or base-catalyzed hydrolysis to yield carboxylic acids.[100]

## 1. Metabolism of Acrylonitrile

Extensive metabolic studies have been reported, which explain in part the bioactivation and degradation of acrylonitrile.

### a. CYANIDE LIBERATION

Increased blood and urine concentrations of thiocyanate in animals were reported after VCN administration.[101] Brieger *et al.*[5] found that acute VCN exposure also produced increased blood concentrations of cyanomethemoglobin. In dogs, which are particularly susceptible to VCN toxicity, the concentration of cyanomethemoglobin increased with length of exposure, so that by the end of the lethal exposure period, most of the methemoglobin present was converted to cyanomethemoglobin.[5]

Acrylonitrile, clearly, is capable of liberating cyanide under biological conditions. However, the percentage of the total urinary excretion of thiocyanate after VCN administration ranges from 4 to 25% of the administered dose.[5,10,96,102]

Gut et al.[8] found that the conversion of VCN to cyanide was dependent on the route of administration and decreased in the following order: oral > intraperitoneal > subcutaneous > intravenous. Thus, the more slowly VCN enters the system (oral administration), the more extensively it is converted to cyanide. This suggests that conversion of VCN to cyanide involves saturable metabolic processes.

Ahmed and Patel[93] studied the metabolism of VCN to cyanide in both rats and mice. In rats, early signs of VCN toxicity were cholinomimetic, which were different from the central nervous system disturbances observed after giving potassium cyanide. In mice, however, the only signs of acrylonitrile toxicity were central nervous system effects; these were identical to those seen after giving potassium cyanide. Treatment of rats and mice with phenobarbital or Aroclor 1254, or fasting, increased blood cyanide concentrations, whereas treatment with cobaltous chloride or SKF 525A resulted in decreased blood cyanide concentrations. These data indicate species differences in VCN toxicity and metabolism and suggest that VCN is metabolized to cyanide by a mixed-function oxidase enzyme system.

*In vitro,* the metabolism of VCN to cyanide was localized in the microsomal fraction of rat liver and required NADPH and $O_2$.[103-105] Metabolism of VCN was increased in microsomes obtained from phenobarbital-, Aroclor 1254-, or 3-methylcholanthrene-treated rats and decreased after cobaltous chloride treatment. Addition of SKF 525A or carbon monoxide to the incubation mixture inhibited VCN metabolism. Addition of the epoxide hydrase inhibitor 1,1,1-trichloropropane 2,3-oxide decreased the formation of cyanide from VCN. The addition of glutathione, cysteine, D-penicillamine, or 2-mercaptoethanol enhanced the release of cyanide from VCN. These findings indicated that VCN is metabolized to cyanide by a cytochrome *P*-450-dependent mixed-function oxidase system.

Earlier investigators believed that the aliphatic nitriles, including VCN, might be direct inhibitors of cytochrome *c* oxidase. The *in vitro* studies in our laboratory[106,107] and studies by Willhite and Smith[108] and Nerudova et al.[109] showed no inhibition of cytochrome *c* oxidase by nitriles. Nerudova et al.[109] reported that the administration of lethal (100 mg/kg) or sublethal doses (40 mg/kg = $LD_{50}$) of VCN to mice inhibited cytochrome *c* oxidase in liver and brain. In rats, after giving $LD_{50}$ doses of VCN, a 50% inhibition of cytochrome *c* oxidase in liver, kidney, and brain was observed.[106,107] Nerudova et al.[109] suggested that after the administration of a lethal as well as $LD_{50}$ dose of VCN, cyanide is present in the organism in a concentration that produces a 50% *in vitro* inhibition of cytochrome *c* oxidase.

b. GLUTATHIONE DEPLETION

Protein and soluble thiol contents are depleted after VCN administration. Hashimoto and Kanai[94] found the decrease in soluble nonprotein thiol concentrations to be most marked in the liver (13% of control), less in brain (50% of control), and least in the blood (83% of control) of guinea pigs 1 hr after giving a $2 \times LD_{50}$ dose of VCN. Dinu[110,111] also measured nonprotein and protein thiol levels 1 hr after giving a $2 \times LD_{50}$ dose of VCN to rats. She found the most marked depletion of soluble nonprotein thiol concentrations in the liver, kidney, and adrenals, and less marked decreases in brain, lung, and testes. However, protein thiol contents were depleted only in the brain and lung. Szabo et al.[112] also reported similar decreases in glutathione contents at earlier time periods.

These changes in glutathione levels after giving VCN likely reflect the formation of an acrylonitrile–glutathione conjugate. The *in vitro* reaction rate of VCN with glutathione S-transferases of rat hepatic cytosol (determined by rate of glutathione depletion) was low compared with other compounds known to form glutathione conjugates.[113] In contrast, VCN was effective *in vivo* in depleting hepatic glutathione contents by a reaction catalyzed by glutathione S-transferase.

Ghanayem and Ahmed[114,115] reported that 27% of administered VCN was excreted in the bile of rats within 6 hr. Treatment of rats with cobaltous chloride or overnight fasting significantly increased biliary excretion, treatment with phenobarbital produced no change, and diethyl maleate treatment significantly decreased biliary excretion of VCN in 6 hr. Four biliary metabolites were identified as glutathione conjugates of VCN, which indicates a significant role for glutathione in VCN biotransformation.

Numerous other metabolic studies have indicated that VCN is metabolized by two pathways as shown in Fig. 5.[93,104,105,114-123] The first major pathway of VCN metabolism is its direct conjugation with glutathione, which is catalyzed by glutathione S-transferases.[113-115] The product S-(2-cyanoethyl)glutathione is further metabolized to N-acetyl-S-(2-cyanoethyl)cysteine. The other pathway involves an epoxide intermediate and is catalyzed by the hepatic microsomal P-450 system.[104,105,116,123] Further metabolism of the epoxide by glutathione epoxide transferase may result in another glutathione conjugate, depending on the site of nucleophilic attack on the epoxide molecule.[114,115] Alternatively, the epoxide intermediate could produce cyanide and other metabolites by rearrangement or by catalysis with epoxide hydrase.[104,105] Further metabolism of glutathione conjugates leads to the excretion of various mercapturic acids.[116,121]

Guengerich et al.[116] have demonstrated that rat liver microsomes or a reconstituted cytochrome P-450 system catalyzed the mixed-function oxi-

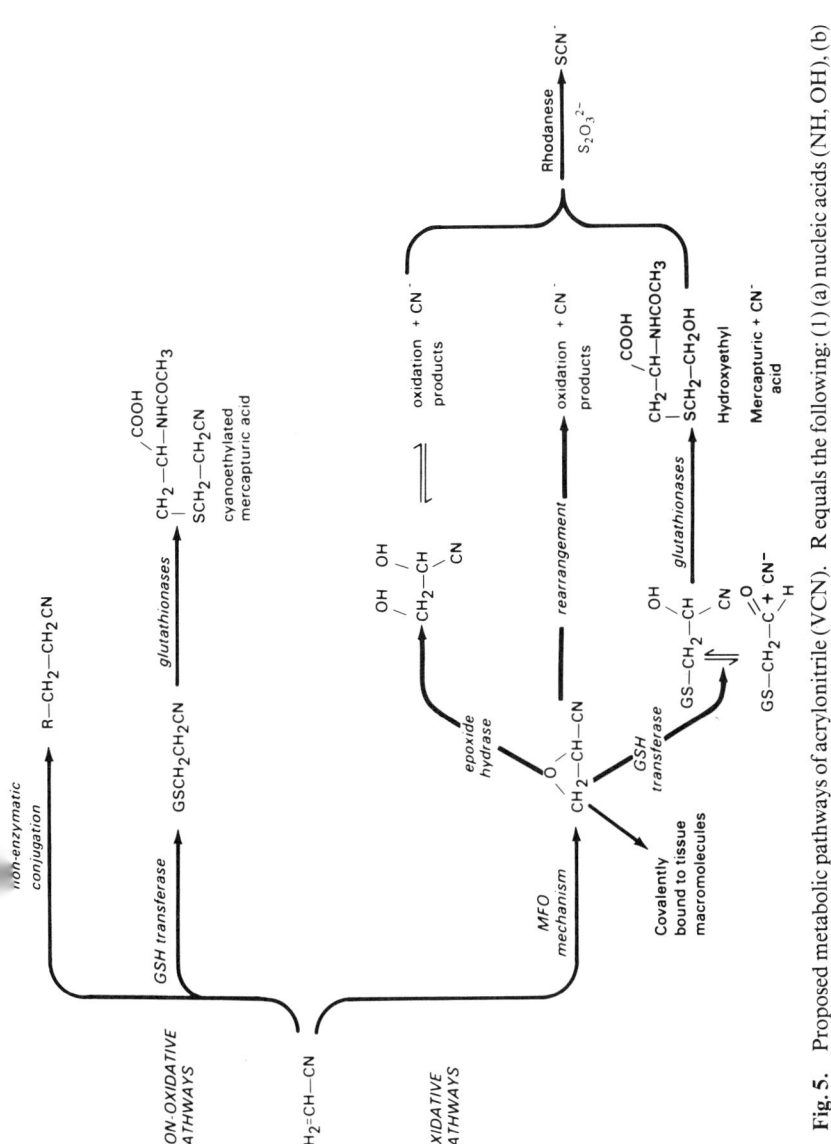

**Fig. 5.** Proposed metabolic pathways of acrylonitrile (VCN). R equals the following: (1) (a) nucleic acids (NH, OH), (b) proteins (NH$_2$, OH, SH), (c) lipids (OH, NH, PO$_4$); (2) biological neurotransmitters: (a) adrenaline and its analogs, (b) serotonin, (c) $\gamma$-aminobutyric acid, (d) histamine; or (3) other nucleophilic components of tissues. (Modified from Ahmed and Trieff.[11])

dation of VCN to 2-cyanoethylene oxide, and this epoxide served as a substrate for microsomal epoxide hydrase.

### 2. Covalent Binding of VCN

In addition to biochemical effects of VCN, several reports have described molecular interaction of VCN with target and nontarget tissues.[124-131] Pharmacokinetic and distribution studies have shown irreversible incorporation of radioactivity from [1-$^{14}$C]VCN into various macromolecules.[124-128] Whole-body autoradiography of rats and monkeys after oral and intravenous administration of [1-$^{14}$C]VCN has shown uptake of radioactivity in the blood, liver, kidney, lung, adrenal cortex, and gastric mucosa.[130] Farooqui and Ahmed[126,127] found that rat erythrocytes retained covalently bound radioactivity for more than 10 days. A maximum of 94% of the $^{14}$C from VCN was covalently bound to cytoplasmic and membrane proteins in erythrocytes. The study indicated that VCN may damage red cells by mechanisms other than release of cyanide. The irreversible reaction of VCN or its reactive metabolites with cellular sulfhydryls or other nucleophiles has been suggested to be of significance for its acute toxicity.[97] Ahmed et al.[125] and Farooqui et al.[128] have shown extensive *in vivo* covalent binding of [1-$^{14}$C]- and [2,3-$^{14}$C]VCN to macromolecules in various target and nontarget tissues. Guengerich et al.[116] have demonstrated the *in vitro* covalent binding of VCN and its metabolite 2-cyanoethylene oxide to proteins and nucleic acids. The *in vivo* studies of Farooqui and Ahmed[129] have shown extensive covalent binding of [2,3$^{14}$C]acrylonitrile to DNA, RNA, and proteins. The covalent binding indices for DNA in liver, stomach, and brain were 6, 52, and 65, respectively.

### 3. Comments on Metabolic Pathways of VCN

Although Fig. 5 proposes four potentially toxic pathways, only the glutathione *S*-transferase-dependent pathway (second from the top) should lead directly to a more stable and excretable product, a cyanoethylated mercapturic acid. Nonenzymatic conjugation (top pathway) by cyanoethylation is possibly the most dangerous pathway because of the numerous biologically important macromolecular nucleophiles that could be attacked. Conjugation of VCN with nucleic acids has carcinogenic potential. In addition, conjugation with neurotransmitters could account for the effects on the central nervous system, and extensive conjugation with vulnerable enzymes of a particular organ could impair organ function. Proteins and other macromolecules containing exposed thiols and primary amines may be particularly vulnerable to attack.

# 17. NITRILES

Activation by a mixed-function oxidase (MFO)-dependent mechanism could lead to a reactive epoxide. This epoxide may be biotransformed by epoxide hydrase or rearranged to a molecular species that liberates cyanide and forms potentially toxic compounds, such as glycoaldehyde or glyoxal. Alternatively, this epoxide could be detoxified by glutathione S-transferase and further biotransformed to excretable products, such as mercapturic acids.

The proposed pathways shown in Fig. 5 account for the known effects of acrylonitrile on the central nervous system, for injury to specific biological systems or organs, for depletion of glutathione, and for the suspected carcinogenicity; they also indicate potential metabolic products.

Figure 6 proposes two potentially toxic pathways for acrylonitrile. Interaction with a biological radical initiator could lead to the formation of an acrylonitrile radical, which must either undergo (1) termination by interaction with a second radical, such as an enzyme component, or (2) a propagation reaction. Polyunsaturated fatty acid moieties of phospholipids are particularly vulnerable to radical propagation reactions because of the labile hydrogen atoms of methylene bridge carbons. Lipid radicals readily react with the diradical oxygen and undergo oxidative degradation to aldehydes and acids. Thus, there is a rational basis for the postulated injury-producing "peroxidative decomposition" reactions of membranous components of cells and tissues.

In summary, acrylonitrile is a demonstrated tumorigen, teratogen, mutagen, and potent central nervous system toxin, and causes injury to a variety of other organs. The underlying chemical mechanisms for these injurious actions are uncertain, in that little is known concerning the biological fate of

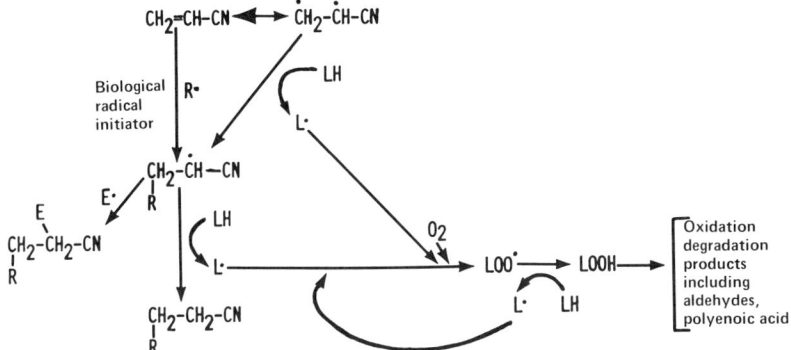

**Fig. 6.** Proposed metabolic pathways of acrylonitrile showing interactions with a biological radical initiator. L, lipid species, such as polyunsaturated fatty acid; E, enzyme component; R, radical or $e^-$ and $H^+$.

acrylonitrile. Two mechanisms must be considered—cyanide liberation and alkylation of tissue nucleophiles.

The liberation of cyanide from acrylonitrile, an enzymatic process, may have a role in the toxic response but cannot entirely account for its acute toxicity, because tests on various animal species have indicated that antidotes against cyanide poisoning are not entirely effective. Studies that have demonstrated modulating effects on acrylonitrile toxicity, including fasting and cysteine administration (both of which affect levels of the soluble antioxidant glutathione), suggest that conjugation of acrylonitrile with glutathione may play a role in detoxication. The studies that demonstrate some enhancing effects of liver enzymes on mutagenic response suggest that enzymatic processes may promote injurious biotransformation reactions of acrylonitrile.

## B. CROTONONITRILE

Crotononitrile ($CH_3CH=CH-C\equiv N$) is metabolized to two mercapturic acids (Fig. 5) in a manner similar to acrylonitrile. However, the proportion of a dose of crotononitrile converted to the hydroxyethylmercapturic acid was much lower than that for acrylonitrile.[123,132] Thus, after giving a dose of 0.05 mmol of the nitriles to rats, the ratios of cyanoethylmercapturic acid to hydroxyethylmercapturic acid were about 2:1 after a dose of acrylonitrile and about 10:1 after a dose of crotononitrile.[115,126]

Another unsaturated nitrile, the cinnammonitrile, will be discussed under alkylaryl cyanides.

## VI. Alkylaryl Nitriles

Phenylalkyl cyanides undergo degradation analogous to the alkyl cyanides, except that conversion into cyanide and thiocyanate occurs more efficiently with these compounds. Benzyl cyanide is chiefly metabolized into mandelonitrile and hence to benzaldehyde and benzoic acid, as shown in Eq. (15).

Benzyl Cyanide → Mandelonitrile → Benzoic Acid → $CN^-$ → $SCN^-$ (15)

## A. AMYGDALIN (LAETRILE)

Amygdalin (D-mandelonitrile-β-D-glucosido-6-β-D-glucoside) [Eq. (16)] is a cyanogenic glycoside found in a variety of plant species, particularly in the seeds of apricots and bitter almonds. Laetrile has been proposed to be effective in both the prevention and the treatment of human cancer[133,134] although many reports do not show a significant effect of amygdalin in experimental tumor systems.[135-137]

Toxicological and pharmacological studies of amygdalin and related cyanogenic glycosides have indicated that the toxicity is related to cyanide liberation.[138] The enzyme α-glucosidase, found predominantly in seeds and nuts containing the glucosides, hydrolyzes amygdalin into two molecules of glucose and one molecule of mandelonitrile.[139] The cyanohydrin is unstable and readily decomposes to benzaldehyde and hydrocyanic acid [Eq. (16)], and this decomposition may also be catalyzed by nitrile lyase, an enzyme found in glycoside-containing plants.[134] In dogs, the toxicity of amygdalin is clearly related to formation of cyanide.[140] High blood concentrations of cyanide and cyanide toxicity have also been reported in rats after the oral administration of amygdalin.[141]

$$\text{Amygdalin} \longrightarrow \text{Glucose} + \text{Glucose} + \text{Mandelonitrile} \longrightarrow \text{Benzoic acid} + \text{HCN} \quad (16)$$

## B. 2-CHLOROBENZYLIDENE MALONONITRILE

Coroson and Stoughton[142] prepared a number of benzylidene nitriles including 2-chlorobenzylidene malononitrile (CS) and observed that some

$$\text{2-Cl-C}_6\text{H}_4\text{-CH=C(CN)}_2$$

of these compounds caused sneezing and irritation. 2-Chlorobenzylidene malononitrile is now the active material in a wide range of devices used against civil disturbances and in war.[143-145] Benzylidene malononitriles are reactive substances and form adducts with amines, n-butanethiol, Grignard reagents, hydrazoic acid, hydrogen cyanide, and other compounds.[146-153] Mercapturic acid formation from CS, although likely by analogy with the metabolism of related nitriles, has not been reported.

## C. BROMOBENZYL CYANIDE

Bromobenzyl cyanide [$C_6H_5CH(Br)C\equiv N$], a strong lachrymator, is used as a war gas because of the high reactivity of the halide group and the high electron deficiency on the benzylic carbon. This compound might act by a progressive reaction with sulfhydryl groups rather than cyanide elimination.[154]

## D. CINNAMONITRILE

Cinnamonitrile is an unsaturated alkylaryl cyanide that undergoes conjugation with glutathione, which is catalyzed by the glutathione S-transferases.[113] van Bladeren et al.[123] and Delbressine[132] have demonstrated that cinnamonitrile metabolized in rats to produce hydroxy- and cyanomercapturic acids in the ratio of 98:2.

$$\text{C}_6\text{H}_5\text{-CH=CH-CN}$$

## VII. Aryl Nitriles

Aromatic nitrile groups, such as benzonitriles, are widely used in manufacture of pesticides, notably dichlobenil (2,6-dichlorobenzonitrile) and bromoxymil (3,5-dibromo-4-hydroxybenzonitrile).[155]

Aromatic nitriles undergo two types of metabolic changes *in vivo;* a major pathway involves aromatic hydroxylation and yields a mixture of isomeric monocyanophenols, and a minor route involves hydrolysis, with ultimate formation of aromatic acid. A modified scheme derived from de Bruin[61] and Harper[155] is shown in Eq. (17) with benzonitrile as an example.

$$\underset{\text{Cyanophenol}}{\text{HO-C}_6\text{H}_3\text{-CN}} \xleftarrow{(O)} \underset{\text{Benzonitrile}}{\text{C}_6\text{H}_5\text{-CN}} \xrightarrow{OH^-} \left[ \underset{}{\text{C}_6\text{H}_5\text{-C(OH)=N}^-} \xrightarrow{H^+} \underset{}{\text{C}_6\text{H}_5\text{-C(OH)=NH}} \right] \longrightarrow \underset{\text{Benzamide}}{\text{C}_6\text{H}_5\text{-CONH}_2} \xrightarrow{H_2O} \underset{\text{Benzoic acid}}{\text{C}_6\text{H}_5\text{-COOH}} + \underset{\text{Ammonia}}{\text{NH}_3} \quad (17)$$

Hook and Robinson[156] and Robinson and Hook[157] reported that a microbial enzyme nitrilase, which has rather restricted substrate specificity, is responsible for nitrile group metabolism. Mimura et al.[158] reported that in *Corynebacterium* species, the metabolism of nitrile yields the corresponding amide. Mahadevan and Thimann[159] have examined a nitrilase that catalyzed the direct conversion of the nitrile into the corresponding acid with no detectable formation of amide. Harper[155] found that in *Nocardia* species, benzonitrile degradation yields benzoic acid and catechol. An enzyme nitrilase was found responsible for catalyzing the conversion of benzonitrile directly to benzoic acid without the intermediate formation of benzamide.

## VIII. Summary

The nitriles, because of their versatile physical and chemical properties, are significant as solvents and as intermediates in polymer, plastic, synthetic fiber, resin, dyestuff, pharmaceutical, vitamin, insecticide, and pesticide industries. Several reports have described fatalities or intoxication in humans by various aliphatic and aromatic nitriles.

Since the early work of Lang[16] and Giacosa[101] indicated that aliphatic nitriles are metabolized to cyanide ions, many investigators have assumed that the toxicity of aliphatic nitriles is related to their ability to liberate cyanide ion,[5-10] and others that the toxicity of aliphatic nitriles is due to the "molecule as a whole" or to metabolites other than cyanide.[96,102,106,107,125,126] The detailed biochemical mechanisms underlying the toxic actions of aliphatic nitriles remain unknown.

Although most of the aliphatic nitriles are potent neurotoxins, their toxic actions on other organ systems differ. Acrylonitrile causes adrenal necrosis,[97,160] and its saturated homolog propionitrile causes duodenal ulcers.[50] Malononitrile produces nuclear changes in neurons and satellite spinal ganglia,[161] but thyroid hyperemia and hyperplasia are seen after exposure to acetonitrile.[162] Kriess et al.[163] have reported neurological bladder dysfunction in workers exposed to dimethylaminopropionitrile.

Aliphatic nitrile biotransformation (Fig. 5) requires activation of the molecule before release of the cyanide group. The instability of the cyanohydrin formed suggests that a nascent form of aldehyde may be formed; it may escape the detoxication mechanisms of the cell and, perhaps, have a role in the mechanism of toxicity of these compounds. Indeed, Hickman[164] and Gescher et al.[165] have reported a role for the toxicity of the nascent aldehyde formed as a result of instability of a carbinolamine. Elucidation of the mechanism of cyanide liberation would contribute to the understanding of the toxic actions of aliphatic nitriles that are known to liberate cyanide. Clarification of the other metabolic fates of aliphatic nitriles is essential to the understanding of the molecular events underlying their toxicity. Studies describing the *in vivo* fate of metabolic products other than cyanide are singularly lacking.

## Acknowledgments

Research in this laboratory on topics related to the subject matter of this chapter is supported by National Institute of Health Grant ES 01871 and R. A. Welch Foundation, Grant H-416. The skillful assistance of Mrs. Ruth Buffington in preparing the manuscript is gratefully acknowledged.

## References

1. Thompson, A. R. (1972). Organic nitrogen compounds. *Chem. Technol.* **4**, 532.
2. U.S. Department of Health, Education and Welfare. (1978). "NIOSH Criteria for a Recommended Standard—Occupational Exposure to Nitriles." U.S.D.H.E.W., Washington, D.C.
3. Trieff, N. M., and Ahmed, A. E. (1979). Acrylonitrile: Mammalian toxicity and human health effects, Environmental Protection Agency. *Fed. Regist.* **44**, 69479–69484.
4. Willhite, C. C. (1979). Studies on the basis of the toxicity of aliphatic organonitriles. Ph.D. Thesis, Dartmouth College, Hanover, New Hampshire.
5. Brieger, H., Rieders, F., and Hode, W. A. (1952). Acrylonitrile: Spectrophotometric determination, acute toxicity and mechanism of action. *Arch. Ind. Hyg. Occup. Med.* **6**, 128–140.

6. Dudley, H. C., and Neal, P. A. (1942). Toxicity of acrylonitrile (vinyl cyanide). I. A study of the acute toxicity. *J. Ind. Hyg. Toxicol.* **24,** 27–36.
7. Fassett, D. W. (1963). Cyanide and nitriles. *In* "Industrial Hygiene and Toxicology" (F. A. Patty, ed.), 2nd rev. ed., pp. 1991–2036. Wiley, New York.
8. Gut, I., Nerudova, J., Kopecky, J., and Holecek, V. (1975). Acrylonitrile biotransformation in rats, mice and Chinese hamsters as influenced by the route of administration and by phenobarbital, SKF 525-A, cysteine, dimercaprol, or thiosulfate. *Arch. Toxicol.* **33,** 151–161.
9. Hashimoto, K., and Kanai, R. (1965). Toxicology of acrylonitrile, metabolism, mode of action, and therapy. *Ind. Health* **3,** 30–46.
10. Paulet, G., Desnos, J., and Battig, J. (1966). The toxicity of acrylonitrile. *Arch. Mal. Prof. Med. Trav. Secur. Soc.* **27,** 849.
11. Ahmed, A. E., and Trieff, N. M. (1982). Aliphatic nitriles: Metabolism and toxicity. *Prog. Drug Metab.* **7,** 229–294.
12. U.S. Department of Health, Education and Welfare. (1976). "NIOSH Criteria for a Recommended Standard—Occupational Exposure to Hydrogen Cyanide and Cyanide Salts [NaCN, KCN and Ca(CN)$_2$]." U.S.D.H.E.W., Washington, D.C.
13. Albaum, H. G., Tepperman, J., and Bodansky, O. (1964). A spectrophotometric study of the competition of methemoglobin and cytochrome oxidase for cyanide *in vitro. J. Biol. Chem.* **163,** 641–647.
14. Ansell, M., and Lewis, F. A. S. (1970). A review of cyanide concentrations found in human organs—a survey of literature concerning cyanide metabolism, "normal," nonfatal, and fatal body cyanide levels. *J. Forensic Med.* **17,** 148–155.
15. Halström, F., and Møller, K. O. (1945). The content of cyanide in human organs from cases of poisoning with cyanide taken by mouth—with a contribution to the toxicology of cyanides. *Acta Pharmacol. Toxicol.* **1,** 18–28.
16. Lang, S. (1894). Über die Umwandlung des Acetonitrils und seiner Homologen im Tierkorper. *Arch. Exp. Pathol. Pharmakol.* **34,** 247–258.
17. Baumann, E. J., Sprinson, D. B., and Metzger, N. (1933). The relation of thyroid to the conversion of cyanides to thiocyanate. *J. Biol. Chem.* **102,** 773–782.
18. Mukerji, B., and Smith, R. G. (1943). Cyanide detoxication in rabbit and dog as measured by urinary thiocyanate excretion. *Ann. Biochem. Exp. Med.* **39,** 23–34.
19. Smith, R. P. (1980). *In* "Toxicology: The Basic Science of Poisons" (J. D. Doull, C. D. Klassen, and M. O. Amdur, eds.), 2nd ed., pp. 328–329. Macmillan, New York.
20. Boxer, G. E., and Rickards, J. C. (1952). Studies on the metabolism of the carbon of cyanide and thiocyanate. *Arch. Biochem. Biophys.* **39,** 7–26.
21. Wood, J. L., Williams, E. F., and Kingsland, N. (1947). The conversion of thiocyanate sulfur to sulfate in the white rat. *J. Biol. Chem.* **170,** 251–259.
22. Goldstein, F., and Reiders, F. (1951). Formation of cyanide in dogs and man following administration of thiocyanate. *Am. J. Physiol.* **167,** 47–59.
23. Goldstein, F., and Reiders, F. (1953). Conversion of thiocyanate to cyanide by an erythrocytic enzyme. *Am. J. Physiol.* **173,** 287–290.
24. Lang, K. (1933). [Rhoadanyl formation in the animal body.] *Biochem. Z.* **259,** 243–256.
25. Williams, R. T. (1959). "Detoxication Mechanisms," 2nd ed., pp. 390–395. Chapman & Hall, London.
26. Sörbo, B. H. (1951). On the properties of rhodanese. *Acta Chem. Scand.* **5,** 724–734.
27. Sörbo, B. H. (1951). On the active group in rhodanese. *Acta Chem. Scand.* **5,** 1218–1219.
28. Sörbo, B. H. (1953). Crystalline rhodanese. *Acta Chem. Scand.* **7,** 238–239.

29. Fromageot, C., and Royer, A. (1945). La présence constante du thiosulfate dans l'urine des animaux supérieurs et sa signification physiologique. *Enzymologia* **11**, 361–372.
30. Villarejo, M., and Westley, J. (1963). Rhodanese-catalzyed reduction of thiosulfate by reduced lipoic acid. *J. Biol. Chem.* **238**, 1135–1136.
31. Wood, J. L., and Cooley, S. L. (1952). Thiocyanate formation from cystine derivatives. *Fed. Proc., Fed. Am. Soc. Exp. Biol.* **4**, 314.
32. Himwich, W. A., and Saunders, J. P. (1948). Enzymatic conversion of cyanide to thiocyanate. *Am. J. Physiol.* **153**, 348–354.
33. Meister, A., Fraser, P. E., and Tice, S. V. (1954). Enzymatic desulfuration of $\beta$-mercaptopyruvate. *J. Biol. Chem.* **246**, 294–301.
34. Wood, J. L., and Cooley, S. L. (1956). Detoxication of cyanide by cystine. *J. Biol. Chem.* **218**, 449–457.
35. Gal, E. M., and Greenberg, D. M. (1953). Metabolism of nitriles. *Fed. Proc., Fed. Am. Soc. Exp. Biol.* **12**, 207.
36. Okhawa, H., Okhawa, R., and Yamamoto, Z. (1972). Enzymatic mechanisms and toxicological significance of hydrogen cyanide liberation from various organothiocyanates and organonitriles in mice and houseflies. *Pestic. Biochem. Physiol.* **2**, 95–112.
37. Amdur, M. L. (1959). Accidental group exposure to acetonitrile — A clinical study. *J. Occup. Med.* **1**, 627–633.
38. Dequidt, J., Furon, D., and Haguenoer, J. M. (1972). [Fatal intoxication by acetonitrile.] *Bull. Soc. Pharm. Lille* **4**, 143–148.
39. Dequidt, J., and Haguenoer, J. M. (1972). [Experimental toxicological study of acetonitrile in rats. I. Acute intraperitoneal intoxication.] *Bull. Soc. Pharm. Lille* **4**, 149–154.
40. Dequidt, J., Furon, D., Wattel, F., Haguenoer, J. M., Scherpereel, P., Gosselein, B., and Ginestet, A. (1974). [Intoxication with acetonitrile with a report on a fatal case.] *Eur. J. Toxicol.* **7**, 91–97.
41. Grabois, B. (1955). Fatal exposure to methyl cyanide. *Mon. Rev. N. Y. State Dep. Labor* **34**, 7–8.
42. Grahl, V. R. (1970). Toxikologie und Wirkungsweise von Acrylonitrile. *Zentralb. Arbeitsmed. Arbeitsschutz* **20**, 369–378.
43. Haguenoer, J. M., Dequidt, J., and Jacquemont, M. C. (1975). [Experimental acetonitrile intoxications. I. Acute intoxications by the intraperitoneal route.] *Eur. J. Toxicol.* **8**, 94–101.
44. Haguenoer, J. M., Dequidt, J., and Jacquemont, M. C. (1975). [Experimental acetonitrile intoxications. II. Acute intoxications by the pulmonary route.] *Eur. J. Toxicol.* **8**, 102–106.
45. Hunt, R. (1923). The acetonitrile test for thyroid and of some alterations of metabolism. *Am. J. Physiol.* **63**, 257–299.
46. Frimin, J. L., and Gray, D. O. (1976). The biochemical pathway for the breakdown of methyl cyanide (acetonitrile) in bacteria. *Biochem. J.* **158**, 223–229.
47. Giampaolo, C., Feldman, D., Reynolds, E. S., Dzau, V. J., and Szabo, S. (1975). Ultrastructural characterization of propionitrile-induced duodenal ulcer in the rat. *Fed. Proc., Fed. Am. Soc. Exp. Biol.* **34**, 227.
48. Haith, L. R., Jr., Szabo, S., and Reynolds, E. S. (1975). Prevention of propionitrile — or cysteamine-induced duodenal ulcer by vagotomy or hypophysectomy in rats. *Clin. Res.* **23**, 577A.
49. Robert, A., Nezamis, J. E., and Lancaster, C. (1975). Duodenal ulcers produced in rats by propionitrile — factors inhibiting and aggravating such ulcers. *Toxicol. Appl. Pharmacol.* **31**, 201–207.
50. Szabo, S., and Selye, H. (1972). Duodenal ulcers produced by propionitrile in rats. *Arch. Pathol.* **93**, 390–391.

51. Szabo, S., Bailey, K. A., Boor, P. J., and Jaeger, R. J. (1977). Acrylonitrile and tissue glutathione: Differential effect of acute and chronic interactions. *Biochem. Biophys. Res. Commun.* **79**, 32–37.
52. Meurice, J. (1900). [Intoxication and detoxification of different nitriles with sodium thiosulfate and metal salts.] *Arch. Int. Pharmacodyn. Ther.* **7**, 11–53.
53. Thiess, A. M., and Hey, W. (1969). [On the toxicity of isobutyronitrile and $\alpha$-hydroxyisobutyronitrile (acetone cyanohydrin).] *Arch. Toxikol.* **24**, 271–282.
54. Zeller, H. V., Hofmann, H. T., Thies, A. M., and Hey, W. (1969). Zur Toxizität der Nitrile. *Zentralbl. Arbeitsmed. Arbeitsschultz* **19**, 225–238.
55. Haguenoer, J. M., Dequidt, J., and Jacquemont, M. C. (1974). [Experimental intoxication by butyronitrile.] *Bull. Soc. Pharm. Lille* **4**, 161–171.
56. Szabo, S., and Reynolds, E. S. (1975). Structure activity relationships for ulcerogenic and adrenocorticolytic effects of alkylnitriles, amines and thiols. *Environ. Health Perspect.* **11**, 1350–1338.
57. Tsurmi, K., and Kawada, K. (1971). [Acute toxicity of isobutyronitrile.] *Gifu Ika Daigaku Kiyo* **18**, 655–664.
58. Pattison, F. L. M., Cott, W. J., Howell, W. C., and White, R. W. (1956). Toxic fluorine compounds. V. $\omega$-Fluoro-nitriles and $\omega$-Fluoro-$\omega'$-nitroalkanes. *J. Am. Chem. Soc.* **78**, 3483–3487.
59. Sunderman, F. W., and Kincaid, J. F. (1953). Toxicity studies of acetone cyanohydrin and ethylene cyanohydrin. *Arch. Ind. Hyg. Occup. Med.* **8**, 371–376.
60. Ghiringhelli, L. (1956). Studio comparitivo sulla tossicita di alcuni nitrilli e di alcune amidi. *Med. Lav.* **47**, 192–199.
61. de Bruin, A. (1976). "Biochemical Toxicology of Environmental Agents," pp. 109–113. Elsevier/North-Holland Biomedical Press, Amsterdam.
62. Wolfsie, J. H. (1960). Glycolonitrile toxicity. *J. Occup. Med.* **2**, 588–590.
63. Shkodich, P. E. (1966). [Experimental substantiation of the maximum permissible concentration of acetone cyanohydrin in bodies of water.] *Gig. Sanit.* **31**, 8–12.
64. Sauerhoff, M. W., Braun, W. H., Ramsey, J. C., Humiston, C. G., and Iersey, G. C. (1976). Toxicological evaluation and pharmacokinetic profile of $\beta$-hydroxypropionitrile in rats. *J. Toxicol. Environ. Health* **2**, 31–44.
65. Barrow, M. W., Simpson, C. F., and Miller, E. J. (1974). Lathyrism: A review. *Q. Rev. Biol.* **49**, 101–128.
66. Levene, C. I. (1961). Structural requirements for lathyrogenic agents. *J. Exp. Med.* **114**, 295–310.
67. Sievert, H. W., Lipton, S. H., and Strong, F. M. (1960). Quantitative determination of cyanoacetic acid as an enzymic product of $\beta$-aminopropionitrile. *Arch. Biochem. Biophys.* **86**, 311–316.
68. Fleisher, J. H., Arem, A. J., Chvapil, M., and Peacock, E. E., Jr. (1976). Metabolic disposition of $\beta$-aminopropionitrile in the rat. *Proc. Soc. Exp. Biol. Med.* **152**, 469–474.
69. Keiser, H. R., and Sjoerdsma, A. (1967). Studies on $\beta$-aminopropionitrile in patients with scleroderma. *Clin. Pharmacol. Ther.* **8**, 593–602.
70. McEwen, C. M., Jr., and Cohen, J. D. (1963). An amine oxidase in normal human serum. *J. Lab. Clin. Med.* **62**, 766–776.
71. Fleisher, J. H., Speer, D., Brendel, K., and Chvapil, M. (1977). Metabolism of $\beta$-aminopropionitrile by rabbits. *Fed. Proc., Fed. Am. Soc. Exp. Biol.* **36**, 1069.
72. Fleisher, J. H., Peacock, E. E., and Chvapil, M. (1978). Urinary excretion of $\beta$-aminopropionitrile and cyanoacetic acid. *Clin. Pharmacol. Ther.* **23**, 520–524.
73. Hyden, H., and Hartelius, H. (1948). Stimulation of the nucleoprotein-production in the nerve cells by malononitrile and its effect on psychic functions in mental disorders. *Acta Psychiatr. Neurol., Suppl.* **48**, 1–117.

74. MacKinnon, I. H., Hock, P. H., Cammer, L., and Waelsch, H. B. (1949). The use of malononitrile in the treatment of mental illnesses. *Am. J. Psychiatry* **105,** 686–688.
75. Meyers, D., Shoemaker, T. E., Adamson, W. C., and Sussman, L. (1950). Effect of "malononitrile" on physical and mental status of schizophrenic patients. *Arch. Neurol. Psychiatry* **63,** 586–592.
76. Stern, J., Weil-Malherbe, H., and Green, R. H. (1952). The effects and the fate of malononitrile and related compounds in animal tissues. *Biochem. J.* **52,** 114–125.
77. Marigo, M., and Pappalardo, G. (1966). [A fatal incident from therapeutic administration of succinonitrile.] *Med. Leg. Assicur.* **14,** 155–185.
78. Contessa, A. R., and Santi, R. (1973). Liberation of cyanide from succinonitrile. *Biochem. Pharmacol.* **22,** 827–832.
79. Cavanna, R., and Pocchiari, R. (1972). Fate of succinonitrile-1-$^{14}$C in the mouse. *Biochem. Pharmacol.* **21,** 2529–2531.
80. Contessa, A. R., Floreani, M., Bonetti, A. C., and Santi, R. (1978). Liberation of cyanide from succinonitrile. 2. The effect of ethanol. *Biochem. Pharmacol.* **27,** 1135–1138.
81. Curry, S. H. (1975). Cumulative excretion of succinonitrile in mice. *Biochem. Pharmacol.* **24,** 351–354.
82. Floreani, M., Caspenedo, F., Santi, R., and Contessa, A. R. (1981). Metabolism of succinonitrile in liver: Studies on the systems involved in cyanide release. *Eur. J. Drug Metab. Pharmacokinet.* **6,** 135–140.
83. Hasger, R. N., and Hulpieu, H. R. (1949). Toxicity of tetramethyl succinonitrile and the antidotal effects of thiosulfate nitrite and barbiturates. *Fed. Proc., Fed. Am. Soc. Exp. Biol.* **8,** 205.
84. Ghiringhelli, H. (1955). [Toxicity of adipic nitrile—clinical picture and mechanism of poisoning.] *Med. Lav.* **46,** 221–228.
85. Maltoni, C. (1977). Carcinogenicity bioassays on rats of acrylonitrile administered by inhalation and by ingestion. *Med. Lav.* **68,** 401–411.
86. O'Berg, M. T (1980). Epidemiologic studies of workers exposed to acrylonitrile. *J. Occup. Med.* **22,** 245–256.
87. Murray, F. J., Schwetz, B. A., Witschke, K. D., John, J. A., Norris, J. M., and Gehring, P. J. (1978). *Food and Cosmetic Toxicol.* **16,** 547–556.
88. Murray, F. J., Schwetz, S. A., Nitscke, K. D., John, J. A., Norris, J. M., and Gehring, P. J. (1978). Teratogenicity of acrylonitrile given to rats by gavage or by inhalation. *Food Cosmet. Toxicol.* **16,** 547–551.
89. DeMeester, C., Duverger-Vanbogaert, M., Lambotte-Vandepaer, M., Roberfroid, M., Poncelet, F., and Mercier, M. (1979). Liver extract mediated mutagenicity of acrylonitrile. *Toxicology* **13,** 7–15.
90. Milvy, P., and Wolff, M. (1977). Mutagenic studies with acrylonitrile. *Mutat. Res.* **48,** 271–278.
91. Rabello-Gay, M. N., and Ahmed, A. E. (1980). Acrylonitrile: *In vivo* cytogenetic studies in mice and rats. *Mutat. Res.* **79,** 249–255.
92. Venitt, S., Bushnell, C. T., and Osborne, M. (1977). Mutagenicity of acrylonitrile (cyanoethylene) in *E. coli. Mutat. Res.* **45,** 283–286.
93. Ahmed, A. E., and Patel, K. (1981). Acrylonitrile *in vivo* metabolism in rats and mice. *Drug Metab. Dispos.* **9,** 219–222.
94. Hashimoto, K., and Kanai, R. (1972). Effect of acrylonitrile on sulfydryls and pyruvate metabolism in tissues. *Biochem. Pharmacol.* **21,** 635–641.
95. Krysiak, B., and Knobloch, K. (1971). Effect of acrylonitrile in the central nervous system. *Med. Pr.* **22,** 601–608.
96. Farooqui, M. Y. H., and Ahmed, A. E. (1981). Effect of acrylonitrile and potassium cyanide on red cell metabolism. *Fed. Proc., Fed. Am. Soc. Exp. Biol.* **40,** 678.

97. Szabo, S., Hunter, I., Kovacs, K., Horvath, E., Szabo, D., and Horner, H. C. (1980). Pathogenesis of experimental adrenal hemorrhagic necrosis: Ultrastructural, biochemical, neuropharmacologic, and blood coagulation studies with acrylonitrile in the rat. *Lab. Invest.* **42**, 533–546.
98. Ralls, J. W. (1959). Unsymmetrical 1,6-additions to conjugated systems. *Chem. Rev.* **59**, 329–344.
99. Musgrave, W. K. R. (1974). Organic halogen compounds. *In* "Encyclopedia Britannica Macropaedia," 11th ed., Vol. 13, (Chisholm, H., ed.), pp. 682–693. Benton, Chicago, Illinois.
100. Curran, D. J., and Siggia, S. (1970). Detection and determination of nitriles. *In* "The Chemistry of the Cyano group" (Z. Rappoport, ed.), pp. 168–170. Interscience, New York.
101. Giacosa, P. (1883). [Metabolism of nitriles.] *Hoppe-Seyler's Z. Physiol. Chem.* **8**, 95–113.
102. Benes, V., and Cerna, V. (1959). Acrylonitrile: Acute toxicity and mechanism of action. *J. Hyg., Epidemiol., Microbiol., Immunol.* **3**, 106–110.
103. Abreu, M. E., and Ahmed, A. E. (1979). Studies on the mechanism of acrylonitrile neurotoxicity. *Toxicol. Appl. Pharmacol.* **48**, A54.
104. Abreu, M. E., and Ahmed, A. E. (1980). Metabolism of acrylonitrile to cyanide *in vitro* studies. *Drug Metab. Dispos.* **8**, 376–379.
105. Ahmed, A. E., and Abreu, M. E. (1982). Microsomal metabolism of acrylonitrile in liver and brain. *In* "Biological Reactive Intermediates" (R. Snyder, D. V. Parke, J. J. Kocsis, D. J. Jollow, C. G. Gibson, and C. M. Witmer, eds.), Vol. II, Part B, 1229–1238. Plenum, New York.
106. Ahmed, A. E., Chieco, P., and Patel, K. (1980). Comparative toxicity of aliphatic nitriles. *Toxicol. Lett., Spec. Issue* p. 174.
107. Ahmed, A. E., and Farooqui, M. Y. H. (1982). Comparative toxicities of aliphatic nitriles. *Toxicol. Lett.* **12**, 157–163.
108. Willhite, C. C., and Smith, R. P. (1981). The role of cyanide liberation in the acute toxicity of aliphatic nitriles. *Toxicol. Appl. Pharmacol.* **8**, 589–598.
109. Nerudova, J., Gut, I., and Kopecky, J. (1981). Cyanide effect in acute acrylonitrile poisoning in mice. *In* "Industrial and Environmental Xenobiotics" (I. Gut, M. Cikrt, and G. L. Plaa, eds.), pp. 245–249. Springer-Verlag, Berlin.
110. Dinu, V. (1975). Activity of glutathione peroxidase and catalase and the concentration of lipid peroxides in acute intoxication with acrylonitrile. *Rev. Roum. Biochim.* **12**, 11–17.
111. Dinu, V. (1975). Intracellular thiol concentration in acrylonitrile intoxication. *Rev. Roum. Biochim.* **12**, 155–158.
112. Szabo, S., Bailey, K. A., Boor, P. J., and Jaeger, R. J. (1977). Acrylonitrile and tissue glutathione, differential effect of acute and chronic interactions. *Biochem. Biophys. Res. Commun.* **79**, 32–37.
113. Boyland, E., and Chasseaud, L. F. (1967). Enzyme-catalyzed conjugations of glutatione with unsaturated compounds. *Biochem. J.* **104**, 95–102.
114. Ghanayem, B., and Ahmed, A. E. (1981). Biotransformation and biliary excretion of 1-$^{14}$C-acrylonitrile in rats. *Toxicologist* **1**, 85.
115. Ghanayem, B. I., and Ahmed, A. E. (1982). *In vivo* biotransformation and biliary excretion of 1-$^{14}$C-acrylonitrile in rats. *Arch. Toxicol.* **50**, 175–183.
116. Guengerich, F. P., Gieger, L. E., Hogy, L. L., and Wright, P. L. (1981). *In vitro* metabolism of acrylonitrile to 1-cyanoethylene oxide, reaction with glutathione, and irreversible binding to proteins and nucleic acids. *Cancer Res.* **41**, 4925–4933.
117. Gut, I., Kopecky, J., and Filip, J. (1981). Acrylonitrile-$^{14}$C metabolism in rats: Effect of

the route of administration on the elimination of thiocyanate and other radioactive metabolites in urine and feces. *J. Hyg., Epidemiol., Microbiol., Immunol.* **25,** 12-16.

118. Holecek, V., and Kopecky, K. (1981). Conjugation of glutathione with acrylonitrile and glycidonitrile. *In* "Industrial and Environmental Xenobiotics" (I. Gut, M. Cikrt, and G. L. Plaa, eds.), pp. 239-244. Springer-Verlag, Berlin.

119. Kopecky, J., Gut, I., Nerudova, J., Zacchardova, D., and Holecek, V. (1981). Metabolic studies on acrylonitrile. *In* "Industrial and Environmental Xenobiotics" (I. Gut, M. Cikrt, and G. L. Plaa, eds.), pp. 221-230. Springer-Verlag, Berlin.

120. Lambotte-Vandepaer, M., Duverger-Van Bogat, M., and de Muster, C. (1981). Identification of two urinary metabolites of rats treated with acrylonitrile: Influence of several inhibitors on the mutagenicity of those urines. *Toxicol. Lett.* **7,** 321-328.

121. Langvardt, P. W., Putzig, C., Braun, W. H., and Young, J. D. (1980). Identification of the major urinary metabolites of acrylonitrile in the rat. *J. Toxicol. Environ. Health* **6,** 273-282.

122. Lin, Y., and Ahmed, A. E. (1979). Analysis of the urinary metabolites of the carcinogen acrylonitrile. *Annu. Meet., Am. Soc. Mass. Spectrosc.* Extended Abstract, p. 613.

123. van Bladeren, P. J., Delbressine, L. P. C., Hoojeterp, J. J., Beaumont, A. H. G. M., Breimer, D. D., Seutter-Berlage, F., and van der Gen, A. (1981). Formation of mercapturic acids from acrylonitrile, crotonitrile, and cinnamonitrile by direct conjugation and via an intermediate oxidation process. *Drug Metab. Dispos.* **9,** 246-249.

124. Ahmed, A. E., and Patel, K. (1979). Pharmacokinetics, distribution and binding of 1-$^{14}$C-acrylonitrile in rats. *Pharmacologist* **21,** 212.

125. Ahmed, A. E., Farooqui, M. Y. H., Upreti, R. K., and Elshabrawy, O. (1982). Distribution and covalent interactions of 1-$^{14}$C-acrylonitrile in the rat. *Toxicology* **23,** 159-176.

126. Farooqui, M. Y. H., and Ahmed, A. E. (1980). Molecular interaction of acrylonitrile and KCN with rat red blood cells. *Fed. Proc., Fed. Am. Soc. Exp. Biol.* **39,** 749.

127. Farooqui, M. Y. H., and Ahmed, A. E. (1982). Molecular interaction of acrylonitrile and potassium cyanide with rat blood. *Chem.-Biol. Interact.* **38,** 145-159.

128. Farooqui, M. Y. H., Elshabrawy, O., Upreti, R. K., and Ahmed, A. E. (1982). Comparative distribution and covalent binding study of 1-$^{14}$C- and 2,2-$^{14}$C-acrylonitrile in rats. *Toxicologist* **2,** 168.

129. Farooqui, M. Y. H., and Ahmed, A. E. (1983). *In vivo* interactions of acrylonitrile with macromolecules in rats. *Chem. Biol. Interact.* **47,** 363-371.

130. Peter, H., and Bolt, H. M. (1981). Irreversible protein binding of acrylonitrile. *Xenobiotica* **11,** 51-56.

131. Sandberg E. C., and Slanina, P. (1980). Distribution of 1-$^{14}$C-acrylonitrile in rat and monkey. *Toxicol. Lett.* **6,** 187-191.

132. Delbressine, L. P. C. (1981). "Metabolic Detoxification of Olefinic Compounds," pp. 70-83. Drukkerij-Uitgeverij Brakkenstein, Nijmegen.

133. O'Brien, P. H. (1973). Unproven cancer remedies. *J. S. C. Med. Assoc.* **69,** 403-404.

134. Kreks, E. T., Jr. (1970). The nitrolisides (vitamin B$_{17}$): Their nature, occurrence and metabolic significance. *J. Appl. Nutr.* **22** (3) and (4).

135. Hill, G. J., Shine, T. E., Hill, H. Z., and Miller, C. (1976). Failure of amygdalin to arrest B16 melanoma and BW5147 AKR leukemia. *Cancer Res.* **36,** 2102-2107.

136. Laster, W. R., Jr., and Schakel, F. M., Jr. (1975). Experimental studies of the antitumor activity of amygdalin (NSC-15780) alone and in combination with $\beta$-glucosidase (NSC-128056). *Cancer Chemother. Rep.* **59,** 951-965.

137. Wodinsky, I., and Swiniarski, J. R. (1975). Antitumor activity of amygdalin MF (NSC-15780) as a single agent and with $\beta$-glucosidase (NSR-128056) on a spectrum of transplantable rodent tumors. *Cancer Chemother. Rep.* **59,** 939-950.

138. Ames, M. M., Moyer, T. P., Kovach, J. S., Moertel, G. C., and Rubin, J. (1981). Pharmacology of amygdalin (laetrile) in cancer patients. *Cancer Chemother. Pharmacol.* **6,** 51–57.
139. Haisman, D. R., and Knight, D. J. (1967). The enzymic hydrolysis of amygdalin. *Biochem. J.* **103,** 528–534.
140. Schmidt, E. S., Newton, G. W., and Sanders, S. M. (1978). Laetrile toxicity studies in dogs. *JAMA, J. Am. Med. Assoc.* **239,** 943–947.
141. Miller, K. W., Anderson, J. L., and Stoewsand, G. S. (1981). Amygdalin metabolism and effect on reproduction of rats fed apricot kernels. *J. Toxicol. Environ. Health* **7,** 457–467.
142. Coroson, B. B., and Stoughton, R. W. (1928). Reaction of $\alpha,\beta$-unsaturated dinitriles. *J. Am. Chem. Soc.* **50,** 2825–2837.
143. Bryant, P. J. R., Owen, A. R., and Scanes, F. S. (1968). U.S. Patent 3,391,036.
144. Finn, D. H., Hogg, M. A. P., and Crichton, D. (1964). British Patent 967,660.
145. White, G. T., and Rothstein, L. R. (1967). U.S. Patent 3,314,835.
146. Punte, C. L., Weimer, J. T., Ballard, T. A., and Wilding, J. L. (1962). Toxicologic studies on o-chlorobenzylidene malononitrile. *Toxicol. Appl. Pharmacol.* **4,** 656–662.
147. Lough, C. E., Currie, D. J., and Holmes, H. L. (1968). Rates of reaction of n-butanethiol with some conjugate heteroenoid compounds. *Can. J. Chem.* **46,** 771–774.
148. Ostroverskhov, V. G., and Shilov, E. A. (1961). Theory of nucleophilic additions. VII. Possibility of addition in the action of weak nucleophiles and molecules with double bonds. *Ukr. Khim. Zh.* **27,** 209–212.
149. Patai, S., and Rappoport, Z. (1962). Nucleophilic attacks on carbon-carbon double bonds. Part II. Cleavage of arylmethylenemalononitrile by water in 95% ethanol. *J. Chem. Soc.* pp. 383–403.
150. Pritchard, W. R., Henderson, W., and Beall, C. W. (1958). Experimental production of dissecting aneurysms in turkeys. *Am. J. Vet. Res.* **19,** 696.
151. Rapporport, Z., and Gertler, S. (1964). Nucleophilic attack on carbon-carbon double bonds. Part IX. The primary (equilibrium) step. The reaction of Tri-n-butylphosphine with arylmethylenemalononitrile in methanol. *J. Chem. Soc.* **2,** 1360–1368.
152. Shrauzer, G. N., and Eichler, S. (1962). Über Komplexe von Kupfer(I)-Halogeniden mit ungesättigten Nitrilen und einigen Olefin. Modelversuch zum Bindings-problem in bis-Acrylnitril-nickel. *Chem. Ber.* **95,** 260–267.
153. Silver, R. F., Kerr, K. A. Frandsen, P. A., Kelley, S. J., and Holmes, H. L. (1967). Synthesis and chemical reactions of some conjugated heteroenoid compounds. *Can. J. Chem.* **45,** 1001–1006.
154. Kinnear, A. M. (1948). Determination of phenylbromoacetonitrile. *J. Soc. Chem. Ind.* **67,** 35–38.
155. Harper, D. B. (1977). Microbiol metabolism of aromatic nitriles. Enzymology of C–N cleavage by *Nocardia* sp. (Rhodorhorus group) N.C.I.B. 11216. *Biochem. J.* **165,** 309–319.
156. Hook, R. H., and Robinson, W. G. (1964). Ricinine nitrilase. II. Purification and properties. *J. Biol. Chem.* **239,** 4263–4267.
157. Robinson, W. G., and Hook, R. H. (1964). Ricinine nitrilase. I. Reaction product and substrate specificity. *J. Biol. Chem.* **239,** 4257–4262.
158. Mimura, A., Kawano, T., and Yamaga, K. (1969). Application of microorganisms to the petrochemical industry. I. Assimilation of nitriles by microorganisms. *J. Ferment. Technol.* **47,** 631–638.
159. Mahadevan, S., and Thimann, K. V. (1964). Nitrilase. II. Substrate specificity and possible mode of action. *Arch. Biochem. Biophys.* **107,** 62–68.

160. Szabo, S., and Selye, H. (1971). Adrenal apoplexy and necrosis produced by acrylonitrile. *Endokrinologie* **57,** 405–408.
161. Van Breeman, V. L., and Hiraoka, J. (1961). Ultrastructure of nerve and satellite cells of spinal ganglia of rats treated with malononitrile. *Am. Zool.* **1,** 473.
162. Spence, A. W., and Marine, D. (1932). Production of thyroid hyperplasia in rats and mice by administration of methyl cyanide. *Proc. Soc. Exp. Biol. Med.* **29,** 967–972.
163. Kreiss, K., Wegman, D. H., Niles, C. A., Siroky, M. B., Krame, R. J., and Feldman, R. G. (1980). Neurological dysfunction of the bladder in workers exposed to dimethylaminopropionitrile. *JAMA, J. Am. Med. Assoc.* **243,** 741–748.
164. Hickman, J. A. (1978). Investigation of the mechanism of action of antitumor dimethyltriazenes. *Biochimie* **60,** 997–1002.
165. Gescher, A., Hickman, J. A., and Stevens, M. F. (1979). Oxidative metabolism of some $N$-methyl-containing xenobiotics can lead to stable progenitors of formaldehyde. *Biochem. Pharmacol.* **28,** 3235–3238.

Chapter 18

# Thiono-Sulfur Compounds

ROBERT A. NEAL

*Chemical Industry Institute of Toxicology*
*Research Triangle Park, North Carolina*

    I. Introduction . . . . . . . . . . . . . . . . . . . . . 519
   II. Mechanisms of Metabolism of Parathion, Carbon Disulfide, and
        Thioacetamide . . . . . . . . . . . . . . . . . . . . 523
        A. Parathion . . . . . . . . . . . . . . . . . . . . . 523
        B. Carbon Disulfide . . . . . . . . . . . . . . . . . 533
        C. Thioacetamide . . . . . . . . . . . . . . . . . . . 535
  III. Summary . . . . . . . . . . . . . . . . . . . . . . 537
        References . . . . . . . . . . . . . . . . . . . . . 538

## I. Introduction

A number of thiono-sulfur compounds, which find use as pesticides, drugs, and in industrial processes, exhibit toxic properties. Figure 1 shows the general structure of thiono-sulfur compounds. The central atom (X) to which the sulfur is bound can be either carbon or phosphorus, and the groups attached to the carbon and phosphorus ($R_1$, $R_2$) can be nitrogen, carbon, or sulfur. Also, in the case of pentavalent phosphorus thiono-sulfur compounds, a third group ($R_3$) would be attached to the X group. The toxic effects of thiono-sulfur compounds include bone marrow depression, liver damage, lung damage, induction of neoplasia, and inhibition of various enzymes, including the cytochrome *P*-450-dependent monooxygenases.

During the past decade, it has become increasingly evident that metabo-

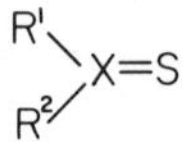

$$R^1_{\phantom{1}}\!\!\diagdown\!\!\!\diagup\!\!R^2\, X=S$$

**Fig. 1.** General structure for thiono-sulfur compounds. $R^1, R^2$ = nitrogen, carbon, sulfur; X = carbon, phosphorus.

lism of the thiono-sulfur group plays a fundamental role in eliciting the toxic effects of thiono-sulfur compounds. The experimental support for this conclusion is based on four different lines of investigation: the effects of inhibitors and inducers of metabolism on toxicity, the comparative toxicity of analogs (especially oxygen analogs), direct toxicity testing of intermediate metabolites of sufficient stability to be isolated, and covalent binding of metabolites of thiono-sulfur compounds *in vitro* and *in vivo*.

Phenobarbital treatment enhances the *in vivo* and *in vitro* desulfuration of representatives of most of the major classes of thiono-sulfur compounds, including parathion,[1,2] carbon disulfide,[3-5] thioacetamide,[6] and α-naphthyl isothiocyanate (ANIT).[7] Consistent with these findings, phenobarbital treatment enhances the liver damage produced by diethyl phenyl phosphorothionate (a parathion analog),[8] and by carbon disulfide,[9-12] thioacetamide,[13] and ANIT.[7] However, in this context, it should be pointed out that the effect of phenobarbital treatment on carbon disulfide toxicity might not be mediated through an increase in the rate of formation of the toxic metabolite, but may result from an increased susceptibility to the action of carbon disulfide itself.[12,14]

The effects of inhibitors of metabolism on the organ damage caused by carbon disulfide, thioacetamide, ANIT, and α-naphthylthiourea (ANTU) are also consistent with a requirement for metabolic activation of the compounds. The inhibitor SKF 525A gives some protection against the toxicity of all four compounds.[7,9,11,13,15] Piperonyl butoxide treatment protects against the toxicity of ANIT[7] and ANTU,[15] and cobaltous chloride treatment blocks the toxicity of ANIT[7] and thioacetamide.[13]

Consistent with the results just described, treatment of rats with small, nonlethal doses of ANTU inhibited the *in vitro* and *in vivo* metabolism of ANTU and protected against the lethality and lung damage of subsequent normally lethal doses.[16]

Studies of mono- and disubstituted thioureas suggest a good correlation between acute toxicity and desulfuration *in vivo*. Thus, the toxic monoarylthioureas (phenyl, *p*-chlorophenyl-, and *p*-tolyl-) were desulfated to a considerable extent (66 – 100%) on *in vivo* administration to rabbits, whereas the relatively nontoxic diarylthioureas (diphenyl- and 4-hydroxydiphenyl-)

were not (<25%). Similarly, *p*-hydroxyphenylthiourea, which was at least five times less toxic to rats and rabbits than phenylthiourea, was found to undergo only a slight degree of desulfuration.[17]

The most compelling evidence that the thiono-sulfur group is required for toxicity is provided by a comparison of the potency of thiono-sulfur compounds with their corresponding oxygen analogs. In many cases, these oxygen analogs are also the major metabolites found *in vitro* or *in vivo*. With the exception of the anticholinesterase activity of the phosphorothionate and phosphonothionate insecticides, which is greatly enhanced on conversion to the oxygen analogs, the toxicity of thiono-sulfur compounds is reduced greatly or abolished on replacement of sulfur by oxygen. For example, the 24-hr $LD_{50}$ for intraperitoneal administration of ANTU to male Sprague-Dawley rats was 10 mg/kg, whereas the value for the oxygen analog α-naphthylurea (ANU) was more than 800 mg/kg. Moreover, at this dose, ANTU caused severe pulmonary edema, whereas none was observed after α-naphthylurea administration.[16] Similarly, replacement of the sulfur atom of ANIT with oxygen abolished the hepatic dysfunction, as judged by the absence of hyperbilirubinemia.[18] The inhibitory activity toward cytochrome *P*-450 *in vivo* or *in vitro* of a number of thiono-sulfur compounds, including thiourea, ethylene thiourea, thiouracil, methylthiouracil, propylthiouracil, phenythiourea, parathion, ANTU, and ANIT, is abolished on replacement of sulfur by oxygen.[10] Finally, substitution of oxygen for sulfur abolishes the carcinogenicity of thiourea and thiouracil and greatly decreases the tumorigenic potency of thioacetamide.

The low order of toxicity of oxygen analogs of thiono-sulfur compounds, on the one hand, and the good correlation between desulfuration *in vivo* and toxicity on the other, are of interest for two reasons. First, they demonstrate that the toxicity of thiono-sulfur compounds is inherent in the thiono-sulfur group. Second, they strongly implicate reactive intermediates formed during the desulfuration process as being responsible for toxicity.

Some thiono-sulfur compounds are metabolized to *S*-oxides of sufficient stability to be chemically synthesized and tested directly for toxicity. Thus, it has been reported that rats exposed by inhalation to the *S*-oxide of the pulmonary toxin thiourea exhibit pathological changes similar to those observed with the parent compound and that the *S*-oxide is at least 10-fold more potent.[19] Similarly, the *S*-oxide of thioacetamide is a more potent hepatotoxin than thioacetamide itself. Thus, at equimolar doses, thioacetamide *S*-oxide produces a more rapid onset and a greater severity of centrilobular necrosis than thioacetamide.[20] However, because the toxicity of thioacetamide *S*-oxide is enhanced by prior treatment of animals with phenobarbital and is inhibited by pyrazole and cobaltous chloride in noninduced animals and by SKF 525A in phenobarbital-induced animals, it appears that thioacetamide S-oxide must undergo further oxidative metabolism to exert its necrotic effect.[20] One study showed thiobenzamide S-oxide to be more hepatotoxic than thiobenzamide, in agreement with the foregoing results.[21]

A number of thiono-sulfur compounds, including parathion,[2,22] carbon disulfide,[5,22] thioacetamide,[6] ANTU,[16] ANIT,[7] and methimazole,[23] have been shown to give rise to metabolites that bind covalently to microsomes when incubated in the presence of NADPH. In some cases (carbon disulfide,[5,22] methimazole,[23] and ANIT[7]), covalent binding is observed even in the absence of NADPH, suggesting some covalent binding of the parent compound. Covalent binding has been observed *in vivo* with parathion,[1] carbon disulfide,[4] thioacetamide,[20] ANIT,[7] and ANTU.[16]

In most cases, a good correlation has been obtained between covalent binding *in vivo* and organ damage after treatments that alter metabolism or after administration of different analogs of thiono-sulfur compounds. Results obtained with ANTU[16] and ANIT[7] illustrate some of the approaches used. Treatment of rats for 5 days with a sublethal dose of ANTU decreased by 60% the covalent binding of $^{35}S$ to lung tissue after a subsequent challenge dose of [$^{35}S$]ANTU and protected against the lung damage and lethality.[16] Consistent with this finding, treatment of rats with diethyl maleate increased by 150% the covalent binding of $^{35}S$ to lung tissue and enhanced the lung damage and lethality. Furthermore, less than 10% as much binding from the relatively nontoxic [$^{14}C$]ANU was observed as from [$^{14}C$]ANTU. In the case of ANIT,[7] treatments that decreased the amount of covalent binding, such as SKF 525A and cobaltous chloride, also protected against hyperbilirubinemia in rats, whereas phenobarbital treatment increased both covalent binding to liver macromolecules and hyperbilirubinemia.

In some cases, the relationship between covalent binding and toxicity is less obvious. Thus, treatment of rats with ipomeanol for 5 days completely protected against the lung damage and lethality of a challenge dose of [$^{35}S$]ANTU sufficient to kill all of the nontreated animals, but only a 15% decrease in covalent binding of $^{35}S$ to lung macromolecules was observed.[16] With ANIT, treatment of rats with 16α-pregnenolone carbonitrile protected against hyperbilirubinemia but caused an increase in covalent binding both *in vivo* and *in vitro*.[7]

Despite the few discrepancies noted, most of the available data strongly suggest a role of reactive intermediates in the toxicity of thiono-sulfur compounds. The remainder of this chapter will review the mechanism of bioactivation of three representative thiono-sulfur compounds: parathion, carbon disulfide, and thioacetamide.

## I. Mechanisms of Metabolism of Parathion, Carbon Disulfide, and Thioacetamide

### A. PARATHION

Parathion is one of a class of phosphorothionate triesters widely used as insecticides. These compounds exert their toxic effects in insects and mammals by inhibiting the enzyme acetylcholinesterase. The phosphorothionates, in general, are poor inhibitors of acetylcholinesterase but are converted by the cytochrome P-450-containing monooxygenase enzyme systems in insects and mammals to the corresponding phosphate triesters, which are potent inhibitors of this enzyme.

The hepatic microsomal metabolism of parathion by the rat and other experimental animals is shown in Fig. 2. There are five major products of parathion metabolism. One of these is the corresponding phosphate triester paraoxon, which is formed in a monooxygenase-catalyzed reaction in which the sulfur atom of parathion is replaced by an oxygen atom.[24] The paraoxon is subject to hydrolysis by esterases present in various tissues to form diethyl phosphate and p-nitrophenol.[25] Parathion is not a substrate for these esterases, presumably because the phosphorus atom is not as electrophilic as that in paraoxon. The difference in the electrophilicity of the phosphorus atom in these two compounds is most likely due to the greater electronegativity of oxygen as compared to sulfur.

Parathion is also metabolized to diethyl phosphorothioic acid and p-nitrophenol in a reaction requiring a cytochrome P-450-containing monooxygenase enzyme system.[26,27] Studies with $^{18}$O-labeled water have indicated that water, in addition to molecular oxygen and NADPH, is required in this reaction.[28] Diethyl phosphate and p-nitrophenol can also be formed from parathion in a monooxygenase-catalyzed reaction.[29]

In an examination of the chemical mechanism of the microsomal metabolism of parathion to paraoxon, two principal questions were posed: first, what was the initial site of attack of the cytochrome P-450-generated oxygen atom on the parathion molecule, and second, was the attacking oxygen atom retained in the product of the reaction, paraoxon? In regard to the question of the retention of the cytochrome P-450-generated oxygen atom in paraoxon, parathion was incubated with rabbit liver microsomes and NADPH in an atmosphere enriched with [$^{18}$O]dioxygen or in a buffer enriched with $^{18}$O-labeled water.[28] The paraoxon formed in these reactions was isolated and examined for an increased content of $^{18}$O using mass spectrometry. No $^{18}$O enrichment was evident in the paraoxon isolated from the incubation carried out in the $^{18}$O-labeled water-enriched buffer. However, as can be

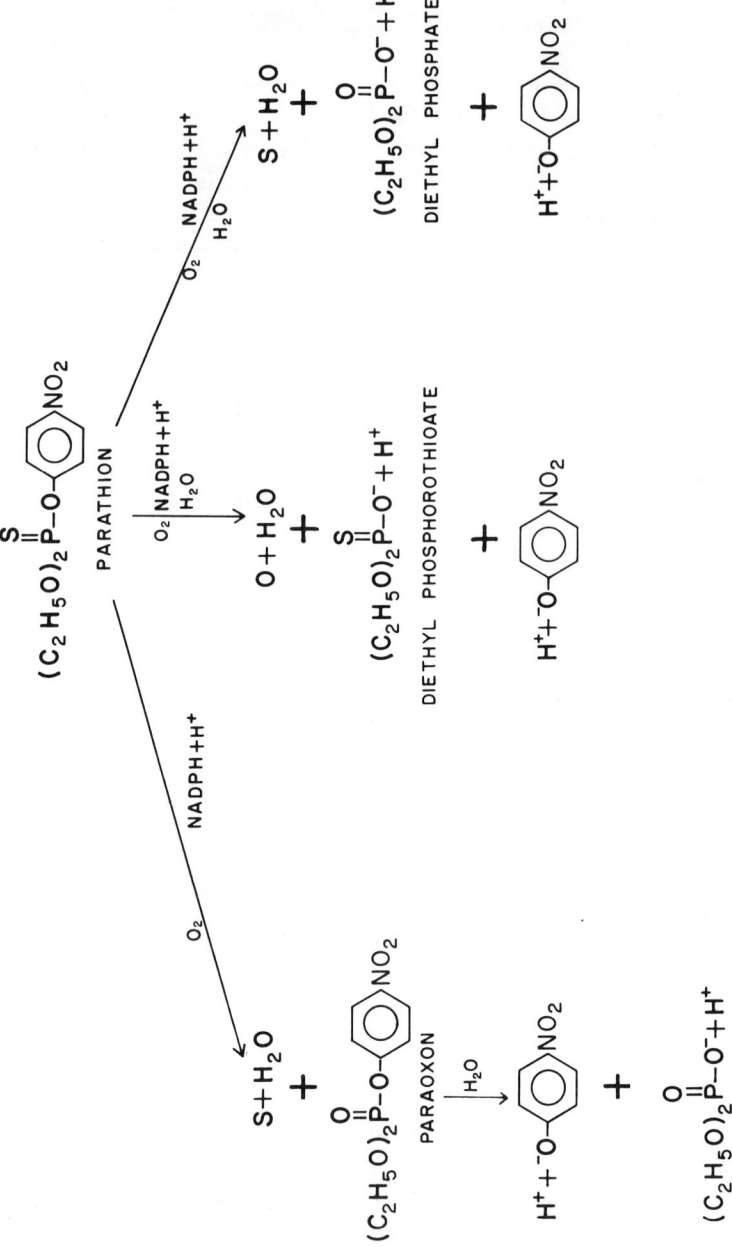

Fig. 2. Scheme for the metabolism of parathion by mammalian hepatic microsomes.

**Table I**

$^{18}O$ Enrichment in Paraoxon after Incubation of Parathion with Rabbit Liver Microsomes[a]

| Experiment | Source of $^{18}O$ (atom % excess) | Observed atom % excess $^{18}O$ in paraoxon |
|---|---|---|
| 1 | $^{18}O$-Labeled water (20) | 0.0 |
| 2 | [$^{18}O$]Dioxygen (46) | 42.0 |

[a] The details of this experiment are described by Ptashne et al.[28]

seen in Table I, when the reaction was carried out in an atmosphere enriched with [$^{18}O$]dioxygen, the paraoxon formed contained $^{18}O$ to about the same atom-percentage excess as the atmosphere in which the incubation took place. These data indicated that the oxygen atom attached to the phosphorus of paraoxon was transferred to parathion by the cytochrome P-450-containing monooxygenase system. Results similar to those just described for parathion were obtained by McBain et al.[30,31] in oxygen-18 studies of the metabolism of the phosphonothionate fonofos (O-ethyl S-phenyl ethylphosphonothiolothionate) with rat liver microsomes.

Additional studies have been carried out to determine the initial site of attack of the cytochrome P-450-generated oxygen atom on parathion. These studies have indicated that the sulfur atom is the initial site of attack. Thus, studies of products of the reaction of parathion[32] and menthylmethylphenylphosphinothioate[33] with peroxy acid model systems for the cytochrome P-450-containing monooxygenase systems indicated that these compounds were readily converted to their oxygen analogs by these model systems. However, no products were seen on incubation of paraoxon or menthylmethylphenylphosphinate with the peroxy acid.[32,33] These results suggested that the attack of the oxygen atom of the peroxy acid was on the sulfur atom rather than on the phosphorus atom. The studies of Herriott[33] were carried out with the two diastereoisomers of menthylmethylphenylphosphinothioate. When the reactions were carried out with m-chloroperbenzoic acid, there was a high degree of retention of configuration about the phosphorus atom, but oxidation with peroxytrifluoroacetic acid occurred with predominant inversion. Thus, with the less acidic m-chloroperbenzoic acid there was retention of configuration. These results again suggest predominant attack of the peroxy acid on the sulfur atom to form an S-oxide followed by closure to a phosphooxathiirane and concomitant or successive loss of sulfur to produce the oxygen analogs of these phosphothionates. The stereospecificity of the metabolism of the chiral isomers of fonofos (O-ethyl S-phenyl ethylphosphonothiolothionate) by mouse liver microsomal monooxygenase systems was examined by Lee et al.[34,35] The oxidative desulfuration of the two isomers of fonofos by mouse liver microsomes proceeded predominantly with retention of configuration. The stereospecific oxidative desulfuration of O-ethyl O-2-nitro-5-methylphenyl N-isopropylphosphoramidothioate to the corresponding oxygen analog by the action of rabbit liver microsomes has also been reported.[36] In this case, the desulfuration also occurred with retention of configuration. Thus, in these experiments with cytochrome P-450-containing monooxygenase systems, the initial attack of the oxygen atom generated in the active site of cytochrome P-450 appeared to be the sulfur atom of the phosphothionates.

From these data, a chemical mechanism for the formation of paraoxon from parathion has been proposed.[29] This mechanism is shown in Fig. 3. It is postulated that a singlet oxygen atom generated in a cytochrome P-450-

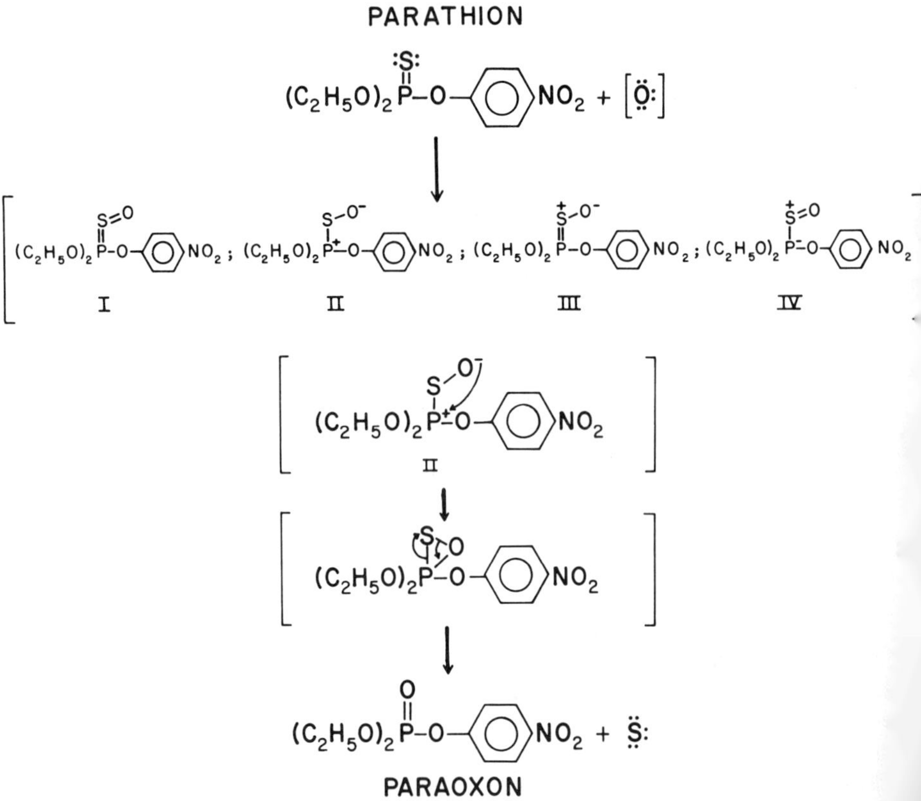

Fig. 3. Chemical mechanism for the metabolism of parathion to paraoxon by the cytochrome P-450-containing monooxygenase system.

catalyzed reaction is donated to the sulfur atom of parathion to yield a compound analogous to the S-oxide formed in the reaction of peracids with thioketones.[37,38] The attacking singlet oxygen atom is shown in brackets to suggest that it is likely to be transferred from cytochrome P-450 to the phosphorothionate sulfur in a concerted reaction. As shown in Fig. 3, there are four different structures that may contribute to the resonance stabilization of the resultant S-oxide. It is proposed that one of these resonance forms (perhaps form II) reacts internally to form a cyclic phosphorus–sulfur–oxygen intermediate analogous to the oxathiirane that has been porposed by numerous investigators to be an intermediate in the reaction pathways of various S-oxides.[39] This resultant phosphooxathiirane then undergoes a cyclic electron shift with the loss of sulfur, forming paraoxon. It is proposed that the sulfur atom that is released is in its singlet form. Although there is no evidence to support this hypothesis, it is probable that, if the attacking oxygen atom is in its singlet state, the departing sulfur atom may also be in its singlet state.

Diethyl phosphorothionate and diethyl phosphate plus p-nitrophenol are also products of the cytochrome P-450 monooxygenase-catalyzed metabolism of parathion. Figure 4 shows the proposed mechanism for the metabolism of parathion to diethyl phosphorothioic acid and to diethyl phosphoric

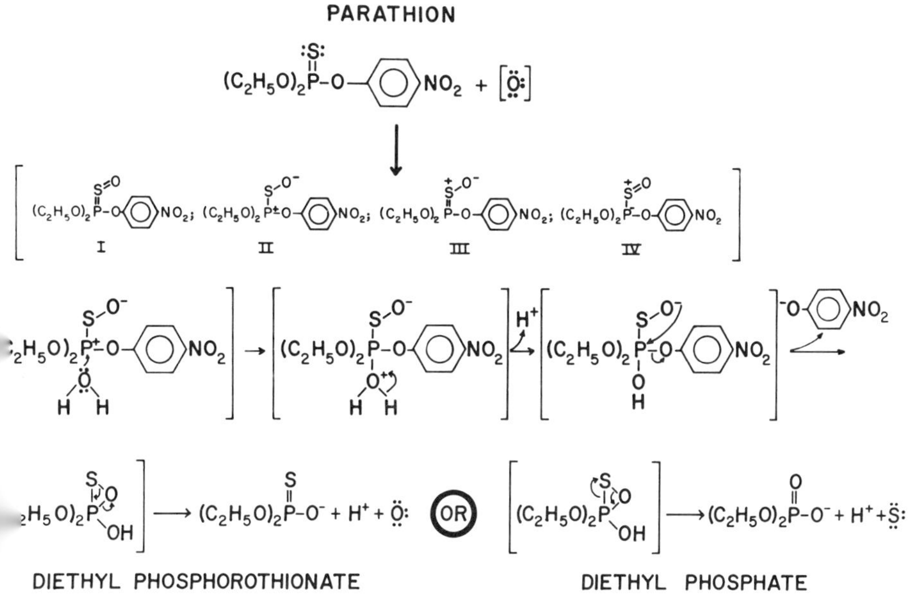

**Fig. 4.** Chemical mechanism for the metabolism of parathion to diethyl phosphorothionate and diethyl phosphate by the cytochrome P-450-containing monooxygenase system.

acid. It is proposed that the initial reaction is again the transfer of a singlet oxygen atom to one of the unshared electron pairs on the sulfur atom to form the same intermediate $S$-oxide proposed in Fig. 3. One of these resonance forms (perhaps form II) may be subject to nucleophilic attack by water, as shown in this scheme, followed by loss of a proton and the $p$-nitrophenol group, forming a phosphooxathiirane derivative that may break down with the loss of an oxygen atom, forming diethyl phosphorothioic acid or, with the loss of a sulfur atom, forming diethyl phosphoric acid. Alternatively, the $S$-oxide may first react internally, as shown in Fig. 3, followed by attack by water. Each of these alternatives is possible.

The majority of the diethyl phosphate and $p$-nitrophenol formed in the mammalian metabolism of parathion is undoubtedly derived by the action of esterases or phosphatases[25] or paraoxon formed from parathion in a cytochrome $P$-450-catalyzed reaction. However, a significant portion of the diethyl phosphate and $p$-nitrophenol must also be the result of the attack of water on the intermediate $S$-oxide of parathion.[29]

The question of what factors control the rate of breakdown of the proposed intermediate $S$-oxide of parathion to the various products is of interest. With all the systems examined so far — whole microsomes, the reconstituted $P$-450 monooxygenase system, and the peroxy acid model systems — the predominant product is paraoxon, followed by diethyl phosphorothioic acid and then by diethyl phosphoric acid. It may be that the rearrangement of the intermediate $S$-oxide to paraoxon is thermodynamically or kinetically more favorable than water attack followed by rearrangement to form diethyl phosphorothioic acid or diethyl phosphoric acid. Alternatively, the limited accessibility of water to the active site of cytochrome $P$-450 to form diethyl phosphorothioic acid or diethyl phosphoric acid may be a controlling factor.

If the sulfur atom released from parathion, as proposed in Fig. 3, is in its singlet state, it would be a highly reactive electrophile that would bind readily to nucleophiles near the site of its release. The thiono-sulfur atom of parathion has been found to be covalently bound to tissue macromolecules after administration of [$^{35}$S]parathion *in vivo*[1] and on incubation with hepatic microsome *in vitro*.[2,27] Table II shows the results of an experiment in which double-labeled [$^{32}$P,$^{35}$S]parathion was incubated with microsomes isolated from the livers of phenobarbital-treated rats.[1] In the absence of NADPH, only a trace of radioactivity could be found bound to the microsomes. However, as shown in Table II, when the incubation was carried out in the presence of NADPH, a substantial amount of sulfur became covalently bound to the microsomes. A small, but significant, amount of the phosphorus-containing portion of the parathion molecule was also covalently bound to the microsomes. The results of this experiment indicate clearly that th

### Table II
$^{35}$S and $^{32}$P Bound to Microsomes after Incubation with Labeled Parathion and Its Metabolites[a]

| Substrate | nmol $^{35}$S bound/mg protein/15 min | nmol $^{32}$P bound/mg protein/15 min |
|---|---|---|
| [$^{32}$P,$^{35}$S]Parathion ($2.5 \times 10^{-4}$ $M$) | $12.93 \pm 0.37$ | $1.12 \pm 0.21$ |
| [$^{32}$P]Paraoxon ($2.5 \times 10^{-5}$ $M$) | — | $0.25 \pm 0.03$ |
| Diethyl [$^{32}$P]phosphate ($4 \times 10^{-5}$ $M$) | — | None |
| Diethyl [$^{32}$P,$^{35}$S]phosphorothioic acid ($4 \times 10^{-5}$ $M$) | None | None |

[a] The details of this experiment are described by Poore and Neal.[1] This experiment was carried out with hepatic microsomes from phenobarbital-treated rats.

majority of the sulfur bound to the microsomes is free of the phosphorus-containing portion of the molecule and, thus, must be atomic sulfur released in the metabolism of parathion to paraoxon. This is further substantiated by the finding that the amount of sulfur bound under these conditions is equivalent to the amount of paraoxon formed in the incubation.[2] Also shown in Table II are the results of an experiment in which an amount of paraoxon, which was approximately five times the amount that would be expected to be formed in the incubation containing double-labeled parathion, was incubated with an equal aliquot of the same preparation of microsomes. As can be seen, a much smaller amount of $^{32}$P was bound with [$^{32}$P]paraoxon than was bound using the double-labeled parathion. Thus, the binding of $^{32}$P in a reaction of paraoxon with nucleophilic sites on the endoplasmic reticulum is responsible for only a small portion of the total binding of $^{32}$P seen with double-labeled parathion. It appears that the greater portion of the $^{32}$P binding in the incubation using double-labeled parathion is likely the result of the reaction of one or more of the intermediate S-oxides shown in Fig. 3 with nucleophiles on the endoplasmic reticulum. In examining the remainder of the data in Table II, it can be seen that the incubation of the other phosphorus-containing metabolites of parathion with microsomes in the presence of NADPH does not lead to the binding of radioactivity.

The binding of sulfur or an activated intermediate of the phosphorus-containing portion of the parathion molecule to the endoplasmic reticulum leads to a decrease in the amount of cytochrome P-450 detectable as its

### Table III
Parathion Metabolism by a Reconstituted Monooxygenase System from Rabbit Liver[a]

| | Product formation (nmol/nmol cytochrome $P$-450/5 min) | | |
|---|---|---|---|
| Conditions | Paraoxon | Diethyl phosphorothioic acid | Diethyl phosphate |
| Complete | 4.510 ± 0.424 | 2.420 ± 0.156 | 0.470 ± 0.107 |
| Minus cytochrome $P$-450 | 0.048 ± 0.002 | 0.040 ± 0.002 | 0.016 ± 0.002 |
| Minus reductase | 0.018 ± 0.001 | 0.070 ± 0.003 | 0 |
| Minus lipid | 1.790 ± 0.021 | 0.816 ± 0.002 | 0.206 ± 0.049 |
| Minus deoxycholate | 3.860 ± 0.679 | 1.990 ± 0.347 | 0.381 ± 0.100 |
| Minus both lipid and deoxycholate | 1.690 ± 0.120 | 0.756 ± 0.008 | 0.245 ± 0.005 |

[a] From Kamataki et al.[29]

carbon monoxide complex and to a decrease in the rate of metabolism of substrates such as benzphetamine.[2] Neither paraoxon nor any other isolatable metabolite of parathion decreases the amount of cytochrome $P$-450 or inhibits the ability of microsomes to metabolize substrates, such as benzphetamine.[2]

We have also examined the metabolism of parathion by purified reconstituted monooxygenase systems isolated from the livers of phenobarbital-treated rabbits and rats. Cytochrome $P$-450 and NADPH–cytochrome $c$ reductase were purified to apparent homogeneity from both species as described previously.[40] The results of an experiment examining the metabolism of parathion by a reconstituted system from rabbit liver are shown in Table III.[40] The cytochrome $P$-450 used in this experiment had a specific content of 18.5 nmol/mg protein. As can be seen, there is a requirement for both cytochrome $P$-450 and the reductase. The activity was also stimulated by addition of lipid (dilauroylphosphatidylcholine). Deoxycholate also had a slight stimulating effect. Another important aspect of the data shown in Table III is that all three of the major phosphorus-containing metabolites of parathion are formed by what appears to be a single species of cytochrome $P$-450. These data and those from the studies with a peroxy acid model system[32] suggest that the monooxygenase system is only involved in the addition of an oxygen atom to the phosphorothionate sulfur (Fig. 3) and that the various products are formed nonenzymatically from a common intermediate. It is believed that this intermediate is one or more of the resonance forms of the intermediate $S$-oxide shown in Fig. 3.

We have examined whether the sulfur that was bound to the proteins of a reconstituted system from the liver of phenobarbital-treated rats was bound to both the reductase and cytochrome P-450.[40] In this experiment, the reconstituted system was incubated with [$^{35}$S]parathion. The reaction mixture was dialyzed and applied to a Sephadex G-25 column to remove unreacted parathion and its noncovalently bound metabolites. The protein fraction from the Sephadex column was reduced in volume and analyzed by SDS–polyacrylamide gel electrophoresis in the absence of either dithiothreitol or mercaptoethanol. The results are shown in Fig. 5. There was considerable protein and radioactivity at the origin. This material at the origin represents an aggregate of reductase and cytochrome P-450 molecules caused by the binding of the sulfur atom. Treatment of the concentrated eluant from the Sephadex column with 100 m$M$ cyanide for 3 hr at room temperature before gel electrophoresis, a treatment that released most of the bound sulfur, prevented the accumulation of both protein and radioactivity at the origin. Little radioactivity was associated with the position of reductase (fP$_2$) on the gel, whereas a significant amount was found in the area of the gel corresponding to cytochrome P-450. This was the case in both the gel of the sample that had been incubated with cyanide before gel electropho-

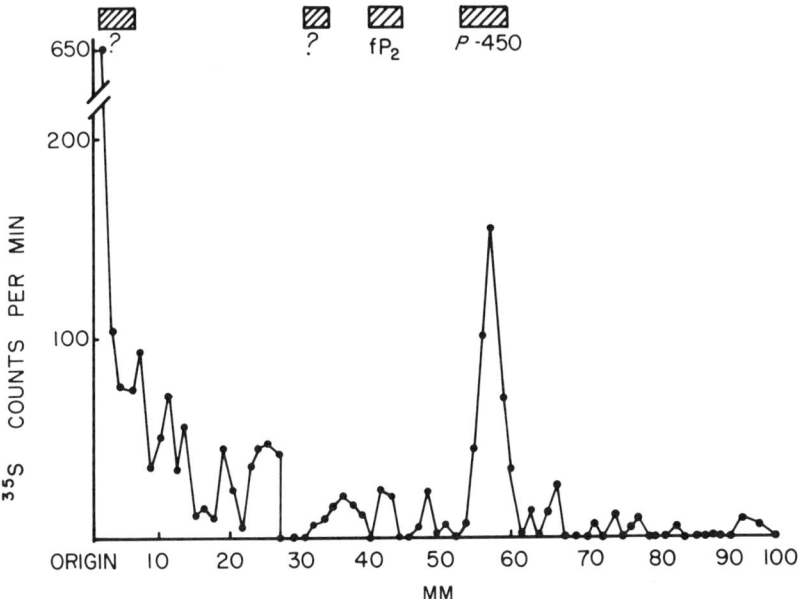

**Fig. 5.** Polyacrylamide gel electrophoresis of a reconstituted monooxygenase system from rat liver that had been labeled with $^{35}$S by incubation with [$^{35}$S]parathion.

resis and that which had not (Fig. 5). These data indicate that the sulfur is bound predominantly or exclusively to cytochrome P-450.

Experiments in this laboratory have employed the reconstituted monooxygenase system from rat liver microsomes to elucidate the mechanism by which parathion causes a loss of cytochrome P-450 detectable as its carbon monoxide complex and a loss of cytochrome P-450-dependent monooxygenase activity. Inhibition of the cytochrome P-450 is accompanied by covalent binding of sulfur to the enzyme and occurs only in the presence of a complete system, where metabolism of parathion takes place.[40] The sulfur that becomes covalently bound to cytochrome P-450 appears to be free of the rest of the parathion molecule, based on the almost negligible amount of covalent binding observed when [ethyl-$^{14}$C]parathion rather than [$^{35}$S]parathion is used.[40]

The cyanide treatment of a reconstituted cytochrome P-450 monooxygenase system labeled with $^{35}$S as a result of incubation with [$^{35}$S]parathion released approximately 50% of the covalently bound sulfur as [$^{35}$S]thiocyanate, suggesting the presence of hydrodisulfide linkages (R-S-$^{35}$SH), apparently formed by attack of $^{35}$S on cysteine residues in the cytochrome P-450 (Fig. 6).[40] In other experiments, dithiothreitol treatment was found to release 75% of the $^{35}$S bound to the proteins of the reconstituted system.[41] The remaining $^{35}$S was stable to treatment with performic acid and to acid hydrolysis and was distributed among at least three different amino acids. Probable mechanisms for the formation of such amino acid adducts are carbon–hydrogen insertion (a known reaction of singlet sulfur) or addition of atomic sulfur across a double bond.[40] There is no evidence, however, that modification of these three amino acids plays any direct role in the loss of cytochrome P-450 or monooxygenase activity.[41]

Metabolism of parathion was found to result in a considerable loss of heme from the cytochrome P-450 of the reconstituted system.[41] The heme loss, which could not be prevented with catalase, was sufficient to account for most of the loss of cytochrome P-450 detectable as its carbon monoxide complex. However, the heme loss could account for only 50% of the loss of cytochrome P-450-dependent monooxygenase activity, suggesting that al-

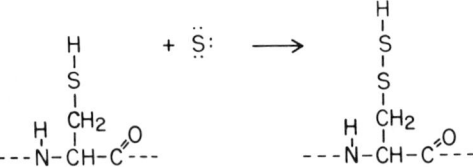

Fig. 6. Scheme for hydrodisulfide formation in a reaction of atomic sulfur with the side chain of a cysteine in the cytochrome P-450 polypeptide chain.

terations of the protein moiety were also in part responsible for the inactivation of the enzyme. However, such modifications do not appear to involve simple derivatization of essential amino acid residues in cytochrome $P$-450, in that removal of 75% of the bound radioactivity regenerates no enzymatic activity. At present, the critical modification of the protein moiety is unclear, although it appears likely that the aggregation accompanying the binding of atomic sulfur to cysteine residues plays a role.

## B. CARBON DISULFIDE

Although the neurological effects of carbon disulfide may be the most prevalent from the standpoint of human exposure, most experimental research using laboratory animals has focused on the effects of carbon disulfide exposure on the liver. A marked decline in the level of cytochrome $P$-450 and cytochrome $P$-450-dependent monooxygenase activity measured *in vitro* is detectable in rats given carbon disulfide orally or by intraperitoneal injection[3-5,9,10,22,42] or exposed by inhalation.[43,44] The administration of carbon disulfide to rats treated with phenobarbital causes an even greater decrease in cytochrome $P$-450 concentrations and associated enzymatic activities,[9,10] and also produces moderate to severe centrilobular hydropic degeneration of the liver.[9-12] Prior administration of SKF 525A to phenobarbital-treated animals decreases the liver damage caused by subsequent exposure to carbon disulfide, suggesting that the hepatic toxicity of carbon disulfide is caused by toxic metabolites produced by the cytochrome $P$-450-dependent monooxygenase system.[3,11] This was substantiated by the findings that phenobarbital treatment increases the *in vivo* binding of both $^{14}C$ and $^{35}S$ from carbon disulfide to liver microsomes[4] and that there is a correlation between the amount of a dose of [$^{14}C$]carbon disulfide excreted as [$^{14}C$]carbon dioxide and the degree of liver damage.[3]

Experimental efforts have been concentrating on elucidating the nature of the metabolites responsible for the inhibition of cytochrome $P$-450 and the liver damage caused by carbon disulfide. It was speculated that carbon disulfide might be metabolized to carbonyl sulfide and then to carbon dioxide in two sequential steps analogous to the conversion of parathion to paraoxon and leading to the release and covalent binding of one or both of the sulfur atoms. *In vitro* experiments with microsomes from untreated and phenobarbital-treated rats and [$^{14}C$]-carbon disulfide or carbon [$^{35}S$]disulfide were carried out to test this hypothesis.[5] In the absence of NADPH, equal amounts of $^{35}S$ and $^{14}C$ became bound to the microsomal proteins, suggesting the direct reactions of carbon disulfide with amino or sulfhydryl groups. In the presence of NADPH, $^{35}S$ binding to hepatic microsomes from untreated rats was increased twofold and to microsomes from phenobarbital-treated rats by sixfold. The stimulation by NADPH of the binding of sulfur from carbon disulfide to microsomes suggested that carbon disulfide as a substrate for cytochrome $P$-450. This was supported by the finding that the amount of sulfur bound in the presence of NADPH was inhibited by carbon monoxide[5] or SKF 525A.[42]

Whereas phenobarbital treatment led to a sixfold increase in $^{35}S$ binding, only a twofold increase in $^{14}C$ binding was observed.[5] This indicated that the majority of the $^{35}S$ bound to the microsomal proteins was free of the carbon atom. Experiments where the $^{35}S$-labeled microsomes were treated with cyanide indicated that approximately 50% of the sulfur that had become bound to the microsomal proteins during the metabolism of carbon disulfide was in the form of a hydrodisulfide,[45] analogous to the results obtained with parathion, which were described in the previous section.

The current data clearly indicate that cytochrome $P$-450 is the major enzyme responsible for converting carbon disulfide to carbonyl sulfide. However, some data indicate that the major enzyme responsible for metabolism of carbonyl sulfide to carbon dioxide is carbonate dehydratase.[46] The carbonate dehydratase activity in rat liver measured with carbonyl sulfide as the substrate was more than adequate to account for the rates of carbonyl sulfide metabolism observed. Carbonate dehydratase was proposed to convert carbonyl sulfide to monothiocarbonic acid ($H_2CO_2S$), which breaks down to carbon dioxide and hydrogen sulfide. On this basis, it was suggested that production of hydrogen sulfide from carbonyl sulfide by carbonate dehydratase might be the mechanism of toxicity of carbonyl sulfide.[47] Thus, it was demonstrated that carbonyl sulfide is converted to hydrogen sulfide *in vivo* and that the blood levels of hydrogen sulfide and the toxicity of carbonyl sulfide are decreased by treatment with acetazolamide.[47] Acetazolamide had no effect on hydrogen sulfide toxicity per se, whereas treatment with sodium nitrite, which protects against hydrogen sulfide toxicity, also protected against carbonyl sulfide toxicity. These results are all consistent with a key role of hydrogen sulfide in the central respiratory arrest caused by carbonyl sulfide. Whether hydrogen sulfide plays a role in other symptoms of the acute or chronic toxicity of carbon disulfide remains unclear.

The proposed pathway of carbon disulfide metabolism in rat liver is shown in Fig. 7. Analogous to the mechanism proposed for parathion, the initial step (reaction i) is the formation of an unstable oxathiirane in a reaction catalyzed by the cytochrome $P$-450-dependent monooxygenase system.[8] This intermediate may break down spontaneously (step ii), releasing carbonyl sulfide and elemental sulfur. Alternatively, it may be subject to attack by water to form monothiocarbonate and elemental sulfur (step iii). Monothiocarbonate, at physiological pH, is the hydrated form of carbonyl

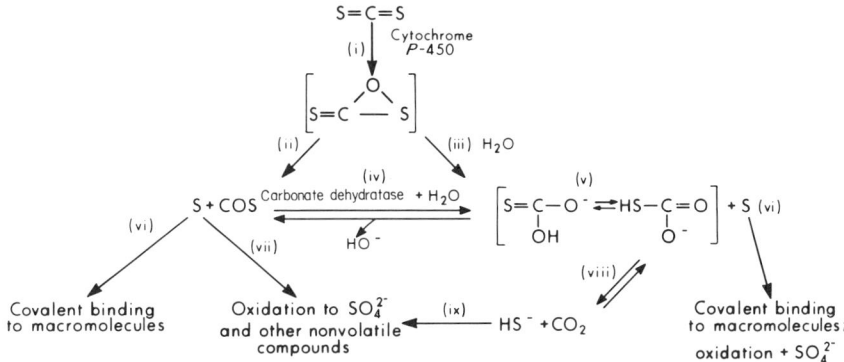

**Fig. 7.** Proposed scheme for the metabolism of carbon disulfide by rat liver cytochrome P-450-dependent monooxygenases.

sulfide, just as bicarbonate is the hydrated form of carbon dioxide. Thus, it is likely that a breakdown of monothiocarbonate to carbonyl sulfide occurs and that the carbonyl sulfide is converted back to thiocarbonic acid by carbonate dehydratase (step iv). This series of reactions (iii and iv) accounts for the findings that the oxygen atom in carbonyl sulfide, formed in the hepatic microsomal metabolism of carbon disulfide, is derived from water rather than from molecular oxygen.[48] The elemental sulfur formed in steps ii and iii binds covalently to available nucleophiles (step vi) and is likely responsible for the inhibition of liver cytochrome P-450 monooxygenase activity seen after administration of carbon disulfide *in vivo* or incubation with hepatic microsomes *in vitro*.[45] The elemental sulfur may also be oxidized to sulfate (step vii). The thiocarbonate also dissociates to carbon dioxide and hydrogen sulfide (step viii). The hydrogen sulfide is, in turn, oxidized to sulfate (step ix).

## C. THIOACETAMIDE

Thioacetamide and a metabolite, thioacetamide $S$-oxide, both produce centrilobular hepatic necrosis in male Sprague–Dawley rats when administered in doses of 1.25 to 5.0 mmol/kg intraperitoneally. Thioacetamide $S$-oxide produces more severe necrosis 24 hr after administration than does an equimolar dose of thioacetamide; additionally, histological changes in liver cells induced by thioacetamide $S$-oxide are detected at earlier times than corresponding changes induced by thioacetamide. The toxicity of both compounds is enhanced by prior treatment of the rats with phenobarbital and is inhibited by pyrazole and cobaltous chloride treatments in noninduced animals and by SKF 525A treatment in phenobarbital-induced animals, suggesting that further metabolic activation is required for either compound to exert its necrotic effect.[13]

On administration of [$^3$H]thioacetamide *in vivo* to rats, maximal covalent binding to liver molecules is observed after 6 hr, and necrosis is detected initially after approximately 12 hr.[20] With [$^3$H]thioacetamide *S*-oxide, the maximal covalent binding, which is about twice that of an equimolar dose of thioacetamide, occurs at approximately 3 hr, and necrosis is detectable at about 7 to 8 hr. In isolated hepatocytes, approximately six times as much binding of $^3$H from thioacetamide *S*-oxide as from thioacetamide is observed.[49] In addition, incubation of hepatocytes with thioacetamide *S*-oxide leads to a significant decrease in the viability of the cells, as measured by trypan blue exclusion, at about 6 hr, and only about 20% survival at 10 hr. On the other hand, with thioacetamide, little decrease in viability is seen at 10 hr. The results of both the *in vivo* and *in vitro* studies strongly suggest that a reactive metabolite of thioacetamide *S*-oxide may be responsible for hepatic necrosis.

Efforts have been directed toward elucidating the nature of the metabolite that becomes covalently bound to liver macromolecules *in vivo*.[20] In one study, thioacetamide *S*-oxide radiolabeled in differing portions of the molecule was administered to rats, and the nature of the radioactivity bound covalently to liver macromolecules was examined.[20] The results of these experiments showed that the amount of binding of [*methyl*-$^3$H]thioacetamide *S*-oxide and [*carbonyl*-$^{14}$C]thioacetamide *S*-oxide was approximately equal. However, little or no covalent binding of $^{35}$S from [$^{35}$S]thioacetamide *S*-oxide was observed. Thus, the bound metabolite appeared to contain the methyl group and the carbonyl carbon, but not the sulfur atom.

In contrast to other thiono-sulfur compounds examined, thioacetamide does not cause an immediate decrease in the amount of cytochrome *P*-450 detectable as its carbon monoxide complex or in cytochrome *P*-450 monooxygenase activity when incubated with hepatic microsomes *in vitro*.[10] The lack of inhibition of this enzyme by thioacetamide could be because thioacetamide, in contrast to other thiono-sulfur compounds examined, forms a stable *S*-oxide. Thus, the rearrangement of the thioacetamide *S*-oxide with the release and covalent binding to cytochrome *P*-450 of atomic sulfur does not occur.

Some data have indicated that thioacetamide is a substrate for both the cytochrome *P*-450-dependent monooxygenase system and the flavoprotein, amine oxidase.[40,50] The proposed mechanism of metabolism of thioacetamide by rat liver is shown in Fig. 8. The initial reaction in the metabolism of thioacetamide to thioacetamide *S*-oxide is a reaction catalyzed by both the cytochrome *P*-450-containing monooxygenase enzymes and amine oxi-

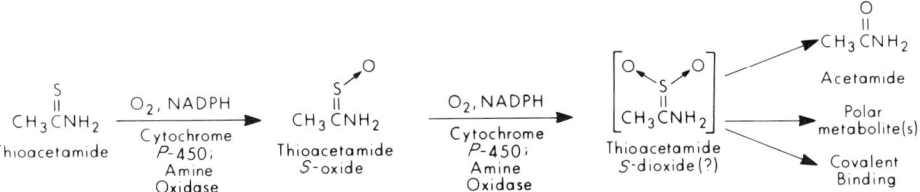

Fig. 8. Proposed scheme for the cytochrome $P$-450-dependent monooxygenase and amine oxidase-catalyzed metabolism of thioacetamide.

dase.[51,52] In contrast to the $S$-oxide of parathion or carbon disulfide, the $S$-oxide of thioacetamide is stable and can be isolated. Thioacetamide $S$-oxide is further metabolized in a reaction also catalyzed by the cytochrome $P$-450 monooxygenase or amine oxidase to a product postulated to be thioacetamide $S$-dioxide. This compound may be subject to attack by water to form acetamide and sulfoxylate ($HSO_2^-$), may break down to unidentified minor metabolites, or may bind covalently to cellular nucleophiles, including the $\epsilon$-amino group of lysine groups present in liver proteins.[50]

### III. Summary

Thiono-sulfur-containing compounds cause a wide variety of toxic effects in mammals. These toxic effects of thiono-sulfur-containing compounds appear to be, at least partially, the result of their metabolism to reactive intermediates by the cytochrome $P$-450-containing monooxygenase enzyme systems. Covalent binding of atomic sulfur released in the cytochrome $P$-450 monooxygenase-catalyzed metabolism of certain thionosulfur compounds appears to be responsible for the inhibition of monooxygenase activity and the loss of cytochrome $P$-450 seen after administration of these thiono-sulfur compounds *in vivo* or incubation with cytochrome $P$-450-dependent monooxygenase enzymes *in vitro*. Liver necrosis and, perhaps, the induction of lung edema and neoplasia as well as other effects of thiono-sulfur-containing compounds are probably the result of the covalent binding of the electrophilic $S$-oxides or $S$-dioxides or carbene derivatives of these $S$-oxides and $S$-dioxides to tissue macromolecules. The rationale for implicating metabolites of thiono-sulfur compounds other than atomic sulfur in these effects derives from the experiments with thioacetamide and the fact that atomic sulfur is highly reactive and appears to bind predominantly or exclusively to cytochrome $P$-450. It is difficult to rationalize why binding to and inhibition of cytochrome $P$-450 would lead to the production of, for example, liver necrosis.

## References

1. Poore, R. E., and Neal, R. A. (1972). Evidence for extrahepatic metabolism of parathion. *Toxicol. Appl. Pharmacol.* **23**, 759–768.
2. Norman, B. J., Poore, R. E., and Neal, R. A. (1974). Studies of the binding of sulfur released in the mixed-function oxidase-catalyzed metabolism of diethyl-*p*-nitrophenyl phosphorothionate (parathion) to diethyl-*p*-nitrophenyl phosphate (paraoxon). *Biochem. Pharmacol.* **23**, 1733–1744.
3. De Matteis, F., and Seawright, A. A. (1973). Oxidative metabolism of carbon disulfide by the rat: Effect of treatments which modify the liver toxicity of carbon disulfide. *Chem.-Biol. Interact.* **7**, 375–388.
4. Jarvisalo, J., Savolainen, H., Elovaara, E., and Vainio, H. (1977). The *in vivo* toxicity of $CS_2$ to liver microsomes: Binding of labelled $CS_2$ and changes of the microsomal enzyme activities. *Acta Pharmacol. Toxicol.* **40**, 329–336.
5. Dalvi, R. R., Poore, R. E., and Neal, R. A. (1974). Studies of the metabolism of carbon disulfide by rat liver microsomes. *Life Sci.* **14**, 1785–1796.
6. Porter, W. R., and Neal, R. A. (1978). Metabolism of thioacetamide and thioacetamide *S*-oxide by rat liver microsomes. *Drug Metab. Dispos.* **6**, 379–388.
7. El-Hawari, A. M., and Plaa, G. L. (1977). α-Naphthylisothiocyanate (ANIT) hepatotoxicity and irreversible binding to rat liver microsomes. *Biochem. Pharmacol.* **26**, 1857–1866.
8. Seawright, A. A., Hrdlicka, J., and De Matteis, F. (1976). The hepatotoxicity of *O,O*-diethyl, *O*-phenyl phosphorothioate ($SV_1$) for the rat. *Br. J. Exp. Pathol.* **57**, 16–22.
9. Bond, E. J., and De Matteis, F. (1969). Biochemical changes in rat liver after administration of carbon disulfide, with particular reference to microsomal changes. *Biochem. Pharmacol.* **18**, 2531–2549.
10. Hunter, A. L., and Neal, R. A. (1975). Inhibition of hepatic mixed-function oxidase activity *in vitro* and *in vivo* by various thiono-sulfur-containing compounds. *Biochem. Pharmacol.* **24**, 2199–2205.
11. Bond, E. J., Butler, W. H., De Matteis, F., and Barnes, J. M. (1969). Effects of carbon disulfide on the liver of rats. *Br. J. Ind. Med.* **26**, 335–337.
12. Magos, L., and Butler, W. H. (1972). Effect of phenobarbitone and starvation on hepatotoxicity in rats exposed to carbon disulfide vapor. *Br. J. Ind. Med.* **29**, 95–98.
13. Hunter, A. L., Holscher, M. A., and Neal, R. A. (1977). Thioacetamide-induced hepatic necrosis. 1. Involvement of the mixed-function oxidase enzyme system. *J. Pharmacol. Exp. Ther.* **200**, 439–448.
14. Magos, L., Butler, W. H., and White, I. N. H. (1973). Hepatotoxicity of $CS_2$ in rats: Relation to postexposure liver weight and pre-exposure cytochrome *P*-450 level. *Biochem. Pharmacol.* **22**, 992–994.
15. van den Brenk, H. A. S., Kelly, H., and Stone, M. G. (1976). Innate and drug-induced resistance to acute lung damage caused in rats by α-naphthylthiourea (ANTU) and related compounds. *Br. J. Exp. Pathol.* **57**, 621–636.
16. Boyd, M. R., and Neal, R. A. (1976). Studies of the mechanism of toxicity and of development of tolerance to the pulmonary toxin, α-naphthylthiourea (ANTU). *Drug Metab. Dispos.* **4**, 314–322.
17. Scheline, R. R., Smith, R. L., and Williams, R. T. (1961). The metabolism of arylthioureas. II. The metabolism of $^{14}$C- and $^{35}$S-labeled 1-phenyl-2-thiourea and its derivatives. *J. Med. Pharm. Chem.* **4**, 109–135.
18. Plaa, G. L., and Priestley, B. G. (1976). Intrahepatic cholestasis induced by drugs and chemicals. *Pharmacol. Rev.* **28**, 207–273.

19. Poulsen, L. L., Hyslop, R. M., and Ziegler, D. M. (1979). S-Oxygenation of N-substituted thioureas catalyzed by the pig liver microsomal FAD-containing monooxygenase. *Arch. Biochem. Biophys.* **198**, 78–88.
20. Porter, W. R., Gudzinowicz, M. J., and Neal, R. A. (1979). Thioacetamide-induced hepatic necrosis. II. Pharmacokinetics of thioacetamide and thioacetamide S-oxide in the rat. *J. Pharmacol. Exp. Ther.* **208**, 386–391.
21. Hanzlik, R. P., Cashman, J. R., and Traiger, G. J. (1980). Relative hepatotoxicity of substituted thiobenzamides and thiobenzamide S-oxides in the rat. *Toxicol. Appl. Pharmacol.* **55**, 260–272.
22. De Matteis, F. (1974). Covalent binding of sulfur to microsomes and loss of cytochrome P-450 during the oxidative desulfuration of several chemicals. *Mol. Pharmacol.* **10**, 849–854.
23. Lee, P. W., and Neal, R. A. (1978). Metabolism of methimazole by rat liver cytochrome P-450-containing monooxygenases. *Drug Metab. Dispos.* **6**, 591–599.
24. Gage, J. C. (1953). A cholinesterase inhibitor derived from $O,O$-diethyl $O$-$p$-nitrophenyl thiophosphate *in vivo*. *Biochem. J.* **54**, 426–430.
25. Aldridge, W. N. (1953). Serum esterases. 2. An enzyme hydrolysing diethyl $p$-nitrophenyl phosphate (E600) and its identity with the A-esterase of mammalian sera. *Biochem. J.* **53**, 117–124.
26. Neal, R. A. (1967). Studies of the metabolism of diethyl 4-nitrophenyl phosphorothionate (parathion) *in vitro*. *Biochem. J.* **103**, 183–191.
27. Nakatsugawa, T., and Dahm, P. (1967). Microsomal metabolism of parathion. *Biochem. Pharmacol.* **16**, 25–38.
28. Ptashne, K. A., Wolcott, R. M., and Neal, R. A. (1971). Oxygen-18 studies on the chemical mechanisms of the mixed-function oxidase-catalyzed desulfuration and dearylation reactions of parathion. *J. Pharmacol. Exp. Ther.* **179**, 380–385.
29. Kamataki, T., Lee Lin, M. C. M., Belcher, D. H., and Neal, R. A. (1976). Studies of the metabolism of parathion with an apparently homogeneous preparation of rabbit liver cytochrome P-450. *Drug Metab. Dispos.* **4**, 180–189.
30. McBain, J. B., Yamamoto, I., and Casida, J. E. (1971). Oxygenated intermediate in peracid and microsomal oxidations of the organophosphonothioate insecticide Dyfonate. *Life Sci.* **10**(2), 1311–1319.
31. McBain, J. B., Yamamoto, I., and Casida, J. E. (1971). Mechanism of activation and deactivation of Dyfonate ($O$-ethyl $S$-phenyl ethylphosphonodithioate) by rat liver microsomes. *Life Sci.* **10**(2), 947–954.
32. Ptashne, K. A., and Neal, R. A. (1972). Reaction of parathion and malathion with peroxytrifluoroacetic acid, a model system for the mixed-function oxidases. *Biochemistry* **11**, 3224–3228.
33. Herriott, A. W. (1971). Peroxy acid oxidation of phosphinothioates, a reversal of stereochemistry. *J. Am. Chem. Soc.* **93**, 3304–3305.
34. Lee, P. W., Allahyari, R., and Fukuto, T. R. (1976). Stereospecificity in the metabolism of the chiral isomers of fonofos by mouse liver microsomal mixed-function oxidase. *Biochem. Pharmacol.* **25**, 2671–2674.
35. Lee, P. W., Allahyari, R., and Fukuto, T. R. (1978). Studies on the chiral isomers of fonofos and fonofos oxon. II. *In vitro* metabolism. *Pestic. Biochem. Physiol.* **8**, 158–164.
36. Ohkawa, H., Mikami, N., and Miyamoto, J. (1976). Stereospecific metabolism of $O$-ethyl $O$-2-nitro-5-methylphenyl $N$-isopropyl phosphoramidothioate by liver microsomal mixed-function oxidase. *Agric. Biol. Chem.* **40**, 2125–2127.
37. Battaglia, A., Dondoni, A., Giorgianni, P., Maccagnani, G., and Mazzanti, G. (1971).

Sulfines. Part III. Kinetics of oxidation of thiobenzophenones with peroxybenzoic acid. *J. Chem. Soc.* **7,** 1547–1550.
38. Strating, J., Thijs, L., and Zwanenburg, B. (1966). Sulfines by oxidation of thioketones. *Tetrahedron Lett.* pp. 65–67.
39. Snyder, J. P. (1974). Oxathiiranes. Differential orbital correlation effects in the electrocyclic formation of sulfur-containing three-membered rings. *J. Am. Chem. Soc.* **96,** 5005–5007.
40. Kamataki, T., and Neal, R. A. (1976). Metabolism of diethyl $p$-nitrophenyl phosphorothionate (parathion) by a reconstituted mixed-function oxidase enzyme system: Studies of the covalent binding of the sulfur atom. *Mol. Pharmacol.* **12,** 933–944.
41. Halpert, J., Hammond, D., and Neal, R. A. (1980). Inactivation of purified rat liver cytochrome $P$-450 during the metabolism of parathion (diethyl $p$-nitrophenyl phosphorothionate). *J. Biol. Chem.* **255,** 1080–1089.
42. Dalvi, R. R., Hunter, A. L., and Neal, R. A. (1975). Toxicological implications of the mixed-function oxidase-catalyzed metabolism of carbon disulfide. *Chem.-Biol. Interact.* **10,** 347–361.
43. Jarvisalo, J., Savolainen, H., and Vainio, H. (1977). Effects of acute $CS_2$ intoxication on liver protein and drug metabolism. *Chem.-Biol. Interact.* **17,** 41–50.
44. Freundt, K. J., and Dreher, W. (1969). Inhibition of drug metabolism by small concentrations of carbon disulfide. *Naunyn-Schmiedebergs Arch. Pharmakol. Exp. Pathol.* **263,** 208–209.
45. Catignani, G. L., and Neal, R. A. (1975). Evidence for the formation of a protein-bound hydrodisulfide resulting from microsomal mixed-function oxidase-catalyzed desulfuration of carbon disulfide. *Biochem. Biophys. Res. Commun.* **65,** 629–636.
46. Chengelis, C. P., and Neal, R. A. (1979). Hepatic carbonyl sulfide metabolism. *Biochem. Biophys. Res. Commun.* **90,** 993–999.
47. Chengelis, C. P., and Neal, R. A. (1980). Studies of carbonyl sulfide toxicity: Metabolism by carbonic anhydrase. *Toxicol. Appl. Pharmacol.* **55,** 198–202.
48. Chengelis, C. P., and Neal, R. A. Unpublished observations.
49. Gudzinowicz, M. J., and Neal, R. A. Unpublished observations.
50. Dyroff, M. C., and Neal, R. A. (1981). Identification of the major protein adduct formed in rat liver after thioacetamide administration. *Cancer Res.* **41,** 3430–3435.
51. Poulsen, L. L., Hyslop, R. M., and Ziegler, D. M. (1974). S-oxidation of thioureylenes catalyzed by a microsomal flavoprotein mixed-function oxidase. *Biochem. Pharmacol.* **23,** 3431–3440.
52. Prough, R. A., and Ziegler, D. M. (1977). The relative participation of liver microsomal amine oxidase and cytochrome $P$-450 in N-demethylation reactions. *Arch. Biochem. Biophys.* **180,** 363–373.

# Index

## A

AAF, see 2-Acetylaminofluorene
AB, see 4-Aminoazobenzene
Acetaldehyde, 494
Acetamide, 366, 457, 473, 492, 537
4-Acetamidoazobenzene, 390
4-Acetamidobiphenyl, 390
2-Acetamidofluorene, 390, 393
2-Acetamidonaphthalene, 390
2-Acetamidophenanthrene, 390, 394
4-Acetamidostilbene, 356, 390
Acetaminophen, 17, 78, 81–82, 92, 265–267, 350, 358–360, 364, 366; see also 4-Hydroxyacetanilide
Acetanilide, 162, 166, 350, 379
Acetazolamide, 534
Acetic acid, 410, 457
Acetone, 494
Acetone cyanohydrin, 486
Acetonitrile, 486, 490–492, 510
$N$-Acetoxy-2-acetylaminofluorene, 12, 14, 17, 392–393
$N$-Acetoxy-$N$-arylacetamides, 388
$N$-Acetoxyarylamides, 364
$N$-Acetoxyarylamines, 388
α-Acetoxynitrosamines, 410
α-Acetoxynitrosopyrrolidine, 420
Acetylacrolein, 251
2-Acetylaminofluorene (AAF), 7, 10–12
$N$-Acetyl-$N$-arylnitrenium ion, 378
$N$-Acetylbenzidine, 394

$N$-Acetyl-$p$-benzoquinoneimine, 265, 359–360
$N$-Acetylcarboxymethylcysteine, 324–325
Acetylcholinesterase, 10, 523
Acetyl CoA:arylamine $N$-acetyltransferase, 387, 389
$N$-Acetyl-$S$-(2-cyanoethyl)cysteine, 502
$N$-Acetylcysteine, 81–83, 444
Acetylenes
  destruction of cytochrome $P$-450, 135–144
  halogenated, 318
  metabolism of, 131–135
  oxidative chemistry of, 122–126
  properties of, 123
  toxicity of, 144–145
$N$-Acetylethanolamine, 457
$S$-Acetylglutathione, 444
Acetylhydrazine, 437, 439, 444
$N$-Acetylisoniazid, 437, 439
Acetylphenylhydrazine, 445
$N$-Acetyltransferase, 437
Acid-base concept, 73
cis-Aconitase, 10
Acrolein, 144
Acrylamide, 112, 116, 494
Acrylonitrile, 80–81, 486, 490, 499–506, 510
Active transport, 33–36
Acylamidase, 437
Acyl chlorides, 319
Acyl halides, 288–289
$N$-Acyloxyarylamines, 389
Acyltransferase, 14
$N,O$-Acyltransferase, 377, 386–391

541

Adenosine 5′-phosphosulfate kinase, 378
Adiponitrile, 499
Administration of foreign compound
  repeated, 41–43
  routes of, 48–54
Adriamycin, 275–276
AF-2, 460
Aflatoxin $B_1$, 15–16, 145
Aflatoxin $B_2$, 15
AIA, see Allylisopropylacetamide
β-Alanine, 80–81
Alclophenac, 129
Alcohol dehydrogenase, 320, 440
Aldehyde dehydrogenase, 320
Aldehydes
  chlorinated, 319
  halogenated, 289–290
Aldrin, 129
Alkanes, 111–119, 444
  bioactivation of, 112–116
  halogenated, see Halogenated alkanes
  toxicity of, 112
Alkenes, 121–145, 297
  halogenated, 317–341
Alkoxides, 170
Alkylaryl nitriles, 506–508
Alkylating agents, 73
Alkyldiazohydroxide, 406–407
Alkyldiazonium ion, 407
7-Alkylguanine, 419
$O^6$-Alkylguanine, 420
Alkylhydrazines, 15
Alkylnitrosamides, 15
Alkylnitrosamines, 15
2-Alkylthiohydroquinone, 272
Alkynes, 121–145
  halogenated, 317–341
$N^6$-Allyladenine, 338
Allyl alcohol, 129, 144
Allyl bromide, 338
Allyl chloride, 337–338
$O^6$-Allylguanine, 338
Allyl halides, 318, 335–339
Allyl iodide, 338
Allylisopropylacetamide (AIA), 135–137, 143–145
Allylisopropyl-2-isopropyl-4-pentenamide, see Allylisopropylacetamide
Amidase, 366
Amidyl radical, 355–356
Amine oxidase, 16, 438

Aminium cation radical, 361
4-Aminoazobenzene (AB), 6, 11, 387–391
Aminoazo dyes, 4–7, 382
p-Aminobenzoic acid, 357
p-Aminobenzoic acid N-acetyltransferase, 389
4-Aminobiphenyl, 387
Amino cation radical, 471
2-Aminofluorene, 387
2-Aminohexanoic acid, 114
2-Aminoimidazole, 454–455, 461
4-Aminoimidazole, 454–455
5-Aminoimidazole, 457
Aminoimidazoles, preparation of, 454–455
3-Amino-1-methyl-5H-pyrido[4,3-b]indole, 377, 394
Aminomisonidazole, 473–474
Aminonitriles, 495–497
Amino oxidase, 417, 536–537
Aminophenols, 357–358
N-Aminopiperidine, 439
β-Aminopropionitrile, 495–496
Aminopyrine, 360
Aminothiols, 80–83
Aminotransferase, 436–437
Ammonia, 509
Amygdalin, 507
Aniline, 8, 350
Anisole, 166
ANIT, see α-Naphthylisothiocyanate
Anthraquinones, 270, 276
Antioxidants, 85–86, 90, 262
ANTU, see α-Naphthylthiourea
ANU, see α-Naphthylurea
Aprobarbital, 136
Arachidonic acid, 470–471
Arene oxides, 159–163, 169–170, 180, 260–261
  from benzo[a]pyrene, 182–185
  reactions of, 163–170
Aroclor 1254, 501
Aromatic amine oxides, 364
Arylamides, 349–366, 376
  chemical properties of, 350–352
  fate of metabolites of, 363–366
  N-oxidation of, 352–363
Arylamines, 349–366, 376
  chemical properties of, 350–352
  fate of metabolites of, 363–366
  N-oxidation of, 352–363
8-Arylaminoguanine, 394
Arylation, 271

# INDEX

Aryl dialkyltriazenes, 15
Arylhydroxamic acids, 364–365, 375–394
 bioactivation by conjugation reactions, 376–394
 O-sulfonation of, 377–381
Arylhydroxyamides, 365
Arylhydroxylamines, 350, 363–366, 375–394
 bioactivation by conjugation reactions, 376–394
Arylnitrenium ion, 386, 392–393
Aryl nitriles, 508–509
Arylnitrosamines, 364
Ascorbate, 72, 75, 86, 94, 469–474
Aspirin, 10
Axon proteins, derivatization of, 117–118
Azide, 170
1,1-Azobis(N,N-dimethylformamide), 444–445
Azomethane, 440
Azomisonidazole, 455
Azomycin, see 2-Nitroimidazole
Azoprocarbazine, 440, 444
Azoxymethane, 440, 445
Azoxymisonidazole, 455

## B

Barban, 137
Barbiturates, 9, 498
Bay region theory, 203–205
Benzaldehyde, 506–507
Benzamide, 509
Benz[a]anthracene, 179–180, 191–192, 204–210, 214–215
Benz[a]anthracene 3,4-dihydrodiol, 216–221
Benz[a]anthracene 8,9-diol 10,11-oxide, 91
Benz[a]anthracene 5,6-oxide, 195
Benzene, 17, 130, 157–170
 bioactivation of, 159–163
 chemical properties of, 158
 reactions of, 163–170
Benzene oxide, 160–161, 164–165, 168–170, 185–186
Benzenes, substituted, 157–170
 binding to macromolecules, 169–170
 bioactivation of, 159–163
 chemical properties of, 158
 reactions of, 163–170
1,2,4-Benzenetriol, 260, 272–273
Benzidine, 358, 387, 394
Benzil, 91

5,6-Benzoflavone, 194
7,8-Benzoflavone, 192
Benzo[b]fluoranthene, 204–205
Benzohydroquinone, 272–273
Benzoic acid, 506–507, 509
Benzoin, 91
Benzonitrile, 166, 508–509
Benzo[c]phenanthrene, 179, 189, 204–209, 212–215
Benzo[c]phenanthrene 3,4-dihydrodiol, 216
Benzo[a]pyrene (B[a]P), 7, 9, 16, 76–77, 87, 90–91, 145, 178–209
 biological activity of, 195–203
 metabolism of, 181–195
 mutagenicity of, 197–199
 tumorigenicity of, 200–203
Benzo[e]pyrene (B[e]P), 178–179, 192, 204–209
Benzo[a]pyrene 4,5-dihydrodiol, 182–187, 200–201, 215
Benzo[a]pyrene 7,8-dihydrodiol, 182–191, 199–203, 211, 213, 215–220
Benzo[a]pyrene 9,10-dihydrodiol, 182–187, 191–193, 200–201, 214–215
Benzo[a]pyrene 11,12-dihydrodiol, 182
Benzo[e]pyrene 9,10-dihydrodiol, 192
Benzo[a]pyrene 7,8-diol 9,10-epoxide, 188–191, 196–205, 213, 216–221
Benzo[a]pyrene 9,10-diol 7,8-epoxide, 192, 198–199
Benzo[a]pyrene 7,10-diol 9,10-epoxide, 198–199
Benzo[a]pyrene-3,6-diphenol, 194
Benzo[a]pyrene $H_4$-7,8-epoxide, 198–199
Benzo[a]pyrene $H_4$-9,10-epoxide, 198–199
Benzo[a]pyrene 4,5-oxide, 182–186, 194–195, 198–203, 212–213
Benzo[a]pyrene 7,8-oxide, 182–186, 194, 199–203, 212–213
Benzo[a]pyrene 9,10-oxide, 182–183, 186, 194, 200–203
Benzo[a]pyrene 11,12-oxide, 194, 200–203
Benzo[a]pyrene-1,6-quinone, 200–201
Benzo[a]pyrene-3,6-quinone, 194, 200–201
Benzo[a]pyrene-6,12-quinone, 200–201
Benzoquinone, 366
o-Benzoquinone, 260, 266–267, 270–272
p-Benzoquinone, 260, 263, 266, 269–275
p-Benzosemiquinone, 275–276
N-Benzoyloxy-N-methyl-4-aminoazobenzene, 12, 17, 388, 394

Benzphetamine, 530
Benzylamine, 441
Benzyl chloride, 336–338
Benzyl cyanide, 490–491, 506
N-Benzylcyclopropylamine, 361–362
Benzylhydrazine, 436
N-Benzyloxy-N-methyl-4-aminoazobenzene, 393
BHA, see 2(3)-tert-Butyl-4-hydroxyanisole
BHT, see Butylated hydroxytoluene
Biphenols, 261–262, 266
Biphenylacetylene, 125–126, 132–133
Biphenyls, 158, 166–167
  chemical properties of, 158–159
Blood, arterial
  drug concentration in, 32–70
Blood concentration vs. time curve, area under, 43–44, 53, 56–57, 67–69
BNF, see β-Naphthoflavone
B[a]P, see Benzo[a]pyrene
B[e]P, see Benzo[e]pyrene
Brallobarbital, 130
Bromoacetaldehyde, 289
Bromobenzene, 78, 87, 89, 96, 163–165, 169
o-Bromobenzonitrile, 169
o-Bromobenzotrifluoride, 169
Bromobenzyl cyanide, 508
2-Bromo-3-chlorobutane, 303–304
1-Bromo-2-chloroethane, 303
1,2-Bromoethane, 305
Bromoform, 288
o-Bromotoluene, 169
Bromotrichloromethane, 291–292, 295
Bromoxymil, 508
Butamoxane, 267
1-Butanol, 411
2-Butanol, 411
tert-Butanol, 410
1-Butene, 123
2-Butene, 303–304
Butylated hydroxytoluene (BHT), 88, 90, 94
tert-Butyldiazotic acid, 410
tert-Butyl hydroperoxide, 92
2(3)-tert-Butyl-4-hydroxyanisole (BHA), 90–91
4-(N-Butylnitrosoamino)-4-hydroxybutyric acid, 416
1-Butyne, 123
Butyronitrile, 493

## C

Calcium, 95–96
Carbenes, 297, 300
Carbinolamines, 362–363, 365
Carbolic acids, see Phenols
Carbonate dehydratase, 534–535
Carbon dioxide, 331
Carbon disulfide, 112, 520, 522, 533–535
Carbonium ion, 168
Carbon monoxide, 289, 300, 330–331
Carbon tetrachloride, 15–16, 78, 95–96, 291–292, 295–301, 498
Carbonyl bromide, 289
Carbonyl halides, 288–289
Carbonyl metabolites, 285–290
Carbonyl sulfide, 533–535
Carboxylic acids, 289, 500
Carboxymethylcysteine, 321, 324–325
Carcinogens, 4–19
  binding to macromolecules, 4–7, 12–14
β-Carotene, 72, 75, 94
Catalase, 72, 75, 357, 443, 445
Catechol estrogen, 264
Catechol O-methyltransferase, 274
Catechols, 181, 260–263, 265, 509
  biosynthesis of, 267
  chemical properties of, 266–267
  metabolism of, 268–269
  toxicity of, 268
Central-peripheral distal axonopathy, 112
Ceruloplasmin, 357
Chloral, 329–330
Chloramphenicol, 289, 299, 303, 459
Chloroacetaldehyde, 320–323
Chloroacetic acid, 320–321, 324–325
Chloroacetyl chloride, 324–325
Chlorobenzene, 161, 166
4-Chlorobenzenesulfenyl chloride, 122–123
2-Chlorobenzylidene malononitrile, 507–508
4-Chlorobiphenyl, 169
1-Chloro-2-butene, 338
2-Chloro-2-butene, 338
1-Chloro-2-cyclohexene, 337
3-Chloro-1-cyclohexene, 338
2-Chloroethanol, 320
S-(2-Chloroethyl)-DL-cysteine, 306
Chloroform, 16, 78, 286, 288, 292, 295–299, 340–341
1-Chloro-3-hydroxyacetone, 339

1-Chloro-2-methyl-2-butene, 338
3-Chloro-2-methyl-1-propene, 338
Chlorooxirane, 321, 323
Chloroperoxidase, 293–294
1-Chloro-1-propene, 338
Chlorotrifluoroethane, 306
2-Chloro-1,1,1-trifluoroethane, 299–300
$S$-(2-Chloro-1,1,2-trifluoroethyl)glutathione, 306
Chlorpromazine, 357
Cholesterol, 118
Chrysene, 179, 191–192, 204–209, 214–215
Chrysene 1,2-dihydrodiol, 216–220
Chrysene 1,2-diol 3,4-epoxide, 219–220
Cinnamonitrile, 508
Clearance, 66–67
  dose-dependent, 69
  organ, 44–48
  total body, 44–48
Cobalamin, 489
Cobaltous chloride, 520–522, 535–536
Coenzyme Q, see Ubiquinone
Compartmentation, cellular, 72, 88–89
Crotononitrile, 506
Cumene, 92
Cyanate, 488
Cyanide, 485–510
Cyanides, inorganic, 486–489
Cyanoacetic acid, 496
Cyanocobalamin, see Vitamin $B_{12}$
2-Cyanoethylene, 504
$S$-(2-Cyanoethyl)glutathione, 502
Cyanogenic glycosides, 507
Cyanohydrins, see Hydroxynitriles
Cyanomethemoglobin, 500
Cyanophenol, 509
Cycasin, 15
Cyclohexane, 127
1,2-Cyclohexanediol, 127
Cyclohexene, 129
4-Cyclohexyl-$N$-hydroxy-$N$-phenylacetamide, 390
Cyclooxygenase, 262
Cyclopropyl amines, 250
Cystathionine pathway, 79
Cysteamine, 83–84, 251
Cysteine, 74, 78–79, 83, 89, 305–306, 444, 488, 506
Cysteine conjugate β-lyase, 305–306
Cystine, 79, 83, 89
Cytochrome $c$ oxidase, 501

Cytochrome $P$-450, 8–11, 16, 87–88, 91, 94
  acetylation of, 289
  catalytic site of, 212–217
  destruction by olefins and acetylenes, 135–144
  isozymes of, 161–163
  in metabolism of
    acetylene, 132–133
    acrylonitrile, 501
    arylamines and arylamides, 351, 354–356, 359–362
    benzenes, 161–163, 169
    carbon disulfide, 533–535
    halogenated alkanes, 291–293, 296–302
    halogenated alkenes and alkynes, 320, 327
    hydrazines, 438–440, 443–444
    nitroimidazoles, 471
    olefins, 124, 128–129
    parathion, 523–533
    phenols, catechols and quinones, 264–268
    polycyclic aromatic hydrocarbons, 180, 208–209
    thiono-sulfur compounds, 521, 536–537
  in nitrosodimethylamine demethylase, 417–419
Cytochrome $P$-450$_I$, 247, 253
Cytochrome $P$-450$_{II}$, 247, 253

# D

DAB, see $N,N$-Dimethyl-4-aminoazobenzene
Daunomycin, 275–276
DDT, see 1,1,1-Trichloro-2,2-bis($p$-chlorophenyl)ethane
Deacetylase, 14
Deacylase, 12
Defense mechanisms, cellular, 71–97, see also Protection
Dehydrohalogenation, oxidative, 285–288
Demethylase, 438
Detoxication, 71–97
Diacetylbenzidine, 390
Dialdehydes, 497
5,5-Diallylbarbituric acid, 135–136
Diamine oxidase, 436
Diarylthioureas, 520–521
Diazenes, 434–444
Diazoalkane, 406–407
Diazomethane, 414

Dibenz[a,h]anthracene, 7, 178–180, 192, 204–209, 214–215
Dibenzo[a,h]pyrene, 204–205
Dibenzo[a,i]pyrene, 204–205
2,3-Dibromobutane, 303–304
1,2-Dibromoethane, 289, 303
3,5-Dibromo-4-hydroxybenzonitrile, 508
Dibromomethane, 286
2,6-Di-*tert*-butyl-4-carboxyphenol, 251
Dichlobenil, 508
Dichloroacetaldehyde, 324, 327–328
Dichloroacetic acid, 327–328
1,3-Dichloroacetone, 339
2′,4′-Dichloroacetophenone, 305
Dichloroacetyl chloride, 329–330
Dichloroacetylene, 340–341
2,6-Dichlorobenzonitrile, *see* Dichlobenil
Dichlorobiphenyls, 159, 161, 163–165
Dichlorocarbene, 292, 301
1,2-Dichloroethane, 303, 306
Dichloroethanol, 327–328
1,2-Dichloroethylene, 78, 327–328
Dichloromethane, 286, 288
1,1-Dichlorooxirane, 324–325
1,3-Dichloropropene, 338
2,3-Dichloro-1-propene, 338-339
S-(1,2-Dichlorovinyl)cysteine, 305
N-Diethyl-4-aminoantipyrine, 8
Diethyl phosphate, 523–524, 527–530
Diethyl phosphorothionate, 523–524, 527–530
Diglutathionyl dithiocarbonate, 288
Dihaloalkanes, 290
1,2-Dihaloethanes, 289–290, 304–305
Dihalomethanes, 289, 302–303
Dihydrodiol dehydrogenase, 91
Dihydrodiols, 180–181, 185–187
1,2-Dihydro-6-ethoxy-2,2,4-trimethylquinoline, 90
15, 16-Dihydro-11-methylcyclopenta[a]phenanthrene-17-one, 204–205
4,5-Dihydroxy-2-isopropylpentanamide, 127
9,10-Dihydroxystearic acid, 127
γ-Diketone hypothesis, 115–119
N-Dimethyl-4-aminoantipyrine, 8
N,N-Dimethyl-4-aminoazobenzene (DAB), 5–6, 365
4-Dimethylaminophenol (DMAP), 82–83
Dimethylaminopropionitrile, 510
N,N-Dimethylaniline, 360

7,12-Dimethylbenz[a]anthracene, 204–205, 210, 212
1,3-Dimethylbenzene, 167–168
1,4-Dimethylbenzene, 170
1,4-Dimethylbenzene oxide, 166, 168
4,5-Dimethylbenzene oxide, 168
3,6-Dimethylcholanthrene, 212
5,11-Dimethylchrysene, 212
7,14-Dimethyldibenz[a,j]anthracene, 212
7,12-Dimethyldibenz[a,h]anthracene, 212
2,5-Dimethylfuran, 114
1,1-Dimethylhydrazine, 436, 439
1,2-Dimethylhydrazine, 440, 445
Dimethylnitrosamine, 7, 12, 16
1,4-Dimethylphenanthrene, 212
4,10-Dimethylphenanthrene, 212
3,3-Dimethyl-1-phenyltriazene, 16
Dinitriles, 497–499
2,4-Dinitrophenol, 262
Diol dehydrogenase, 181
Diphenols, 265
1,2-Diphenyl-1,2-ethanediol, 127
1,3-Diphenylpropane, 130
Dipiperidyltetrazene, 439
Dithiocarbamates, 83
Dithioglycolic acid, 324–325
Dithionite, 462
Dithiothreitol, 83, 251
Divinyl ether, 145
DMAP, *see* 4-Dimethylaminophenol
DNA polymerase, 15, 274
DT-diaphorase, 275

**E**

Elagic acid, 77
Electron transport, 274–275
Electrophilic reactants
 formation of, 11–14
 protection against, 72–74
 role of, 15–17
Endoplasmic reticulum, 7–10
Enedial, 251
Enflurane, 288
Enolase, 116
Enol ethers, 123
Enzyme induction, 90–92
Epichlorohydrin, 332–333
Epinephrine, 469
Epoxidation, of olefins, 126–131, 317–341

Epoxide hydrase, 502–505
Epoxide hydratase, 16
Epoxide hydrolase, 9, 72, 76–77, 89–91, 126–128, 160–161, 164, 180, 183–187, 209, 267
1,2-Epoxybutane, 332–333
Epoxydiols, 260
Epoxyphenols, 260, 265, 267
Esterase, 528
Estragole, 15
Estrogens, 264–265
Estrogen semiquinone, 264–265
Ethacrynic acid, 89
Ethanolamine, 457
Ethchlorvynol, 134–135, 138
Ethinamate, 137
17α-Ethinylestradiol, 133–134, 136
4′-Ethinyl-2-fluorobiphenylacetylene, 132–133
2-Ethinylnitrobenzene, 131–132
17α-Ethinyl sterols, 133–134, 136, 145
Ethionine, 7
Ethoxyquin, see 1,2-Dihydro-6-ethoxy-2,2,4-trimethylquinoline
Ethyl 3-furoate, 244
4-Ethyl-3,5-bis(ethoxycarbonyl)-2,6-dimethyl-1,4-dihydropyridine, 361
Ethyl carbamate, 15
2-(N-Ethylcarbamoylhydroxymethyl)furan, 246–247, 252
Ethylcyanide, see Propionitrile
Ethylene, 122–123, 135–136, 138–139
Ethylenes, halogenated, 130–131, 318–335
Ethylene thiourea, 521
4-Ethyl-1-hexene, 135–136
Ethylhydrazine, 438
Ethylmethanesulfonate, 96
N-Ethyl-N-nitrosoureas, 73–74
Ethyl radical, 438

## F

2-FAA, see 2-Fluoroenylacetamide
FANFT, 460
Fatty acid, 505
Fatty acid nitroxide, 470–471
Fick's Law, 35
First-pass organ, 61–67
Flagl, see Metronidazole
4′-Fluoro-4-acetamidobiphenyl, 394
Fluoracetic acid, 10, 493

Fluoroacetonitrile, 493
6-Fluorobenzo[a]pyrene, 206–209, 211
7-Fluorobenzo[a]pyrene, 211
8-Fluorobenzo[a]pyrene, 211
9-Fluorobenzo[a]pyrene, 211
10-Fluorobenzo[a]pyrene, 211
2-Fluorobiphenyl, 159
2-Fluoroenylacetamide (2-FAA), 350, 366
4-Fluorohexanonitrile, 493
Fluoronitrile, 493
Fluorooxirane, 323
3-Fluoropropionitrile, 493
Fluorotrichloromethane, 299
Fluroxene, 135–139, 142–143, 145
Fonofos, 525–526
Formaldehyde, 302–303, 406, 410–411, 418, 439, 494
Formaldehyde dehydrogenase, 303
Formic acid, 303, 330–331, 488–489
S-Formylglutathione hydrolase, 303
Formyl halide, 289
Free radicals, protection against, 74–75
Fructose-6-phosphate kinase, 116
Furamide, 246
Furan hydroperoxide, 251–252
Furans, 17, 243–254
  binding to macromolecules, 247–249, 252–253
  bioactivation of, 246–252
  occurrence of, 245
  toxicity of, 252–253
Furfural, 245
Furosemide, 17, 87, 245–251

## G

gem-Dihaloalkanes, 302–303
gem-Halohydrin, 285–286
Geraniol, 129
Gluconeogenesis, 437
α-Glucosidase, 507
Glucuronic acid, 383–385
Glucuronidation
  of arylhydroxamic acids, 383–385
  of arylhydroxylamines, 385
γ-Glutamyltransferase, 77–78, 83
Glutathionase, 503
Glutathione, 74–81, 87–96, 168, 194–195, 253, 302–306, 393, 444, 502–506
Glutathione disulfide reductase, 90

Glutathione S-epoxide transferase, 168, 502
Glutathione hydroperoxides, 94
Glutathione peroxidase, 72, 75, 92–95
Glutathione reductase, 72, 75, 93, 95
Glutathione S-transferase, 72, 77, 79, 81, 89–91, 97, 126, 180, 185–187, 289–290, 302–306, 502–505, 508
Glutathionuria, 78
Glyceraldehyde-3-phosphate dehydrogenase, 116–117
Glycine, 457
Gycoaldehyde, 505
Glycol aldehyde, 320
Glycolic acid, 331
Glycolysis, 116–117
Glyconitrile, 494
N-Glycoylethanolamine, 457
Glyoxal, 505
Glyoxylic acid, 330–331
N-(Guanosine-8-yl)-2-acetamidofluorene, 393

# H

Halobenzenes, 17
Halocarbon radicals, oxygenation of, 290–296
S-(2-Haloethyl)glutathione, 304–305
Halogenated alkanes, 283–306, 317–318, 339–340
 glutathione-dependent metabolism of, 302–306
 reductive-oxygenation of, 290–296
 reductive reactions of, 296–302
 toxicity of, 295–296
Halogenated alkenes, 317–341
Halogenated alkynes, 317–341
Halogenated biphenyls, 158
Halogen-carbon bond, chemistry of, 284–285
α-Haloketones, 305
Halothane, 288, 294, 299–301
Heinz body, 442
Heme, 135–144, 351, 361–362, 442–444, 532
Hemoproteins, 442–444
Henderson–Hasselbalch equation, 35
Heptachlor, 126
2,5-Heptanedione, 115
1-Heptene, 135–136
Heterocyclic N-hydroxy compounds, O-sulfonation of, 382–383
Hexachlorobutadiene, 305–306, 340–341

Hexachloroethane, 294–295, 298–301
1-Hexadecene, 129
Hexamethylmelamine, 365
n-Hexane, 111–115
2,5-Hexanediol, 114–115
2,5-Hexanedione, 113–119
2-Hexanol, 113, 115
2-Hexene, 124
Hexobarbital, 129
Hydralazine, 437
Hydrazides, 434–435, 437, 439
Hydrazines, 405, 433–445
 bioactivation of, 436–440
 chemical properties of, 434–436
 fate of reactive intermediates from, 441–445
Hydrazones, 434–441
Hydrocyanic acid, 507
Hydrogen cyanide, 486, 494–495
Hydrogen peroxide, 94, 97, 170, 355, 363, 442–443
Hydrogen sulfide, 534–535
Hydrohalic acids, 289
Hydronitroxides, 469–470
Hydroperoxidase, 358
Hydroperoxides, 94, 97
4a-(Hydroperoxy)flavins, 353
Hydroquinones, 161, 260–263, 271, 275
 biosynthesis of, 267
 chemical properties of, 266–267
 metabolism of, 268–269
 toxicity of, 268
Hydrosulfide, 534
Hydroxamic acids, 355–356, 360
N-Hydroxy-4-acetamidoazobenzene, 380
3-Hydroxy-4-acetamidobiphenyl, 390
N-Hydroxy-2-acetamidobiphenyl, 391
N-Hydroxy-4-acetamidobiphenyl, 380, 384, 390
N-Hydroxy-2-acetamidofluorene, 377–394
N-Hydroxy-3-acetamidofluorene, 381, 391
N-Hydroxy-1-acetamidonaphthalene, 391
N-Hydroxy-2-acetamidonaphthalene, 380–381
N-Hydroxy-2-acetamidophenanthrene, 380, 384, 394
N-Hydroxy-4-acetamidostilbene, 380, 384
3-Hydroxyacetaminophen, 267
4-Hydroxyacetanilide, 379; see also Acetaminophen
N-Hydroxyacetanilide, 380–381
N-Hydroxy-2-acetylaminofluorene, 11–12, 471

N-Hydroxyamine reductase, 471
N-Hydroxy-4-aminoazobenzene, 377, 382
N-Hydroxy-4-aminobiphenyl, 382, 385, 388, 390
N-Hydroxy-2-aminofluorene, 12, 377, 382, 385, 391, 394
4-Hydroxyaminoquinoline 1-oxide, 14, 16, 377
N-Hydroxy-4-aminostilbene, 382
N-Hydroxyaniline, 382
1-Hydroxybenzo[a]pyrene, 193, 198–201
2-Hydroxybenzo[a]pyrene, 199–203, 263
3-Hydroxybenzo[a]pyrene, 183, 193–195, 198–201
4-Hydroxybenzo[a]pyrene, 198–201
6-Hydroxybenzo[a]pyrene, 193, 198–199, 202–203
7-Hydroxybenzo[a]pyrene, 193, 198–201
9-Hydroxybenzo[a]pyrene, 193–194, 198–201
10-Hydroxybenzo[a]pyrene, 200–201
11-Hydroxybenzo[a]pyrene, 200–202
12-Hydroxybenzo[a]pyrene, 198, 200–201
2-Hydroxy-p-benzoquinone, 260
N-Hydroxy-4-chloroacetanilide, 384–385
N-Hydroxy-p-chloroacetanilide, 380–381
$O^6$-7-(1-Hydroxyethano)deoxyguanine, 322
N-Hydroxy-N-ethyl-4-aminoazobenzene, 382
S-(2-Hydroxyethyl)cysteine, 321
1,2-Hydroxyethyl-2-methyl-5-nitroimidazole, see Metronidazole
N-(2-Hydroxyethyl)oxamic acid, 457, 473
N-Hydroxy-2-fluorenylacetamide, 355, 364, 366
5-Hydroxy-2-hexanone, 113, 115
α-Hydroxyisobutyronitrile, 494
Hydroxylamines, 353, 355, 457, 469
Hydroxylaminomisonidazole, 458
Hydroxyl radical, 94–95, 355
(2-Hydroxy-3-methoxypropyl)guanidine, 474
N-Hydroxy-N-methyl-4-aminoazobenzene, 382, 392
6-Hydroxymethyl-benzo[a]pyrene, 181, 199–202
N-Hydroxy-N-methyl-N-benzylamine, 382
1-Hydroxy-3-methylcholanthrene, 210–211
2-Hydroxy-3-methylcholanthrene, 210–211
3-Hydroxy-3-methylglutaryl-CoA reductase, 118
7-Hydroxymethyl-12-methylbenz[a]anthracene, 195, 211

N-Hydroxy-1-naphthylamine, 382
N-Hydroxy-2-naphthylamine, 382, 385
Hydroxynitriles, 490, 493–495
α-Hydroxynitrosamines, 406–407, 410
β-Hydroxynitrosamines, 413–414
β-Hydroxynitrosodialkylamines, 405
5-Hydroxypentanal, 408
N-Hydroxyphenacetin, 350, 366, 380–381, 390
N-Hydroxy-N-phenylacetamide, 390
p-Hydroxyphenylthiourea, 521
β-Hydroxypropionitrile, 494–495
2-Hydroxytetrahydrofuran, 408
3-Hydroxyxanthine, 16, 382–383
Hypochlorous acid, 296

## I

Iminium ion, 361–363
2-Iminothiacolidine-4-carboxylic acid, 489
1,2-Indanediol, 127
Indene, 127, 129
Induction, of enzymes, 90–92
Initiation, 18, 200–202, 393–394
Insecticide, 136
Intraaortic administration of foreign compound, 48–50
Intramuscular administration of foreign compound, 50
Intravenous administration of foreign compound, 50, 53, 68
Ionophore A23187, 95–96
4-Ipomeanol, 244–247, 250, 252–254, 522
1,4-Ipomeanol, 253
Iproniazid, 17, 439, 443
Iron, 74
Isobutyronitrile, 493
Isoniazid, 17, 387, 437
Isopropylhydrazine, 438–439, 444
2-Isopropyl-4-pentenamide, 127
Isoproterenol, 469

## K

Ketenes, 132–133, 145
α-Ketoacids, 437
β-Ketonitrosamines, 413
Kynureninase, 436

## L

Lactonitrile, 494
Laetrile, see Amygdalin
Lathyrism, 495
Lethal synthesis, 10
Lipid peroxidation, 84–85, 89, 92–96, 295–298, 505

## M

MAB, see N-Methyl-4-aminoazobenzene
Malononitrile, 486, 497, 510
Mandelonitrile, 490–491, 506–507
Menadione, 270, 275
Menthylmethylphenylphosphinothioate, 525
Mercaptoimidazoles, 83
β-Mercaptopropionic acid, 80–81
β-Mercaptopyruvic acid, 488
Mercapturic acids, 289, 302, 502–506, 508
Mesitylene, 166
Metabolite
  chemically reactive, 30
  covalent binding of, 86–88
  first-generation, 62–67
  intermediate-lived, 31, 57–60
  long-lived, 31–32, 60–69
  second-generation, 56, 62–69
  short-lived, 30–32, 54–57
  ultralong-lived, 31–32
  ultrashort-lived, 30
Metal compounds, 15
Methane, 440
Methanol, 406, 410–411
Methemoglobin, 363, 376, 442
Methenamide, 10
Methimazole, 522
Methionine, 79
N-Methoxy-2-acetamidofluorene, 389–390
N-Methylacetaminophen, 359
N-Methyl-4-aminoazobenzene (MAB), 5, 11–12, 392–394
N-Methylaminoazo dyes, 8–9
Methylazoxymethanol, 440, 445
7-Methylbenz[a]anthracene, 204–205
1-Methylbenzene oxide, 168
3-Methylbenzene oxide, 168
4-Methylbenzene oxide, 168
6-Methylbenzo[a]pyrene, 199–202
11-Methylbenzo[a]pyrene, 212

2-Methylbutenedial, 251
Methyl n-butyl ketone, 112–116
3-Methylcholanthrene, 9, 11, 87, 91, 169, 179, 185, 189, 191, 204–210, 501
5-Methylchrysene, 204–205, 210, 212
11-Methylcyclopentaphenanthrene-17-one, 212
α-Methyldopa, 268
Methyl ethyl ketone, 115
2-Methylfuran, 244–245, 251
3-Methylfuran, 246, 251
7-Methylguanine, 406, 412–413, 445
$O^6$-Methylguanine, 413, 445
3-Methylhistidine, 406
Methylhydrazine, 439
Methyl isobutyl ketone, 115
Methylketones, 305
1-Methylmercapto-2-acetamidofluorene, 379
3-Methylmercapto-2-acetamidofluorene, 379
3-Methylmercapto-N-methyl-4-aminobenzene, 12
8-Methylmercaptoxanthine, 383
1-Methyl-2-nitroimidazole, 461
4-(N-Methyl-N-nitrosamino)-1(3-pyridyl)-1-butanone, 408
N-Methyl-N-nitrosourea, 73–74, 88
Methyl phenyldiazenecarboxylate, 444
p-Methylphenyl-1,2-ethanediol, 127
Methyl phenyl sulfone, 166
p-Methylstyrene, 127
Methylthioacetylaminoethanol, 325–326
Methylthiourea, 521
Metronidazole, 452–457, 460–468, 472–473
Microsomes, see Endoplasmic reticulum
Microtubules, 274, 276
Misonidazole, 94, 453, 455–462, 468, 473–475
Mitochondria, 88–89, 274–275
Mitomycin C, 275–276
Mixed-function oxidase, 8–11, 161, 246–252, 492, 498, 505
Monoamine oxidase, 84, 417, 438, 441, 496
Monoarylthioureas, 520
Monochloroacetyl chloride, 324–325
Monochlorobutenes, 337–338
Monochloroglycol, 320
Monocyanophenol, 508–509
Monohaloacetaldehyde, 289
Monohaloalkanes, 290
Mono-input organs, 44, 50, 61
Monooxygenase, 83, 94, 113, 209–210, 263–265, 352–354, 439, 490

# INDEX

Monothiocarbonic acid, 534–535
Muconic acid, 267
Mutagenesis, 17–18, 197–199, 219–222

## N

NADPH-cytochrome $c$ reductase, 275
NADPH cytochrome $P$-450 reductase, 93, 95
NADPH:quinone reductase, 90
Naphthalene, 180
Naphthalene 1,2-oxide, 164, 186
β-Naphthoflavone (BNF), 163–164
1-Naphthol, 262
1,2-Naphthoquinone, 270, 275
1,4-Naphthoquinone, 270
Naphthoquinones, 271–272
$N$-2-Naphthoylhydroxylamine $O$-acetate, 390
2-Naphthylamine, 16, 350, 386, 394
α-Naphthylisothiocyanate (ANIT), 520, 522
α-Naphthylthiourea (ANTU), 520–522
α-Naphthylurea (ANU), 521–522
Neurotoxicity, 112–119
Neurotransmitter, 503–504
Ngaione, 245–246
Nifuradene, 457
Nifurtimox, 460
Niridazole, 463
Nitramine, 405
Nitrenium ion, 364–365, 392–393, 471
Nitrile lyase, 507
Nitriles, 485–510
  saturated aliphatic, 491–499
  unsaturated aliphatic, 499–506
Nitro anion radical, 462–469
Nitrobenzene, 166, 457, 459
$p$-Nitrobenzoate, 459, 463
Nitro compounds
  enzymatic reduction of, 458–461
  free-radical metabolites of, 462–472
Nitrodimethyl-$d_6$-amine, 407
Nitrofurans, 245–246, 452, 467–468
Nitrofurantoin, 459, 462, 464
Nitrofurazone, 457, 459–460, 462–466
Nitrogen mustard, 7
Nitroimidazoles, 451–475
  bacterial metabolism of, 472–473
  reactions with macromolecules, 462–463
  reduction of, 461–462
    enzymatic, 456–458
    zinc, 455–456

Nitrones, 353, 364
4-Nitrophenol, 379
$p$-Nitrophenol, 523, 527–529
3-(2-Nitrophenyl)propynoic acid, 131–132
4-Nitroquinoline $N$-oxide, 469
Nitro radical anions, 93
Nitroreductase, 453, 458–460, 463–469, 475
Nitrosamines, 403–420
  bioactivation of, 405–420
  chemical properties of, 404–405
  fate of reactive intermediates from, 419–420
Nitroso(α-acetoxyalkyl)alkylamines, 410
Nitroso(α-acetoxymethyl)benzylamine, 410–411
Nitroso(α-acetoxymethyl)-$tert$-butylamine, 410
$N$-Nitroso-$N$-acetoxymethyl-$N$-2-oxopropylamine, 414
Nitrosoalkyl-2-carboxyethylamine, 416
Nitrosoalkyl-3-carboxypropylamine, 416
Nitrosoalkyl-4-hydroxybutylamine, 416
$N$-Nitroso-α-amino acids, 405
Nitrosobenzene, 469
Nitrosobutane, 469–472
Nitrosobutylamine, 415
Nitrosobutyl-3-carboxy-2-hydroxypropylamine, 415
Nitrosobutylcarboxymethylamine, 415
Nitrosobutyl-3-carboxypropylamine, 415
Nitrosobutyl-4-hydroxybutylamine, 415
Nitrosobutyl(α-hydroxymethyl)amine, 411
Nitrosobutyl-2-oxopropylamine, 415
Nitroso compounds, 363–364
Nitrosodiallylamine, 407
Nitrosodibenzylamine, 409
Nitrosodibutylamine, 407, 412, 415
Nitrosodicylohexylamine, 409
Nitrosodiethylamine, 407
Nitrosodimethylamine, 403–406, 409–413, 417–420
Nitrosodimethylamine demethylase, 417–418
Nitrosodipropylamine, 407, 412–413, 419
Nitrosoethylethylamine, 408
Nitrosofluorene, 14, 470
Nitrosohexamethyleneimine, 408, 420
Nitroso(α-hydroxymethyl)amine, 411
Nitroso-2-hydroxypropylpropylamine, 412–413, 419
Nitrosoisopropylamine, 409
Nitrosomethyl(α-acetoxybenzyl)amine, 410–411
Nitrosomethyl(α-acetoxymethyl)amine, 410

Nitrosomethylalkylamines, 412–413
Nitroso-N-methylaniline, 411
Nitrosomethylbenzylamine, 408, 410–411
Nitrosomethylbutylamine, 408, 410, 420
Nitrosomethyl-2-carboxyethylamine, 416
Nitrosomethyl-3-carboxypropylamine, 416
Nitrosomethyldodecylamine, 416
Nitrosomethylethylamine, 409
Nitrosomethyl-2-hydroxyalkylamines, 413–414
Nitrosomethyl(2-phenylethyl)amine, 408
Nitrosomethylpropylamine, 413
Nitrosomorpholine, 408–409, 420
Nitrosonornicotine, 408
Nitroso-2-oxopropylpropylamine, 412–414
Nitrosopiperidine, 408, 419–420
Nitrosoproline ethyl ester, 409
Nitrosopropylamine, 413
Nitrosopyrrolidine, 408, 419–420
Nitroxides, 469–471
Nonfirst-pass organ, 62–67
Norbornane, 250
Norethisterone, 134, 136, 138
Norgestrel, 133–134
Normenthofuran, 244
Novonal, 130
Nucleic acids, reactions with
　acrylonitrile, 503–504
　allyl halides, 336–338
　arylamine and arylamide metabolites, 366
　benzene derivatives, 169
　benzo[a]pyrene derivatives, 196–197
　carcinogens, 4–7, 12–14, 86–88
　furans, 247–249, 252–253
　haloacetaldehydes, 288–289
　hydrazines, 445
　nitroimidazoles, 462–463
　nitrosamines, 406–407, 419–420
　trichloroethylene metabolites, 332
　vinyl chloride metabolites, 321–323
　vinylidene chloride, 326
Nucleophiles, 72–73
Nucleophilic substitution reactions, 73

## O

1,2-Octanediol, 127
4,5-Octanediol, 127
3,6-Octanedione, 115
Octene, 138–139

1-Octene, 127–129, 139, 142
2-Octene, 124
4-Octene, 126–129
Octyne, 138–139
Olefins
　destruction of cytochrome $P$-450, 135–144
　metabolism of, 126–131
　oxidative chemistry of, 122–126
　properties of, 123
　toxicity of, 144–145
Oleic acid, 127–128
Oral administration of foreign compound, 49–53
Organosulfides, 250
Oxadiazoline, 414
Oxadiazolium ion, 413–414
Oxalic acid, 331, 335
Oxaziranes, 355–356
Oxepin, 169
Oxidative phosphorylation, 262, 275
Oxidocyclohexane, 128
Oxiranes, 318–335
Oxirenes, 125, 143–145
7-Oxoethylguanine, 322
2-Oxopropyldiazotate, 413–414
$N$-(2-Oxopropyl)-$N$-nitrosourea, 414
2-Oxothiazolidine-4-carboxylic acid, 288
6-Oxybenzo[a]pyrene, 181
Oxygen-mediated toxicity, 75

## P

PAH, see Polycyclic aromatic hydrocarbons
Parallel-tube model, 45–48
Paraoxon, 523–533
Paraquat, 86, 464
Parathion, 10, 520–533
Pargyline, 137
Passive diffusion, 33–36
PBB, see Polybrominated biphenyls
PCB, see Polychlorinated biphenyls
$S$-(1,1,2,3,4-Pentachloro-1,3-butadienyl)cysteine, 306
$S$-(1,1,2,3,4-Pentachloro-1,3-butadienyl)glutathione, 306
Pentachloroethane, 295, 300–301
Pentachloroethylperoxyl radical, 295
Pentachlorophenol, 379
Pentanoic acid, 114
Perilla ketone, 245, 250

Peroxidase, 265–266, 268, 357–358, 443, 445, 471
Peroxyflavins, 353
Pharmacokinetics, 29–70
Phenacetin, 10, 82, 350, 356, 360, 366, 384–385
Phenanthrene, 179–180, 191, 206–207, 209, 214–215
Phenanthrene 1,2-dihydrodiol, 216–217
Phenanthrene 9,10-oxide, 77, 195
Phenelzine, 437–438
Phenobarbital, 87, 91, 163–164, 169, 189, 208, 246, 289, 498, 501, 519–522, 533–536
Phenols, 160–161, 165, 168, 180–181, 259–260, 267
  chemical properties of, 261–262
  metabolism of, 263–266
  toxicity of, 262–263
Phenylacetylene, 123, 131
Phenylalkyl cyanides, 506
Phenylarsonic acids, 10
Phenyldiazene, 436, 441
Phenylenediamines, 269, 357
Phenyl-1,2-ethanediol, 127
2-Phenylethyl radical, 438
Phenylhydrazine, 437, 441–443
Phenylhydronitroxide, 469
Phenylhydroxylamine, 365, 471
2-Phenyl-4-pentenamide, 136
Phenylthiourea, 521
Phosgene, 288–293, 296, 340–341
Phosphatase, 528
3′-Phosphoadenosine 5′-phosphosulfate, 378–381
Phospholipid, 505
Phosphonothionate, 521
Phosphooxathiirane, 527
Phosphorothionate, 521
Piperonyl butoxide, 520
Plastoquinone, 270, 274–275
Polybrominated biphenyls (PBB), 158
Polychlorinated biphenyls (PCB), 158–159, 161, 169
Polycyclic aromatic hydrocarbons (PAH), 4–9, 15, 76–77, 90–91, 177–222
  biological activity of, 219–222
  metabolism of, 180–181, 206–212
    stereoselectivity in, 212–218
Poly-input organs, 45, 50
Porphyrins, 138–144, 443

16α-Pregnenolone carbonitrile, 522
Procarbazine, 438–440
Procarcinogen, 10–11, 18
Prodrug, 10
Prolyl-tRNA synthetase, 377
Promotion, 18, 200–202
Prontosil, 10
Propanol, 407
Propene, 138–139, 144–145
Propionaldehyde, 407, 414, 492
Propionitrile, 492, 510
$n$-Propyldiazonium ion, 407
Propyl gallate, 251
7-$n$-Propylguanine, 407
Propylthiourea, 521
Propyne, 138–139
Prostaglandin endoperoxidase, 9, 266
Prostaglandin hydroperoxidase, 471
Prostaglandin synthetase, 191, 358
Protection, 71–97
  against electrophiles, 72–74
  against free radicals, 74–75
  cellular aspects of, 88–89
  covalent binding of metabolites, 86–88
  types of, 75–88
Proteins, reactions with
  acrylonitrile, 503–504
  allyl halides, 336–338
  arylamine and arylamide metabolites, 366
  benzene derivatives, 169
  benzo[a]pyrene derivatives, 196–197
  carcinogens, 4–7, 12–14, 86–88
  furans, 247–249, 252–253
  hydrazines, 445
  hydroxylamines and hydroxamic acids, 377–394
  nitroimidazoles, 462–463
  trichloroethylene metabolites, 332
  vinyl chloride metabolites, 321–323
  vinylidene chloride, 326
Purines, 383
Pyrazole, 521–522, 535–536
Pyrene 4,5-oxide, 195
Pyridoxal phosphate, 436, 441
Pyrocatechase, 267
Pyrrole, 244
Pyrrolizidine alkaloids, 15–16

# Q

Quinine, 129

Quinones, 94, 260, 268
  biosynthesis of, 269–270
  chemical properties of, 271–273
  toxicity of, 273–276

## R

Radiosensitizer, 452–453
Rate vs. time curve, area under, 56–57, 67–68
Receptor site, 33–36, 40
Redox cycling, 93–97
Reductases, 364, 376
Rhodanese, 487–488, 490, 503
RO 07-0582, see Misonidazole
Ronidazole, 464

## S

Safrole, 15–16
Secobarbital, 127, 129, 135–136
Secobarbital diol, 127
Selenium, 92
Selenocysteine, 92
Semiquinone radical, 94, 260
Seryl-tRNA synthetase, 16, 377
SKF 525A, 501, 520, 522, 533–534, 536
Sodium thiophenoxide, 170
Squalene, 128–129
Steady-state concentration, of parent compound in blood, 33–36
Sterols, synthesis of, 118–119
cis-Stilbene, 124, 127–129
trans-Stilbene, 124, 127
trans-Stilbene oxide, 91
Styrene, 127, 129–131, 144–145
Styrene 7,8-oxide, 89
Subcutaneous administration of foreign compound, 50
Succinate:quinone reductase, 275
Succinonitrile, 486, 498
Suicide enzyme inhibitor, 30
Sulfamethazine, 387
Sulfamethazine N-acetyltransferase, 389
O-Sulfonation, of arylhydroxylamines, 382
N-Sulfooxy-acetylaminofluorene, 17
Sulfotransferase, 14, 16, 181, 195, 377–379, 382–383, 488
Sulfur mustard, 7

Superoxide anion, 161, 442–443, 469
Superoxide anion radical, 93–95, 292, 464–469
Superoxide dismutase, 72, 75, 93, 469
Systemic lupus erythematosus, 437

## T

Tamoxifen, 131
TCDD, see 2,3,7,8-Tetrachlorodibenzo-$p$-dioxin
Terpenoids, 245
Testicular toxicity, 112
Testosterone, 379
2,5,2',5'-Tetrachlorobiphenyl, 161, 167
2,3,7,8-Tetrachlorodibenzo-$p$-dioxin (TCDD), 87, 91
2,3,7,8-Tetrachloro-$p$-dioxin, 19
Tetrachloroethene, 295
Tetrachloroethylene, 333–335
2,2,3,3-Tetrachlorooxirane, 334–335
Tetrachloroquinone, 270
Tetradymol, 245
2,6,2',6'-Tetrafluorobiphenyl, 159
$\Delta^9$-Tetrahydrocannabinol, 129
Tetramethylsuccinonitrile, 498–499
Thioacetamide, 78, 520–522, 535–537
Thioacetamide $S$-oxide, 521–522, 535–537
Thioacetate, 170
Thiobenzamide, 522
Thiocarbamides, 83
Thiocarbonic acid, 535
Thiocyanate, 486–501, 506, 532
Thiocyanate oxidase, 487
Thiodiglycolic acid, 324–325
Thioethanol, 170
Thioglycolic acid, 324–325
Thiolate anion, 97
Thiol–disulfide exchange reaction, 84
Thiols, 72, 74
  endogenous, 77–81
  exogenous, 81–86
  protein, 88–89
  reactions with hydrazines, 444–445
Thiono-sulfur compounds, 519–537
Thiophene, 17, 244
Thiosulfate, 487–488, 497, 499
Thiouracil, 521
Thiourea, 521
Thiyl radical, 82, 444

Three-compartment system, 33
Thyroxine, 261
α-Tocopherol, 75, 85–86, 94
β-Tocopherol, 85–86
δ-Tocopherol, 85–86
γ-Tocopherol, 85–86
Toluene, 166
Toluene 1,2-oxide, 167
Toluene 2,3-oxide, 167
Toluene 3,4-oxide, 164
Toxic cell death, 95
Transacetylase, 14
Transaminase, 488
Tremorine, 137
Tricarboxylic acid cycle, 437, 492–493
Trichloracetic acid, 328, 334–335
2,2′,4′-Trichloroacetophenone, 305
Trichloroacetyl chloride, 334–335, 340–341
Trichloroacryloyl chloride, 340–341
1,1,1-Trichloro-2,2-bis(p-chlorophenyl)ethane (DDT), 299
Trichloroethanol, 328
Trichloroethylene, 318, 328–333, 335
Trichloromethyl hypochlorite, 292–293
Trichloromethylperoxyl radical, 290–298
Trichloromethyl radical, 290–292, 297–300
3,3,3-Trichloro-1,2-oxidopropane, 128
2,2,3-Trichlorooxirane, 328–332
1,1,1-Trichloropropane 2,3-oxide, 501
Trifluoroacetic acid, 300
Trifluoromethylcarbene, 301
Trimethylamine, 441–442
Trimethylamine dehydrogenase, 441–442
7,11,12-Trimethylbenz[a]anthracene, 212
Triorthocresylphosphate, 112
Tumorigenicity, 219–222
Two-compartment system, 32–33, 36–40
  repeated administration of foreign compound, 41–43
Tyramine, 442
Tyrosinase, 263, 267–269

## U

Ubiquinone, 270, 273, 275
UDPglucuronosyltransferase, 9, 181–182, 195, 377, 383–385
Uric acid, 75

## V

α-Valerolactone, 114
vic-Dihaloalkanes, 303–304
vic-Diols, 319
Vinyl bromide, 322–323
Vinyl chloride, 16, 320–323
Vinyl cyanide, see Acrylonitrile
4-Vinylcyclohexene, 129
Vinyl fluoride, 123, 138–139, 142–143, 323
Vinyl halides, 250
Vinylidene chloride, 323–326
Vinylidene fluoride, 326
Vitamin A, 72
Vitamin $B_6$, see Pyridoxal phosphate
Vitamin $B_{12}$, 489
Vitamin C, 72, 75, 86, 94
Vitamin E, 72, 74, 85–86, 89, 273–274
Vitamin K reductase, 275

## W

Warfarin, 167
Water, in cellular protection, 72, 75–77
Well-stirred model, 45–48

## X

Xanthine oxidase, 456–458, 460
Xylene, 166
p-Xylene 3,4-oxide, 164